QUATERNARY CARBONATE AND EVAPORITE SEDIMENTARY FACIES AND THEIR ANCIENT ANALOGUES

Other publications of the International Association of Sedimentologists

Special Publication Number 43 of the International
Association of Sedimentologists

Quaternary carbonate and evaporite sedimentary facies and their ancient analogues

A Tribute to Douglas James Shearman

Edited by

Christopher G. St.C. Kendall
Earth and Ocean Sciences, University of South Carolina, Columbia, SC 29208 USA

Abdulrahman S. Alsharhan
Faculty of Science, UAE University, P.O. Box 17551, Al-Ain United Arab Emirates

SERIES EDITOR

Ian Jarvis
*School of Geography, Geology and the Environment
Centre for Earth and Environmental Science Research
Kingston University London
Penrhyn Road
Kingston upon Thames KT1 2EE
UK*

SERIES CO-EDITOR

Tom Stevens
*Department of Geography
Royal Holloway, University of London
Egham, Surrey
TW20 0EX
UK*

A John Wiley & Sons, Ltd., Publication

This edition first published 2011 © 2011 by International Association of Sedimentologists

Blackwell Publishing was acquired by John Wiley & Sons in February 2007. Blackwells publishing program has been merged with Wileys global Scientific, Technical and Medical business to form Wiley-Blackwell.

Registered office: John Wiley & Sons Ltd, The Atrium, Southern Gate, Chichester, West Sussex, PO19 8SQ, UK

Editorial offices: 9600 Garsington Road, Oxford, OX4 2DQ, UK

For details of our global editorial offices, for customer services and for information about how to apply for permission to reuse the copyright material in this book please see our website at www.wiley.com/wiley-blackwell.

Library of Congress Cataloging-in-Publication Data

Quaternary carbonate and evaporite sedimentary facies and their ancient analogues : a tribute to Douglas James Shearman / edited by Christopher G.
  St. C. Kendall and Abdulrahman Alsharhan.
      p. cm. (Special publication ; no. 43)
  Includes bibliographical references and index.
  ISBN 978-1-4443-3910-9 (hardback)
 1. Carbonate rocks—Persian Gulf Coast (Persian Gulf States)   2. Evaporites—Persian Gulf Coast (Persian Gulf States)
3. Facies (Geology)—Persian Gulf Coast (Persian Gulf States)   4. Geology, Stratigraphic—Quaternary. I. Shearman, Douglas James, 1918–2003. II. Kendall, Christopher G. St. C. III. Alsharhan, A. S.
  QE471.15.C3Q38 2010
  552'.58–dc22
2010040513

A catalogue record for this book is available from the British Library.

This book is published in the following electronic formats: ePDF 9781444392302; ePub 9781444392319

Set in 10/12pt Melior by Thomson Digital, Noida, India
Printed and bound in Malaysia by Vivar Printing SdnBhd

1 2011

Douglas James Shearman (1918–2003): Father of the sabkha model

Contents

Int. Assoc. Sedimentol. Spec. Publ. (2011) **43**, 9–14

Douglas James Shearman (1918–2003): "The Father of the sabkha model"

This volume is dedicated to the late Professor Douglas Shearman, who is best remembered for his outstanding research contributions to unravelling the nature and origin of the sabkha evaporites of the United Arab Emirates, and his recognition that these were probably analogues for many ancient evaporite sequences.

Douglas James Shearman (Figs. 1 and 2) was an amazingly perceptive and innovative geologist who showed remarkable courage and determination to proselytise his views, both to the geological community and to any others who crossed his path. He was renowned for his all-embracing bursts of enthusiasm for the latest subject that had caught his attention. He would drag academics, students and even cleaners into his room and lecture them on problems they never knew existed.

Douglas was born in Isleworth on 2 July 1918. Before the Second World War, he trained and worked as a post-office engineer. His war service was as a radio operator on a minesweeper on the Arctic convoys to Russia with the British Royal Navy. He waxed eloquently on epic hours chipping ice off radio aerials. His interest in geology was

Fig. 2. Douglas Shearman recounting one of his many adventures to, as always, an attentive and amused audience. Christopher G. St.C. Kendall fourth, and Abdullah Al-Zamel fifth from left. Kuwait 1986 (photograph courtesy of Professor Claudio Vita-Finzi).

triggered by meeting a soldier and a terebratulid in the Dover NAAFI canteen. After the war he graduated from Chelsea Polytechnic, that nursery of many famous geologists, and obtained a post as Lecturer in Sedimentology in the Geology department of Imperial College, London, where he was to spend the rest of his professional life.

Douglas spent his first eleven years in academe busily studying the structural geology and geomorphology of North Devon, UK. His research demonstrated that the *en echelon* offset of the West Country granites was the result of series of lateral crustal tears. This was to have been submitted for a PhD, but the research was never completed. It was typical of Shearman that this work was never written up, neither as a thesis nor for wider publication; until a short paper appeared in 1967. This habit of solving one geological problem and leaving it unpublished to move on to another typified his career. He worked as a consultant geologist a great deal, particularly in south-east England but also in Africa, where he worked on the Kariba dam site and in other remote locations. In addition, he

Fig. 1. Douglas Shearman examining a section revealing anhydrite showing a dish-form just beneath the surface of the sabkha, Abu Dhabi, 1964 (photograph courtesy of Christopher G. St.C. Kendall).

supervised research students, unofficially, as he did not have a PhD and this contravened the rules of the University of London for supervisors at that time. These students' research work was on diverse topics including: Carboniferous limestones and turbidites, Jurassic limestones, Cretaceous sands, Holocene intertidal-flat deposits and carbonates.

During these early years, he had little interest in carbonates and evaporites. In fact he never lectured on the latter in his courses to undergraduates, and the teaching collection at Imperial contained only a few drawers of rather dusty samples of rock salt and gypsum crystals. Douglas' interest in carbonates started with the arrival in the department of Derek Price, a mature student who worked on the Yoredale carbonates of Yorkshire. It developed further when Shearman became involved in the supervision of some students engaged in a study of the Mesozoic carbonates of the Jura Mountains, France, which had been initiated and organized by Derek Ager. This led to a series of papers on dedolomitization, a process that had hitherto not been recorded as occurring in sedimentary rocks (Shearman *et al.*, 1961; Shearman & Shirmohammadi, 1969).

The study of carbonate rocks took up more and more of his time with the initiation of a programme of research into the Quaternary deposits of the Trucial Coast (now the United Arab Emirates, UAE) in the early 1960s by the sedimentology section of the Geology Department at Imperial. At that time, the site now occupied by Abu Dhabi City was a scorching coastal salt flat that extended for several kilometres from land to sea, and the airport was a bamboo "lean to" with a windsock. The opportunity to investigate modern carbonates allowed him and his co-workers to develop, and at times to confirm, suggestions that they had made in earlier papers, on the basis of studies of ancient carbonates, on limestone diagenesis (Skipwith & Shearman, 1965; Evamy & Shearman, 1969; Shearman *et al.*, 1970).

His interest in evaporites, one which was to occupy every moment of his later life, started when he realized that some scattered crystals in the wash of a sample collected by David Kinsman from the edge of the saline coastal plain (the sabkha) were anhydrite. He immediately recognized the significance of this discovery, and in the following years pursued this discovery and its significance (Shearman, 1965, 1966). He and his co-workers P. R. Bush, G. Butler, G. Evans, D.J.J. Kinsman, C.G. St.C. Kendall and Sir P.A. d'Estoteville Skipwith,

produced several papers describing the now famous coastal sabkha evaporites of Abu Dhabi.

This stimulated Douglas into an investigation of various ancient evaporite deposits: firstly, those of the Purbeck of the Warlingham borehole in southeast England (Shearman, 1966), later with studies in Canada with J. Fuller on the Middle Devonian Winnepegosis Formation (Shearman & Fuller, 1969), and then more cursory studies on other famous evaporite deposits of the geological column. He never made a detailed study of any particular deposit except the Winnepegosis. Instead, he merely picked the plums which supported his growing conviction (which later mellowed a little!) that all evaporites were probably of a coastal plain origin! His lack of willingness to provide detailed comprehensive descriptions of any succession led to the remark of one of his assessors, when Douglas attempted to obtain promotion to the status of Reader in Geology in the University of London: "*the trouble is, Shearman, you have not produced a single major paper for the Quarterly Journal of the Geological Society of London*". Shearman had to wait several years for his promotion!

It is interesting to note that the discovery of modern anhydrite produced little interest in the older establishment of the British oil industry. One of these asked the writer: "*Why are you chaps wasting the Government's money studying the sabkha and these coastal areas?*" However, younger members of the geological fraternity were enthused by the discovery and its implications, as were researchers in the Shell Oil Company, The Netherlands. Shell had its own carbonate research group which included Bruce Purser, A. Wells, V. Colter, P. Zeigler and Brian Evamy (one of Shearman's former research students), as well as a group of eminent American carbonate and evaporite specialists, including James Wilson, E.A. "Gene" Shinn and Raymond Murray. These scientists were very supportive, and after a series of lectures by Shearman and the writer, gave the Imperial group what was in those days, a handsome donation to their research budget. The Trucial Coast project had been funded originally by the Department of Science and Industrial Research (now NERC) and Shell oil company and was later supported by Socony-Mobil Dallas (Henry Nelson and Volkmar Schmidt of that company participated in the drilling of the first line across the sabkha plain that revealed the history of the development of this interesting feature).

In 1965, Shearman astounded the geological fraternity and produced cries of disbelief from the gurus of the evaporites, with his audacious claims that they had all misinterpreted ancient evaporite deposits, and that the hypothesized evaporating-dish origin of these was incorrect. Instead, he suggested that the classical evaporite sequence, together with its frequent replacement textures, was the result of the deposition of a pile of carbonate-gypsum/anhydrite sabkha sediments, and the subsequent upward expulsion of the connate brines to produce mineralogical changes whilst leaving the lower part unaltered. He developed this thesis in his most famous paper (Shearman, 1966, and discussion, 1967) in which he introduced the term "sabkha measures". After a visit to Baja California, this was extended to include layered halite deposits and their chevron-type habit; these, he showed, were not indicative of deep water, thus supporting the earlier conclusions of Holliday (1967). In addition, he argued that individual salt layers did not always represent annual layers, thus supporting the earlier denial of this by Phleger (1969).

Douglas became intrigued, together with two Canadian research students, by the evidence provided by gypsum veins of hydraulic fracturing in the subsurface (Shearman *et al.*, 1972, 1973), and later by the changes of gypsum fabrics during the exhumation of evaporitic rocks (Mossop & Shearman, 1973). Acceptance of his theories was followed by academic recognition. Shearman was awarded a DSc, and then a personal Chair by the University of London (1978). One of his final papers on sabkha evaporites was written with his old friend and colleague Amanda Gunatilaka, when he was a visiting professor in the University of Kuwait. They described some laminated gypsum-carbonate deposits from an ephemeral channel on the sabkha (Gunatilaka & Shearman, 1988). It was an attempt to answer some of his critics who claimed that he had never demonstrated the presence of laminated evaporites in an intertidal-supratidal setting.

Later, Douglas became fascinated by the curious mineral ikaite ($CaCO_3 \cdot 6H_2O$) and supervised work on this in the Greenland fjords, as well as writing a review of the occurrences of this mineral and its variously named pseudomorphs (Shearman & Smith, 1985). Subsequently, he and a colleague carried out some laboratory studies on the formation of this mineral (Shaikh & Shearman, 1986). His last episode of fieldwork was in the USA to examine the tufa mounds of the Lake Lahotan and Mono Lake Basins, which he and his co-workers claimed were originally formed of ikaite but had later been replaced by calcite (Shearman *et al.*, 1989).

Whilst Shearman was not always the originator, rather sometimes reviving interest in earlier ignored or forgotten views, and was occasionally incorrect in some of his suggestions on the origin of evaporites, he will always be remembered for his persuasive skill as a teacher and lecturer, with his beautifully drafted figures, and as the initiator of the re-examination of these deposits. His impact was considerable, and led to him being appointed as the first Professor of Sedimentology at Imperial College in 1978, and being awarded the Lyell Fund (1967), the Lyell Medal (1984) and the Wollaston Medal (1997) of the Geological Society of London, the Matson Award (1971) of the American Association of Petroleum Geologists, and the Geological Association of Canada Medal (1999). Above all, he will always be remembered by former colleagues, students and acquaintances for his unbounded enthusiasm and skill as an articulate exponent of his views, and his bottomless fund of entertaining stories of his escapades in the Royal Navy during the Second World War, and others from his earlier travels.

As his old student Dick Selley remarked "*an account of Shearman's academic achievements tells little of his character. He was unforgettable, eccentric, charming, frustrating to administrators who required forms filled in. Generations of students worshipped him. He had a special soft spot for overseas students. Shearman had a puckish sense of humour, a delight in practical jokes, and an ability to attract unforeseen events. When wrongly convicted of some naval misdemeanour he was set to paint the ship's funnel. This he did - complete with black swastika. The Israeli army shot up his tent when he was working for the UN looking for water in Jordan. Fortunately he was out at the time.*"

Further and more detailed information on Douglas Shearman's life and people's appreciation of him can be found in Bush (2003), Evans (2003, 2005) and Selley (2003).

Graham Evans

National Oceanography Centre, Southampton, UK, and formerly of Imperial College, London, UK.

REFERENCES

Blyth, F. and **Shearman, D.J.** (1962) A conglomeratic grit at Knowle Wood, near Woolley, South Devon. *Geol. Mag.*, **99**, 30–32.

Bush, P.R. (2003) Professor Douglas Shearman. *The Independent*, 23 May.

Butler, G., Evans, G., Kinsman, D.J., Shearman, D.J. and **Skipwith, P.A.** (1964) Discussion of I.M. West: Evaporite diagenesis in the Lower Purbeck beds of Dorset. *Proc. Yorks. Geol. Soc.*, **34**, 315–330.

Butler, G., Kendall, C.G., Kinsman, D.J.J., Shearman, D.J. and **Skipwith, P.A.** (1965) Recent evaporite deposits along the Trucial Coast of the Arabian Gulf. *Proc. Geol. Soc. London*, **1623**, 91.

Curtis, R., Evans, G., Kinsman, D.J.J. and **Shearman, D.J.** (1963) On the association of dolomite and anhydrite in the Recent sediments of the Persian Gulf. *Nature*, **197**, 679.

Evamy, B.D. and **Shearman, D.J.** (1962) The application of chemical staining techniques to the study of diagenesis of limestones. *Proc. Geol. Soc. London*, **1599**, 102.

Evamy, B.D. and **Shearman, D.J.** (1965) The development of overgrowths from echinoderm fragments. *Sedimentology*, **5**, 211–233.

Evamy, B.D. and **Shearman, D.J.** (1969) Early stages in development of overgrowths on echinoderm fragments in limestones. *Sedimentology*, **12**, 317–322.

Evans, G. (2003) Douglas James Shearman, 1918–2003. *Carbonates and Evaporites*, **18**, 171–174.

Evans, G. (2005) Douglas Shearman, 1918–2003. *Proc. Geol. Assoc.*, **116**, 191–205.

Evans, G. and **Shearman, D.J.** (1964) Recent celestite from the sediments of the Trucial Coast of the Persian Gulf. *Nature*, **202**, 385–386.

Evans, G., Kinsman, D.J.J. and **Shearman, D.J.** (1964) A reconnaissance survey of the environment of Recent carbonate sedimentation along the Trucial Coast, Persian Gulf. In: *Deltaic and Shallow Marine Deposits* (Ed. L.M.J.U. van Straaten), *Dev. Sedimentol.*, **1**, 129–135.

Gunatilaka, A. and **Shearman, D.J.** (1988) Gypsum-carbonate laminites in a recent sabkha Kuwait. *Carbonates Evaporites*, **3**, 67–73.

Gunatilaka, A., Al-Zamel, A., Shearman, D.J. and **Reda, A.** (1987) A spherulitic fabric in selectively dolomitized siliciclastic crustacean burrows, northern Kuwait. *J. Sed. Petrol.*, **57**, 922–927.

Holliday, D.W. (1967) Contribution to discussion of Shearman, D.J., 1966. Origin of marine evaporites by diagenesis. *Trans. Inst. Min. Metall.*, **B76**, 179–180.

Mossop, G. and **Shearman, D.J.** (1973) Origins of secondary gypsum rocks. *Trans. Inst. Min. Metall.*, **B82**, 147–154.

Phleger, F.B. (1969) A modern evaporite deposit in Mexico. *AAPG Bull.*, **53**, 824–829.

Pitcher, W.S., Shearman, D.J. and **Pugh, D.C.** (1954) The loëss of Pegwell Bay, Kent, and its associated frost soils. *Geol. Mag.*, **91**, 308–314.

Pugh, E.E. and **Shearman, D.J.** (1967) Crytoturbation structures at the south end of the Isle of Portland. *Proc. Geol. Assoc.*, **78**, 463–471.

Selley, R. (2003) Professor Douglas Shearman. *The Times*, 21 May.

Selley, R.C. and **Shearman, D.J.** (1962) The experimental production of sedimentary structures in quicksands. *Proc. Geol. Soc. London*, **1599**, 101.

Selley, R.C., Shearman, D.J., Sutton, J. and **Watson, J.** (1963) Some underwater disturbances in the Torridonian of Skye and Raasay. *Geol. Mag.*, **100**, 224–243.

Shaikh, A. and **Shearman, D.J.** (1986) On ikaite and the morphology of its pseudomorphs. *Proc. Int. Meeting on Geochemistry of the Earth Surface and Processes of Mineral Formation, Granada, Spain.* C.I.S.C., Madrid, 79–83.

Shearman, D.J. (1963) Recent anhydrite, gypsum, dolomite and halite from the coastal flats of the Arabian shore of the Persian Gulf. *Proc. Geol. Soc. London*, **1607**, 63.

Shearman, D.J. (1964) On the penecontemperaneous disturbance of bedding by quicksand movement in the Devonian rocks of North Devon. In: *Deltaic and Shallow Marine Deposits* (Ed. L.M.J.U. van Straaten), *Dev. Sedimentol.*, **1**, 368–370.

Shearman, D.J. (1965) Discussion. In: *Salt Basins Around Africa* (Ed. Anon), pp. 75-76. Institute of Petroleum, London.

Shearman, D.J. (1966) Origin of marine evaporites by diagenesis. *Trans. Inst. Min. Metall.*, **B75**, 208–215 (see also Discussion in **B76**, 82-86).

Shearman, D.J. (1967) On the Tertiary fault movements in North Devonshire. *Proc. Geol. Assoc.*, **78**, 555–566.

Shearman, D.J. (1970) Recent halite rock. *Baja, California. Trans. Inst. Min. Metall.*, **B79**, 127–136.

Shearman, D.J. (1971) Discussion on paper by Beales and Onasick on Mississippi Valley ore deposits. *Trans. Inst. Min. Metall.*, **80**, 50–52.

Shearman, D.J. (1976) The geological evolution of southern Iran. *Geogr. J.*, **142**, 393–410.

Shearman, D.J. (1978) Evaporites of coastal sabkhas. In: *Marine Evaporites* (Eds W.E. Dean and B.C. Schreiber), *SEPM Short Course Notes*, **4**, 6-42.

Shearman, D.J. (1979) A field test for the identification of gypsum in soils and sediments. *Q. J. Eng. Geol.*, **12**, 51.

Shearman, D.J. (1980) Sabkha facies evaporates. In: *Evaporite Deposits. Chambre sindicate de la Recherche et de la Prediction du Petrol et du Gaz Naturel.* Editions Techniq., **27**, 96-109, Paris, France.

Shearman, D.J. (1981) Displacement of sand grains in sandy gypsum crystals. *Geol. Mag.*, **118**, 303–316.

Shearman, D.J. and **Fuller, J.G.C.M.** (1969) Anhydrite diagenesis, calcitization and organic laminites. Winnipegosis Middle Devonian, Saskatchewan. *Bull. Can. Petrol. Geol.*, **17**, 496–525.

Shearman, D.J., Khouri, J. and **Taha, M.S.** (1961) On the replacement of dolomite by calcite in some Mesozoic limestones from the French Jura. *Proc. Geol. Assoc.*, **72**, 1–12.

Shearman, D.J., Mossop, G., Dunsmore, H. and **Martin, M.** (1972) Origin of gypsum veins by hydraulic fracture. *Trans. Inst. Min. Metall.*, **B81**, 149–155.

Shearman, D.J., Mossop, G., Dunsmore, H. and **Martin, M.** (1973) Discussion and contribution to above paper. *Trans. Inst. Min. Metall.*, **B82**, 66–67.

Shearman, D.J. and **Orti Cabo, F.** (1978) Upper Miocene gypsum: San Miguel de Salinas, southeast Spain. *Mem. Soc. Geol. Ital.*, **16**, 327–339.

Shearman, D.J. and **Shirmohammadi, N.H.** (1969) Distribution of strontium in dedolomites from the French Jura. *Nature*, **223**, 606–608.

Shearman, D.J., Twyman, J. and **Zand-Karimi, M.** (1970) The genesis and diagenesis of oolites. *Proc. Geol. Assoc.*, **81**, 561–575.

Shearman, D.J. and **Smith, A.J.** (1985) Ikaite, the parent mineral jarrowite-type pseudomorphs. *Proc. Geol. Assoc.*, **96**, 305–394.

Shearman, D.J., McGugan, A., Stein, C. and **Smith, A.J.** (1989) Ikaite, $CaCO_3 \cdot 6H_2O$, precursor of the thinolites in the Quaternary tufas and tufa mounds of the Lahontan and Mono Lake Basins, western USA. *Geol. Soc. Am. Bull.*, **101**, 913–917.

Skipwith, P.A. and **Shearman, D.J.** (1965) Organic matter in Recent and Ancient limestones and its role in their diagenesis. *Nature*, **208**, 1310.

West, I.M., Shearman, D.J. and **Pugh, E.E.** (1969) Whitsun Field Meeting in Weymouth area, *1966. Proc. Geol. Assoc.*, **80**, 331–340.

Int. Assoc. Sedimentol. Spec. Publ. (2011) **43**, 1–10

Introduction to Quaternary carbonate and evaporite sedimentary facies and their ancient analogues

ABDULRAHMAN S. ALSHARHAN* and CHRISTOPHER G. St. C. Kendall[†]

Faculty of Science, United Arab Emirates University, PO Box 17551, Al Ain, UAE
(E-mail: sharhana@emirates.net.ae)
[†]*Department of Geological Sciences, University of South Carolina, Columbia, South Carolina 29208, USA*
(E-mail: Kendall@geol.sc.edu)

ABSTRACT

Douglas James Shearman (1918–2003) used his imagination to extend our understanding of the sabkha evaporites of the Arabian Gulf, and their use as analogues for evaporites that are now associated worldwide with hydrocarbon exploration and exploitation. His work on the Holocene carbonates and evaporites of the southern coast of the Arabian Gulf has meant that these are now the most frequently cited examples of type-analogues for assemblages of shallow-water carbonates, evaporites and siliciclastics found throughout the geological record. Striking examples from the United Arab Emirates (UAE) and nearby regions include those found in the Tertiary and Mesozoic sedimentary rocks of the immediate subsurface. Other analogues of this setting include the Palaeozoic carbonates of the western USA, Europe and Asia, and the Mesozoic carbonates of the Gulf of Mexico and Europe.

INTRODUCTION

The Arabic word for salt flat, "sabkha", was coined by Shearman in the field in 1961 to differentiate the facies he first saw exposed on the then Trucial Coast (now the United Arab Emirates, UAE). Since then, the word sabkha has been incorporated into the evaporite literature to describe both continental and near-coastal evaporative sediments that accumulate close to the sedimentary surface. The subaerial boundary of this flat is a geomorphic surface whose level is dictated by, and is in equilibrium with, the local water table. Sabkhas are occasionally covered by ephemeral shallow water, but for most of the time, they are subaerial mud and sand flats. The term "marine sabkha" normally denotes a near-coastal salt flat dominated by marine-derived brines and processes, though the character of the adjacent continental water-table often influences it. A "continental sabkha", in contrast, is an inland salt flat dominated by continental brines and processes. Both settings receive water via either subsurface or overland (storm-induced) flow, and may be associated with aeolian dune fields and intermittently submerged interdunal corridors. All these settings were observed and described more than half century ago by Shearman and his colleagues from the Arabian Gulf.

Until the late 1950s there were no well-documented modern evaporite analogues to ancient evaporite settings. However, this changed with research into the carbonates and evaporites of the Arabian Gulf, and particularly Abu Dhabi, in the late 1950s and early 1960s with the initial work of Emery (1956) and Houbolt (1957). The pace of this research was stepped up when the Imperial College of London research team, led by Graham Evans and Douglas Shearman, arrived in Abu Dhabi. As a result of carrying out extensive research along the Abu Dhabi coast through 1961–1970, this group became a leader in the study of modern carbonates and evaporites. Initiated by Evans and Shearman, the research observations of the group were published in a mix of many PhD dissertations, which included those of David Kinsman, Christopher Kendall and Patrick Skipwith, and articles in professional journals and books. These all documented the Holocene carbonate (ooids, grapestones, pellets, mud and cyanobacterial flats) and evaporite (halite, anhydrite and gypsum) sediments and various other associated sedimentary facies that had their counter points in the subsurface of ancient sedimentary sections.

Shell Research of the Netherlands followed the Imperial College team and carried out regional surveys along the coast of the Arabian Gulf, from

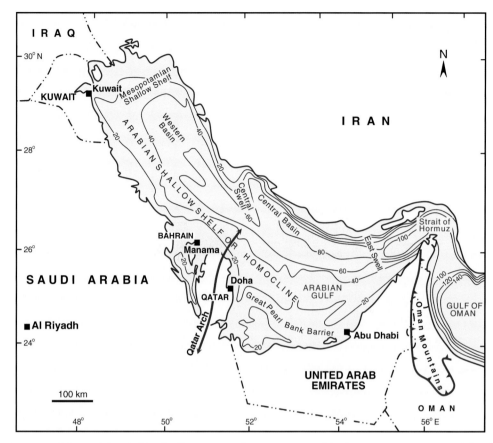

Fig. 1. Location map of the Arabian Gulf and adjacent area showing principal bathymetric provinces and depth of water (in metres) (modified from Kassler, 1973; Alsharhan & Kendall, 2003).

Qatar to the UAE. Bruce Purser directed this research, working with many now famous geologists that included Gene Shinn, Leslie Illing and J.C.M. Taylor. Later, Purser and his team carried out fieldwork sponsored by the National Museum of Natural History of Paris and the TOTAL Oil Company. Ken Hsu and his graduate students (who included Godfrey Butler and Judith Mckenzie) arrived from Riverside California and then switched to the Polytechnic Institute of Zurich, where they made significant contributions to our understanding of the recent carbonates and evaporites of Abu Dhabi in the early 1970s. These workers were followed by Stjepko Golubic and colleagues from Boston, who studied the cyano-bacteria of the protected tidal flats of the UAE in the late 1970s, as well as R.J. Patterson, R.K. Park and D.J. Kinsman who came from Princeton in the late 1970s and 1980s. Also in the 1980s, Gunatilaka worked with Shearman in Kuwait, while G. Walkden, and A. Williams from Aberdeen investigated the southern margin of the Gulf. A direct consequence of all these studies of the

southern coast of the Arabian Gulf (Fig. 1) is that this region is now one of the best, if not the best, documented modern sea-margin of Holocene carbonate-evaporites and dolomites.

The Holocene marine and coastal sediments of the southern Arabian Gulf coast represent the arid equivalent of the shelf sediments of the Yucatan, British Honduras and Florida, and the isolated platform of the Bahamas. The sediments of the Holocene of the UAE exhibit a wide range of sedimentary facies (Fig. 2; Table 1) that include: (a) offshore bivalve sands mixed with lime and argillaceous mud; (b) bivalve-rich sediment in the deeper tidal channels between the barrier island lagoons and deeper portions of the Khor al Bazam; (c) coral reefs and coralgal sands of coastal margins to the west; (d) oolite shoals that accumulate on the tidal deltas of channels that debouch from between barrier islands to the east; (e) grapestones that occur on exposed coastal terraces of the western Khor al Bazam and to the lee of the reefs and oolite shoals in eastern Abu Dhabi; (f) pelleted lime muds that accumulate in the lagoons of

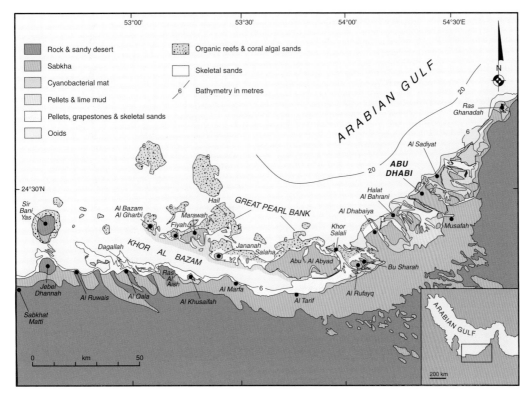

Fig. 2. General sedimentary facies along the coastal areas of Abu Dhabi Emirate.

Table 1. Types and characteristics of tidal flats along the United Arab Emirates coast.

Types	Characteristics	Occurrence
Barrier island-lagoon sabkha	Sabkhas prograde into barrier lagoons that form behind a series of Pleistocene barrier islands. These sabkhas have complexities such as buried beach ridges, abundant siliciclastic matrices and facies, mosaic distributions of anhydrite-gypsum-dolomite.	Coast of Abu Dhabi
Mainland beach-dune sabkha	Lack low-energy lagoonal sediments. Evaporites form in the mixing zone between marine and continental influenced groundwaters. They infill depressions behind extensive Holocene mainland beach-dunes.	NE of Ras Ghanada in Abu Dhabi all the way to Dubai
Continental interdunal sabkha	These interdunal sabkhas are filled with late Pleistocene/Holocene evaporitic sediments. The evaporites are displacive sulphates (gypsum and anhydrite) mixed in with displacive and bedded halite in a siliciclastic matrix. There appears to be a much higher proportion of halite preserved in these continental sabkhas compared to the coastal sabkha.	In many of the interdunal corridors there are currently elongate interdunal sabkhas. Sabkhat Matti is the largest in the area.
Fan delta sabkha	The alluvial fans are fed by ephemeral streams from the mountains and pass directly into the waters of the Gulf. On the distal portions of these fan deltas, sabkhas have nodular and bedded gypsum, and displacive halite grows in a matrix of siliciclastic fan delta sediments.	Northern part of Ras Al Khaimah in Al Rams area

the eastern Abu Dhabi; (g) cyanobacterial mats and mangrove swamps lining the inner shores of the protected lagoons of Abu Dhabi and the east Khor al Bazam; and (h) supratidal salt flats (sabkhas) where evaporite minerals accumulate along the inner shoreline. The coastal sediments pass landward into continental aeolian facies (Alsharhan & Kendall, 2003).

Plio–Pleistocene tectonic events have dominated the evolving morphology of the Arabian Gulf though Quaternary erosion and deposition, and have modified the relief of the resulting structures (Kassler, 1973). For example a sea-level fall of around 120 m during the Pleistocene left the Arabian Gulf entirely exposed, with rivers channeling into its flanks and moving down its axis. During this maximum fall and during the subsequent rise, a series of platforms were cut into the pre-existing surface (Kassler, 1973; Weijermars, 1999). The Late Pleistocene and Holocene sea-level changes were also associated with dramatic climate change, and there is a broad coincidence between the deduced sea-level and temperature curves of this time period. Currently, the Arabian Gulf is situated in the low latitudes of the tropics, and the distribution of its sediment is controlled by many factors that include an arid climate, a mix of the influence of low and high wave energy, a coastal orientation that is a response to northwesterly Shamal winds and the presence or absence of offshore barriers (Wagner & Van der Togt, 1973).

ANCIENT ANALOGUES AND THE SIGNIFICANCE OF EVAPORITES FOR PETROLEUM EXPLORATION

Similar associations of shallow-water carbonate-evaporite facies to those of the coastal areas of the southern Arabian Gulf occur in the subsurface and form good reservoirs in many parts of the world (Table 2). The Arabian Gulf examples include the Permian Khuff Formation, the Upper Jurassic Arab Formation and Hith Anhydrite, and the Tertiary Umm Er Rhaduma Formation and Fars Group. Other similar associations from other parts of the world include the Ordovician Red River Formation of the Williston Basin, and the Ordovician Bauman Fjord Formation of Ellsmere Island, the Devonian of Western Canada and Western Australia, the Pennsylvanian of the Paradox Basin of Utah, the Permian of West Texas and the Zechstein Sea in the North Sea area, the Jurassic sedimentary rocks

Table 2. Ancient sabkha reservoirs.

I. Marine and continental sabkha
1. Ordovician Red River Formation, Williston Basin, USA.
2. Lower Clear Fork Formation (Permian), Texas.
3. Upper Minnelusa Formation, Wyoming.
4. Jurassic Smackover and Lower Buckner Formations, South Texas.
5. Rift Basins: Gulf of Suez and Red Sea.

II. Shallow-water evaporite
1. Ferry Lake Anhydrite, Fairway Field, East Texas.
2. Jurassic Todilto Formation, New Mexico and Colorado.
3. San Andres Formation, NW Shelf of Permian Basin, West Texas.
4. Upper Part of the Arab Formation, Upper Jurassic, Arabian Gulf.
5. Khuff Formation, Upper Permian, Arabian Gulf.
6. Permian Guadalupian, West Texas and SE New Mexico.

III. Deep-water evaporites
1. Permian Delware Basin, Texas.
2. Permian Zechstein of Europe.
3. Upper Silurian of the Michigan Basin.

of the Gulf of Mexico, and parts of the Lower Cretaceous of south-eastern Texas.

Cycles of ancient sabkha facies were distinguished in the Upper Jurassic Arab Formation of Abu Dhabi by Wood & Wolf (1969). In fact, the present UAE coastline appears to match some of the evaporite settings of the Upper Jurassic Hith and is a recognizable analogue of the Hith (Alsharhan & Kendall, 1994). The Upper Jurassic evaporitic sulphates and minor chlorides of Arabia accumulated within a complex of giant playas and sabkhas (72 million km^2) along the southern margin of the Tethys Ocean. These evaporites now form excellent seals to some of the world's most prolific oil reservoirs (Murris, 1980; Ayres *et al.*, 1982; Alsharhan & Kendall, 1986). Examples of this association include the grain carbonates of the Arab Formation of the eastern Arabian Peninsula and the Hith Anhydrite. The Hith forms an excellent seal in Saudi Arabia, Qatar, Bahrain, and western Abu Dhabi. It prevents the upward movement of oil generated in the Jurassic source rocks (Alsharhan & Kendall, 1994). This seal is dominated by the playa facies but eastward, where the Hith facies change to a dominantly sabkha and peritidal carbonate, oil escapes upward and is found in the Lower Cretaceous reservoirs of eastern Abu Dhabi. In contrast, the series of Holocene algal flats associated

Table 3. Some carbonate hydrocarbon reservoirs associated with tidal-flat evaporites and algal flats.

Field(s)	Formation	Age	Location	Reference(s)
Puckett	Ellenberger	Ordovician	Texas, USA	Loucks and Anderson (1985)
Cabin Creek and Pennel	Red River	Ordovician	Montana, USA	Clement (1985); Ruzyla & Friedman (1985)
Cabin Creek	Intertake	Silurian	Montana, USA	Roehl (1985)
Rainbow	Keg River	Middle Devonian	Alberta, Canada	Schmidt *et al.* (1985)
Northwest Lisbon	Leadville	Mississippian and Pennsylvanian	Paradox Basin, USA	Miller (1985)
Little Knife	Mission Canyon	Mississippian	North Dakota, USA	Lindsay and Kendall (1985)
Tarchaly, Rybaki, and Sulecin	Zechstein	Early Permian	Poland	Depowski & Peryt (1985)
Qatif	Arab	Late Jurassic	Saudi Arabia	Wilson (1985)
Ghawar	Arab	Late Jurassic	Saudi Arabia	Mitchell *et al.* (1988)
Umm Shaif	Arab	Late Jurassic	Abu Dhabi	Alsharhan (1989)
Dukhan	Arab/Qatar	Late Jurassic	Qatar	Qatar General Petroleum Corp. & Amoco Petroleum Co. (1991)
Chatom	Smackover	Late Jurassic	Southeast Arkansas, USA	Feazel (1985)
Mt. Vernon	Smackover	Late Jurassic	South Arkansas, USA	Druckman & Moore (1985)
Sunniland	Sunniland Limestone	Early Cretaceous	South Florida, USA	Halley (1985)
Gachsaran and Bibi Hakimeh	Asmari	Oligocene and EarlyMiocene	Southwest Iran	McQillan (1985)

with high concentrations of organic matter inter-bedded with carbonates and evaporites in the region is not common to the Hith. However, hydrocarbon traps related to evaporite and carbonate cycles are quite common in the geological record and support these authors' contention that the association between evaporites and carbonates seen in Abu Dhabi is not unique to the Holocene. In fact, such associations may represent potential source rocks for some ancient carbonate petroleum reservoirs (Table 3).

Thick repetitions of dolomite-anhydrite in the geological column are interpreted in terms of arid coastal-plain accretionary processes, forming not only the seal but also the reservoir and the source rocks. Shoaling-upward sequences of carbonates associated with evaporites are extremely common as hydrocarbon traps. In the United States, similar carbonate traps are associated with major oil fields of the Central Basin platform and on the northwest shelf of the Permian basin of Texas and New Mexico, where shelf carbonates inter-finger with updip evaporites and clastics (Ward *et al.*, 1986). This association also occurs within fields of the Williston Basin, where the Madison Limestone/Charles evaporites (Lindsay & Kendall, 1985) and the Ordovician Red River Formation carbonate/evaporite sequence trap and form

hydrocarbon reservoirs (Roehl, 1985). Similar rocks can also be found in the Western Canadian Basin, where the Devonian shoaling-upward carbonates and evaporites (Schmidt *et al.*, 1985) are associated with sequences that are similar to those seen in Abu Dhabi and in Lake MacLeod (Alsharhan & Kendall, 1994). Kenig *et al.* (1989) demonstrated from their studies that this association has some bearing on the occurrence of petroleum in similar sequences in the Middle East and the western United States and Canada (see also Table 3).

PAPERS IN THIS VOLUME

The Abu Dhabi International Conference on "Evaporite Stratigraphy, Structure and Geochemistry, and their Role in Hydrocarbon Exploration and Exploitation" was first held in Abu Dhabi on 12–13 October 2004 and a second conference was held in Abu Dhabi on 7–8 November 2006.

The first conference in 2004 honoured Professor Douglas Shearman's outstanding research contributions to unravelling the causes of the character of the sabkha evaporites of the United Arab Emirates and his recognition that these were probably the analogues for many similar ancient evaporite sequences. The second conference honoured

Professor Bruce Purser's outstanding research contributions to our current understanding of the geology of the Holocene evaporites and carbonates of the Arabian Gulf.

TOPICS

The topics reflecting the theme of the two conferences have been divided into six parts:

(1) "An Abu Dhabi retrospective on contributions from the 1960s and 1970s, their impact on our current thinking on evaporites and their association with hydrocarbon exploration and exploitation."
(2) "Evaporite stratigraphic signals of base-level change in the geological record" with contributions relating these to hydrocarbon exploration and exploitation in the Infracambrian, Permian, Jurassic and Tertiary of the Arabian Plate, the North Sea Zechstein, the Mediterranean Messinian salinity crisis, the Western Canadian Devonian, the Permian Basin of West Texas and New Mexico, the Paradox Basin of the Four Corners area, the Sverdrup Basin, and the Mesozoic evaporites of the early Atlantic break up.
(3) "Tectonic response of evaporites to burial and lateral compression" with contributions related to hydrocarbon exploration and exploitation in the Middle East, the Gulf of Suez, offshore Gabon, Equatorial Guinea, Brazil, the Gulf of Mexico, offshore Senegal through Morocco, and Pakistan.
(4) "Geochemical controls on evaporites and associated dolomitization of carbonates" as related to their contribution to the prediction of porosity and its role in hydrocarbon exploration and exploitation.
(5) "Effect of evaporites on reservoir quality and fluid flow" with contributions related to oil exploration and exploitation associated with salt plug and diapir examples of the Arabian Gulf, and world-wide.
(6) "Evaporite controversies: evolution of evaporites in time and space" as related to their contribution to hydrocarbon exploration and exploitation.

The papers presented in this volume are dedicated to our friend the late Professor Douglas Shearman, who is best remembered for his work on evaporites from the Holocene to the ancient. Shearman was an enthusiastic scientist who inspired many geologists to examine the natural world around them and establish the origins of the evaporites they observed.

OVERVIEW OF CONTRIBUTIONS

The papers in this volume are from the first and second evaporites conferences held in Abu Dhabi in 2004 and 2006 respectively. Selected papers from these two meetings were chosen to reflect the themes of the two conferences. In the dedication, *Evans* enthusiastically recalls the contribution of Douglas Shearman to the field of evaporite geology. Shearman's life is described and much of his bibliography is listed, providing a valuable resource for future evaporite geologists.

This volume is divided into three parts, with the first focused on the Holocene carbonate-evaporite sequences and associated sediments, the second deals with geochemistry, and the third part examines their ancient analogues. The reader will note that the individual papers are broadly categorized based on the themes of the two conferences. A brief summary of the highlights from individual papers is as follows:

Kendall & Alsharhan review the geomorphology of the sedimentary settings formed along the coast of the UAE. These are traced from the high-energy seaward reefs that rim the Pearl Banks to the ooid barrier island shoals, to inner shelf coastal terraces with grapestones and cyanobacterial mats. It is described how these sediments relate to the evaporite-rich protected coast. In the second contribution, *Evans* tracks the evolution of the Quaternary geological record of the Arabian Gulf region. Emphasis is placed on the influence of sea-level variations on the occurrence of aeolian sediments in the area. *Park* ties the updip sabkha cycle of Abu Dhabi to the cyclic character of many shoaling upward tidal flat cycles. This cyclicity is related to variations in sea-level that cause the carbonate/evaporites to onlap onto the updip portions of a ramp.

Gunatilaka describes how stable isotopes can be used to track the evolution of the sabkha evaporites of the Arabian Gulf region, from Kuwait to the UAE. It is shown how the sabkha sedimentary cycles are a relatively recent phenomenon, related to an increase in the aridity of the local climate in the last 5000 years and its close association with the sea. Earlier sedimentary cycles reflect a cooler wetter climate when fresher waters flushed the marine brines seaward.

Shinn describes carbonate beaches, spits and channelized sabkhas nurtured by longshore currents on the coast of Qatar that favour evaporite and carbonate sedimentation. Coastal spits are shown to have accreted at a remarkable rate since the mid-1960s by periodic beach/sabkha complex "jumps". Rapid beachrock formation is noted on the low-energy landward side of exposed linear ridges, while the seaward side of the cheniers mix accretion and erosion. It is suggested that this sedimentary complex could serve as a model for exploration, and guide production drilling where similar ancient sabkha and evaporitic settings occur. *Strohmenger et al.* provide a detailed description of the assemblages of shallow-water carbonates, evaporites and siliciclastics from a traverse of the sabkha near Al Rufayq Island in Abu Dhabi. Numerous spectacular illustrations of trenches cut in this coastal sabkha demonstrate the shoaling-upward character of the stacking patterns.

Mettraux et al. describe the formation of the Bar Al Hikman sabkhas on the east coast of Oman in an extremely arid climate with high evaporation. These sabkhas accumulate over a complex geological substrate in which recent sea-level falls and structural movement have caused the emergence of marine to lagoonal carbonates in which evaporites form. Here, the sabkhas are grouped into "coastal sabkhas", "continental sabkhas" and older sabkha deposits preserved as terraces. The movement of groundwater along faults and fractures is linked by the authors to early lithification of Quaternary sediments. *Kendall & Alsharhan* record the occurrence of cyanobacteria coating and infesting carbonate grains along the coast of the UAE. It is argued that these cyanobacteria are a major cause of micritization of the carbonate grains.

Epps recognizes the current building trends on Abu Dhabi Island and its vicinity, and describes how engineers have been testing the physical properties and strength of the sediments in these coastal areas. The geotechnical properties summarised in this paper will serve as an important reference for construction engineers working in the region.

Wood reviews current models of dolomitization and considers a new mechanism that is based on regional groundwater models. These models are used to explain the current high magnesium content of the groundwaters of the Eastern Arabia and how this impacts on the coastal dolomitization. *Qabazard et al.* describe the organic matter associated with the shallow-water carbonates that accumulate on the areas behind the Al Khiran barrier island of Kuwait. The preservation of this organic matter is explained by its occurrence in a protected reducing environment.

Kirkham offers an unconventional explanation for the occurrence of bands of carbonate and evaporite found on the sabkha surface of Abu Dhabi. It is proposed that these are the products of deposition from lagoonal waters that have flooded over the sabkha surface of Abu Dhabi. This in contrast to most previous authors who have ascribed storm washover onto a supratidal surface as the source of widespread halite precipitation, but it has not generally been regarded as the source of anhydrite banding.

Kenig describes the occurrence of cyanobacterial mats at the margin of the inner coast of the United Arab Emirates. A series of man-made excavations that form canals in Abu Dhabi have been used to trace the mats from their formation on the early Holocene transgressive surface, to where they currently occur within the regressive coastal sediments. The descriptions provided will assist sedimentologists in their interpretation of similar sediments that occur in the geological record, while the associated data should help establish the source rock potential of carbonate tidal-flat sediments.

Wright & Kirkham explain the occurrence of enigmatic carbonate fabrics that exhibit a variety of exotic forms. These patterned fabrics are ascribed to the replacement of evaporites by carbonates. *Warren* considers how periodic influxes of freshwater into brine water-bodies induce blooms of cyanobacteria in the water column. It is argued that, should this organic matter be preserved, it has a good chance of maturing to become the source rock for some hydrocarbon fields.

Costa et al. describe the extension and compression associated with the salt tectonics of passive margin basins. A series of physical models are reviewed that explain these features. These examples should serve to help geologists making interpretations of the fabrics formed by the massive halite often associated with the initial phases of continental pull apart. *Hafid et al.*

describe the tectonic features found in the disturbed and distorted salt of the Essaouira Basin of western Morocco. These features show many of the fabrics found in other salt basins, while their character within the near-surface seismic sections provides excellent analogues for the interpretation of similar features found elsewhere in the world.

Al-Suwaidi et al. describe the Arab sabkha cycles found in the offshore Jurassic petroleum reservoir facies using seismic cross-sections and well logs. It is shown how the sediments of these cycles change their character from west to east and seaward, and how the hydrocarbons are trapped at the updip pinch-out of the grain carbonates beneath sabkha evaporites. *Orti* describes gypsum crystals that form in many settings. The occurrence and character of gypsum is reviewed, with a particular focus on the selenite fabrics of Sicily, Italy, Spain and Poland. The information presented should help both the novitiate and evaporite expert to better understand other selenite fabrics in the geological record.

ACKNOWLEDGEMENTS

We would like to thank Abu Dhabi National Oil Company (ADNOC), Abu Dhabi Marine Operating Company (ADMA-OPCO), Abu Dhabi Company for Onshore Oil Operations (ADCO) and Zakum Development Company (ZADCO) for their support and encouragement for the conferences of 2004 and 2006. We extend our sincere thanks to the authors of the papers in this volume for their interest and timely contributions to this special publication. We would like to thank all reviewers, whose critical revisions of the manuscripts assured us of the high quality and comprehensive compilations expressed by this volume. We greatly appreciate the effort of Mr M. Shahid, who processed the chapters for this volume from inception to final completion, incorporated the authors' changes and handled all correspondences with authors and reviewers.

REFERENCES

Alsharhan, A.S. (1989) Petroleum geology of the United Arab Emirates. *J. Petrol. Geol.*, **12**, 253–288.

Alsharhan, A.S. and Kendall, C.G.St.C. (1986) Precambrian to Jurassic rocks of the Arabian Gulf and adjacent areas: their facies, depositional setting and hydrocarbon habitat. *AAPG Bull.*, **70**, 977–1002.

Alsharhan, A.S. and Kendall, C.G.St.C. (1994) Depositional setting of the Upper Jurassic Hith Anhydrite of the Arabian Gulf: An analog to Holocene evaporites of the United Arab Emirates and Lake MacLeod of Western Australia. *AAPG Bull.*, **78**, 1075–1096.

Alsharhan, A.S. and Kendall, C.G.St.C. (2003) Holocene coastal carbonates and evaporites of the southern Arabian Gulf and their ancient analogues. *Earth-Sci. Rev.*, **61**, 191–243.

Ayres, M.G., Bilal, M., Jones, L.R.W., Slenz, W., Tartir, M. and Wilson, A.O. (1982) Hydrocarbon habitat in main producing areas, Saudi Arabia. *AAPG Bull.*, **66**, 1–9.

Clement, J.H. (1985) Depositional sequences and characteristics of Ordovician Red River reservoirs, Pennel field, Williston basin, Montana. In: *Carbonate Petroleum Reservoirs* (Eds P.O. Roehl and P.W. Choquette), pp. 71–84. Springer-Verlag, New York.

Depowski, S. and Peryt, T.M. (1985) Carbonate petroleum reservoirs in the Permian dolomites of the Zechstein, Fore-Sudetic area, western Poland. In: *Carbonate Petroleum Reservoirs* (Eds P.O. Roehl and P.W. Choquette), pp. 251–264. Springer-Verlag, New York.

Druckman, Y. and Moore, C.H., Jr., (1985) Late subsurface secondary porosity in a Jurassic grainstone reservoir, Smackover Formation, Mt. Vernon field, southern Arkansas. In: *Carbonate Petroleum Reservoirs* (Eds P.O. Roehl and P.W. Choquette), pp. 369–384. Springer-Verlag, New York.

Emery, K.O. (1956) Sediments and water of the Persian Gulf. *AAPG Bull.*, **40**, 2354–2383.

Feazel, C.T. (1985) Diagenesis of Jurassic grainstone reservoirs in the Smackover Formation Chatoim field, Alabama. In: *Carbonate Petroleum Reservoirs* (Eds P.O. Roehl and P.W. Choquette), pp. 357–368. Springer-Verlag, New York.

Halley, R.B. (1985) Setting and geologic summary of the Lower Cretaceous, Sunniland field, southern Florida. In: *Carbonate Petroleum Reservoirs* (Eds P.O. Roehl and P.W. Choquette), pp. 443–454. Springer-Verlag, New York.

Houbolt, J.J.H.C. (1957) *Surface Sediments of the Persian Gulf near the Qatar Peninsula.* Mouton & Co., The Hague, *113 pp*

Kassler, P. (1973) The structural and geomorphic evolution of the Persian Gulf. In: *The Persian Gulf: Holocene Carbonate Sedimentation and Diagenesis in a Shallow Epicontinental Sea* (Ed. B.H. Purser), pp. 11–32. Springer-Verlag, Berlin.

Kenig, F., Huc, A.Y., Purser, B.H. and Oudin, J.L. (1989) Sedimentation, distribution and diagenesis of organic matter in a Recent carbonate environment Abu Dhabi, U.A.E. *Org. Geochem.*, **16**, 735–745.

Lindsay, R.F. and Kendall, C.G.St.C. (1985) Depositional facies, diagenesis, and reservoir character of Mississippian cyclic carbonates in the Mission Canyon Formation, Little Knife field, Williston Basin, North Dakota. In: *Carbonate Petroleum Reservoirs* (Eds P.O. Roehl and P.W. Choquette), pp. 175–190. Springer-Verlag, New York.

Loucks, R.G. and Anderson, J.H. (1985) Depositional facies, diagenetic terranes and porosity development in Lower Ordovician Ellenberger Dolomite, Puckett field, West Texas. In: *Carbonate Petroleum Reservoirs* (Eds P.O.

Roehl and P.W. Choquette), pp. 19–38. Springer-Verlag, New York.

McQillan, H. (1985) Fracture controlled production from the Oligo-Miocene Asmari Formation in Gachsaran and Bibi Hakimeh fields, southwest Iran. In: *Carbonate Petroleum Reservoirs* (Eds P.O. Roehl and P.W. Choquette), pp. 161–174. Springer-Verlag, New York.

Miller, J.A. (1985) Depositional and reservoir facies of the Mississippian Leadville Formation, Northwest Lisbon field, Utah. In: *Carbonate Petroleum Reservoirs* (Eds P.O. Roehl and P.W. Choquette), pp. 161–174. Springer-Verlag, New York.

Mitchell, J.C., Lehmann, P.J., Cantrell, I.A., Al-Jallal, I.A. and **Al Thagafy, M.A.R.** (1988) Lithofacies, diagenesis and depositional sequence; Arab-D Member, Ghawar Field, Saudi Arabia. In: *Giant Oil and Gasfields* (Eds A.J. Lomando and RM. Harris). *SEPM Core Workshop*, **12**, 459–514.

Murris, R.J. (1980) Middle East: Stratigraphic evolution and oil habitat. *AAPG Bull.*, **64**, 597–618.

Qatar General Petroleum, Corp., and **Amoco-Qatar Petroleum, Co.,** (1991) Dukhan Field - Qatar, Arabian Platform. In: *Structural Traps V.* (Eds N.H. Foster and E.A. Beaumont) *AAPG Treatise Petrol. Geol., Atlas of Oil and Gas Fields*, pp. 103–120.

Roehl, P.O. (1985) Depositional and diagenetic controls on reservoir rock development and petrophysics in Silurian tidalites, interlake formation, Cabin Creek field area, Montana. In: *Carbonate Petroleum Reservoirs* (Eds P.O. Roehl and P.W. Choquette), pp. 85–105. Springer-Verlag, New York.

Ruzyla, K. and **Friedman, G.M.** (1985) Factors controlling porosity in dolomite reservoirs of the Ordovician Red River Formation, Cabin Creek field, Montana. In: *Carbonate Petroleum Reservoirs* (Eds P.O. Roehl and P.W. Choquette), pp. 39–58. Springer-Verlag, New York.

Schmidt, V., McIlreath, I.A. and **Budwill, A.E.** (1985) Origin and diagenesis of Middle Devonian pinnacle reefs encased in evaporites, "A" and "E" pools, Rainbow field, Alberta. In: *Carbonate Petroleum Reservoirs* (Eds P.O. Roehl and P.W. Choquette), pp. 141–160. Springer-Verlag, New York.

Wagner, C.W. and **Van der Togt, C.** (1973) Holocene sediment types and their distribution in the southern Persian Gulf. In: *The Persian Gulf: Holocene Carbonate Sedimentation and Diagenesis in a Shallow Epicontinental Sea* (Ed. B.H. Purser), pp. 123–156. Springer-Verlag, New York, Heidelberg, Berlin.

Ward, R.F., Kendall, C.G.St.C. and **Harris, P.M.** (1986) Late Permian (Guadalupian) facies and their association with hydrocarbons. *AAPG Bull.*, **70**, 239–262.

Weijermars, R. (1999) Quaternary evolution of Dawahat Zulum (Half Moon Bay) region of Eastern Province, *Saudi Arabia. GeoArabia*, **4**, 71–90.

Wilson, A.O. (1985) Depositional and diagenetic facies in the Jurassic Arab-C and -D reservoirs, Qatif Field, Saudi Arabia. In: *Carbonate Petroleum Reservoirs* (Eds P.O. Roehl and P.W. Choquette), pp. 319–340. Springer-Verlag, New York.

Wood, G.V. and **Wolfe, M.H.** (1969) Sabkha cycles in the Arab/Darb Formation off Trucial Coast of Arabia. *Sedimentology*, **12**, 165–192.

Part 1

Recent carbonate and evaporite sediments

Int. Assoc. Sedimentol. Spec. Publ. (2011) **43**, 11–44

An historical review of the Quaternary sedimentology of the Gulf (Arabian/Persian Gulf) and its geological impact

GRAHAM EVANS

Department of Ocean and Earth Sciences, National Oceanography Centre, University of Southampton Waterfront Campus, European Way, Southampton, SO14 3ZH (E-mail: grahamandrosemary@googlemail.com)

ABSTRACT

The Arabian Gulf is a foreland basin that lies between the Arabian Shield and the Zagros fold belt. Today, it is being infilled at its northern head by the Tigris-Euphrates-Karun delta, receives further detrital sediment from Iran on its NE flank, and along its Arabian flank is the site of carbonate and evaporite sedimentation. Other than a few preliminary surveys by officers of the Indian Geological Survey and other minor geological studies associated with archaeological excavations on the Mesopotamian Plains, prior to the 1950s the Quaternary deposits of the region received little attention. However, since the publication of pioneering sedimentological studies in the late 1950s and early 1960s, these sediments have attracted considerable interest. Today, the Arabian Gulf is quoted as a model for foreland basin sedimentation, carbonate evaporite-dune-fan associations and carbonate ramps. It has become a classic area for open-water spontaneous precipitation of calcium carbonate (whitings), formation of hardgrounds, ooid production, dolomite precipitation and, most famously, for the development of shallow-water carbonate-evaporite associations. The surprise discovery of anhydrite forming in the supratidal coastal plain sediments and not in the subaqueous environment of the Gulf had a profound effect on interpretations of ancient evaporites. This discovery caused considerable re-examination and reconsideration of these deposits.

Keywords: Arabian Gulf, carbonates, dolomitization, evaporites, anhydrite, sabkha.

INTRODUCTION

Until the 1950s the Gulf, otherwise known as the Persian Gulf or the Arabian Gulf, was an area which was almost always discussed with reference to its oil-rich rocks. However, a few geologists of the Indian Geological Survey (e.g. Pilgrim, 1908), who worked in this area, studied the Quaternary deposits; notably the wind-blown carbonates (the so-called "miliolite"), and a series of marine carbonates that either cap the former or lie on erosional terraces cut into the older rocks of the Arabian and Iranian shorelines and the intervening islands. Others carried out rather cursory examinations of the delta of the Tigris-Euphrates-Karun rivers at the head of the Gulf, where they were particularly concerned with the historical development of the Mesopotamian plains and their relationship to the famous archaeological sites such as those of Ur and Eridu. Except for the results of bottom soundings and sampling by the British Royal Navy and some fishery studies (Belvgad, 1944), little was known about the floor of the Gulf and its coastal environment and their associated deposits.

In the early 1950s and 1960s, several geologists (Emery, 1956; Houbolt, 1957; Sugden, 1963a,b 1966) began work on the modern deposits of the Gulf. This was either as a result of ancillary studies to their company work, when an opportunity arose to sample the seafloor in areas of marine petroleum exploration, or when a reasonable cover of samples were made available by hydrographic surveys of various navies. Emery (1956) published the first significant paper dealing with the oceanography, offshore topography and sediments. He stressed the importance of climate in the development of basinal sediments and compared the contemporary deposits with earlier oil-bearing deposits of the area. He also made the very perceptive observation that:

"Study of the present conditions of sedimentation in the Arabian Gulf thus may be useful in interpreting the older sediments of the geosyncline" (Emery, 1956, p. 2354).

This indeed proved to be true, as witnessed by subsequent studies that bear out this relationship. The paper initially seemed to have aroused little attention, except for a discussion correcting minor geological details and agreeing that perhaps the modern deposits have more similarities to those of the middle Fars, or even younger rocks (mid-Pliocene), than to the Mesozoic and Tertiary deposits in which abundant source rocks of the region's extensive oil deposits occur (Elder, 1957). Subsequent work has shown that Emery's predictions were correct, particularly as regards the shallow-water carbonates.

The work of Emery (1956), Houbolt (1957) and Sugden (1963a,b, 1966) aroused a great deal of interest in the area, undoubtedly spurred on by the wealth of fascinating new information on carbonate sediments that was emerging from the Shell Oil Company research group studies of Ginsburg and others in Florida (Ginsburg, 1953, 1956, 1957), and the pioneering work of Illing on the Bahamas (Illing, 1954). This interest gained momentum over the years and even today, more than fifty years after the original studies, there are numerous workers pursuing valuable studies on the Quaternary sedimentation of the Arabian Gulf that are of widespread geological significance, as well as others that are of importance for understanding the evolution of the landscape and archaeology of the Mesopotamian plains.

An attempt is made in this contribution to review the progress of this work whilst stressing the important discoveries and their significance. This is best done by dealing in turn with the various aspects of these studies rather than giving a general chronological account. General reviews of the Quaternary sedimentation in the Arabian Gulf have been given by Emery (1956), Evans (1966, 1995), Seibold & Vollbrecht (1969), Purser & Seibold (1973), Seibold et al. (1973), Gunatilaka (1986), Uchupi et al. (1996, 1999), and particularly of the UAE coastline by Evans (1995), Friedman (1995), Kirkham (1998b), Alsharhan & Kendall (2003) and Evans & Kirkham (2005). A collection of papers compiled by Purser (1973a) provided a very comprehensive survey of many aspects of Quaternary sedimentation in the Arabian Gulf, as was known up to that date. Lambeck (1996) gave a detailed modern review of sea-level changes in the area, in which a model of the glacio-hydro-isostatic effects was compared with previous observations and shoreline reconstructions. The ecosystems of the Arabian Gulf, including a discussion of the oceanography and faunas, were summarised by Khan et al. (2002).

THE TIGRIS-EUPHRATES-KARUN DELTA

Earlier writers, including Becke (1835) and Loftus (1855), discussed the progradation of the Tigris-Euphrates-Karun delta into the Gulf. Becke (1835), basing his conclusions on historical work, suggested that the delta had prograded approximately 640 km into the Gulf since late prehistoric times (i.e. since the famous drowning of the plains of Mesopotamia during "the Deluge" or "Flood" of Sumerian legend). This was disputed by Carter (1834, 1835) who, in contrast, considered that there had been little change in the location of the coastline during the historical period. However, Ainsworth (1838), who carried out the first geological reconnaissance work in the area, suggested that there had been approximately 112 km of progradation since "the Deluge", i.e. between approximately 5000–4000 years ago.

Later, de Morgan (1900), using historical texts and the earlier literature, presented a map showing a former shoreline that lay approximately 200 km landward of the present coastline (Anon, 1944). This latter view seems to have been accepted by most archaeologists and others interested in the delta (de Morgan 1900). However, Lees & Falcon (1952) astonished the archaeologists and historians by claiming, using evidence from boreholes, topographic maps, aerial photos and some fieldwork, that the coastline at the head of the Arabian Gulf had hardly changed its position during the Holocene. Instead, they suggested that subsidence had accommodated the sediment delivered by the Tigris, Euphrates and Karun rivers. Due to the eminence of these two authors, this view was generally accepted. For example, Mitchell (1957) seemed to agree with their surprising conclusions. Hudson et al. (1957) described, with some lithological details, the fauna of sediments from some boreholes, and also claimed that their results agreed with the conclusions of their superiors, Lees & Falcon (1952). Later, Larsen (1975), Larsen & Evans (1978) and Evans (1979), whilst accepting the importance of local tectonics, refuted the analysis of Lees & Falcon (1952) and argued that a marine transgression, the Holocene transgression, had resulted in the inundation of the Mesopotamian plains. Subsequent progradation of approximately 150–200 km, with possibly some aggradation, had deposited a wedge of sediment, the so-called Hammar Formation, which underlies much of the lower Mesopotamian plain (Fig. 1A).

Larsen (1975), Larsen & Evans (1978) and Evans (1979) drew attention to a radiocarbon date produced from a freshwater peat in a borehole at Fao (Godwin *et al.*,1958; Godwin & Willis, 1959) that falls on the general eustatic sea-level curve, rather than below, as is common with dates from

subsiding areas (Fig. 1B). If it is accepted that the peat was unlikely to have formed much above contemporary sea level, it indicates that the area shows little evidence of subsidence. Furthermore, Kassler (1973), in a review of the tectonics and Quaternary evolution of the Arabian Gulf, concluded that there had been neither tectonic activity in the proximity of the delta nor along the Arabian Coast, except for local changes around Bahrain, other islands and parts of the Saudi Arabian coast,

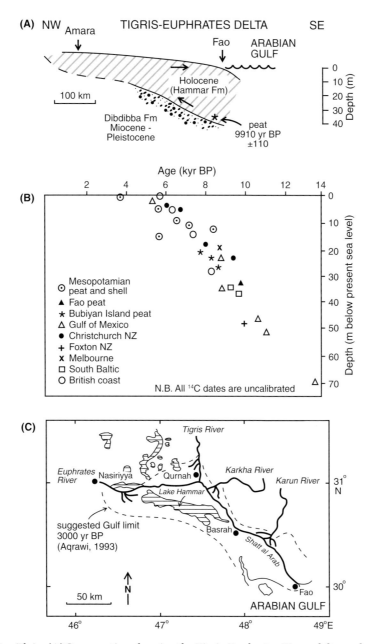

Fig. 1. The Mesopotamian Plain. (A) Cross section showing the Tigris-Euphrates-Karun delta wedge (modified from Larsen & Evans, 1978). (B) Radiocarbon dates from Al-Zamel (1983) showing the local eustatic sea-level curve, with additions from Aqrawi (2001b). (C) Proposed former extent of the Arabian Gulf (after Aqrawi, 1993b).

a view supported by McClure & Vita-Finzi (1982). This stability of the Arabian coast, contrasted markedly with the changes on the Iranian coast illustrated by Vita-Finzi (1982).

MacFayden & Vita-Finzi (1978) re-evaluated the fauna found in the boreholes of Hudson *et al.* (1957) and agreed that there was little evidence for any major subsidence in the area. Cooke (1987) gave a reconstruction of the Mesopotamian shoreline over the Holocene and suggested that the coastline around 5000 yrs ago lay approximately 400 km inland of its present position. Sanlaville (1989) presented a map giving a similar landward maximum position at the shoreline. Later, Aqrawi (1993b) produced a map of the former position of the shoreline and suggested that it lay approximately 200 km from the present coast around 3000 years ago (Fig. 1C).

Very little has been written about the submarine parts of the Tigris-Euphrates-Karun delta except for a paper by Off (1963), who described a series of ridges that appear to be formed by strong tidal currents and are similar to such tidal ridges in the North Sea. A short note on the bar at the river mouth was also written by Wilson (1925). Subsequent studies of the geomorphology, hydrology and a reconnaissance of the sediments of the Mesopotamian delta plain were made by Rzoska (1980), Baltzer *et al.* (1982), Purser *et al.* (1982) and Baltzer & Purser (1990). Al-Zamel (1983) described the sediments of Bubiyan Island on the southwest flanks of the delta and Hilmy *et al.* (1971), Khalaf & Ala (1980), Khalaf *et al.* (1982, 1984), Al-Ghadban (1990) and Aqrawi (1994) described the marly sediments of Kuwait Bay, a delta flank depression, whose composition proved to be similar to that of the load of the Tigris-Euphrates except for the addition of high contents of the diagenetic clay palygorskite. Later, Aqrawi & Evans (1994) described the sediments of Lake Hammar (Fig. 1C) and remarked on the poor preservation of organic matter in the sediments despite the abundance of reed beds and the high organic productivity of the area. This was attributed to the abundant sunlight and high temperatures that prevented the formation of peat described from the earlier Holocene sediments. However, even these are better described as organic-rich muds rather than true peats (Aqrawi, 2001a).

Aqrawi (1993a, 1995b) suggested that palygorskite, as well as being partly supplied by the rivers and wind, also grew authigenically in the sediments together with gypsum and dolomite. It was suggested that authigenic (low-calcium) dolomite formed in hypersaline settings, high-Mg calcite in brackish-marine environments, and low-Mg calcite in lacustrine settings on the delta plain. It was claimed that the association of dolomite and pyrite indicate that sulphate reduction may have been the important factor in the formation of the former. Later, Aqrawi (1995a) discussed the importance of compaction in the consideration of rates of sedimentation of the Holocene sediments, and also the stratigraphic signature of Holocene climate changes in the deltaic sediments (Aqrawi, 2001b).

Early estimates of the fluvial supply to the Gulf were made by Rawlinson (reported in Pilgrim, 1908), Ionides (1937, 1955), Phillip (1968) and Berry *et al.* (1970). The last of these authors showed that the suspended sediment found in the Shatt al Arab, where the Tigris and Euphrates join one another below Qurna (Fig. 1C), is dominated by the sediment delivered by the Tigris. Milliman & Meade (1983), in their review of the sediment loads of the world's rivers, estimated that the combined Tigris-Euphrates sediment discharge to the Gulf is approximately $50-100 \times 10^6$ tons each year.

The large contribution of sediment by wind to the delta, and indeed to the adjacent Gulf, was noticed by earlier workers (e.g. Emery, 1956) and was studied in detail by Kukal & Saadallah (1970, 1973), Khalaf *et al.* (1979), Khalaf & Al-Hashash (1983) and Foda *et al.* (1985). The mean monthly dust fall out in 1970–1975 in neighbouring Kuwait was at a maximum in summer (May, 224.7 tonnes km^{-2}) and a minimum in winter (February, 20.2 tonnes km^{-2}). In 1979–1980, dust falls as high 1002.7 tonnes km^{-2} were recorded in July. Unfortunately, there are at present no available estimates of the total annual supply of aeolian sediments to the Arabian Gulf. Details of the mineralogy of the fluvial sediments were given by Phillip (1968), Berry *et al.* (1970) and Aqrawi (1993a), while that of the aeolian sediment by Khalaf *et al.* (1979) and Aqrawi (1993a). All of these studies showed that both fluvial and aeolian sediment are rich in both calcite and dolomite.

The possible relationship between the rise of sea-level during the Holocene transgression and the so called "Noah's Flood" or "the Deluge" of Sumerian legend has been discussed by various writers (Cooke, 1987; Lambeck, 1996; Teller *et al.*, 2000). A very useful and comprehensive review of the geological and archaeological development of Mesopotamia was given by Kennet & Kennet (2006). Clearly, the low fluvial plains of Mesopotamia must have been affected by the rise of Holocene sea-level to a point slightly higher than that of the present at around 5000–4000 yrs ago, which has been so widely recorded in the Gulf. It is tempting to correlate the transgression, and the associated river floods over the adjacent fluvial plains that would have accompanied it, with the floods of the Sumerian legends. However, until more detailed sedimentological and palaeontological studies are undertaken on the deltaic plain, particularly in the vicinity of Ur, it remains nothing more than an interesting speculation. Surprisingly, there appears to have been no geological study of Woolley's "flood layer" (Woolley, 1929); for example, whether it is a marine or a fluvial crevasse deposit possibly induced by a rise in sea level.

THE IRANIAN TROUGH AND ITS SEDIMENTS

Emery (1956) showed that marly sediments dominate the deeper parts of the Arabian Gulf bordering Iran (Fig. 2). Further details of these sediments were provided by Evans (1966), who also remarked on the thinness of the sediment cover and the presence of relict shallow-water sediments, such as oolites, with small admixtures of marl (i.e. which would now more accurately be termed palimpsest sediments). Pilkey & Noble (1966) showed that the carbonate fraction is dominated by low-Mg calcite and detrital wind-blown dolomite, and that the distribution parallels the trend of the Gulf. Additionally, it was demonstrated that the clay minerals consist of illite, montmorillonite mixed-layered clays, chlorite and kaolinite, and indicate that the Iranian coast is the dominant source area, with some particular point sources producing lobes into deeper waters. A more complete description of the sediments was given by Wagner & Van der Togt (1973).

A detailed sampling programme was undertaken in 1960 by a German group from Kiel under the leadership of Eugene Seibold, using the German

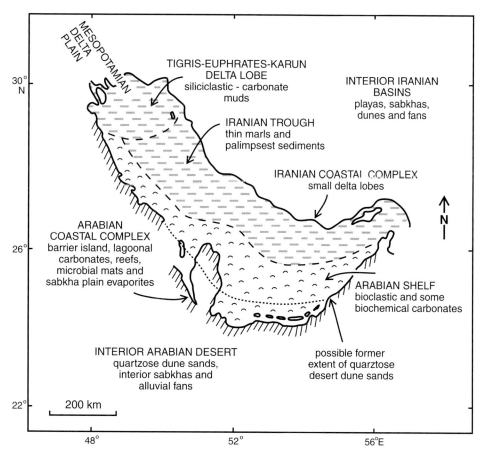

Fig. 2. Schematic sketch showing the distribution of the main sedimentary facies of the Arabian Gulf (modified from Evans, 1995).

research ship Meteor II (Dietrich *et al.*,1966). A review of the sedimentology and evolution of the area, as known at that time, was presented by Seibold & Volbrecht (1969) and Seibold & Ulrich (1970). Martini (1967) showed that oceanic nanno-plankton extended as far north as the tip of the Qatar peninsula and that reworked microfauna from the Cretaceous and Tertiary rocks were abundant in the marly sediments. The presence of glauconite in the relict sediments was discussed by Lange & Sarnthein (1970).

The detailed composition of the Gulf marls was described by Sarnthein (1970, 1971) and later by Lange (1970), Melguen (1971, 1973) and Sarnthein & Walger (1973), who supported the conclusion of Pilkey & Noble (1966) that Iranian rivers were the main suppliers of detrital sediment to the area, except in the extreme northwest, where the Tigris-Euphrates were dominant (Fig. 3). Esteoule *et al.* (1970) showed that palygorskite was supplied by Iranian rivers, as had been described from the Mesopotamian rivers, and that it was the load

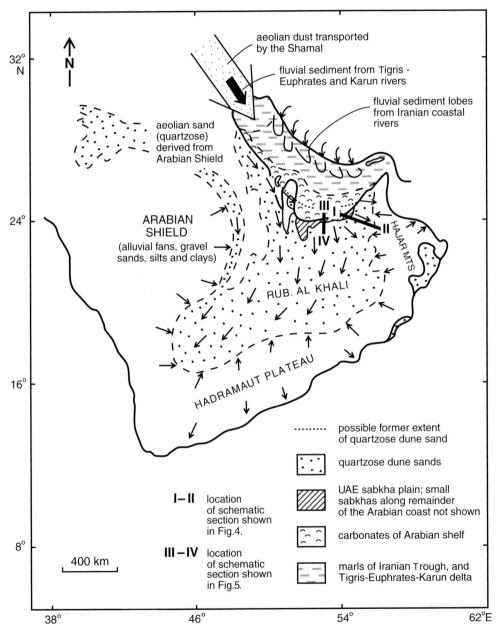

Fig. 3. Schematic sketch showing the pattern of sediment dispersal in the Arabian Gulf and adjacent land areas. Locations of cross-sections in Figs Fig. 4 (I–II) and Fig. 5 (III–IV) are shown.

of the latter rivers plus organic matter that together influenced the geochemistry of the sediments. Studies were made of the larger foraminifera by Lutze *et al.* (1971). Hoefs & Sarnthein (1971) showed that the oxygen stable-isotope ratios of pteropods faithfully reflected the average salinity and temperature conditions of the Gulf waters.

Sarnthein (1970) was able to demonstrate, using the coarse fraction of the sediments, that the effective wave base in the Gulf was 40 m near the islands and on the flanks of the shoals. Diester-Hass (1971, 1972a,b, 1973) studied the vertical distribution of the coarse fraction of the marls in cores and distinguished several horizons representative of different climate zones during the Holocene since 9000 years ago. Sarnthein (1971, 1972) confirmed the presence of relict sediments (i.e. palimpsest sediments) and reiterated the thinness of the marl, as noted by earlier writers (Emery, 1956; Evans, 1966). A first attempt was made to reconstruct the conditions during the Holocene transgression (between 18,000–7000 years ago) and it was suggested that this was intermittent, with periods of still-stand at levels around 64–61 m, 53–40 m and possibly at 30 m below present sea level. Seibold *et al.* (1973) gave an excellent review of the German work and suggested that the deposition of the sediments of the trough along the Iranian side of the Gulf was a useful model for marine marly molasse (see also Seibold, 1970). Sarnthein's studies (*op. cit.*) were amplified later by Stoffers & Ross (1979) of the Woods Hole Institute, USA, who produced a more detailed discussion of the late Pleistocene and Holocene history of the deeper parts of the Gulf using longer cores.

Uchupi *et al.* (1996), using high-resolution echo-soundings, described in detail the micro-topography of the Gulf floor and discussed the presence of a network of pockmarks considered to have been produced by the escape of water or of thermogenic gas. The latter was thought more likely, and the authors quoted examples of abundant gas seeps in the Gulf (Hoveland & Judd, 1988). However, water seepages are also widespread and were used at one point by pearl divers, in former days, as a source of drinking water. Further work (Uchupi *et al.* 1999) indicated that prior to approximately 22,800 years ago both carbonates and terrigenous muds from the Iranian hinterland accumulated on the floor of the deeper parts of the Gulf. These were later buried during the Holocene transgression by more mud and then by shallow-water carbonates including aragonitic muds. Finally, with a change in climate

after 9000 years ago, the carbonates were cloaked by lobes of terrigenous mud extending out from the Iranian coast that, together with some aeolian sediment and the autochthonous carbonate, has produced the present-day discontinuous cover of marl. It was suggested that this fluvial supply is now rather limited in comparison with that of the earlier Holocene (Uchupi *et al.* 1999).

THE ARABIAN SHELF

Emery's (1956) general survey was refined by Houbolt's (1957) study of the shelf region offshore of Qatar using samples collected over the Shell Offshore Concession. Houbolt (1957) gave a detailed discussion of the shallow shelf carbonates and stressed how the Shamal (the strong NW wind) produced waves which spread shallow-water detritus from isolated island highs to distances of up to 20 km in the downwind direction. Detailed studies by the Shell group extended the work of Houbolt (1957). It was shown that the Arabian Shelf was covered with a complicated patchwork of skeletal carbonate sands, muddy sands and muds, reflecting the variations in the underlying topography, with the coarser sediments on the highs and the muddier sediment infilling the depressions (Wagner & Van der Togt, 1973; Hughes-Clarke & Keij, 1973). Sedimentation on the highs was described in detail by Purser (1973) who again stressed the importance of the Shamal-generated waves on the dispersal of sediments, as well as on the distribution of sediment facies on the highs themselves.

Peculiar clouds of milky water had been reported in the 19th-century by Royal Navy hydrographic vessels in both the West Indies and off the Saudi Arabian coast. The first documented sighting was in 1873 off Dharhan (Commander Desmond Scott, Hydrographic Officer, pers. comm.). Similar features from the Bahaman Banks were described later by Cloud (1962), known as "whitings", who suggested that these were formed by spontaneous precipitation of calcium carbonate from seawater. Wells & Illing (1964) investigated similar clouds in the lagoons and off the coastal areas of Qatar. These appeared to have been produced by spontaneous precipitation of aragonite needles in the water and were considered to have been triggered by diatom blooms extracting CO_2 from the surface water. It was suggested that this process was important in the production of carbonate mud on the shelf and in the near-shore environments over much of the Arabian shelf. By contrast, Uchupi et al. (1999) proposed that aragonite precipitation could have been triggered by the release of CO_2 or CH_4 from the seafloor, but this seems unlikely.

de Groot (1965), following up the work of Wells & Illing (1964), showed that, whereas the composition of the mud derived from the whitings was similar to that of the bottom mud, it was different from that produced by artificially-induced precipitation from Arabian Gulf waters. Furthermore, it was shown that the rate of crystallization of aragonite was slow when induced experimentally from Arabian Gulf waters, and did not explain the rapid precipitation reported from the field studies in the Arabian Gulf. There is the possibility that such blooms could have been triggered by the deposition of iron minerals from aeolian dusts, but their patchiness makes this unlikely. Ellis & Milliman (1985), as a result of a study of suspended sediment in the waters of the Gulf, concluded that the material in suspension in the mid- and northern Gulf waters is of aeolian and biogenic origin, or is produced by re-suspension of bottom sediments.

Perhaps the most significant discovery of the offshore work by the Shell group was the evidence of contemporary submarine cementation of the surface sediments on the seafloor down to depths of 30 m, and over an estimated area of $70,000 \, km^2$ (Shinn, 1969, 1971). Generally, there had been considerable scepticism until the 1970s about the reality of submarine cementation, and it had been assumed that cementation had occurred mainly after exposure of sediments to the action of continental groundwaters. However, around that time numerous examples of submarine cementation were being discovered in a variety of environments and depths (Bricker, 1971).

The process of submarine cementation appears to be favoured by relatively low rates of sedimentation and sediment stability, as well as a high permeability of the sediment. The cement is composed of aragonite and high-Mg calcite produced by local inversion from aragonite. There are often several crusts in the superficial sediment overlying soft unconsolidated sediment, and these are obviously of recent origin as they enclose pottery and show radiocarbon ages that range from 460–8000 ^{14}C years BP (uncalibrated). The cemented layers or crusts are often ruptured and showed metre scale (10–40 m) "tepee structures", and are commonly extensively bored and colonized by epifauna. The surfaces of these crusts are modern analogues of the hardgrounds so common in many ancient neritic carbonates, and represent diastems or longer term hiatuses.

COASTAL GEOMORPHOLOGY AND FACIES PATTERNS

Very few details, other than descriptions in the British Admiralty pilots, were available about the bottom sediments and physiography of the nearshore and coastal areas of the Arabian Gulf until the 1960s. Even now the information available for the Saudi Arabian coast is sketchy and fragmentary, and only a few studies have been made of the Iranian shoreline (Baltzer & Purser, 1990).

Evans et al. (1964b) described the general physiography of the Abu Dhabi coastline and Kinsman (1964a,b) partly described the general sediment pattern determined during the initial Imperial College field season undertaken in 1961 by Evans and Kinsman. The coastal complex is remarkably similar in many ways to the barrier-lagoon complex of the Dutch coast (Evans, 1970). On the Abu Dhabi coast, nearshore skeletal sands pass landward into oolitic sands on the beaches and dunes of the barrier islands, and particularly the tidal-deltas where the waters of the lagoons debouche into the open Gulf. These tidal-deltas pass landward into a complex of channels and banks covered by pelletal carbonate sands with muds on the inner ends of the channels. Further landward, broad subtidal and intertidal terraces pass landward into cyanobacterial mats bordered by a relatively featureless coastal plain – the sabkha (Fig. 4). In the Dutch Wadden Sea, siliciclastic-shelf sands pass landward into a barrier and tidal-delta complex composed of siliciclastic sand, and then into a complex of channels and banks and intertidal flats covered with sands and muds, and finally these pass into salt marshes with brackish/fresh marshes and swamps (which now cannot be seen as they have been reclaimed) that border a broad featureless coastal plain (Evans et al. 1964b, 1973; Evans, 1970; Evans & Bush, 1970; Purser & Evans, 1973, Edwards et al., 1986).

Kendall (1966), Skipwith (1966) and Kendall & Skipwith (1969a,b) extended the original Imperial College survey around Abu Dhabi and described the rather different elongated island-capped barrier dominated by reefs and skeletal sands, with local oolitic sands in the west. This barrier

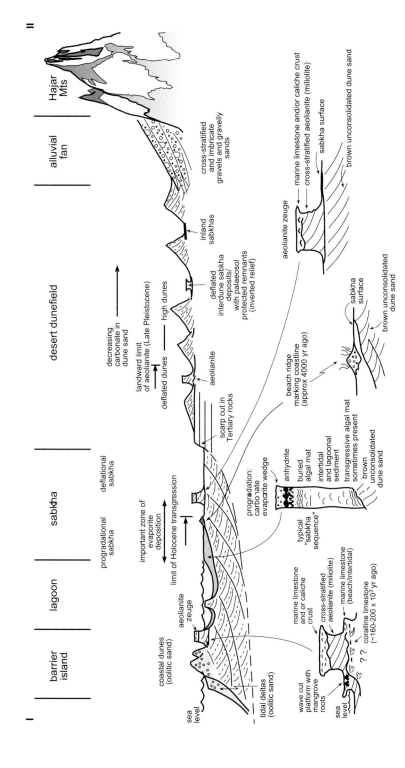

Fig. 4. Schematic cross-section from the UAE coastline to the Hajar Mountains (modified from Evans & Kirkham, 2005). Location of cross-section I–II is shown in Fig. 3.

encloses the lagoon of Khor Al Bazm, which is infilled with fine-grained carbonate sediments in the lee of the barrier, that grade into skeletal sands over the open lagoonal floor. Along the mainland coast of the lagoon, these in-turn pass into broad pelletal- and grapestone-covered intertidal flats and cyanobacterial mats that are succeeded landward by beach ridges and the supratidal zone of the salt-covered sabkha plain (Evans *et al.*,1964a).

Whereas much of the sediment of the Abu Dhabi coastal complex has been produced by the breakdown of skeletal debris, there is here, as in other carbonate provinces, the problem of the origin of the ooids and of the carbonate muds. Shearman *et al.* (1970) described the internal structure of the ooids as comprising interlayers of microcrystalline organic-rich layers and others of tangentially orientated aragonite crystals. Later, Loreau & Purser (1973) and Loreau (1982) provided a detailed description of the structure and ultra-structure of ooids. It was shown that those of the tidal deltas have a complex origin. It was suggested that the aragonite cortex grows as radiating crystals in periods of quiescence (in depressions on the tidal deltas), and that these crystals are then reoriented to a mainly tangential orientation when the grains are agitated by waves and currents on the crests of the various bedforms. They thus produce a fabric that is statistically tangential with respect to the laminations (i.e. with 80–90% of the crystals deviating by <30°). Generally, ooids with radially orientated fabrics are characteristic of sheltered lagoonal embayments.

Kinsman (1969a) showed that the mud (and mud in pellets) of the lagoons has a strontium content which suggests that it is of a non-skeletal origin and has probably been precipitated, although there are problems in ascribing this to direct chemical precipitation, and it may involve organic complexing. Codiacean algae, whose disintegration produce much of the carbonate muds of the Florida and Bahaman lagoons, are rare in the Arabian Gulf and are unlikely to contribute much to the production of mud.

Dalongeville *et al.* (1993) and Bernier *et al.* (1995) described the coastline and its development in the northeast of the UAE, where the alluvial fans from the Oman Mountains reach the coast. Goudie *et al.* (2000) showed, using achival material, old maps and air photographs, that the coast near Ras Al Khaimah has changed considerably in historic times by coastal breaching and by spit growth and decay. Other detailed studies have been made on various parts of the Arabian coast, but large areas still remain undescribed. Shinn (1973b) documented a complex of elongate spits and lagoons on the east coast of Qatar and the coast near Umm Said where, in contrast to the coast of the UAE, the dominant wind is offshore, which leads to a supply of aeolian quartzose sand to the near-shore zone where it mixes with the locally produced carbonates (Shinn 1973a). A description of the coastline of Bahrain and its Quaternary history was given by Evans *et al.* (1980); more recently, Bush *et al.* (2005) described the Holocene evolution of the Tubli inlet, leading to the famous archaeological site of Saar, Bahrain.

In the Dharhan area of the Saudi Arabian coast, the coastal morphology and sedimentology is dominated, as is that of southwestern Qatar, by seaward prograding quartzose sands mixing with local carbonate sediments (Fryberger *et al.*, 1983). Lomando (1999) stressed the importance of a structural control on the development of the coast. It was shown along the coast of Saudi Arabia between Dharhan and the Kuwait border, that cuspate spits formed on the structural highs and the intervening embayments were infilled by beach ridge and sabkha sediments. Picha (1978) documented a similar pattern of barrier islands with tidal deltas enclosing lagoons fringed by sabkha plains along the southern coast of Kuwait. Gunatilaka (1986) gave a detailed description of the area and its Quaternary development, together with a comprehensive review of earlier work. The northern coast of Kuwait and its siliciclastic-dominated sabkhas were described by Saleh *et al.* (1999). Their deposits are similar to the siliciclastic sabkha deposits documented by Glennie & Evans (1976) from the Ranns of Kutch, India.

Baltzer *et al.* (1982) and Baltzer & Purser (1990) described the Iranian coastline and showed that for much of its length it is dominated by the development of small fan-deltas formed by intermittent streams. These streams act as point sources for the supply of detrital sediment to the north-eastern Gulf while, locally, the fluvially discharged sediment forms mud banks in the nearshore area.

Cyanobacterial mats

The Arabian coast, particularly the area fringing the barrier island-lagoon complex of Abu Dhabi has some of the best developed cyanobacterial mats in the world. These were first described in detail by Kendall & Skipwith (1968) who classified the intertidal mats on the basis of their form and position in the intertidal zone. This early work was refined by Park (1976) and Kinsman & Park (1976) who gave more detail on the species, and showed the relationship between the various growth forms and their degree of exposure. The distribution and character of similar mats in the Al Khiran area of Kuwait were described by Saleh (1978). Park (1977) also documented the presence of strange "thrombolytic-like" growths in the Gulf of Salwa between Qatar and Saudi Arabia that have a general resemblance to the well-known stromatolites of Shark Bay, Australia.

Organic matter

Carbonate rocks are important sources of petroleum deposits, both as reservoir rocks and possibly also as source rocks. The original work of Emery (1956) and Evans (1966) showed that organic carbon content in the modern deposits in the Gulf range from 0.5–1.5 wt%. Later, Vita-Finzi & Phethean (1980) reported higher values, between 0.15–3.34 wt%, of muds in the inlets of the Musandam Peninsula near the entrance of the Gulf. Shearman & Skipwith (1965) suggested that the mucilaginous jackets which surround most shallow-water carbonate grains may indeed be the source of much of the kerogen in carbonate rocks. Later, Ferguson & Ibe (1981) analysed the light hydrocarbons in the shallow-water oolites off the coast of Abu Dhabi, and surprised earlier workers when they obtained much higher results than those reported by Evans & Bush (1970) from the tidal delta sediments (average values 0.27 wt%). However, these higher values seem to have been produced by doubtful analytical techniques, as pointed out by Kenig *et al.* (1989, 1991).

Kenig *et al.* (1989, 1991), together with Baltzer *et al.* (1994), gave a detailed analysis of the organic matter in shallow-water lagoonal and intertidal sediments of Abu Dhabi. Three different organo-sedimentary facies were distinguished, with distinct organic signatures that persist after burial and which are related to the rate of change of sea-level, substrate morphology and rates of sedimentation. Both Kenig *et al.* (1991), and later Kendall *et al.* (2002), pointed out the source potential of these microbial mats (see also Evans, 1989).

Carbonate cementation

During the 1950s and 1960s there were numerous brief reports of modern contemporary cementation of carbonate sediments in a variety of environmental settings (see reviews in Bricker, 1971): Sugden (1963b) described aragonitic cements and the development of composite grains, "grapestones" of Illing (1954), and the penecontemporaeous cementation of intertidal flat sands; Taylor & Illing (1969, 1971a,b,c), Evamy (1973), Evans *et al.* (1973), Shinn (1973), Kendall *et al.* (1994), Holail *et al.* (2004) and Whittle *et al.* (1998), described modern cementation in the shallow-water intertidal and subtidal sediments of the UAE and Qatar.

Aragonite occurs in the form of both lithified cryptocrystalline mud and as precipitated fringe cements, and high-Mg calcite (> 4 mol% $MgCO_3$) is common, occurring both as a precipitate and as a replacement of aragonite by a dissolution and re-precipitation process, which affects the enveloping aragonite druses and associated pellets. Shinn (1969, 1971) documented the considerable areal extent of submarine cementation by aragonite and high-Mg calcite. de Groot (1969) explained the importance of the degree of super-saturation in controlling the mineralogy of the cement and suggested that high-Mg calcite, the normal stable form, precipitates so slowly that it is swamped by aragonite. Purser & Loreau (1973) described interesting coatings of aragonite on intertidal and splash-zone rocks that resemble stromatolites, but which were considered to be quite distinct and are produced solely by physio-chemical processes (the terms "coniatolite" and "pelagosite" were proposed to describe these tufa-like deposits).

The sabkha plain

The term "sabkha" was hardly mentioned in the geological literature until the 1960s, although Tricart (1954) wrote about this feature in North Africa, and Holm (1960) described similar features in Arabia. However, these descriptions had aroused little interest. Much later, Purser (1985) presented a comprehensive review of sabkhas and other coastal evaporite systems. Evans *et al.* (1964a) produced the first description of the sabkha plain of the Abu Dhabi coast (UAE) and compared this low, relatively featureless plain fringing the coastal lagoons with coastal progradational plains elsewhere, such as the chenier plain of Louisiana and the reclaimed intertidal flat plains of Holland and other parts of northern Europe (Fig. 4). They incorrectly assumed that the whole plain had been produced by coastal progradation of intertidal and supratidal flats, having been impressed by the cliff of Miocene rocks that formed its inner margin.

Subsequent studies, firstly by Butler (1965, 1969), and later by Evans *et al.* (1969), Patterson & Kinsman (1977) and Kirkham 1997, 1998b, showed that the outer part of the sabkha plain was indeed the result of coastal progradation of intertidal and supratidal flats into the adjacent lagoons, followed by accretion of aeolian sediments and interstitial deposition of evaporitic minerals to produce a distinctive vertical sequence (Fig. 4; see also Lokier & Steuber (2010) and Strohmenger *et al.*, this volume, pp. xx–xx). However, in contrast, the inner part of this plain was produced by deflation of aeolian dune sand, down to the level of the capillary zone. The sabkha thus has a dual origin resulting from: (1) the drowning of the coastal deserts of Arabia during the Holocene transgression, followed by a regression at the outer part produced by coastal progradation from an apparent highstand 5000–4000 years ago; and (2) the deflation of the inner part down to the capillary zone by the onshore winds, to produce a Stoke's surface (Evans *et al.*, 1969; Evans & Kirkham, 2002; Fig. 5).

Goodall (1995) described the rather different extensive salt-covered sabkha plains of the Sabkha Matti, which are separated by alluvial gravel plains

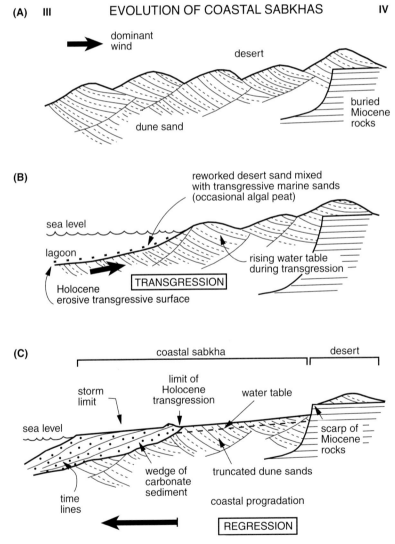

Fig. 5. Episodes in the Quaternary evolution of the UAE coastal sabkha. (A) Initial situation. (B) After culmination of the Holocene rise in sea level. (C) Contemporary situation (after Evans, 1995). Location of cross-section III–IV is shown in Fig. 3.

and aeolian sand sheets, and have unusually thick gypsum crusts in the outer parts. It was shown that the inner part of this huge sabkha is formed by deflated fluvial sediments, in contrast to the dominant aeolian sediments of the coastal sabkhas and those to the east. Further details on Arabian sabkhas were provided by Gunatilaka (1986), Al-Reda (1987); Gunatilaka & Mwango (1987) and Saleh *et al.* (1999) who described the sabkhas of Kuwait.

Water movement in the sabkha

The early workers who described the initial diagenetic changes in the sabkha generally seemed to think that these had been produced either by lateral movement of lagoon water into the intertidal sediments, or by lagoon water sinking into the surface after flooding by high tides or storms (Wells, 1962; Curtis *et al.* 1963; Fig. 6A). Little attention was given to the problems of the permeability of the deposits and the inherent difficulties this presented. Later, Butler (1969) coined the term "flood recharge" to describe this surface flooding of the outer sabkha surface. It was suggested that the most important water movements were flooding of the surface of the outer sabkha and subsequent downward infiltration into the sediments of part of this water, together with runoff of the remainder as a return flow to the adjacent lagoon. The importance of seepage (i.e. down-gradient lateral flow) of continental water into the sabkha sediments from its landward margins was emphasized. Bush (1973) also stressed the downward and seaward movement of water following surface flooding by marine water (Fig. 6B).

These studies were followed by the first serious detailed attempt to determine the origin of the sabkha groundwater, carried out by Patterson & Kinsman (1977, 1981). They discovered that Na^+/Cl^- and Cl^-/K^+ ionic ratios were not, as originally hoped, satisfactory indices of the origin of the sabkha brines. On the other hand, they showed that Cl^-/Br^- and K^+/Br^- ratios were useful. Whereas Cl^-/Br^- ratios <1000 were characteristic of marine waters, ratios 1000–5000 were characteristics of mixed continental and marine waters, and ratios >5000 were indicative of continental waters. It appeared that K^+/Br^- ratios <25 were indicative of marine water, ratios of 25–150 were characteristic of mixed continental and marine waters, and ratios of >150 of continental waters. Interestingly, the term "reflux" was employed, apparently in the same sense as used by earlier authors concerned with the origin of dolomitizing fluids. However, Butler (1969) had stated that this latter process was different from his idea of flood recharge. Patterson & Kinsman (1971, 1981) claimed that the sabkha groundwater system was dominated by a slow regional seaward flow of water, with a seaward movement of a narrow zone of mixed continental and marine waters, which indicated a groundwater flow of 0.3–$1.0\,\text{m}\,\text{yr}^{-1}$ over the last 5000 years. These pioneering papers offer an interesting discussion of the likely preservation of the evaporite minerals under both falling and rising sea levels.

Further detailed groundwater studies were made in the 1970s by the group from ETH Zurich, Switzerland, led by Kenneth Hsü and employing oxygen stable-isotopes. This work refined earlier ideas of Butler (1969) and others, and introduced the term "evaporative pumping" (Hsü & Siegenthaler, 1969; Hsü & Schneider, 1973). This proposed an essentially upward "Darcy flow" of subsurface waters under evaporitic conditions. These authors suggested that the water movement in the outer part of the sabkha (i.e. that progradational part in which the evaporitic minerals are concentrated) is best considered as occurring in three distinct phases: "flood recharge", as had been suggested by Butler (1969), followed by capillary evaporation from the sediment as the surface dried, and finally by "evaporative pumping" (Fig. 6C). It was stressed that lateral movement of water in the outer progradational wedge was unimportant. A very important result of this work was the demonstration that the contents of non-stable tritium indicated that the interstitial waters of the outer sabkha were young, <20 years old, indicating the introduction of considerable volumes of contemporary water. However, Kendall *et al.* (1998) questioned this mechanism of evaporative pumping, and suggested that in some cases flood recharge was the most important process.

In the late 1990s, the United States Geological Survey (USGS) was commissioned to study the sabkha and its waters to assess the possible exploitation of magnesite from this extensive feature. This group, led by Warren Wood, astounded those working in the area, by claiming that approximately 90% of the groundwater of the sabkha originated as rainwater, with <10% being derived from lateral landward flow from the adjoining Tertiary rocks and its cover, or by upward movement from the rocks underlying the sabkha. Again, lateral seepage of water from the lagoons into the sediments was argued to be unimportant (Fig. 6D).

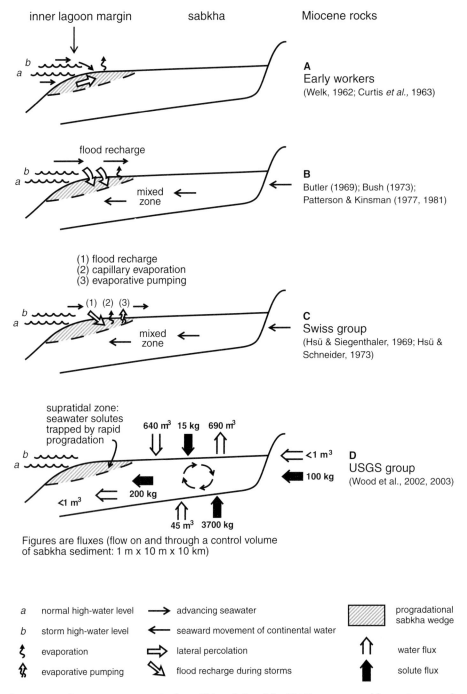

inner lagoon margin   sabkha   Miocene rocks

A
Early workers
(Welk, 1962; Curtis *et al.,* 1963)

flood recharge

mixed zone

B
Butler (1969); Bush (1973);
Patterson & Kinsman (1977, 1981)

(1) flood recharge
(2) capillary evaporation
(3) evaporative pumping

(1) (2) (3)

mixed zone

C
Swiss group
(Hsü & Siegenthaler, 1969; Hsü & Schneider, 1973)

supratidal zone:
seawater solutes
trapped by rapid
progradation

640 m³   15 kg   690 m³

<1 m³

100 kg

200 kg

<1 m³

45 m³   3700 kg

D
USGS group
(Wood et al., 2002, 2003)

Figures are fluxes (flow on and through a control volume
of sabkha sediment: 1 m x 10 m x 10 km)

a   normal high-water level   ⟶   advancing seawater   progradational
                                                                                       sabkha wedge
b   storm high-water level   ⟵   seaward movement of continental water

      evaporation   ⟹   lateral percolation   water flux

      evaporative pumping   ⟹   flood recharge during storms   solute flux

Fig. 6. Suggested patterns of water movement in the sabkha plain of the UAE as proposed by various workers. (A) Original views of Wells (1962) and Curtis *et al.* (1963). (B) Views of Butler (1969), Bush (1973), Patterson & Kinsman 1977, 1981. (C) View of Hsü & Siegenthaler (1969) and Hsü & Schneider (1973). (D) The balance of the flow of water and solutes proposed by Wood *et al.* (2002, 2003).

It was claimed that, in contrast to the supply of water, 95% of the solutes are derived from upward movement from the underlying rocks, with merely small additions from the adjacent rocks and from rainfall (Fig. 6D). It was proposed that the chemical changes in the progradational wedge are due to exchange between water trapped in the sediment and the enclosing sediment (Wood & Imes, 1995; Sandford & Wood, 2001; Wood & Sandford, 2002; Wood *et al.* 2002, 2003; Yechieli & Wood, 2002).

Elsewhere in the Gulf, Robinson *et al.* (1990) and Robinson & Gunatilaka (1991) showed, using isotopic studies, that in the Kuwait sabkhas the major part of the interstitial water and precipitated sulphate minerals are derived from continental groundwater, and that marine water contributed very little of either of these, except in restricted intertidal areas. That the sources of solutes are the underlying rock formations and not the adjacent sea agrees with earlier studies made on the sabkhas of Libya by Imperial College students (Rouse & Sherif, 1980).

Modern dolomite, magnesite and celestite

Undoubtedly, the discovery of what appeared to be recent penecontemporaneously-formed dolomite in the intertidal sediments of the Qatar peninsula by the Shell research group (consisting of A. Wells, V. Coulter and others working in conjunction with L. Illing, J.C.M. Taylor and R.W. McQueen of the V. C. Illing Partnership, London), led to a great deal of interest in the Arabian Gulf. The dolomite was found associated with gypsum in the carbonate sediments occurring between normal diurnal tidal and spring tidal levels (Wells, 1962). It was not clear whether the dolomite was primary or replacive. This was soon followed by the discovery of a similar occurrence in the sabkha sediments in Abu Dhabi (Curtis *et al.* 1963). These latter authors noted also that the dolomite was a calcium-rich type; it was associated with anhydrite and gypsum and appeared to be replacive. Later, Illing *et al.* (1965) proved that the dolomite was indeed modern and had ages of 3310–2670 years BP (uncalibrated) and they further showed petrographically that, whereas the dolomite replaces aragonitic mud, the coarser skeletal debris appears to be unaffected. The geochemistry of the groundwaters associated with the dolomitization was discussed by de Groot (1973).

In contrast to the intertidal flat setting in Qatar and Abu Dhabi, Gunatilaka *et al.* (1984, 1987) discovered calcium-rich dolomites forming by direct precipitation subtidally in a shallow but occasionally drying lagoon in Kuwait. It was suggested that the reduced level of dissolved sulphate in the pore fluids due to microbial activity, was more important than the moderate Mg/Ca ratios in controlling the formation of the dolomite.

McKenzie *et al.* (1980, 1981) confirmed the penecontemporaneous nature of the Gulf coastal dolomite and, using oxygen-isotope data, showed

that the dolomite had developed in isotopic equilibrium with the porewaters at temperatures of between 34–49 °C. Later, Patterson & Kinsman (1982) suggested that the process may be even more complicated than originally thought, including a possible two periods of dolomitization, and that the process was related to the presence of buried creek deposits under the sabkha surface acting as water conduits. Furthermore, it was proposed that the dolomitization was a subsurface phenomenon in contrast to the near-surface dolomitization of the Bahaman and Florida sediments.

Gunatilaka (1987, 1989, 1991) concluded that dolomite in the Kuwait sabkhas had also developed at different times: first during the early transgressive phase; and later during coastal progradation. Dolomite, with a spherulitic fabric, in a burrowed siliciclastic sand was also described by Gunatilaka *et al.* (1987) in Kuwait. Later, Müller *et al.* (1990) using Sr isotopes, confirmed earlier ideas that the dolomite was produced by reaction between earlier carbonate sediments and marine-derived water.

A group from the University of Paris, Orsay, led by Bruce Purser, discovered dolomite associated with fossilized algae and mangrove roots in the Abu Dhabi coastal area (Kenig *et al.*,1991). This occurred in the transitional zone between areas of oxic and anoxic conditions. Decay of organic matter of vegetation in the palaeosols had produced anoxic and slightly acidic conditions that aided in the dissolution of the aragonitic host sediment, again, as suggested by workers in other areas, pointing to the importance of a reduced sulphate content in the porewaters (Baltzer et al. 1994).

Illing & Taylor (1993) confirmed the early suggestion by Klement (1894, 1895) and later that of Kinsman (1965) that, as the dolomitization proceeded, the enclosed brines in the sediments lost magnesium and sulphate at a ratio of 1:1, to produce almost equal amounts of dolomite and gypsum; i.e. two moles of $CaCO_3$ are replaced by one mole of dolomite and one mole of gypsum:

$$2CaCO_3 + Mg + SO_4 + 2H_2O$$

$$\rightarrow CaMg(CO_3)_2 + CaSO_4(2H_2O) \quad (1)$$

These early studies also revealed the development of other minerals during the process of early diagenesis of sabkha sediments. Evans & Shearman (1964) showed that celestite ($SrSO_4$) occurred in the dolomitized sediments and appeared to have been produced by Sr released from the aragonitic sediment reacting with the sulphate of the connate water (definite confirmation was provided by Dr Robin Walls and his assistant at the BP Research Laboratory, Sunbury-on-Thames, UK, as the Imperial College x-ray equipment had broken down). Continued reaction between the Mg-rich connate waters and the diagenetic dolomite led to the production of magnesite ($MgCO_3$) (Kinsman, 1965; Bush, 1973).

Anhydrite, gypsum, bassanite and other evaporite minerals

Considerable excitement was generated by the discovery of modern penecontemporaneous anhydrite ($CaSO_4$) within the sediments of the sabkha plain, and not in subaqueous deposits. As noted earlier, the mineral was first noticed by Douglas Shearman in a wash of a sample from a core collected by David Kinsman from the sabkha just landward of Abu Dhabi Island. Shearman saw the prismatic crystals and immediately recognized their significance.

He sent frantic cables to the author, who together with Christopher G. St. C. Kendall and Sir Patrick d'Esotoville Skipwith, was in Abu Dhabi to investigate the sabkha. Subsequent pits dug into the sabkha surface (Fig. 7) revealed for the first time the crenulated seams and nodules of anhydrite in the near-surface supratidal sediments of the sabkha plain (Curtis *et al.* 1963; Shearman, 1963). Later, more extensive pitting revealed a wide range of structures, including diapirs, enterolithic folds and chicken-wire mosaics (Kinsman, 1966; Shearman, 1966; Butler, 1970; Butler *et al.* 1965, 1982).

There is field evidence for replacement of gypsum ($CaSO_4 \cdot 2H_2O$) by anhydrite (Kinsman, 1969b; Bush, 1972, 1973). Cuff (1973) was able to dehydrate gypsum in the laboratory to form anhydrite (which had a disorganized structure, as did the naturally occurring sabkha anhydrite) using brines similar in composition to those of the sabkha, supporting an earlier suggestion of such an origin (Cuff, 1969). However, much of the anhydrite of the sabkha plain is believed to have formed as a primary precipitate in the capillary zone, perhaps being seeded by earlier replacement of gypsum. Kirkham (1998b; this volume, pp. 265–76), by contrast, has repeatedly asserted that the anhydrite seams are probably formed after gypsum that crystallized from ponded seawater in salinas during previously higher Holocene sea levels. Butler (1966, 1969) showed that, in the inner parts of the Abu Dhabi sabkha, reaction between continental groundwater, seeping from the Miocene rocks, and anhydrite causes the reversion or replacement of anhydrite by gypsum.

Fig. 7. Digging the first pit which revealed nodular anhydrite in the UAE, autumn 1962 (from left to right: Rosemary Evans; Graham Evans; Christopher G.St.C. Kendall; and guard). Photograph courtesy of Sir Patrick A.d'E. Skipwith.

Anhydrite has been described from sabkha sediments in Bahrain (Evans *et al.*, 1980), and has been observed in similar sediments by the author in Saudi Arabia near Dharhan, and as far north as Kuwait by Gunatilaka *et al.* (1980) and Gunatilaka (1990, 1991). However, so far, it has not been discovered in the sabkha sediments of Qatar and occurs rarely on the offshore barrier island sediments of Abu Dhabi. Interestingly, northeast of Ras Al Khaimah there is little gypsum or anhydrite in the sabkha sediments. This is presumably due to the incursion of abundant fresh/brackish water from the adjacent alluvial fans.

Following the early discoveries, Kinsman (1965, 1966, 1969a,b), Butler (1965, 1966, 1969, 1970, 1973), Butler *et al.* (1973) and Bush (1972, 1973) discussed the physical and chemical conditions of the formation of anhydrite in the sabkha plain sediments of Abu Dhabi. Later Gunatilaka and colleagues (Gunatilaka *et al.* 1980; Gunatilaka, 1991) showed that anhydrite in the Kuwait sabkhas was formed by direct dehydration of gypsum around halophytic plants and that it reverted to gypsum on their death. Gunatilaka (1991) further discovered that in some very shallow brine-filled channels, prismatic gypsum is sometimes altered to anhydrite subaqueously, but reverts to gypsum when the salinity of the channels decreases after rain. Furthermore, it was shown that in the Arabian Gulf, the latitudinal range of anhydrite is 23°–28°N; whilst, in contrast, it is found in latitudes as low as 18°N in the Red Sea (Gunatilaka, 1989, 1990, 1991).

Kirkham (1997) demonstrated that as well as anhydrite developed as a cap to the deposits of the regressive sabkha cycle, as described by Evans *et al.* (1969) and others, an earlier anhydrite had developed at the height of the Holocene transgression. This developed in a zone landward of the main ancient beach-ridge complexes in some parts of the sabkha, but had not been seen in the areas to the east studied by Evans *et al.* (1969), Butler (1969) and Patterson & Kinsman (1977, 1982). This earlier anhydrite marks the innermost parts of the sabkha affected by marine waters and has developed in the brown aeolian sand which underlies the sabkha. It is overlain by a gypsum mush and intertidal-subtidal muds with their microbial layers, and then by the usual supratidal regressive anhydrite. It is clearly transgressive and is correlatable with the transgressive intertidal cyanobacterial deposits described by Butler (1969), Patterson & Kinsman (1977), McKenzie *et al.* (1980) and Kenig *et al.*

(1989), although these were missed by Evans *et al.* (1969), probably due to the crudeness of their drilling techniques. However, there is no transgressive anhydrite in the Kuwait sabkhas (Gunatilaka, pers comm. 2006). The transgressive anhydrite is thus the arid zone equivalent of the transgressive peat that underlies transgressive lagoonal deposits in more humid areas (Evans, 1970). However, it appears to differ from the latter as it is only found on the most landward limits of the transgression and has not been found to underlie the whole of the transgressive-regressive wedge of Holocene sediment lying to seaward.

Although extensive, the more common gypsum has received rather less attention than anhydrite. Bramkamp & Powers (1955) described subaqueous gypsum from a small lagoon, and M.W. Hughes Clarke and E.A. Shinn (reported in Loreau & Purser, 1973, p. 289) observed subaqueous gypsum in the Khor Odaid on the eastern coast of Qatar. Subaqueous gypsum occurs associated with halite behind coastal spits near Dhahran, Saudi Arabia (C.G.St.C. Kendall, pers. comm. 2006). In contrast, primary subaqueous precipitation of gypsum has not been observed anywhere in Abu Dhabi, except in large brine-filled artificial shot holes left by seismic crews when it forms encrustations of beautiful selenite twins on the surrounding walls.

Elsewhere, Gunatilaka & Shearman (1988) described gypsum laminites (balatino type) and gypsum-carbonate/organic couplets from shallow ephemeral pools on the sabkha of Kuwait. Its first appearance is normally in the intertidal-supratidal deposits beneath the microbial mat and then as a mush of small crystals which overlie the latter. Scattered larger crystals (desert-rose type) are common in the wind-blown sands exposed on the inner parts of the sabkha (Kinsman, 1966; Butler, 1970; Bush, 1973; Shearman, 1978, 1980, 1981). Other gypsum crystals, often turbid with inclusions, occur beneath and within the intertidal microbial mats; the latter, as pointed out by Illing (1963), appear to be secondary and are bi-products of the dolomitization of aragonitic mud (Illing *et al.*, 1965; Illing & Taylor, 1993).

Other authigenic evaporitic minerals are found in the sabkha sediments as well as the abundant gypsum and anhydrite. Bassanite ($CaSO_4 \cdot 0.5H_2O$) has been described by various workers as an ephemeral coating on exposed gypsum crystals (Kinsman, 1965, 1967; Butler, 1969; Gunatilaka *et al.*, 1985) and rarer minerals such as huntite ($CaMg_3(CO_3)_4$) were recorded by Kinsman (1965) and Butler (1969). Later Baltzer & Purser (1990) described thenardite (Na_2SO_4) crystals up to 5 cm in length growing in sabkha sediments in Khor Al Logait. Although the sabkha has a thin halite (NaCl) crust on its surface and is often covered, particularly on its inner parts, by sheets of glistening white halite with occasional cubic halite just below the salt crust, none of this is preserved within the sediments. This lack of persistence of salt in the area is probably due to high humidity (Kinsman, 1976).

Fauna and flora

The Gulf contains an abundant but lower diversity fauna and flora than the adjacent Indian Ocean (Melville & Stadden, 1901; Belvgad, 1944; Hughes-Clarke & Keij, 1973; Bradford, 1975; Basson et al., 1977; Jones, 1985). Early studies of the marine biology of the Arabian Gulf were made during fishery investigations, and later ecological studies were undertaken attempts to monitor environmental changes, including studies of the intertidal and nearshore faunas and floras of the Saudi Arabian coast by Basson et al. (1977) and of the Kuwait coastal areas by Jones (1985).

During the later studies, a great deal of attention focused on the distribution of organisms whose remains contributed to the cover of carbonate sediments, as well as the use of these as environmental, i.e. facies, indicators. Hughes-Clarke & Keij (1973) described the general distribution of the major faunal-floral groups. Members of the Kiel group described the distribution of benthic and planktonic foraminifera and other groups in the sediments of the Iranian parts of the Gulf (Haake, 1970; Diester, 1971; Sarnthein, 1971, 1972; Sarnthein & Walger, 1973), and Lutze et al. (1971) studied the importance of larger foraminifera as sediment producers. Martini (1967) showed that both oceanic nannoplankton and planktonic foraminifera penetrated into the Gulf only as far as the tip of the Qatar peninsula.

The fauna and flora of the nearshore and lagoons of the UAE were studied by members of the Imperial College group. Kinsman (1964c) showed that corals (particularly Porites sp.) were able to live under higher salinity and temperature conditions in the Arabian Gulf than had previously been thought. Murray provided a comprehensive survey of the foraminifera of the coastal-nearshore and lagoons of the UAE (Murray, 1965a,b, 1966a,b,c, 1970a,b) and the ostracods of these shallow-water

environments were described by Bate (1971). A general review of the fauna and flora of the coastal-nearshore and lagoons of the area was given by Evans et al. (1973).

Surprisingly, in spite of the great amount of engineering works, such as the dredging and deepening of existing channels and the excavation of new channels, along the UAE coastline, the foraminiferal faunas of the nearshore and lagoonal environments showed little difference between the 1960s and 1990s (J. Murray, pers. comm. 2006). However, as in many other areas, George & John (1999) have shown that higher temperatures have caused coral bleaching, the death of corals, and the widespread covering of reefs with macro-algae. Recent reviews of the fauna and flora of the nearshore areas and coastal lagoons of the UAE have been given by Boer & Aspinall (1975, mangrove swamps), George (2005, marine invertebrates), George & John (2005, marine habitats), John & George (2005, the shore and shallow seas); John (2005, marine plants), and other ecological features were described in Hellyer & Aspinall (2005).

Pleistocene deposits

Geologists of the Indian Geological Survey in the 19th century, during the days of the British Raj, described the Pleistocene rocks of the Arabian coastlines. These consist mainly of a cross-bedded aeolianite, "the miliolite", capped by marine limestones (Carter, 1849, 1859; Pilgrim, 1908). The name "miliolite" originated from similar deposits described in western India which contained the remains of miliolid foraminifera (Theobald, 1878; Chapman, 1900; Evans, 1900). These deposits form the cores of most of the islands of the UAE, and occur as scattered zeugen on the coastal plain and the islands (Fig. 4). They were referred to fleetingly by Evans et al. 1964b, 1973 and many others (Williams & Walkden, 2002), who all agreed on an aeolian origin for the miliolite and suggested that the marine cap, often modified by the development of caliche, found on most zeugen was deposited at the peak of the last interglacial (marine oxygen-isotope stage 5, $\sim$120 ka). Evans et al. (2002), Evans & Kirkham (2005) and Kirkham & Evans (2010) supported this age assignment, based on stratigraphic grounds and comparisons to global relative sea-level variations.

Attempts to ^{14}C date the marine cap on the miliolite have yielded ages around 30 ka (Williams & Walkden, 2002; i.e. close to the limit of radiocarbon dating) but these were regarded as being unreliable and probably affected by diagenesis of the limestone. More recently, Wood *et al.* (2006) dated the marine carbonates at between 24.2–28.8 ka using optically stimulated luminescence techniques on quartz grains and calibrated radiocarbon dating of shells. It was claimed that this indicates that the UAE land surface has been uplifted approximately 80 m in the last 25,000 years ($\sim$3 mm yr^{-1}). However, this seems unlikely and requires further investigation. Scholle & Kinsman (1974) discovered that some of the later caliche crusts unusually contained aragonite and high-Mg calcite and not the usual low-Mg calcite found in most other areas.

Kirkham (1998a) concluded that the aeolianites represent barchan carbonate-sand dunes which had been reworked later into seif dunes by north-westerly winds. It was concluded that the subsequent Holocene marine transgression had drowned a dunefield: first eroding the aeolianites; and then capping them with a marine skeletal limestone. Hadley *et al.* (1998) proposed the name Gayathi Formation for the aeolianites and the Aradah Formation for continental sabkha-type deposits that they found overlying these inland in the UAE. Walkden & Williams (1998) made a brief reference to these deposits and later Williams & Walkden (2001, 2002) gave a detailed description of these Pleistocene limestones. They accepted the terminology of Hadley *et al.* (1998) and, in addition, proposed the name Fuwayrit Formation for the marine limestones which commonly overlie them in the coastal areas.

The Fuwayrit Formation is divided into two members: the Futaisi Member; overlain by the Dabb'iya Member and separated from it by a pitted surface infilled with the sediments of the latter. It has been suggested that the lowermost part of the Dabb'iya Member is possibly a tsunami deposit and that the pits are of karstic origin linked to the former presence of large plants (Williams & Walkden, 2001, 2002). However, it seems more likely that these were produced by burrowing crabs or some other bottom-feeding organism (Kirkham & Evans, 2010). In contrast to Abu Dhabi, the Dabb'iya Member in Qatar is overlain by an oolitic aeolianite unit, which was named the Al Wusayi Member. Boreholes in offshore Abu Dhabi have revealed at least six Quaternary sequences consisting of marine carbonates and sabkha deposits within the top 50 m of sediment (Williams & Walkden, 2001, 2002).

Evans *et al.* (2002) described what appears to be the most complete Quaternary sequence in the UAE on Marawah Island off Abu Dhabi. A coralline limestone cropping out near present sea-level has been dated as 160–200 ka (Evans & Kirkham, 2005; Fig. 4). This is overlain by a flat-bedded marine limestone, which in-turn is succeeded by an aeolianite (miliolite), so commonly seen elsewhere, covering an erosion surface. Also, as elsewhere, this aeolianite is succeeded by a bioturbated marine limestone and on Marawah Island it is capped by an oolitic limestone. As in other investigations, no meaningful dates have been obtained from the capping limestone, but it is thought to be of last interglacial age.

Marawah and other islands formed by outcropping limestones in the lagoon, are usually surrounded by a wave-cut platform whose surface is just above normal high-tide level, and one metre above the modern low-tide terrace. However, the platform is covered by a thin sheet of water at high-water spring tides and during Shamals. On its surface is a network of mangrove rhizoliths and, although its exact age has not been determined, it is thought to be late Holocene: probably formed contemporaneously with the inner beach ridges of the sabkha plain (i.e. 5000–4000 uncalibrated ^{14}C years BP; Evans *et al.*, 2002; Evans & Kirkham, 2005). New exposures in the banks of canals at the landward part of Abu Dhabi have revealed vuggy carbonate overlain by evaporites, which in-turn are capped by miliolite. The basal carbonate horizon together with the overlying evaporite layer is probably the back-barrier equivalent of the coralline limestone that occurs on Marawah Island (Evans & Kirkham, 2005).

Similar aeolianites and marine limestones have been described from Saudi Arabia (Felber *et al.*, 1978). According to Holm (1960), Steineke wanted to name them the Bahr Formation after their occurrence at Jabal Bahr near Al Jubayl, but the name was never adopted. Picha (1978) described the development of a series of Pleistocene beach-dune ridges composed of oolitic-quartzose sands which enclose coastal sabkhas in Kuwait. Miliolite was described by Pilgrim (1908), Cooke & Goudie (1980) and Evans *et al.* (1980) on the island of Bahrain. Abu Zeid *et al.* (1999, 2001), Holail (1999) and Hussain (2006) described the petrology of similar Quaternary carbonates in Qatar.

The petrology and geochemistry of some Quaternary limestones recovered from drilling for oilfield installations offshore of Abu Dhabi were documented by Khantaprab (1968). Darwish & Conley (1989) described a sequence of Pleistocene-Holocene sediments from boreholes between Saudi Arabia and Bahrain. These consisted of a lower dolomitic unit overlain by skeletal limestones and several quartzose sand layers produced by aeolian transport of quartzose sand across the Gulf of Salwa at lower sea levels, which link the dune sands of Bahrain with those of Saudi Arabia. Supposed Pleistocene carbonates, including dolomites together with gypsum, have been described offshore of Saudi Arabia (Chafetz et al., 1988; Chafetz & Rush, 1994) and have been interpreted as Pleistocene sabkha deposits.

The Arabian desert

Holm (1960) gave the first comprehensive summary of the geomorphology and the sediments of the Arabian Peninsula and its deserts. These are composed mainly of quartzose sands, contrasting markedly with the carbonate sediments of the coastline. Since this early work, the main advances in the study of the area have been made by Kenneth Glennie and his students and associates (Glennie & Evamy, 1968; Glennie, 1970, 1972, 1987, 1994, 1997, 2005; Teller et al., 2000; Glennie & Singhvi, 2002; Glennie et al., 2002). An early general review by Glennie (1970) concentrated on the recognition of desert sediments. Later, he and his co-workers demonstrated that the mobile sand cover has been moulded into various dune forms in the UAE. This consists dominantly of spectacular linear dunes which pass southwards into the barchanoid megadunes of the Al Liwa, or in the northeast into the bordering fans (Glennie & Singhvi, 2002).

Modern dune patterns are dominated by the Shamal wind, blowing from a northerly quarter, but during cold phases of the Quaternary, WNW rather than the present NNW Shamal was the dominant wind direction (Glennie & Singhvi, 2002), a response to the movement of climate belts. Teller et al. (2000) claimed that at least six generations of dunes can be identified from satellite and aerial photographs, and these have been constantly modified by later periods of aeolian activity. These workers showed that older cemented dunes (aeolianites) underlie the modern mobile cover of sand.

Studies of the composition of the aeolian sands in the UAE were made by Pugh (1997), Hadley et al. (1998), Ahmed et al. (1998), Alsharhan et al. (1998) and El-Sayed (1999). These authors showed that most of the sands were originally derived from the rocks of the Arabian Shield. Pugh (1997) claimed that the immediate sources of the sand were the Pliocene and younger alluvial sediments. Intermixed with the siliciclastic, mainly quartzose, sand is carbonate sand derived from the exposed Gulf floor during periods of lower sea-level. Generally, the carbonate content of the dune sand decreases inland, as does that of the older cemented aeolianites, from >70% near the coast, to <30% between 70–100 km from the coast (Pugh, 1997). A similar decrease was reported by Teller et al. (2000) who found <10% carbonate 40–50 km from the coast. These workers suggested that this may mark the landward limit of marine-derived aeolian sand, which spread inland when the coastal dunes had been flooded by the rise of sea-level during the Holocene. This cut off the supply of sand for large areas of the coast. Subsequent deflation of the area immediately landward of the Holocene beach ridges produced the inner deflational part of the sabkha plains and provided some additional sand, although only for a short period as this supply must soon have been exhausted (Evans et al.,1969).

Today, modern carbonate sand is transported landwards along the Saudi Arabian coast and the NE UAE to intermix with older sands. However, in much of the UAE the carbonate dunes occur on the islands and are separated from the mainland shore by coastal lagoons and sabkhas. Thus sand cannot reach the mainland except in the area between the NE of Ras Ghanada and Dubai where no lagoons fringe the coastline today, although these may have existed in the past and are now infilled.

Changing sea levels throughout the Quaternary have controlled the sand supply to the deserts of the UAE in particular. At times of lower sea level, sand derived originally from the Arabian Shield continued to be transported along the Saudi Arabian shoreline and coastal deserts (Fryberger et al., 1983), and crossed the Gulf of Salwa, where Darwish & Conley (1989) have shown the presence of two horizons of quartzose sands interbedded with marine carbonates. Al-Hinai et al. (1987) described what may be relict dune forms on the floor of the Gulf of Salwa. However, these may merely be cemented tidally-produced bedforms that are common in the area (Cloet, 1954). At lower sea-levels, aeolian sands reached Bahrain Island, where today all that remains are large areas of deflated dune sand (Evans et al.,1980). Finally, the sand reached and crossed the Qatar peninsula, where the remnants of the once extensive cover are now building into the sea on the east coast of the peninsula, south of Umm Said (Shinn, 1973a). This sand-sheet was presumably formerly connected with the deserts of the UAE across the southern Gulf (Fig. 3). However, curiously, no aeolian sand has been identified from this offshore area. Potentially it was either removed by deflation during periods of lower sea level or its remains have been buried beneath later marine sediments.

Details of the sedimentary structures of the aeolian sands and aeolianites were described by Glennie (1970, 1972, 1987), Fryberger *et al.* (1983), Gunatilaka & Mwango (1989), Kirkham (1998a), and Williams & Walkden (2001, 2002). Glennie & Evamy (1968) also described fossil plant root struc-tures ("dikaka") from the dune sands of the UAE. Later, Bristow *et al.* (1996) revealed the detailed internal structure of some of the large dunes of the Liwa area of Abu Dhabi by trenching and using ground-penetrating radar. Stokes & Bray (2005) also described the internal structures of the dunes, provided optically stimulated luminescence dat-ing results of core and sand-quarry samples and presented an account of the late Pleistocene aeo-lian history of the Liwa (UAE).

The aeolian sediments pass landward into de-posits of the alluvial fans, which form a wide bajada (Glennie, 1998; Rodgers & Gunatilaka, 2002) that borders the mountainous rim of the Rubal Khali (Fig. 4). These fans are still relatively poorly understood. In general, the fans in Oman, which have been studied more fully, are thought to be mainly Miocene-Pleistocene in origin (Maizels, 1988). Although they are still affected by modern floods, they must have formed during a wetter climate than that of today. They are crossed by a complex of channels which have changed their courses frequently (Maizels & McBean, 1990; Glen-nie & Singhvi, 2002; Al-Farraj & Harvey, 2005), and are composed mainly of gravels and sands rich in debris from the limestones and ultrabasic rocks derived from the adjacent highlands. Diagenetic changes between the relatively unstable ferromag-nesian-mineral-rich rocks derived from the Oman Mountains and the groundwater have produced a dolomitic alteration product, named "barzamanite" by Maizels (1988). Similar fans and alluvial sedi-ments which formed in the late Tertiary and during the Quaternary have been described from Sabkha Matti (Goodall, 1995), and as far north as the fringes of the Tigris-Euphrates-Karun delta (Purser *et al.*, 1982; Baltzer & Purser, 1990) indicating that, in the past, considerable fluvial sediment reached the Gulf from the Arabian Peninsula.

Inland sabkhas, such as the Umm al Samim, occur 200 km away from the Hajar Mountains at the foot of the alluvial fans or in inter-dune areas elsewhere, although these have received relatively sparse attention. In one such small inland sabkha in Abu Dhabi, a cm-thick horizon of palygorskite has developed at what appears to be the top of the water table. It is thought to have been produced by reaction between aeolian dust and the local saline groundwater (Evans & Kirkam, 2005). Elsewhere, in various depressions, lakes developed during the earlier wetter periods of the Quaternary (Glennie *et al.*, 1998). McClure (1976, 1978, 1984, 1988) described lake deposits including *Chara* marls, which contain ostracods and fresh-water molluscs from the Rub al Khali.

Glennie and his colleagues (Juyal *et al.* 1998; Glennie *et al.*,1998, 1999; Teller *et al.*, 2000; Glennie *et al.*, 2002, Glennie & Singhvi, 2002), using radiocarbon and optically stimulated lumi-nescence dating of aeolian sands, as well as iso-topic dating, reconstructed an event stratigraphy for the southern Arabian peninsula that has im-proved understanding of the evolution of the area during the Quaternary.

SUMMARY OF THE SIGNIFICANCE OF THE ARABIAN GULF STUDIES

Studies of Quaternary sedimentation in the Arabian Gulf since the pioneering work of Emery (1956) have yielded some interesting results. Some of them have been applied to help in the interpre-tation of rock sequences in the Middle East (Alsharhan & Kendall, 1994; Alsharhan & Whittle, 1995; Alsharhan & Kendall, 2002), but also far afield in many parts of the world (e.g. Kendall, 1978; Achauer, 1982; Shinn, 1983).

The Gulf is a typical foreland basin bordered on one side by the Arabian Shield and on the other by the Iranian foldbelt. It is a marginal sea of the Indian Ocean whose waters show greater extremes of salinity and temperature than the open ocean. It is also a high productivity area but is colonized by a depleted Indian Ocean flora and fauna with plank-tonic foraminifera only penetrating into the Gulf as far as approximately the latitude of the tip of the Qatar peninsula.

The Gulf has an asymmetric supply of terrigeous sediment from the Iranian flank and is being in-filled longitudinally by the deposits of the Tigris, Euphrates and Karun rivers which form the deltaic plains of Mesopotamia (i.e. it is a typical A-zone delta of Audley-Charles *et al.*, 1987; Fig. 3). The situation is unusual because due to the strong northwesterly wind, the Shamal, which blows down the Gulf, the delta receives, in addition to its fluvial load, an unusually large supply of aeolian sediment from the same hinterland (up to 43% of the total supply; A. Gunatilaka, pers. comm., 2006). The delta is also unusual in that it is building into a sea dominated by carbonate sediment and is flanked by a desert with a carbonate sediment-dominated coastal plain (Gunatilaka 1986, Evans, 1989).

The south-western flank of the Gulf was considered by Ahr (1973) to be a good modern example of a carbonate ramp, although the validity of this suggestion was questioned by Walkden & Williams (1998). Marls in the northeast sector along the Iranian side of the Gulf, pass gradually southwestwards into a mosaic of neritic skeletal carbonate muds and sands, which in-turn pass into a barrier complex of oolitic and skeletal carbonate sands and fringing reefs (Fig. 4). Landward of the barrier are lagoons in which pelletal carbonate muds and sands accumulate. Extensive microbial mats (stromatolites) occur along the mainland shoreline with scattered groves of black mangrove (*Avicennia marina*), which although common on the coastline of the southern Gulf and along parts of the Saudi Arabian coast, do not reach as far north as Kuwait (A. Gunatilaka, pers. comm., 2006).

Two phenomena which occur in the lagoons and offshore of the Arabian coast have aroused a great deal of interest. Firstly, the apparent spontaneous precipitation of aragonite: the "whitings". These have been claimed to be initiated by diatom blooms, but they are problematic because of geochemical difficulties in explaining their formation and due to the presence within them of considerable non-aragonitic material. Secondly, the widespread contemporary subaqueous cementation of sediments at depths ranging from tens of metres in the offshore, to several metres on the tidal deltas and on the terraces of the lagoons, as well as those of the intertidal flats and beaches. Although submarine cementation has been found elsewhere, it is on a very large scale in the Arabian Gulf. This early cementation, with its development of cemented crusts, bored surfaces and colonizing epifauna, is obviously analogous to the hardgrounds of ancient carbonates and helps in the interpretation of these. Although well known and already studied by several workers, it is clear that these hardgrounds are the result of slow or non-deposition. However, the exact reason for their multiple development and what cyclical changes in the water column produce a succession of these crusts is not yet understood. Although the hardgrounds may be contemporaneous over large areas on the shelf, they appear to be diachronous in the lagoons, where they have advanced seaward over the in-filled lagoon sediment and are capped by the intertidal and sabkha plain sediment.

The primary reason that the Gulf has attracted considerable attention is because of the coastal plain landward of the inner lagoonal shoreline (the "sabkha"). The outer part of this has been constructed by coastal progradation of intertidal-supratidal sediments and is not only the site of dolomitization, as are similar intertidal and supratidal deposits in the Bahamas and Florida, but it is also the site of precipitation of extensive gypsum, and more importantly, anhydrite. Before the studies in the Gulf, the latter mineral was unknown in the modern environment, and conventionally it was assumed that the extensive anhydrite deposits in the stratigraphical column had either been precipitated subaqueously or had resulted from diagenetic changes of precipitated gypsum.

Anhydrite in the sabkha plain sediments of the Gulf has all the textural varieties seen in ancient evaporites, but clearly forms above the water table in supratidal sediments. This discovery has led to the re-examination and reinterpretation of many ancient carbonate-evaporite sequences and has produced fierce debate in the sedimentological community following the very persuasive writings and lectures of Shearman (1965, 1966, 1971, 1978, 1980). Undoubtedly, following this there was initially a tendency to interpret all anhydrite deposits as of sabkha origin by some over-enthusiastic workers, and indeed all evaporites were sometimes claimed to have a sabkha origin. However, this is clearly not the explanation for the origin of many (Warren & Kendall, 1985), although the discovery of anhydrite in the sabkha sediments undoubtedly led to a renewed interest in evaporites and a re-investigation of many ancient deposits.

The lateral passage from sabkha evaporites, through aeolian dune sands, and into the gravels of the alluvial fans has also attracted considerable interest. This is clearly seen on the south-western flank of the Arabian Gulf and has been used as a model for the understanding of the Permo-Triassic deposits of the North Sea and surrounding areas, as well as marine carbonate to terrestrial red-bed sequences elsewhere (Glennie, 1972, 1987, 1994).

In addition to the interest in the sedimentological work and its application to the understanding of ancient rock sequences, the gradual evolution of ideas on the Holocene rise of sea-level and its impingement on the surrounding land has begun to provide some possible answers to the intriguing historical problem of the meaning of the Flood of Sumerian history, upon which the story of "Noah's Flood" was apparently based (Lambeck, 1996; Glennie, 1997, 1998; Teller *et al.,*2000; Kennet & Kennet, 2006). However, there is still a relatively sketchy understanding of the Quaternary history of the Mesopotamian plains.

ACKNOWLEDGEMENTS

The writer would like to record his thanks to Professor Ananda Gunatilaka, Professor Christopher G.St.C. Kendall and especially Dr Anthony Kirkham for their careful reading and numerous corrections and useful suggestions in the improvement of the original manuscript. Miss Emma Bennet, Mrs Linda Kirkham, Dr Anthony Kirkham and Mrs Anna Messervy-Evans are thanked for typing the various versions of the manuscript, as well as Mrs Kate Davis who drafted the figures. The author would like to take the opportunity to record his and his colleagues at Imperial College London, thanks to: Captain H. Hatfield and the crew of HMS Dalrymple of the Hydrographic Service; to the late Sir Hugh Boustead and the staff of the British Agency in Abu Dhabi, especially Mr Douglas Gordon; the late William Clark OBE, the Ruler's Secretary; Mr & Mrs Christopher Willy (ADMA), John Wilkinson (PDTC), Mr & Mrs I. Kirkbride (PDTC), Mr & Mrs Ian McGaskill (Gray-McKenzie Ltd.); and other members of the Abu Dhabi "ex-pat" community who helped and extended their hospitality to various members of the Imperial College party between 1961–1964. NERC UK, Shell Oil Company The Hague, and Socony Mobile Dallas financed the Imperial College research. Peter Hellyer is thanked for his assistance to the author and to Dr Anthony Kirkham, which has allowed us to pursue our studies in Abu Dhabi over the last few years. Thanks are here recorded to the late H.H. Sheikh Shakhbut and the late H.H. Sheikh Zayed for permission to work in Abu Dhabi, and their hospitality, and finally, the help given to all by many of the local inhabitants.

REFERENCES

Abu Zeid, M.M., Abd El-Monem, T., Abd El-Hameed, T. and Al-Kuwari, A.J. (1999) Petrology, mineralogy and sedimentation of the coastal Quaternary carbonates, Qatar, Arabian Gulf. *Carbonates Evaporites*, **14**, 209–224.

Abu Zeid, M.M., Abd El-Monem, T., Abd El-Hameed, T. and Al-Kuwari, A. J. (2001) Diagenesis in the coastal Quaternary carbonates in Qatar, Arabian Gulf. *Carbonates Evaporites*, **16**, 26–36.

Achauer, C.W. (1982) Sabkha anhydrite, the supratidal facies of cyclic deposition in the upper Minnelusa Formation (Permian), Rozet fields area, Powder River basin, Wyoming. In: *Depositional and Diagenetic Spectra of Evaporites: a Core Workshop. SEPM Core Workshop*, **3**, 193–209. Calgary.

Ahmed, E.A., Soliman, M.A., Alsharhan, A.S. and Tamer, S. (1998) Mineralogical characteristics of the dunes in the eastern province of Abu Dhabi, United Arab Emirates. In: *Quaternary Deserts and Climatic Change* (Eds A.S. Alsharhan, K.W. Glennie, G.L. Whittle and C.G.St.C. Kendall), pp. 85–90. Balkema, Rotterdam.

Ahr, W.M. (1973) The carbonate ramp: an alternative to the shelf model. *Trans. Gulf Coast Assoc. Geol. Soc.*, **23**, 221–115.

Ainsworth, W. (1838) *Researches in Assyria, Babylonia and Chaldea*. John W. Parker, London.

Al-Farraj and Harvey, A.M. (2005) Morphology and depositional style of late Pleistocene alluvial fans, Wadi Al-Bih, northern U.A.E. and Oman. In: *Alluvial Fans: Geomorphology, Sedimentology and Dynamics* (Eds. A.M. Harvey, A.E. Mather and M. Stokes). *Geol. Soc. London Spec. Publ.*, **241**, 187–206.

Al-Ghadban, A.N. (1990) Holocene sediments in a shallow bay, southern coast of Kuwait, Arabian Gulf. *Mar. Geol.*, **92**, 237–354.

Al-Hinai, K.G., Moore, J.M. and Bush, P.R. (1987) Landsat image enhancement study of possible submerged sand dunes in the Arabian Gulf. *Int. J. Remote Sens.*, **8**, 251–258.

Al-Reda, A. (1987) *The Clastic Sabkha of the Bahra-Sabbiyah Area of Northern Kuwait*. Unpubl. MSc thesis, University of Kuwait. Kuwait.

Al-Zamel, A.Z. (1983) *Geology and Oceanography of Recent sediments Jazirat Bubiyan and Ras Al-Sabiyah, Kuwait, Arabian Gulf*. Unpubl. PhD thesis, University of Sheffield, UK.

Alsharhan, A.S. and Kendall, C.G.St.C. (1994) Depositional setting of the Upper Jurassic Hith anhydrite of the Arabian Gulf an analogue to Holocene evaporite of the United Arab Emirates and Lake Macleod of western Australia. *AAPG Bull.*, **78**, 1075–1096.

Alsharhan, A.S. and Kendall, C.G.St.C. (2002) Holocene carbonate/evaporites of Abu Dhabi, and their Jurassic ancient analogs. In: *Sabkha Ecosystems* (Eds H.J. Barth and B.B. Boer), pp. 187–202. Kluwer Academic Publishers, The Netherlands.

Alsharhan, A.S. and Kendall, C.G.St.C. (2003) Holocene coastal carbonates and evaporites of the southern Arabian Gulf and their ancient analogues. *Earth Sci. Rev.*, **61**, 191–243.

Alsharhan, A.S. and Whittle, G.L. (1995) Carbonate-evaporite sequences of the Late Jurassic, southern and southwestern Arabian Gulf. *AAPG Bull.*, **79**, 1608–1630.

Alsharhan, A.S., Ahmed, E.A. and **Tamer, S.** (1998) Textural characteristics of Quaternary sand dunes in the eastern province of Abu Dhabi, United Arab Emirates. In: *Quaternary Deserts and Climatic Change* (Eds A.S. Alsharhan, K.W. Glennie, G.L. Whittle and C.G.St.C. Kendall), pp. 91–108. Balkema, Rotterdam.

Al-Thukair, A.A. and **Golubic, S.** (1991) Five new *Hyella* species from the Arabian Gulf. *Algol. Stud.*, **64**, 167–197.

Anon (1944) *Geographical Handbook. Iraq and the Persian Gulf.* Naval Intelligence Division, London.

Aqrawi, A.A.M. (1993a) Palygorskite in the fluvio-lacustrine and deltaic sediments of southern Mesopotamia. *Clay Min. Bull.*, **28**, 153–159.

Aqrawi, A.A.M. (1993b) Implications of sea-level fluctuations, sedimentation and neotectonics for the evolution of the marshlands (Ahwar) of southern Mesopotamia. In: *International Symposium on Neotectonics; Recent Advances* (Eds L. A. Owen, I. Stewart and C. Vita-Finzi), 6–7 June, Burlington House, London. *Quat. Proc.*, **33**, 21–31.

Aqrawi, A.A.M. (1994) Petrography and mineral content of sea-floor sediments of the Tigris-Euphrates Delta, NW Arabian Gulf, Iraq. *Estuar. Coast Shelf Sci.*, **38**, 569–582.

Aqrawi, A.A.M. (1995a) Correction of Holocene sedimentation rates for near-surface compaction; the Tigris-Euphrates Delta. *Mar. Petrol. Geol.*, **12**, 409–416.

Aqrawi, A.A.M. (1995b) Brackish-water and evaporitic Ca-Mg carbonates in the Holocene lacustrine/deltaic deposits of southern Mesopotamia. *J. Geol. Soc. London*, **152**, 259–268.

Aqrawi, A.A.M. (2001a) Nature and preservation of organic matters in the lacustrine/deltaic sediments of southern Mesopotamia. *J. Petrol. Geol.*, **21**, 69–90.

Aqrawi, A.A.M. (2001b) Stratigraphic signatures of climatic change during the Holocene evolution of the Tigris-Euphrates delta, lower Mesopotamia. *Global Planet. Change*, **28**, 267–283.

Aqrawi, A.A.M. and **Evans, G.** (1994) Sedimentation in the lakes and marshes (Ahwar) of the Tigris-Euphrates delta, southern Arabia. *Sedimentology*, **41**, 755–776.

Audley-Charles, M., Curray, J. and **Evans, G.** (1987) Location of major deltas. *Geology*, **5**, 341–344.

Baltzer, F. and **Purser, B.H.** (1990) Modern alluvial fan and deltaic sedimentation in a foreland tectonic setting, Lower Mesopotamian plain and the Arabian Gulf. *Sed. Geol.*, **67**, 175–197.

Baltzer, F., Conchon, O., Freytet, P. and **Purser, B.H.** (1982) Un complexe fluvio-deltaique sursalé et son. Contexte: originalité au Mehran (SE Iran). *Mém. Soc. géol. Fr. Neue Ser.*, **144**, 207–216.

Baltzer, F., Kenig, F., Boichard, R., Plaziat, J.C. and **Purser, B.H.** (1994) Organic matter distribution, water circulation and dolomitization beneath the Abu Dhabi Sabkha (United Arab Emirates). In: *Dolomites: a Volume in Honour of Dolomieu* (Eds B.H. Purser, M.R. Tucker and D. Zenger), *Int. Assoc. Sedimentol. Spec. Publ.*, **21**, 409–428.

Basson, P.W., Burchard, Jr., J. E., Hardy, J.T. and **Price, A.R. G.** (1977) *Biotopes of the Western Arabian Gulf.* Arabian American Oil Company, pp. 284.

Bate, R.H. (1971) The distribution of Recent Ostracoda in the Abu Dhabi Lagoon, Persian Gulf. In: *Paleoecologie Ostracodes* (Ed. H.J. Oerte). *Bull. Centre Rech. Pau-SNPA*, **5**, 239–256.

Becke, C.T. (1835) On the geological evidence of the advance of the land at the head of the Persian Gulf. *Phil. Mag.*, **7**, 40–46.

Bernier, P., Dalongeville, R., Depuis, B. and **Medwecki, V.** (1995) Holocene shoreline variations in the Persian Gulf. Example of the Umm Al-Qowayn lagoon, U.A.E. *Quat. Int.*, **29/30**, 95–103.

Berry, R. W., Brophy, G. P. and **Naqash, A.** (1970) Mineralogy of the suspended sediments in the Tigris, Euphrates, and Shatt al-Arab rivers of Iraq, and the recent history of the Mesopotamian plain. *J. Sed. Petrol.*, **40**, 131–139.

Belvgad, H. (1944) *Fishes of the Iranian Gulf. Danish Scientific Expedition in Iran (1935–1938)*, **3**, 1–247.

Boer, B. and **Aspinall, S.** (1975) Life in the mangroves. In: *The Emirates: a Natural History* (Eds P. Hellyer and S. Aspinall), pp. 133–136. Trident Press, London.

Bradford, M.R. (1975) New dinoflagellates cyst genera from the recent sediments of the Persian Gulf. *Can. J. Bot.*, **53**, 3064–3074.

Bramkamp, R.A. and **Powers, R. W.** (1955) Two Persian Gulf lagoons. *J. Sed. Petrol.*, **25**, 139–140.

Bricker, O.P. (Ed.) (1971) *Carbonate Cements.* Johns Hopkins Press, Baltimore and London, pp. 376.

Bristow, C., Pugh, J. and **Goodall, T.** (1996) Internal structure of aeolian dunes in Abu Dhabi determined using ground penetrating radar. *Sedimentology*, **43**, 99–1003.

Bush, P.R. (1972) *Sedimentology and Groundwater Chemistry of some Recent Coastal Plain Sediments, Abu Dhabi, Trucial Coast, Persian Gulf.* Unpubl. PhD thesis, Imperial College, London, UK.

Bush, P.R. (1973) Some aspects of the diagenetic history of the sabkha in Abu Dhabi Persian Gulf. In: *The Persian Gulf: Holocene Carbonate Sedimentation and Diagenesis in a Shallow Epicontinental Sea* (Ed. B.H. Purser), pp. 395–408. Springer-Verlag, Berlin.

Bush, P.R., Evans, G. and **Glover, E.** (2005) Geological investigations. In: *The Early Dilmun Settlement at Saar* (Eds R. Killick and J. Moon), pp. 339–347. Archaeology International Ltd, Ludlow, UK.

Butler, G.P. (1965) Brines and their evaporites in the Trucial Coast. *Int. Assoc. Sedimentol. Meeting, Reading*, 1–4.

Butler, G.P. (1966) *Early Diagenesis in the Recent Sediments of the Trucial Coast of the Persian Gulf.* Unpubl. MSc thesis, Imperial College, London.

Butler, G.P. (1969) Modern evaporite deposition and geochemistry of co-existing brines, the sabkha, Trucial coast, Arabian Gulf. *J. Sed. Petrol.*, **39**, 70–89.

Butler, G.P. (1970) Recent gypsum and anhydrite of the Abu Dhabi sabkha, Trucial coast: an alternative explanation of origin. In: *Third Symposium on Salt*, **1**. Geol. Soc. N. Ohio, pp. 120–152.

Butler, G.P. (1973) Strontium geochemistry of modern and ancient calcium sulphate minerals. In: *The Persian Gulf: Holocene Carbonate Sedimentation and Diagenesis in a Shallow Epicontinental Sea* (Ed. B.H. Purser), pp. 423–452. Springer-Verlag, Berlin.

Butler, G.P., Kendall, C.G.St.C., Kinsman, D.J.J., Shearman, D.J. and **Skipwith, Sir Patrick A.d'E.**

(1965) Recent anhydrite from the Trucial coast of the Arabian Gulf. *Geol. Soc. London Circ.*, **120**, 3.

Butler, G.P., Krouse, R.H. and Mitchell, R. (1973) Sulphur-isotope geochemistry of an arid supratidal evaporite environment Trucial coast. In: *The Persian Gulf: Holocene Carbonate Sedimentation and Diagenesis in a Shallow Epicontinental Sea* (Ed. B.H. Purser), pp. 453–463. Springer-Verlag, Berlin.

Butler, G.P., Kendall, C.G.St.C. and Harris, P.M. (1982) Recent evaporites from the Abu Dhabi coastal flats. In: *Depositional and Diagenetic Spectra of Evaporites* (Eds G.R. Handford, R.G. Loucks and G.R. Davies), *SEPM Core Workshop*, **3**, 33–64.

Carter, H. J. (1849) On the foraminifera, their existence in a fossilized state in Arabia, Sindh, Kutch and Kattyawar. *J. Roy. Asiatic Soc. Bombay*, **3**, 158.

Carter, H. J. (1859) Report on geological specimens from the Persian Gulf collected by Lieutenant C.G. Constable, 1.N. *J. Roy. Asiatic Soc. Bombay*, **28**, 41–48; **29**, 359–365.

Carter, W.G. (1834) Remarks on Beke's papers on the Gopher Wood and the former extension of the Persian Gulf. *Phil. Mag. Ser.* **3**, 244–252.

Carter, W.G. (1835) On the ancient and modern formation of the Delta by the Euphrates and Tigris. *Phil. Mag. Ser.* **3**, 192–202, 250–256.

Chafetz, H.S. and Rush, P.F. (1994) Diagenetically altered sabkha-type Pleistocene dolomite from the Arabian Gulf. *Sedimentology*, **41**, 409–421.

Chafetz, H.S., McIntosh, A.G. and Rush, P.F. (1988) Freshwater phreatic diagenesis in the marine realm of recent Arabian Gulf carbonates. *J. Sed. Petrol.*, **57**, 433–440.

Chapman, F. (1900) Notes on consolidated aeolian sands of Kathiawar. *Q. J. Geol. Soc. London*, **LVI**, 584–589.

Cloet, R.L. (1954) Sand waves in the southern North Sea and in the Persian Gulf. *J. Inst. Navig., VII*, **3**, 272–279.

Cloud, P.E. (1962) Environment of calcium carbonate deposition west of Andros Island, Bahamas. *US Geol. Surv. Prof. Pap.*, **350**, 138.

Cooke, G.A. (1987) Reconstruction of the Holocene coastline of Mesopotamia. *Geoarchaeology*, **2**, 15–28.

Cooke, R.U. and Goudie, A.S. (1980) Aeolian landforms and deposits: the aeolianite. In: *Geology, Geomorphology and Pedology of Bahrain* (Eds J.C. Doornkamp, D.C. Brunsden and D.K.C. Jones), pp. 216–219. GeoAbstracts Ltd, Norwich.

Cuff, C. (1969) Lattice disorder in Recent anhydrite and its geological implications. *Geol. Soc. London Proc.*, **1659**, 326–330.

Cuff, C. (1973) *The Crystallography and Geochemistry of Authigenic Calcium Sulphate and Carbonate Minerals of the Recent Sediments of the Trucial Coast.* Unpubl. PhD thesis, Imperial College, London, UK.

Curtis, R., Evans, G., Kinsman, D.J.J. and Shearman, D.J. (1963) Association of dolomite and anhydrite in the Recent sediments of the Persian Gulf. *Nature*, **197**, 679–680.

Dalongeville, R., Bernier, P., Dupuis, B. and de Medwicki, V. (1993) Les variations récentes de la ligne di rivage dans le Golfe Persique: l'exemple de la lagune d'Umm Al-Qowayn (Emirates Arabes Unis). *Bull. Inst. Géol. Bassin Aquitaine*, **53**, 179–192.

Darwish, A.H. and Conley, C.D. (1989) Pleistocene-Holocene sedimentation and diagenesis along the King Fahd causeway between Saudi Arabia and Bahrain. *J. King Fahd Univ. Earth Sci.*, **3**, 63–79.

de Groot, K. (1965) Inorganic precipitation of calcium carbonate from sea water. *Nature*, **207**, 404–405.

de Groot, K. (1969) The chemistry of submarine cement formation at Dohat Hussain in the Persian Gulf. *Sedimentology*, **12**, 63–68.

de Groot, K. (1973) Geochemistry of tidal flat brines at Umm Said, SE Qatar, Persian Gulf. In: *The Persian Gulf: Holocene Carbonate Sedimentation and Diagenesis in a Shallow Epicontinental Sea* (Ed. B.H. Purser), pp. 377–395. Springer-Verlag, Berlin.

de Morgan, J. (1900) *Mémoires de la Délégation en Perse, Paris*, MDP, **1**, 4–48.

Diester, L. (1971) *Grobfraktionsanalyse von Sedimentkernen aus dem Persischen Golf.* Unpubl. PhD thesis, University of Kiel, Germany.

Diester, L. (1972a) Zur spätpleistozänen und holozänen Sedimentation im zentralen und östlichen Persischen Golf. *Meteor Forschungsergeb., Reihe C.*, **8**, 37–83.

Diester, L. (1972b) Grobfraktionsanalyse von Sedimentkernen aus dem Persischen Golf. *Meteor Forschungsergeb., Reihe C.*, **8**, 87–103.

Diester, L. (1973) Holocene climate in the Persian Gulf as deduced from grain-size pteropod distribution. *Mar. Geol.*, **14**, 207–223.

Dietrich, G., Krause, G., Seibold, E. and Vollbrecht, K. (1966) Reisebericht der Indischen Ozean Expedition mit Forschungsschiff "METEOR" 1964–1965. *Meteor Forschungsergeb., Reihe A.*, **1**, 1–52.

Edwards, A., Bush, P.R. and Evans, G. (1986) Salt and contaminant flux in the Abu Dhabi lagoon. *Rapp. Réunion Conserv. Int. Explor.*, **186**, 412–422.

El-Sayed, M.I. (1999) Sedimentological characteristics and morphology of the aeolian sand dunes in the eastern part of the U.A.E. a case study from Ar' Rub al Khali. *Sed. Geol.*, **124**, 219–238.

Elder, S. (1957) Discussion of K.O. Emery's paper: the sediments and water of the Persian Gulf. *AAPG Bull.*, **41**, 332–333.

Ellis, J.P. and Milliman, J.D. (1985) Calcium carbonate suspended in Arabian Gulf and Red Sea waters: biogenic and detrital, not "chemigenic". *J. Sed. Petrol.*, **55**, 805–808.

Emery, K.O. (1956) Sediments and water of the Persian Gulf. *AAPG Bull.*, **40**, 2354–2383.

Esteoule, J., Esteoule-Choux, J., Melguen, M. and Seibold, E. (1970) Sur la présence d'attapulgite dans des sédiments récents du Nord-Est du Golfe Persique. *Centre Rech. Acad. Sci., Paris*, **271**, 1153–1156.

Evamy, B.D. (1973) The precipitation of aragonite and its alteration to calcite on the Trucial coast of the Persian Gulf. In: *The Persian Gulf: Holocene Carbonate Sedimentation and Diagenesis in a Shallow Epicontinental Sea* (Ed. B.H. Purser), pp. 329–342. Springer-Verlag, Berlin.

Evans, G. (1966) The Recent sedimentary facies of the Persian Gulf region. *Phil. Trans. Roy. Soc. London Ser. A.*, **259**, 291–298.

Evans, G. (1970) Coastal and nearshore sedimentation: a comparison of clastic and carbonate deposition. *Proc. Geol. Assoc. London*, **81**, 493–508.

Evans, G. (1979) A discussion of: the Mesopotamian delta in the first millennium BC by Hansman J. F. *Geogr. J.*, **144**, 529–531.

Evans, G. (1989) The control of source and climate on coastal facies and their associated economic deposits: is there a rationale? *Bull. Soc. Géol. Fr.*, **8**, 105–1064.

Evans, G. (1995) The Arabian Gulf: a modern carbonate-evaporite factory: a review. *Cuad. Geol. Ibérica*, **19**, 61–96.

Evans, G. and Bush, P.R. (1970) Some sedimentological and oceanographical observations on a Persian Gulf lagoon. *Proc. UNESCO Conf. Coastal Lagoons, Mexico City, November 1967* 155–169.

Evans, G. and Kirkham, A. (2002) The Abu Dhabi sabkha. In: *Sabkha Ecosystems* (Eds H.J. Barth and B.B. Böer), pp. 7–20. Kluwer Academic Publishers, Dordrecht, The Netherlands.

Evans, G. and Kirkham, A. (2005) The Quaternary deposits. In: *The Emirates: A Natural History* (Eds P. Hellyer and S. Aspinall), pp. 65–78. Trident Press, London.

Evans, G. and Kirkham, A. (2009) A Geological Description of Belghelam Island, North-east Abu Dhabi, U.A.E. *Tribulus*, **18**, 4–9.

Evans, G. and Kirkham, A. (2009) A Report on a Geological Reconnaissance of Al Aryam Island, March 2004, Abu Dhabi. *Tribulus*, **18**, 10–17.

Evans, G. and Shearman, D.J. (1964) Recent celestite from the sediments of the Trucial coast of the Persian Gulf. *Nature*, **202**, 385–386.

Evans, G., Kendall, C.G.St.C. and Skipwith, P.A.d'E. (1964a) Origin of the coastal flats, the sabkha, of the Trucial coast, Persian Gulf. *Nature*, **202**, 579–600.

Evans, G., Kinsman, D.J.J. and Shearman, D.J. (1964b) A reconnaissance survey of the environment of Recent carbonate sedimentation along the Trucial coast, Persian Gulf. In: *Developments in Sedimentology, 1, Deltaic and Shallow Marine Deposits* (Ed. L.M.J.U. van Straaten), pp. 129–135. Elsevier, Amsterdam.

Evans, G., Schmidt, V., Bush, P. and Nelson, H. (1969) Stratigraphy and geologic history of the sabkha Abu Dhabi, Persian Gulf. *Sedimentology*, **12**, 145–159.

Evans, G., Murray, J.W., Biggs, H.E.J., Bate, R. and Bush, P.R. (1973) The oceanography, ecology, sedimentology and geomorphology of parts of the Trucial coast barrier island complex, Persian Gulf. In: *The Persian Gulf: Holocene Carbonate Sedimentation and Diagenesis in a Shallow Epicontinental Sea* (Ed B.H. Purser), pp. 233–278. Springer-Verlag, Berlin.

Evans, G., Bush, P.R. and Temple, P. (1980) The coastal zone of Bahrain. In: *Geology, Geomorphology and Pedology of Bahrain* (Eds J.C. Doornkamp, D. Brundsden, and D.K.C. Jones), pp. 269–327. GeoAbstracts Ltd, Norwich.

Evans, G., Kirkham, A. and Carter, R.A. (2002) Quaternary development of the United Arab Emirates coast: new evidence from Marawah Island, Abu Dhabi. *GeoArabia*, **7**, 441–458.

Evans, J.W. (1900) Mechanically formed limestones from Jungarh (Kathiawar) and other localities. *Q. J. Geol. So. London*, **LXVI**, 559–583.

Felber, H., Hötzl, H., Maurin, V., Moser, H., Rauert, W. and Zötl, J.G. (1978) Sea level fluctuations during the Quaternary period. In: *Quaternary Period in Saudi Arabia* (Eds S.A. Al-Sayari and J.G. Zötl), pp. 50–57. Springer-Verlag, Vienna.

Ferguson, J. and Ibe, A.C. (1981) Origin of light hydrocarbons in carbonate ooliths. *J. Petrol. Geol.*, **4**, 103–107.

Foda, M.A., Khalaf, F.I. and Al-Khadi, A.S. (1985) Estimation of dust fallout and rates in the northern Arabian Gulf. *Sedimentology*, **32**, 595–603.

Friedman, G.M. (1995) The arid peritidal complex of Abu Dhabi: a historical perspective. *Carbonates Evaporites*, **10**, 2–7.

Fryberger, S.G., Al Sari, A.M. and Clisham, T.J. (1983) Eolian dune, interdune sand sheet and siliciclastic sabkha sediments of an offshore prograding sand sea, Dharhan, Saudi Arabia. *AAPG Bull.*, **67**, 280–312.

George, D. (2005) Marine invertebrates. In: *The Emirates: a Natural History* (Eds P. Hellyer and S. Aspinall), pp. 197–220.

George, D. and John, D. (1999) High sea temperatures along the coast of Abu Dhabi, U.A.E., Arabian Gulf – their impact upon corals and macroalgae. *Newsl. Int. Soc. Reef Stud., Reef Encounter*, **25**, 21–23.

George. D. and John, D. (2005) The marine environment. In: *The Emirates: a Natural History* (Eds P. Hellyer and S. Aspinall), pp. 111–115.

Ginsburg, R.N. (1953) Beachrock in south Florida. *J. Sed. Petrol.*, **23**, 85–92.

Ginsburg, R.N. (1956) Environmental relationships of grain size and constituent particles in some south Florida carbonate sediments. *AAPG Bull.*, **40**, 2384–2427.

Ginsburg, R.N. (1957) Early diagenesis and lithification of shallow water carbonate sediments in south Florida. In: *Regional Aspects of Carbonate Deposition* (Eds R.J. Le Blanc and J.G. Breeding). *SEPM Spec. Publ.*, **5**, 80–100.

Glennie, K.W. (1970) *Desert Sedimentary Environments. Dev. Sedimentol.*, **14**, 222 pp. Elsevier, Amsterdam,.

Glennie, K.W. (1972) Permian Rotliegendes of northwestern Europe interpreted in the light of modern desert sedimentation studies. *AAPG Bull.*, **5**, 1048–1071.

Glennie, K.W. (1987) Desert sedimentary environments, present and past – a summary. *Sed. Geol.*, **50**, 135–165.

Glennie, K.W. (1994) Quaternary dunes of SE Arabia and Permian (Rotliegendes) dunes of NW Europe: some comparisons. *Zbl. Geol. Paläontol.*, **1**, 1199–1215.

Glennie, K.W. (1997) Evolution of the Emirates' land surface: an introduction. In: *Perspectives of the United Arab Emirates* (Eds J.G. Ghareeb and I. Al Abed), pp. 17–35. Trident Press, London.

Glennie, K.W. (1998) The desert of southeast Arabia: a product of Quaternary climatic change. In: *Quaternary Deserts and Climatic Change* (Eds A.S Alsharhan, K.W. Glennie, G.L. Whittle and C.G.St.C. Kendall), pp. 279–291. Balkema, Rotterdam.

Glennie, K.W. (1999) Dunes as indicators of climatic change. In: (Eds A.K. Singhvi and E. Derbyshire), pp. 153–174. *Paleoenvironmental Reconstruction in Arid Lands*. Oxford & IBH, New Delhi.

Glennie, K.W. (2005) *The Deserts of Southeast Arabia*. Gulf Petrolink, Manama, Bahrain.

Glennie, K.W. and Evamy, B.D. (1968) Dikaka: plants and plant-root structures associated with aeolian sand. *Palaeogeogr. Palaeoclimatol. Palaeoecol.*, **4**, 77–87.

Glennie, K.W. and Evans, G. (1976) A reconnaissance of the Recent sediments of the Ranns of Kutch, India. *Sedimentology*, **23**, 625–647.

Glennie, K.W. and Singhvi, A. K. (2002) Event stratigraphy, palaeoenvironment and chronology of SE Arabian Deserts. *Quat. Sci. Rev., Spec. Issue*, **21**, 853–869.

Glennie, K.W., Al-Belushi, J. and Al-Maskery, S. (1998) The inland sabkhas of the Huqf, Oman: a product of past extremes of humidity and aridity. In: *Quaternary Deserts and Climatic Change* (Eds A.S. Alsharhan, K.W. Glennie, G.L. Whittle and C.G.St.C. Kendall), pp. 117–122. Balkema, Rotterdam.

Glennie, K.W., Lancaster, N., Singhvi, A.K. and Teller, J.T. (1999) Responses of dune systems to climatic and sea level changes: United Arab Emirates. *Poster, XV INQUA Congress, The environmental background to Hominid evolution in Africa, 3–11 August 1999, Durban, South Africa*

Glennie, K. W., Singhvi, A. K., Lancaster, N. and Teller, J. T. (2002) Quaternary climatic changes over southern Arabia and the Thar desert, India. In: *The Tectonic and Climatic Evolution of the Arabian Sea Region* (Eds P. Clift, D. Kroon, C. Gaedicke and J. Craig). *Geol. Soc. London Spec. Publ.*, **195**, 301–316.

Godwin, H. and Willis, E.H. (1959) *Am. J. Sci. Radiocarbon Suppl.*, **1**, 63–75.

Godwin, H., Suggate, R.P. and Willis, E.H. (1958) Radiocarbon dating of the eustatic rise in ocean level. *Nature*, **181**, 1518–1519.

Goodall, T.M. (1995) *The Geology and Geomorphology of the Sabkhat Matti Region (United Arab Emirates): a Modern Analogue for Ancient Desert Sediments from North-west Europe.* Unpubl. PhD thesis, University of Aberdeen, UK.

Goudie, A.S., Colls, A., Stokes, S., Parker, A., White, K. and Al-Farraj, A. (2000) Latest Pleistocene and Holocene dune construction at the north-eastern edge of the Rub Al Khali, United Arab Emirates. *Sedimentology*, **47**, 1011–1021.

Goudie, A.S., Parker, A.G. and Al-Farraj, A. (2000) Coastal changes in Ras Al Khaimah (United Arab Emirates), a cartographic analysis. *Geogr. J.*, **166**, 14–25.

Gunatilaka, A. (1986) Kuwait and the northern Arabian Gulf: a study in Quaternary sedimentation. *Episodes*, **9**, 223–231.

Gunatilaka, A. (1987) The "dolomite problem" in the light of recent studies. *Modern Geol.*, **11**, 311–324.

Gunatilaka, A. (1989) Low latitude anhydrite from the Red Sea coast: palaeoclimatic and palaeohydrological implications. *Terra Nova*, **1**, 280–283.

Gunatilaka, A. (1990) Anhydrite diagenesis in a vegetated sabkha, Al Khiran, Kuwait, Arabian Gulf. *Sed. Geol.*, **69**, 95–116.

Gunatilaka, A. (1991) Dolomite formation in coastal Al-Khiran, Kuwait, Arabian Gulf. A re-evaluation of the sabkha model. *Sed. Geol.*, **72**, 35–53.

Gunatilaka, A. and Mwango, S. (1987) Continental sabkha pans and associated nebkhas in southern Kuwait, Arabian Gulf. In: *Desert Sediments Ancient and Modern* (Eds L. Frostick, and I. Reid). *Geol. Soc. London Spec. Publ.*, **35**, 187–203.

Gunatilaka, A. and Mwango, S. (1989) Flow separation and the internal structure of shadow dunes. *Sed. Geol.*, **61**, 125–134.

Gunatilaka, A. and Shearman, D.J. (1988) Gypsum-carbonate laminites in a recent sabkha, Kuwait. *Carbonates Evaporites*, **3**, 67–73.

Gunatilaka, A., Saleh, A. and Al-Temeemi, A. (1980) Plant controlled supratidal anhydrite from Al Khiran, Kuwait. *Nature*, **288**, 257–260.

Gunatilaka, A., Saleh, A. and Al-Temeemi, A. (1984) Occurrence of subtidal dolomite in a shallow hypersaline lagoon, Kuwait. *Nature*, **311**, 450–452.

Gunatilaka, A., Saleh, A., Al-Temeemi, A. and Nassar, N. (1984) Occurrence of subtidal dolomite in a hypersaline lagoon, Kuwait, Arabian Gulf. *Nature*, **311**, 450–452.

Gunatilaka, A., Al-Temeemi, A., Saleh, A. and Nassar, N. (1985) A new occurrence of basanite in recent evaporite environments, Kuwait, Arabian Gulf. *J. Univ. Kuwai, Sci.*, **12**, 157–166.

Gunatilaka, A, Al-Zamel, A., Shearman, D.J. and Reda, A. (1987) A spherulitic fabric in selectively dolomitised siliciclastic crutacean burrows, Northern Kuwait. *J. Sed. Petrol.*, **57**, 922–927.

Gunatilaka, A., Saleh, A., Al-Temeemi, A. and Nassar, N. (1987) Calcium-poor dolomites from the sabkhas of Kuwait, Arabian Gulf. *Sedimentology*, **34**, 999–1006.

Haake, F.W. (1970) Zur Tiefenverteilung von Miliolinen (Foram.) im Persischen Golf. *Paläontol. Zool.*, **44**, 196–200.

Hadley, D.G., Brouwers, E.M. and Bown, T.M. (1998) Quaternary paleodunes, Arabian Gulf coast, Abu Dhabi Emirate: age and paleoenvironmental evolution. In: *Quaternary Deserts and Climatic Change* (Eds A.S. Al-Sharhan, K.W. Glennie, G.L. Whittle and C.G.St.C. Kendall), pp. 123–139. Balkema, Rotterdam.

Hellyer, P. and Aspinall, S. (Eds) (2005) *The Emirates, A Natural History.* Trident Press, London. 424 pp

Hilmy, M.E., Slansky, E. and Khalaf, F.I. (1971) Opaque minerals in Recent beach sediments of Kuwait. *Neues Jb. Geol. Paläontol.*, **6**, 340–344.

Hoefs, J. and Sarnthein, M. (1971) $^{18}O/^{16}O$ ratios and related temperatures of Recent Pteropod shells (*Cavolinia longirostris* LESUEUR) from the Persian Gulf. *Mar. Geol.*, **10**, 11–16.

Holail, H.M. (1999) The isotopic composition and diagenetic history of Pleistocene carbonates, W. Qatar. *Carbonates Evaporites*, **14** 41–55.

Holail, H.M., Shaabon, M.N. and Mansour, A.S. (2004) Cementation of Holocene beachrocks in the Aqaba and Arabian Gulfs, a comparative study. *Carbonates Evaporites*, **19**, 142–150.

Holm, D.A. (1960) Desert geomorphology in the Arabian Peninsula. *Science*, **132** 1369–1379.

Houbolt, J.J.H.C. (1957) *Surface Sediments of the Persian Gulf near the Qatar Peninsula.* PhD thesis, University of Utrecht. Den Haag: Mouton and Co., The Netherlands.

Hoveland, M. and Judd, A.G. (1988) *Seabed Pockmarks and Seepages: Impact on Geology, Biology and the Marine Envronment.* Graham and Trotman, London.

Hsü, K.J. and Schneider, J. (1973) Progress report on dolomitization-hydrology of Abu Dhabi sabkhas, Arabian Gulf. In: *The Persian Gulf: Holocene Carbonate Sedimentation and Diagenesis in a Shallow Epicontinental Sea* (Ed. B.H. Purser), pp. 409–422. Springer-Verlag, Berlin.

Hsü, K.J. and Siegenthaler, C. (1969) Preliminary experiments on hydrodynamic movement induced by evaporation and their bearing on the dolomite problem. *Sedimentology*, **12**, 11–25.

Hudson, R.G.S., Eames, F.B. and Wilkins, G.L. (1957) The fauna of some recent marine deposits near Basrah, Iraq. *Geol. Mag.*, **94**, 393–401.

Hughes-Clarke, M.W. and **Keij, A.J.** (1973) Organisms as producers of carbonate sediment and indicators of environment in the southern Persian Gulf. In: *The Persian Gulf: Holocene Carbonate Sedimentation and Diagenesis in a Shallow Epicontinental Sea* (Ed. B.H. Purser), pp. 33–56. Springer-Verlag, Berlin.

Hussain, M. (2006) Recognised attributes and oomoldic porosity development in eolianite in a semi-arid setting. An example fom the Quaternary Gulf coastline, Saudi Arabia. *Carbonates Evaporites*, **21**, 124–132.

Illing, L.V. (1954) Bahamian calcareous sands. *AAPG Bull.*, **38**, 1–95.

Illing, L.V. (1963) Discussion of Shearman: Recent anhydrite, gypsum, dolomite and halite from the coastal flats of the Arabian shore of the Persian Gulf. *Proc. Soc. London*, **1607**, 4.

Illing, L.V. and **Taylor, J. C. M.** (1993) Penecontemporaneous dolomitization in Sabkha Faishakh, Qatar: evidence from changes in the chemistry of the interstitial breaks. *J. Sed. Petrol.*, **63**, 1042–1048.

Ionides, M.G. (1937) *The Regime of the Rivers Euphrates and Tigris.* Spon Ltd, London, 278 pp.

Ionides, M.G. (1955) A reply to Lees and Falcon, 1952. *Geogr. J.*, **121**, 394–395.

John, D. (2005) Marine plants. In: *The Emirates: A Natural History* (Eds P. Hellyer and S. Aspinall), pp. 161–167. Trident Press, London.

John, D. and **George, D.** (2005) The shore and shallow seas. In: *The Emirates: A Natural History* (Eds P. Hellyer and S. Aspinall), pp. 123–128. Trident Press, London.

Jones, D.A. (1985) The biological characteristics of the marine habitats found within the ROPME. *Proc. ROPME Symposium, Al Ain, Dec. 1985, ROPME/GC-4/2* 71–89.

Juyal, N., Glennie, K.W. and **Singhvi, A.K.** (1998) Chronology and paleoenvironmental significance of Quaternary desert sediments in SE Arabia. In: *Quaternary Deserts and Climatic Change* (Eds A.S. Alsharhan, K. W. Glennie, G.L. Whittle and C.G.St.C. Kendall), pp. 315–325. Balkema, Rotterdam.

Kassler, P. (1973) Geomorphic evolution of the Persian Gulf. In: *The Persian Gulf: Holocene Carbonate Sedimentation and Diagenesis in a Shallow Epicontinental Sea* (Ed. B.H. Purser), pp. 11–32. Springer-Verlag, New York.

Kendall, A.C. (1978) Facies models, 11; continental and supratidal evaporites. *Geosci. Canada*, **5**, 66–78.

Kendall, C.G.St.C. (1966) *Recent Sediments of the Western Khor al Bazam, Abu Dhabi, Trucial Coast.* Unpubl. PhD thesis, University of London, UK.

Kendall, C.G.St.C. and **Skipwith, P.A.d'E.** (1966) Recent algal stromatolites of the Khor al Bazam, southwest Persian Gulf. *Geol. Soc. Am. Spec. Pap. Abstr.*, p. 108.

Kendall, C.G.St.C. and **Skipwith, P.A.d'E.** (1968) Recent algal mats of a Persian Gulf Lagoon. *J. Sed. Petrol.*, **38**, 1040–1058.

Kendall, C.G.St.C. and **Skipwith P.A.d'E.** (1969a) Geomorphology of a recent shallow water carbonate province: Khor Al Bazam, Trucial Coast, southwest Persian Gulf. *Geol. Soc. Am. Bull.*, **80**, 865–892.

Kendall, C.G.St.C. and **Skipwith, P.A.d'E** (1969b) Holocene shallow-water carbonate and evaporite sediments of Khor al Bazam, Abu Dhabi, southwest Persian Gulf. *AAPG Bull.*, **53**, 841–869.

Kendall, C.G.St.C., Sadd, J.L. and **Alsharhan, A.S.** (1994) Holocene marine cement of beachrocks of the Abu Dhabi coastline (United Arab Emirates): analogs for cement fabrics in ancient limestones. *Carbonates Evaporites*, **9**, 119–131.

Kendall, C.GSt.C., Alsharhan, A.S. and **Whittle, G.L.** (1998) The flood re-charge sabkha model supported by recent inversions of anhydrite to gypsum in the U.A.E. sabkhas. In: *Quaternary Deserts and Climatic Change* (Eds A.S. Alsharhan, K.W. Glennie, G.L. Whittle and C.G.St.C. Kendall), pp. 29–42. Balkema, Rotterdam.

Kendall, C.G.St.C., Alsharhan, A.S. and **Cohen, A.** (2002) The Holocene tidal flat complex of the Arabian Gulf Coast of Abu Dhabi. In: *Sabkha Ecosystems* (Eds H.J. Barth and B.B. Böer), pp. 21–35. Kluwer Academic Publishers, Dordrecht, Holland.

Kenig, F., Hue, A.Y., Purser, B.H. and **Oudin, J.L.** (1989) Sedimentation, distribution and diagenesis of organic matter in a recent carbonate environment, Abu Dhabi, U.A.E. *Adv. Org. Geochem.*, **16**, 735–749.

Kenig, F., Boichard, R. and **Purser, B.H.** (1991) Holocene organo-sedimentary sequence of the Abu Dhabi coastal carbonate environment, United Arab Emirates. In: *Dolomieu Conference on Carbonate Platforms and Dolomitization* (Abstract Book) (Eds A. Bosellini, R. Brander, E. Flügel, B.H. Purser, W. Schlager, M. Tucker and D. Zenger), pp. 134–140. Int. Assoc. Sedimentol., Oxford, U.K.

Kennet, O.J. and **Kennet, J.P.** (2006) Early slate formation in southern Mesopotamian sea levels, shorelines and climatic change. *J. Isl. Coast Archaeol.*, **1**, 67–99.

Khalaf, F. and **Ala, M.** (1980) Mineralogy of the recent intertidal muddy sediments of Kuwait-Arabian Gulf. *Mar. Geol.*, **359**, 331–342.

Khalaf, F.I. and **Al-Hashash, M.** (1983) Aeolian sedimentation in the north-western parts of the Arabian Gulf. *J. Arid Environ.*, **6**, 319–332.

Khalaf, F.I., Al-Saleh, S., Al-Houty, F., Ansari, L. and **Shublaq, W.** (1979) Mineralogy and grain size distribution of dust fall-out in Kuwait. *Atmos. Environ.*, **13**, 1719–1723.

Khalaf, F.I., Al-Ghadban, A., Al-Saleh, S. and **Al-Omran, L.** (1982) Sedimentology and mineralogy of Kuwait Bay bottom sediments, Kuwait-Arabian Gulf. *Mar. Geol.*, **46**, 71–99.

Khalaf, F., Al-Bakri, D. and **Al-Ghadban, A.** (1984) Sedimentological characteristics of the surficial sediments of the Kuwaiti marine environment, northern Arabian Gulf. *Sedimentology*, **31**, 531–545.

Khan, N.Y., Munawar, M. and **Price, A.R.G.** (2002) *The Gulf Ecosystem: Health and Sustainability.* Backhuys Publishers, Leiden, The Netherlands.

Khantaprab, C. (1968) *Petrology and Geochemistry of some Quaternary Limestones from the Persian Gulf.* Unpubl. PhD thesis, Imperial College, London, UK.

Kinsman, D.J.J. (1964a) *Recent Carbonate Sedimentation near Abu Dhabi, Trucial Coast, Persian Gulf.* Unpubl. Ph. D thesis, Imperial College, London, UK.

Kinsman, D.J.J. (1964b) The recent carbonate sediments near Halat el Bahrani, Trucial coast, Persian Gulf. In: *Deltaic and Shallow Marine Deposits* (Ed. L.M.J.U. van Straaten), pp. 185–192. *Dev. Sedimentol.*, **1**, Elsevier, Amsterdam.

Kinsman, D.J.J. (1964c) Reef coral tolerance of high temperatures and salinities. *Nature*, 1280–1282.

Kinsman, D.J.J. (1965) Experimental studies and recent occurrence of gypsum, basanite and anhydrite. *Int. Assoc. Sedimentol. Meeting, Reading*, 1–4.

Kinsman, D.J.J. (1966) Gypsum and anhydrite of Recent age, Trucial coast, Persian Gulf. *Proc. 2nd Salt Symposium, 1. Northern Ohio Geological Society Cleveland, Ohio*, 302–326.

Kinsman, D.J.J. (1967) Huntite from a carbonate – evaporite environment. *Am. Mineral.*, **52**, 1332–1340.

Kinsman, D.J.J. (1969a) Interpretation of Sr^{++} concentrations in carbonate minerals and rocks. *J. Sed. Petrol.*, **39**, 486–508.

Kinsman, D.J.J. (1969b) Modes of formation, sedimentary associations and diagnostic features of shallow-water and supratidal evaporites. *AAPG Bull.*, **53**, 830–840.

Kinsman, D.J.J. (1976) Evaporites: relative humidity control of primary mineral facies. *J. Sed. Petrol.*, **46**, 273–279.

Kinsman, D.J.J. and Park, R.K. (1976) Algal belt and coastal sabkha evolution, Trucial coast, Persian Gulf. In: *Stromatolites* (Ed. M.N. Walter). *Dev. Sedimentol.*, **26**, 421–433.

Kirkham, A. (1997) Shoreline evolution, aeolian deflation and anhydrite distribution of the Holocene, Abu Dhabi. *GeoArabia*, **2**, 403–416.

Kirkham, A. (1998a) Pleistocene carbonate seif dunes and their role in the development of complex past and present coastlines of the U.A.E. *GeoArabia*, **3**, 19–32.

Kirkham, A. (1998b) A Quaternary proximal foreland ramp and its continental fringe, Arabian Gulf, U.A.E. In: *Carbonate Ramps* (Eds V.P. Wright and T.P. Burchette). *Geol. Soc. Londo, Spec. Publ.*, **149**, 15–41.

Kirkham, A. and Evans, G. (2008) Giant burrows in the Quaternary Limestones of Futaysi Island and Al Dabb'iya, Abu Dhabi Emirate. *Palaeogeogr. Palaeclimatol. Palaeoecol.*, **270**, 324–331.

Kirkham, A. and Evans, G. (2010) Quantification of carbonate ramp sedimentation and progradation rates for the late Holocene Abu Dhabi coastline: Discussion. *J. Sed. Res.*, **80**, 300–301.

Klement, M.C. (1894) Sur la formacion de la dolomie: communication preliminaire. *Bull. Soc. Belge géol. (Mém.)*, **8**, 219–224.

Klement, M.C. (1895) Sur l'origine de la dolomie dans les formations sedimentaires. *Bull. Soc. Belge géol. (Mém.)*, **9**, 23.

Kukal, Z. and Saadallah, A. (1970) Composition and rate of deposition of the Recent dust storm sediments in Iraq. *Cas. Mineral. Geol.*, **15**, 227–234.

Kukal, Z. and Saadallah, A. (1973) Aeolian admixtures in the sediments of the northern Persian Gulf. In: *The Persian Gulf: Holocene Carbonate Sedimentation and Diagenesis in a Shallow Epicontinental Sea* (Ed. B.H. Purser), pp. 115–122. Springer-Verlag, Berlin.

Lambeck, K. (1996) Shoreline reconstructions for the Persian Gulf since the last glacial maximum. *Earth Planet. Sci. Lett.*, **142**, 43–57.

Lange, J. (1970) Geochemische Untersuchungen an Sedimenten des Persischen Golf. *Contrib. Mineral. Petrol.*, **28**, 288–305.

Lange, J.H. von Sarnthein, M. (1970) Glauconitkörner in rezenten sedimenten des Persischen Golfs. *Geol. Rundsch.*, **60**, 256–264.

Larsen, C.E. (1975) The Mesopotamian Delta Region: a reconstruction of Lees and Falcon. *J. Am. Orient. Soc.*, **95**, 43–57.

Larsen, C.E. and Evans, G. (1978) The Holocene geological history of the Tigris-Euphrates-Karun delta. In: *The Environmental History of the Near and Middle East* (Ed. W.C. Brice), pp. 227–244. Academic Press, London.

Lees, G.M. and Falcon, N.L. (1952) The geographical history of the Mesopotamian plains. *Geogr. J.*, **116**, 24–39.

Loftus, W.K. (1855) On the geology of portions of the Turko-Persian frontier and of the districts adjoining. *Q. J. Geol. Soc. London*, **11**, 247–344.

Lokier, S. and Steuber, T. (2008) Quantification of carbonate ramp sedimentation and progradation role for the late Holocene, Abu Dhabi shoreline. *J. Sed. Res.*, **78**, 423–431.

Lokier, S. and Steuber, T. (2009) Large scale intertidal polygonal features of the Abu Dhabi coastline. *Sedimentology*, **56**, 609–621.

Lokier, S. and Steuber, T. (2010) Quantification of carbonate ramp sedimentation and progradation rates for the late Holocene Abu Dhabi coastline: Reply. *J. Sed. Res.*, **80**, 302.

Lomando, A.J. (1999) Structural influences on facies trends of carbonate inner ramp systems, examples from the Kuwait-Saudi Arabian coast of the Arabian Gulf and the northern Yucatan, Mexico. *GeoArabia*, **4**, 339–360.

Loreau, J.-P. (1982) Sediments aragonitiques et leur genése. *Mém. Musée Nat. Hist. Sér. C, Géol.*, **XLVII**, p. 312.

Loreau, J.-P. and Purser, B.H. (1973) Distribution and ultrastructure of Holocene ooids in the Persian Gulf. In: *The Persian Gulf: Holocene Carbonate Sedimentation and Diagenesis in a Shallow Epicontinental Sea* (Ed. B.H. Purser), pp. 279–328. Springer-Verlag, Berlin.

Lutze, G.F., Grabert, B. and Seibold, E. (1971) Lebendbeobachtungen an Groß-Foraminiferen (Heterostegina) aus dem Persischen Golf. *Meteor Forschungsergeb., Reihe C.*, **6**, 21–40.

MacFadyen, W.A. and Vita-Finzi, C. (1978) Mesopotamia: the Tigris-Euphrates delta and its Holocene Hammar fauna. *Geol. Mag.*, **115**, 287–300.

Maizels, J. (1988) Palaeochannels: Plio-Pleistocene raised channel systems of the western Sharqiyah. In: *The Scientific Results of the Royal Geographical Society's Oman Wahiba Project 1985–1987* (Ed. R.W. Dutton). *J. Oman Stud. Spec. Rep.*, **3**, 95–112.

Maizels, J. and McBean, C. (1990) Cenozoic alluvial fan system of interior Oman: palaeoenvironmental reconstruction based on discrimination of palaeochannels using remotely sensed data. In: *The Geology and Tectonics of the Oman Region* (Eds A.H.F. Robertson, M.P. Searle and A.C. Ries). *Geol. Soc. Lond. Spec. Publ.*, **49**, 565–582.

Martini, E. (1967) Nannoplankton und Umlagerungserscheinungen im Persischen Golf und im nördlichen Arabischen Meer. *Neues Jb. Geol. Paläontol.*, **59**, 597–607.

McClure, H.A. (1976) Radiocarbon chronology of late Quaternary lakes in the Arabian Desert. *Nature*, **263**, 755–756.

McClure, H.A. (1978) The Rub al-Khali. In: *Quaternary Period in Saudi Arabia* (Eds S.A. Al-Sayari and J.G. Zotl), pp. 252–263. Springer-Verlag, Vienna.

McClure, H.A. (1984) *Late Quaternary Palaeoenvironments of the Rub' Al Khali.* Unpubl. PhD thesis, University of London.

McClure, H.A. (1988) Late Quaternary palaeogeography and landscape evolution of the Rub'Al Khali. In: *Araby the Blest* (Ed. D.T. Potts), pp. 9–13. Museum Tusculanum Press, University of Copenhagen.

McClure, H.A. and **Vita-Finzi, C.** (1982) Holocene shorelines and tectonic movements in eastern Saudi Arabia. *Tectonophysics,* **85,** T37–T43.

McKenzie, J.A., Hsü, K. and **Schneider, J.F.** (1980) Movement of subsurface waters under the sabkha, Abu Dhabi, U.A.E. and its relation to evaporite dolomite genesis. In: *Concepts and Models of Dolomitization* (Eds D.H. Zenger, J.B. Dunham and R.L. Ethington). *SEPM Spec. Pub.,* **28,** 11–30.

McKenzie, J.A., Hsü, K. and **Schneider, J.F.** (1981) Holocene dolomitization of calcium carbonate sediments from the coastal sabkhas of Abu Dhabi, U.A.E.: a stable isotope study. *J. Geol.,* **89,** 185–197.

Melguen, M. (1971) *Etude de Sédiments Pleistocène-Holocène au Nordouest du Golfe Persique. Analyse de Faciès par Ordinateur.* Unpubl. PhD thesis, University of Rennes, France.

Melguen, M. (1973) Corresponding analysis for recognition of facies in homogeneous sediments off an Iranian River mouth. In: *The Persian Gulf: Holocene Carbonate Sedimentation and Diagenesis in a Shallow Epicontinental Sea* (Ed. B.H. Purser), pp. 99–114. Springer-Verlag, Berlin.

Melville, J.C. and **Stadden, R.** (1901) The molluscs of the Persian Gulf, Gulf of Oman and Arabian Sea. *Proc. Zool. Soc.,* **2,** 327–346.

Milliman, J. D. and **Meade, R.H.** (1983) World delivery of river sediment to the oceans. *J. Geol.,* **1,** 1–21.

Mitchell, R.C. (1957) Recent tectonic movements in the Mesopotamian plain. *Geogr. J.,* **123,** 569–571.

Müller, D.W., McKenzie, J.A. and **Mueller, P.A.** (1990) Abu Dhabi sabkha, Persian Gulf, revisited: Application of strontium isotopes to test an early dolomitization model. *Geology,* **18,** 618–621.

Murray, J.W. (1965a) The Foraminiferida of the Persian Gulf. Part 1. *Rosalina adhaerens* sp. *Ann. Mag. Nat. Hist. Ser.,* **13,** 8, 77–79.

Murray, J.W. (1965b) The Foraminiferida of the Persian Gulf. Part 2. The Abu Dhabi region. *Palaeogeogr. Palaeoclimatol. Palaeoecol.,* **1,** 307–332.

Murray, J.W. (1966a) The Foraminiferida of the Persian Gulf. Part 3. The Halat al Bahrani region. *Palaeogeogr. Palaeoclimatol. Palaeoecol.,* **2,** 59–68.

Murray, J.W. (1966b) The Foraminiferida of the Persian Gulf. Part 4. Khor al Bazam. *Palaeogeogr. Palaeoclimatol. Palaeoecol.,* **2,** 153–169.

Murray, J.W. (1966c). The Foraminiferida of the Persian Gulf. Part 5. The shelf of the Trucial Coast. *Palaeogeogr. Palaeoclimatol. Palaeoecol.,* **2,** 267–278.

Murray, J.W. (1970a) The Foraminiferida of the Persian Gulf. Part 6. Living forms in the Abu Dhabi region. *J. Nat. Hist.,* **4,** 55–67.

Murray, J.W. (1970b) The Foraminifera of the hypersaline Abu Dhabi Lagoon, Persian Gulf. *Lethaia,* **3,** 51–68.

Off, T. (1963) Rhythmic linear sand bodies caused by tidal currents. *AAPG Bull.,* **47,** 324–341.

Park, R.K. (1976) A note on the significance of lamination in stromatolites. *Sedimentology,* **23,** 379–393.

Park, R.K. (1977) The preservation potential of some recent stromatolites. *Sedimentology,* **24,** 485–506.

Patterson, R.J. and **Kinsman, D.J.J.** (1977) Marine and continental ground-water source in a Persian Gulf coastal sabkha in reefs and related carbonates-ecology and sedimentology. *AAPG Bull.,* **4,** 381–397.

Patterson, R.J. and **Kinsman, D.J.J.** (1981) Hydrologic framework of a sabkha along the Arabian Gulf. *AAPG Bull.,* **65,** 1457–1475.

Patterson, R.J. and **Kinsman, D.J.J.** (1982) Formation of diagenetic dolomite in a coastal sabkha along the Arabian (Persian) Gulf. *AAPG Bull.,* **66,** 28–43.

Phillip, G. (1968) Mineralogy of Recent sediments of Tigris and Euphrates rivers and some of the older detrital deposits. *J. Sed. Petrol.,* **38,** 35–44.

Picha, F. (1978) Depositional and diagenetic history of Pleistocene and Holocene oolitic sediments and sabkhas in Kuwait, Persian Gulf. *Sedimentology,* **25,** 427–429.

Pilgrim, G.E. (1908) Geology of the Persian Gulf and adjoining portions of Persia and Arabia. *Geol. Surv. India Mem.,* **34,** 1–179.

Pilkey, O.H. and **Noble, D.** (1966) Carbonate and clay mineralogy of the Persian Gulf. *Deep-Sea Res.,* **13,** 1–16.

Pugh, J.M. (1997) T*he Quaternary Desert Sediments of the Al Liwa Area, Abu Dhabi.* Unpubl. PhD thesis, University of Aberdeen UK.

Purser, B.H. (Ed.) (1973a) *The Persian Gulf: Holocene Carbonate Sedimentation and Diagenesis in a Shallow Epicontinental Sea.* Springer-Verlag, Berlin, 471 pp.

Purser, B.H. (1973b) Sedimentation around bathymetric highs in the southern Persian Gulf. In: *The Persian Gulf: Holocene Carbonate Sedimentation and Diagenesis in a Shallow Epicontinental Sea* (Ed. B.H. Purser), pp. 157–178. Springer-Verlag, Berlin.

Purser, B.H. (1985) Coastal evaporite systems. In: *Hypersaline Ecosystems* (Eds G.M. Friedman and W.E. Krumbein), pp. 72–102. Springer-Verlag, Berlin.

Purser, B.H. and **Evans, G.** (1973) Regional sedimentation along the Trucial coast, SE Persian Gulf. In: *The Persian Gulf: Holocene Carbonate Sedimentation and Diagenesis in a Shallow Epicontinental Sea* (Ed. B.H. Purser), pp. 211–231. Springer-Verlag, Berlin.

Purser, B.H. and **Loreau, J.-P.** (1973) Aragonitic supratidal encrustations on the Trucial coast, Persian Gulf. In: *The Persian Gulf: Holocene Carbonate Sedimentation and Diagenesis in a Shallow Epicontinental Sea* (Ed. B.H. Purser), pp. 343–376. Springer-Verlag, Berlin.

Purser, B.H. and **Seibold, E.** (1973) The principal environmental factors influencing Holocene sedimentation and diagenesis in the Persian Gulf. In: *The Persian Gulf: Holocene Carbonate Sedimentation and Diagenesis in a Shallow Epicontinental Sea* (Ed. B.H. Purser), pp. 1–10. Springer-Verlag, Berlin.

Purser, B.H., Al-Azzawi, M., Al-Hassani, N.H., Baltzer, F., Hassan, K.M., Orszag-Sperber, F., Plaziat, J.C., Yacoub, S.Y. and **Younis, W.R.** (1982) Caractères et évolution du complex deltaique Tigre Euphrate. *Géol. Soc. Fr. Mém., NS,* **144,** 207–216.

Robinson, B.W. and Gunatilaka, A. (1991) Stable isotope studies and the hydrological framework of sabkhas, southern Kuwait, Arabian Gulf. *Sed. Geol.*, **72**, 35–53.

Robinson, B.W. Stewart, M.K. and Gunatilaka, A. (1990) Use of sulphur and other stable isotopes in environmental studies of regional groundwater flow and sulphate mineral formation in Kuwait. In: *Use of Isotopes in Sulphur-Cycle Studies* (Eds H. Krouse and V. Grineko), pp. 361–371. John Wiley & Sons, New York.

Rodgers, D.W. and Gunatilaka, A. (2002) Bajada formation by monsoonal erosion of a subaerial forebulge, Sultanate of Oman. *Sed. Geol.*, **154**, 127–146.

Rouse, G.E. and Sherif, N. (1980) Major evaporite deposition from groundwater remobilized salts. *Nature*, **285**, 470–472.

Rzoska, J. (1980) Euphrates and Tigris: Mesopotamian ecology and density. In: *Monographiae Biologicae* (Ed. J. Illiiested), **38**, 122. Dr W/Junk Bv., The Hague.

Saleh, A. (1978) *Algal Mats and Stromatolites of Al-Khiran, Kuwait.* Unpubl. PhD thesis, Imperial College, London, UK.

Saleh, A., Al-Ruwaih, F., Al-Reda, A. and Gunatilaka, A. (1999) A reconnaissance study of a clastic coastal sabkha in Northern Kuwait, Arabian Gulf. *J. Arid Environ.*, **43**, 1–19.

Sandford, W.E. and Wood, W.W. (2001) Hydrology of the coastal sabkhat of Abu Dhabi, United Arab Emirates. *Hydrogeol. J.*, **9**, 358–366.

Sanlaville, P. (1989) Considérations sur l'évolution de la basse Mésopotamie aucours des derniers millénaires. *Paléontol.*, **15**, 5–27.

Sarnthein, M. (1970) Sedimentologische Merkmale für die Untergrenze der Wellenwirkung im Persischen Golf. *Geol. Rundsch.*, **59**, 649–666.

Sarnthein, M. (1971) Oberflächensedimente im Persischen Golf und Golf von Oman. II. Quantitative Komponenten Analys der Grobfraktion. *Meteor Forschungsergeb., Reihe C.*, **5**, 113 pp

Sarnthein, M. (1972) Sediments and history of the post glacial transgression in the Persian Gulf and northwestern Gulf of Oman. *Mar. Geol.*, **12**, 245–266.

Sarnthein, M. and Walger, E. (1973). Classification of modern marl sediments in the Persian Gulf by factor analysis. In: *The Persian Gulf: Holocene Carbonate Sedimentation and Diagenesis in a Shallow Epicontinental Sea* (Ed. B.H. Purser), pp. 81–98. Springer-Verlag, Berlin.

Scholle, P.A. and Kinsman, D.J.J. (1974) Aragonitic and high-magnesium caliche from the Persian Gulf – a modern analogue for the Permian of Texas and New Mexico. *J. Sed. Petrol.*, **44**, 904–916.

Seibold, E. (1970) Nebenmeere im humiden und ariden Klimabereich. *Geol. Rundsch.*, **60**, 73–105.

Seibold, E. and Ulrich, J. (1970) Zur Bodengestalt des nordwestlichen Golfs von Oman. *Meteor Forschungsergeb., Reihe C.*, **3**, 1–14.

Seibold, E. and Vollbrecht, K. (1969) Die Bodengestalt des Persischen Golfs. *Meteor Forschungsergeb., Reihe C*, **2**, 29–56.

Seibold, E., Diester, L, Fuetterer, D., Lange, H., Muller, D. and Werner, F. (1973) Holocene sediments and sedimentary processes in the Iranian part of the Persian Gulf. In: *The Persian Gulf: Holocene Carbonate Sedimentation and Diagenesis in a Shallow Epicontinental Sea* (Ed. B.H. Purser), pp. 57–80. Springer-Verlag, Berlin.

Shearman, D.J. (1963) Recent anhydrite, gypsum, dolomite and halite from the coastal flats of the Arabian shore of the Persian Gulf. *Proc. Geol. Soc. London*, **1607**, 63.

Shearman, D.J. (1965) Discussion. In: *Salt Basins around Africa* (Ed. Anon), pp. 75–76. Institute of Petroleum, London.

Shearman, D.J. (1966) The origin of marine evaporites by diagenesis. *Trans. Inst. Min. and Metall.*, **75**, B208–215.

Shearman, D.J. (1971) Marine evaporites: the calcium sulphate facies. *AAPG Seminar, University of Calgary, Alberta*, 65.

Shearman, D.J. (1978) Evaporites of coastal sabkhas. In: *Marine Evaporites. SEPM Short Course Notes* **4**, 6–42.

Shearman, D.J. (1980) Sabkha facies evaporites. In: *Evaporite Deposits*, **27**, pp. 96–109. Chambre syndicate de la recherche et de la prediction du pétrole et du gaz naturel. Editions Technip, Paris, France.

Shearman, D.J. (1981) Displacement of sand grains in sandy gypsum crystals. *Geol. Mag.*, **118**, 303–316.

Shearman, D.J. and Skipwith, P.A. (1965) Organic matter in Recent and ancient limestones and its role in their diagenesis. *Nature*, **208**, 1310–1311.

Shearman, D.J., Twyman, J. and Zand Karimi, M. (1970) The genesis and diagenesis of oolites. *Proc. Geol. Assoc.*, **81**, 561–575.

Shinn, E.A. (1969) Submarine lithification of Holocene carbonate sediments in the Persian Gulf. *Sedimentology*, **12**, 109–144.

Shinn, E.A. (1971) Holocene submarine cementation in the Persian Gulf. In: *Carbonate Cements* (Ed. O.P. Bricker), pp. 63–65. Johns Hopkins Press, Baltimore and London.

Shinn, E.A. (1973a) Sedimentary accretion along the leeward southeastern coast of the Qatar peninsula, Persian Gulf. In: *The Persian Gulf: Holocene Carbonate Sedimentation and Diagenesis in a Shallow Epicontinental Sea* (Ed. B.H. Purser), pp. 199–210. Springer-Verlag, Berlin.

Shinn, E.A. (1073b) Recent intertidal and nearshore carbonate sedimentation around rock highs E. Qatar, Persian Gulf. In: *The Persian Gulf: Holocene Carbonate Sedimentation and Diagenesis in a Shallow Epicontinental Sea* (Ed. B.H. Purser), pp. 193–196. Springer-Verlag, Berlin.

Shinn, E.A. (1973c) Carbonate coastal accretion in an area of longshore transport, NE Qatar, Persian Gulf. In: *The Persian Gulf: Holocene Carbonate Sedimentation and Diagenesis in a Shallow Epicontinental Sea* (Ed. B.H. Purser), pp. 179–192. Springer-Verlag, Berlin.

Shinn, E.A. (1983) Tidal flat environment. In: *Carbonate Depositional Environments* (Eds P.A. Scholle, D.G. Bebout and C.H. Moore). *AAPG Mem.*, **33**, 171–210.

Skipwith, P.A.d'E (1966) *Recent Carbonate Sediments of Eastern Khor Al Bazam, Abu Dhabi, Trucial coast.* Unpubl. PhD thesis, Imperial College, London.

Stoffers, D. and Ross, D.A. (1979) Late Pleistocene and Holocene sedimentation in the Persian Gulf – Gulf of Oman. *Sed. Geol.*, **23**, 181–208.

Stokes, S. and Bray, H.E. (2005) Late Pleistocene aeolian history of the Liwa Region, Arabian Peninsular. *Geol. Soc. Am. Bull.*, **117**, 1464–1480.

Sugden, W. (1963a) The hydrology of the Persian Gulf and its significance in respect to evaporite deposition. *Am. J. Sci.*, **261**, 741–755.

Sugden, W. (1963b) Some aspects of sedimentation in the Persian Gulf. *J. Sed. Petrol.*, **33**, 355–364.

Sugden, W. (1966) Pyrite staining of pellety debris in carbonate sediments from the Middle East and elsewhere. *Geol. Mag.*, **103**, 250–256.

Taylor, J.C.M. and Illing, L.V. (1969) Holocene intertidal calcium carbonate cementation, Qatar, Persian Gulf. *Sedimentology*, **12**, 69–107.

Taylor, J.C.M. and Illing, L.V. (1971a) Development of Recent cemented layers within intertidal sand flats, Qatar, Persian Gulf. In: *Carbonate Cements* (Ed. O.P. Bricker), pp. 27–31. Johns Hopkins Press, Baltimore and London.

Taylor, J.C.M. and Illing, L.V. (1971b) Variations in Recent beachrock cements, Qatar, Persian Gulf. In: *Carbonate Cements* (Ed. O.P. Bricker), pp. 32–35. Johns Hopkins Press, Baltimore and London.

Taylor, J.C.M. and Illing, L.V. (1971c) Alteration of Recent aragonite to magnesium calcite cement, Qatar, Persian Gulf. In: *Carbonate Cements* (Ed. O.P. Bricker), pp. 36–39. Johns Hopkins Press, Baltimore and London.

Teller, J.T., Glennie, K.W., Lancaster, N. and Singhvi, A.K. (2000) Calcareous dunes of the United Arab Emirates and Noah's Flood: the post glacial reflooding of the Persian (Arabian) Gulf. *Quat. Int.*, **68–71**, 297–308.

Theobald, W. (1878) The Geology of Guzeral (1–14), The Geology of Kathiawar peninsula. In: *The Geology of Guzeral. Geol. Soc. India Mem.*, **XXI**, 73–136.

Tricart, J. (1954) Une forme de relief climatique: les sebkhas. *Rev. Géomorphol. Dynam.*, **5**, 97–101.

Uchupi, E., Swift, S.A. and Ross, D.A. (1996) Gas venting and late Quaternary sedimentation in the Persian (Arabian) Gulf. *Mar. Geol.*, **129**, 237–269.

Uchupi, E., Swift, S.A. and Ross, D.A. (1999) Late Quaternary stratigraphy, palaeoclimate and neotectonism of the Persian (Arabian) Gulf region. *Mar. Geol.*, **160**, 1–23.

Vita-Finzi, C. (1982) Recent coastal deformation near the Straits of Hormuz. *Proc. Roy. Soc. London*, **A382** 441–457.

Vita-Finzi, C. and Phethean, S.J. (1980) Recent inshore sediments in Musandam, Oman. *Mar. Geol.*, **36**, 241–251.

Wagner, C.W. and Van der Togt, C. (1973) Holocene sediment types and their distribution in the southern Persian Gulf. In: *The Persian Gulf: Holocene Carbonate Sedimentation and Diagenesis in a Shallow Epicontinental Sea* (Ed. B.H. Purser), pp. 123–156. Springer-Verlag, Berlin.

Walkden, G. and Williams, A. (1998) Carbonate ramps and the Pleistocene–Recent depositional systems of the Arabian Gulf. In: *Carbonate Ramps* (Eds V.P. Wright and T.P. Burchette). *Geol. Soc. London Spec. Publ.*, **149**, 43–53.

Warren, J.K. and Kendall, C.G.St.C. (1985) Comparison of sequences formed in marine sabkha (subaerial) and salina (subaqueous) settings: modern and ancient. *AAPG Bull.*, **69**, 1013–1023.

Wells, A.J. (1962) Recent dolomite in the Persian Gulf. *Nature*, **194**, 274–275.

Wells, A. J. and Illing, L.V. (1964) Present day precipitation of calcium carbonate in the Persian Gulf. In: *Deltaic and Shallow Marine Deposits* (Ed. L.M.J.U. van Straaten). *Dev. Sedimentol.*, **1**, 429–435.

Whittle, G.L., Alsharhan, A.S. and Kendall, C.G.St.C. (1998) Petrography of Holocene beachrock and hardgrounds, Abu Dhabi, United Arab Emirates. In: *Quaternary Deserts and Climatic Change* (Eds A.S. Alsharhan, K. W. Glennie, G.L. Whittle, and C.G.St.C. Kendall), pp. 57–68. Balkema, Rotterdam.

Williams, A.H. and Walkden, G.M. (2001) Carbonate eolianites from an eustatically influenced ramp-like setting: the Quaternary of the southern Arabian Gulf. In: *Modern and Ancient Carbonate Eolianites: Sedimentology, Sequence Stratigraphy and Diagenesis* (Eds F.E. Abegg, P.M. Harris and D.B. Loope). *SEPM Spec. Publ.*, **71**, 77–92.

Williams, A.H. and Walkden, G.M. (2002) Late Quaternary highstand deposits of the southern Arabian Gulf: a record of sea-level and climatic change. In: *The Tectonics and Climatic Evolution of the Arabian Sea Region* (Eds P.D. Clift, D. Kroon, C. Gaedick and J. Craig). *Geol. Soc. London Spec. Publ.*, **195**, 371–386.

Wilson, A.T. (1925) The delta of the Shatt-al-Arab and proposals for dredging the bar. *Geogr. J.*, **65**, 225–229.

Wood, W.W. and Imes, J.L. (1995) How wet is wet? Precipitation constraints on late Quaternary climate in the southern Arabian Peninsula. *J. Hydrol.*, **164**, 263–268.

Wood, W.W., Sandford, W.E. and Al-Habashi, A.R. (2002) Source of solutes to the coastal sabkha of Abu Dhabi. *Geol. Soc. Am. Bull.*, **114**, 259–268.

Wood, W.W. and Sandford, W.E. (2002) Hydrology and solute chemistry of the coastal-sabkha aquifer in the Emirate of Abu Dhabi. In: *Sabkha Ecosystems* (Eds H.J. Barth and B.B. Böer), pp. 173–186. Kluwer Academic Publisher, The Netherlands.

Wood, W.W., Rizk, Z.S. and Alsharhan, A.S. (2003) Timing of recharge and origin, evolution and distribution of solutes in a hyperarid aquifer system. In: *Water Resources Perspectives: Evaluation, Management and Policy* (Eds. A.S. Alsharhan and W.W. Wood), pp. 295–312. *Developments in Water Science*, **50**. Elsevier.

Wood, W.W., Stokes, S., Brandt, D., Kraemer, T.F. and Imes. J.L. (2006) Rapid rise (~3mm/year) of coastal Abu Dhabi. *Geol. Soc. Am. Ann. Meeting and Exposition, Abstracts with Programs*, **38**, 238.

Woolley, C.L. (1929) *Ur of the Chaldees*. London, 151 pp.

Yechieli, Y. and Wood, W.W. (2002) Hydrogeologic processes in saline systems: playas, sabkhas and saline lakes. *Earth Sci. Rev.*, **58**, 343–365.

Int. Assoc. Sedimentol. Spec. Publ. (2011) **43**, 45–88

Holocene geomorphology and recent carbonate-evaporite sedimentation of the coastal region of Abu Dhabi, United Arab Emirates

CHRISTOPHER G. St.C. KENDALL* and ABDULRAHMAN S. ALSHARHAN[†]

*Department of Geological Sciences, University of South Carolina, Columbia, South Carolina 29208, USA
(E-mail: Kendall@geol.sc.edu)
[†]Faculty of Science, United Arab Emirates University, Post Box 17551, Al Ain, United Arab Emirates*

ABSTRACT

Seaward reefs, barrier islands and tidal flats characterize the Holocene shallow-water carbonate and supratidal evaporite tract that lines the United Arab Emirates (UAE) coastal embayment of the southern Arabian Gulf coast. The character of the sediments of the entire province has constantly changed. An offshore bank is accreting seaward through the agency of coral growth and tidal delta formation. South of these banks, supratidal flats are infilling the coastal lagoons with shallow subtidal sediments, beach ridges and algal flats.

The sediments of the UAE coast mark the transition between landward continental facies and seaward basinal facies. To the east, the Holocene coast of Abu Dhabi trends NE–SW and is composed of a microtidal barrier/lagoon complex which narrows northward. To the west, in central Abu Dhabi, the protecting barrier islands are more widely spaced than those to the east. They are nucleated on extensive carbonate shoals and coral banks cut by tidal channels. South of the barrier, the Khor Al Bazam lagoon forms a continuous open body of water. The western end of this lagoon is connected to the Arabian Gulf so circulation is less restricted than in the lagoons to the NE.

The character of the Holocene sediments of central Abu Dhabi is tied to the physiography of this coast. Coral reefs grow along the seaward side of most of the offshore banks. This contrasts to the east, where ooids accumulate on inter-island tidal deltas and coral reefs are restricted to small patches. In the central area, to the west of the Al Dhabaiyah Peninsula, carbonate muds are being deposited in a narrow belt south of the offshore bank. Grapestones and skeletal debris are the dominant components of the inner western coastal terraces. Eastwards, extensive barrier islands protect the lagoon, while instead of a coastal terrace a widespread intertidal cyanobacterial mat is aggrading seaward of supratidal flats. These flats are the site of evaporite accumulation. In eastern Abu Dhabi, protected lagoons occur to the lee of the barrier islands and are sites of carbonate mud and pellet accumulation.

Keywords: Carbonates, reefs, tidal deltas, tidal flats, tidal bars, barrier islands, beaches, mangrove swamps, sabkhas, ooids, grapestones, pellets, bioclastic sediments, lime muds, cyanobacterial mats, anhydrite, gypsum

INTRODUCTION

The United Arab Emirates (UAE) is located on the eastern side of the Arabian Peninsula, between latitudes 22°40′N and 26°00′N, and longitudes 51°00E′ and 56°00′E (Fig. 1). It overlies the subsurface interior platform of the Arabian shelf. The UAE parallels the approximately 600 km long linear southeastern coast of the Arabian Gulf between Qatar and the northern end of the Musandam Peninsula of Oman. Relatively pure carbonate sediments with minor siliciclastic components accumulate here. These sediments vary laterally in character and have had different depositional and diagenetic histories. The Qatar Peninsula provides protection from heavy seas coming from the west. In contrast, the central coast is protected from wave action by the 60 m deep broad, structurally

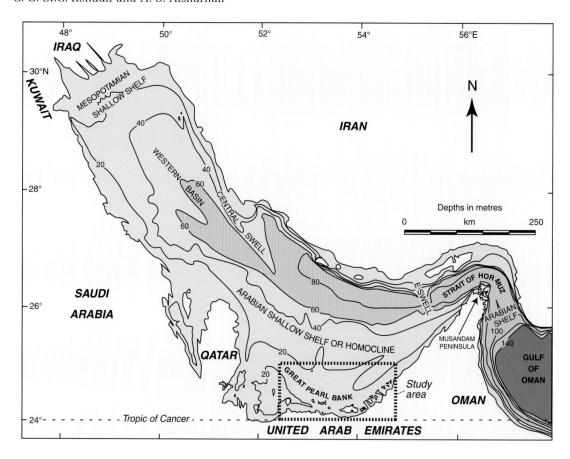

Fig. 1. Map showing the location of the study area along the northern coast of the UAE. Note the bathymetry of the Arabian Gulf and general geomorphology of the marine provinces of this enclosed sea (modified from Purser, 1973).

controlled, shallow shelf of the Great Pearl Bank (Houbolt, 1957). The NE section is unprotected (Fig. 1), and it faces directly towards the wind fetch of the entire length of the Arabian Gulf. The variation in coastal protection means that the detailed morphological features of the UAE coast (Fig. 2A and B) are more complex than those typical of an essentially linear coastline. The associated carbonate and evaporite depositional settings are the products of an eclectic mix of tides, wave, water chemistry, and temperatures.

It is the intention here to provide an overview of the Holocene geomorphology and sediments of coastal Abu Dhabi in the UAE. This shore forms an extensive system of lagoons and barrier islands stretching from Ras Ghanadha to the east, to Jebel Dhannah peninsula to the west (Figs. 2 and 3). The paper is dedicated to the late Douglas Shearman of Imperial College London for the inspiration he gave to the geological world and the authors. In remembrance of Shearman's work of the early sixties this paper updates the work he inspired

that was undertaken by Kendall & Skipwith (1966, 1968, 1969a, b). The present paper quotes this earlier work directly, but to better illustrate and describe of the Holocene carbonates and evaporites distributions, makes extensive use of new high-quality Landsat TM, and SPOT Panchromatic satellite images (Fig. 2A and B), and others from Google Earth. The figured photographs capture most of the morphological features described in this paper. The reader is referred to the high-resolution satellite images of the UAE coastline displayed on Google Earth for other examples of the same features (*http://earth.google.com/download-earth.html*).

The Holocene sediments of the Abu Dhabi coastal flats provide one of the best known Holocene analogues to ancient shallow-water carbonate and supratidal evaporite sequences (Curtis *et al.*, 1963, 1964; Sugden, 1963; Kinsman, 1964a, b, c; Butler *et al.*, 1965; Murray, 1966; Kendall, 1969; Wood & Wolfe, 1969; Purser, 1973; Purser & Evans, 1973; Purser & Loreau, 1973; Wagner & Van der Togt (1973); Shearman, 1978; McKenzie *et al.*, 1980;

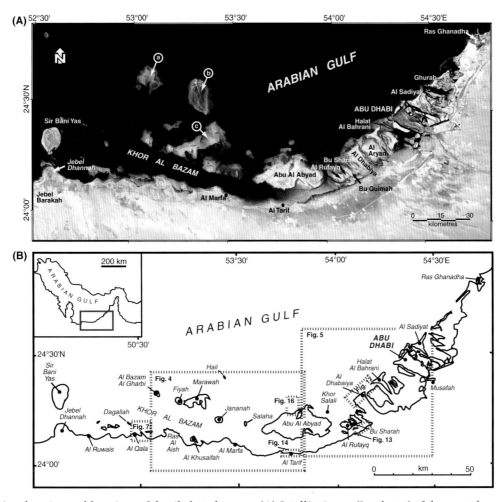

Fig. 2. Regional setting and locations of detailed study areas. (A) Satellite image (Landsat 7) of the coastal area of western UAE. (a) Bu Tini bathymetric high, and (b) Mubarras bathymetric high are the sites of coral reef shoals while (Cc) is an offshore barrier shoal area capped local with sabkha islands. This barrier extends roughly parallel to the coastline and defines the northern margin of the Khor Al Bazam. (B) Location on satellite image of different morphological features exhibited by the shoals and sabkhas of the Abu Dhabi coastal area.

Patterson & Kinsman, 1981; Butler *et al.*, 1982; Warren & Kendall, 1985; Swart *et al.*, 1987; Kenig *et al.*, 1990; Muller *et al.*, 1991; Baltzer *et al.*, 1994; Kenig *et al.*, 1995; Hughes, 1997; Kirkham, 1997; Kendall *et al.*, 1998; Alsharhan & Kendall, 2003; Kendall *et al.*, 2007). Several distinct facies belts have been identified within this region of abundant islands and shoals. These include: (1) extensive shallow-water carbonates of (a) coral and coralline algae, (b) oolitic sand, (c) grapestones and pellets, (d) carbonate mud and pellets, (e) molluscan sand, (f) cyanobacterial mat facies; (2) a supratidal evaporite tract; and (3) aeolian sands. These sedimentary facies are described below in terms of their distribution, structures, biota and features. Their character reflects their response to the processes active in the area.

The seaward carbonate barrier islands, tidal channels and deltas, and shallow lagoons of coastal Abu Dhabi are sites for the accumulation of oolitic sands and finer mud-sized carbonates. Supratidal islands protect the inner coastal lagoons that are rimmed southward by a cover of intertidal to supratidal cyanobacterial flats. These in turn have been partly covered by prograding supratidal salt flats or sabkhas (see also Kendall & Skipwith, 1968a, b; Purser, 1973a; Butler *et al.*, 1982; Alsharhan & Kendall, 2003). Traced from the seaward margin to a line of early Holocene beach sediments just landward of the present shoreline, the surface of the intertidal to supratidal sediments of this protected area has a variable character. In order, the sediment surface includes: a series of distinctive cyanobacterial mats; a moist carbonate-

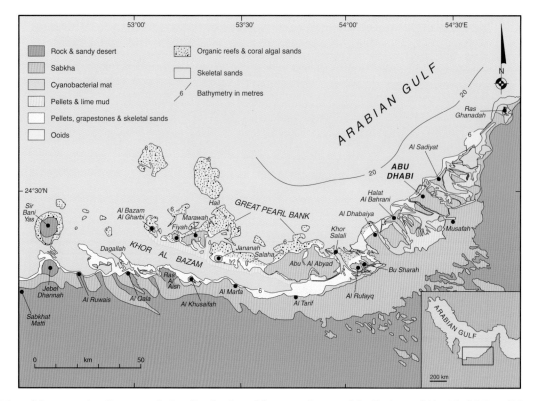

Fig. 3. Map of the general sedimentary facies distribution of the coastal areas of the Emirate of Abu Dhabi. Localities referred to in the text are indicated (modified from Kendall & Skipwith, 1969a, b and Alsharhan & Kendall, 2003).

mud gypsum mush; and a convolute halite crust passing into a thin polygonally-cracked halite surface. These variations in surface sediment type and morphology reflect the topography and a relationship to tidal range. The sabkha slope is disturbed by beach ridges, tidal channel depressions, and by tepee structures formed by elevated hardgrounds just below the high-water mark of the tidal flats.

Below the late Holocene sequence, lie Pleistocene and early Holocene sediments that consist of outwash fans of mixed siliciclastic and carbonate deposits, windblown and marine carbonates, cyanobacterial sediments, and supratidal flat deposits. The base of the Holocene varies from marine sediments seaward, to more restricted evaporative sediments landwards, together reworked with sediments of the early Holocene (Wood *et al.*, 2002). The sediments of the coastal region pass landward into continental dunes and wadis, and seaward into a basinal mud-sized carbonate and clay-rich facies. The surface is expressed by rock outcrops, desert sands, and coastal carbonate and evaporite sediments as seen on the generalized lithofacies map of Abu Dhabi (Fig. 3).

GEOMORPHOLOGICAL FEATURES OF THE ABU DHABI COASTAL AREA

From west to east, the general coastal complex of the Abu Dhabi Emirate can be subdivided into three major geomorphological provinces: (1) the open Khor Al Bazam lagoon and seaward bank (Figs. 2 and 3); (2) the Al Rafayq - Bu Sharah Shelf and channel region; (3) the Abu Dhabi barrier island and protected lagoon. Each of these provinces is composed of major and minor geomorphological features, that are summarized in Tables 1–3.

(1) The open Khor Al Bazam lagoon and seaward barrier province of the Great Pearl Bank (Fig. 4). Traced from seaward (north) to landward (south) this consists of:
(a) A northern offshore bank whose northern flank is rimmed by coral reefs that pass southward to shoals (commonly capped by supratidal islands nucleated about Pleistocene dune sediments). These are incised by tidal channels that terminate in flood deltas on the southern flank of the

Table 1. Major geomorphological features of Abu Dhabi coastal complex.

Major features and units	Relation to water level	Subenvironment
Shelf and shoals	Features below LWM	Coral reefs
	Intertidal features	Sand flats
	Dry land	Islands of cemented limestone
Offshore bank, shoal and channel complex	Features below LWM	Coral reefs
	Features below and above LWM	Tidal channels and deltas
	Intertidal features	Seaweed patches
		Sand flats
		Mangrove areas
		Cyanobacterial flats
Lagoon	Supratidal	Sabkha barrier islands
	Dry land	Islands of cemented rock
	Features below LWM	Open lagoon with flat floor
		Open lagoon with dissected floor
Coastal terrace	Features below LWM	Coral reefs
	Features below and above LWM	Tidal channels and deltas
	Intertidal features	Weed patches
		Sand flats
		Cyanobacterial flats
Mainland coastal plain	Supratidal and dry land	Sabkha plain, hills and alluvial fans

LWM, low water mark

bank. The sediments of this province are largely derived from the reef system and are bioclastic in origin.

(b) A central open lagoon, the Khor Al Bazam, that deepens westward and shallows eastward. This feature appears to represent a submerged wadi valley system formed during Pleistocene sea-level lowstands.

Table 2. Genesis of major features of offshore Abu Dhabi.

Environment	Primary agent(s) of genesis	Major feature(s)
Marine	Waves and currents	Intertidal sand flat
	Currents	Tidal channels and deltas
		Open lagoon with flat floor
		Open lagoon with dissected floor
	Biological	Coral reefs
	Wind, waves and currents	Seaweed covered areas
		Cyanobacterial flats
		Mangrove area
		Sabkha barrier islands
		Mainland sabkha
Marine and subaerial	Wind and biological	Coastal dunes
	Wind, fluvial, waves and currents	Hill and alluvial fans

Bioclastic sediments are currently filling the lagoon. South and east of Abu Al Abyad this lagoon terminates in the Al Rafayq - Bu Sharah Shelf channel system (Fig. 5).

(c) A southern coastal terrace that is commonly rimmed by calcareous algae and corals. It is capped on its seaward margin by a wave-truncated cemented carbonate hardground. To the west, the terrace is covered by the prograding coastal complex of carbonate spits and megaripples. To the east this cover changes to extensive intertidal cyanobacterial mats. This accumulation is protected from waves by the low gradient of the intertidal zone and the presence of the Abu Abyad barrier island to the

Table 3. Minor features of the offshore, Abu Dhabi.

Primary agents of genesis	Features	Occurrence
Waves	Break-point bars	Lower part of the littoral zone to below LWM
	Runnels and bars	
	Intertidal spits and beaches	Sand flats
		Upper part of littoral zone on sand flats
Currents	Submarine spits	Variable
	Megaripples	Variable
	Ebb gullies	Littoral zone

LWM, low water mark

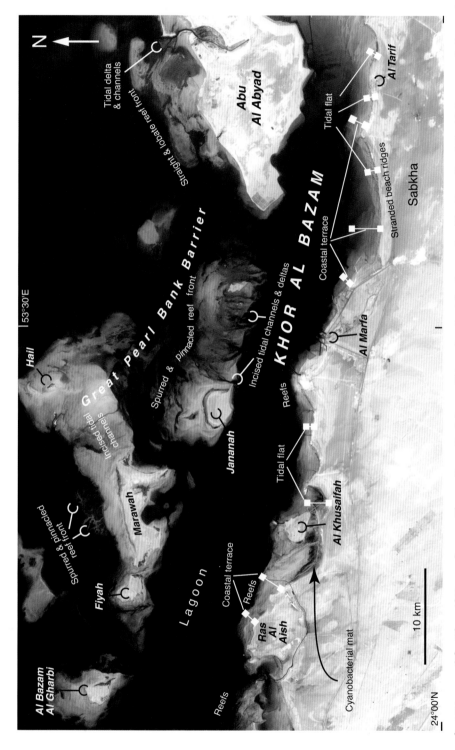

Fig. 4. Landsat 7 image of the Khor Al Bazam lagoon. The northern flank is an offshore barrier island complex (the Great Pearl Bank) that extends from Al Bazam Al Gharbi to Abu Al Abyad island. The southern coast of this lagoon is formed by a series of irregular coastal terraces capped by supratidal flats that include cyanobacterial mats and salt flats (sabkhas). For location see Fig. 2B.

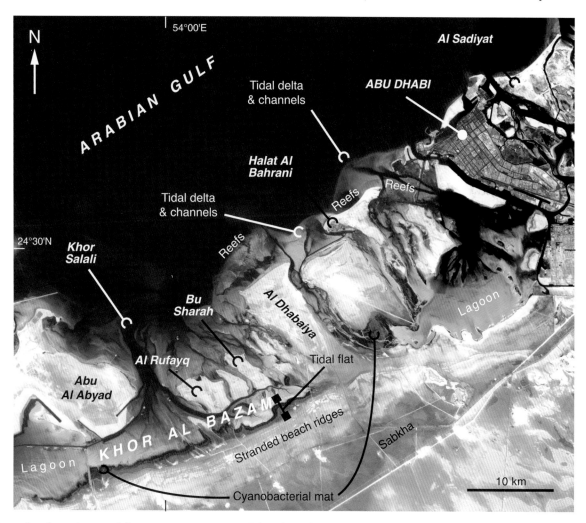

Fig. 5. Landsat 7 image of the eastern termination of the Khor Al Bazam lagoon where it is encroached on by beach ridges, cyanobacterial mat and coastal sabkha along the coast. To the east of the Al Dhabaiya peninsula are the enclosed protected lagoons to the lee of the barrier islands of Halat Al Bahrani, Abu Dhabi and Al Sadayat. For location see Fig. 2B.

north (Fig. 5). These tidal flat sediments and associated supratidal evaporites are flanked by reworked sediments of the early Holocene. Seaward of these are commonly stranded high-energy beach ridges that accumulated at the beginning of the late Holocene still-stand of 3000 BC.

(2) The Al Rafayq - Bu Sharah Shelf and channel province (Fig. 5) is bounded to the west by the Khor Salil (a wide N–S oriented deep tidal channel situated on the eastern margin of Abu Al Abyad). The eastern boundary of this province is the Al Dhabaiya peninsula. The shelf and channel province is formed by an area of shoals that are dissected by a mix of smaller N–S tidal channels and a major E–W channel. The province hosts a corresponding system of numerous small N–S linear islands oriented perpendicular to the coast that are nucleated seaward about Pleistocene dune sediments and are flanked by mangrove swamps. The tidal range here is 1 to 2 metres. The sediments of this channeled region are a mix of bioclastic material derived from more seaward areas, together with pellets and lime mud. To the south of these islands, in the shallowest portion (here mostly intertidal) of the Khor Al Bazam, are sediments composed of pelleted carbonate and bioclasts. These pass landward into intertidal to supratidal cyanobacterial and evaporative flats. As with the sediments of the sabkha flats to the west, these flats are flanked to the south by a high-energy beach ridge system that marks the margin to the sabkha formed

by reworked early Holocene aeolian and outwash sediments.

(3) The third province consists of the Abu Dhabi barrier islands and protected microtidal lagoons (Fig. 5). This area is characterized by system of "drum-stick" barrier islands that have long leeward trailing tails. Tidal deltas and channels are developed between these islands; coral reefs form their seaward centres, and are growing along the tidal channels. The tidal deltas are sites for the accumulation of ooids, and the channels are rich in skeletal debris. The seaward barrier islands protect carbonate-mud prone lagoons and mangrove swamps that are flanked to the south by protected intertidal cyanobacterial flats and supratidal sabkha evaporites. Landward are stranded high-energy beach ridges and reworked early Holocene aeolian sands and outwash fans.

MARINE GEOMORPHOLOGICAL FEATURES

The geomorphology of the coastline (Fig. 6) is controlled by the interaction between waves, currents and biological processes.

Major features formed by waves and currents

Intertidal sand flats

Intertidal sand flats are present on the eroded surface of ancient reef flats and wave-cut benches of pre-Holocene rocks (Kendall & Skipwith, 1969b; Alsharhan & Kendall, 2003). These intertidal flats are ubiquitous to the region and are found on shoals, the offshore bank, and coastal terrace (Figs. 4, 5 and 7, Table 1). They have gradients of approximately 1:5000 and vary in width from 0.5 km to 8 km. They are generally covered by varying thicknesses of unconsolidated sand which may be underlain by alternating layers of poorly cemented limestone (1–3 cm thick) and unconsolidated carbonate sand (3–6 cm thick). The surface of the intertidal sand flat is ripple-marked or covered by growing seaweed, break-point bars, and runnels and bars (Figs. 6, 8 and 9), all of which are interpreted to have been controlled in their distribution by the direction of wave approach. Although observed in the field, on aerial photographs, and matched to analogues in the literature, it has not been possible to make direct field observation of these features forming. Their landward edge is delimited by intertidal spits, beaches or by cyanobacterial flats or mangrove clumps, or both.

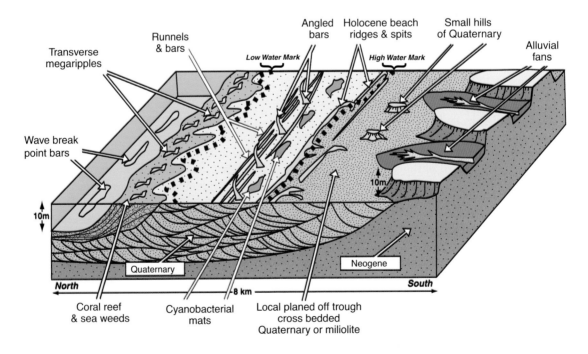

Fig. 6. Block diagram of the general character to the coastal terraces of the western Khor Al Bazam and the associated sedimentary settings.

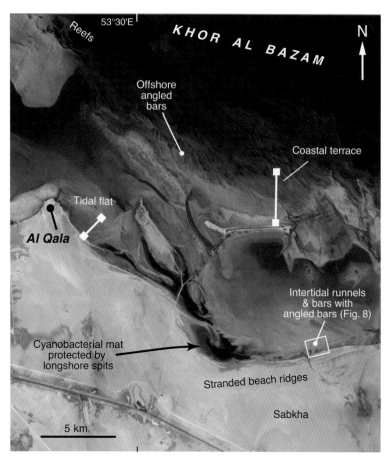

Fig. 7. Google Earth image (September 2008) in the vicinity of Al Qala headland flanked to the north by coastal terraces with reefs and offshore angled bars often covered by weeds. To the south are tidal flats marked by runnels and bars, cyanobacterial mats protected by longshore spits, beach ridges and halite encrusted sabkha flats. For location see Fig. 2B.

The intertidal sand flats may be divided on the basis of fauna into *Cerithium* flats in the lower area and crab flats in the upper part (Kinsman, 1964c). *Scopimera* sp., the crab whose activity characterizes the latter zone, creates radial patterns of feeding balls around its burrows (Fig. 10). Crabs, worms, and molluscs burrow the sediments of the flats extensively so that the primary sedimentary structures are destroyed. Ginsburg (1957) reported similar characteristics for the carbonate sediments of Florida. Commonly cemented layers occur that are associated with the loose surface sediment. These have flat upper surfaces and irregular lower surfaces. Some layers, as with beachrock, are cemented as the unconsolidated sediment of the sand flat accumulates. The cements are composed of aragonite probably precipitated by cyanobacteria (Nesteroff, 1956) and from the evaporation of capillary water at low tide (Ginsburg, 1953b). This latter process is likely especially active during the

height of summer. Commonly before additional sediment accumulates on these flats the upper surface of the cemented sediment is truncated by storm wave erosion. Sugden (1963) and Purser & Loreau (1973) have described similar crusts from western Abu Dhabi.

Not all of the cemented layers found in the Khor Al Bazam region are contemporaneous, and some have different compositions from the loose sand overlying them. These latter layers commonly occur a few centimetres below high-water mark and are commonly covered by cyanobacteria and beach sediments. These crusts form megapolygons on the intertidal sand flats south of Rafayq and Bu Sharah.

Other cemented horizons occur in the subtidal zone some 2 m below low-water mark. For instance, in the east Khor Al Bazam they form megapolygons tens of metres across (Assereto & Kendall, 1977). Most of these later layers are

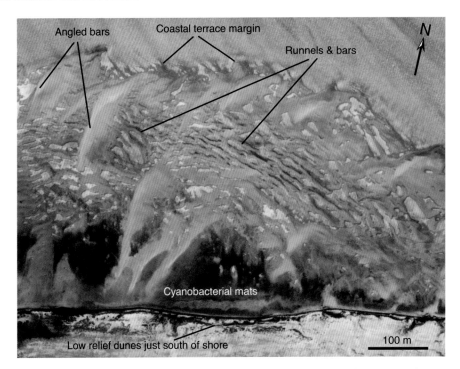

Fig. 8. Google Earth image (September 2008) east of Al Qala headland (located in white rectangle inset on Fig. 7). Note the tidal flats marked by runnels and bars, cyanobacterial mats protected by angled bars, and small beach ridges just south of the shoreline.

cementing in the submarine today, as do those recorded by Shinn (1969) offshore and just to the east of Qatar. Elsewhere, subtidal limestones may represent sediments that accumulated as the sea transgressed across the shelf during the last Holocene sea-level rise. These were probably cemented as beachrock contemporaneous with this change in sea level.

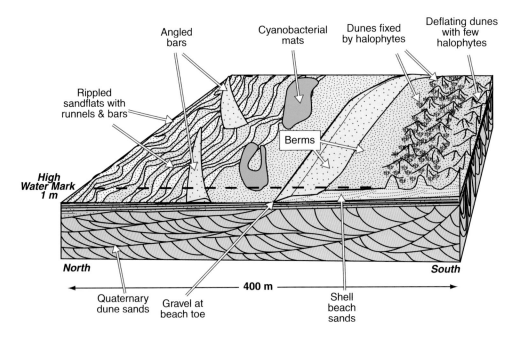

Fig. 9. Schematic block diagram of intertidal sand flats, and cyanobacterial flats. Landward beaches and dunes are part of a migrating intertidal to supratidal spit. The dunes are fixed by halophytes.

Fig. 10. *Scopimera* crab burrows on carbonate sand tidal flat, surrounded by their feeding balls. The pen is 15 cm long.

Subtidal crusts were recorded by Assereto & Kendall (1977) as forming polygons that averaged 400 m across at the eastern end of the Khor Al Bazam. Here they were marked by seaweed growth but have since been destroyed by dredging. The polygonal outlines were believed to be the surface expression of cracks in the hardened surface that lay beneath the veneer of sediment (Assereto & Kendall, 1977). Lines of seaweed preferentially colonized the loose sediment that filled the cracks. Small polygonal cracks were seen where strong currents keep the bottom clear of loose sediment. Polygon formation may have been connected with cementation of the sea floor penecontemperaneous with submarine cementation or from the release of gases accumulated in the sediment beneath the hard layer. Shinn (1969) recognized similar polygonal patterns in submarine crusts forming offshore Qatar. These were ascribed to modern submarine cementation, citing the occurrence of modern man-made artifacts cemented into the crusts as providing evidence of the penecontemporaneous character of the crusts. The size of the polygons probably reflects the thickness and competency of the layers. The polygons and the tepees of the back-reef facies in the Upper Permian of the Guadalupe Mountains may have the same form and origin (Kendall & Skipwith, 1969b; Alsharhan & Kendall, 2003).

Minor features formed by waves

Break-Point bars

Sand ridges lying parallel to the shore and covered by the highest tides are common along the seaward margins of the offshore bank and coastal terrace (Kendall & Skipwith, 1969b). These correspond to the break-point bars of King (1959). The aerial photographs of Khor Al Bazam support the interpretation that these features as the product of plunging breakers, which are thought to transit the edge of the offshore shoal areas and coastal terrace (Figs. 6 and 8) during storms. The spill of the breakers carries sand from the seaward margin across the top of the reef fronts onto the landward terrace.

The characteristics of these break-point bars enable them to be subdivided into: (a) single break-point bars; (b) lunate bars; and (c) multiple break-point bars. The single break-point bars are well developed along the coastal terraces east

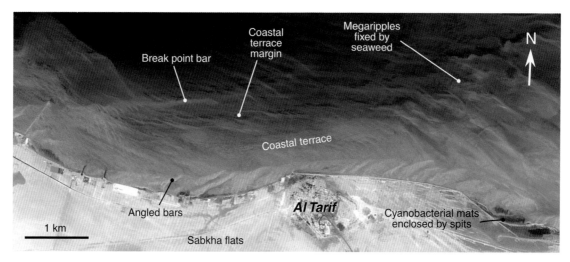

Fig. 11. Google Earth image (September 2008) of the southern Khor al Bazam shoreline at Al Tarif just south of Abu Al Abyad. Here offshore sand ribbons occur in the shallow 1–2 m of water just seaward of the sandy beaches. Locally these ribbons have been modified by waves to form break-point bars. Note the intertidal angled bars and the cyanobacterial mats enclosed by migrating spits along the shore. For location see Fig. 2B.

of Al Tarif (Fig. 11), west of Al Marfa and off the Ras Al Aish peninsula. They occur north of the offshore bank where coral reefs are trapping sediment. They also exist to the east of the island of Al Bazam Al Garbi (Fig. 4), in an area free of coral, just north of the bank. In general, the bars have an axial length that ranges from 0.5 km to 2 km, have a width of 60 m, and are 1 m high in 5 m of water.

Kendall & Skipwith (1969b) reported finding lunate bars just north of the small bay at the tip of Ras al Aish. These have since been destroyed by dredging. Similar features have been called lunate bars by Shepard (1952), and crescentic bars by King (1959) and Williams (1960). They probably formed where there is interaction of waves from two directions. Multiple break-point bars are developed at the entrance and to the west of the Dagallah lagoon. Shepard (1952) and King (1959) discussed similar features and suggested that they are the product of different sized waves. The deeper outermost line of bars formed at the plunge line of large waves and the inner lines are formed by smaller similar waves.

Runnels and bars

Parallel ridges of sand separated by "lows" within the intertidal zone correspond to the "balls and lows" of Johnson (1919), the "runnels and ridges" of King (1959), and the "runnels and bars" of Williams (1960). The last of these terms, as

described by Kendall & Skipwith (1969a), is used here (Figs. 6–8).

In the Khor Al Bazam, these structures take three forms: (a) runnels and bars with frequent cross-cutting channels; (b) runnels and bars with few cross-cutting channels; and (c) angled runnels and bars. Williams (1960) suggested that all three varieties form by the same mechanism, and the close proximity of these structures to one another in the Khor Al Bazam supports this idea (Figs. 6–8). The interpretation here is that the bars probably form from the swash of breaking waves, while the cross-cutting channels are the product of backwash and rip-current outlets.

Runnels and bars with frequent cross-cutting channels are well developed in the Khor al Bazam to the west of the Dagallah lagoon entrance. In cross section the bars are similar to Imbrie & Buchanan's (1965) asymmetric avalanche and accretion ripples. The front of the bars has a rounded slope while their landward face may be steep but is commonly curvilinear. The bars are approximately 10 m wide and 20–40 cm high. Except where there is wave refraction around an obstacle, they lie parallel to the coast. Shallow channels, generally perpendicular to the coast, frequently cut them.

Runnels and bars with few cross-cutting channels occur as sand ridges of some 10 to 20 m in wavelength and <1 m in amplitude. They are common on the upper and middle intertidal flats of the offshore barrier and coastal terraces. Near Al Marfa, these bars reach an axial length of over

1 km. They form parallel to the shore in evenly spaced groups of 2 to 20 individual bars; for example, west of Dagallah lagoon entrance. In the Khor Al Bazam these features are commonly cemented or partially covered by other features. This suggests that the conditions necessary for their formation only occurred rarely. Storm-driven waves are the most probable agents. On the basis of aerial photographs covering close to 60 years, it would appear that once formed, the bars remain as long-lived to permanent features.

Angled runnels and bars form on the upper intertidal flats of the offshore bank and coastal terrace at variable angles to the coast (Fig. 6). They are clearly seen at Dagallah mouth, west of Dagallah mouth, Qala west of Ras Al Aish (Fig. 8) and in front of the island of Marawah. They are between 3 to 6 m across and 20 to 40 cm high (approximately the same dimensions as the other runnels and bars). In cross section their frontal edges are gently curved and they have steep shoreward faces, as are the asymmetric ripples described by Imbrie & Buchanan (1965). In plan view, the frontal seaward edge is usually forms an undulating line, whereas the lee or backside forms a series of interlocking crescents. The bars are broadest closer to the shore and taper seaward. Their seaward tip is hooked back toward the shore. The surfaces of these bars are loose and fresh, suggesting continuous reworking. The angled runnels and bars are similar to features described by King (1959) off the Lancashire coast and by Davies (1962) at Gibraltar Point, but the Khor Al Bazam bars have some differences. These are the low gradient (1:5000) of the intertidal flats, the small tidal range (1 m) and the fact that some bars lie normal to the shore.

King (1959) suggested that angled bars form during the incoming tide and are almost perpendicular to the waves that produce them. The orientation of the bars within the Khor Al Bazam is consistent with this interpretation. They are interpreted to have formed in response to small waves that cross wide stretches of shallow intertidal flat without breaking. These waves apparently do not refract at the edges of the coastal terrace or barrier. The angles of the bars suggest that the waves did not necessarily come from the dominant wind direction, but are generated by occasional winds from the direction of greatest local fetch. After each wave breaks, the direction in which the water runs out from between the bars is thought to be modified by the angle between the bars and the shore. This can be seen in the lagoons of Dagallah, near

Al Marfa, Al Khusaifa and seaward of Marawah. The surfaces of these angled bars are commonly reworked by *Scopimera* sp. crabs (Fig. 10).

Somewhat larger angled bars occur in the eastern Khor Al Bazam. These are thought to form from waves that are not refracted by the bottom topography and are generated by large storms from the northwest. The long axes of the bars appear to parallel the wave crests and the resultant angles to the coast expressed by the bars depend very much on the coastal trend. The differences in trend of the angled bars to the east and west of Al Tarif are interpreted to be a response to the waves refracted around headlands.

Larger angled bars are commonly broad and indistinct, being separated from each other only by narrow outwash channels. The size of the bars is interpreted to depend upon the strength of the waves that form them and on the amount of material available. Although smaller and more distinct bars do exist, they are usually about 1000 m long, while their breadth varies from 50 to 600 m. Commonly, these bars are modified and covered by runnels and bars, and wave ripples.

The larger angled bars do not form in areas that are protected from wave action, such as the lee of headlands. Even exposed areas require that the terraced platform be wide enough so that constructive waves are able to produce these bars. When the coastline is traced from protected to exposed areas, there is a progression from small angled bars dominated by runnels and bars, to large angled bars that are superimposed on the runnels and bars.

Intertidal spits and beaches

Intertidal spits are some of the commonest features associated with beach formation in the west Khor Al Bazam (Figs. 6, 7, 9 and 11). They also delineate the upper limit of the intertidal flats (Kendall & Skipwith, 1969b). They exhibit a west to east trend that reflects the longshore drift. The seaward side of each spit has a beach face slope with a gradient of 1:6 that contrasts with the slope of the adjacent seaward intertidal flat with its gradient of approximately 1:5000. A line of drifted seaweed normally marks the bottom of the slope, or toe of the beach. Above the beach face, there is at least one berm backed by a line of hummocky dunes (Fig. 9). In cross section, the landward-facing slope of the spit is convex and curves down to the approximate level of the old intertidal flat upon which it rests.

The sediments show the sheet-like relationships described by Imbrie & Buchanan (1965). The spits increase in width seaward, unless a further spit develops seaward of them.

Spits nucleate along lines of drifted seaweed that accumulate on the tidal flat. Two stages of development occur. The first stage is exemplified by the spits developed just off Al Tarif (Fig. 11). Here, the most recent more northward sandy ridge is nucleated on a line of seaweed to form a lunate or cusped ridge of sand. In the second stage that can be seen just south of this where accretion has joined a sand ridge to the land to form a spit. The initiation of new spits probably occurs during stormy periods (Evans *et al.*, 1964a, b).

The entire coastline of the west Khor Al Bazam is lined with compound forms of spits that are now stranded inland by coastal accretion (Kirkham, 1997). These features are similar to the stranded quartzose beaches of the Louisiana chenier plains (Byrne *et al.*, 1959; Gould & McFarlan, 1959) and parts of the Niger delta (Allen, 1965) but are composed of carbonate sands. Spits within each compound feature of the inner portions of the UAE coast are commonly hooked and in the extreme case formed into looped bars (Evans, 1942). Other spits drape headlands to form winged headlands (Thornbury, 1954); for example, Ras Al Aish, Marawah and west of Al Marfa. Kendall & Skipwith (1969b) reported an unusual spit at right angles to the coast east of Al Tarif. This feature was interpreted to have formed when a N–S trending sand ribbon became attached to the shoreline. Wave action built up the ribbon above normal high water and longshore drift produced a series of southerly facing hooks. This has been destroyed by coastal construction filling the lagoon.

Major geomorphological features produced by currents (tidal channels)

Tidal channels are common on the offshore banks north of the Khor Al Bazam, the bank flanking the Dagallah lagoons on the southern margin of the Khor Al Bazam, in the shoal and channel area to the east around Al Rafayq and Bu Sharhah (Kendall & Skipwith, 1969b), and between the barrier islands of Halat Al Bahrani, Abu Dhabi, and Sadiyat (Figs. 2–5 and 12). The following criteria, originally developed by Van Veen (1953) and

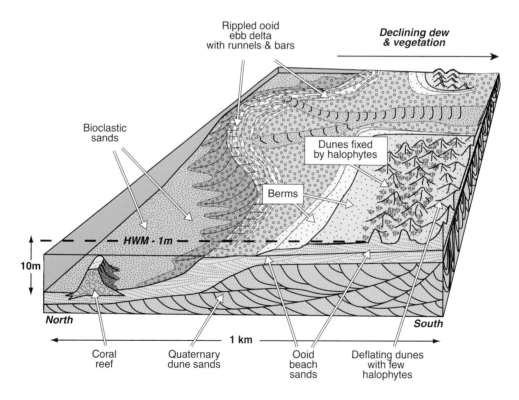

Fig. 12. Schematic block diagram showing a representative morphology of the ooid tidal deltas formed between the drumstick barrier islands of Halat Al Bahrani, Abu Dhabi and Sadiyat. Notice the runnels on these bars and the aeolian dunes located on the adjacent barrier islands. HWM, high water mark.

Price (1963), were used to determine the direction of current flow:

(a) the bifurcated end of a straight channel is directed toward its lower reaches;
(b) dendritic patterns converge in the direction of ebb drainage, away from elevated embayed areas;
(c) spits are aligned away from the major current direction;
(d) linguoid ripples and deltas are convex downstream.

The channels that cut through the offshore barrier match the "tidal inlets" of Price (1963) since they channel water from the open "sea" into a lagoon and vice-versa.

On the bank north of the Khor Al Bazam there are two areas of tidal channels that are of note. One is a set of flood-tide channels SW of Hail and the other is a set of flood and ebb channels east of Jananah (Fig. 4). Both sets have deltas or spill-over lobes (Ball, 1967a, b) on their southern ends. This implies that flood-tide currents cut the channels. The Hail group deltas are markedly displaced westward by the westward-flowing flood tide of the major channel. The Jananah set is also displaced westward, but here the effect is not so pronounced. In the eastern channels of the Jananah (Fig. 4), the flow direction of the ebb current in the eastern channels is the same as the flood current. The orientation of linguoid megaripples at their northern ends and the orientation of sand tails behind coral pinnacles on the bank support this. Many of the channels near Jananah exhibit sand-ribbons and rhomboid sand-streaks. On the bank to the west of Hail, two multiple deltas have formed at the mouths of a series of short shallow channels. Along these channels the tidal movement is from east to west. The fronts of the deltas are deflected by waves from the dominant wind direction. Maxwell *et al.*, (1964) suggested that currents and waves working in opposition produce such deltas.

The tidal channels of the shoal and channel area are similar to those described for the Dutch "wadden" or tidal flats, since they carry the tidal waters in circulation around the shoal areas. Flow patterns and the distinction between flood and ebb channels have been described by van Veen (1953) and can be recognized in the area under discussion. Many channels in the shoal and channel area have levees on their northeast side. They commonly have an ebb delta at their seaward ends

even though flood tidal movement and waves dominate this area, with ebb tides being dominant in the inner reaches. These deltas have a variety of sizes and shapes. The main delta is a large complex covered with bars and is fed by an axial channel. This delta is similar in size and shape to those found between the islands of Halat Al Bahrani, Abu Dhabi, and Sadiyat (Evans *et al.*, 1964a, 1964b; Figs. 2, 3, 5 and 12). An unusual delta is found off the mouth of the Khor Salali (Fig. 5). It is U-shaped and skewed eastward in the direction of the long-shore ebb current. It is similar to the river bars of the Niger delta (Allen, 1965).

A most beautiful multiple ebb delta occurs at the exit of the mangrove inlet on the north side of Abu Al Abyad. Most ebb deltas have their apex at the distal end; this one has its apex at the channel mouth and is more like a classical fluvial delta.

Minor geomorphological features formed by currents

Submarine spits

Submarine spits are attached to both land and shoals. Tidal currents form most submarine spits of the Khor Al Bazam area (Kendall & Skipwith, 1969a). Numerous underwater spits occur in the major channel south of the Dagallah Bank and are formed by the flood tide. In the mouth of west Dagallah lagoon, spits are produced by both ebb and flood tides, as for example just south of Dagallah.

Megaripples

In the Khor Al Bazam, transverse, diagonal and longitudinal ripples occur. These are current ripples that have wavelengths greater than 1 m (van Straaten, 1953a). They lie either transverse to, or parallel with, the current direction. Their shapes depend on four factors:

(a) supply and grain size of sediment; (b) depth of water; (c) velocity of current; (d) bottom topography.

A series of "megaripples" form at the eastern end of the Khor Al Bazam (Fig. 11). Here, longitudinal megaripples and sand ribbons are formed by fast-moving currents in the very shallow water at the head of the lagoon and the adjacent terrace. As the lagoon waters deepen toward the west, larger megaripples or sand waves occur that extend into long thin lines of alternating sand and seaweed parallel to the current direction. The velocities of the ebb current are reduced in deeper water and are counteracted by flood currents, so that closely bunched sand waves occur.

Transverse megaripples

Three distinct different types of megaripple features can be seen in the Khor Al Bazam (Kendall & Skipwith, 1969b). One matches that described by van Straaten (1950) which occurs in tidal channels. It is formed by ebb currents and tends to lie along the sides of channels. This ripple forms by the downstream migration of ordinary current ripples when there is continuous water flow, rather than a burst of fast current movement at the peak of the tidal change (van Straaten, 1950). Ripples ranges in wavelength from 1 to 200 m and have heights from 25 cm to more than 1 m (Fig. 11).

A second type of megaripple is most abundant on the seaweed-covered coastal terrace platforms (Fig. 6). Examples occur just northeast of Fiyah, north of Bazam Al Garbi, the northern part of Marawah, in the lagoons southwest of Dagallah, west of Al Tarif, and on the northwest coast of Abu Al Abyad. Ripple height is between 20 and 50 cm and width is between 30 and 200 m; the length of the ripple crest can be more than 500 m. The ripples are consistently aligned NW–SE in plan, and their NE faces are gently scalloped, while their SW faces are far more irregular. Lines of seaweed commonly grow between the crests. They are very similar to the barchan-like "underwater dunes" observed by Newell & Rigby (1957) in the Bahamas. These structures in the Khor Al Bazam have consistent alignment and fresh surfaces, but interestingly their shapes have not altered between 1958 and 2008. Their orientation is obviously unrelated to adjacent wave-formed features. Two possible origins are suggested. Either they are formed by tidal currents, or they are the result of waves and currents generated by large storms from the NNE. If the latter is true, then the crests of the ripples are probably kept free of weeds and their surfaces are reworked by normal wave action. The depressions between the ripples are protected from waves, forming a sheltered setting for the growth of abundant seaweed.

Cornish (1914) called this kind of transverse megaripple a sand wave. Since then they have been described under a number of names: Stride (1963) and Gilbert (1919) called them dunes; Kindle (1917) referred to them as large sand ripples;

and van Straaten (1953a) used the term "transverse megaripples". However, these features are larger than the transverse megaripples that van Straaten (1950) described forming in channels.

In the Khor Al Bazam transverse megaripples occur either to the side of, or between the sand and weed lines or streaks. Although similar in shape to subaerial barchan dunes, the sand waves have their steep side on their convex slope. They are some 400 m in width at their maximum, decreasing in size in both shallower and deeper waters. They form in response to the ebb currents that drain the lagoon.

The term "sand cloud" is used here to describe an amalgamation of a series of sand waves. These sand clouds occur west of the sand waves and in deeper water. They are essentially formed by the heaping up of sand waves in response to the easterly flowing flood tides where they counteract and overcome the strength of the ebb currents. Individual waves that form in the amalgamation tend to be small.

Diagonal megaripples

Two kinds of diagonal megaripples occur in the Khor Al Bazam: (1) linguoid megaripples; and (2) rhomboid megaripples.

The linguoid megaripple is a common feature of shallow tidal channels (Kendall & Skipwith, 1969b). Ball (1967a, b) referred to these as spillover lobes where they occur as ribbons of tidal deltas in the Bahamas. Such ripples were suggested by van Straaten (1953a) to have formed in the same way as transverse megaripples but in shallower water.

Rhomboid megaripples and sand streaks are usually found in shoal areas crossed by sheets of fast-moving tidal water. The wide stretches of shoal water found in the Khor Al Bazam are ideal for their formation. Good examples can be seen on the banks to the east of Jananah, to the NE of Marawah and SW of Fiyah. The ratio of bare sand to weed is approximately 1:1. The ripples disappear in the deeper water of the southernmost channels of the Jananah-Salaha Bank. The rhomboid pattern seen in all of these areas may have been formed in one of two ways: either by the accretion of rhomb-shaped tongues of sand (pointed down current) on the edges of weed patches; or by the erosion of rhomb-shaped scours (pointed up current) into the sand below on the edge of the weed patches. The relief of the features has not been determined, but similar features on the Norfolk coast of England are only 5–10 cm in height and have the same areal extent as those of the Khor Al Bazam. Such patterns or ripples can range from a few square metres to several square kilometres in extent.

There is a certain amount of ambiguity in interpreting the direction of the currents that form these diagonal megaripples. To the NW of Salaha, the linguoid megaripple marks and the sand tails behind small coral pinnacles suggest a northward current. The weed patches associated with the rhomboid ripples have serrated edges pointing into the current. At Al Khusaifah, the serrated edges of the boundary of the Al Khusaifah cyanobacterial flat are clearly erosional scours and point down current. However, just north of this edge there are rhomboid sand streaks and weed patches pointed toward the cyanobacterial flat. If these features are accretionary then they indicate a reverse current direction to that off the Al Khusaifah cyanobacterial flat. Rhomboid megaripples are fairly stable features.

Longitudinal megaripples

Longitudinal megaripples were recognized by van Straaten (1950, 1953a, b) in the Dutch tidal flats or Wadden. These current-formed features lie with their long axes parallel to the current. According to van Straaten (1953a), their wavelength varies from 1 to 60 m and their height from 1 to 20 cm. They may attain lengths of more than 1 km. The ripple is usually symmetrical in cross section with equally rounded troughs and crests. Van Straaten (1950, 1953a, b) also distinguished between deep-water and shallow-water varieties and showed they depend on the tide and the smoothness of the bottom. The deep-water variety is more regular and simpler. Imbrie & Buchanan (1965) recognized similar structures in the Bahamas but preferred to call them "large-scale current lineations".

In the Khor Al Bazam, longitudinal megaripples were subdivided by Kendall & Skipwith (1969b) into: (1) longitudinal megaripples; (2) sand ribbons; (3) sand streaks.

The term longitudinal megaripple is used where the elongate sand bars form a regular series. They do not have equally rounded troughs and crests, but are closely spaced sand ribbons of nearly equal dimensions and are not always symmetrical in cross section. They are found in water less than 4 m deep and parallel to both flood and ebb tidal currents. Their wavelengths average 100 m, which is greater than those described by van Straaten (1950, 1953a, b), while their axial lengths vary between 200 m and 1.5 km. Their amplitude is never more than 1 m. It is believed that the depth of water, bottom topography and current strength are all critical factors in their development but, as van Straaten (1950, 1953a) pointed out, their mode of formation is not clear. Imbrie & Buchanan (1965) believed that wind-driven currents related to hurricanes produce them.

Sand ribbons occur in the shallow water at the head of the Khor Al Bazam and are believed to have a similar mode of formation to those described by Stride (1963). Apart from the fact that they are not all evenly spaced, they resemble longitudinal megaripples. In the Khor Al Bazam, the ribbons are probably formed by both ebb and flood tide current and are elongate parallel to the direction of the currents of formation. They commonly start as a narrow ribbon and then splay out toward their distal ends.

Sand streaks are similar to sand ribbons but are much smaller, very closely spaced and commonly joined laterally. These linear features are found at the southern ends of the flood channels of Khor Salali and are usually less than 100 m long (Fig. 5). The width of the area between individual streaks differs from that of ribbons. The area between sand streaks is less than the width of the sand body, while the area between sand ribbons is wider than the ribbons.

Near the head of the Khor Al Bazam, sand and seaweed streaks are found. These are slightly different from the sand streaks described above, since the seaweed controls the distribution of the sand. The seaweed runs in N–S sinuous bands cut by E–W sand streaks of 0.5 km length and less than 10 m width. This is about half the width of the seaweed streaks. The origin of these features is not known, but appears to be related to currents.

Occurring in a similar setting to that exhibiting longitudinal megaripples are long, wide, erosive "striae" caused by ebb currents in shoal areas. The best examples are seen on the east bank of Khor Salali. These are over 1 km long, 0.25 km wide and are less than 1 m deep. They are also found along the sides of many of the channels and commonly run into the centre of the shoal bank. Striae also occur on the coastal terrace west of Al Khusaifah. Sand bodies between the striae are completely irregular in shape and elongate parallel to the "rills." Aerial photographs indicate that the sand bodies are positive features lying above the general surface of the shoal. Their shapes have been modified subsequently by rill erosion.

Ebb gullies or creeks

Ebb gullies are drainage channels that periodically carry water. They have steep sides relative to their width and are formed by the strong erosive effect of shallow water turbulence during ebb tides. The flow of the water, and hence the axis of the gullies, is usually at right angles to the main tidal channels into which they drain. According to Price (1963), the flood deposits of the Dutch tidal flats are reworked by ebb creeks. The gullies of the Khor Al Bazam are similar to "creeks" as described by Evans (1965) in the Wash and by van Straaten (1950) in the Wadden Sea. They drain the southern part of the shoal and channel area and occur in large numbers around the southern end of the Khor Salali (Fig. 5; Kendall & Skipwith, 1969a) and north of the main south eastern cyanobacterial flats of the Khor Al Bazam (Fig. 13). They also occur extensively in the cyanobacterial flats and in mangrove areas.

Other gullies are found along the south coast of Abu Al Abayad, but they are more like small tidal creeks than drainage gullies.

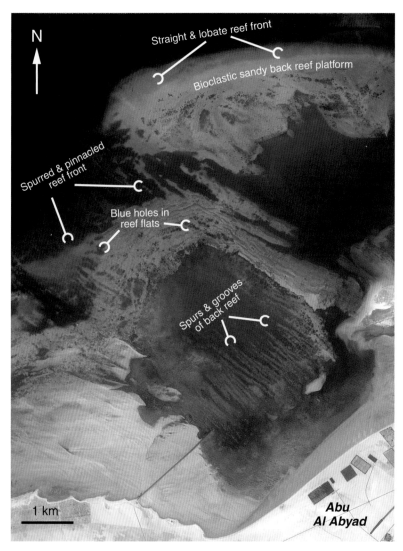

Fig. 13. Google Earth image (September 2008) of the reef off the northern shore of Abu Al Abyad showing a protected lagoon with spurs and grooves, pinnacles and blue holes. For location see Fig. 2B.

MAJOR MARINE GEOMORPHOLOGICAL FEATURES WITH BIOLOGICAL AFFINITIES

Coral reefs

Corals grow profusely on the offshore shoals, along the northern or windward edge of the offshore bank and the seaward edge of the coastal terrace (Figs. 3–7; Kendall & Skipwith, 1969b). Small reefs also line tidal channels and parts of the lagoon edge. *Acropora* sp., *Porites* sp. and several species of brain coral in conjunction with calcareous algae have been found to be the main reef builders. Kinsman (1964a) found similar corals abundant in the Abu Dhabi reefs. As in the Bahamas (Newell & Rigby, 1957), much of the coral of the Khor Al Bazam is being colonized and bored by a variety of sponges, bivalves and other organisms.

The coral reefs of the Khor Al Bazam are subdivided into coral reef front and reef platform.

Coral reef front

The coral reef front is at the seaward edge of a coral embankment where coral growth is most prolific. The slope of the reef front is dependent on the rate of coral growth on the initial bottom slope and the intensity of waves and currents. The reef front may take several forms: (a) straight and lobate vertical fronts; (b) spurred and pinnacled front; and (c) terraced front (Kendall & Skipwith, 1969b). None of these reefs exhibit the *Lithothamnion* ridge or marked rubble zone of Pacific reefs.

Straight and lobate vertical fronts

These are common and form a vertical wall of corals and calcareous algae. They occur around the northeast edge and northern tip of the Hail Bank and north of Abu Al Abyad (Figs. 4 and 13). Kendall & Skipwith (1969b) have shown that poorly developed examples of this front partially rim the coastal terrace on the southern side of the Khor Al Bazam.

Straight and lobate fronts are related to the relative intensities of the cross currents and the incoming waves. The straight front forms where there are strong crosscurrents; a spurred front where wave-driven currents flow perpendicular to the front; and a lobate front is formed in intermediate conditions. All forms can be seen in the reefs lying in front of Abu Al Abayad (Fig. 13) and Marawah. The slope of the face of the reef front is probably nearly the same as the surface on which the reef was initiated. This is believed to be a cliff surface formed by the immediate basement of Pleistocene to early Holocene windblown and marine carbonate (Fig. 6), known throughout the Gulf area as "miliolite". This rock is so called on the basis of the common occurrence in the rock of the many-chambered microscopic shells of *Miliola* benthic foraminifera (Pilgrim, 1908).

Vertical fronts undoubtedly once extended along the length of the northern edge of the coastal terrace of the southern margin of the Khor Al Bazam. Older coral reefs probably underlie part of this terrace. Coral growth here has been inhibited, as the offshore bank has become a more effective barrier to water circulation. Salinities and temperatures in the present lagoonal waters are probably too high and turbulence too low for the development of a flourishing reef. As these conditions become more extreme, the corals can be expected to cease to grow. The terrace is thus a dying fringing reef. Kinsman (1964a) recognized the effect of these high salinities in the enclosed lagoons to the east, between Halat Al Bahraini and Abu Dhabi.

Spurred and pinnacled fronts

Coral pinnacles are sharply projecting vertical colonies of corals (Shepard, 1948). Similar structures have been referred to as coral heads (single corals) and coral stacks, which are upright structures of many corals (Storr, 1964). Coral pinnacles occur north of the offshore bank near Salaha, northwest of Abu Al Abyad (Fig. 13) and opposite Marawah (Fig. 4). They indicate that pinnacles are also present in the shelter of the back-reef lagoons, as at Al Dhabaiyah (Fig. 14) where they are associated with micro-atolls. These pinnacles are usually about 10 to 20 m in diameter and may coalesce to form a coral ridge or small coral embankment. Micro-atolls (Krempf, 1927) are a halfway stage in the outward growth of coral pinnacles and the formation of coral ridges and small coral banks. The atolls are needle-shaped pillars, rimmed entirely by living coral, with the exception of the top where the central corals are deprived of nutrients, smothered by sediment, and exposed at low tide so that they die leaving a bald patch. The atolls are best developed in the back-reef off the Al Dhabaiyah Peninsula (Fig. 14).

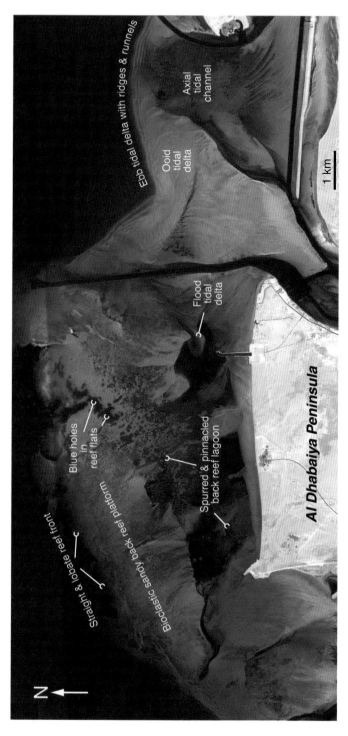

Fig. 14. Google Earth image (September 2008) of the reef and lagoon off the northern windward shore of the Al Dhabaiya Peninsula and the ebb tidal delta to the west of Halat al Barani. Note the micro-atolls in the lagoon and the tidal delta to the NE debouching from the tidal channel to the east of the Al Dhabaiya Peninsula. Note the ridges and runnels on the ebb delta. For location see Fig. 2B.

Spurred and grooved fronts are associated with pinnacles. Shinn (1963), reviewing the development of spurs and grooves, collected evidence from the Florida reef tract, which suggested that there the spurs are growing seaward but are periodically trimmed back by waves generated by severe storms. Detritus is removed from between the spurs by storms, preventing the infilling of the grooves. The spurs grow seaward since only those coral heads that point directly into the heavy seas are resistant to destruction. Newell *et al.* (1951) discussed similar features in the Bahamas, and Cloud (1959) studied them in Saipan.

In the Khor Al Bazam, three forms of spurs and grooves occur and are distinguished by the size and position on the reef:

(1) Seaward of the reef front, spurs can be as much as 2 km long, 400 m wide and between 5 to 10 m high. Grooves may be 800 m wide. Some striking examples of these spurs and grooves lie just north-northwest of Abu Al Abyad (Fig. 13) and north of Marawah (Fig. 4). These spurs and the long axes of most of the pinnacles are aligned parallel to the dominant wind direction. Once the pinnacles begin to grow on the sea floor, they trap sediment on their leeward side. The pinnacles expand radially to form micro-atolls, while coral and calcareous algae fix their sediment tails. These tails gradually extend toward the reef front, which they join to form a spur. An oriented colony protected in the lee of another colony grows seaward to connect to it; this increases the linearity of the growth centres (Shinn, 1963). Once initiated, spurs grow slowly seaward and grooves between spurs are maintained by wave and tidal scour. Spurs could not develop on the reef front where strong cross-currents occur. Some of the large spurs and grooves may have been enhanced by solution along joint planes during a Pleistocene sea-level low; this contention is supported by the fact that joint directions and present dominant wind directions coincide.

(2) Spurs and grooves on the reef front range between 3 and 5 m high, are sometimes over 16 m wide, and can be between 400 and 800 m long. Grooves are between 3 and 6 m wide. These features resemble the spurs and grooves of Florida and Spain (Shinn, 1963). They are believed to be initiated as patch reefs that developed tails that became fixed by coral. Any gaps between adjacent patch reefs are kept clear by heavy storms (Shinn, 1963) and are maintained as the spurs grow forward. The best example of these features can be seen northeast of Marawah (Fig. 4).

(3) On the reef flat behind the reef front, spurs can be as much as 1 km long, 0.5 to 2 m wide, and 0.5 to 1 m high. The grooves are 2–6 m wide. These lines of coral occur to the NNW of Abu Al Abyad (Fig. 13) and NW of Salaha (Kendall & Skipwith, 1969b). Like pinnacles and patch reefs, individual coral colonies act as sediment traps. Tails of sediment develop behind the coral and are fixed by coral growth. Once the coral line is initiated, it grows shoreward. Cloud (1959) suggested that similar features in the back-reef lagoon of Saipan Mariana Islands and on the Campech Bank Mexico are produced by movement of sediment trains that scour deeply into the floor of the lagoon and reef. Potholes within grooves observed in Saipan and on the Alacran reef complex, (Campeach Bank, Mexico; Kornicker & Boyd, 1962) have not been observed in the Khor Al Bazam region of the Arabian Gulf.

Terraced fronts

These consist of two or more levels of coral growth and occur along the southeastern edge of the Hail Bank and in front of the Jananah-Salaha Bank Each terrace is believed to have a similar origin and to have formed on offshore bars parallel to the offshore bank. The different levels indicate different periods of coral growth and the lowest tidal level limits their upward growth. The lower level terraces and the seaward spurs and pinnacles occur at the same depth. Thus, they probably began to grow at the same time.

Coral fronts may begin to grow on bars, since these bars have some permanence on aerial photographs. Cloud (1959) has shown that although the sand may be in constant motion, corals may colonize its surface. In the Khor Al Bazam, *Acropora* was observed growing on shells lying on banks of moving sand.

Reef platform

A reef platform forms at the top of any reef structure that is fairly flat (Storr, 1964). Most of the offshore bank north of Khor Al Bazam and the shelves north of Abu Al Abyad and north of Al Dhabaiyah may therefore be considered reef platform (Kendall & Skipwith, 1969b). The platforms are largely covered by coral/coralline algae sand and are underlain by Pleistocene or early Holocene miliolite along the coastal areas (Fig. 6). This matches the Florida reef platform (Ginsburg, 1956) where sand accumulates as megaripples that are just awash at low tide. Similarly parallel ribbons of coral and seaweed occur to the south of Hail, where they trend NNW and SSE. Between Jananah and Salaha they trend E–W. The ribbons are perpendicular to both the expected direction of waves refracted against the reef front and to the ebb and flow of the tidal currents. They extend almost to the southern edge of the Jananah-Salaha section of the offshore bank. Here, tidal channels cutting across the bank have been unable to erode the coral but have deepened the areas between the ribbons.

Hollows or depressions in the reef platform and reef fronts range from 2 to 10 m deep and can be from 0.4 to 2 km wide. They occur southeast of Hail or on the seaward side of the Jananah-Salaha Bank, and to the north of the Al Dhabaiyah Peninsula (Fig. 13) and Abu Al Abyad (Fig. 16). They have vertical walls that are rimmed by actively growing coral and are being filled by reef debris washed in by tidal currents and waves. It is unlikely that all of the hollows on the reef have the same origin. Some are depressions remain as the reef front advances. Large hollows southeast of Hail are of this type and represent old fluvial channels cut into the miliolite at low Pleistocene sea-level, now partially overgrown by coral. Still other hollows in reefs to the north of Salaha are remnant features left by random growth of coral spurs and ridges. Other hollows are erosional features such as abandoned tidal channels. Two hollows just north of Salaha may well be this type of feature. The last possibility is that the hollows represent solution at low Pleistocene sea-levels similar to the "blue holes" of the Bahamas (Newell & Rigby, 1957).

Seaweed-Covered areas

The seaweed growing in the Khor Al Bazam includes bladder wrack, green filamentous algae, small broad-leafed seaweed and *Halodule (Diplanthera) uninervis*. Areas of seaweed growth are considered geomorphological features because they profoundly modify wave and current movement and act as sediment traps (Ginsburg & Lowenstam, 1958).

Sargassum or bladder wrack areas

Sargassum algae or bladder wrack (brown algae) areas commonly occur as large patches along the edge of the coastal terrace on the Khor Al Bazam floor (where they lie below 3 m and are not visible on aerial photographs) and in patches on the Shelha Al Bazam (Kendall & Skipwith, 1969b). John (2005) identified two forms of these algae as *Sargassum latifolium* (Turner) C. Agardh, forming dense growths on the rocky platforms in shallow water (0.5–3 m), and *Sargassum decurrens* (Brown ex Turner) C. Agardh, growing, in conjunction with *Hormophysa cuneiformis*, along the coast on rocks partially covered by sand.

In the shallow-water regions where direct field observations could be made, the *Sargassum* is seen to be growing on hard cemented surfaces. These range from cemented portions of the coastal terraces, fragments of colonial coral or individual small boulders. Sugden (1963) recorded that layers of aragonite encrust the stems of bladder wrack off Qatar. He suggested that photosynthesis of the rich benthic floras promote aragonite precipitation. This process would account for not only the veneer of aragonite but also the cemented sediment around the bases of the weeds. This same process formed the rock platform beneath these cemented bases. An alternative but unlikely scenario is that it may represent beachrock developed at lower sea levels.

The seaweed does not prevent the formation of transverse megaripples. The bladder wrack also grows in conjunction with corals on the coastal terrace edge, some tidal channels and on the offshore bank, such as on the reef northwest of Bazam Al Garbi and on the reef lines north of Salaha. The *Sargassum* commonly breaks free from the sea floor and forms great rafts of seaweed that float on the Khor Al Bazam, accumulate on beaches and are incorporated into the sediments.

Areas of green filamentous algae

These areas are not as widespread as the *Sargassum*. They grow on shallow intertidal rock platforms free of violent wave action, between patches of *Sargassum* weed. Their location includes parts of the southern edge of the Shelha Al Bazam, the tidal flats just north of the Dagallah beach barrier, the bank north of Al Qala and Thimairyan, and the bank between Bu Shiyarar and Ras Al Aish.

Areas of halophila sea grass

John (2005) described how *Halophila ovalis* (R. Brown) Hook, *Halophila stipulacea* (Forskål) Ascherson and *Halodule uninervis* (Forskål) Ascherson occur in shallow sandy areas just below the low-water mark down to depths of about 8 metres along the coasts of the UAE. In the Khor Al Bazam they extend along the floors and sides of deeper channels, and along the lee slopes of the offshore bank, up onto the offshore bank, the Dagallah Bank, and the coastal terrace. In many cases they are either too deep or too sheltered to be affected by violent wave movement. Thus, these sea grasses are able to bind the sediment together, to drape the banks and form centres about which mud mounds accumulate. These mounds are aligned along major current directions and are probably very similar to the mud mounds described by Ginsburg (1956) in the Florida embayment, where they are fixed by *Halodule* sp. and *Thalasia* sp. As with the *Thalasia* of Florida Bay, these sea grasses of the Khor Al Bazam grew essentially in what is described as a "back reef" area by Ginsburg, (1956). Undoubtedly these grasses act as a sediment traps. Their epibiont colonies are centres of much animal life, with the subsequent accumulations of untransported death assemblages. Murray (1966) agreed with this observation as it relates to the foraminifera assemblages.

Cyanobacterial flats

Extensive flats of laminated cyanobacteria are forming on the protected intertidal and supratidal flats of the Khor Al Bazam (Figs. 4–8 and 15–17; Kendall & Skipwith, 1966). They have an average width of approximately 2 km and a thickness of at least 30 cm. At the eastern end of the lagoon, the largest flat is 42 km long (Fig. 5), and to the west, another flat is 9 km long (Fig. 4) at Al Khusaifah. In some areas the flats extend landward in the subsurface more than 2 km beneath a thin cover of evaporites and wind-blown sediments. Smaller flats occur in the shelter of islands, headlands and swash bar (Figs. 6–9 and 11).

The larger cyanobacterial flats are divided by surface morphology into four geographical belts (Fig. 16). From the low-water mark moving landward these are: (i) Cinder Zone (2 in Fig. 12), a warty black cyanobacterial surface, the colour and size of the raised bumps resembling a weathered volcanic cinder layer (Fig. 15). These bumps are

Fig. 15. A cinder-like cyanobacterial mat (right) marks the seaward margin of the algal flats and often overlies a cemented crust (left), as it does here. Note the abundance of cerithid gastropods on the crust surface. The coin is 2.5 cm across.

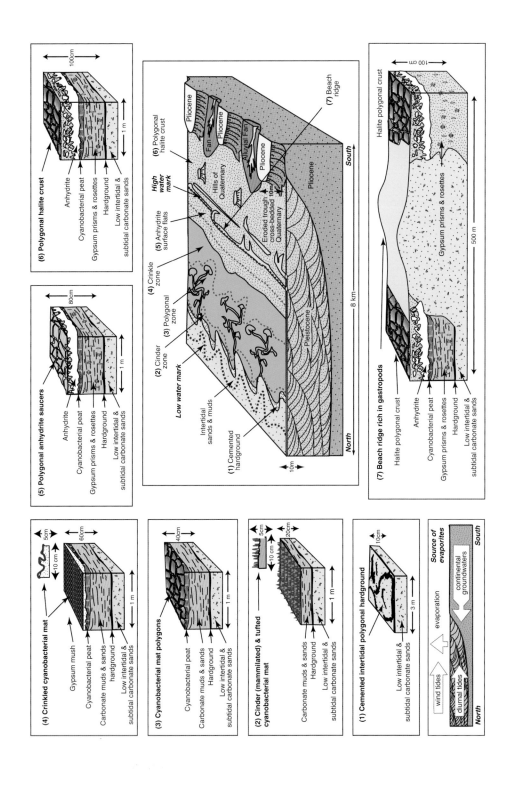

Fig. 16. Schematic composite block diagrams showing conceptual distribution of sabkha and cyanobacterial sediments of the Khor Al Bazam (modified from Butler *et al.*, 1982). (1) Seaward lagoonal carbonate sands and/or muds capped by carbonate hardgrounds; (2) Poorly laminated lower tidal-flat cinder-like cyanobacterial peat; (3) Upper tidal-flat cyanobacterial mat; commonly laminated and capped by desiccation polygons formed by growing cyanobacterial mat; (4) High intertidal to supratidal crenulated cyanobacterial mat capping a mush of gypsum crystals; (5) Cap of supratidal anhydrite polygons with storm washover and windblown carbonate and quartz. Below, anhydrite is interlayered with storm washover carbonate and replaces the supratidal mush of gypsum crystals. Note the large gypsum crystals (prismatic or lenticular) that form within the cyanobacterial peat, and carbonate and overlie the hardground layers; (6) Halite crust formed into compressional polygons over the anhydrite layers, algal peat, hardgrounds and intertidal and lagoonal carbonate; (7) Cerithid-rich beach ridge flanked by cyanobacterial mats and marine sabkha seaward (left) and storm-washover carbonates and supratidal evaporites landward (right).

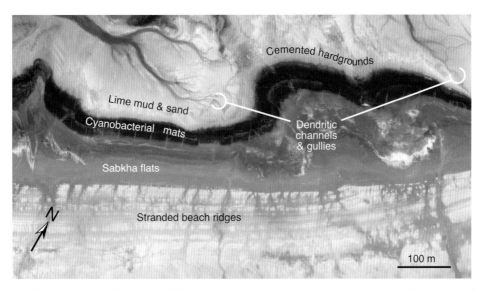

Fig. 17. Google Earth image (September 2008) of the eastern Khor Al Bazam shore line showing distribution of cyanobacterial mats and sabkha sediments just south of Al Rafayq and Bu Shara. Note the tidal channels incised into the lagoon and lower tidal flat carbonate sands and lime muds flanked by lower intertidal hardgrounds that locally are formed into compressional megapolygons two to five metres across. For location see Fig. 2B.

shaped like small pustules, 2 to 3 cm in diameter and contain an unlaminated cyanobacterial and sediment peat. Smaller cyanobacterial flats have one or more of these zonal belts; (ii) Polygonal Zone (3 in Fig. 16), cyanobacterial mat separated into desiccation polygons from a few cm to 2 m in diameter which cover laminated cyanobacterial peat; sediment fills the cracks between the polygons (Figs. 18 and 19); (iii) Crinkle Zone (4 in Fig. 16), leathery cyanobacterial skin forming a

Fig. 18. Polygonal cyanobacterial mat along the coast. Note the upturned edges and discolouration toward the centre of the mat. Quite commonly this is a pinkish colour. Sandals are 30 cm long.

Fig. 19. Megapolygons in the cyanobacterial mat along a small tidal creeks dissecting the tidal flats. Note the quite common pinkish colour of this mat.

blistered surface over gypsum and carbonate mush (Fig. 19); and (iv) Flat Zone, firm, smooth cyano-bacterial mat with no topographic relief, overlying quartz-rich carbonate sand and evaporites (Fig. 20).

The cyanobacterial growth and structures appear to be determined by the frequency and duration of subaerial exposure and the salinity of the tidal waters. They are only related to wave energy where they are limited by wave and tidal scour at the edge of the Cinder Zone and along ebb channels.

Mangrove areas

Only the Grey Mangrove, *Avicennia marina*, with its characteristic system of protruding roots or "pneumatophores" (Crumbie, 1987), has been found in the Khor Al Bazam and near-by islands (Kendall & Skipwith, 1969b). This was recognized as *Avicennia marina* but misnamed Black Man-grove by Kendall & Skipwith (1969b). It occurs as small bushes and trees, either in narrow strips parallel to the shore at the very top of the intertidal flats (Figs. 21–23), or in areas lining the edges of creeks draining cyanobacterial flats (Fig. 24). These areas are protected from vigorous wave action

and undergo frequent tidal interchange so that the mangrove roots and lower trunk are normally cov-ered at high water.

Tanner *et al.* (1963) observed that good tidal interchange is important for mangrove growth. As a result, larger mangroves grow on the edges of channels and diminish in height and distribu-tion inland as they do in the marshes and swamps of the Niger delta (Allen, 1965). In the Khor Al Bazam, when these channels become filled with carbonate silt, the area is colonized by cyanobac-terial mat and the mangroves die. The cyanobac-terial mats tend to grow landward of the mangroves (Fig. 24). They sometimes grow on creek banks, however, where they take the form of a cinder-like blue-green cyanobacterial mat. This is commonly cemented to form a beachrock. Inland, the cyano-bacterial mat is marked by a superficial polygonal pattern. Dunes covered by halophytes such as *Arthrocnemum glaucum* back this area. Carbonate mud accumulates around the mangroves of the Khor Al Bazam, as it does in Florida (Vaughan, 1909) and in the Bahamas (Newell *et al.*, 1951). This is commonly highly burrowed by crabs (Fig. 25). Some of the mud probably precipitates *in situ*.

Fig. 20. Crenulated or crinkle mat of the upper, intertidal zone. Note gypsum mush and cyanobacterial peat and the cross section of a polygonal saucer below. The knife is 25 cm long.

MARINE AND SUBAERIALLY FORMED GEOMORPHOLOGICAL FEATURES

Sabkhas are supratidal coastal regions composed largely of unconsolidated carbonate sediment and are the site of deposition of various evaporite minerals (Evans *et al.*, 1964a, b). Sabkhas form the subaerial portions of the surface area of barrier islands and the coastal plains in the UAE (Figs. 3–5 and 16).

Sabkha barrier islands

Sabkhas may evolve in the protection of offshore beaches to form barrier islands. These beaches are commonly cusped and lie on the landward side of the offshore bank bordering a lagoon, as at Dagallah. The sand that makes up the beach barrier is derived from the bank it borders. The beach slopes steeply into the deeper water and has a smooth curving profile. It is probably produced by waves refracted around the banks that interact with tidal flow across the bank, in a similar way to that described by Maxwell *et al.* (1964), Steers (1929) and Stoddart (1962). The barrier beaches act as a nucleus for barrier island development and are commonly cemented to form beachrock. In the early stages of formation, the beaches are often breached by tides and waves.

The early stages of barrier island formation are found near Dagallah. Here there is a complex of lunate beaches, spits and cyanobacterial flats. The final stage of development takes the form of an island of accreting cyanobacterial flat and chenier-like beach ridges, which can be seen at Jananah and Marawah (Figs. 4 and 5). A different form of barrier island occurs where miliolite acts as a core around which initial barrier beaches develop. This is followed by supratidal accretion in the form of mangrove stands and cyanobacterial flats. Two good examples of this kind of barrier island are those of Marawah and Abu Al Abyad, while the islands of Al Bazam Al Garbi and Salaha are smaller versions.

Mainland sabkha plain

The mainland "sabkha" plain represents the Holocene accretion of supratidal, intertidal and lagoonal sediments. This evaporite facies has been interpreted by Kirkham (1997) to have been initiated during the Flandrian transgression that deposited sediments and minerals in a discontinuous coastal

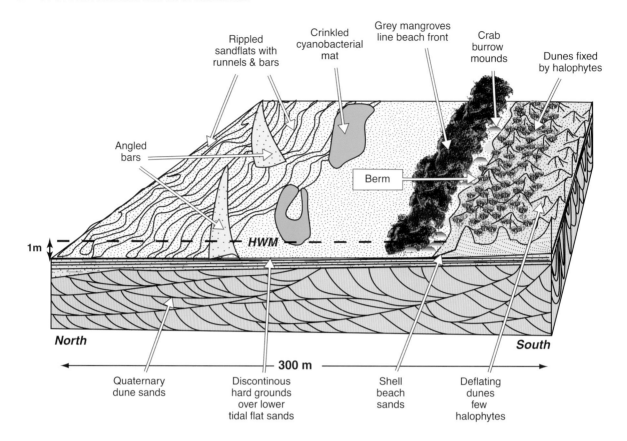

Fig. 21. Schematic block diagram of the intertidal sand flats beside which mangroves grow in stands parallel to shell-rich beaches with adjacent dunes fixed by halophytes. These growths occur along the western Abu Dhabi coast on the islands on the bank north of the Khor Al Bazam, particularly in the Al Dhabaiya-Al Rafayq area. HWM, high water mark.

lagoon just to the lee of the cerithid beach ridges. Locally, instead of carbonate sand cyanobacterial sediments are developed. Evans *et al.* 1964a, 1964b described how the mainland sabkha stretches "from Ras Ghanada almost to the Qatar peninsula in the west, a total distance of almost 320 kilometres". The width of the sabkha in the Khor Al Bazam varies and can be as much as 32 km wide (Fig. 3). Except where hills of Tertiary and Quaternary rocks jut out as peninsulas, it backs the intertidal flats of the coastal terrace. In places, the sabkha surface lies flush with eroded Quaternary rocks.

The surface of the sabkha can take several forms. It can be formed by ancient beach ridges; as a sandy surface between and behind beach ridges; or a salt-encrusted cyanobacterial flat. The most conspicuous features of the sabkha are the ancient beach ridges, identified by their distinctive appearance in the field, or directly from aerial photographs (Figs. 4, 5 and 17). As with the "cheniers" of Louisiana they have linear shapes with a smooth

seaward margin and an irregular landward outline (Byrne *et al.*, 1959). The beach ridges normally consist of well-sorted, coarse skeletal sands, and mark various stages of the seaward accretion of the sabkha in the form of intertidal spits. These spits drape headlands (for example, Ras Al Aish) and cross embayments (for example Al Marfa and Al Khusaifa). It is likely that some of the beach ridges represent barrier beaches that formed seaward of reworked earlier sediment and wadi outwash of the early Holocene (Wood *et al.*, 2002) and local small lagoons.

The beach ridge belt is some two to three kilometres wide. It is thought to represent a shoreline formed some 4000 years BP (Kirkham, 1997) when the Khor Al Bazam was more open and deeper. South of these ridges the sediment surface is very sandy and probably represents reworked early Holocene and earlier sediment and local small lagoon infill. Cross sections cut in some of the sandy sediment revealed cross bedding very similar to that shown by much of the miliolite,

Fig. 22. Mangroves occurring just seaward of and beside intertidal beachrock on an island to the west of Al Dhabaiya peninsula.

matching similar sediments of Quaternary age. Other sediments of the sandy surface show crude horizontal laminae, with well-sorted horizons of foraminifera-rich sand that probably accumulated during storm flooding. These beds are commonly contorted due to the crystallization of halite or the rotting of *Sargassum* and other seaweed washed inland with the foraminifera.

Salt encrusted cyanobacterial flats occur between beach ridges, and to the north and seaward of them. Storm washover sediments cover them as on the sabkha, south of the large eastern cyanobacterial flat. Trenches dug at least 2 km inland from the present high-water mark reveal that cyanobacterial laminae are still preserved above lagoonal sediments and a beachrock crust (Fig. 20).

The marine groundwater of the sabkha shows a progressive landward increase in salinity as a result of evaporation (Kinsman, 1964b; Butler, 1969; Butler *et al.*, 1982; Wood *et al.*, 2002; Alsharhan & Kendall, 2003). This produces four parallel gradational belts of distinct evaporite mineral assemblages and associated structures (Fig. 16):

(1) Upper intertidal: gypsum and celestite crystals (Evans & Shearman, 1964) and dolomitized

calcium carbonate within the capillary zone (Fig. 20).

(2) High water: calcium sulphate hemihydrate (Skipwith, 1966), anhydrite nodules and dolomite accompanying solution of gypsum in the capillary zone; dolomite and large "sand crystals" of gypsum below the water table; halite precipitated by the evaporation of stranded tidal waters and capillary water at the air/sediment interface.

(3) Above high water: anhydrite polygons (which are festoon in cross section) (Fig. 26) and diapirs within the capillary zone; gypsum and dolomite below the water table.

(4) Adjacent to outwash fans: anhydrite converted to gypsum by the influx of less saline ground water.

The occurrence of these mineralogical belts is variable and one or more is commonly absent. The typical sabkha cycle is shown in Fig. 16. Traced landward, gypsum normally appears in the capillary zone of the algal flats as a mush of lenticular crystals. Each crystal is about 0.5 mm in diameter and is flattened perpendicular to its c-axis (Masson, 1955). Inland from the intertidal zone, the layer of crystals thickens to 20 cm. A thin

Fig. 23. Large grey mangroves (*Avacennia marina*) growing just to the seaward of the beach face on the island of Marawah.

cyanobacterial mat generally covers the gypsum (Fig. 20). At the landward edge of this mat, small blebs of anhydrite are included in the surface layer. Some of the gypsum crystals in this zone show signs of solution and calcium sulphate hemihydrate ($CaSO_4.0.02H_2O$) may be present (Skipwith, 1966). Just seaward of the first occurrence of anhydrite, buried cyanobacterial and lagoonal sediments contain larger flattened gypsum crystals up to 15 cm in diameter. This size is thought to be the result of slow growth within the water table (Masson, 1955). Where this gypsum is found in sandy lagoonal sediments, it contains many inclusions. However, in cyanobacterial sediments gypsum characteristically contains few impurities.

Large gypsum crystals may commonly protrude through the sabkha surface landward of the beach ridges. This occurs where the water table has fallen and the surface sediments, no longer bound together by capillary water, are removed by deflation. The gypsum crystals are fragmented on exposure and this is explained as being the result of diurnal thermal expansion and contraction enhanced by the formation of small halite crystals within cleavage cracks.

Anhydrite first forms as small nodules about 0.5–1 mm in diameter (Butler, 1969). They lie within the sediment surface and are of a soft, white cheese-like texture. They are thixotropic. Like the gypsum mush, the anhydrite forms in the capillary zone. Traced into areas of higher salinity, the

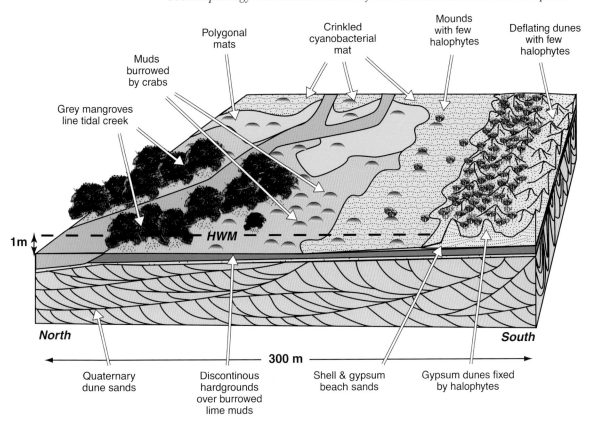

Fig. 24. Schematic block diagram illustrating the occurrence of mangroves growing in and beside tidal creeks that are flanked by cyanobacterial mats and dunes fixed by halophytes on the west coast of Al Dhabaiya and into the Al Rafayq area.

nodules become more abundant and larger, in places as much as 4–6 cm across. They form in aeolian and storm washover sediments that begin to accumulate on the upper algal flats and extend back onto parts of the sabkha. Landward of the nodules a surface layer of anhydrite develops into a series of interlocking saucer-shaped structures that are polygonal in outline. Inland, as the sediments overlying the cyanobacterial mat thicken further, the anhydrite also thickens and forms layers at depth, near, and at the surface. The lower enterolithic layers commonly are contorted and they form small crenulations and tight folds similar to those of ptygmatic quartz veins. In the Khor Al Bazam, as in the sabkha inland from Abu Dhabi, some of the enterolithic anhydrite layers thicken to form layers more than 20 cm thick. West of Al Tarif, antiform structures more than 20 m in diameter occur.

As Butler (1969) first reported near Abu Dhabi, the sequence of anhydrite development in the Khor Al Bazam ends with the influx of groundwater into the sabkha via outwash fans. Anhydrite is hydrated to gypsum, forming white, coarse, elongate, tooth-like monoclinic crystals about 2–3 cm long and 0.5 cm in diameter. The hydration process disturbs the surface of the sabkha so much that it resembles a ploughed field. The blistered surface contains large quantities of halite. Elongate, wispy crystals of rock salt are common in surface sediments of the sabkha at the top of the capillary zone. After flash floods and marine incursions, surface waters evaporate and leave a thin white crust of halite or, alternatively, sandy salt blisters. The salt is removed by wind and may be replaced by evaporation of capillary water. Where depressions in the sabkha retain pools of floodwater for any length of time, "hopper" crystals develop.

X-ray analysis of core samples from cyanobacterial flats and sabkha west of Ras Al Aish indicates the presence of dolomite (Butler, 1969). This dolomite forms contemporaneously with the gypsum and anhydrite. The percentage of dolomite appears to be a function of grain size. The proportion of dolomite in calcareous mud is generally greater than that in calcareous sand. One sabkha mud from west of Ras Al Aish contained more than 80% dolomite.

Fig. 25. Cemented crab burrows (approx. 20 cm across) being formed on tidal creek margins where small salt tolerant bushes and larger grey mangroves (*Avacennia marina*) are growing (see Fig. 24).

The surface of the sabkha is modified by both marine and aeolian erosion. For instance, on beach ridges, wind carries away only the finer grades of sand, leaving behind a lag deposit of gastropod shells. Aeolian erosion is limited by where sands are dampened by capillary water from the water table, but marine erosion can extend deeper. In parts of the sabkha where it has been particularly effective, marine erosion exposes the water table. A thin, dry, halite crust usually covers such areas, gypsum protrudes as vertical sand crystals, and the polygonal forms of anhydrite may be exposed. Marine flooding also breaches old beach lines transporting their sediment onto the sabkha behind. The outgoing water re-breaches the beach ridges to produce small deltas (for example, west of Al Khusaifa and west of Al Qala). The sheets of floodwater are driven about the sabkha by strong winds. The evaporites have a mix of at least three origins (Wood *et al.*, 2002; Alsharhan & Kendall, 2003): (1) regional upwelling of the waters of brine-dominated local aquifers (Wood *et al.*, 2002); (2) local rainfall; and (3) marine waters from flood recharge and capillary upwelling from the adjacent water table.

SUBAERIAL GEOMORPHOLOGICAL FEATURES

Coastal dunes

In the Khor Al Bazam, halophytes like *Arthrocnemum glaucum* line the shore in areas above the high-water mark and on intertidal spits (Fig. 9). They trap sand blown off the beach at low tide and help to stabilize coastal dunes. Despite low rainfall, the plants survive on heavy nightly dews and moisture retained in the beach sediments by seaweed humus. Adjacent and parallel to the beach, foredunes form mounds of up to 2.5 m high; behind them there may be one or two beach ridges. Approximately 30 m from the shore the dunes begin to lose plant cover, probably because the amount of dew precipitated on the plants decreases rapidly with increasing distance from the sea. Without the plant cover, the dunes are blown away, leaving behind beach ridges.

Hills and alluvial fans

Quaternary (miliolite) and Tertiary rocks crop out close to the coast as low mesas and buttes

Fig. 26. Trench cut in the surface sabkha sediments in the coastal region near Al Marfa in 1963. The surface is marked by anhydrite polygons that exceed ten centimetres in thickness and below which are nodular and enterolithic anhydrite, interbedded with stormwashover carbonates and overlying subtidal carbonate deposits. Douglas Shearman provides the scale.

surrounded by alluvial fans (Figs. 6 and 27). These hills are remnants of a much more continuous cover of almost horizontal rock, eroded by the combined action of flash floods and wind (Kirkham, 1997). Where the miliolite projects as peninsulas into the sea, seawater, boring algae, and animal activity etch the rock surface, in much the same way as the cay rock of the Bahamas (Newell *et al.*, 1960). Commonly in inland areas, wind and marine flooding have eroded the miliolite flush with the water table of the surrounding sabkha (Kirkham, 1997).

SEDIMENT SIZE, COMPOSITION AND DISTRIBUTION ALONG THE COAST

Kendall & Skipwith (1969a) and Hughes Clarke & Keij (1973) provided a general description of the principal carbonate-producing and environmentally significant organisms of the Abu Dhabi coastline. The bulk of the skeletal material that constitutes most of the Holocene sediments of the Arabian Gulf represent a relatively small number of organisms that include, in order of decreasing abundance, molluscs, calcareous algae, corals, foraminifera, bryozoa and echinoids. The carbonate sediments of the southern Arabian Gulf reflect the broad ecological limitations of the setting on their organism assemblages, with elevated salinity having the clearest restrictive effects (Hughes Clarke & Keij (1973)). Further details are to be found in a general survey of the distribution and character of sediment types of the southern Arabian Gulf by Wagner & Van der Togt (1973). They described the textures and grain types which they related to the sedimentary and diagenetic processes that formed them.

Grain-size distribution

The pattern of grain-size distribution in central Abu Dhabi and in particular in the vicinity of the Khor Al Bazam is associated with changes in relief of the sea floor. Sediments of the shelf areas generally are well sorted and lagoonal sediments are poorly sorted. Most of the Khor Al Bazam sediments are coarse to fine sand and are positively skewed about a mode in the medium-sand grade.

Fig. 27. Example of a butte of Pleistocene cross-bedded miliolite sediment cropping out in the Abu Dhabi coastal sabkha that extends from Ras Ghanada to Jebel Dhanna.

However, in the lee of the offshore bank, at the western end of the Khor Al Bazam opposite Bazam Al Garbi, and in mangrove swamps, the silt-and mud-sized content of the sediments in many places exceeds 30%. These fine constituents are probably the product of transport and deposition from the offshore bank, as well as from direct precipitation (Kendall & Skipwith, 1969a).

Ginsburg (1956) used the percentage of material <0.125 mm as the most critical environmental indicator in Florida Bay and the adjacent reef tract. The percentage of material <0.625 mm has been found to be much more sensitive to environmental conditions in the Khor Al Bazam (Kendall & Skipwith, 1969a). This distribution coincides with an area of mud mounds fixed by *Halodule*, much as with the mounds described in Florida Bay by Ginsburg (1956).

Sediments

Most of the sediments found in central Abu Dhabi in the Khor Al Bazam area are bioclastic (Kendall & Skipwith, 1969a). Many species, but few phyla are represented. The skeletal material is composed of either aragonite or high- or low-magnesium calcite. Non-skeletal carbonates are composed of

aragonite, which probably is derived biochemically, but not from direct secretion by biota (Kinsman, 1964a). Using strontium content as an indicator, Kinsman (1964a) found that the aragonite mud in a lagoon southwest of Abu Dhabi was not abraded skeletal material. However, it was not possible to establish whether this aragonite was precipitated by evaporation and concentration of marine waters through physicochemical means or was precipitated within plant microenvironments during photosynthesis. Kendall & Skipwith (1969a) suggested that in the Khor Al Bazam, cyanobacteria are probably the major agency of precipitation of non-skeletal aragonite. Recent observations of "whiting" blooms of precipitating carbonate in the Gulf suggest carbonate is precipitated intermittently within the eastern Gulf throughout the year but particularly during the summer (Kendall *et al.*, 2007).

Components

The major components in the sediments from the study area of the Khor Al Bazam are shown in Table 4. The sediment components are similar to those of the Bahamas (Illing, 1954) and Florida Bay (Ginsburg, 1956). There are exceptions. Bryozoan

Table 4. Major components identified in the sediments of the Khor Al Bazam, Abu Dhabi.

Components constituting >1% of the sand fraction	Components constituting <1% of the sand fraction
Carbonate grains	Serpulid tubes
Mollusc shells and fragments	Sponge spicules
Ostracod valves	Echinoid fragments
Foraminifera tests	Fish bones
Calcareous algae	Crustacean fragments
Coral fragments	
Bryozoan skeletal fragments	
Faecal pellets	
Ooids	
Unidentified particles	
Non-carbonate grains:	
Quartz	
Heavy-mineral grains	
Unidentifiable sugary brown grains	
Evaporite minerals	
Flocculent material	
Mixed non-carbonate and carbonate grains:	
Aggregates	
Unidentified particles	

Table 5. Sediment components and their equivalent geomorphological unit in the Khor Al Bazam of Abu Dhabi.

Sediment component	Geomorphological unit
Molluscan sand	Lagoon and open sea
Oolitic sand	Coastal terrace
Pellet aggregate and pellet sand	Coastal terrace and south edge of offshore bank
Coral and coralline algal sand	Seaward shoals and outer edge of offshore bank
Faecal pellet sand	Southern shoal and channel area
Carbonate mud and pellets	Protected lagoon and mangrove swamp
Cyanobacterial mat	Intertidal
Evaporites	Supratidal

skeletal fragments are present in greater amounts than in Florida and the Bahamas. In addition, detrital grains of quartz, other silicates and gypsum and anhydrite crystals are present in the UAE coastal regime but are not found in Florida or the Bahamas.

In the descriptions that follow, each sediment component is related to the major lithofacies units of the Khor Al Bazam. The units are separated on the basis of component content and texture. Their equivalent geomorphological unit is shown in Table 5. Each of the lithofacies is related to the hydrodynamics and water chemistry of the depositional setting. Flora and fauna, though affected by the same physical agents, also control lithofacies.

Molluscs

The list of molluscs in Table 6 is compiled for the Khor Al Bazam and represents a collection made by Kendall & Skipwith (1969a) that was identified by J. Biggs of the Natural History Museum, London. The heavy, robust forms of molluscs, mainly gastropods, are mainly on found the coastal terraces and bank. In contrast, thin-shelled bivalves, commonly still articulated, dominate the lagoonal sediments. Their shells are commonly uncoloured or lightly coloured, as are those of the interior of the Bahama Banks (Illing, 1954).

Many of the shells are in the process of being destroyed by boring and crushing organisms and by waves. Boring organisms include small molluscs, worms, sponges and cyanobacteria. Crushing organisms include a variety of crustaceans that roam the whole of the Khor Al Bazam and offshore bank. Crabs are particularly active on rock platforms covered by green algae. Kendall & Skipwith (1969a) reported that at low tide, the sound of shells being crunched is common. Breakage caused by waves is confined largely to the edges of the offshore bank and the coastal terrace.

Mollusc shells normally form a high percentage of the sediments on, and directly in the lee of, the offshore bank, in the centre of the lagoon, and on the coastal terrace. On the bank and terrace the coarse shells and fragments are probably *in situ* and are not being carried into the lagoon. Similarly, the thin-shelled molluscs of the lagoon are probably *in situ*. Articulated shells and lack of abrasion support this interpretation. The most clearly defined setting for mollusc shell accumulation is to the lee of the offshore bank, where it coincides with the higher percentages of carbonate mud (Fig. 3). Near the head of the eastern Khor Al Bazam and beyond the offshore bank where currents are active, the shells are better sorted and free of flocculent material. The material may be transported over short distances in these areas.

From Al Khusaifa to the west, the same general pattern continues. It is affected by changes in relief of the lagoon floor; cyanobacteria are most active on areas of high relief in the mid-lagoon and the percentage of molluscs decreases over those areas. The highest concentration of shells is found on

Table 6. Molluscs found in the Khor Al Bazam of Abu Dhabi.

Bivalvia	Gastropoda
Angulus (Fabulina) rhomboides Quoy and Gaimard	*Atys cylindrica* Jan
	Ancilla cinnamomea Sowerby
Angulus (Fabulina) immaculata Philippi	*Ancilla eburnea* Deshayes
Arca plicata (Chemn) Dillwyn	*Bullaria ampulla* Linne
Arca lacerata Linne	*Calyptraea pellucida* Reeve
Arca tortuosa Linne	*Cerithidea cingullatus* Gmelin
Arcopagia robttsta Hanley	*Cerithiurn morus* Lamarck
Brachidontes variabilis Krauss	*Cerithium scabridum* Philippi
Cardium sueziensis Issel	*Cerithium petrosum* Wood
Chlamys ruschenbergeri Tryon	*Clava (Clava) fasciata* (Bruguiere)
Circe corrugata Dillwyn	*Columbella* sp.
Circe scripta Linne	*Cypraea caurica* Linne
Codakia fischeriana Issel	*Cytherea* sp.
Corbula acutangula Issel	*Diodora funiculata* Reeve
Corbula subquadrata Melvill	*Drupa margariticola* Broderip
Crenella adamisiana Melvill and Standen	*Euchelus bicinctus* Philippi
Diplodonta ravayensis Sturany	*Finella pupoides* A. Adams
Diplodonta sp.	*Finella scabra* A. Adams
Dosinia histrio Gmelin	*Laemodonta (Laemodonta) bicolour* (Pfr.)
Dosinia sp.	*Minolia gradata* Sowerby
Grastrochaena cuneiformis Spengler	*Minolia holdsworthiana* Nevill
Glycymeris hoylei Melvill	*Monilea obscura* Wood
Glycymelis pectunculus Linne	*Murex (Chicoreus) anguliferus* Lamarck
Glycymeris spurcus Reeve	*Mitrella (Mitrella) blanda* Sowerby
Lima tenuis A. Adams	*Perrinia stellata* A. Adams
Lima (Limatula) leptocarya Melvill	*Persicula* sp.
Lioconcha pitta Lamarck *Laevicardium papyraceum* Bruguiere	*Phasianella nivosa* Reeve *Phasianella* sp.
Lucinae dentula Linne	*Pterigia* sp.
Mactra cf. olorina Philippi	*Pyrene (Seminella) phaula* Melvill
Macoma jeanae Dance and Eames	*Retusa omanensis* Melvill
Malleus regula Forskal	*Ringicula propinquans* Hinds
Meretrix sp.	*Rissoina sismondiana* Issel
Motirus irus Linne	*Rissoina distans* (Anton)
Nuculana confusa Hinds	*Rissoina savignyi* Jousseaume
Phacoides semperianus Issel	*Scaliola arenosa* A. Adams
Pinna sp.	*Scaliola elata* Semper

Table 6. (*Continued*)

Bivalvia	Gastropoda
Pinctada radiata Leach	*Strombus* sp.
Pitaria spp. (incl. *hagenhowi*) Dunker	*Tricolia foridlana* Pilsby
Quadrans pristis Lamarck	*Trochus (Infundibulops) erythraeus* Brocchi
Solenocurtus strigullatus (Linne)	*Turbo radiatus* Gmelin
Spondylus exilis Sowerby	*Turbo coronatus* Gemelin
Sunetta effosa Hanley	*Turitella aurocincta v.* Martens
Tapes undulata Born	*Umbonium vesticzrium* Linne
Tellina pygmaea Loven	*Xenophora caperata* Philippi
Tellydora pellyana A. Adams	
Timoclea sp.	
Trachycardium lacunosum Reeve	
Trachycardium maculosum (Wood)	
Venus (Callanaitis) calophylla Hanley	

certain beaches, where they comprise nearly 100% of the <5 mm fraction. Common to most beaches are cuttlefish skeletons. Blackened shells are common in the central parts of the west Khor Al Bazam.

Ostracods

Ostracods generally are articulated in the lagoon samples, but in shelf samples they are disarticulated and bored (Kendall & Skipwith, 1969a). The ostracods appear to be living in the central lagoon and in the muddy areas in the lee of the bank.

Foraminifera

The foraminifera collected in the Khor Al Bazam were identified by Murray (1966). Very few living individuals were found in the samples. Representatives of the superfamily Miliolacea make up most of the populations, the dominant forms being species of Quinqueloculina, Triloculina and Peneroplidae. Murray (1966) illustrated their areal distribution and suggested that they are mostly untransported death assemblages. It was concluded that the foraminifera must have been living on the seaweed, which had not been sampled.

The foraminifera generally are spread evenly over the lagoon. However, foraminifera are transported from the bank and are concentrated just to the lee side, causing a small increase in their percentage at this location. In the intertidal areas of the coastal terrace and certain beaches, the foraminifera percentages are high, but they decrease landward over the cyanobacterial flats. In the shoal areas around the margins of the Khor Al Bazam, foraminifera tests may be partially or wholly altered by the activities of cyanobacteria (Kendall & Skipwith, 1969a).

Calcareous algae

Extensive encrustations of robust red calcareous algae, including *Lithothamnium* sp. and *Archaeolithothamnium* sp., are living along the reef front of the offshore bank and parts of the edge of the coastal terrace. These organic remains undergo rapid alteration, becoming indistinguishable from the microcrystalline grains a short distance from their source areas. Illing (pers. comm., 1965) reported that the calcareous algae *Goniolithon* is common in reefs in Qatar and this algae is widely present in the reefs of the west Khor Al Bazam. The delicate red calcareous alga, *Jania*, is a more important component, particularly in finer grades of sand. Over much of the coastal terrace it grows in small colonies that break up to form rod-like grains. The influence of the red calcareous algae detritus begins at the headland west of Al Marfa and continues to the terrace just south of Bazam Al Garbi, a distance of 40 km. The percentage of algal fragments decreases abruptly lagoonward and increases shoreward. Like other red algae, rods of *Jania* are subjected to micritization caused by cyanobacterial infestation, but remain recognizable because of their distinctive shape.

Coral fragments

It was very difficult to identify coral fragments with the stereoscopic microscope, partly because of the rapid alteration of the grains. Where alteration has not occurred, the grains abrade readily and thus their characteristic surface texture is destroyed. Many of the components, which were classified by Kendall & Skipwith (1969a) as unidentifiable on the offshore bank and coastal terrace, may be of coral origin. Offshore bank samples have high percentages of coral when traced across the bank from the reefs.

Bryozoan skeletal structures

Mrs Patricia Cook of the Natural History Museum, London identified the following specimens of bryozoa for Kendall & Skipwith (1969a): (1) Cheilostomata, Anasca: *Thalamoporella gothica* var. *indica* Hincks; Ctenostomata, Carnosa: *Sundanella sibogae* Harmer. *Thalamoporella gothica* is very common and grows as encrustations on the stems of the "bladder wrack" seaweed. It forms a very lightweight, cellular structure that accumulates as detritus in the areas of maximum development of bladder wrack. It is found particularly on the terrace and southern part of the west Khor Al Bazam between Al Marfa and Al Khusaifa. When the bladder wrack breaks free from the sea floor, it drifts as "*Sargassum*-like" rafts that eventually wash up on the shore. The strandline of the beaches is lined by accumulations of this weed, which becomes incorporated in beach ridges as they build seaward. The weed rots in beach ridges and leaves *Thalamoporella gothica* to form lenses 20 cm or more thick.

Pellets

Three types of pellets were identified by Kendall & Skipwith (1969a) to be forming in the Khor Al Bazam. These grains range in size from 0.05 to 1 mm and may be cylindrical and ellipsoid, spherical, or irregular. Dissolution of any of the pellets with HCl leaves a mucilaginous sludge that commonly contains cells of cyanobacteria. The process by which the pellets are hardened is interpreted to be related to this organic sludge, particularly if it includes cyanobacteria.

Cemented ellipsoid and cylindrical pellets of aragonite mud are rare in most Khor Al Bazam samples, but were found by Kendall & Skipwith (1969a) to be numerous just seaward of cyanobacterial mats. Pellets have been found in the guts of some worms. This suggests that many of the pellets are of faecal origin and were formed by the myriad of worms that burrow in the sediment. Alternatively, some are likely excreted by molluscs. Kornicker & Purdy (1957) and Folk and Robles (1964) found that the gastropod *Battilaria minima*, a species of cerithium, produced pellets similar in shape to the ones observed in the Khor Al Bazam by Kendall & Skipwith (1969a).

The pellets are initially very soft. If they are subjected to violent wave or current movement, they are easily disrupted, but if conditions are quiet, the pellets become cemented and harden. In the Bahamas, Illing (1954) showed this hardening to be very rapid. The fact that only a few soft pellets are present in the Khor Al Bazam suggests that the hydrodynamic setting is too violent to allow many to be preserved. The surfaces of some of the grains suggest that they were abraded during cementation. Kinsman (1964b) recorded a similar occurrence of pellets in the "crab flats" near Abu Dhabi.

Spherical pellets are distinguished from the faecal pellets by their high sphericity, and are more common than faecal pellets in the Khor Al Bazam. Kendall & Skipwith (1969a) found them to be normally confined to shoal areas like that of Al Qala embayment and the broad shelf west of Al Marfa. They have four probable origins: (1) material altered by cyanobacteria (Dalrymple, 1965); (2) derived from broken faecal pellet fragments; (3) are comparable to friable aggregates (Illing, 1954); (4) are quiet-water ooids (Freeman, 1962).

More irregular and, consequently, less readily identified pellets also are present in the lagoon. In the quieter waters of the coastal terrace, these grains commonly are surrounded by flocculent material.

Aggregates (grapestones, botryoidal grains, and encrusted lumps)

Kendall & Skipwith (1969a) described these aggregates forming on the lagoon floor and on some of the adjacent shoal areas (Table 7). Material from the shoal areas is not transported far into the lagoon. As Illing (1954) recognized for the Bahamas, several sand-sized accretionary grains, including friable aggregates, grapestones, botryoidal grains, encrusted lumps, and shell infillings occur in the Khor Al Bazam region. Samples taken across the cyanobacterial mats show that, although aggregates are forming in front of them, the aggregates are not transported across the mats. The percentage of aggregates decreases sharply landward, but the older buried, frontal algal cyanobacterial mats contain a higher proportion of aggregates.

Friable aggregates composed of discrete silt-sized particles. They are generally lightly bound together by threads of seaweed, cyanobacteria or some other organic tissue (Illing, 1954). Component particles of aggregates in the Khor Al Bazam consist of nearly equal proportions of quartz and carbonate. In contrast, those described by Illing (1954) are exclusively carbonate. The complete sequence illustrated by Illing (1954) of progressive cementation of these grains to form ovoid grains, was not found in the Khor Al Bazam, although the two end members were found. The highest percentage of uncemented friable aggregates is present in sea grass areas of the lagoon floor and the ovoid grains are found both in sea-grass and on the coastal terrace.

Grapestones are aggregates composed of sand-sized particles bound together by aragonite cement. Component grains protrude, giving the appearance of a bunch of grapes. The components are predominantly aragonite, with up to 20% quartz. Illing's (1954) description of the Bahamian examples fits the Khor Al Bazam grapestone. Worm tubes and shell infillings were recognized in the Khor Al Bazam.

The cement that joins the grains is finely divided aragonite of varying texture. It forms first around the points of contact, and is friable and chalky white, in contrast to the greasy or matt textured

Table 7. Aggregate grains from the Khor Al Bazam of Abu Dhabi.

Grain type	Composition	Environment of occurrence
Friable aggregates	Discrete silt-sized particles bound by organic fibres	Lagoonal grass area
Lumps; grapestones	Discrete sand-sized particles bound by, and protruding from, an aragonite cement	Lagoon floor, coastal terrace and offshore bank
Botryoidal lumps	Similar to grapestones but with polished aragonite envelope disguising components	Coastal terrace
Encrusted lumps	Similar to grapestones but with surface so bored by cyanobacteria and altered, as to be unidentifiable except in thin section	Lagoon floor, coastal floor, coastal terrace and offshore bank
Worm tubes	Discrete sand-sized particles bound by aragonite cement to form tubes	Coastal terrace and offshore bank
Shell infillings	Discrete sand-sized particles bound by aragonite cement and enclosed or partially enclosed by mollusc shells	Lagoon, coastal terrace and offshore bank

grains. It can easily be scraped off with a pin, and is composed of aggregated particles of mud size. From its occurrence, it is clear that it is being precipitated from seawater, yet the particles show no recognizable crystalline shape and are similar to the material that forms the matrix of the grains themselves. Externally, the chalky white cement is restricted to the crevices between the protruding grains. As cementation proceeds, the cement beneath this surface layer becomes firmer and mat textured. Additional grains or small lumps may be joined by the same sequence of stages. However, the forces of mechanical disintegration prevent unlimited growth and finally all traces of chalky white texture are lost.

In the Khor Al Bazam, grapestones form on the grassy lagoon floor and shoal areas. During periods of minimum wave activity in the lagoon, surface sediment is bound initially by an organic slime forming a thin crust (Nesteroff, 1956). On intertidal shoals, the initial binding agent is either organic cyanobacterial mucilage or incipient aragonite beachrock cement; the latter is precipitated at low tide during hot summer months (Ginsburg, 1953a,b). In both areas, if the cementation process continues long enough, photosynthesis by cyanobacteria precipitates aragonite. It cements and fuses the carbonate grains and is indistinguishable from the grains. The acicular crystals of beachrock cement are altered and the whole grain ultimately becomes homogeneous. The longer the surface is undisturbed by wave action, the better developed the binding.

Once formed, the surface crust may be broken into small lumps by wave action. The size of the lumps depends on the strength of the cement and the force of the waves. With a heavy sea and lightest of bindings, the crust may be separated into its constituent grains, but with gentle action and strong cementation, aggregates are produced. Thus on the average, aggregates are larger and more irregular in the protected grass areas of the lagoon than those of the exposed shoal regions.

In the intertidal region of Al Qala Bay, Kendall & Skipwith (1969a) have found that aggregates are composed of 20% quartz, 60% aragonite pellets and 20% aragonite cement. The sediment in which they are present contains a ratio of aggregate to quartz to pellet of approximately 3:3:1. If the sediment were cemented unselectively, the aggregates would be expected to contain more quartz than they do. The low percentage of quartz in aggregates may be explained in three ways:

(1) The percentage of quartz in the sediment is not usual. It is abnormally high because the samples were collected only a short time after a strong onshore gale. The excess quartz was carried in from the headlands.

(2) When the partly cemented surface crust is broken by waves, it generally fractures along the boundaries of the quartz grains, rather than along the fused boundaries of carbonate grains. The separated quartz grains may retain patches of the cement. This could account for the fact that many of the quartz grains have carbonate jackets, some even with one or more hemispheroidal aragonite shapes protruding from the surface. The latter are fused aragonite pellets.

(3) Another possible process of aggregation involves flocculent blebs of cyanobacteria and their investing mucilage. The cyanobacteria float free in sediment, and act as centres of aragonite precipitation to form pellets. In their early stages these pellets have "sticky" surfaces that adhere easily to other grains. Thus aggregates, quartz grains, and pellets quickly acquire aragonitic coatings. The loosely bound aggregates become cemented by cyanobacterially precipitated aragonite. Pellets formed by this process are unlikely to be unconsolidated because of their adhesive qualities. Although cyanobacteria and their mucilage are present on all carbonate grains, there is no evidence that cyanobacteria colonize non-carbonate grains. If they do not, or only do so poorly, this ecological factor also may influence the ratio of quartz to carbonate in the final aggregate.

Kendall & Skipwith (1969a) found that in thin section, the initial cement that binds grapestones is of darker colour and is apparently coarser grained than the final cement. As cementation proceeds, the textures of the cement and the component grains become indistinguishable. Purdy (1963a,b) observed a similar paragenesis in the Bahamas.

In the Khor Al Bazam, the presence of acicular aragonite crystals as a cement in the voids of grapestones is rare. However, development of some crystals occurs below the intertidal level, and Kinsman (1964a) found them in the base of corals from the Abu Dhabi area. A portion of the component carbonate particles commonly are blackened, but are bound by white aragonite cement. In places, the cement itself is blackened. As with the sediment components previously described, blackening is largely confined to the lagoon samples and follows the pattern shown by Murray (1966).

Botryoidal lumps are like grapestones in that they are composed of discrete sand-sized particles. However, they differ in being covered by a sheath of polished aragonite. As with grapestones, the mamillated surface of these grains is grape-like. In the Khor Al Bazam, few botryoidal lumps grow larger than 0.54 mm in diameter. The smaller the grain, the more complete the cover and the better the polish. Botryoidal lumps represent grapestones that, by a process similar to ooid formation, acquire successive coatings of aragonite. These coatings eventually disguise the component grains and fill the cracks between them. The lumps form in shallow and intertidal waters, the highest energy environments of the Khor Al Bazam.

Encrusted lumps are grapestones whose surfaces have been modified by intense cyanobacterial activity. The process of alteration changes the surfaces, to such an extent that it is impossible to distinguish the lumps from other amorphous grains (Illing, 1954). They are most abundant in the areas of greatest development of bladder wrack seaweed.

Worm tubes are the cemented walls of worm burrows. When fresh, they are easily distinguished by their shape, but if they are broken and abraded, it becomes increasingly difficult to distinguish them from grapestones. Normally the wall of the worm tube is no more than three layers of sand grains in thickness. Other types of tubes may have walls more than 2 cm thick. The thicker tubes probably are formed by bivalves or arthropods. Tubes broken free from reworked intertidal sand commonly litter the beach. Shell infillings are cemented aggregates that accumulate within the protection of shells. These, like the worm tubes, can be an important source of grapestones. When fresh, they are easy to identify, but if broken, abraded, or cyanobacterially altered, it is difficult to establish their origin.

Ooids

Ooids have been recognized in thin sections from one or two localities on the shallow coastal shelf near Ras Al Aish. At Jebel Dhannah and east of Al Marfa, where they are in the form of perfect spheres of the fine- to medium-grained sand. They probably formed on the swash bars of the area. Although they exhibit a series of concentric mucilaginous layers that contain algal cells, few of these ooids are bored by cyanobacteria, as are the ooids in the lee of the Abu Dhabi tidal deltas. This lack of boring is the result of either the turbulence of the setting or its water chemistry.

Unidentifiable grains

There are several unidentifiable grains of extremely variable appearance and probably diverse origins. Most of them are products of alteration of other carbonate grains. An exception is detritus in the fine-sand and silt grades that may represent disaggregated or poorly formed pellets. Unidentifiable material is commonly blackened, particularly in the western part of the Khor Al Bazam.

Unidentifiable components constitute a high percentage of the sediment throughout the lagoon and decrease in abundance on the coastal terrace. In coarse sand, the unidentifiable material is concentrated on the edges of the shoals and in areas of high relief in the lagoon. Exceptions occur in areas where local conditions encourage algal alteration, as indicated by intense pitting of the grain surface.

Quartz grains

Although some quartz grains may have an aeolian origin from the south, they are derived largely from outcrops of Tertiary and Quaternary rocks. Most of the identified source outcrops lie above low-water mark on the coastal strip and on the offshore bank. At Al Marfa, the sand commonly contains as much as 50% quartz. Similarly, at the peninsula of Al Qala, adjacent beaches and intertidal flats are composed of nearly 100% quartz sand that drowns any carbonate material present. Bayward, the percentage of quartz decreases and carbonate coated grains increase. Many of the jackets show one or more projecting hemispheroids of carbonate. Kendall & Skipwith (1969a) have shown that in the Khor Al Bazam the quartz percentage decreases markedly away from shoal areas, particularly in the coarser grades. Finer grades have an even spread and hence may be transported more easily.

Quartz grains are generally in the fine-sand grade and are subround in shape. Surfaces of many quartz grains have a partial carbonate coating similar to that of the quiet-water ooids (Freeman, 1962). Sand grains in a carbonate province accrete until an optimum size is reached (Illing, 1954). This process is controlled by the hydrodynamics of the environment. The Khor Al Bazam grains generally have thicker carbonate jackets in shoal areas and thinner ones in the lagoonal areas.

Saccharoidal brown grains composed of stained quartz, calcite and feldspar are widespread throughout the area, but are not abundant. They are unimportant except where they parallel quartz in distribution; suggesting that many of these grains have the same source as the quartz and are largely derived from outcrops of Quaternary and Tertiary rock.

Evaporite minerals

Unlike previously described grains, and as described in the earlier section on the mainland sabkah plain, evaporite minerals are forming after the deposition of intertidal and sabkha (supratidal) sediments.

Flocculent matter

Flocculent matter is a mixture of carbonate mud and a high percentage of organic material. It is very common in the lagoon samples (Kendall & Skipwith, 1969a), but forms a low percentage of all except the most sheltered intertidal sediments. One sheltered area is the Al Qala embayment. Like pellets, flocculent matter is a good indication of an environment protected from vigorous water movement.

Serpulid nodules

Serpulids form accretionary nodules up to 30 cm or more in diameter on intertidal flats and shoals. Gastropods, bivalves and barnacles colonize the nodules.

Sponge spicules

Sponges grow in abundance in parts of the bladder wrack weed zone. Spicules released by decay of sponges are present in the finer grades of sand and silt.

Echinoid fragments

Common echinoids are the irregular, flattened sand dollars in the lagoons and the regular urchins found in the reefs. Their fragmented skeletons are dispersed so quickly that they do not form an important constituent of the sand.

Fish bones and teeth

In spite of the high salinity of the Khor Al Bazam, there is an extensive nekton, and notable are mackerel, rays, sharks, sea snakes, turtles, dolphins, and manatees, all of which may be found in the lagoon or offshore. The coral banks support a reef nekton that includes groupers and parrot fish. Cloud (1959) and Kaulback *et al.* (1962) discussed their geological importance. A few fish bones and teeth were found in one or two samples.

Crustacea

Although living crustacea are common, their skeletal remains are quickly fragmented and dispersed, and they form a minor constituent of the sediments. Barnacles colonize boulders of serpulid tubes but otherwise are rare. They include *Chthamalus* cf. C. sp of Hoelc and *Banalus amphitrite* Darwin.

Flora

Cyanobacterial mats, mangroves (*Avicennia marina*) and the bushy halophyte (*Arthrocnemum glaucum*) are found in protected areas. Bladder wrack grows on a hard substrate in waters where currents and waves stir the bottom. In quieter waters, large areas are covered by *Halodule (Diplanthera) uninervis*, a grass-like weed that is extremely effective in binding sediment and in trapping silt- and mud-sized particles (Kendall & Skipwith, 1968). Its role is equivalent to that of the *Thalassiu* of the Bahamas (Illing, 1954; Purdy, 1963a,b). In the most protected waters of the lagoon where the sediment is not affected by current action, a weed grows which has small, rounded "cress-like" leaves; in the lee of the offshore bank where muddy sediments accumulate, a weed grows that has elongate leaves reminiscent of "mint".

CONCLUSIONS

Seaward reefs, barrier islands and tidal flats line the Holocene shallow-water carbonate and supratidal evaporite tract of the Khor Al Bazam. In this part of Abu Dhabi, the barrier islands protecting the inner coast are more widely spaced than those to the east. They occur on extensive carbonate shoals and coral banks cut by tidal channels. The lagoon south of the barrier forms a continuous open body of water. It has a less restricted circulation than the lagoons to the northeast and at its western end it is connected to the Arabian Gulf.

The distribution of the Holocene sediments here is linked to the local character of the coast. For instance, coral reefs have developed on the seaward margins of most of the offshore banks to the west, which contrasts with the ooid shoals to the east. To the west of the Al Dhabaiyah Peninsula carbonate muds accumulate only in a narrow belt south of the offshore bank. Grapestones and skeletal debris are the dominant components of the inner western coastal terraces, but some ooid shoals occur on the more exposed beaches and channel margins. Extensive barrier islands protect the lagoons so that instead of a coastal terrace, an extensive intertidal cyanobacterial mat occurs seaward of supratidal flats that are the site of evaporite accumulation. The offshore bank is a coastline accreting seaward through the agency of coral growth and tidal delta formation. South of these banks, supratidal flats are encroaching on the lagoons with the development of beach ridges and cyananobacterial flats.

ACKNOWLEDGEMENTS

We gratefully acknowledge the support of the United Arab Emirates University, and to all of our scientific colleagues who have made contributions and suggestions that improved the text. Much of the material described here directly quotes the earlier work of Kendall & Skipwith (1966, 1968, 1969a) but has been upgraded using modern satellite images.

REFERENCES

Alsharhan, A.S. and Kendall, C.G.St.C. (2003) Holocene coastal carbonates and evaporites of the Southern Arabian Gulf and their ancient analogues. *Earth-Sci. Rev.*, **61**, 191–243.

Allen, J.R.L. (1965) Coastal geomorphology of Eastern Nigeria: beach ridges, barrier islands and vegetated tidal flats. *Geol. Mijnbouw*, **44**, 1–21.

Assereto, R.L.A.M. and Kendall, C.G.St.C. (1977) Nature, origin and classification of peritidal tepee structures and related breccias. *Sedimentology*, **24**, 153–210.

Ball, M.M. (1967a) Tectonic control of configuration of the Bahama Banks. *Trans. Gulf Coast Assoc. Geol. Soc.*, **17**, 265–267.

Ball, M.M. (1967b) Carbonate sand bodies of Florida and the Bahamas. *J. Sed. Petrol.*, **37**, 556–591.

Baltzer, F., Kenig, F., Boichard, R., Plaziat, J.-C. and Purser, B.H. (1994) Organic matter distribution, water

circulation and dolomitization beneath the Abu Dhabi sabkha (United Arab Emirates). *Int. Assoc. Sedimentol. Spec. Publ.*, **21**, 409–427.

Butler, G.P. (1969) Modern evaporite deposition and geochemistry of co-existing brines, the Sabkha, Trucial Coast, *Arabian Gulf. J. Sed. Petrol.*, **39**, 70–89.

Butler, G.P., Kendall, C.G.St.C., Kinsman, D.J.J., Shearman, D.J. and Skipwith, P.A. (1965) Recent evaporite deposits on the Trucial Coast of the Arabian Gulf. *Proc. Geol. Soc. London*, **1623**, 91–92.

Butler, G.P., Harris, P.M. and Kendall, C.G.St.C. (1982) Recent evaporites from Abu Dhabi Coastal Flats. In: *Depositional and Diagenetic Spectra of Evaporites* (Eds C.R. Handford, G.L. Robert and R.D. Graham), pp. 33–64. SEPM, Core Workshop no. 3. Tulsa.

Byrne, J.V., LeRoy, D.O. and Riley, C.M. (1959) The Chenier plain and its stratigraphy, southwestern Louisiana. *Trans. Gulf Coast Assoc. Geol. Soc.*, **9**, 237–260.

Cloud, P.E., Jr. (1959) Geology of Saipan Mariana Islands. Part 4: Submarine topography and shoal water ecology. *US Geol. Surv. Prof. Pap.*, **280**, 361–445.

Cornish, V. (Ed.) (1914) *Waves of Sand and Snow and the Eddies Which Make Them*. Fisher Unwin, London.

Crumbie, M.C. (1987) *Avicennia marina*: the Grey Mangrove: general notes and observations. *Bull. Emirates Nat. Hist. Gr. (Abu Dhabi)*, **32**, 2–13.

Curtis, R., Evans, G., Kinsman, D.J.J. and Shearman, D.J. (1963) Association of dolomite and anhydrite in the recent sediments of the Persian Gulf. *Nature*, **197**, 679–680.

Curtis, R., Evans, G., Kinsman, D.J.J. and Shearman, D.J. (1964) A reconnaissance survey of the environment of Recent carbonate sedimentation along the Trucial coast, Persian Gulf. In: *Deltaic and Shallow Marine Deposits* (Ed. L.M.J.U. van Straaten), **1**, pp. 129–135. Elsevier, Amsterdam.

Dalrymple, D.W. (1965) Calcium carbonate deposition associated with blue-green algal mats, Baffin Bay, *Texas. Inst. Mar. Sci. Univ. Tex. Publ.*, **10**, 187–200.

Davies, W. (1962) *Sediments of Gibraltar Point Area, Lincolnshire*. Unpubl. PhD Thesis, London University.

Evans, G. (1965) Intertidal flat sediments and their environments of deposition in the Wash. *Q. J. Geol. Soc. London*, **121**, 209–245.

Evans, G. and Shearman, D.J. (1964) Recent celestite from the sediments of the Trucial coast of the Persian Gulf. *Nature*, **202**, 385–386.

Evans, G., Kinsman, D.J.J. and Shearman, D.J. (1964a) A reconnaissance survey of the environment of Recent carbonate sedimentation along the Trucial Coast, Persian Gulf. In: *Deltaic and Shallow Marine Deposits* (Ed. L.M.J.U. van Straaten), **1**, pp. 129–135. Elsevier, Amsterdam.

Evans, G., Kendall, C.G.St.C. and Skipwith, P.A. (1964b) Origin of the coastal flats, the sabkha, of the Trucial Coast. Persian Gulf. *Nature*, **202**, 759–761.

Evans, O.F. (1942) The origin of spits, bars and related structures. *J. Geol.*, **50**, 846–865.

Folk, R.L. and Robles, R. (1964) Carbonate sands of Isla Perez, Alacran reef complex, *Yucatan. J. Geol.*, **72**, 255–293.

Freeman, T. (1962) Quiet water oolites from Laguna Madre, Texas. *J. Sed. Petrol.*, **32**, 475–483.

Gilbert, G.K. (1919) The transportation of debris by running water. *US Geol. Surv. Prof. Pap.*, **86**, 1–263.

Ginsburg, R.N. (1953a) Beachrock in South Florida. *J. Sed. Petrol.*, **23**, 85–92.

Ginsburg, R.N. (1953b) Intertidal Erosion in the Florida Keys. *Bull. Mar. Sci.*, **3**, 55–69.

Ginsburg, R.N. (1956) Environmental relationship of grain size and constituent particles in some south Florida carbonate sediments. *AAPG Bull.*, **40**, 2384–2427.

Ginsburg, R.N. (1957) Early diagenesis and lithification of shallow water carbonate sediments in south Florida. In: *Regional Aspects of Carbonate Deposition* (Eds R.J. Le Blanc and J.G. Breeding). *SEPM Spec. Publ.*, **5**, 80–100.

Ginsburg, R.N. and Lowenstam, H.A. (1958) The influence of marine bottom communities on the depositional environment of sediments. *J. Geol.*, **66**, 310–318.

Gould, H.R. and McFarlan, E. Jr. (1959) Geologic history of the Chenier Plain, southwestern Louisiana. *Trans. Gulf Coast Assoc. Geol. Soc.*, **9**, 261–270.

Houbolt, J.J.H.C. (1957) Surface Sediments of the Persian Gulf near the Qatar Peninsula. Unpubl. PhD Thesis Utrecht. Univ., Moutan and Co., Den Haag, 113 pp.

Hughes G.W. (1997) The Great Pearl Bank Barrier of the Arabian Gulf as a possible Shu'aiba analogue. *GeoArabia*, **2**, 279–304.

Hughes Clarke, M.W. and Keij, A.J. (1973) *Organisms as producers of carbonate sediment and indicators of environment in the southern Persian Gulf.* In: The Persian Gulf: Holocene Carbonate Sedimentation and Diagenesis in a Shallow Epicontinental Sea (Ed. B.H. Purser), pp. 33–56. *Springer Verlag*, Berlin, New York.

Illing, L.V. (1954) Bahamian calcareous sands. *AAPG Bull.*, **38**, 1–95.

Imbrie, H. and Buchanan, H. (1965) Sedimentary structures in modern carbonate sands of the Bahamas. In: *Primary Sedimentary Structures and their Hydrodynamic Interpretations* (Ed. G. V. Middleton). *SEPM Spec. Publ.*, **12**, 149–172.

John, D., (2005) Marine Plants. In: *Emirates Natural History*, pp. 161–167. *Trident Press Ltd*, London, UK.

Johnson, J.W. (Ed.) (1919) *Shore Processes and Shoreline Development.* Wiley, New York.

Kaulback, J.A., Kendall, C.G.St.C., and Skipwith P.A. (1962) *Cyclothems on the Islands of Kharg and Kharg, Persian Gulf.* Unpubl. Rep. Iranian Oil Exploration and Producing Companies, 19 pp.

Kendall, C.G.St.C. (1969) An environmental re-interpretation of the Permian evaporite/carbonate shelf sediments of the Guadalupe Mountains. *Geol. Soc. Am. Bull.*, **80**, 2503–2526.

Kendall, C.G.St.C. and Skipwith, P.A. (1966) Recent algal stromatolites of the Khor Al Bazam, Abu Dhabi, southwest Persian Gulf (abstract). *Geol. Soc. Am., Spec. Pap.* 108.

Kendall, C.G.St.C. and Skipwith, P.A. (1968) Recent algal mats of a Persian Gulf lagoon. *J. Sed. Petrol.*, **38**, 1040–1058.

Kendall, C.G.St.C. and Skipwith, P.A. (1969a) Holocene Shallow Water Carbonate and evaporite sediments of Khor Al Bazam, Abu Dhabi, Southwest Persian Gulf. AAPG Bull., **53**, 841–869.

Kendall, C.G.St.C. and Skipwith, P.A. (1969b). Geomorphology of a Recent shallow water carbonate province: Khor Al Bazam, Trucial Coats, southwest Persian Gulf. *Geol. Soc. Am. Bull.*, **80**, 865–891.

Kendall, C.G.St.C., Alsharhan, A.S. and Whittle, G.L. (1998) The flood re-charge sabkha model supported by recent inversions of anhydrite to gypsum in the UAE sabkhas. In: *Quaternary Deserts and Climatic Change* (Eds A.S. Alsharhan, K.W. Glennie, G.L. Whittle and C.G.St.C. Kendall), pp. 29–42. A.A. Balkema, Rotterdam.

Kendall, C.G.St.C., Shinn, G. and Janson, X. (2007) Holocene cyanobacterial mats and lime muds: links to Middle East carbonate source rock potential. Abstract, *Proc. AAPG, SEPM Annual Meeting, Long Beach.*

Kenig F., Huc A.Y., Purser B.H.,and Oudin J.L. (1990) Sedimentation, distribution and diagenesis of organic matter, in a carbonate hypersaline environment, Abu Dhabi (UAE). *Org. Geochem.* **16**, 735–747.

Kenig F., Sinninghe Damste J., de Leeuw J.W. and Huc, A.Y. (1995) Occurrence and origin of mono-, di- and trimethylalkanes in modern and Holocene microbial mats from Abu Dhabi (UAE). *Geochim. Cosmochim. Acta* **59**, 2999–3015.

Kindle, E.M. (1917) Recent and fossil ripple marks. *Bull. Can. Geol. Surv.*, **25**, 1–121.

King, C.A.M. (Ed.) (1959) *Beaches and Coasts.* Edward Arnold, London.

Kinsman, D.J.J. (1964a) Reef coral tolerance of high temperatures and salinities. *Nature*, **202**, 1280–1282.

Kinsman, D.J.J. (1964b) *Recent Carbonate Sedimentation near Abu Dhabi, Trucial Coast, Persian Gulf.* Unpubl. PhD Thesis, University of London.

Kinsman, D.J.J. (1964c) The recent carbonate sediments near Halat Al Bahraini, Trucial Coast, Persian Gulf. In: *Deltaic and Shallow Marine Deposits* (Ed. L.M.J.U. van Straaten), **1**, pp. 189–192. *Elsevier*, Amsterdam.

Kirkham, A. (1997) Shoreline evolution, aeolian deflation and anhydrite distribution of the Holocene, Abu Dhabi. *GeoArabia*, **4**, 403–416.

Kornicker, L.S. and Purdy, E.G. (1957) A Bahamian faecalpellet sediment. *J. Sed. Petrol.*, **27**, 126–128.

Kornicker, L.S. and Boyd, D.W. (1962) Shallow-water geology and environments of Alacran Reef complex, Campech Bank, *Mexico. AAPG Bull.*, **46**, 640–673.

Krempf, A. (1927) La forme des récif coralliens et le régime des vents alternants. *Trav. Serv. Océanog. Pêches de l'Indochine, Mem.*, **2**, 3–33.

Masson, P.H. (1955) An occurrence of gypsum in southwest Texas. *J. Sed. Petrol.*, **25**, 1–25.

Maxwell, W.G.H., Jell, J.S. and McKeller, R.G. (1964) Differentiation of carbonate sediments in Heron Island Reef. *J. Sed. Petrol.*, **34**, 294–308.

McKenzie, J.A., Hsu, K.J. and Schneider, J.F. (1980) Movement of subsurface waters under the sabkha, Abu Dhabi, UAE, and its relation to evaporative dolomite genesis. In: *Concepts and Models of Dolomitization* (Eds D.H. Zenger, J.B. Dunham and R.L. Ethington). *SEPM Spec. Publ.*, **28**, 11–30.

Muller, D.W., McKenzie, J.A., Kendall, C.G.St.C. and Alsharhan, A.S. (1991) Application of strontium isotopes to the study of brine genesis during sabkha dolomitization (abstract). In: *Dolomieu Conference on Carbonate Platforms and Dolomitization* (Eds A. Bosellini, R. Brander, E. Flugel, B. Purser, W. Schlager, M. Tucker and D. Zenger), p. 181.

Murray, J.W. (1966) The foraminiferids of the Persian Gulf, 4. Khor Al Bazam. *Palaeogeogr. Palaeoclimatol. Palaeoecol.*, **2**, 153–169.

Nesteroff, W.D. (1956) La substratum organique dans les dépots calcaires, sa signification. *Bull. Soc. Géol. Fr.*, **6**, 381–389.

Newell, N.D. and **Rigby, J.K.** (1957) Geological studies on the Great Bahama Bank. In: *Regional Aspects of Carbonate Deposition* (Eds R.J. LeBlanc and J.G. Breadway). *SEPM Spec. Publ.*, **5**, 15–72.

Newell, N.D., Rigby, J.K., Whiteman, A.J. and **Bradley, J.S.** (1951) Shoal water geology and environments, eastern Andros Island, *Bahamas. Bull. Am. Mus. Nat. Hist.*, **97**, 7–26.

Newell, N.D., Purdy, E.G. and **Imbrie, J.** (1960) Bahamian oolitic sand. *J. Geol.*, **68**, 481–497.

Patterson, R.J. and **Kinsman, D.J.J.** (1981) Hydrologic framework of a sabkha along Arabian Gulf. *AAPG Bull.*, **65**, 1457–1475.

Pilgrim, G.E. (1908) Geology of Persian Gulf and adjoining portions of the Persia and Arabia. *Geol. Surv. India Mem.* **34**, 1–179.

Price, W.A. (1963) Patterns of flow and channeling in tidal waters. *J. Sed. Petrol.*, **33**, 279–290.

Purdy, E.G. (1963a) Holocene calcium carbonate facies of the Great Bahama Bank, part. 1, Petrography and reaction groups. *J. Geol.*, **71**, 334–355.

Purdy, E.G. (1963b) Holocene calcium carbonate facies of the Great Bahama Bank: part. 2, Sedimentary facies. *J. Geol.*, **71**, 472–297.

Purser, B.H. (Ed.) (1973) *The Persian Gulf: Holocene Carbonate Sedimentation in a Shallow Epicontinental Sea.* Springer-Verlag, New York, 471 pp.

Purser, B.H. and **Evans, G.** (1973) Regional sedimentation along the Trucial Coast, SE Persian Gulf. In: *The Persian Gulf: Holocene Carbonate Sedimentation and Diagenesis in a Shallow Epicontinental Sea* (Ed. B.H. Purser), pp. 211–232. Springer-Verlag, New York.

Purser, B.H. and **Loreau, J.P.** (1973) Aragonitic supratidal encrustations on the Trucial Coast, Persian Gulf. In: *The Persian Gulf: Holocene Carbonate Sedimentation in a Shallow Epicontinental Sea* (Ed. B.H. Purser), pp. 343–376. Springer-Verlag, New York.

Shearman, D.J. (1978) Evaporites of coastal sabkhas. In: *Marine Evaporites* (Eds W.E. Dean and B.C. Schreiber). *SEPM Short Course*, **4**, 6–42.

Shepard, F.P. (Ed.) (1948). *Submarine Geology.* Harper, New York.

Shepard, F.P. (1952) Revised nomenclature for coastal features. *AAPG Bull.*, **36**, 1902–1912.

Shinn, E.A. (1963) Spur and groove formation on the Florida reef tract. *J. Sed. Petrol.*, **33**, 291–303.

Shinn, E.A. (1969) Submarine lithification of Holocene carbonate sediments in the Persian Gulf. *Sedimentology*, **12**, 109–144.

Skipwith, P.A.d'E. (1966), *Recent carbonate sediments of the Eastern Khor al Bazam, Abu Dhabi, Trucial Coast.* Unpubl. PhD Thesis, University of London, 477 pp.

Steers, J.A. (1929) The Queensland coast and Great Barrier reefs. *J. Geogr.*, **74**, 232–257.

Stoddart, D.R. (1962) Three Caribbean atolls: Turnette Islands, Lighthouse Reef, and Glovers Reef, British Honduras. *Atoll Res. Bull.*, **87**, 1–151.

Storr, J.F. (1964) Ecology and oceanography of the coral reef tract, Abaco Island, Bahamas. *Geol. Soc. Am. Spec. Pap.*, **79**, 1–98.

Stride, A.H. (1963) Current-swept sea floors near the southern half of Great Britain. *Q. J. Geol. Soc. London*, **119**, 175–201.

Sugden, W. (1963) Some aspects of sedimentation in the Persian Gulf. *J. Sed. Petrol.*, **33**, 355–364.

Swart, P.K., Shinn, E.A., Mckenzie, J.A., Kendall, C.G.St.C. and **Hajari, S.E.** (1987) Spontaneous dolomite precipitation in brines from Umm Said sabkha, Qatar: (abstract) *Geol. Soc. Am. Ann. Meet.*, p. 862.

Tanner, W.F., Evans, R.G. and **Holmes, C.W.** (1963) Low energy coast near Cape Ramano, *Florida. J. Sed. Petrol.*, **33**, 713–722.

Thornbury, W.D. (Ed.) (1954) *Principles of Geomorphology.* Wiley, New York.

van Straaten, L.M.J.U. (1950) Giant ripples in tidal channels In: *Overdruck van de Mei-Afleuing 1950: Wadden symposium. K. Ned. Aardrijksk. Genoot. Tijdschr.*, 76–81.

van Straaten, L.M.J.U. (1953a) Megaripples in the Dutch Wadden sea and in the basin of Arcachon (France). *Geol. Mijnbouw*, **15**, 1–11.

van Straaten, L.M.J.U. (1953b) Rhythmic patterns in Dutch North Sea beaches. *Geol. Mijnbouw*, **15**, 31–43.

van Veen, J. (1953) Eb en vloedschaar systemen in de Nederlandse Getijwateren, In: *Overdruk van de Mei-Afleveuing 1950. Wadden symposium. K. Ned. Aardrijksk. Genoot. Tijdschr.* 43–66.

Vaughan, T.W. (1909) Geology of the Florida Keys and the marine bottom deposits and Holocene corals of southern Florida. *Carnegie Inst. Wash. Yb.*, **7**, 131–136.

Wagner, C.W. and **Van der Togt, C.** (1973) Holocene sediment types and their distribution in the southern Persian Gulf. In: *The Persian Gulf: Holocene Carbonate Sedimentation and Diagenesis in a Shallow Epicontinental Sea* (Ed. B.H. Purser), pp. 123–156. Springer-Verlag, Berlin, New York.

Warren, J.K. and **Kendall, C.G.St.C.** (1985) Comparison of marine (subaerial) and salina (subaqueous) evaporites, modern and ancient. *AAPG Bull.*, **69**, 1013–1023.

Williams, W.W. (Ed.) (1960) *Coastal Changes.* Routledge and Kegan Paul, London.

Wood, G.V. and **Wolfe, M.J.** (1969) Sabkha cycles in the Arab/Darb Formation off Trucial Coast of Arabia. *Sedimentology*, **12**, 165–192.

Wood, W.W., Sanford, W.E. and **Al Habschi, A.R.S.** (2002) The source of solutes in the coastal sabkha of Abu Dhabi. *Geol. Soc Am. Bull.*, **114**, 259–268.

Int. Assoc. Sedimentol. Spec. Publ. (2011) **43**, 89–112

The impact of sea-level change on ramp margin deposition: lessons from the Holocene sabkhas of Abu Dhabi, United Arab Emirates

ROBERT K. PARK

Kodeco Energy Co. Ltd, Indonesian Stock Exchange Building, Twr1 23fl, Jln Jend. Sudirman, kav 52, Jakarta 12190, Indonesia (E-mail: robert_park@kodeco.co.id),

ABSTRACT

Around the globe, the Holocene is replete with examples of the rapid accumulation and outbuilding of sediment wedges in response to, and associated with, a rise in sea level. It is particularly evident among the much-studied shallow-marine carbonate tidal flat association and arguably, nowhere better displayed than by the sabkha cycle developed along the coastal margin of the UAE. This cycle of rapid transgression and steady outward progradation throughout the last 10,000 years is only the most recent of a number of similar cycles that have been a characteristic of ramp margin sedimentation and a feature of stratigraphic associations in the Arabian Basin since the late Palaeozoic. It is not coincidental that these successions lying on top of earlier rifted basin sequences, many salt filled, are home to huge reserves of hydrocarbons.

The sabkha cycle, as represented by the Holocene transgressive – highstand sequence found along the northern coast of the Arabian Platform is barely 2 m thick. Its sedimentary facies, associated mineralogical suites and early diagenesis are well documented. The gently sloping ramp margin setting is one that is extremely sensitive to even minor fluctuations in sea level and thin high-frequency sediment cycles are its hallmark. Fluctuations in sea level, however small, produce dramatic changes in sedimentation and groundwater chemistry in this environment, all of which can and do impact the rates of growth and destruction of the sequence. This may occur via increased erosion and ablation, leaching, alteration and replacement of existing minerals and precipitation of new ones, to aerobic destruction of organic material.

Coastal flats are situated at the interface between groundwater, seawater and meteoric water which, especially in a sabkha setting, exposes the sequence to a multiplicity of diagenetic overprints. These, through their impact on porosity and permeability properties of potential reservoirs and seals, are of vital interest to the petroleum industry. The preservation potential, or lack thereof, in such tidal flat facies should be recognised as an inherent feature of high-frequency cyclicity. Despite this, relics and indicators of high frequency cycles can be observed in cores and wireline logs and offer some insights into the history of formation and preservation of porosity and hydrocarbon migration.

Keywords: Carbonate tidal flat, sabkha, transgression, progradation, ramp margin, erosion, leaching, groundwater, seawater, meteoric water, high-frequency cyclicity, porosity, hydrocarbon migration.

INTRODUCTION

Modern environments are commonly studied with a view to documenting the products and understanding the processes that contribute to their formation. Physiographic setting and climate set the stage and, to a degree, select the cast of players who will act out this unfolding story. Taking a complex piece of machinery apart can be an effective way of gaining insights as to how it works and, not infrequently, raise a host of new questions in the process. During the late 1950s–60s, sedimentology, though an integral part of stratigraphy in the classical sense, became established as a major sub-discipline in its own right within the field of geology. At that time the focus was very much on *process* as a means of taking apart some of the geological complexities of the time.

Ginsburg and Shinn, using the Florida-Bahamas platform as their laboratory, sought to unravel and thereby identify some of the key elements of carbonate depositional systems. In particular they directed their attention on the main elements of the intertidal and supratidal facies and sequence associations, culminating in their seminal paper on the anatomy of a tidal flat (Shinn *et al.*, 1977). Subsequently, this was extended to incorporate a whole gamut of tidal deposits (Ginsburg, 1975). Parallel studies in cool-water carbonates in western Ireland were being undertaken by Lees and his colleagues at the University of Reading (Lees & Buller, 1972; Gunatilaka, 1976); while the Imperial College London group, under Shearman and Evans, sought the warmer climes of the Arabian Gulf (Evans *et al.*, 1964a,b; Evans, 1966, 1969). That work in these areas continues even now, is testimony to the number of new questions spawned through this dismantling process.

While the initial focus among each of the study groups was to document and understand facies associations, the discovery of recent dolomite, as a supratidal crust in Florida (Shinn *et al.*, 1965), in Queensland (Cook, 1973) and the Coorong (Alderman & Skinner, 1957; Alderman & von der Borch, 1961, 1963), and in the Arabian Gulf (Wells, 1962; Shearman, 1966; Curtis *et al.*, 1963), and of "beach rock" in the upper intertidal reaches of Connemara's beaches (Buller & Scott, pers. comm., 1970), saw the emphasis begin to shift towards processes involving groundwater hydrology and geochemistry and their impact on diagenesis (Kinsman & Park, 1969). The intertidal-supratidal environment is the point of contact between marine waters, groundwaters and meteoric waters. Combine this with atmospheric and oceanographic circulation patterns, temperature and rainfall variations, and the stage is set for a drama of Shakespearean proportions with myriad plots and subplots.

During the 1960s and 1970s, dolomite, and dolomitization was of particular interest to the petroleum industry, dolomite reservoirs being widely regarded as attractive if yet little understood exploration targets. Nowhere was this better exemplified than among the giant reservoirs of the Gulf region, where dolomite-evaporite associations were particularly prevalent. The recognition of a modern analogue and the chance to study its workings received much attention (and funding; Illing & Wells, 1964; Hsu, 1966; Purser, 1973). Dolomitization models proliferated at that time,

from evaporative pumping and flooding (Patterson & Kinsman, 1982), alongside seepage reflux models (Deffeyes *et al.*, 1964; Hanshaw *et al.*, 1971;Lucia & Major, 1994) to Dorag mixing-zone scenarios (Badiozamani, 1973) and deep-basin brine stratification (Schmalz, 1969). Dolomite and dolomitization and their apparent linkage with evaporitic systems had become subjects of intense scrutiny because of their impact on porosity development and evolution, of obvious interest to the petroleum industry (Pray & Murray, 1965; Perkins *et al.*, 1994; Purser *et al.*, 1994).

With the totally unexpected discovery of anhydrite within the supratidal sabkhas of Abu Dhabi (Shearman, 1963; Kinsman, 1964) it became apparent that these tidal and supratidal sabkha flat settings also offered a challenge and an opportunity to understand something of the geochemical cycle and the kinetics involved, not only in dolomite and evaporite associations, but of a much wider range of diagenetic reactions, processes and products. Thus, the subsequent work of Kinsman and Patterson in the late 1960s–early 1970s was particularly focused on the hydrology and hydrological dynamics of the sabkha (Kinsman, 1969; Patterson, 1972; Patterson & Kinsman, 1981).

THE GULF REGION TODAY

Present-day climatic conditions of the southern Gulf area are well documented (Kinsman, 1969; Patterson, 1972; Patterson & Kinsman, 1981, 1982). In summary, it is subtropical arid, with average annual air temperatures of 20 °C, ranging from winter low of 17 °C, to July maxima of around 45 °C. The protected shallow-water lagoons behind the barrier island chain also exhibit temperatures well above the adjacent Gulf waters, ranging from 15–40 °C. Rainfall is $<5\,\mathrm{cm\,yr^{-1}}$, hence net evaporative conditions would be anticipated, although these are significantly offset by the high humidity levels that prevail along the coastal areas, resulting in annual evaporation losses of only 5–$6\,\mathrm{cm\,yr^{-1}}$, and these may have been even lower during the early Holocene. The onset of more extreme arid conditions during the mid-Holocene was noted by Gunatilaka (*this volume*) with reference to the demise of several major cultures at that time.

The modern Arabian Gulf evolved and developed as a foreland basin situated between the Arabian Shield and Eurasia, its asymmetry being accentuated by the rising Zagros Mountain chain in the north. The gently dipping northern margin of the shield provided a stable ramp setting that has supported a long and relatively uniform depositional style along the southern flank. Evans (1966) provided an early summary and review of the major facies distribution patterns within the Gulf and in particular the presence of a carbonate ramp sequence along the Emirates coast. Significant volumes of siliciclastics continue to be discharged into the Gulf by the Tigris-Euphrates rivers as the primary steady source of supply, much of it being deposited in the associated delta complex. The prevailing anti–clockwise gyre of the tidal currents, particularly in the more northern reaches of the Gulf, further ensure that suspended fines are transported along the northern shore and well away from the southern coast. It is this isolation from siliciclastics that has resulted in the carbonate-evaporite system that operates so effectively today (Fig. 1).

Such siliciclastics as do reach the coastal sabkhas of the UAE are primarily airborne, transported from the northern coast by the Shamal winds. Farther west and north these winds are more commonly offshore, and transport sands via migrating dunes into the Gulf (Shinn, 1973). Slightly higher annual rainfall ($10–20 \, cm \, yr^{-1}$) is experienced by the Oman Mountain region to the south, even so, there is little active surface run off, save for sporadic flash-flood discharge, little if any of which impinges on the modern coastal plain environment. Fluvial discharge from the Oman Mountains is confined to their immediate vicinity and does not extend as far as the modern sabkha plains; the gravels encountered at the inner sabkha margins and hinterland derive from an earlier more humid phase, before the onset of severe aridity in the mid-Holocene at around 6500 BP.

The main elements present on the southern Gulf margin are the sabkha coastal plain notable for the prominence of evaporites, an intertidal zone dominated in its upper part by microbial mats and in its lower part by gastropod-bearing pellet sands, that pass seaward into a lagoonal facies characterised by muddy skeletal sands rich in foraminifera and calcareous green algae. A chain of islands, many having relict Pleistocene cores, acts as an energy buffer, shielding the lagoons from all but the strongest storm activity. The straits between the islands

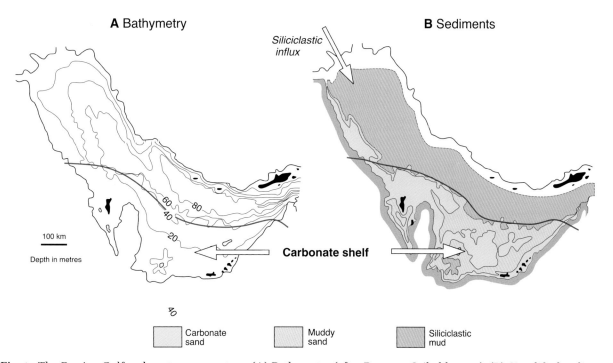

Fig. 1. The Persian Gulf carbonate ramp system. (A) Bathymetry (after Purser & Seibold, 1973). (B) Simplified sediment composition and distribution map (after Wagner & Togt, 1973). The bathymetry reveals an asymmetric basin with its axial trough close to the Iran coastline to the north. The southern margin comprises a broad shallow shelf dominated by carbonate sands and muds and is largely free of siliciclastic sediments whose distribution is focused and confined to the northern half of the Gulf.

have become factories for ooid formation and the focused tidal and wave energy has created some well-developed ebb and flood oolite-dominated deltas. Patchy coral growth is restricted to lower energy shadows in front and at the centre of the islands (Fig. 2A and B).

Salinities in the Gulf are slightly above normal seawater, raising $CaCO_3$ supersaturation levels and triggering the widespread phenomenon of whitings that occur when aragonite is periodically instantaneously precipitated. The high water temperatures and salinities also apply environmental stress to the biota, and in particular to the coral communities. Glynn (2000) has discussed at some length the potential negative impact that high water temperatures and attendant knock-on effects might have on coral communities. It was noted that corals show a tendency to thrive during periods of general warming and sea-level highstands, as is demonstrable from the Holocene record from the Indo-Pacific Caribbean and Arabian Gulf regions, and potentially from a number of intervals throughout the geological record.

The upper limit of temperature tolerance of corals may not have been definitively established, but it is evident that at higher temperatures, such as those experienced in the present-day Arabian Gulf or brought on by El Niño, reefs may continue to be well developed but community diversity tends to become much reduced. Glynn's (2000) observations are supported by anecdotal evidence from recent reef community surveys and personal observation on islands of the Indonesian archipelago and from the Arabian Gulf (Lamondo pers. comm., 2005) although some well-developed coral reefs are found farther offshore. Commonly, other elements of the biota also suffer, including restricted populations of foraminifera and molluscs (Evans et al., 1973; Murray, 1965, 1970).

Marine marginal areas are noteworthy for their near total lack of vegetation, save for isolated halophytic shrubs and sparse mangroves. However, slightly more moderate conditions at the northern end of the Gulf allow for more extensive halophyte growth, though mangroves are notable by their absence. Oolite shoals and tidal deltas are a feature of the Abu Dhabi coastal region and farther north they form prominent beach ridges. Here, marine lagoons are absent and algal/microbial mat development is restricted to a bare 100 m wide fringe along some tidal creeks. In addition, it is of note that the groundwaters of these northern sabkhas have a strong continental source, in contrast to the more marine-influenced groundwaters of the southern margin (Gunatilaka, this volume). This is in striking contrast to the mangrove swamps and forests bordering similar coastal topography in equatorial regions that are the product of rapid fluvial deposition and marine reworking, such as the east coast of Sumatra and south Kalimantan (Fig. 3), and to a lesser degree their physiographic relatives in the northern Gulf region, along the Queensland coast, or in Florida-Bahama settings.

In the absence of competition, the intertidal zone of the Abu Dhabi lagoons is dominated by broad expanses, up to 2 km wide in places, of microbial mats comprising a rich assemblage of blue-green cyanophyte algae and bacteria (Park, 1973, 1976, 1977; Golubic, 2000). Beyond this, lies the barren tract of evaporite-bearing aeolian sands that is the sabkha (Fig. 4). Though recognised elsewhere (Mawson, 1929; Monty, 1967; Gebelein, 1969), rarely do such mats exhibited so rich an array of morphological expression and community variation (Kendall & Skipwith, 1968; Kinsman & Park, 1969; Park, 1973, 1977), although the Gulf mats do not quite attain the spectacular columnar morphologies of their cousins in Shark Bay, Western Australia (Davies, 1970b; Playford & Cockbain, 1976).

The importance of the microbial community as part of the dynamic of the geochemical cycle in supratidal-intertidal diagenesis has gained increasing recognition over the years: from Walter et al.'s (1973) records of hydromagnesite and aragonite in an association with stromatolites in the Coorong, to selective dolomitization of crustacean burrows in northern Kuwait (Gunatilaka et al., 1987), to mineralization in Brazilian lagoons (Vasconclos & McKenzie, 1997). Such reactions are not limited to carbonate shorelines, as can be seen in the weathering profiles and mineralization found throughout equatorial coastal mangrove flats. The coastal plain and adjacent intertidal reaches provide a setting where the hydrology is dynamic and the associated geochemical reactions and attendant diagenesis often aggressive.

The UAE coastal margin represents an archetypal homoclinal ramp situated in a tropical arid setting. The nearshore area is bypassed by siliciclastic discharge from the Tigris–Euphrates outflow and the coastal plains (sabkhas) are backed by a desert hinterland virtually devoid of rainfall and subject to near negligible fluvial outwash. Sediment accumulation is largely windblown and the surface, devoid of vegetation, is controlled

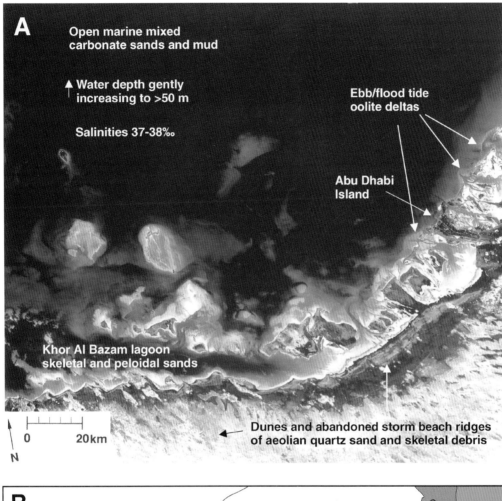

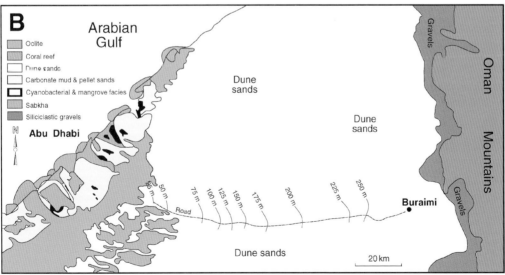

Fig. 2. Depositional environments and facies of the Abu Dhabi coastline, UAE. (A) Satellite image showing the ramp characteristics of the shoreline sequence, transiting from the open marine Gulf waters, past a string of barrier islands interrupted by oolitic tidal exchange deltas and patch coral growth, via peloidal sandy lagoons and fringing microbial mats, to the evaporite-bearing sabkha plains beyond. On this surface can be seen preserved shell ridges marking the steady seaward accretion of the sabkha, reinforced by migrating aeolian dunes (after Harris & Kowalik, 1994). (B) Summary map showing the distribution of major facies (Kendall & Skipwith, 1969a,b).

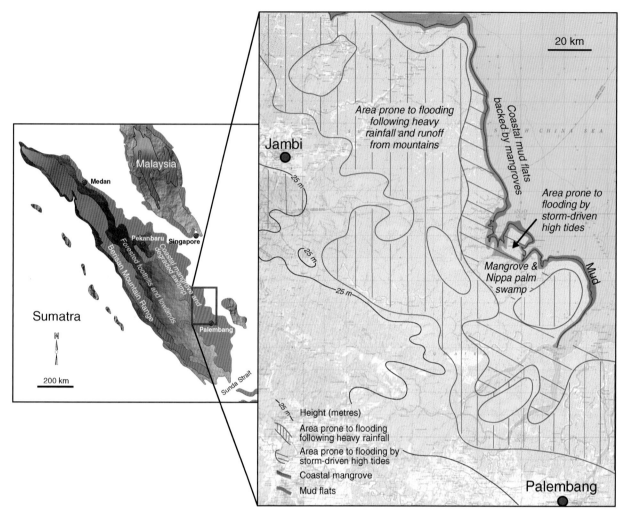

Fig. 3. Coastal environments on the eastern margin of Sumatra: a broad low-angle prograding coastal plain, not dissimilar in scale to that found in the UAE. The coastal margin comprises a mangrove–grass marsh terrain traversed by meandering streams that are perennially highly charged with fine sediment derived from the rising hinterland beyond. High rainfall results in regular flooding while seasonal easterly storms can also cause flooding of the coastal marsh areas. High temperatures and high rainfall combine to promote aggressive weathering and high sediment discharge, while the marsh peats will become the coals of tomorrow. Just as the carbonate-evaporite association has a long history in the Gulf area, so too do the peat-rich marshes of Sumatra and Kalimantan, which dominate the Tertiary stratigraphy of Indonesia and are a primary source of hydrocarbons, much as the microbial mats in the Gulf have been. Here, however, collision tectonics has played a major role in the development of many of the high-frequency events.

almost solely by wind deflation. The warm waters of the Gulf, with their slightly higher salinities, inhibit some of the major skeletal carbonate producers yet the sheer volume of carbonate, mostly aragonite, is testimony to the existence of a very effective carbonate factory.

Ooid formation is widespread in the high-energy loci of inter-island straits, while in the interior lagoons aragonite is reworked into a wide array of pellets and peloids. The intertidal zone represents a transition between marine net carbonate production and the net evaporative conditions of

the sabkha where sulphate mineralization prevails. It is also an area where diurnal exposure triggers cementation, generating surficial crusts with aragonite, high-Mg calcite, dolomite and even magnesite as the primary cementing agents. Within and below the upper intertidal zone surface the microbial community is actively involved in dolomite formation, and which becomes intimately associated in time and space with the appearance of significant amounts gypsum. Under the more extreme conditions of the supratidal setting of the sabkha, gypsum is transformed to and

Fig. 4. Sabkha carbonate-evaporite facies associations. (A) Lagoonal facies with peloidal foraminifera-rich sands. (B) Intertidal area with sparse halophytic shrubs, abundant crab activity locally; the mounds are cemented by dolomite and aragonite. (C) Broad expanse of microbial mats developed around the innermost protected margins of the lagoons. (D) The broad low-relief low-angle wind deflation surface that is the sabkha, with a light sprinkling of surficial halite.

supplemented by additional anhydrite precipitation and, where conditions of ponding arise, halite accumulation.

What makes the UAE ramp margin system unique is the presence of evaporites produced within the sabkha coastal plains. Kinsman (1969) defined the term to refer to "a salt flat that is inundated only occasionally". He further distinguished "continental" and "coastal" variants. The former are dominated by continental sediments and physiographically are wind deflation surfaces; the latter are dominated by marine sediments and are a product of regressive offlap. Both represent equilibrium surfaces under the control of the groundwater table. These definitions and the recognition of the underlying controls, set the stage for the work that was to follow in modelling the hydrological dynamics and the essential elements of sabkha geochemistry (Kinsman, 1969, 1976; Kinsman & Park, 1969; Patterson, 1972; Park, 1973; Patterson & Kinsman, 1981, 1982). These models, however, are specific to the coastal sabkha plains of the Emirates and are not necessarily representative

of the Arabian Gulf coastal areas as a whole. To the north, for instance, the classic marine sabkha sequences are less well developed, and siliciclastics and oolitic chenier sands are more prominent, while biological activity is much more constrained (Saleh *et al.*, 1999; Gunatilaka, pers. comm. 2006). Some generalisations may, however, be applicable to other areas of similar physiography but with groundwater chemistry the key to early diagenetic reactions, climatic conditions will have a key influence in: (a) moderating evaporative conditions; and (b) encouraging colonisation by flora. In the opposing extreme setting of a tropical equatorial region, this produces a heavily vegetated mangrove-grass plain (Fig. 3), a source of many of the coals and not insignificant amounts of hydrocarbons found throughout the Indonesian archipelago for instance.

The Holocene, then, is studied primarily with a view to understanding sedimentological, biological, atmospheric, and oceanographic processes and their dependence/inter-dependence on and within the physical, climatic, and chemical world. Linking the two together can be fraught with difficulties, given that Holocene studies tend to focus on what is accumulating now, but not necessarily on what has the potential to be preserved in the geological record. While the interpreter of cores, logs and seismic is faced with a record full of gaps, missing links and the distinct possibility, even probability, that there is more of the record missing than is present.

On the issue of preservation, much can be learned of the dynamics and interdependence of the many factors involved in the creation, accumulation and transformation of any Holocene sediment wedge. However, it is an equally legitimate question to ask whether, at a macro-scale, any of it will ever be preserved and find its way into the rock record. Observation of day-to-day processes and inter-relationships can offer a mass of information not all of which will have a bearing on assessing the potential for removal/preservation in response to single or multiple fall(s) in sea level. Many details of the intricate coexistence and co-dependence of interstitial chemistry, fauna and microflora are lost, although isotopic signatures may remain to help us unravel something of the origins and changes that have taken place (e.g. McKenzie, 1981). While some components of the microbial community may not survive into the record, the dolomite, in whose formation it is instrumental, both at the initiation stage and ongoing, almost certainly will (Vasconclos & McKenzie, 1997). Cored records, spanning a much longer period of time will potentially reveal trends in facies evolution and preservation that can in-turn be evaluated in terms of a response to a shift in the global climatic regime, such as the increased run-off and erosion brought on by the passage into more humid conditions. The celebrated and much studied Holocene of the UAE sabkhas offers insights into some of these issues.

Kinsman (1969) recognised the importance of physiography, in conjunction with local oceanographic and climatic conditions, as a controlling factor in so far as it could control the rates and patterns of flow into, around and out of lagoons and onto the supratidal flats, where evaporation becomes a dominant factor. Recognition of the multifaceted interactive relationship of a variety of natural processes that include tectonics, the atmosphere and the oceans is now widely recognised and essential to explaining changes in everything from basin geometries, current patterns,

climate shifts and ultimately to local sedimentary and diagenetic facies associations (Clift & Molnar, 2003).

The physiography and geomorphology of broad coastal plains, delta plains and adjacent ramp settings anywhere are, by their very nature, susceptible to small shifts in relative sea level. Hence the highly charged concerns over the impact of global warming on coastal and island communities through a potential sea-level rise. It is recognised that carbonate/evaporite productivity is the product of a linkage between the physical realm of winds, tides, climate and physiography with the biogenic realm. This is reflected in diversity and growth rates of individuals and communities that contribute to and are integral parts of the carbonate environment, a fundamental distinguishing feature of carbonate sequence evolution compared to their siliciclastic counterparts, and thence to diagenesis, with its clear link to geochemical and hydrological controls. All of which is, in reality, an embellishment of the early principles of orbital forcing established by Croll (1875 in Fischer et al., 2004), who was able to link orbital eccentricity and axial obliquity to changing intensity of seasonal variations, the albedo effects of increased ice caps, their impact on atmospheric zonation and, by extension, trade wind and oceanic circulation.

If the warm and saline waters of the Gulf and lagoons are considered to constitute the local carbonate factory, it is equally appropriate to consider the supratidal sabkha flats as an evaporite factory. The presence and prominence of evaporite minerals, notably gypsum and anhydrite, provide what is the essence of the sabkha environment. Within the upper part of the sediment wedge, the distinction between what is diagenetic and what may be a primary precipitate can become blurred. The gamut of minerals reported from the sabkha environment, from the abundant sulphates, gypsum and anhydrite and, more rarely, celestite and huntite, to ongoing carbonate mineral production, especially of aragonite, dolomite and to a lesser extent magnesite, are all essentially contemporaneous elements and integral to the classic sabkha sequence (Fig. 5). Similar mineralogical extremes have also been recorded from the evaporitic Coorong lagoons (Alderman & von der Borch, 1961).

Frequent flushing and evaporation throughout the intertidal zone is responsible for extensive surface cementation and crust formation, some aragonitic, some dolomitic, and in a few instances

The Classic Sabkha Section

E Enterolithic layers of hard to pasty anhydrite within aeolian sand wedge

D Cottage cheese anhydrite replacement after gypsum

C Upper intertidal microbial laminites (locally dolomitic) with interlayers of gypsum

B Middle intertidal microbial laminites with scattered large poikilotopic discoidal gypsum

A Lower intertidal - lagoonal muddy cerithid-rich peloidal sands

Fig. 5. The classic sabkha sequence. (1) Subtidal to lower intertidal muddy pellet/skeletal sands, predominantly aragonitic, within which large poikilotopic discoidal gypsum crystals are found (A – B). (2) Mid- to upper intertidal section dominated by a laminite sequence generated by microbial mats (C). (3) Upper intertidal facies comprises a mush of gypsum crystals, which initially seed between the microbial laminae and eventually contribute to their partial destruction. As the sabkha progrades, it is gradually buried under an advancing aeolian sand wedge. The gypsum mush is eventually replaced by anhydrite having a distinctive "cottage cheese" fabric (D). (4) A subaerial supratidal section comprised of brown aeolian sands in which large-scale dune cross bedding may be recognised, and within which contorted layers of pasty anhydrite and more discontinuous layers of hard anhydrite nodules occur (E). The sabkha mineral assemblage consists of: (1) Sulphates, with the nucleation and growth of gypsum and its eventual replacement by anhydrite, along with the primary precipitation of anhydrite, plus intermittent rehydration back to gypsum. These record a story of progressive and increasingly intense evaporation interrupted by brief episodic wetting. Gypsum-cemented crusts occasionally form a layer at the base of the intertidal sequence. (2) Carbonates, where aragonite is the dominant carbonate mineral, but evaporation, coupled with the role of the microbial community encourages localised precipitation of dolomite within the laminites and as surficial dolomitic and magnesitic crusts. (3) Halite is notable by its absence, occurring only as ephemeral crusts in the aftermath of episodic storm floods.

huntite- and magnesite-bearing (Kinsman, 1967). These pass into the subsurface where they are preserved to varying degrees but are rarely prominent, except for aragonite-cemented crusts within the mid- to lower intertidal reaches (Taylor & Illing, 1969). Gypsum crystals appear initially as ephemeral precipitates in microponds created by convolutions in the intertidal microbial mat surface. Meanwhile, a more permanent mush of crystals begins to develop beneath the surface of the mat. Contemporaneously, larger, discoidal and poikilotopic gypsum crystals become commonplace at depth within the buried lower intertidal

and lagoonal facies. In a few areas, a continuous gypsum-cemented crust may be encountered within this zone, at or just below the water table. Towards the high-water mark, which defines the limit of development of the microbial mat cover, the mush of crystals is well established and has effectively severed all connection between the surface mat cover and the accumulated peats below. Here, prolonged exposure and desiccation inhibits colonisation and regeneration as the mat shrivels and decays (Fig. 6A–C). Many of the intricacies of microbial mat morphology are thus lost and will rarely, if ever, be preserved.

Ponded areas and shallow rills, with their more long-lasting and persistent water film, support and encourage the development of thick peats and associated anaerobic conditions help ensure their preservation. Sediment-packed polygon cracks from such an environment are commonplace in the subsurface. Associated channel margins however, by virtue of even a few centimetres height above that permanent water film, are raised into an aerobic environment, under which regime almost all vestiges of the mat cover are lost through decay. The high-water mark also marks the seaward limit of the aeolian wedge, which is what forms much of the sabkha surface. This salt-encrusted surface has a 1:2500 slope that is controlled by wind deflation and the capillary forces above the water table. As a result, the present-day water table is nowhere in excess of 1 m below the surface (Fig. 7; Kinsman, 1969; Park 1973). Only on occasions, when the Shamal onshore winds reinforce high tides, do the flood waters extend any distance inland, typically following the barely discernible path of pre-existing tidal channels (Fig. 8).

Surface flooding, documented by the annual fluctuations in groundwater chemistry and reported on in some detail by Patterson (1972) and Patterson & Kinsman (1981), was perceived as an important recharge mechanism. Furthermore, it was presented as a mechanism for recent diagenetic dolomite formation within a sabkha setting (Patterson & Kinsman, 1982). Patterson & Kinsman (1981) presented a schematic flow net that showed the pattern of interference between groundwater flow, evaporation and surface recharge but without

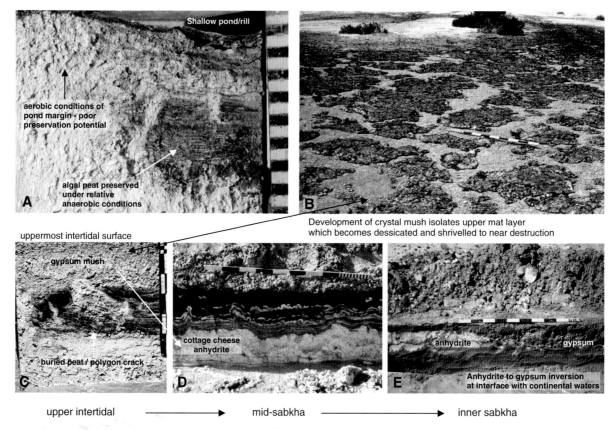

Fig. 6. Preservation of sabkha sediments. The variety of intricate surface morphologies have a low potential for preservation (Park, 1973). (A) Ponded areas offer some protection, the relatively anoxic conditions aiding mat preservation while oxic conditions in adjacent highs encourage bacterial decay. (B, C) With the development of the gypsum mush (C) a significant portion of the upper mat section is physically destroyed, while the thin layer stranded on top (B), becomes desiccated, shrivels, is torn apart and lost. (D) Under conditions of net evaporation, gypsum is replaced by anhydrite beyond the high-water mark. (E) It does not to take much water influx via groundwater seepage from the interior to initiate a reversal of this process. In some inland areas, large blades of gypsum, whose whitish cores are found to be of the hemi-hydrate bassanite, locally break the sabkha surface. This is an on-going dynamic diagenetic flux and the changes will continue into the subsurface.

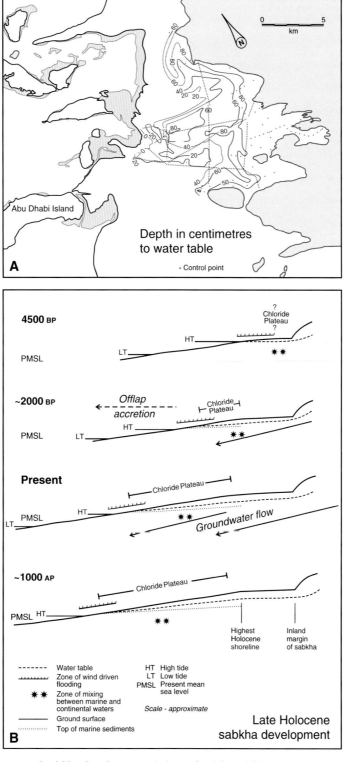

Fig. 7. Water table distribution and sabkha development. (A) Depth of the sabkha water table: this is rarely more than 80 cm below the surface, as confirmed by extensive pit sampling. The flooding pattern (Fig. 8) indicates areas of the sabkha that regularly and repeatedly receive a fresh influx of seawater via wind-assisted high-tide flooding. (B) In a regional sense, the sabkha surface is controlled largely by wind deflation and capillary forces within the near-surface section; the slope of the water table is only slightly less than the slope of the sabkha surface itself (after Patterson & Kinsman, 1981).

Fig. 8. Aerial view of the Abu Dhabi sabkha. The absence of vegetation and the low gradient of the sabkha surface allow wind-assisted storm floods to advance "uphill" well beyond the normal high-tide level. In a microcosm, it also illustrates how rapidly and how far facies belts could and do shift in response to even minor changes in relative sea level.

reference as to how this may impact diagenesis. However, Wood (*this volume*) has noted that surface recharge volumes are small and that, based on the ionic budgets and hydrostatic gradients now documented, it is possible to infer that the most significant and long-lasting source of fluids for diagenesis may be deeper aquifers.

Beyond the high-water mark, net evaporative conditions prevail and the first gypsum transformation to anhydrite occurs, with the replacement of the gypsum mush by the distinctive "cottage cheese" anhydrite fabric. Within the overlying aeolian sands larger nodules of anhydrite begin to grow displacively, some moderately firm and hard, some soft like toothpaste. These develop as layers a few cm thick, some continuous, some less so with varying amounts of sediment between. Heavy rains recharge and freshen the near surface groundwaters resulting in the rehydration of the anhydrite to gypsum (Fig. 6C–E) which, subject to renewed conditions of prolonged drought and evaporation, may undergo dewatering via

bassanite back to anhydrite. This recurring cycle, part seasonal, is a reminder of how fragile and potentially ephemeral this mineral assemblage is.

Saline waters and an interface with freshwaters from the hinterland and/or from below, coupled with high temperatures and net evaporative conditions, provide a setting in which aggressive and ongoing diagenesis is the norm. This reaction encapsulates one of the more complex exchange systems that drive diagenesis in the sabkha environment. To this end what can or will be preserved of the diagenetic history, if indeed anything, may be something of a lottery and not necessarily readily predicted. Even yet, it is only partly understood and difficult to quantify but manifestly a ubiquitous and integral part of the sabkha environment.

Sabkha water chemistry and diagenetic reactions have received much attention over the years. Although initially studied in isolation or as part of a diagenetic series, it has been increasingly recognised that the microbial community may be a key catalyst in many of these reactions. This, in association with bacterially induced sulphate reduction under anoxic conditions, has been recognised as a key element in replacement dolomitization of the shallow subsurface offshore Bahamas by Whitaker *et al.* (1994) and in Brazilian coastal lagoons by Vasconclos & McKenzie (1997). Diagenesis, by its very definition, dictates that some components have undergone change, and in this environment these changes occur and continue to occur, before, as well as during and post burial. To the extent that some of these changes can be documented and, with the aid of isotopes and fluid inclusions, inferred from buried and ancient sequences, it can enable the recognition of past sabkha-like conditions (McKenzie, 1981; Schreiber, 1988; Logan, 1987; Pufahl & Wefer, 2001).

THE HOLOCENE OF ABU DHABI IN A SEQUENCE CONTEXT

The essentially homoclinal surface of the Arabian platform margin has become the archetypal model for carbonate ramp sedimentation as defined by Ahr (1998). The gentle slope of the sabkha coastal plain allows ready access to flood-tide waters driven far onshore and "upslope" by strong supporting onshore winds (Fig. 8). This form of periodic flooding was called upon by Patterson & Kinsman (1982) to provide a recurring source of magnesium for dolomite formation within the sabkha section. Additional support for this was found in the pattern of distribution of dolomite across the sabkha, which appeared to show a close relationship to old residual channel-ways.

More recently, focus has shifted to the role of the cyanobacterial community as the primary agents in the formation of dolomite. Given that the thickest peats tend to accumulate in ponds and shallow channels (Fig. 9) it is perhaps then understandable that dolomite distribution across the sabkha also tends to follow old channel ways (Fig. 10). The biotic agents still require a steady source of magnesium, and within the modern sediment wedge periodic flooding still offers an attractive and appealing source. Surface flooding and near-surface recharge do have an immediate impact on near-surface penecontemporaneous diagenesis, although more recent mass-balance calculations would seem to indicate that, long

term, it is deeper source waters that have the most lasting and permanent effect on what enters the geological record (Wood, *this volume*). Of interest in the photograph shown in Fig. 8 is the broad swath of freshly deposited carbonate in the middle of the "algal belt". This produces a gentle clinoform and is the active face of seaward accretion (cf. Fig. 7B).

Fischer *et al.* (2004) noted that geologists and natural historians have long had a perennial fascination with the phenomenon of cyclicity, which has been recognised in some form or other throughout the geological record. Milankovitch (1941) recognised 100,000 year ice age cycles within the Quaternary, which he attributed to the combined impact of equinoxial precession, obliquity of Earth's axis, and changes in orbital eccentricity on solar radiation and insolation. More recently, Lisiecki & Raymo (2005), amongst others, have reinforced this with the documentation of as many as 57 cycles, extending back into the Pliocene, as reflected in changes in $\delta^{18}O$. The sabkha cycle found presently along the coastal margin the Trucial Coast represents the single parasequence (~3–4 m thick) deposited since the end of the last glacial period, some 12,500 years ago. This is directly underlain by lithified Pleistocene aeolianites, and it is arguable as to whether there has been sufficient time to allow the preservation of any Pleistocene or older sabkhas formed during earlier glacial-interglacial cycles.

While evaporites may be lost, porosity is often created and preserved, and stacked porosity cycles are a feature of many platform carbonate reservoirs. At the parasequence scale, as represented by the Holocene sabkha wedge and similar coastal plain sequences developed on shallow ramp margins, sequences are developed whose facies assemblages and distribution patterns are highly sensitive to minor fluctuations in sea level. These are commonly recognisable in cores if not at seismic scale (Yang *et al.*, 1993, 1994).

Following Vine & Matthews' (1963) report on magnetic striping, opportunities arose for geophysics to enter the debate, ultimately giving rise to the development of seismic stratigraphy by Vail and colleagues (Vail *et al.*, 1977; Vail, 1987; Haq *et al.*, 1988). They recognised the existence of a hinge mechanism that effectively controlled the relative position of sea level in relation to the shelf and where it lay relative to the fundamental interface between marine and non-marine

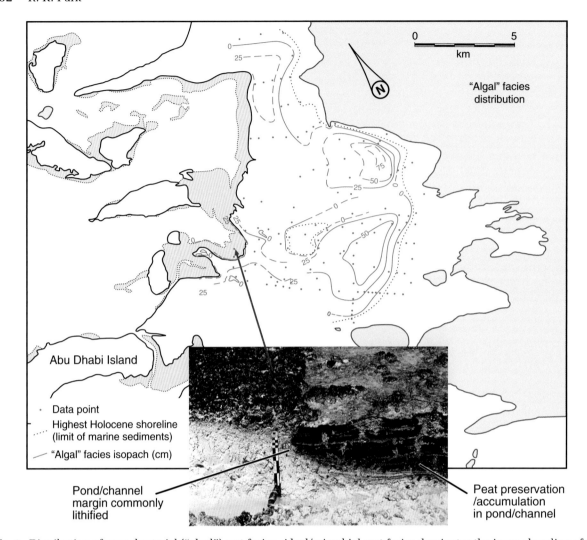

Fig. 9. Distribution of cyanobacterial ("algal") mat facies. Algal/microbial mat facies dominates the inner shoreline of the lagoon but preservation appears largely confined to traces of buried tidal channels, evident from the subtle relief revealed by the air photograph in Fig. 8. The inset photograph (bottom), taken near the inner margin of mat development, shows just how rapidly peat preservation is lost via aerobic destruction along raised pond and channel margins.

conditions. Pittman (1978) and others (e.g. Dewey & Pittman, 1998) disagreed with precisely where that hinge lay, which in-turn influenced their choice of a preferred means by which such changes were effected. Either way, the evolving science (or art) of sequence stratigraphy provided a practical tool for the interpretation of seismic data within a dynamic regional setting, and the basic tenets of sequence stratigraphy have been a guiding principle to stratigraphers and seismic interpreters alike for a generation.

The low-gradient sabkha surface, barren of any vegetation cover, offers little resistance to flooding, which is rapid and far reaching. This mimics, in essence, the rapid transgression that would ensue

from a sea-level rise. The stratigraphic section along the UAE coast records a long history of such rapid transgressions, followed by more measured regression, progradation and infilling of the available accommodation space. In any ramp setting, high-frequency transgressive-regressive events are likely to be the norm. Under such a regime, a multiplicity of associated high-frequency diagenetic events will follow and in this evaporitic environment, dominated as it is by metastable mineralogies, mineral transformations will be commonplace. Even with the aid of the tools now available, the full story of such cycles and their overprints will be difficult to fully resolve. The pore systems within these sequences may thus experience several phases of

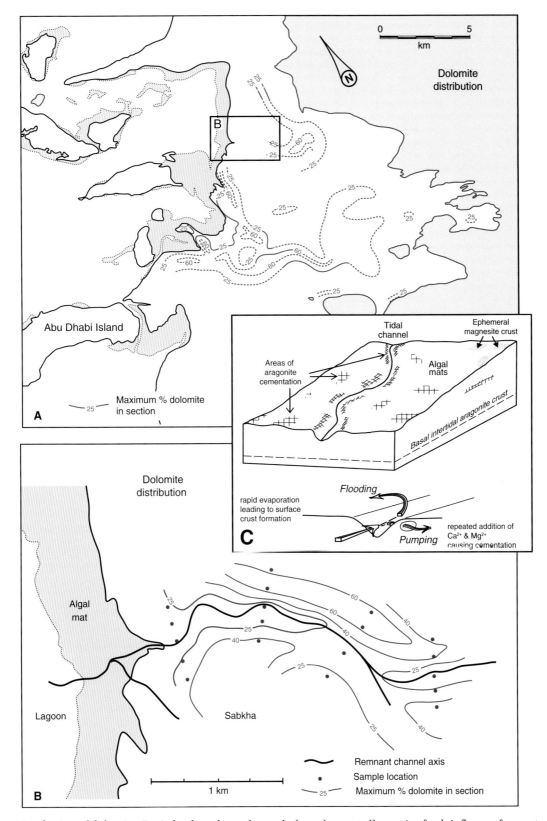

Fig. 10. Distribution of dolomite. Buried palaeochannels regularly and repeatedly receive fresh influxes of seawater. (A, B) The distribution of dolomite appears to follow old tidal drainage channels and ponds. (C) Dolomitization is likely promoted by tidal pumping and evaporation which raises the Mg^{2+}/Ca^{2+} ratio of sabkha porewaters and triggers dolomite precipitation and crust formation along the channel margins (Park, 1973; Patterson & Kinsman 1981).

enhancement and occlusion, long before any migrating hydrocarbons come onto the scene.

During any regression, the metastable and unstable mineral assemblages will be exposed to waters of a different chemistry resulting in a fresh round of diagenesis. This is precisely the scenario recognised by Ricken (1994) who postulated not only a direct link between 3rd-order transgression-regression cycles and carbonate productivity, but also a timing relationship with respect to diagenesis and its impact on cementation, organic content, bed thickness and overall preservation potential. A key driver here will be associated climatic changes, which in the first instance will affect runoff and sediment discharge rates and more subtly impact redox conditions in the adjacent pelagic settings of the deeper shelf. Under highstand conditions carbonate productivity will increase, and in addition will be the locus of additional carbonate cementation supplied via redistribution of carbonate dissolved from adjacent marly sediments. The cementation patterns described by Yang & Baumfalk (1994) from the Lower Permian Upper Rotleigende Group fit well with such a model.

The Holocene transgressive-regressive wedge (TST–HST) of sediment that borders the UAE coast is a record of the last 10,000–12,000 years of depositional history. At its thickest, the wedge is less than 10 m thick, and over much of the sabkha plain less than 3 m thick, a time and thickness association that is consistent with it being equated to a parasequence.

The sequence concept was promulgated by Sloss (1963), derived from studies of the expansive shallow epeiric marine deposits that dominated the North American craton during the Carboniferous. It was absorbed and adapted by geophysics into the realm of seismic stratigraphy as expounded by Vail and his colleagues (Vail *et al.*, 1977), and latterly reclaimed by geologists in the pursuit of sequence stratigraphy (Galloway, 1989; Brink *et al.*, 1993). The interplay of tectonism and tectonism-induced subsidence and rates of uplift with erosion and sediment supply rates to form seismically resolvable stratigraphic events and features was reaffirmed by Brink *et al.* (1993). Gischler & Lomando (2000) also demonstrated that local attributes, such as underlying and inherited topography, orientation and disposition relative to waves and currents, can often override or mask global eustatic trends in sea-level fluctuation.

The sequence stratigraphic debate, as Weimar & Posmentier (1993) succinctly stated, revolves around matters of scale; scales of observation and scales of resolution, or to use their analogy, looking at the forest versus the component trees. Studies of the Holocene, with their focus inevitably on process, clearly fall in the latter category. Most seismic-based studies deal with 3rd- and 4th-order cycles, while field–based and core studies will inevitably focus on much higher frequency 5th-, 6th- and even 7th-order events and attendant responses. This scale is well within the window of Milankovitch astronomical cycles. The recognition of various hierarchies of cycles and sequences in the field were also evident. The petroleum industry was quick to realise the potential of adapting this concept to the interpretation of cores and wireline logs and tying back to seismic. Studies of the Rotliegende Group by Yang & Baumfalk (1994), in addition to numerous others pertaining to the Palaeozoic–Mesozoic sections of the Arabian shelf itself, demonstrated the relevance of the Holocene of the Arabian Gulf for developing sequence stratigraphic models (Alsharhan & Scott, 2000).

Whether seismic or sequence stratigraphy is being discussed, the terminology employed is useful in providing a framework for the description and identification of the stratal packages and disconformities recognised in seismic sections, and more so with the enhanced resolution capabilities now afforded by 3-D seismic data. So where does the classic Holocene sabkha sequence fit in such discussions? It is after all, no more than a parasequence, barely a scratch on the canvas of geological history. Its importance is not in the thickness of the cycle *per se* but in the reaffirmation of the value of uniformitarian principles that incorporate not simply litho- and biofacies associations but just as importantly, the hydrological framework and dynamics and the diagenetic scenarios that they set up. The latter may have far more long-lasting effects, particularly in terms of say, porosity development and hydrocarbon reservoir evolution and preservation (Yang & Nio, 1993).

Shearman's observation that the sabkha is but the arid equivalent of the temperate fenlands of East Anglia, or mangrove marshlands of tropical areas such as Sumatra (Fig. 3), has been reiterated by Evans (*this volume*). In the sabkha environment, "coal measures" are replaced by a carbonate-evaporite assemblage, where net evaporation promotes precipitation, versus the aggressive leaching associated with humic acids in the latter. In the former, the environment is largely bereft of fauna and flora, save for empire-building microbial communities that form a mesh of cyanophytes and bacteria, best developed around the more protected inner margins of lagoons, and capable of producing a thick (>40 cm) peat. Each of these coastal margin environments is host to very different interstitial water chemistry regimes. This obviously will impact the immediate style and rates of chemical reactions and likely products in a given setting. These "now" reactions, however, may well be overwhelmed and supplanted by the dynamics of the longer cycle overprint imposed by deeper basin hydrology, as suggested Wood (*this volume*), the latter offering additional support and a rationale for Kendall's (1989) thesis of regional dolomitization by brine mixing.

The UAE ramp setting hosts two intimately linked sediment factories, one carbonate, the other evaporite (Fig. 11). Both produce minerals that are at best metastable under present Earth conditions and thus subject, and demonstrably prone, to alteration and reformation at the slightest whiff of a drop of rain or flood recharge. This evaporite/carbonate association can be shown to have persisted since the late Proterozoic. It is therefore not unreasonable to infer that cause-and-effect relationships proposed for the relatively limited area and unique setting of the Emirates coastline, have some resonance in the much more extensive evaporite-bearing sequences of an earlier Arabian platform margin that was perhaps one order of magnitude larger (Alsharhan & Scott, 2000). Furthermore, the associated microbial mat sequences such as those of the uppermost Precambrian-Cambrian Huqf Group, are considered to be a major source of petroleum in the region (Pollastro, 1999), just as the mangrove swamps are for the equatorial Indonesian basins. The association of evaporites and hydrocarbons is a topic that has long been championed by Schreiber (1988).

It is no coincidence that the Arabian platform succession is host to a significant proportion of the world's oil and gas reserves, and accounts for why the sabkhas of Abu Dhabi have commanded such interest, not to mention celebrity status. While details of the frequency, timing and multiplicity of diagenetic overprints remain to be resolved and documented in full, there is little doubt that gross patterns of facies assemblages and fabrics, notably the evaporite fabrics, provide testimony to the staying power of this sequence. For all its vulnerabilities, the 5th-order parasequence represented by the present-day Holocene wedge does appear to have the ability to survive into the geological record, at least as far as major facies assemblages and sequence relationships are concerned. Some of the details may get lost in the process, but there seems little question that the fundamental characteristics and associations of the evaporite-carbonate ramp setting can be preserved. This is clearly evident from the host of sabkha-like sections that have been recognised from various parts of the geological column, not least on the Arabian platform itself, which has a long history of persistent sabkha-like conditions associated with recurring proximal and distal ramp settings (Alsharhan & Scott, 2000). Throughout the Phanerozoic, where new rift basins and epeiric seas have evolved, such conditions are repeated.

Perhaps Shearman's most celebrated contribution (Shearman, 1966) is his comparison of the Purbeck of Dorset, where high-frequency 1 m scale cycles have recently been interpreted with reference to Milankovitch-type orbital forcing (Anderson, 2004), with the trenches dug in the sabkha flats of Abu Dhabi. Elsewhere, the evaporite-bearing sections bordering the early proto-Atlantic (Fig. 12) have been described by Evans *et al.* (1977), Park (1983) and Cocozza & Gandin (1990), while the Miocene Messinian deposits in the Mediterranean Basin has been documented at length through the work of Schreiber (1976) and Schreiber *et al.* (1977). In all cases high-frequency transgressive-regressive cycles are typical. Davies & Montaggioni (1985) documented high-frequency cyclicity within the Great Barrier Reef and this is perhaps echoed in Preto *et al.*'s (2004) interpretation of orbitally forced cyclic sequences in the Middle Triassic of the Italian Dolomites. What may not be preserved, or be difficult to discern in evaporitic sequences, are details of the complex diagenesis and paragenesis, not least because of the metastable minerals originally present. The carbonate-evaporite ramp setting, with its vulnerability to high-frequency short-term changes in near-surface interstitial chemistry, can readily impact reservoir compartmentalisation and require detailed analysis and modelling (Meyer *et al.*, 2004).

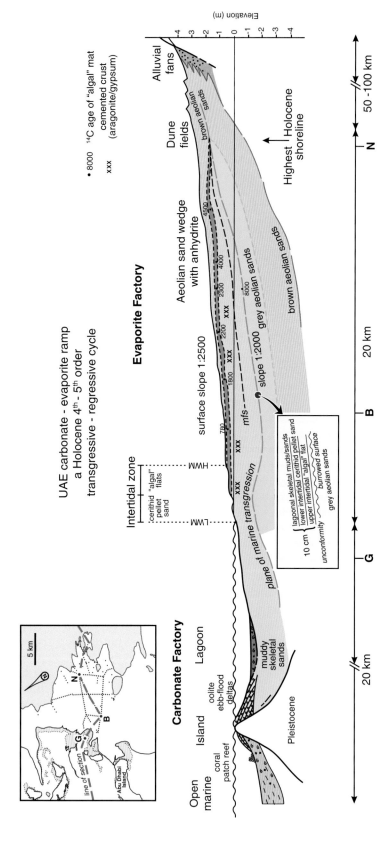

Fig. 11. Transect showing the intimate relationship of the two factories, one carbonate, the other evaporite, that are responsible for the evaporite-carbonate facies associations and diagenetic scenarios found in this modern ramp setting (modified from Park 1973, Kinsman & Park, 1976).

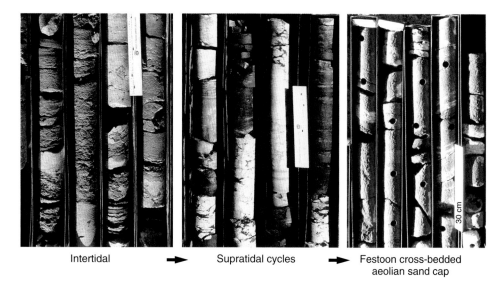

Intertidal　➡　Supratidal cycles　➡　Festoon cross-bedded aeolian sand cap

Fig. 12. Lower Jurassic carbonate-evaporite ramp sediments, Morocco. The geological record contains numerous examples of sabkha-type sequences and profiles. In addition to the long history of such sediments that has been documented in the Arabian Basin, examples from the early proto-Atlantic, such as this succession from the Lias of Morocco, are also well documented (Park, 1983), as are the Miocene Messinian sequences of the Mediterranean region. The high frequency 1–2 m scale cycles illustrated are typical of sedimentation on a gently dipping shelf margin subject to low-amplitude, high-frequency, flooding-evaporation cycles.

CONCLUSION

The interplay of tectonism and tectonism-induced subsidence and uplift rates with erosion and sediment supply rates to form seismically resolvable stratigraphic events and features was reaffirmed by Brink *et al.* (1993). In the carbonate realm, Sarg (1988) had developed this concept further, stating that "controls on carbonate productivity and platform development tend to be short-term eustatic changes superimposed on longer term tectonic changes". Sarg (1988) also noted that the rates of relative change in sea level will also affect the formation, the location and *in situ* longevity of meteoric lenses, and hence impact diagenetic processes, cementation, leaching and dolomitization. These are the events which may have far more long–lasting effects, particularly in terms of porosity development and hydrocarbon reservoir evolution and preservation.

Those forces that directly impact ocean basin geometry also exert a primary control on low-frequency modulations in sea level. This was recognised by Pittman (1978) and Dewey & Pittman (1998) and was recently reiterated by Immenhauser & Matthews (2004). Since the Jurassic, the geometry of the proto-Arabian Gulf has been determined by the subsidence history imposed by the collision of Arabia against the buttress of the Zagros Mountains. Higher frequency events on the other hand, such as 5th-order and higher periodicity, are likely to result from orbital forcing in so far as they impact events and processes especially within a shallow-water epeiric sea environment such as the Gulf and its adjacent land areas. All combine to determine the resulting sequence stratigraphic record (Immenhauser & Matthews, 2004).

The Holocene wedge of the Abu Dhabi sabkha sequence appears to record a single transgressive-regressive facies cycle produced during a 4th- to 5th-order cycle of relative sea-level change brought on by post-glacial melting of the polar icecaps and subsequent cooling. Fluctuations in this cycle, as reported elsewhere, are not readily evident within the facies patterns and associations preserved within the Abu Dhabi sabkha, where there are no obvious interruptions, discontinuities or other indications of such higher frequency events and perturbations affecting the stratigraphic sequence.

Chemical signatures and detailed documentation of the diagenetic history may tell us otherwise. Indeed a possible expression of such induced events may be the stacked array of crusts through the more seaward reaches of the sabkha and intertidal sections, as reported by Strohmenger (*this volume*). Recognition of such events farther inland

and their spatial distribution will require a greater sampling density and event correlation than currently exists. Here too, interpretation of results my be compounded by the mixed signal of long cycle low-frequency input from the deeper aquifer (Wood, *this volume*) and short-term high-frequency flood recharge and evaporative cycles that exist within the Holocene wedge. Yang & Baumfalk (1994) documented high-frequency cycles within a 2nd-order sequence within the Permian Upper Rotliegende Group, in which they calculated accumulation rates of 9–11 cm kyr^{-1}. Compare this to accumulation rates for the Holocene of Abu Dhabi, where a 2–3 m-thick regressive wedge has developed in less than 10 kyr.

Siliciclastic supply is very much controlled by climate, and increased rainfall promotes increased run off and dilution of carbonate and loss of evaporites. No evaporites occur within the Tertiary of Indonesia, for example, where high temperatures were accompanied by high rainfall and humidity and high-volume fines removal and redistribution to form coastal plains and deltas. Rates of accumulation also influence bed thickness. In the adjacent deeper water facies, carbonate may be dissolved and redistributed generally towards the tops of cycles (Ricken, 1994).

Biofacies and lithofacies associations within a Waltherian temporal and spatial framework are now routinely integrated with and supplemented by isotopic and palaeomagnetic data, as in the studies by Zuhlke (2004) and Sacchi & Muller (2004). Furthermore, situated as the sabkha is, at the interface between meteoric water, seawater and groundwater, the stage is set for complex diagenetic scenarios both penecontemporaneous and later. Combined, they can have a profound effect on potential hydrocarbon reservoir development and preservation.

ACKNOWLEDGEMENTS

This has very much been a trip down memory lane, so to speak, and so I would like to take this opportunity to express my thanks and gratitude to David Kinsman for inviting me to join in his continuing research in Abu Dhabi and to Ron Patterson for his companionship in the field and stimulating discussions, not forgetting the wonderful bread making prowess of his wife Carol, and to Steve Golubic for introducing me to the world of cyanophytes. And then there was Doug, the ever willing teacher, friend and mentor, who encouraged and cajoled us all to pursue every avenue, to strive to satisfy our curiosities and, above all, showed us how to exact the maximum enjoyment out of a lifetime of questioning the natural world around us – how we shall miss him. My thanks to Kodeco, notably Supardan, for their support and last but not least to Tony Lamando and Ana Gunatilaka for helpful comments in completing this manuscript.

REFERENCES

Ahr, W. (1998) Carbonate Ramps, 1973–1996: A Historical Review. In: *Carbonate Ramps* (Eds V.P. Wright and T.P. Burchette). *Geol. Soc. London, Spec. Publ.*, **149**, 7–14.

Alderman, A.R. and **Skinner, H.C.W.** (1957) Dolomite sedimentation in the southwest of S. Australia. *Am. J. Sci.*, **255**, 561–567.

Alderman, A.R. and **von der Borch, C.C.** (1961) Occurrence of magnesite-dolomite in South Australia. *Nature*, **192** (4805), 861.

Alderman A.R. and **von der Borch, C.C.** (1963) A dolomite reaction series. *Nature*, **198** (4879), 465–466.

Alsharhan, A.S. and **Scott, R.W.** (Eds) (2000) Middle East Models of Jurassic/Cretaceous Carbonate Systems. *SEPM Spec. Publ.*, **69**, 364 pp.

Anderson, E.J. (2004) Facies patterns that define orbitally forced third-, fourth- and fifth-order sequences of sixth-order cycles and their relationship to ostacod faunicycles: the Purbeckian (Berriasian) of Dorset, England. In: *Cyclostratigraphy: Approaches and Case Histories* (Eds B. D'Argenio, A.G. Fischer, I.P. Silva, H. Weissert and V. Ferreri). *SEPM Spec. Publ.*, **81**, 245–260.

Badiozamani, K. (1973) The Dorag dolomitization model - application to the Middle Ordovician of Wisconsin. *J. Sed. Petrol.*, **34**, 965–984.

Brink, G.J., **Keenan, J.H.G.** and **Brown Jr., L.F** (1993) Deposition of 4th order, post-rift sequences and sequence sets, Lower Cretaceous (Lower Valanginian to Lower Aptian), Pletmos Basin, southern offshore South Africa. In: *Siliciclastic Sequence Stratigraphy* (Eds P. Weimar and H.W. Posamentier). *AAPG Mem.*, **58**, 43–70.

Clift, P. and **Molnar, P.** (2003) An IODP Workshop for drilling of the Indian Ocean fan systems *JOI/USSAC Workshop on "Continent-Ocean Interactions in the East Asian Marginal Seas"*, 6.

Cocozza, T. and **Gandin, A.** (1990) Carbonate deposition during early rifting: the Cambrian of Sardinia and the Triassic-Jurassic of Tuscany, Italy. In: *Carbonate Platforms: Facies, Sequences and Evolution* (Eds M. Tucker, J.L. Wilson, P.D. Crevello, J.R. Sag and J.F. Read). *IAS Spec. Publ.*, **9**, 9–37.

Cook, P.J. (1973) Supratidal environment and geochemistry of some recent dolomite concretions, Broad Sound, Queensland, *Australia. J. Sed. Petrol.*, **43**, 998–1011.

Curtis, R., **Evans, G.**, **Kinsman, D.J.J.** and **Shearman, D.J.** (1963) Association of dolomite and anhydrite in the Recent sediments of the Persian Gulf. *Nature*, **197** (4868), 670–680.

Davies, G.R. (1970b) Algal laminated sediments, eastern Shark Bay, *Western Australia. AAPG Mem.*, **13**, 85–168.

Davies, P.J. and Montaggioni, L. (1985) Reef growth and sea-level change: the environmental signature. *Proc. 5th Internat. Coral Reef Conf. Tahiti*, **3**, 477–511.

Deffeyes, K.S., Lucia, F.J. and Weyl, P.K. (1964) Dolomitization of Recent and Plio-Pleistocene sediments by marine evaporite waters on Bonaire, Netherlands Antilles. In: *Dolomitization and Limestone Diagenesis, A Symposium* (Eds L.C. Pray and R.C. Murray). *SEPM Spec. Publ.*, **13**, 71–88.

Dewey, J.F. and Pittman III, W.C. (1998) Sea-level changes: mechanisms, magnitudes and rates. In: *Paleogeographic Evolution and Non-glacial Eustasy* (Eds J.L. Pindell and C. Drake). *SEPM Spec. Publ.*, **58**, 2–16.

Evans, G. (1966) The Recent sedimentary facies of the Persian Gulf region. *Phil. Trans. Roy. Soc. London Ser. A*, **259**, 291–209.

Evans, G. (1969) Stratigraphy and geologic history of the sabkha, Abu Dhabi, *Persian Gulf. Sedimentology*, **12**, 145–159.

Evans, G., Kendall, C.G.St.C. and Skipwith, P.A.d'E. (1964a) Origin of the coastal flats, the sabkha of the Trucial coast, Persian Gulf. *Nature*, **202** (4934), 759–761.

Evans, G., Kinsman, D.J.J. and Shearman, D.J. (1964b) A reconnaissance survey of the environment of Recent carbonate sedimentation along the Trucial Coast, Persian Gulf. In: *Deltaic and Shallow Marine Deposits* (Ed. L.M.J.U. van Straaten), pp. 129–135. Elsevier, The Netherlands.

Evans, G., Murrat, J.W., Biggs, H.E.J., Bate, R. and Bush, P.R. (1973) The oceanography, ecology, sedimentology and geomorphology of parts of the Trucial Coast barrier island complex, Persian Gulf. In: *The Persian Gulf, Holocene Carbonate Sedimentation and Diagenesis in a Shallow Epicontinental Sea* (Ed. B.H. Purser), pp. 233–277. Springer Verlag, New York.

Evans, G., Kendall, C.G.St.C. and Butler, J. (1977) Genesis of Liassic shallow and deepwater rhythms, central High Atlas Mountains, Morocco. *J. Sed. Petrol.*, **47**, 120–128.

Fischer, A., D'Argenio, B., Silva, I.P., Weissert, H. and Ferreri, V. (2004) Cyclostratigraphic approach to Earth's history: an introduction. In: Cyclostratigraphy: Approaches and Case Histories (Eds *D'Argenio, A.G. Fischer, I.P. Silva, H. Weissert* and *V. Ferreri). SEPM Spec. Publ.*, **81**, 5–13.

Galloway, W.E. (1989) Genetic stratigraphic sequences in basin analysis. I. Architecture and genesis of flooding–surface bounded depositional units. *AAPG Bull.*, **73**, 125–142.

Gebelein, C.D. (1969) Distribution, morphology and accretion rate of Recent subtidal algal stromatolites, Bermuda. *J. Sed. Petrol.*, **39**, 49–69.

Ginsburg, R.N. (Ed.) (1975) *Tidal Deposits, A Casebook of Recent Examples and Fossil Counterparts*. Springer-Verlag, New York.

Gischler, E. and Lomando, A.J. (2000) Isolated carbonate platforms of Belize, Central America: sedimentary facies, late Quaternary history and controlling factors. In: *Carbonate Platform Systems: Components and Interactions* (Eds E. Insalaco, P.W. Skelton and T.J. Palmer). *Geol. Soc. London, Spec. Publ.*, **178**, 135–146.

Glynn, P.W. (2000) El Nino-Southern Oscillation mass mortalities of reef corals: a model of high temperature marine extinctions? In: *Carbonate Platform Systems: Components and Interactions* (Eds E. Insalaco, P.W. Skelton and T.J. Palmer). *Geol. Soc. London, Spec. Publ.*, **178**, 117–133.

Golubic, S. (2000) Microbial landscapes: Abu Dhabi and Shark Bay: environmental evolution: effects of the origin and evolution of life on planet. *Earth Edition*, **2**, 117–139.

Gunatilaka, A. (1976) Thallophyte boring and micritization within skeletal sands from Connemara, western Ireland. *J. Sed Petrol.*, **46**, 548–554.

Gunatilaka, A., Al–Zamel, A., Shearman, D.J. and Reda, A. (1987) A spherulitic fabric in selectively dolomitised siliciclastic crustacean burrows, northern Kuwait. *J. Sed. Petrol.*, **57**, 922–927.

Harris, P.M. and Kowalik, W.S. (1994) Satellite Images of Carbonate Depositional Settings: Examples of Reservoir and Exploration Scale Geologic Facies Variation. *AAPG Methods Explor. Ser.*, **11**, 167 pp.

Hanshaw, B.B., Back, W. and Deike, R.G. (1971) A geochemical hypothesis for dolomitization by ground water. *Econ. Geol.*, **66**, 710–724.

Haq, B.U., Hardenbol, J. and Vail, P. (1988) Mezozoic and Cenozoic chronostrratigraphy and cycles of sea-level change. In: *Sea Level Changes: An Integrated Approach* (Eds C.K. Wilgus, B.K. Hastings, H. Posamentier, J. Van Waggoner, C.A. Ross and G.G.St.C. Kendall). *SEPM Spec. Publ.*, **42**, 71–108.

Hsu, K.J. (1966) Origin of dolomite in sedimentary sequences: a critical analysis. *Mineral. Deposita*, **2**, 133–138.

Illing, L.V. and Wells, A.J. (1964) Penecontempory dolomite in the Persian Gulf. *AAPG. Bull.*, **448**, 532–533.

Immenhauser, A. and Matthews, R.K. (2004) Albian sea-level cycles in Oman: the "Rosetta Stone" approach. *Geoarabia*, **9**, 11–46.

Kendall, A.C. (1989) Brine mixing in the Middle Devonian of western Canada and its possible significance in regional dolomitization. *Sed. Geol.*, **64**, 271–285.

Kendall, C.G.StC. and Skipwith, P.A.d'E. (1968) Recent algal mats of a Persian Gulf lagoon. *J. Sed. Petrol.*, **38**, 1040–1058.

Kendall, C.G.St.C. and Skipwith, P.A.d'E. (1969a) Holocene shallow-water carbonate and evaporite sediments of Khor al Bazam, Abu Dhabi, south-east Persian Gulf. *AAPG Bull.*, **53**, 841–869.

Kendall, C.M.St.G. and Skipwith, P.A.d'E. (1969b). Geomorphology of a Recent shallow water carbonate province, Khor al Bazam, Trucial Coast, Persian Gulf. *Bull. Geol. Soc. Am.*, **80**, 865–891.

Kinsman, D.J.J. (1964) *Recent carbonate sediments near Abu Dhabi, Trucial Coast, Persian Gulf.* Unpubl. PhD Thesis, Imperial College London, UK.

Kinsman, D.J.J. (1967) Huntite from a carbonate-evaporite environment. *Am. Mineral.*, **52**, 1332–1340.

Kinsman, D.J.J. (1969) Modes of formation, sedimentary associations, and diagnostic features of shallow-water and supratidal evaporites. *AAPG Bull.*, **53**, 830–840.

Kinsman, D.J.J. (1976) Sabkha facies nodular anhydrite and paleotemperature determination. *AAPG Meeting Abstract*, 688.

Kinsman, D.J.J. and **Park, R.K.** (1969) Studies in recent sedimentology and early diagenesis, Trucial Coast, Arabian Gulf. *Second Regional Techn. Symp., Soc. Petrol. Eng. AIME, Saudi Arabia Section.*

Kinsman, D.J.J. and **Park, R.K.** (1976) Algal belt and coastal sabkha evolution, Trucial Coast, Persian Gulf. In: *Stromatolites* (Ed. M.R. Walter), *Developments in Sedimentology,* **20**, 421–434. Elsevier, The Netherlands.

Kinsman, D.J.J., **Park, R.K.** and **Patterson, R.J.** (1971). Sabkhas: studies in Recent carbonate sedimentation and diagenesis, Persian Gulf. *Abs. Geol. Soc. Am. Ann. Meetings,* 1971.

Lees, A. and **Buller, A.T.** (1972) Modern temperate water and warm water shelf carbonate sediments contrasted. *Mar. Geol.,* **13**, 1767–1773.

Lisiecki, L.E. and **Raymo, M.E.** (2005) A Pliocene-Pleistocene stack of 57 globally distributed benthic δ^{18}O records. *Paleoceanography,* **20**, PA 1003, doi: 10.1029/2004PA001071

Logan B.W. (1987) The Macleod evaporite basin, Western Australia. *AAPG Mem.,* **44**, 140 pp.

Lucia, F.J. and **Major, R.P.** (1994) Porosity evolution through hypersaline reflux dolomitisation. In: *Dolomites A Volume in Honour of Dolomieu* (Eds. B.H. Purser, M. Tucker and D. Zenger). *Int. Assoc. Sedimentol. Spec. Publ.,* **21**, 325–344.

Mawson, D. (1929) Some South Australian algal limestones in the process of formation. *Q. J. Geol. Soc. London,* **85**, 613–623.

McKenzie, J.A. (1981) Holocene dolomitisation of calcium carbonate sediments from coastal sabkhas of Abu Dhabi, UAE. A stable isotope study. *J. Geol.,* **89**, 185–198.

Meyer, A. Boichard, R. Azzam, I. and **Al-Amoudi, A.** (2004) The Upper Khuff Formation, sedimentology and static core rock type approach – comparison of two offshore Abu Dhabi Fields. SPE 88794 11th ADIPEC Conference, Abu Dhabi, UAE, 3.

Milankovitch, M. (1941) Kanon der Erdbestrahlung und seine Anwendung auf das Eiszeitenproblem. *Akad. Roy. Serbe.,* **133**, 633 pp.

Monty, C.L.V. (1967) Distribution and structure of Recent stromatolitic algal mats, eastern Andros Island, *Bahamas. Ann. Soc. Géol. Belg.,* **90**, 55–100.

Murray, J.W. (1965). The Foraminiferida of the Persian Gulf. 2. The Abu Dhabi region. *Palaeogeogr. Palaeoclimatol. Palaeoecol.,* **1**, 307–332.

Murray, J.W. (1970). The Foraminiferida of the hypersaline Abu Dhabi Lagoon, *Persian Gulf. Lethaia,* **3**, 51–68.

Park, R.K. (1973) *Algal belt sedimentation, the Trucial Coast, Arabian Gulf.* Unpubl. PhD Thesis, University of Reading, UK, 440 pp.

Park, R.K. (1976) A note on the significance of lamination in stromatolites. *Sedimentology,* **23**, 379–393.

Park, R.K. (1977) The preservation potential of some recent stromatolites. *Sedimentology,* **24**, 485–506.

Park, R.K. (1983) Lower Jurassic carbonate buildups and associated facies, central and western Morocco. In: *Carbonate Buildups - A Core Workshop* (Ed. P. M. Harris), SEPM Core Workshop, 4, Dallas, 328–365.

Patterson, R.J. (1972) *Hydrology and carbonate diagenesis of a coastal sabkha in the Persian Gulf.* Unpubl. PhD Thesis, Princeton University, USA, 469 pp.

Patterson R.J. and **Kinsman, D.J.J.** (1981) Hydrologic framework of a sabkha along the Arabian Gulf. *AAPG Bull.,* **65**, 1457–1475.

Patterson R.J. and **Kinsman D.J.J.** (1982) Formation of diagenetic dolomite in coastal sabkhas along the Arabian (Persian) Gulf. *AAPG Bull.,* **66**, 28–43.

Perkins, R.D., **Dwyer, G.S.**, **Rosoff, D.B.**, **Fuller, P.A.** and **Lloyd, R.M.** (1994) Salina sedimentation and diagenesis: West Caicos Island, British West Indies. In: *Dolomites A Volume in Honour of Dolomieu* (Eds. B.H. Purser, M. Tucker and D. Zenger). *Int. Assoc. Sedimentol. Spec. Publ.,* **21**, 37–54.

Pittman III, W.C. (1978) Relationship between eustasy and stratigraphic sequences of passive margins. *Geol. Soc. Am. Bull.,* **80**, 1389–1403.

Playford, P.E. and **Cockbain, A.E.** (1976) Modern algal stromatolites at Hamelin Pool, a hypersaline barred basin in Shark Bay, Western Australia. In: *Stromatolites* (Ed. M.R. Walter), *Dev. Sedimentol.,* **20**, 389–411. Elsevier, The Netherlands.

Pollastro, R.M. (1999) Ghaba salt basin province and Fahud salt basin province, Oman - geological overview and total petroleum systems. *USGS Open File Rep.,* **99-50-D**, 46 pp.

Pray L.C. and **Murray, R.C.** (Eds) (1965) *Dolomitization and Limestone Diagenesis – A Symposium.* SEPM Spec. Publ., 13, 180 pp.

Preto, N., **Hinnov, L.A.**, **Zanche, V.**, **Mietto, P.** and **Hardie, L.A.** (2004) The Milankovitch Iinterpretation of the Latemar platform cycles (Dolomites, Italy): implications for geochronology, biostratigraphy and Middle Triassic carbonate platform accumulation. In: *Cyclostratigraphy: Approaches and Case Histories* (Eds B. D'Argenio, A. G. Fischer, I.P. Silva, H. Weissert and V. Ferreri). *SEPM Spec. Publ.,* **81**, 167–182.

Pufahl, P.K. and **Wefer, G.** (2001) Data report: petrographic, cathodoluminescent and compositional characteristics of organogenic dolomites from southwest Africa. In: *Proc. ODP Sci. Results* (Eds G. Wefer, W.H. Berger and C. Richter), **175**, 1–17.

Purser, B.H. (Ed) (1973) *The Persian Gulf: Holocene Carbonate Sedimentation and Diagenesis in a Shallow Epicontinental Sea.* Springer Verlag, New York, 471 pp.

Purser, B.H. and **Seibold, E.** (1973) The principal environmental factors influencing Holocene sedimentation and diagenesis. In: *The Persian Gulf, Holocene Carbonate Sedimentation and Diagenesis in a Shallow Epicontinental Sea* (Ed. Purser, B.H), pp. 1–9. Springer Verlag, New York.

Purser, B.H., **Brown, A.** and **Aissaoui, D.M.** (1994) Nature, origins and evolution of porosity in dolomites. In: *Dolomites A Volume in Honour of Dolomieu* (Eds. B.H. Purser, M. Tucker and D. Zenger). *Int. Assoc. Sedimentol. Spec. Publ.,* **21**, 283–324.

Ricken, W. (1994) Complex rhythmic sedimentation related to third order sea level variations: Upper Cretaceous, Western Interior Basin, USA. In: *Orbital Forcing and Cyclic Sequences* (Eds P.L. de Boer and D.G. Smith). *Int. Assoc. Sedimentol. Spec. Publ.,* **19**, 167–193.

Sacchi, M and **Muller, P.** (2004) Orbital cyclicity and astronomicval calibration of the Upper Miocene continental succession cored at the Iharosbereny–1 well site, Western Pannonian Basin, Hungary. In: *Cyclostratigraphy:*

Approaches and Case Histories (Eds B. D'Argenio, A.G. Fischer, I.P. Silva, H. Weissert and V. Ferreri). *SEPM Spec. Publ.*, **81**, 275–294.

Saleh, A., Al-Ruwaih, F., Al-Reda, A. and Gunatilaka, A. (1999) A reconnaissance study of a clastic coastal sabkha in northern Kuwait, *Arabian Gulf. J. Arid Environ.*, **43**, 1–19.

Sarg, J.F. (1988) Carbonate sequence stratigraphy. In: *Sea Level Changes: An Integrated Approach* (Eds C.K. Wilgus, B.K. Hastings, H. Posamentier, J. Van Waggoner, C.A. Ross and G.G.St.C. Kendall). *SEPM Spec. Publ.*, **42**, 155–181.

Schmalz, R.F. (1969) Deep–water evaporite deposition: a genetic model. *AAPG Bull.*, **53**, 789–823.

Schreiber, B.C. (Ed.) (1988) *Evaporites and Hydrocarbons.* Columbia University Press, New York, 475 pp.

Schreiber, B.C., Friedman, G.M., Decima, A. and Schreiber, E. (1976) The depositional environments of the upper Miocene (Messinian) evaporite deposits of the Sicilian basin. *Sedimentology*, **23**, 469–479.

Schreiber, B.C., Catalano, R. and Schreiber, E. (1977) An evaporite lithofacies continuum: latest Miocene (Messinian) deposits of Salemi Basin (Sicily) and a modern analog. In: *Reefs and Evaporites – Concepts and Depositional Models* (Ed. J.H. Fisher). *AAPG Stud. Geol.*, **5**, 169–180.

Shearman, D.J. (1963). Recent anhydrite gypsum, dolomite and halite from the coastal flats of the Arabian shore of the Persian Gulf. *Proc. Geol. Soc. London*, **1607**, 63.

Shearman, D.J. (1966) Origin of marine evaporites by diagenesis. *Trans. Inst. Min. Metall. (Sect. B)*, **79**, 155–162.

Shinn, E.A., (1973) Sedimentary accretion along the leeward, SE coast of Qatar Peninsula, Persian Gulf. In: *The Persian Gulf: Holocene Carbonate Sedimentation and Diagenesis in a Shallow Epicontinental Sea* (Ed. B.H. Purser), pp. 199–209. Springer Verlag, Berlin, Heidelberg, New York.

Shinn, E.A., Lloyd, R.M. and Ginsburg, R.N. (1965) Recent supratidal dolomite from Andros Island, Bahamas. In: *Dolomitization and Limestone Diagenesis, A Symposium*, (Eds L.C. Pray and R.C. Murray). *SEPM Spec. Publ.*, **13**, 112–123.

Shinn, E.A., Lloyd, R.M. and Ginsburg, R.N. (1977) Anatomy of a modern carbonate tidal-flat, Andros island, *Bahamas. J. Sed. Petrol.*, **39**, 1202–1228.

Sloss, L.L. (1963) Sequences in the cratonic interior of North America. *Geol. Soc. Am. Bull.*, **64**, 93–113.

Taylor, J.C.M. and Illing, L.V. (1969) Holocene intertidal calcium carbonate cementation, Qatar, Persian Gulf. *Sedimentology*, **12**, 69–107.

Vail, P.R. (1987) Seismic stratigraphy interpretation using sequence stratigraphy. Part 1: seismic stratigraphy interpretation procedure. In: *Atlas of Seismic Stratigraphy 1* (Ed. A.W. Bally). *AAPG Stud. Geol.*, **27**, 1–10.

Vail, P.R., Mitchum, R.M. and Thompson III, S. (1977) Seismic stratigraphy and global changes of sea level, part 3: relative changes of sea level from coastal onlap. In: *Seismic Stratigraphy Applications to Hydrocarbon Exploration* (Ed. C.W. Payton). *AAPG Mem.*, **26**, 63–97.

Vasconclos, C. and McKenzie, J.A. (1997) Microbial mediation of modern dolomite precipitation and diagenesis under anoxic conditions (Lagoa Vermelha, Rio de Janeiro, Brazil). *J.Sed. Res.*, **67**, 578–390.

Vine, F.J. and Matthews, D.H. (1963) Magnetic anomalies over oceanic ridges, *Nature*, **199**, 947–949.

Wagner, C.W. and van der Togt, C. (1973) Holocene sediment types and their distribution in the southern Persian Gulf. In: *The Persian Gulf, Holocene Carbonate Sedimentation and Diagenesis in a Shallow Epicontinental Sea* (Ed. B.H. Purser), pp. 123–155. Springer Verlag, New York.

Walter, M. R., Golubic, S. and Preiss, W.V. (1973) Recent stromatolites from hydromagnesite and aragonite depositing lakes near the Coorong Lagoon, South Australia. *J. Sed. Petrol.*, **43**, 1021–1030.

Weimer, P. and Posamentier, H.W. (1993) Recent development and applications in siliciclastic sequence stratigraphy. In: *Siliciclastic Sequence Stratigraphy* (Eds P. Weimer and H.W. Posamentier). *AAPG Mem.*, **58**, 3–12.

Wells, A.J. (1962) Recent dolomite in the Persian Gulf. *Nature*, **194** (4825), 274–275.

Whitaker, F.F., Smart, P.L., Vahrenkamp, V.C., Nicholson, H. and Wogelius, R.A. (1994) Dolomitisation by near normal seawater? Field evidence from the Bahamas. In: *Dolomites A Volume in Honour of Dolomieu* (Eds. B.H. Purser, M. Tucker and D. Zenger). *Int. Assoc. Sedimentol. Spec. Publ.*, **21**, 111–132.

Yang, C-S and Baumfalk, Y.A. (1994) Milankovitch cyclicity in the Upper Rotleigende Group of the Netherlands offshore. In: *Orbital Forcing and Cyclic Sequences* (Eds P.L. de Boer and D.G. Smith). *Int. Assoc. Sedimentol. Spec. Publ.*, **19**, 47–62.

Yang, C-S. and Nio, S-D. (1993) Application of high resolution sequence stratigraphy to the Upper Rotleigende in the Netherlands offshore. In: *Siliciclastic Sequence Stratigraphy* (Eds P. Weimer and H.W. Posamentier). *AAPG Mem.*, **58**, 285–316.

Zuhlke, R. (2004) Integrated cyclostratigraphy of a model Mesozoic carbonate platform - the Latemar (Middle Triassic, Italy). In: *Cyclostratigraphy: Approaches and Case Histories* (Eds B. D'Argenio, A.G. Fischer, I.P. Silva, H. Weissert and V. Ferreri). *SEPM Spec. Publ.*, **81**, 183–211.

Int. Assoc. Sedimentol. Spec. Publ. (2011) **43**, 113–132

Holocene evolution of Arabian coastal sabkhas: a re-evaluation based on stable-isotope analysis, forty years after Shearman's first view of the sabkha

ANANDA GUNATILAKA

Geological Consultant, 10 Thumbovila, Piliyandala, Sri Lanka (E-mail: gunat@asianet.lk)

ABSTRACT

The location of Arabian Gulf sabkhas at the interface between seawaters, continental groundwaters and meteoric waters results in characteristic geochemical signatures that have been acquired over a single transgressive–regressive cycle of ~10,000 years duration. The stable-isotope compositions (especially those of deuterium, sulphur and oxygen) of diagenetic evaporite minerals and coexisting sabkha groundwaters, when interpreted with respect to coastal geomorphological controls, give perhaps the best description of the hydrology and evolution of sabkhas with time. The hydrological framework of marine sabkhas in the southern Arabian Gulf (UAE) is controlled mainly by marine flood-recharge and reflux with a dominant marine isotopic signature in the sulphate evaporites. By contrast, in the northern Arabian Gulf (Kuwait) the marine sabkha sulphates show dominantly continental water-derived isotopic signatures. In the latter region, continental groundwaters have virtually reached the coast, displacing the marine waters in the sabkhas. As the coastline progrades, a given marine sabkha can have sulphate minerals with both isotopic signatures, but formed at different times. Similarly, the volumes of dolomite formed in sabkhas are dependent on their respective hydrological and geomorphological (i.e. open vs barriered coasts) frameworks. Low dissolved sulphate contents and organic matter availability in the environment are associated with dolomite precipitation. The current (late Quaternary) 100,000-year orbitally driven climatic cycle with its glacial–interglacial contrasts and resulting sea/base-level changes, and accompanying groundwater-table fluctuations in the sabkhas, indicate that their preservation potential is extremely poor. This accounts for the general absence of Pleistocene sabkha deposits. The Arabian Gulf and Peninsula region, which earlier was under the influence of the Indian Ocean Monsoon rainfall belt, went into a climatically hyper-arid phase just over 6000 years ago, more than a thousand years before sabkhas started forming (~4800 BP), thus priming the sabkhas for diagenetic evaporite mineral precipitation, which requires high groundwater salinities. Anhydrite, an index evaporite mineral, is limited to latitudes of 23° to 28° N in the Gulf. Climate and hydrology in conjunction with palaeogeography have important implications when discussing ancient sabkha analogues, which are known since the early Archaean, and were of a different order of magnitude from those forming today (e.g. the giant sabkhas of the Arabian Jurassic).

Keywords: Sabkhas, isotopes, hydrological, climatic, eustatic, geomorphological interactions

INTRODUCTION AND HISTORICAL PERSPECTIVE

Sabkhas (with their myriad meanings) have had a chequered but short history in sedimentology. The term itself has connotations with the Arabian Gulf. The pioneers of sabkha sedimentology originally referred to it as a flat, salt-encrusted geomorphological surface or coastal flat (Shearman, 1963; Evans *et al.*, 1964; Kinsman, 1969a). Sabkhas soon became one of the best-understood modern evaporite sedimentary environments associated with ramp-margin carbonate deposition. However, by the 1990s, the marine sabkha was vanishing from the literature. The reality was that there is a dearth of other modern "sabkhas" to be investigated in the world, where the optimum combination of climate, oceanography and hydrology are rarely developed today.

Based on the Arabian Gulf successions, a Holocene sabkha depositional/diagenetic parasequence is typically associated with a single 4th- to 5th order transgressive–regressive (T–R) cycle of ~10 kyr duration, with the sabkha forming essentially during the progradational phase of sedimentary offlap and current sea-level stabilization (mainly the last 5000 years). The ~2 m thick stratigraphic wedge (the sabkha-cycle) of 2–8 km width and generated in a brief instant of geological time has become the classic model in efforts to understand the giant sabkhas of geological history. This was uniformitarianism at its best, but also in the extreme. It was also serendipitous that both recent and ancient sabkha analogues occurred in Arabia. Three critical evaporite minerals (dolomite, gypsum and anhydrite) were also discovered forming in a modern environment. It was but a short step for the pioneers of evaporite sedimentology and petroleum geologists to recognise that here was a modern analogue for the carbonate reservoirs and tight cap-rocks of many petroleum provinces.

The pioneering work on Arabian Gulf sabkhas was initiated in the early 1960s by the Koninklijke Shell Company, V.C. Illing and Partners and the Imperial College Research Group in the Trucial Coast, Arabian Gulf (now the United Arab Emirates), jointly led by Douglas Shearman and Graham Evans (Fig. 1). In a series of classic publications and several doctoral theses, this area became the mecca for modern marine carbonate–evaporite sedimentation studies in a hyper-arid ramp-margin environment: the so-called arid zone carbonate–evaporite factory. The most important papers published on the area are now part of sabkha history (Wells, 1962; Shearman, 1963–1985; Curtis et al.,1963; Evans & Shearman, 1964; Evans et al., 1964–1973; Butler et al., 1965, 1973, 1982; Kinsman, 1964–1969a, 1974; Evans, 1966, 1995; Illing et al., 1965; Kendall & Skipwith, 1968, 1969; Butler, 1969, 1970; Bush, 1973; Purser, 1973; De-Groot, 1973; Hsu & Schneider, 1973; Patterson & Kinsman, 1977–1982; McKenzie et al., 1980; Wood et al., 2003). A recent update and review of sabkha ecosystems was given in Barth & Boer (2002), and the impact of sea-level changes on ramp-margin sedimentation and the generation of sabkhas over

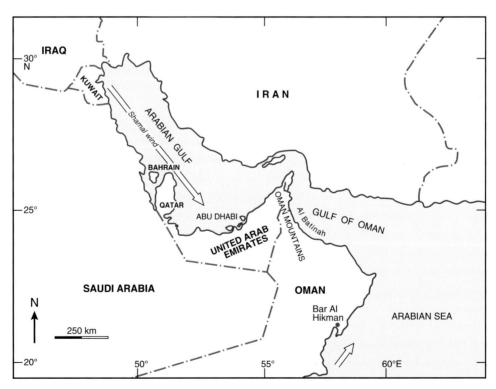

Fig. 1. Locations of Arabian Gulf and Peninsula region coastal sabkhas referred to in the text. The prevailing wind vector (arrow) of the *Shamal* in the Gulf is 300°–320°. The prevailing wind in Oman is from the southwest Indian Ocean.

geological time-scales has been discussed by Park (this volume, pp. 89–112).

The Arabian Gulf coastline is >1000 km long, yet, <100 km of the United Arab Emirates coastline (Trucial Coast) became representative of the entire Arabian (Persian) Gulf and the modern analogue for many ancient marine sabkha environments. The northern Arabian Gulf was hardly examined, except for two studies from Kuwait on the carbonates (Picha, 1978; Saleh, 1978). The evaporites there were an unknown entity. In 1979, the present author initiated a long-term research programme of the northern sabkhas (1979–1989) from the Department of Geology, Kuwait University, Kuwait (in the northern Arabian Gulf). Starting from the Kuwait-Saudi Arabia Border Zone, this study extended all the way to the Kuwait-Iraq border (Fig. 2). Two representative areas were targeted for detailed work: the Al Khiran region of southern

Kuwait (a classic carbonate–evaporite environment), and the Bahra-Sabbiyah region in northern Kuwait, a clastic marine sabkha influenced by the Shatt Al Arab delta and fluvial tidal system (Fig. 2). The results of this research were published over several years (Gunatilaka *et al.*, 1980–1987b; Robinson *et al.*, 1983; Gunatilaka, 1984–1991; Mwango, 1986; Reda, 1986; Gunatilaka & Mwango, 1987; Robinson, 1987; Gunatilaka & Mwango, 1988; Gunatilaka & Shearman, 1988; Robinson & Gunatilaka, 1991; Robinson *et al.*, 1991; Saleh *et al.*, 1999). Other studies have been published by Al Zamel (1983) and Cherif *et al.* (1993). Early carbonate sediments (mostly ooid-rich) underlying the Bahra-Sabbiyah region indicate that siliciclastics came to dominate the sedimentation pattern in the northern Gulf with the stabilization of Holocene sea levels and the role played by the Euphrates-Tigris delta fluvial tidal system (Saleh

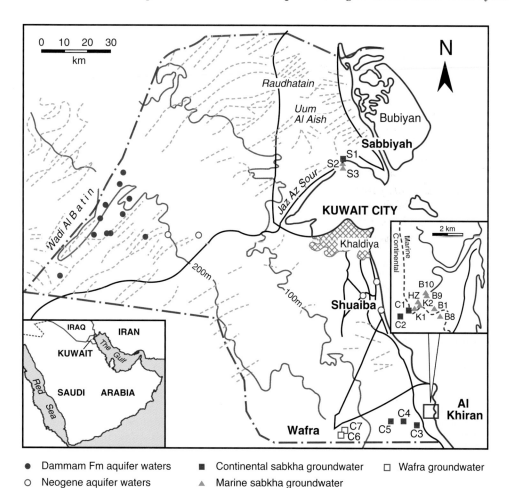

● Dammam Fm aquifer waters   ■ Continental sabkha groundwater   □ Wafra groundwater
○ Neogene aquifer waters   ▲ Marine sabkha groundwater

Fig. 2. Study areas in southern Kuwait (Al Khiran) and northern Kuwait (Bahra-Sabbiyah). Kuwait topography and drainage together with the locations of the aquifer and sabkha groundwaters/brine samples (see Table 1). Bubiyan Island is to the east of Sabbiyah and is separated by a tidal channel.

et al., 1999). It is probably the only modern A-zone delta (Gunatilaka, 1986a) prograding into a shallow, hypersaline, semi-enclosed marine basin and bordering a desert.

Beyond the Gulf and on to the coast of the Gulf of Oman are a series of very small patch sabkhas along the Al Batinah alluvial plain (Fig. 1). Facing the Arabian Sea, on the southeastern part of the Sultanate of Oman is a very unique sabkha called the Barr Al Hikman (Fig. 1). It is the only marine sabkha known that is located within a monsoon climatic belt today (influenced by the Indian Ocean Southwest Monsoon). All Oman sabkhas are siliciclastic dominated with only minor gypsum precipitation, and no anhydrite has been detected. The Barr Al Hikman sabkha is in the direct path of the SW-monsoon track. Downwind to the north, the sabkha merges with the large erg of the Wahiba Sands, which in turn has received sediments from the very long and large alluvial outwash fans (especially the Adam, Halfayn and Andam) coming down the southern Al Hajar range of the Oman Mountains (see Rodgers & Gunatilaka, 2002). This author has reconnoitred these areas. They represent a vast alluvial fan – desert sand/erg – clastic sabkha – marine association.

A near analogue to the Barr Al Hikman area is the Laguna Ojo de Liebre-San Felipe sabkha complex in Baja California. However, there is one major difference: the tidal range in the Arabian Sea is <2 m, while in Baja California it is ~7 m. Consequently, the marine flooding of the Hikman sabkha is limited and is at a maximum only during the SW-monsoon season (May to August). There are also sabkha environments on the Red Sea coast, whose tectonic/climatic setting is different from that of the Arabian Gulf. Here, the critical evaporite mineral anhydrite extends (albeit in very small quantities) to latitudes as low as 18°N (Gunatilaka, 1989b), while it is restricted to 28°N to 23°N in the Arabian Gulf (Gunatilaka, 1986a).

In this paper, the interactions between coastal geomorphology, sabkha hydrology, late Quaternary sea-level and climate change, and the generation of sabkhas are explored. The discussions that follow are based on 12 years of sabkha trekking by this author on the Arabian Peninsula, which have enabled a comparative overview of sabkha dynamics to be presented here. Unfortunately, most of these sabkhas have lost their pristine nature, as much of the Gulf coastline has been rapidly transformed by recent urbanisation.

Table 1. The main cultural phases (ages in kyr BP) in the Mesopotamian-Arabian region during the past 6000 years, after mid-Holocene aridity had set in

Bahrain (Dilmun)	Mesopotamia (Iraq)	Abu Dhabi (UAE)
Ubaid 4 (6.3-5.9)	Ubaid (7.5-5.5)	
Jemdet Nasr (5.9-5.1)	Uruk (5.5-5)	
Barbar I (4.2-3.9)	Sumer (4.5- 4.0)	Umm Al Nar (ca.4.5-4)
Barbar II (3.9-3.7)	Akkadian (4.3-4)	Wadi Suq (4.0-3.3)
Kassites (3.7-2.1)	Dynastic (after 4.0)	

Data from: Sanlaville & Passkoff (1986); Crawford (1991; Algaze (1993); Weiss et al. (1993); Cullen et al. (2000).

Several major civilizations in the region such as the Uruk (5500–5000 BP), the Sumer (4500–4000 BP) and the Akkadian Empire (4300–4000 BP), all reached their zenith and collapsed during the regional hyper-arid climatic phase of the mid-Holocene (Table 1). This was also the period of major sabkha formation, when the monsoon rainfall belt moved south and away from the Arabian Peninsula (Sirocko *et al.*, 1993; Gasse & van Campo, 1994). There is, therefore, a close correlation between eustatic sea-level changes, climatic changes and the rise and fall of cultural phases in the region (Larsen & Evans, 1978; Sanlaville & Paskoff, 1986; Gunatilaka, 2000), evidenced by an unparalleled 7000 year record of cultural developments in this single geographical area, commonly regarded to be the cradle of modern civilisation.

A COMPARISON OF THE ARABIAN GULF SABKHAS

The main differences between the northern and southern Arabian Gulf sabkhas can be related to two or three interdependent factors: (a) the orientation of the coastline with respect to the prevailing northwest (*Shamal*) winds (Fig. 1); (b) the geomorphological framework in the two regions; and (c) climatic-hydrological factors. The sabkha sedimentary facies and hydrodynamics of sabkha groundwater flow and the evaporite mineral assemblages generated are dependent directly or indirectly on these factors.

In the southern Gulf (formerly Trucial Coast), the coastline is oriented normal to the prevailing northwesterly wind vector (~300°) and there are no coastal barriers (beach-dune ridges or cheniers) to restrict the marine flooding of the sabkha by Gulf seawaters, and hence the flood-recharge-reflux of brines, which controls the volume of evaporite minerals generated (especially dolomite, gypsum and anhydrite). Mangrove vegetation is also present in the lagoons. The inland-to-coast topographic gradient is gentler (1:2500) than in the northern Gulf. Consequently, a much wider area of the sabkha is (or was) flooded (up to 8 km). The climate is more arid than in the northern Gulf, with a rainfall of only 3–4 cm yr^{-1} and an average summer temperature is 45 °C, but ranges up to 60 °C (Patterson & Kinsman, 1981). The dolomite content in the sabkha is greater and the supratidal anhydrite layer is thicker. Coastal progradation rates are 1–2 km kyr^{-1} during the past 5000 years.

The network of offshore islands, shoals and reefs that protect the shorelines of the southern Arabian Gulf (Kendall & Skipwith, 1969) are absent in the northern Gulf. In southern Kuwait, the prevailing wind is oblique to the shoreline, the topographic gradient is much steeper (1:400), and a series of five oolitic beach-dune ridges or "cheniers" formed between 5500–2190 BP trails the sabkha, thus progressively restricting marine flooding and reflux of the sabkha brines (Gunatilaka, 1986a). Mangrove vegetation is absent; instead there are wide zones of halophytic vegetation in the sabkha, which play an important role in converting gypsum to anhydrite (Gunatilaka *et al.*, 1980). The geomorphological and stratigraphic setting of the Al Khiran sabkha is shown in Fig. 3. Flooding of the sabkha occurs through two tidal channels, which cut through the barrier ridges. Marine flooding

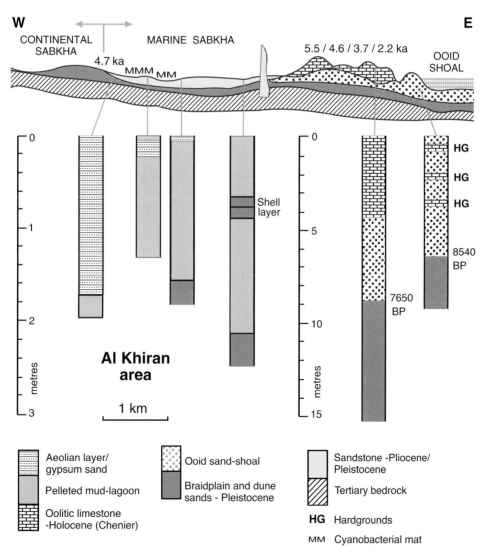

Fig. 3. Geomorphological setting and generalized stratigraphy of the Al Khiran sabkha, southern Kuwait. The line of section is E–W (see Fig. 2). The dated (age BP) chenier ridges are also indicated.

is thus confined mainly to the intertidal supratidal zone of the tidal channels. About 80% of the ~3–5 km wide "marine" sabkha (underlain by marine carbonate sands/muds) remains outside the flooding zone today. Consequently, the amount of dolomite and anhydrite precipitated is volumetrically small (Gunatilaka, 1990, 1991). In the restricted and small Al Khiran lagoon in Kuwait, subtidal dolomite has precipitated from waters with reduced sulphate content (Gunatilaka *et al.*, 1984, 1985b). The amount of wind-borne *materials* deposited in this sabkha can be as high as 40%. Al Khiran, at 28°N latitude, is the northernmost limit of sabkha anhydrite in the Arabian Gulf. The coastal progradation rate is $\sim 1\,km\,kyr^{-1}$.

In the Al Sabbiyah clastic sabkha (~ 5 km wide) near to the Iraq border (Fig. 2), the coastal orientation is different from both Al Khiran and the UAE region (Fig. 1). Wind plays a different role here as it blows from the land to the sea, minimizing the flooding of the sabkhas by tidal waters. Marine flooding is entirely dependent on the tidal cycle and flow (generally a <1 km wide zone), except in the very rare instance of the wind blowing from the southeast (<5% of the time), when a wider area can be flooded especially if the wind coincides with high-water spring tides. Reda (1986) and Saleh *et al.* (1999) referred to it as a wind tidal flat. Continental groundwaters dominate in this region, except for the narrow strip flooded by the daily tides. Only gypsum is found here; no anhydrite has been detected. The sediment source for this sabkha is mainly the reworked fine-grained suspended material derived from the Euphrates-Tigris delta by counter-clockwise tidal currents at the head of the Gulf and the tidal island of Bubiyan, while the coarser material is derived from the small alluvial fans descending onto the sabkha from the Jal Az Sour escarpment to the north (Fig. 2). The coastal progradation rate is ~1.5 km kyr^{-1} (Reda, 1986).

In summary, the sabkhas change from purely siliciclastic dominated in the north, to varying mixed siliciclastic-carbonate southward towards the southern Kuwait–Saudi Arabia border (Fig. 2), and finally to mostly carbonate dominated in the UAE. Hence, it is clear that the UAE sabkhas are not representative of the entire >1000 km long Gulf coastline. Significant variations in sedimentation styles exist, which need to be kept in mind when discussing ancient analogues. The above differences have also had another effect. As the sabkha progrades (1–1.5 km kyr^{-1}), the continental groundwaters draining to the coasts are gradually displacing the marine-derived brines. In the Kuwait sabkhas (both in Al Khiran and Al Sabbiyah), these groundwaters have displaced most of the marine-derived brines (Robinson & Gunatilaka, 1991), while in the UAE there is (was) still a substantial "transitional zone" of continental and marine-derived brines, which continues to shift seawards with time (Patterson & Kinsman, 1981). These relationships are well expressed in the Kuwait sabkhas from the stable-isotope analyses of groundwaters (see below). Comparable isotopic studies are not available for the UAE sabkhas (Butler *et al.*, 1973).

The hydrodynamics of groundwaters and brines in the sabkha zones resulting from the above controls have a direct bearing on their composition, and consequently the volume, composition and distribution of coexisting evaporite minerals, which precipitated from these evolving groundwaters. In other words, sulphate evaporite minerals *in the marine sabkha* may have precipitated from marine brines, continental groundwater-derived brines, or mixed brines. Standard (solute) geochemistry will not bring out these relationships. Only stable isotopes of sulphate minerals and groundwaters (deuterium, oxygen and sulphur) can do so, as demonstrated later. Further, it will be shown that the differences between the northern and southern Arabian Gulf sabkhas have a significant climatic overprint too.

CLIMATIC FRAMEWORK OF SABKHAS IN THE ARABIAN GULF

Climatic controls need to be considered in the context of a single Milankovitch (orbitally-driven) glacial–interglacial cycle of approximately 10^5 years duration, with accompanying sea-level fluctuations and related transgressions and regressions. Quaternary sea-level and climate history have played an important role in the evolution of the sabkha.

Quaternary sea level changes in the Gulf Basin

During the last glacial maximum (~18,000 years ago), sea levels decreased globally by as much as 120 m (Shackleton, 1987; Fairbanks, 1989). The Arabian (Persian) Gulf (maximum depth 80 m) was drained during each of the last eight to nine glacial cycles. The Ur-Schatt River (or Shatt Al Arab, the combined channel of the Euphrates, Tigris and Karun) and floodplain extended beyond the Gulf, and regional shorelines were in the Gulf of Oman (Lambeck, 1996). The Gulf Basin started filling up around 13,500 BP, long before the Holocene period commenced, pushing back the great river. Estuarine development followed for the next ~4000 years. Marine transgression began in the UAE coastline at approximately 7000 BP and reached a maximum between 5000–4000 BP (up to 8 km inland from present shoreline) and progradation started soon after (Evans *et al.*, 1969; Patterson & Kinsman, 1981). In southern Kuwait, the transgression began between 8540–7650 ± 70 BP, and reached a maximum between 4730–4530 ± 60 BP (~ 3–5 km to the west of the present-day shoreline). Since then, intertidal and suptratidal sediments have prograded to produce the sabkhas of this region (Fig. 3).

The transgressive surface in Bubiyan Island is dated at approximately 8500 BP; Al Zamel (1983) dated a freshwater peat just below the transgressive surface at 8490–7870 ± 90 BP. It was demonstrated that Bubiyan Island had emerged by 4500–3500 BP and on its southern shoreline, facing the Gulf, multiple shelly beach-ridges were dated at 3500–2800 BP, and it was estimated that the coast is here prograding at a rate of 1400–1600 m kyr⁻¹. In the Al Sabbiyah sabkha in northern Kuwait, at the head of the Gulf, coastal progradation commenced around 3040 BP at an annual rate of 1.5–2 km kyr⁻¹. In all of the Gulf regions, therefore, the sabkha is less than 5 kyr old (<5 % time of a single glacial interglacial cycle and ~40% of the Holocene transgressive–regressive cycle).

Quaternary climatic history in the Arabian peninsula

The Arabian Peninsula was from time to time under the influence of the Indian Ocean Monsoon, with the Intertropical Convergence Zone (ITCZ) penetrating far to the north of the Peninsula (Gasse & van Campo, 1994; Fleitmann *et al.*, 2003). The best available records are from Oman and the Oman margin (ODP Leg 117). The terrestrial climate record is mainly from lake deposits and isotopic analysis of speleothem deposits and groundwaters (McClure, 1976; Burns *et al.*, 2001; Preusser *et al.*, 2002; Fleitmann *et al.*, 2003). A record of continental pluvial events (Fleitmann *et al.*, 2003) over the last 330 kyr indicates that five pluvial periods occurred in southern Arabia (from

6000–10,500, 78,000–82,000, 120,000–135,000, 180,000–200,000 and 300,000–330,000 years ago). These are equivalent to the early to middle Holocene, Marine Oxygen Isotopes Stages (MIS) 5a, 5e, 7a and 9. A strong monsoon period has also been recorded around 510–525 kyr (Rossignol-Strick *et al.*, 1998). The intervening periods were generally very arid. Isotopic studies ($\delta^{18}O$ and δD) indicate that groundwaters were predominantly recharged by the Indian Ocean Monsoon rainfall, when the monsoon rainfall belt moved northward and reached the Arabian Peninsula during each of these periods. The pluvial event of interest here is the last one (6.2–10.5 ka). Since 6200 years ago, the peninsula was not affected by the Indian Ocean Monsoon, because the ITCZ and associated rainfall belt remained generally to the south of the Arabian Peninsula (Hasternath, 1985; Sirocko *et al.*, 1993; Burns *et al.*, 2001).

During the Holocene transgression, the climate in southern Arabia was humid, as the monsoon rainfall belt extended to the region. However, during the progradational phase when the sabkha and the evaporite mineral belt started forming, mid-Holocene aridity had set in, with the rainfall belt having retreated to the south, away from the peninsula, where it remains today. The climate became hyperarid. There are three different sources of precipitation in the region: the northwestern Shamal winds, which bring rainfall from the Mediterranean frontal systems (December to March); local thunderstorms; and tropical cyclones originating in the Bay of Bengal and the Arabian Sea. It is known that for almost 80% of the time of a single glacial interglacial cycle, the climate was arid-semiarid with lower sea levels and therefore low groundwater tables.

The last 5000 years were not only arid, but were also characterised by high sea levels. The sabkha is an equilibrium deflation surface, whose topography is controlled by the depth of the groundwater table. If, for most of a climatic cycle, low groundwater tables prevail (due to lowering base-levels), then deflation will remove the sabkha sediments and its preservation potential becomes very poor. Although sabkhas would have formed during the previous Quaternary climate cycles, there is no evidence of their preservation in the Gulf region. It is proposed here that since the present 100 kyr glacial interglacial cycles were established about 800 ka, it is unlikely that any previous sabkhas would have been preserved. Prior to 0.8 Ma, obliquity (41 kyr periodicity) dominated glacial interglacial cycles. The resulting transgressions and regressions were probably of smaller amplitude, and nothing much is known of the resulting deposits. Hence, there was a major change in sea-level history with the onset of northern hemisphere glaciations. If any sabkhas were preserved, it would be due to local tectonic uplift or rapid drowning only. In Oman, Pleistocene uplift has been reported from coastal zones, but a preserved sabkha has not been detected (Rodgers & Gunatilaka, 2002).

Climate has a regional or latitudinal effect on sabkha development that can be of use in palaeo-geographic reconstructions and studying ancient sabkha analogues, which range in age from the early Archaean Warawoona Group (in western Australia) to the well-known Miocene examples. Their enormous extent in both space and time is exemplified by several well-known evaporite basins in many continents (e.g. the Elk Point Basin in Canada; Williston, Salina, Gulf Coast, Paradox and Delaware Basins in the USA; Cheshire, Zech-stein, Siberian and Moscow Basins in Europe; McArthur and Amadeus Basins in Australia; Miocene Basins of the Middle East and Mediterranean regions; Arabian Basin, Cambrian Hormuz Basin in Iran; Permian Andes evaporites; late Proterozoic Zambian Copperbelt; Upper Proterozoic Ara evaporites in Oman), all of which have recognizable sabkha facies in their stratigraphic sequences. However, the nature of the ancient climate regimes prevailing remains speculative, in contrast to the precisely known Quaternary orbital cycles. The Arabian Gulf depositional diagenetic parasequence encompassing a time span of <10,000 years, pales into insignificance in comparison with parasequences of the ancient examples. Yet, this is the only modern-day analogue we have.

THE HYDROLOGICAL FRAMEWORK OF THE SABKHAS AND EVAPORITE MINERAL DIAGENESIS: STABLE ISOTOPE ANALYSIS

Several workers have investigated the hydrological framework of sabkhas over the years and the broad aspects of groundwater flow have been discussed extensively (Butler, 1969; Bush, 1973; DeGroot, 1973; Hsu & Schneider, 1973; Shampine *et al.*, 1978; McKenzie *et al.*, 1980; Patterson & Kinsman, 1981; Robinson *et al.*, 1983, 1987; Robinson & Gunatilaka, 1991; Wood *et al.*, 2003). A

precise understanding of sabkha dynamics and evaporite mineral formation has been made possible by stable-isotopic studies of sulphate evaporites and coexisting sabkha brines and groundwaters. The $\delta^{18}O$, $\delta^{34}S$ of sulphate minerals and groundwaters and δ^2H (deuterium) of representative groundwaters are given in Table 2. Unfortunately, detailed stable-isotope data are available only for the Kuwait sabkhas. Yet, some predictions can be made based on these data for other regions of the Gulf if a knowledge of the coastal geomorphology and groundwater-flow data are available. The isotopic compositions are dependent on whether marine- or continental derived groundwaters/brines dominate the sabkha and the extent of their mixing (if any).

Isotopic composition of the Kuwait sabkhas

Groundwaters ($\delta^{18}O$ and δD)

The question arises as to what type of precipitation recharged groundwaters during the present interglacial period. Weyhenmeyer *et al.* (2000) showed that precipitation in the southern Arabian Peninsula from southern moisture sources (Arabian Sea and Bay of Bengal) and northern moisture sources (Mediterranean frontal systems) differ greatly in their isotopic compositions (Fig. 4). The two sources define separate Local Meteoric Water Lines: a northern (N-LMWL; $\delta D = 5.0\ \delta^{18}O + 10.7$) and a southern (S-LMWL; $\delta D = 7.1\ \delta^{18}O + 1.1$). A northern Oman Groundwater Line (N-OGL; $\delta D = 5.3\ \delta^{18}O + 2.7$) has also been distinguished that plots between the two meteoric water lines.

Figure 4 plots precipitation and modern groundwater data from freshwater fields in Kuwait (Table 2), which lie along the Kuwait Meteoric Water Line (K-LMWL; $\delta D = 8.0\ \delta^{18}O + 15$), which is close to the N-LMWL and N-OGL lines. The two sets of data can be considered to represent the northern and southern regions of the Gulf. They show that a northern moisture source has recharged the groundwater in both areas in recent times (~ last 5000 years). However, between 6–10.5 ka (the Holocene monsoon maximum) and prior to this, southern (Indian Ocean) moisture sources dominated the pluvial phases in Arabia (Fleitmann *et al.*, 2003). When sabkha progradation started at around ~4800 years ago, the southern monsoonal rainfall source had ceased and a northern moisture source had taken over groundwater recharge. Consequently, during the

Table 2. Stable-isotope compositions of Kuwait waters and sulphate minerals from Kuwait sabkhas

Date	Sample	Water		Sulphate	
		$\delta^{18}O$ ‰ (VSMOW)	δD ‰ (VSMOW)	$\delta^{34}S$ ‰ (VCDT)	$\delta^{18}O$ ‰ (VSMOW)
Rainwater					
Winter 1982/83	Wadi Al Batin	−0.5	10.3	nd	nd
Winter 1982/83	Khaldiya	−2.9	−6.0	nd	nd
Winter 1982/83	Al Khiran	−2.2	−7.9	nd	nd
October 1982	South Al Khiran	−0.9	2.9	nd	nd
Freshwater wells (0.4-2‰ tds; SO_4^{2-}; Cl^-; HCO_3^-)					
Raudhatain well					
November 1981	R1	−3.5	−9.5	16.0	14.6
November 1981	R3	−3.2	−8.1	15.7	15.1
November 1981	R5	−3.6	−14.4	15.3	14.4
November 1981	R6	−3.2	−9.3	15.6	14.3
November 1981	R9	−3.4	−10.5	15.3	14.9
November 1981	R16	−3.3	−11.5	15.8	15.0
November 1981	R20	−4.1	−11.1	15.7	14.5
November 1981	R27	−3.7	−12.4	15.5	15.8
November 1981	R58	−3.6	−11.5	15.5	14.0
November 1981	R61	−3.2	−11.4	15.8	15.2
November 1981	R64	−3.3	−10.6	15.4	14.0
Umm Al Aish well					
November 1981	U20	−3.3	−13.0	15.4	13.6
November 1981	U22	−3.0	−10.5	15.3	13.7
November 1981	U54	−2.5	−6.1	15.5	13.8
Wafra brackish waterwells (4–9 ‰ tds;Cl^-; SO_4^{2-})					
April 1982	C6	−1.9	−13.9	14.9	14.5
April 1982	C7	−1.9	−14.1	14.6	14.3
Gulf seawater					
45 ‰ tds		2.4	14.7	20.9	10.4
Pits in sabkha (0.5–2 m) 20–300 ‰ tds					
Sabbiyah sabkha					
May 1982	S1	0.8	−11.1	14.9	13.2
May 1982	S2*	6.5	21.0	19.4	0.8
May 1982	S3*	6.2	21.0	19.7	8.5
Al Khiran sabkha brines					
December 1981	B1	1.2	−8.8	17.6	14.3
October 1982	B1	−0.6	−9.1	17.2	13.8
December 1981	B10*	3.1	16.1	21.3	11.0
December 1981	B9 *	6.1	23.1	20.2	10.1
October 1982	B9 *	6.4	21.2	18.9	10.8
December 1981	B10 *	1.2	−8.8	17.6	14.3
October 1982	B10 *	0.6	−9.1	17.2	13.8
December 1981	B8	0.2	−15.0	15.8	13.6
October 1982	B8	0.7	−12.4	14.6	13.6
December 1982	HZ *	4.2	4.6	17.0	10.5
December 1982	K2	0.8	−7.5	13.8	12.8
December 1982	K1	0.8	−9.1	14.0	13.1
December 1982	K1	0.3	−7.6	nd	nd
April 1982	C1	−0.5	−16.7	13.4	13.4
October 1982	C1	−0.8	−15.2	13.7	14.6
April 1982	C2	−0.9	−16.2	13.8	13.3
April 1982	C3	−1.0	−16.1	13.5	13.7
October 1982	C3	−0.8	−14.8	13.8	13.9
April 1982	C4	0.9	−3.3	12.9	12.9
April 1982	C5	−0.3	−8.2	13.3	13.4

(*Continued*)

Table 2. (*Continued*)

Date	Sample	Water		Sulphate	
		$\delta^{18}O$ ‰ (VSMOW)	δD ‰ (VSMOW)	$\delta^{34}S$ ‰ (VCDT)	$\delta^{18}O$ ‰ (VSMOW)
Sulphate minerals					
December 1982	Rus Formation (Eocene) anhydrite	nd	nd	18.6	14.2
December 1982 Al Khiran sabkha	Wadi Al Batin getch	nd	nd	15.6	15.6
December 1980	primary discoid gypsum	nd	nd	15.8	14.9
December 1980	cottage cheese anhydrite (B9)	nd	nd	17.4	13.9
December 1980	bassanite after gypsum	nd	nd	12.1	14.6
December 1980	alabastrine gypsum	nd	nd	15.3	14.5
December 1980	anhydrite	nd	nd	15.8	14.2
December 1980	anhydrite	nd	nd	14.0	13.8
December 1980	cemented gypsum crust	nd	nd	16.2	15.6

tds = total dissolved solids, nd = not determined, * = flood recharge.

current peak interglacial period, both northern and southern moisture sources made their contributions.

In considering the evolution of the hydrological system, it is essential to emphasize the role that Pleistocene pluvial events and eustatic sea-level changes have played in the region (Gunatilaka, 1986a). The regional groundwater network was most likely established during previous pluvial episodes when the Arabian Peninsula had a much

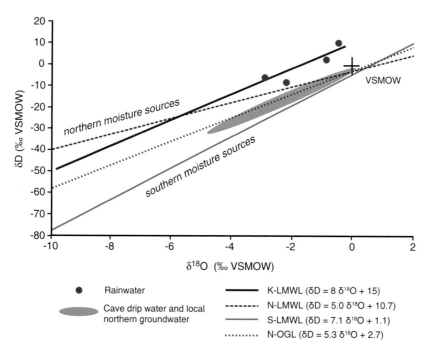

Fig. 4. Isotopic composition (δD and $\delta^{18}O$) of rainwater and groundwaters from Kuwait and southern Arabia. Note that present-day groundwaters are derived from a northern moisture source. LMWL, local meteoric water line; N, northern moisture sources; S, southern moisture sources; K, Kuwait meteoric water; N-OGL, northern Oman groundwater line.

milder and wetter climate than today. These pluvial episodes have now been identified from the speleothem records in northern Oman (Fleitmann *et al.*, 2003). In the continental regions of Kuwait, the main aquifers are in rocks and sediments of the Kuwait Group (Pliocene–Pleistocene) and Hasa Group (Palaeocene–Miocene). All aquifers are interconnected to each other and to the karstified and fissured main Dammam Limestone Formation (middle Eocene) aquifer (brackish water) below. The lower Eocene Rus Formation (anhydrite) is an aquiclude and is hydrologically separated from the Dammam and Umm er Radhuma (Palaeocene-lower Eocene) aquifers (Fig. 5). The entire hydrological complex originates in interior Arabia.

A plot of δD vs δ^{18}O of all groundwaters and brines from the Kuwait sabkhas, Kuwait Group and Dammam Formation waters from western Kuwait, and Dammam Formation and Neogene aquifer waters from Saudi Arabia are shown in Fig. 6. There appears to be very little mixing of freshwaters, formation waters and sabkha waters. The formation waters are isotopically much lighter than the freshwaters, which is consistent with a cooler (?) or wetter climate than today. Previous studies (^{14}C and tritium analyses) place these waters at between 8000 and 40,000 years

(Shampine *et al.*, 1978; Abusada, 1982). Dissolved noble gases in northern Oman groundwaters indicate that the average ground temperature during the late Pleistocene (15,000–24,000 BP) was 5° to 6 °C lower than that of today (Weyhenmeyer *et al.*, 2000). It is suggested here as a working hypothesis that the wetter climate record is probably related to the last one or two pluvial events (6–10.5 ka; 78–82 ka) in Arabia (Fleitmann *et al.*, 2003). It is presumed that these waters fell in interior Saudi Arabia onto the Dammam limestone aquifer and have been slowly penetrating and moving eastwards towards the Gulf, where they can be tapped by drilling today (Fig. 5). Their isotopic composition was probably around δ^{18}O -6‰ and δD -40‰. All pluvial events had a southern moisture source (Indian Ocean Monsoon rainfall).

Coastal Gulf water exhibits an evaporation trend from open ocean seawater (VSMOW). Sabkha samples taken from areas of occasional marine flooding (B9, B10, HZ) represent a higher concentration of D and ^{18}O over Gulf seawater due to evaporation (Fig. 6; Table 2). As the waters become concentrated in Ca, Mg, K and Na, the heavy isotope content of the residual brine increases. However, this evaporation has a limit and the trend reverses as the salinity increases (from B9 to HZ with a salinity of 160–180

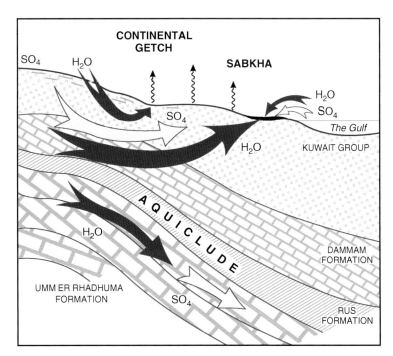

Fig. 5. Schematic stratigraphic cross-section of Kuwait showing the regional groundwater flow pattern and sources of water and sulphate.

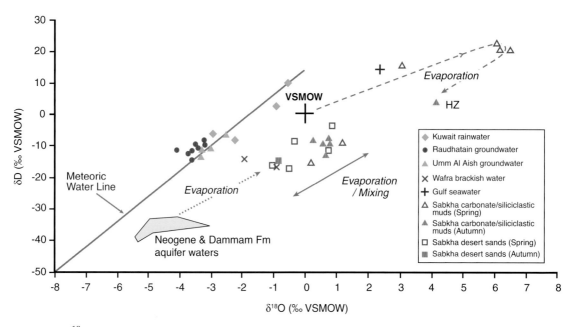

Fig. 6. A plot of $\delta^{18}O$ vs δD for sabkha and related groundwaters in Kuwait and Neogene and Dammam aquifer formation waters. The Kuwait meteoric water line is $\delta D = 8\,\delta^{18}O + 15$. The fine dashed line represents evaporation of formation water to produce Ca- and K-type sabkha waters (see Table 1), which may have undergone minor mixing(?) with HZ-type waters. The long, dashed and looped curve represents evaporation of seawater (VSMOW) and evaporated Gulf seawater. High-salinity samples (B9, B10, HZ) have their δD plotted as activities to compare to this curve. Other lower salinity samples are plotted as δD concentrations, as the difference between activity and concentration is within the error of measurement. See Table 2 for analytical data.

‰). The long dashed curve in Fig. 6 is taken from Pierre *et al.* (1984) and represents evaporation of marine brines from Ojo de Liebre, Baja California. The Al Khiran, Kuwait, marine sabkha samples B10, B9 and HZ lie exactly on this trend. The remainder of the samples taken along an E–W traverse (Fig. 2), were collected from pits in the marine sabkha (B and K samples) and continental zones (C1 to C7); these appear as a broad linear group lying between highly evaporated seawater and the Neogene and Dammam Formation aquifer waters (Fig. 6; Table 2). Sample sites C3 to C5 are located progressively farther from the sea, but the data still show an increase in the heavy isotopes (just as in HZ). This is easily explained without invoking seawater mixing: an evaporation line with a slope of 5 (Fig. 6) from the old formation waters shows that C-type samples can be derived by evaporation of these aquifer waters.

Figure 6 shows two evaporation trends: (i) Gulf seawater (similar to open-ocean water VSMOW), which through enhanced evaporation (e.g. sample B10) or due to capillary pumping in the carbonate muds, produces heavier isotopic compositions; (ii) evaporation of old formation waters through

the water table in desert sand areas and in marine sabkha zones, which are not subjected to marine flooding today. The latter represents the bulk of the waters draining into the marine sabkha zones (K, other B and continental zone samples). Samples B1 and B8 may represent a minor amount of mixing of the two types of waters. Samples C6 and C7 from Wafra area (over 50 km to the west of the coast) may represent aquifer water plus recharge from recent irrigation and rainfall. Similarly, for the Sabbiyah sabkha, samples S2 and S3 from the zone of marine flooding are similar in isotopic composition to B9 and suggest that they are a mixture of evaporated old formation water and evaporated seawater. Sample S1 is from the uppermost supratidal zone, which is beyond marine flooding and has a continental isotopic signature. The old formation waters show a marked isotopic shift from the present-day meteoric water line and may be closer to a line of $\delta D = 8\,\delta^{18}O + 8$. Consequently, variable evaporation of the old formation waters along a slope of 5 could produce the observed trends.

Data from the UAE marine sabkhas give a different story (Butler *et al.*, 1973; McKenzie *et al.*, 1980;

Patterson & Kinsman, 1981). In these sabkhas, which are open to unrestricted marine flooding, the data indicate a marine isotopic signature for a much wider area of the sabkha, except in the innermost zone of the marine sabkha where continental influences can be seen. The presence or absence of coastal barriers will eventually determine which types of water are dominant in a given sabkha. In Kuwait, old formation waters are dominant in the sabkhas, while marine-derived waters have dominated the system in the UAE. With continued progradation (at rates of 1–1.5 km kyr^{-1}) the marine-derived brines will eventually be displaced by continental derived waters. In fact, around Shuaiba, 25 km north of Al Khiran (Fig. 2), the groundwater salinities ~200 m inland from the shoreline were between 35–41‰ (less than the salinity of coastal waters there) in the marine sabkha, now completely cut off from Gulf seawaters by 4–8 m high cemented beach-dune ridges. Here the continental waters have already reached the coast.

Isotopic composition of the sulphate

Dissolved sulphate in groundwaters and sulphate minerals

The same samples represented in Fig. 6 were used for sulphate analyses (Fig. 7). Gulf seawater and brine samples from the Al Khiran sabkha (e.g. B9,

B10, HZ; Table 2) group around normal isotopic values for seawater sulphate. The lower δ^{34}S in HZ and B9 (by up to ~3‰) is due to mineral precipitation and/or sulphate reduction (Butler *et al.*, 1973). Sample B9 is from an area of highly reducing cynobacterial algal mats. All other samples have more positive δ^{18}O and more negative δ^{34}S values. This sulphate is of continental origin. The samples plotting close to the Rus Formation anhydrite (samples B1 and B8) could represent some mixing of evaporated seawater and evaporated old formation waters as discussed previously. Sulphate samples from the Neogene aquifers have δ^{34}S values of around 12 ± 1‰ and the Dammam aquifers 12 ± 4‰ (Shampine *et al.*, 1978). These values are a little lower than the Kuwait samples and will be discussed later.

A plot of δ^{18}O vs δ^{34}S in mineral sulphate is shown in Fig. 8. When sulphates precipitate from brines, it leads to gypsum with heavier δ^{34}S and δ^{18}O than the source waters (by ~ 3–4‰ and 1–2‰ respectively) due to selective fractionation of the precipitated sulphate (Krouse, 1987). This trend would lie on a line of slope 2. This is seen in the Baja California sabkhas (Pierre *et al.*, 1984). No clear trend is visible in the Kuwait samples, although a line of slope 2 could describe the freshwater and continental sabkha sulphate samples. The continental and marine sulphate fields are clearly defined. The anhydrite and bassanite show a wide spread in δ^{34}S values, although δ^{18}O is

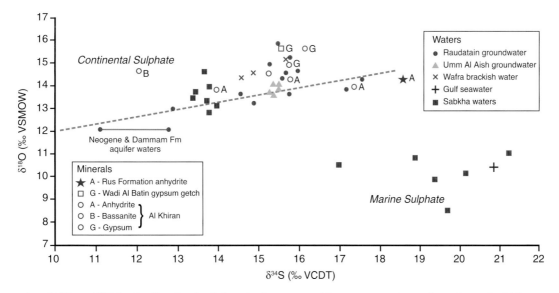

Fig. 7. A plot of δ^{18}O vs δ^{34}S for the dissolved sulphate in Neogene aquifer waters, marine and continental sabkha waters and Kuwait freshwaters (Raudhatain and Umm Al Aish groundwater). Note that only a negligible amount of mixing between marine and continental sulphate has taken place. See Table 2 for analytical data.

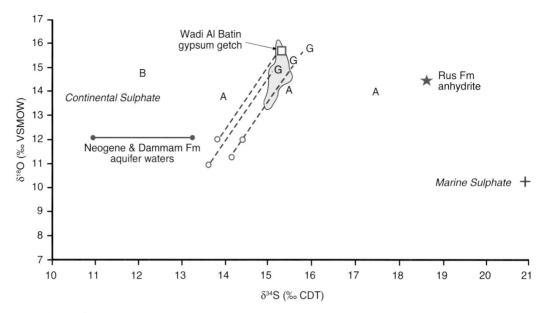

Fig. 8. Plot of $\delta^{18}O$ vs $\delta^{34}S$ to show the derivation of sulphate minerals in the sabkha mostly from continental groundwaters and freshwaters (yellow area). The dashed lines link the precipitated minerals to the calculated isotopic compositions of dissolved sulphate (circles) in the waters from which they were derived. The composition of Dammam and Neogene aquifer formation waters is also indicated. Note that most sulphate mineral samples plot in the continental groundwaters field. A, anhydrite; G, gypsum; B, bassanite ($CaSO_4 \cdot \frac{1}{2}H_2O$). See text for discussion.

comparable. This is partly due to the frequent mineral transformations that occur from gypsum to anhydrite to gypsum, and gypsum to bassanite to gypsum, and the role played by halophytes in these reactions (Gunatilaka *et al.*, 1980; Gunatilaka, 1990; Robinson *et al.*, 1991). Note that except for one sample (collected from the area of marine flooding in the supratidal zone), all sulphates plot in the *continental sulphate* field, and the isotopic compositions of waters from which they have precipitated are closely related to that of the Neogene aquifer waters with "lighter" sulphur isotopic compositions (Fig. 8).

Clearly, the majority of the mineral sulphate samples (>90%) in the marine sabkhas of Al Khiran were derived from a continental groundwater source. In contrast, the $\delta^{34}S$ values of groundwater and precipitated sulphate in the UAE marine sabkhas have a dominant marine source (Butler *et al.*, 1973), which again reflects the open sabkhas of the UAE coast where unrestricted flood-recharge and reflux occur. These differences are also responsible for the contrasting volumes of anhydrite and dolomite deposited in the respective areas. Marine flood-recharge and reflux in the UAE sabkhas keep the salinities high enough to form large volumes of dolomite and thick layers of diagenetic anhydrite after gypsum, whereas in Kuwait the absence of flood-recharge caused by a

barriered chenier-coastline, and the influx of dilute continental waters into the marine sabkha, keep salinities low and prevent the formation of large volumes of anhydrite or dolomite.

Diagenetic alteration of aragonite is the proposed mechanism of dolomitisation in the UAE sabkhas (Patterson & Kinsman, 1981). In Kuwait, this mechanism can be invoked only in the narrow intertidal to supratidal zones bordering the extremities of tidal creeks of Al Khiran where the required high salinities are reached (Gunatilaka, 1991). Further, in both Kuwait and UAE sabkhas, the volume of dolomite formed bears a distinct relationship to the distribution of cyanobacterial mats in the sabkha. Dolomite content decreases substantially away from the organogenic mat zones. In the hypersaline coastal Al Khiran lagoon (and also most likely in the small infilled lagoon sabkhas between the oolite beach ridges), direct precipitation of micrometre-sized dolomite crystals (generally <3–4%) has occurred under low-sulphate concentrations and reducing conditions of the interstitial waters (Gunatilaka *et al.*,1984, 1986b). These differences have been discussed in detail by Gunatilaka (1991) and Robinson & Gunatilaka (1991). The question also arises as to whether the flood-recharge/reflux mechanism provides a sufficient volume of fluid throughput to cause high volume dolomite formation. It has to be emphasized that the above data are only applicable in a Holocene context for high-frequency flooding events. For the longer term, low-frequency transgressive–regressive cycles (2nd and 3rd order parasequences) and basinal-scale mechanisms of fluid flow need to be invoked for dolomitisation.

Origin of the sulphate

A comprehensive review of the data on sulphate isotope studies of sabkhas from the southern Mediterranean to the Arabian Gulf was made by Robinson (1987). Data from Libya, Egypt, Kuwait and the UAE all show a seawater influence only in the outer sabkha areas, whereas on the landward side, the δ^{34}S of sabkha groundwater sulphate decreases to +13–16‰. In Kuwait, the continental gypsum "getch" shows δ^{34}S values of ~ +15.6 ‰ and slightly lower in other areas mentioned above. The lower Eocene Rus Formation anhydrite, being an aquiclude, prevents older water and sulphate from entering the sabkhas (Fig. 5). So, throughout the vast areas mentioned, there appear to be a pervasive and homogeneous sulphur source with δ^{34}S = +13–16‰ (average ~ +14‰) that supplies sulphate to the groundwaters. Gypsum "getch" (gypcrete) and crusts are exposed over vast areas in the continental zones of Arabia and Africa (Watson, 1979). It is proposed that particulate gypsum circulated in dust storms, derived by aeolian erosion of these gypcretes and crusts is the source of this sulphur. The origin of this continental sulphate is rock/soil sulphate plus some rainwater contribution, which may have become concentrated in the getch layers of the upper part of the Kuwait Group. Its final sulphur isotope composition would represent a mixture of airborne seawater sulphate plus continental oxidized H_2S, on top of which is imprinted a heavy isotope enrichment due to bacterial reduction. For comparison, the average deposition of sulphate in arid zones of the USSR is ~ 1.4 g S m^2 yr^{-1} over 3×10^6 km^2 (Zverev, 1968). In Kuwait, the average dust storm carries 3–7% gypsum (Al Bakri *et al.*, 1986).

It follows that sulphates in *marine* sabkhas can have several origins. They may form from marine, continental or mixed waters (Fig. 8). It is notable that the Eocene Rus anhydrite does not have a marine isotopic signature. Early supratidal gypsum and anhydrite (i.e. prior to formation of coastal barriers) may have a marine isotopic signature, while later crystals formed at the same location may have a continental signature as

progradation continued and with barrier restrictions developed (i.e. marine waters no longer reach the area). In this sense, it is incorrect to treat marine and continental sabkhas as separate entities. Rather, they are both parts of a continuum as groundwaters evolve with time, and continental groundwaters effectively displace marine-derived brines with continued sabkha progradation. In some areas, the continental waters have already reached the Gulf. This must be so for ancient analogues too.

DISCUSSION AND CONCLUSIONS

Considerable progress has been made during the last forty years in understanding the interactions of the various processes leading to the generation of Arabian sabkhas. The giant sabkhas of the geological record can be accounted for by extrapolating present-day rates of coastal progradation. A giant sabkha can be formed in <10^5 years. However, the tectonic, climatic, eustatic and hydrological frameworks around vast, shallow epicontinental seas that generated these sabkhas (e.g. the Jurassic Arab Formation) were quite different from those of the present day.

The UAE sabkhas are not representative of the entire 1000 km long Gulf coastline – indeed no single geographical area(s) exemplify the whole coastline. Significant variation in sedimentation styles exist from the extreme north (siliciclastic dominated) to the south (carbonate dominated). Coastal geomorphology (open vs barriered) in conjunction with wind vectors and connection to a continental groundwater source, play a primary role in the evolution of the hydrological framework of sabkhas and hence the volume and composition of the sulphate-facies minerals and dolomite. In Kuwait, the continental groundwater system has virtually completely taken over the hydrological framework in the marine sabkhas. This is, perhaps, partly due to the steeper inland to coast topographical and groundwater table gradients compared to the southern Gulf. Some workers have reported springs issuing from the shallow sea bottom of the Gulf Basin.

The stable-isotope composition of diagenetic sulphate evaporite minerals and the groundwater brines from which they are derived are a reasonably accurate fingerprint of the above controls on sabkha evolution. The modern sabkha is more than a geomorphological surface. It is a highly dynamic environment where geomorphological, hydrological, biotic, eustatic and climatic factors have interacted over a short time-scale to produce the sabkha facies, whose preservation potential is questionable, especially in the context of Quaternary orbital cycles with their attendant climatic and sea-level changes. Nonetheless, the sabkha facies has been preserved in the geological record since early Archaean times. The 4th- and 5th- order Holocene transgressive–regressive cycle is only a guide for reconstructing ancient sabkha analogues. Extrapolation to higher order cycles and parasequences must be undertaken with caution. This in no way diminishes its relevance in environmental reconstructions. In the context of sequence stratigraphy, the arguments centre on problems of scale of the parasequences.

The chronology of the monsoon-controlled Quaternary climatic framework and pluvial events of the region became available only during the last three to four years. The sea-level and climate chronologies have greatly improved the stratigraphic history of the Holocene sabkha complexes (4th- to 5th-order, highstand, regressive interglacial deposits). The onset of glacial–interglacial cycles resulted in the draining of the shallow Arabian (Persian) Gulf every 100,000 years during the last 0.8 Myr; prior to this when obliquity cycles (ca. 41 kyr duration) dominated climate and sea-level changes (Shackleton, 1987), at least partial draining of the Gulf must have occurred. The resulting drops in base levels and water tables have impacted on the preservation potential of sabkha environments and account for the general scarcity of Pleistocene sabkha deposits. Sabkhas are equilibrium geomorphological surfaces whose levels are maintained by the depth of the groundwater table. A drop of a few metres will result in deflation.

Uniquely in the Arabian Gulf region, the sabkha progradational phase (last ~5000 years) coincided with the onset of extreme aridity during the mid-Holocene, which has lasted until the present. This is due to the ITCZ and the associated monsoon rainfall belt moving south, away from the Arabian Peninsula. Coincidentally, several major and successful civilizations (e.g. the Uruk, Sumer and Akkadian from 5500–4000 BP) reached a peak in their development and collapsed during this mid-Holocene climatic deterioration, which in-turn set off a chain of disastrous political, economic and demographic changes resulting in much instability in the region.

ACKNOWLEDGEMENTS

I am honoured to have been invited to make a contribution to this volume and to the memory of Prof. D.J. Shearman, the doyen of evaporite sedimentologists, who had been a colleague and friend for over thirty years. Thanks are due to Professor Graham Evans (formerly of Imperial College, London) who over the years constantly encouraged the author to "put it all together" and write up the story of the Arabian Gulf sabkhas; to Dr. D. Harkness of NERC Radiocarbon Laboratory, East Kilbride, Scotland, UK for the diligent work on ^{14}C dating. The stable-isotope analyses were carried out at the Institute of Nuclear Sciences, DSIR, Lower Hutt, New Zealand. To the editors of this volume, Professors C.G.St.C. Kendall and A. Al Sharhan for their patience and unenviable task of making this volume a reality. Finally, I thank Prof. Andrew Goudie of Oxford University, one of the reviewers, for his suggestions on improving the manuscript.

REFERENCES

Abusada, S.M. (1982) Interpretation of environmental isotope data of Kuwait groundwater. *J. Gulf Arab. Penin. Stud.*, **2**, 80–92.

Al Bakri, D., Khalaf, F. and **Al Ghadban, A.** (1986) Mineralogy, genesis and sources of surficial sediments in the Kuwait marine environment, northern Arabian Gulf. *J. Sed. Petrol.*, **54**, 1266–1279.

Algaze, C. (1993) *The Uruk World System: Dynamics of Expansion of Early Mesopotamian Civilization*. University of Chicago Press, Chicago.

Al Zamel, A. (1983) *The Geology and Oceanography of Recent Sediments of Jazirat Bubiyan and Ras As-Sabbiyah, Kuwait*. Unpubl. PhD thesis, University of Sheffield, UK.

Barth, H-J. and **Boer, B.** (Eds) (2002) *Sabkha Ecosystems, 1 – The Sabkhas of the Arabian Peninsula and Adjacent Countries*. Kluwer Academic Publishers, Dordrecht, 354 p.

Burns, S.J., Fleitmann, D., Matter, A., Neff, U. and **Mangini, A.** (2001) Speleothem evidence from Oman for continental pluvial events during interglacial periods. *Geology*, **29**, 623–626.

Bush, P.R. (1973) Some aspects of the diagenetic history of the sabkha in Abu Dhabi, Persian Gulf. In: *The Persian Gulf: Holocene Carbonate Sedimentation and Diagenesis in a Shallow Epicontinental Sea* (Ed. B.H. Purser pp. 395–407. Springer Verlag, Berlin.

Butler, G., Kendall, C.G., Kinsman, D.J.J., Shearman, D.J. and **Skipwith, P.A.** (1965) Recent evaporite deposits along the Trucial Coasts of the Arabian Gulf. *Proc. Geol. Soc. London*, **1623**, 91 pp.

Butler, G.P. (1969) Modern evaporite deposition and geochemistry of coexisting brines, the sabkha, Trucial Coast, Arabian Gulf. *J. Sed. Petrol.*, **39**, 70–89.

Butler, G.P. (1970) Holocene gypsum and anhydrite of the Abu Dhabi sabkha, Trucial Coast: An alternative explanation of origin. In: *Third Symposium on Salt* (Eds J.L. Rau and L.F. Dellwig) pp. 120–152. *North Ohio Geol. Soc.* Cleveland, Ohio.

Butler, G.P., Krouse, R.H. and Mitchell, R. (1973) Sulphur isotope geochemistry of an arid, supra-tidal evaporite environment, Trucial Coast. In: *The Persian Gulf: Holocene Carbonate Sedimentation and Diagenesis in a Shallow Epicontinental Sea* (Ed. B.H. Purser) pp. 453–462. *Springer-Verlag*, Berlin.

Butler, G.P., Harris, M. and Kendall, C.G.St.C. (1982) Recent evaporites from the Abu Dhabi coastal flats. In: Depositional and diagenetic spectra of evaporites (Eds C.R. Handford, R.G. Loucks and G.R. Davies) pp. 33–64. SEPM Core Workshop 3, Calgary, Alberta.

Cherif, O.H., Al Rifaiy, I.A. and Al Zamel, A. (1993) Sedimentary facies of tidal creeks of Khor Al Mufateh and Khor Al Mamlaha, Khiran, *Kuwait. J. Univ. Kuwait*, 21, 87–108.

Crawford, H. (1991) *Sumer and the Sumerians.* Cambridge University Press. New York.

Cullen, H.M., deMenocal, P.B., Hemming, S., Hemming, G., Brown, F.H., Guilderson, T. and Sirocko, F. (2000) Climate change and the collapse of the Akkadian Empire. Evidence from the deep sea. *Geology*, 28, 379–382.

Curtis, R., Evans, G., Kinsman, D.J.J. and Shearman, D.J. (1963) Association of dolomite and anhydrite in the Recent sediments of the Persian Gulf. *Nature*, 197, 679–680.

DeGroot, K. (1973) Geochemistry of tidal flat brines at Umm Said, S.E. Qatar, Persian Gulf. In: *The Persian-Gulf: Holocene Carbonate Sedimentation and Diagenesis in a Shallow Epicontinental Sea* (Ed. B.H. Purser) pp. 377–394. *Springer-Verlag*, Heidelberg and Berlin.

Evans, G. (1966) The recent sedimentary facies of the Persian Gulf region. *Phil. Trans. Roy. Soc. London, Ser. A*, 259, 291–298.

Evans, G. (1995) The Arabian Gulf: a modern carbonate evaporite factory – a review. *Cuad. Geol. Ibérica*, 19, 61–96.

Evans, G. and Shearman, D.J. (1964) Recent celestite from the sediments of the Trucial Coast of the Persian Gulf. *Nature*, 202, 759–761.

Evans, G., Kendall, C.G. St.C. and Skipwith, P. (1964) Origin of the coastal flats, the sabkha of the Trucial Coast, Persian Gulf. *Nature*, 202, 759–761.

Evans, G., Schimdt, V., Bush, P. and Nelson, H. (1969) Stratigraphy and geologic history of the sabkha, Abu Dhabi, Persian Gulf: *Sedimentology*, 12, 145–159.

Evans, G., Murray, J.W., Biggs, H.E., Bate, R. and Bush, P.R. (1973) The oceanography, ecology and sedimentology and geomorphology of parts of the Trucial Coast barrier island complex, Persian Gulf. In: *The Persian-Gulf: Holocene Carbonate Sedimentation and Diagenesis in a Shallow Epicontinental Sea* (Ed. B.H.Purser) pp. 233–278. Springer-Verlag, New York.

Fairbanks, R.G. (1989) A 17,000 year glacio-eustatic sea level record: influence of glacial melting rates on the Younger Dryas event and deep ocean circulation. *Nature*, 342, 637–642.

Fleitmann, D., Burns, S.J., Neff, U., Mangini, A. and Matter, A. (2003) Changing moisture sources over the last

330,000 years in Northern Oman from fluid inclusion evidence in speleothems. *Quatern. Res.*, 60, 223–232.

Gasse, F. and van Campo, E. (1994) Abrupt post-glacial climate events in West Asia and North Africa monsoon domains. *Earth Planet. Sci. Lett.*, 126, 435–466.

Gunatilaka, A. (1984) A new carbonate-evaporite province from Al Khiran, Kuwait, Arabian Gulf. Paper presented at the 7th Meeting of Carbonate Sedimentologists, Liverpool, UK. (Abstract).

Gunatilaka, A. (1986a) Kuwait and the northern Arabian Gulf – A study in Quaternary sedimentation. *Episodes*, 4, 223–231.

Gunatilaka, A. (1986b) The Quaternary Carbonate-Evaporite Complex in Al Khiran, Southern Kuwait. Guidebook for International Conference and Field Excursions on Quaternary Sedimentation in the Northern Arabian Gulf, Kuwait, 7th–13th February 1986. 23 pp.

Gunatilaka, A. (1987) The "Dolomite Problem" in the light of recent studies. *Modern Geol.*, 11, 311–324.

Gunatilaka, A. (1989a) Spheroidal Dolomites – Origin by hydrocarbon seepage? *Sedimentology*, 36, 701–710.

Gunatilaka, A. (1989b) Low latitude anhydrite from the Red Sea coast – palaeoclimatic and palaeohydrological implications. *Terra Nova*, 1, 280–283.

Gunatilaka, A. (1990) Anhydrite diagenesis in vegetated sabkhas, southern Kuwait, Arabian Gulf. *Sed. Geol.*, 69, 95–116.

Gunatilaka, A. (1991) Dolomite formation in coastal Al Khiran, Kuwait, Arabian Gulf – A reevaluation of the sabkha model. *Sed. Geol.*, 72, 35–53.

Gunatilaka, A. (2000) Sea levels as historical time-markers in prehistoric studies. *J. Roy. Asia. Soc. (Sri Lanka Branch), New Ser.*, XLV, 19–34.

Gunatilaka, A. and Mwango, S.B. (1987) Continental sabkha pans and associated nebkhas in Southern Kuwait, Arabian Gulf. In: *Desert Sediments – Ancient and Modern.* (Eds L. Frostick, and I. Reid) J. Geol. Soc. London Spec. Publ,. 35, 187–203.

Gunatilaka, A. and Mwango, S.B. (1988) Flow separation and the internal structure of shadow dunes. *Sed. Geol.*, 1/2, 125–134.

Gunatilaka, A. and Shearman, D.J. (1988) Gypsum-carbonate laminites in a recent sabkha, Kuwait. *Carbonates Evaporites*, 3, 67–73.

Gunatilaka, A., Saleh, A. and Al Temeemi, A. (1980) Plant controlled supratidal anhydrite from Al Khiran, Kuwait, Arabian Gulf. *Nature*, 288, 257–260.

Gunatilaka, A., Saleh, A., Al Temeemi, A. and Nassar, N. (1984) Occurrence of subtidal dolomite in a hypersaline lagoon, Kuwait, Arabian Gulf. *Nature*, 311, 450–452.

Gunatilaka, A., Saleh, A.A., Al Temeemi, A. and Nassar, N. (1985a) A new occurrence of bassanite from evaporitic environments in Kuwait. *Kuwait Univ. J. Sci.*, 12, 159–168.

Gunatilaka, A., Saleh, A. and Al Temeemi, A. (1985b) Sulphate reduction and dolomitization in a Holocene lagoon, Kuwait. Paper read at SEPM Meeting, Colorado School of Mines, Golden, Colorado, USA. Abstracts.

Gunatilaka, A., Al Zamel, A., Shearman, D.J. and Reda, A. (1987a) A spherulitic fabric in selectively dolomitized siliciclastic crustacean burrows, northern Kuwait. *J. Sed. Petrol.*, 57, 922–927.

Gunatilaka, A., Saleh, A., Al Temeemi, A. and **Nassar, N.** (1987b) Ca-poor dolomites from the sabkhas of Kuwait, Arabian Gulf. *Sedimentology*, **34**, 999–1006.

Hasternath, S. (1985) *Climate and Circulation of the Tropics. Reidel*, Boston, Mass., 433 pp.

Hsu, K.J. and **Schneider, J.** (1973) Progress report on dolomitization and hydrology of the Abu Dhabi sabkhas, Arabian Gulf. In: *The Persian-Gulf: Holocene Carbonate Sedimentation and Diagenesis in a Shallow Epicontinental Sea* (Ed. B.H.Purser) pp. 409–422. Springer-Verlag, Berlin.

Illing, L.V., Wells, A.J. and **Taylor, J.C.M.** (1965) Penecontemporaneous dolomite in the Persian Gulf. In: Dolomitization and limestone genesis: a Symposium. (Eds L.C. Pray and P.C. Murray) *SEPM Spec. Publ.*, **13**, 89–111.

Kendall, C.G.St.C. and **Skipwith, P.A. d'E.** (1968). Recent algal mats of a Persian Gulf lagoon. *J. Sed. Petrol.*, **38**, 1040–1058.

Kendall, C.G.St.C. and **Skipwith, P.A. d'E.** (1969). Geomorphology of a recent shallow water carbonate province, Khor Al Bazam, Trucial Coast, southwest Persian Gulf. *Geol. Soc. Am. Bull.*, **8**, 865–891.

Kinsman, D.J.J. (1964) Reef coral tolerance of high temperatures and salinities. *Nature*, **202**, 1280–1282.

Kinsman, D.J.J. (1966) Gypsum and anhydrite of Recent age, Trucial Coast, Persian Gulf. In: *Second Symposium on Salt.* North. Ohio. Geol. Soc., **1**, 302–326.

Kinsman, D.J.J. (1969a) Modes of formation, sedimentary associations and diagenetic features of shallow water and supratidal evaporates. *AAPG Bull.*, **53**, 830–840.

Kinsman, D.J.J. (1969b) Interpretation Sr^{++} concentrations in carbonate minerals and rocks. *J. Sed. Petrol.*, **39**, 486–508.

Kinsman, D.J.J. (1974) Calcium sulphate minerals of evaporite deposits: their primary mineralogy. *Fourth Symposium on Salt.* North. Ohio Geol. Soc., **3**, 343–348.

Krouse, H.R. (1987) Relationships between the sulphur and oxygen isotopic composition of dissolved sulphate. In: *Studies of Sulphur Isotope Variations in Nature*, pp. 19–29. IAEA, Vienna.

Lambeck, K. (1996) Shoreline reconstructions for the Persian Gulf since the last glacial maximum. *Earth Planet. Sci. Lett.*, **142**, 43–57.

Larsen, C.E. and **Evans, G.** (1978) The Holocene geological history of the Euphrates-Tigris-Karun delta. In: *The Environmental History of the Near and Middle-East since the last Ice Age.* Academic Press, London. (Ed. W.C.Brice) 384 pp.

McClure, H. A. (1976) Radiocarbon chronology of late Quaternary lakes in the Arabian Peninsula. *Nature*, **263**, 755–756.

McKenzie, J.A., Hsu, K.J. and **Schneider, J.F.** (1980) Movement of subsurface waters under the sabkha, Abu Dhabi, UAE and its relation to evaporite dolomite genesis. In: Concepts and Models of Dolomitization. Dolomitization (Eds D.H. Zenger, J. B. Dunham and R.L. Ethington) *SEPM Spec. Publ.*, **28**, 11–30.

Mwango, S.B. (1986) *Sedimentological Studies of Aeolian Terrains in Southern Kuwait.* Unpubl. MSc thesis, Kuwait University. 168 pp.

Patterson, R.J. and **Kinsman, D.J.J.** (1977) Marine and continental groundwater sources in a Persian Gulf coastal sabkha. In: Reefs and Related Carbonates - Ecology and Sedimentology. *AAPG Stud. Geol.*, **4**, 381–397.

Patterson, R.J. and **Kinsman, D.J.J.** (1981) Hydrologic framework of a sabkha along Arabian Gulf. *AAPG Bull.*, **65**, 1457–1476.

Patterson, R.J. and **Kinsman, D.J.J.** (1982) Formation of diagenetic dolomite in a coastal sabkha along Arabian (Persian Gulf). *AAPG Bull.*, **66**, 28–43.

Picha, F. (1978) Depositional and diagenetic history of Pleistocene and Holocene oolitic sediments and sabkhas in Kuwait, Persian Gulf. *Sedimentology*, **25**, 427–450.

Pierre, C., Ortlieb, L. and **Person, A.** (1984). Supratidal evaporitic dolomite at Ojo de Liebre lagoon: Mineralogy and isotopic arguments for primary crystallization. *J. Sed. Petrol.*, **54**, 1049–1061.

Preusser, F., Radies, D. and **Matter, A.** (2002) A 160,000 year record of dune development and atmospheric circulation in southern Arabia. *Science*, **296**, 2018–2020.

Purser, B.H. (1973) *The Persian Gulf: Holocene Carbonate Sedimentation and Diagenesis in a Shallow, Epicontinental Sea. Springer-Verlag*, Heidelberg and Berlin. 473 pp.

Reda, A.A. (1986) A Reconnaissance Study of a Clastic Coastal Sabkha, Bahra Area, Northern Kuwait. Unpubl. MSc thesis, Kuwait University. 245 pp.

Robinson, B. (1987) Sulphur and oxygen isotopic compositions of groundwater and sabkha sulphate in the Middle East. In: *Sulphur Isotopic Variations in Nature*, pp. 77–83. IAEA Vienna.

Robinson, B. and **Gunatilaka, A.** (1991) Stable isotope studies and the hydrological regime of sabkhas in southern Kuwait, Arabian Gulf. *Sed. Geol.*, **73**, 141–159.

Robinson, B., Stewart, M.K. and **Gunatilaka, A.** (1983) Sulphate and water stable isotopes and the origin of ground and sabkha waters in Kuwait. *Proc Fourth Internat. Symp. Water-Rock Interaction, Misasa, Japan* (extended abstracts).

Robinson, B., Stewart, M.K. and **Gunatilaka, A.** (1991). Use of sulphur and other stable isotopes in environmental studies of regional groundwater flow and sulphate mineral formation in Kuwait. In: *Use of Isotopes in Sulphur-Cycle Studies. Stable Isotopes: Scope 43*, (Eds H.K Rouse and V. Grienenko) pp. 361–371. John Wiley & Sons, New York, 431 pp.

Rodgers, D.W. and **Gunatilaka, A.** (2002) Bajada formation by monsoonal erosion of a subaerial forebulge, Sultanate of Oman. *Sed. Geol.*, **154**, 127–146.

Rossignol-Strick, M., Paterne, M., Bassinot, F.C., Emels, K.C. and **De Lange, G.J.** (1998) An unusual mid-Pleistocene monsoon period over Africa and Asia. *Nature*, **392**, 269–272.

Saleh, A.A. (1978) *Recent Algal mats and Quaternary Carbonate Sediments of Al Khiran, Kuwait.* Unpubl. PhD thesis, Imperial College, University of London., 242 pp.

Saleh, A., Al Ruwaih, F., Reda, A. and **Gunatilaka, A.** (1999) A reconnaissance study of a clastic coastal sabkha in northern Kuwait, Arabian Gulf. *J. Arid Environ.*, **4**, 1–19.

Sanlaville, P. and **Paskoff, R.** (1986) Shoreline changes in Bahrain since the beginning of human occupation. In: *Bahrain through the Ages.* KBI Bahrain.

Shackleton, N. (1987) Oxygen isotopes, ice volume and sea level. *Quatern. Sci. Rev.*, **6**, 183–190.

Shampine, W.J., Dincer, T. and **Noory, M.** (1978) An evaluation of isotope concentrations in the groundwater of Saudi Arabia. *Isotope Hydrol.*, **2**, 443–463.

Shearman, D.J. (1963). Recent anhydrite, gypsum, dolomite and halite from the coastal flats of the Arabian shore of the Persian Gulf. *Proc. Geol. Assoc. London*, **1607**, 63–65.

Shearman, D.J. (1966) Origin of marine evaporates by diagenesis. *Trans. Inst. Min. Metall. B.*, **75**, 208–215.

Shearman, D.J. (1971) Marine Evaporites: the calcium sulphate facies. *AAPG Seminar, University of Calgary,* Canada, 65 pp.

Shearman, D.J. (1978) Evaporites of coastal sabkhas. In: *Marine Evaporites* (Eds W.E. Dean and B.C. Schreiber) pp. 6–42. *SEPM Short Course* **4**.

Shearman, D.J. (1985) Syndepositional and late diagenetic alteration of primary gypsum. In: *Sixth Salt Symposium.* (Ed. B.C. Schreiber) North. Ohio Geol. Soc. Cleveland, OH.

Sirocko, F., Sarnthein, M., Erlenkeusser, H., Lange, H., Arnold, M. and **Duplessy, J.C.** (1993) Century scale events in monsoonal climate over the past 24,000 years. *Nature*, **364**, 322–324.

Watson, A. (1979) Gypsum crusts in deserts. *J. Arid. Environ.*, **2**, 3–20.

Weiss, H., Courty, M-A, Wetterstrom, W., Guichard, F., Senior, L., Meadow, R. and **Curnow, A.** (1993) The genesis and collapse of the 3rd millennium north Mesopotamian civilization. *Science*, **26**, 995–1003.

Wells, A.J. (1962) Recent dolomite in the Persian Gulf. *Nature*, **194**, 274–275.

Weyhenmeyer, C. E., Burns, S.J., Waber, H., Aeschbach-Hertig, W., Kipfer, R., Beyerle, R., Loosli, H. and **Matter, A.** (2000) Cool glacial temperatures recorded by noble gases in a groundwater study from northern Oman. *Science*, **287**, 842–845.

Wood, W.W., Sandford, W.E. and **Al Habschi, A.R.S.** (2003) The source of solutes in the coastal sabkha of Abu Dhabi. *Geol. Soc. Am. Bull*, **114**, 259–268.

Zverev, V.P. (1968) Role of precipitation in the cycling of chemical elements between the atmosphere and lithosphere. *Dokl. Akad. Nauk. SSR*, **181**, 716–719.

Int. Assoc. Sedimentol. Spec. Publ. (2011) **43**, 133–148

Interplay between Holocene sedimentation and diagenesis, and implications for hydrocarbon exploitation: return to the sabkha of Ras Umm Said, Qatar

EUGENE A. SHINN

College of Marine Science, University of South Florida, St. Petersburg, FL 33701, USA
(e-mail: eshinn@marine.usf.edu)

ABSTRACT

Carbonate sand beaches and spits nurtured by longshore currents in the Arabian Gulf commonly form sheltered areas favourable for evaporite and carbonate sedimentation. Promontories and embayments created by pre-Holocene topography favour such spit formation, especially in areas where contours are perpendicular to prevailing winds and longshore currents. Ras Umm Said on the NE coast of the Qatar Peninsula provides a useful model for this style of sedimentation. In the mid-1960s, three 12 km long chenier-like Holocene beaches had developed en echelon at Ras Umm Said. Each chenier had several landward-projected carbonate sand bodies that originated as terminal fish-hook-shaped spits during southward migration. The flat-lying channelized sabkha and intertidal areas that formed landward of, and between, the beaches provided low-energy areas suitable for fine-grained carbonate sedimentation and local evaporite deposition. These combined cheniers and channelized sabkhas have continued to grow at a remarkable rate over the last 40 years. Each chenier, with its recurved spits, reaches a stable phase and is then protected from wave action when a new chenier forms on its seaward side.

The chenier beach and recurved spit create conditions favourable for rapid beachrock formation on their low-energy landward side; cementation is apparent within 1 year. The narrow beachrock then becomes exposed as linear ridges on the seaward side of the cheniers as they are eroded and migrate landward, creating a characteristic set of beach laminations. Wave erosion of these exposed rock ridges creates locally abundant lithoclasts that are incorporated into the beach as it transgresses landward. Storm tides and currents break through the curved landward-directed spits allowing flooding and extension of tidal channels into the sabkha and intertidal zones between the cheniers. Periodically, the entire beach/sabkha complex "jumps" and accretes seaward over offshore areas of submarine-cemented sediment. A new "jump" is beginning and growing today.

The bored submarine-cemented layers (hardgrounds) that become buried under the intertidal and sabkha sediments often constrain the depth to which later tidal channels may erode. Such bored surfaces could easily be misidentified as flooding surfaces during sequence stratigraphic analysis. The cemented layers also serve as impermeable seals, preventing or retarding vertical movement of fluids; such early formation of aquitards may explain the distribution of fluids (water or oil) in ancient analogue settings. Creation of mainly low-angle landward-dipping beach bedding, overlying a basal, more steeply dipping, set of landward-dipping beds, is indicative of the migrating chenier sedimentary model. The spits will eventually seal off the lagoon, which will then evolve into a large sabkha possibly composed of wind-blown sand. Chenier migration and marine cementation have considerable significance for coastal development and civil engineers projects in the region.

Keywords: Diagenesis, carbonate cheniers, sabkha, evaporites, beachrock, submarine cementation, hardground, channel migration, sequence stratigraphy, Arabian Gulf, Qatar

INTRODUCTION

The purpose of this review paper is to highlight some relationships between diagenesis, specifically submarine and intertidal cementation, and sedimentary processes, with particular emphasis on tidal-channel downcutting, lateral beach migration, and sequence stratigraphy. These processes may have important implications for the formation of evaporitic minerals and the interpretation of palaeoenvironments. This review draws mainly on earlier research on the sabkhas, intertidal flats, and tidal channels protected by chenier beaches near Ras Umm Said (Fig. 1), in the northeast corner of the Qatar Peninsula (Shinn, 1973). This early work, beginning in 1965, was based mainly on coring, trenching, and aerial photography. Since 1967, the study area has been revisited on four occasions (1983, 1987, 2005, 2006). During this 41-year period, new information has been obtained

with new geological implications (e.g. sequence stratigraphy, a relatively new concept at the time (Vail & Mitchum, 1977), had not come into vogue during the early stages of this research), and this has modified and extended the significance of some of the earlier observations.

Modern carbonate deposition in arid environments was the primary research focus of the Shell Group working under the direction of Bruce Purser from 1964 to the early 1970s. While the Shell Group concentrated mainly on the sedimentation of the entire Arabian Gulf and smaller, localized areas around the Qatar Peninsula, the Shearman and Evans group from Imperial College London concentrated mainly on the widespread sabkhas of Abu Dhabi (Kendall & Skipworth, 1968, 1969). In the mid- to late 1960s, the geological profession first began to appreciate the presence and importance of marine cementation, a process overshadowed by the vadose and phreatic freshwater cementation paradigm espoused by Dunham (1969). The initial discovery of widespread submarine cementation was made in the Arabian Gulf during the mid-1960s (Shinn, 1969). Earlier, Sugden (1963) had recognized submarine cementation over a limited area off the west coast of Qatar, but its wide extent was initially proprietary information within the Shell Group. Nonetheless, Milliman (1966) described marine lithification of fine-grained carbonate sediment on deep-water guyots in the Atlantic, Gevirtz & Friedman (1966) recovered thin aragonite-cemented crusts from deep water in the Red Sea, and Ginsburg *et al.* (1971) described cementation within the so-called boiler reefs of Bermuda. Over the next 20 years, many areas of cementation were documented.

Due to its paradigm-breaking and controversial nature, the discovery phase (within the Shell Group) of marine early cementation was directed mainly toward verifying its existence and distribution, together with the development of petrographic and isotopic criteria for its recognition. It was soon realized that many of the petrographic features observed were similar to those presumed to be indicative of vadose cementation. The next phase was to determine whether this marine cementation had been a viable process throughout geological time and, more importantly, did it have implications for petroleum exploration. Notably, Pray (1965) had already deduced the presence of early submarine cementation in Mississippian bioherms, and Longman (1980) later catalogued

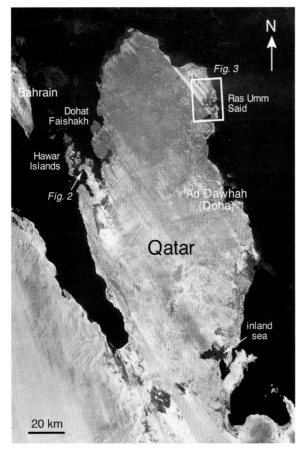

Fig. 1. Satellite image of the Qatar Peninsula showing the location of the Ras Umm Said study area.

petrographic criteria for its recognition. By the 1970s, the questions of how, when, and where, had been mostly determined, but the detailed chemistry remained a mystery. Even without a chemical explanation (initially, some chemists denied that marine cementation existed), the process of marine cementation/lithification could be seen to be constrained to certain geological settings recognizable by sedimentary structures and their sequence characteristics.

As exemplified by the present-day Arabian Gulf, high salinity and evaporation accelerate lithification, while at the same time the process is controlled by rates of sedimentation. In general, pulses of rapid sedimentation, probably caused by storms, followed by variable periods of exposure to seawater and tidal pumping, apparently constrain marine cementation in the Arabian Gulf. The combination of sedimentary processes and diagenesis invariably leads to the creation of hard, relatively impermeable seafloor crusts (hardgrounds). Crusts may become stacked or remain as thin layers sandwiched between non-cemented sediment (Fig. 2). Individual crusts may become bored and/or corroded, or they may blend with overlying uncemented sediment if they retain a sediment veneer during cementation. Given sufficient time (long periods of time with stable sea levels, not available during the Holocene), the processes might potentially produce crusts of unlimited thickness that would be massive limestone with no particular bedding features.

Since the early studies of the Arabian Gulf, the process of submarine cementation has been recognized in many reef settings, especially near shelf margins and breaks in slope where impinging oceanic waves and tidal pumping are thought to force huge volumes of seawater (10,000 to 50,000 pore volumes; Dunham, 1969) through an initially porous framework, thus driving cementation. This diagenetic process produces surfaces (hardgrounds) suitable for hardbottom communities

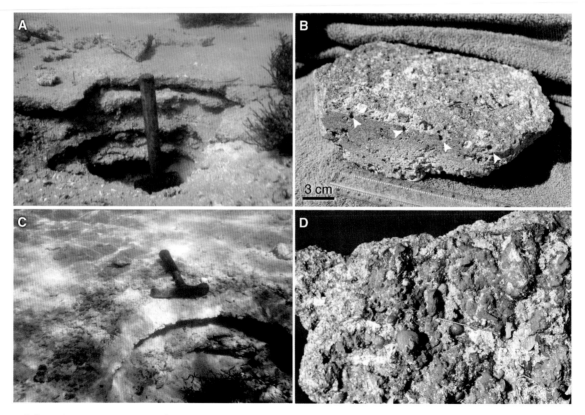

Fig. 2. Submarine cementation on the western Qatar Peninsula. (A) Underwater photograph of a small excavation showing four separate submarine-cemented (hardground) grainstone layers. Hammer shaft is 36 cm long. (B) Detail of upper layer from (A) that has been sawed to reveal pholad borings (arrowed) in its flat surface. (C) View of an excavation in layers that is actively forming. Hammer head is 17.5 cm long. (D) Detail of the upper surface of the actively forming submarine-cemented layer showing fresh carbonate grains being cemented in the brown actively precipitating aragonite layer. Pin is 2.5 cm long.

that can thus completely change the texture and fabric of the resulting accumulation. The process of marine cementation (diagenesis) can control the distribution of fossils (biofacies) that require hard bottoms including the distribution of reef-building corals. In addition, implications for sedimentation rates and sea level can be deduced from the recognition of the marine-cemented surfaces.

STUDY AREA

A series of carbonate chenier beaches nourished by southward-moving longshore currents characterizes the Ras Umm Said area south of Ras Laffan on the northeast corner of the Qatar Peninsula (Fig. 3; Shinn, 1973). At Ras Umm Said (called Ras Umm Sa by Shinn, 1973) the periodic generation of chenier beaches during the later part of the Holocene transgression has produced a series of beaches and hook-shaped spits (Fig. 4). Growth of these cheniers has smoothed and partially blocked off an irregular indentation in the pre-Holocene shoreline. The process is ongoing, and fishing harbours south of Ras Umm Said at Dakhirah and Al Khawr will eventually be blocked and converted to sabkhas if present processes are left to continue.

Each chenier with its hook-shaped spits creates a protected low-energy lagoonal area suitable for the accumulation of fine-grained sediment (Shinn, 1973). Accumulation of fine-grained sediments and periodic flooding by sediment-laden water have led to the classic succession of sediments found in all tidal-flat areas. Facies range upward from grey, reduced, H_2S-rich, subtidal lime mud to oxidized, cream-coloured, intertidal sediment (that may or may not include cyanobacterial mats) overlain by oxidized and evaporitic supratidal/sabkha sediments (Illing & Wells, 1965; Kendall & Skipwith, 1969; Shinn, 1969, 1983). Tidal channels lined with coarser-grained sediments are interlaced through the generally fine-grained sequence. In the tidal channels, coarser particles are often mixed black and white "salt-and-pepper" grains (Shinn, 1983). A relatively stable tidally influenced water table combined with wind deflation of dry sediment above the capillary zone maintain a smooth and level sedimentary surface. The deflation process is a characteristic of arid-climate sabkhas the world over. By contrast, on humid-climate tidal flats the upper surface is controlled more by the height of storm tides that deliver sediments from adjacent marine areas (Shinn *et al.*, 1969).

At Ras Umm Said, the sabkha sequences have been repeatedly created with development of new cheniers on the eastern seaward side of the sedimentary complex. Old aerial photographs and the satellite image shown in Fig. 4 demonstrate that the first chenier to form created the largest of the sabkha sequences. Since the formation of the first chenier, the complex association of cheniers and sabkhas has continued to accrete seaward and southward. This process has proceeded sporadically and invariably leads to accretion of the entire sedimentary complex over extensive submarine-cemented layers that are forming farther offshore.

An offshore cemented layer east of the study area extends seaward many kilometres and out to depths as great as 25 m. This older continuously forming rock is, for the most part, the hard substrate that defines the famous Arabian Gulf pearl beds (Fig. 5). Similar, but much thinner and younger, layers are forming locally within some Qatar lagoons (Fig. 2). However, the fine-grained highly mobile and less permeable subtidal sediment within the Ras Umm Said lagoon tends to retard cementation. The distribution of cemented layers along the eastern Qatar Peninsula is sporadic and tends to be associated with tidal channels and/or sand banks where bottom sediment is coarser. Trenching through abandoned infilled tidal channels on the Ras Umm Said sabkha has demonstrated this association. The formation of cemented layers associated with the early Holocene marine transgression has recently been found in construction excavations along the western side of Doha Bay farther south (D. Puls, pers. comm., 2006).

In other, much larger sabkha areas of the Arabian Gulf, especially around Abu Dhabi, the beaches and sabkhas are nearly perpendicular to the prevailing Shamal winds, and sediment tends to be transported directly onto the sabkhas (Kendall & Skipwith, 1968, 1969). Sabkha accretion rates there have been much more rapid, and the sabkha is many kilometres wider than at Ras Umm Said. The southern Gulf also has chenier-like barrier islands that are wider than at Ras Umm Said and they are separated from the main intertidal and sabkha flats by a broad shallow lagoonal area. Longshore currents there are less strong and hook-shaped spits are generally less obvious than at Ras Umm Said. However, well-developed hook-shaped spits that are much larger and better developed than those at Ras Umm Said occur at Umm al

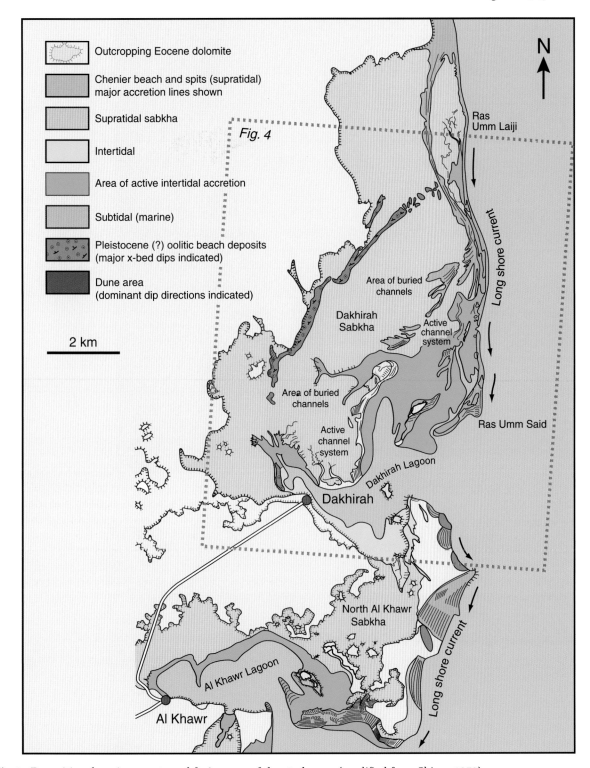

Fig. 3. Depositional environments and facies map of the study area (modified from Shinn, 1973).

Quwain, east of Dubai (Fig. 6). This area has not been thoroughly studied, but its porous chenier beaches would provide an excellent exploration model for the chenier/sabkha accretion model.

In addition to its larger size, the Trucial or Emirates coastal area is more arid than at Ras Umm Said. The increased aridity results in the precipitation of gypsum and anhydrite along with modern

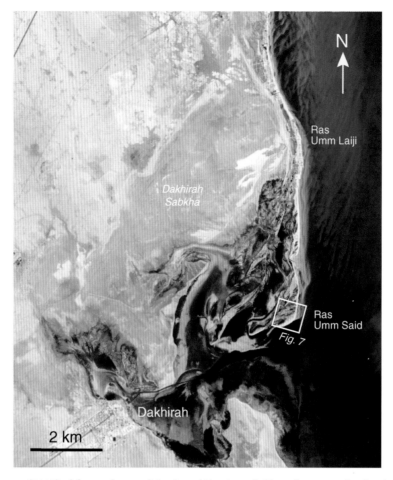

Fig. 4. Google Earth image (2006) of the study area (N 25°46.6′ E 51°34.3′). Note the new spits that have formed since 1973 (compare Fig. 3).

Fig. 5. Underwater photograph of a thick submarine cemented-layer forming in the deeper water of the pearling grounds east of the study area. Note the pholad borings (black arrows) and lighter coloured internal sediments (white arrows). This rock is the substrate that controls the distribution of the pearl-oyster area that provided the basis of the former pearling industry in the Gulf. Hammer head is 17.5 cm long.

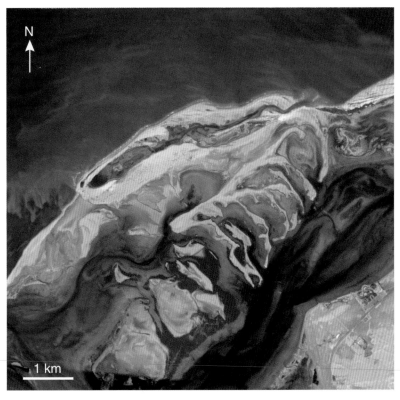

Fig. 6. Google Earth image (2006) showing large hook-shaped spits at Umm al Quwain, 400 km east of the Qatar study area, along the Trucial Coast east of Dubai (N 25°36.3′ E 55°34.5′).

dolomite. The Trucial/Emirates area also has extensive belts of laminated cyanobacterial mats and cemented grainstone crusts in the intertidal zone (Kendall & Skipwith, 1968; Purser, 1973; Strohmenger, 2006). Cyanobacterial mats and evaporites are rare at Ras Umm Said and are mostly restricted to intertidal areas of infilled tidal channels. The largest area of intertidal cyanobacterial mats in Qatar is on the west side of the peninsula at Dohat Faishakh (Illing *et al.*, 1965) and along the inland sea to the southeast of the peninsula (Fig. 1).

New observations have been made through the examination of Google Earth images of the Ras Umm Said area. One such observation, ground-truthed in 2005, is the breakthrough of hook-shaped spits by storm currents. For example, Fig. 7 shows the newest spit at the southern end of Ras Umm Said. Tidal channels have transected several other older spits that had formed since the original study in the mid-1960s. Aerial images show that the new spits curve while growing and that they subsequently merge with the previous chenier beach, effectively blocking off small bodies of water. Storm tides then break through the spit. The results of this process can clearly be seen in Fig. 7.

Once this process was recognized, then many examples in the older cheniers and spits, now encased in older sabkha deposits, became apparent. The formation of hook-shaped spits, such as that shown in Fig. 7, leads to yet another important diagenetic process: the formation of beachrock.

BEACHROCK

The same chemical and sedimentological processes that promote and/or constrain submarine cementation are accelerated by daily tidal fluctuations, and wetting and drying in the intertidal and supratidal zones. The resulting rock, universally known as "beachrock" since the 1800s (Gischler, 2007), dries between tides, leading to diagnostic voids (fenestral structures) related to the presence of air. In beachrock, these features have also been called keystone vugs (Dunham, 1970). Fibrous aragonite is the most common cementing agent in Arabian Gulf beachrock and the precipitated cement is often thicker on the undersides of individual sedimentary grains. This so-called "pendant cement", like keystone vugs, is a result

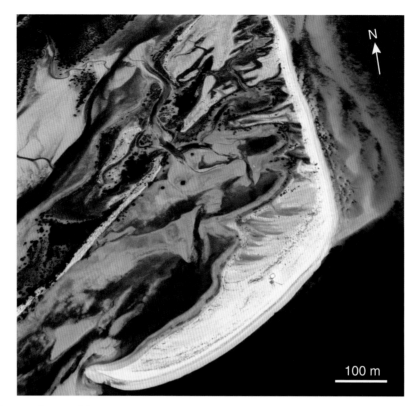

Fig. 7. Google Earth image (2006) of the most recent accretionary spit at Ras Umm Said (N 25°45.5′ E 51°35.5′). Note that tidal currents have transected older spits that formed since the area was studied in the mid-1960s (compare Fig. 3). See Fig. 4 for location.

of drainage, gravity, and the presence of interstitial air and meniscus effects during low tides (Purser & Loreau, 1973).

Beachrock at Ras Umm Said forms in two distinct areas. One is on the outer bends of tidal channels, where the sediment tends to be coarser than on the inner bends; these crusts are periodically eroded. Such crusts are not traditionally referred to as beachrock, but the process of cementation and formation of internal structures is essentially the same. The second site of beachrock formation is on the landward sides of the individual chenier beaches, especially the back sides of hook-shaped spits. In both cases, the cementation processes, discussed below, tend to be geographically localized and controlled by sedimentation.

Tidal currents, which are stronger along the outer bends of tidal channels, form crescent-shaped areas in the upper intertidal zone as a result of spillover during flooding. The outer bends of channels are also the areas of most active erosion during channel migration. Thus, the beachrock-like layers that form in the coarser grained sediments in this area break into slabs that constantly fall into the

channel and become incorporated as large intraclasts within the coarse channel-lag sediments (Fig. 8). As in most of the tidal-flat channels, the coarser outer-bend channel sediments generally are thicker and have the appearance of salt-and-pepper grains (Shinn, 1973, 1983; Shinn & Lidz, 1988): new grains tend to be light in colour, whereas those eroded from reducing muds during channel migration are commonly black.

In contrast, the beachrock associated with chenier beaches is mainly limited to the inner, more protected margin of hook-shaped spits. Such rock is continually forming as the chenier beaches, with their hook-shaped spits, accrete southward. This form of beachrock is lower in elevation, or closer to low tide, than the rock on the outer bends of the channels, and generally extends down to or even slightly below low-tide level. Where the rock is below low tide, it is petrographically identical to the submarine-cemented sediments discussed earlier. When wave-driven alongshore currents erode cheniers they move sediment southward and expose beachrock on the shoreface, which was previously formed on the back sides of the

Fig. 8. (A) The outer bend of a tidal channel in the intertidal to supratidal zone at Ras Umm Said. Note the sloping burrowed intertidal point bar in the foreground under shovel, and supratidal cemented crusts (beachrock) falling into the channel on the right. Channel width is approximately 6 m. (B) Caught in the act! Excavation through a tidal channel near Abu Dhabi showing slabs of cemented crusts (arrowed) "falling" into a channel at left of view. Section is approximately 2.5 m high.

hook-shaped spit. This exposed beachrock breaks into slabs and pebbles to become redeposited as intraclasts within the chenier.

Repeated visits to Ras Umm Said since the original mid-1960s study and the use of serial aerial photographs and Google Earth images show that the formation of beachrock at Ras Umm Said is an extremely rapid process when compared to beachrock formation in other areas in the world (e.g. Shinn, 2009). Figure 9 shows beachrock on the most recently formed spit at Ras Umm Said; this beachrock probably formed in less than one year. Examination of beachrock by progressive sampling

of older hook-shaped spits (Fig. 7) reveals increasing hardness and thickness on the older spits.

Chenier backstepping and intraclast formation

The continual erosion and backstepping of cheniers in the Ras Umm Said area result in the beachrock forming on the leeward side of spits, eventually becoming exposed on the seaward side of the beach (Fig. 10A). Once exposed, the rock is easily undermined and eroded by waves, broken, and redeposited into the beach accumulation as intraclasts. As a result of periodic overwash,

Fig. 9. Intertidal beachrock forming on the concave side of the most recent spit at Ras Umm Said. The house in the background can be seen in Fig. 7. The beach is approximately 15 m wide.

most of the chenier-beach laminations, including intraclasts, dip landward at about 5° (Fig. 10B; Shinn, 1973). Thus, a combination of diagenetic and primary sedimentary processes is continually producing beaches, laminations, and intraclasts. The sedimentation-cementation-erosion process accompanying chenier backstepping produces a distinctly different geometry and intraclast distribution to the process of intraclast formation-redeposition that occurs on the outer banks of migrating tidal channels. The seaward-dipping beach sediments containing intraclasts are preserved when the next chenier marches down the coast to provide armour for the previously formed chenier.

The vertical sedimentary sequences and sedimentary structures produced by the two processes, beachrock on outer bends of channels versus beachrock on hook-shaped spits, therefore, provide the best clues for differentiating between the two styles of porous sedimentary facies accumulation.

Channel downcutting

One of the more interesting and significant effects of submarine cementation is its effect on tidal-channel depth. Trenches dug in the Ras Umm Said sabkha in the 1960s demonstrated that the depth of channels could be limited by the presence of a cemented layer only a few centimetres thick. Not only do such layers retard and/or prevent

downward erosion, they are also relatively impermeable and prevent, or greatly reduce, vertical fluid movement. This was vividly demonstrated when a crust broken during trenching caused subsurface tidal waters to gush up and fill the trench. Thus submarine or intertidal crusts can prevent vertical migration of fluids whether they are evaporative or continental fresh groundwater (Fig. 11).

Control of vertical fluid movement likely has a strong influence on evaporite formation. Less-saline subsurface waters are prevented from rising and diluting evaporating floodwaters closer to the sediment surface, where evaporites generally form. On the other hand, the impermeable layers may also prevent upward movement of magnesium-rich capillary waters necessary for evaporative dolomite formation. Distribution of the submarine layers at Ras Umm Said is sporadic but may become more uniform in the future as the sedimentary complex accretes seaward over the more extensive submarine layers found farther offshore.

In some areas of the Arabian Gulf, tidal channels have actually broken through underlying marine crusts. For example, in 1966 while the author was diving in the Dubai Channel adjacent to the first major westward channel bend, a rock ledge at about 3 m was observed below the water surface (Fig. 12). The exposed submarine-cemented ledge had a heavily bored upper surface and the bed was approximately 30 cm thick. It had been undermined by currents, and cobbles and large slabs had

Fig. 10. (A) Beachrock formed on the back side of the hook-shaped spit is now exposed on the shoreface as the beach erodes. Intraclasts eroded from the band of beachrock become reworked into the beach deposits. Location is about 1 km north of the pit shown in Fig. 7. (B) Trench through the chenier beach shows washover laminations containing intraclasts and coral debris. Light meter (arrowed), about 5 cm in length, is resting on a layer of coarse material near the top of the exposure. Note steep bedding at the base of the trench. Intertidal carbonate mud was exposed below the steeper bedding beneath the short length of timber in the bottom of the trench.

broken free and fallen into the channel. The layer is thought to be continuous with an extensive area of marine cementation, then visible offshore from the air and examined while diving, less than 1 km offshore. A study of borehole data offshore of Dubai could be used to trace the distribution and depth of the cemented layer where it has not been excavated for building purposes.

In areas lacking laterally extensive submarine cementation, such as the Bahamas, the depth to which tidal channels cut is commonly limited by the depth of the underlying Pleistocene limestone that had been exposed during the last glacial period. Tidal fluctuation in the Bahamas is less than half that in the Arabian Gulf, especially in the tidal-flat areas west of Andros Island. Yet channels off Andros have easily eroded down 3 to 4 m, and would probably extend much deeper if Holocene sediments were thicker. The main factor limiting channel cutting depth is the depth of underlying

Fig. 11. Abandoned tidal channels. (A) Trench through a meandering sand-filled tidal channel on the sabkha made visible by varying moisture conditions. (B) Trenching through a wider sand-filled tidal channel. A 5 cm thick cemented grainstone layer underlay this channel. When the layer was broken, highly saline tidal water filled the lower third of the trench.

rock, whether it is Pleistocene bedrock or a Holocene submarine-cemented layer.

Obvious and deep tidal channels are extremely rare in ancient tidal-flat carbonates (Ginsburg, 1975). This repeated observation in ancient tidal flats indicates that either syndepositional marine and intertidal cementation was common and limited tidal channel downcutting, or that ancient tidal flats formed predominantly in areas lacking significant lunar tides. The latter seems unlikely. An abundance of syndepositional cementation, especially during Permian times, is the favoured hypothesis. Knowing that there may have been far more extensive areas of syndepositional

cementation and impermeable crusts in the past may have implications for petroleum production engineers and hydrologists exploring ancient limestones. Such knowledge certainly has implications for modern construction engineers. Good examples can be observed in excavations for large buildings in Doha (Ad Dawhar), Qatar. At the time of writing, construction is occurring on a spoil area of land that constituted the bay bottom in the mid-1960s when the author was mapping Doha Bay. Figure 13 shows three separate Holocene submarine-cemented layers that were observed in an excavation in Doha in November 2006. David Puls and his associates at RasGas Company Ltd are

Fig. 12. Submarine hardgrounds. (A) Underwater view of a 30 cm thick submarine-cemented layer exposed in the Dubai tidal channel in 1966. (B) Close up of the rock layer in the Dubai channel. The layer is thought to be continuous with a submarine-cemented layer identified by diving about 1 km offshore.

presently mapping and dating these layers and their associated marine sediments as new excavations appear and then become the foundations of skyscrapers in downtown Doha.

DISCUSSION AND CONCLUSIONS

As indicated at the start of this paper, the relatively new paradigm of sequence stratigraphy (Vail & Mitchum, 1977) had not been developed when the author first began his research in the Arabian Gulf. The subject of marine cementation was controversial, and most effort was devoted to documenting

the process and its diagnostic criteria. A critical observation made during that phase of study was that rock-boring pholads (*Lithophaga* sp.) were present in rock that was continually being formed on the present-day seafloor. In other words, sea-level lowstands did not control cementation and rock boring: these are entirely marine processes operating during the present Holocene highstand. The marine cemented layers found today in the Doha onshore pits are at and just above current sea level, implying a slightly higher level approximately 5000 years ago when the layers formed (D. Puls pers. comm., 2007). The same organisms can also produce borings in older rocks as the rocks

Fig. 13. Three submarine-cemented layers of Holocene sediment (near present sea-level) exposed in an excavation in Doha city. Note pholads in 2 cm long borings near finger of left hand. The three 15 cm thick cemented layers are overlain and underlain by loosely consolidated sediment.

are inundated by rising sea level. The two different processes could easily be confused. The product of the entirely synsedimentary process can look similar to a true marine flooding surface. Geologists heavily constrained by the sequence-stratigraphy paradigm should be on guard against the misinterpretation of bored or corroded hardground surfaces.

The general lack of clearly recognizable, closely spaced tidal channels in ancient tidal-flat sequences is likely due to the intensity of syndepositional cementation. The absence of well-defined channels in outcrops of the well-known Permian tidal flats of West Texas in the Guadalupe Mountains, for example, may be best explained by early syndepositional cementation. The Permian of the Midland Basin with its evaporites has long been thought to be analogous to the present-day sabkhas of the Arabian Gulf. The Permian tidal flats of this area are often associated with prolific oil production, and production engineers commonly discover vertically impermeable layers that separate producing strata. In addition, tidal-flat strata associated with hydrocarbons could contain porous and permeable chenier-beach accumulations, but such accretions would be difficult to predict. However, any single chenier or spit may connect laterally with other cheniers and spits not detected due to well spacing. Thus, it is important to

recognize the potential presence of such curving and linear accumulations, and distinguish them from beaches or channel deposits during the exploitation phase of drilling and production. With new developments, it may soon be possible to recognize and distinguish such structures by their geometry using ever-evolving 3-D seismic acquisition and processing.

The Qatar model, on various scales, is repeated along parts of the Arabian Gulf Trucial Coast, and very likely, along the southern shore of the Mediterranean Sea. Longshore drift and a steady supply of organically produced sand-sized sediment in any area where there is sedimentation similar to that at Ras Umm Said will result in net off-lap accretion. Given an additional few thousand years, and a continual supply of sediment, the sedimentary complex will increase in area and thickness. If present in ancient sabkha and evaporitic settings, this sedimentary complex could serve as a model for exploration and a guide for production drilling. The model also has various societal implications. For example, at Ran Laffan, a large 1-km-long LNG and GTL tanker loading terminal and wharf have been constructed perpendicular to the shore. The wharf is up-current from the Ras Umm Said area and will potentially cut off or retard southward movement of sediment. Chenier beach and spit formation may be terminated. However, the positive benefit is that the reduced supply of sediment may prevent the closing-off of Dakhirah and Al Khawr lagoons by growing sand spits. Continued access to these lagoons is vital to the fishing fleets and other boats at Dakhirah and Al Khawr city. Such knowledge can be applied to other future shoreline developments in the area.

ACKNOWLEDGEMENTS

The author wishes to thank many people who contributed to this study. First and foremost, Bruce Purser, who was instrumental in bringing a skinny young aspiring geologist from the US to the Shell Research Laboratory in Reisjick, The Netherlands, and later sending him and his family to live in Doha, Qatar where he provided him with the freedom to pursue where field observations led. I wish to thank Rashad Al-Mohanadi of Al Khawr, Qatar, my loyal assistant during the early work, and his son Dr Mohamad Rasad Al-Mohanadi of Qatar University, for his long friendship and for transporting the author by boat to Ras Umm Said in 1987. I thank AAPG for sending Peter Scholle, Bob Halley, Paul Harris, and me to the area in 1983 for production of the AAPG film *Arid Coastlines*, and for the long-lasting friendship and discussions that I have had with these co-actors in the film.

I thank David Puls of RasGas Ltd for transporting the author to the cheniers of Ras Umm Said and providing an eye-opening tour of construction/excavation sites in Doha, Qatar. He repeated the tour again in 2006 and reviewed this manuscript in 2007. Finally I thank Ahmad Al-Suwaidi of ADMA for arranging and supporting the 2006 visit to Abu Dhabi and the 12th ADIPEC Conference. Over the past 25 years I have had useful discussion with Abdulrahman Alsharhan of Dubai. Little of this would have happened without the aid and companionship of Christopher Kendall. Barbara Lidz of USGS edited an early draft and Betsy Boynton of USGS prepared the illustrations.

REFERENCES

Dunham, R.J. (1969) Early vadose silt in Townsend Mound, New Mexico. In: *Depositional Environments in Carbonate Rocks: A Symposium* (Ed. G.M. Friedman). *SEPM Spec. Publ.*, **14**, 139–181.

Dunham, R.J. (1970) Keystone vugs in carbonate beach deposits. *AAPG Bull.*, **54**, 845.

Evans, G. (1966) The Recent sedimentary facies of the Persian Gulf region. *Phil. Trans. Roy. Soc. London, Ser. A*, **259**, 291–298.

Evans, G. (1969) Stratigraphy and geological history of the sabkha, Abu Dhabi, Persian Gulf. *Sedimentology*, **12**, 145–159.

Gevirtz, J.L. and **Friedman, G.M.** (1966) Deep-sea carbonate sediments of the Red Sea and their implication on marine lithification. *J. Sed. Petrol.*, **36**, 143–151.

Ginsburg, R.N. (1975) *Tidal Deposits. A Casebook of Recent Examples and Fossil Counterparts.* Springer Verlag, New York, 428 pp.

Ginsburg, R.N., Schroeder, J.H. and **Shinn, E.A.** (1971) Recent synsedimentary cementation in subtidal Bermuda reefs. In: *Carbonate Cements* (Ed. O.P. Bricker). *John Hopkins Univ. Stud. Geol.*, **19**, 54–58.

Gischler, E. (2007) Beachrock and intertidal precipitates. In: *Geochemical Sediments and Landscapes* (Eds D.J. Nash and S.J. McLaren), pp. 365–390. Blackwell, Oxford.

Illing, L.V., Wells, A.J. and **Taylor, J.C.M.** (1965) Penecontemporary dolomite in the Persian Gulf. In: *Dolomitization and Limestone Diagenesis* (Eds L.C. Pray and P.C. Murray). SEPM Spec. Publ., **13**, 89–111.

Kendall, C.G.St.C. and **Skipwith, P.A.d'E.** (1968) Recent algal mats of a Persian Gulf lagoon. *J. Sed. Petrol.*, **38**, 1040–1058.

Kendall, C.G.St.C. and **Skipwith, P.A.d'E.** (1969) Holocene shallow-water carbonate and evaporite sediments of Khor al Bazam, Abu Dhabi, southwest Persian Gulf. *AAPG Bull.*, **53**, 841–869.

Longman, M.E. (1980) Carbonate diagenetic textures from near-surface diagenetic environments. *AAPG Bull.*, **64**, 461–487.

Milliman, J.D. (1966) Submarine lithification of deep-water carbonate sediments. *Science*, **153**, 994.

Pray, L.C. (1965) Clastic limestone dikes and marine cementation, Mississippian bioherms, New Mexico: Permian Basin section. *SEPM Progr. Abstr. Ann. Meeting*, 21–22.

Purser, B.H. and **Evans, G.** (1973) Regional sedimentation along the Trucial Coast, SE Persian Gulf. In: *The Persian Gulf: Holocene Carbonate Sedimentation and Diagenesis in a Shallow Epicontinental Sea* (Ed. B.H. Purser), pp. 211–213. Springer Verlag, Heidelberg, Berlin.

Purser, B.H. and **Loreau, J.P.** (1973) Aragonitic, supratidal encrustations on the Trucial Coast, Persian Gulf. In: *The Persian Gulf: Holocene Carbonate Sedimentation and Diagenesis in a Shallow Epicontinental Sea* (Ed. B.H. Purser), pp. 343–376. Springer Verlag, Heidelberg, Berlin.

Shinn, E.A. (1969) Submarine lithification of Holocene carbonate sediments in the Persian Gulf. *Sedimentology*, **12**, 109–144.

Shinn, E.A. (1971) Submarine cementation in the Persian Gulf. In: *Atlas of Carbonate Cements* (Ed. O. Bricker), pp. 63–65. John Hopkins Univ. Press, Maryland.

Shinn, E.A. (1973) Coastal accretion in an area of longshore transport, Persian Gulf. In: *The Persian Gulf: Holocene Carbonate Sedimentation and Diagenesis in a Shallow Epicontinental Sea* (Ed. B.H. Purser), pp. 179–191. Springer Verlag, New York.

Shinn, E.A. (1983) Tidal flat environment. In: *Carbonate Depositional Environments* (Eds P.A. Scholle, D.G. Bebout and C.H. Moore). *AAPG Mem.*, **33**, 171–210.

Shinn, E.A. (1986) Modern carbonate tidal flats: their diagnostic features. In: *Carbonate Depositional Environments, Modern and Ancient. Part 3, Tidal Flats* (Eds L.A. Hardie and E.A. Shinn). *Colorado School Mines Q.*, **81**, 7–35.

Shinn, E.A. (2009) The mystique of beachrock. In: *Perspectives in Carbonate Geology: a Tribute to the Career of Robert Nathan Ginsburg* (Eds P.K. Swart, G.P. Eberli and J.A. McKenzie). *Int. Assoc. Sedimentol. Spec. Publ.*, **41**, 19–28.

Shinn, E.A. and **Lidz, B.H.** (1988) Blackened limestone pebbles: fire at subaerial unconformities. In: *Paleokarst* (Eds N.P. James and P.W. Choquette), pp. 117–131. Springer Verlag, New York.

Shinn, E.A., Lloyd, R.M. and **Ginsburg, R.N.** (1969) Anatomy of a modern carbonate tidal flat, Andros Island, Bahamas. *J. Sed. Petrol.*, **39**, 1202–1228.

Strohmenger, C.J. (2006) *Modern Carbonate-Evaporite Depositional Environments of Abu Dhabi (UAE)*. Field Guide 12th ADIPEC 2nd Internat. Evaporite Conf., 7-8 November 2006, Abu Dhabi, 61 pp.

Sugden, W. (1963) The hydrology of the Persian Gulf and its significance in respect to evaporite deposition. *Am. J. Sci.*, **26**, 741–755.

Vale, P.R. and **Mitchum, R.M., Jr.** (1977) Seismic stratigraphy and global changes of sea level. In: *Seismic Stratigraphy - Applications to Hydrocarbon Exploration* (Ed. C.E. Payton). *AAPG Mem.*, **26**, 83–97.

Wells, A. (1962) Primary dolomitization in the Persian Gulf. *Nature*, **194**, 274–275.

Int. Assoc. Sedimentol. Spec. Publ. (2011) **43**, 149–182

Facies stacking patterns in a modern arid environment: a case study of the Abu Dhabi sabkha in the vicinity of Al-Qanatir Island, United Arab Emirates

CHRISTIAN J. STROHMENGER[*1], HESHAM SHEBL[†], ABDULLA AL-MANSOORI[*], KHALIL AL-MEHSIN[‡], OMAR AL-JEELANI[*], ISMAIL AL-HOSANI[*], ALI AL-SHAMRY[*] and SALEM AL-BAKER[*]

[*]*Abu Dhabi Company for Onshore Oil Operations (ADCO), P.O. Box 270, Abu Dhabi, UAE*
(E-mail: christian.j.strohmenger@exxonmobil.com)
[†]*Zakum Development Company (ZADCO), Abu Dhabi, UAE*
[‡]*Abu Dhabi National Oil Company (ADNOC), Abu Dhabi, UAE*

ABSTRACT

Supratidal (sabkha) to intertidal (microbial mat), and lowermost intertidal to shallow-subtidal (peloid-skeletal tidal flat) environments were studied along the Abu Dhabi coastline in the vicinity of Al-Qanatir (Al-Rufayq) Island. A transect from land to sea displays the following classic examples of supratidal to shallow-subtidal facies belts:

(1) Inner, upper sabkha (upper supratidal), buckled polygonal halite crust displaying teepee structures; (2) Stranded beach ridges, forming low-relief topographic highs paralleling the coastline that are mainly composed of cerithid gastropods; (3) Outer, upper sabkha (upper supratidal), buckled polygonal halite crust displaying teepee structures; (4) Middle sabkha (middle supratidal), whitish anhydrite polygons on surface; (5) Lower sabkha (lower supratidal), soft, shiny surface due to sparkling gypsum crystals (gypsum mush); (6) Upper intertidal, thin, leathery, crinkled or crenulated microbial mat; (7) Middle intertidal, blistered microbial mat and pinnacle or domed microbial mat; (8) Lower intertidal, thick, smooth polygonal microbial mat and tufted or cinder-like microbial mat; (9) Lowermost intertidal to shallow-subtidal, peloid-skeletal tidal flat (lagoonal and shallow tidal-channel/tidal-creek deposits) displaying cerithid and littorinid gastropod grazing-traces, *Skolithos*-type burrows, and eroded wave ripples.

The stenohaline bryozoan species *Disporella* sp., which has not been recorded previously from the United Arab Emirates, was found within a thin channel lag deposit of the outer, upper sabkha environment. Significant amounts of dolomite were found within a subsurface crinkly-laminated microbial mat (middle sabkha environment). The fine-crystalline dolomite displays subhedral to euhedral dolomite rhombs embedded in an organic matrix. The formation of dolomite is interpreted to be related to sulphate reducing microbial organisms which form the widespread microbial mat along the Abu Dhabi coastline.

Radiocarbon dating of 15 samples (10 hardground samples, 3 microbial mat samples, and 2 samples from anhydrite-dominated layers) show an age range from ca 3500 uncalibrated ^{14}C yr BP (outer, upper sabkha environment: subsurface hardground, seaward of stranded beach ridges) to ca 900 uncalibrated ^{14}C yr BP (intertidal environment: subsurface microbial mat); thereby supporting the seaward progradation of the facies belts since the last Holocene sea-level highstand (formation of cerithid gastropod stranded beach ridges). An anhydrite layer within aeolian deposits (inner, upper sabkha environment: landward of stranded beach ridges) showed a radiocarbon age of the host sediment of ca 12,900 uncalibrated ^{14}C yr BP, corresponding to Pleistocene dune deposits, pre-dating the Holocene flooding event. The distribution of radiocarbon ages indicates a complex stratigraphic history in which chronostratigraphic time lines clearly cross-cut depositional lithofacies and diagenetic boundaries. This is significant in that depositional lithofacies and diagenetic facies are commonly used in ramp settings to correlate continuous sedimentary

---

[1] Present address: ExxonMobil Oil Indonesia (EMOI), Wisma GKBI, Jl. Jend. Sudirman 28, Jakarta 10210, Indonesia.

packages. Careful attention to chronostratigraphic relationships elucidates complex stratal geometries that may previously have been missed.

Keywords: Sabkha, evaporites, dolomite, microbial mat, sea-level history, radiocarbon dating, bryozoan species *Disporella*.

INTRODUCTION

The giant coastal sabkhas of Abu Dhabi (United Arab Emirates, UAE) are one of the few areas of the world where a geoscientist can observe the complex interplay between siliciclastic, carbonate, and evaporite sedimentation. Coast parallel tidal flats and sabkha are widest and best developed approximately 80 km southwest of Abu Dhabi City where they extend more than 15 km inland to Pleistocene-Holocene dunes (Fig. 1A). Coastal sabkhas have been actively established since 4500 years ago (Evans *et al.*, 1969, 2002) and, concomitantly, significant thicknesses of sabkha sediments have been deposited during this time interval. The widespread sabkha distribution indicates that they are worthy of distinction as a major depositional environment associated with carbonate ramp systems. Until recently, most sabkha work has focused on descriptions of the broad facies tracts, carbonate and evaporite mineralogy and distribution, as well as on the hydrology and chemistry of the coastal sabkha aquifer (Curtis, 1963; Kendall & Skipwith, 1968; Butler, 1969, 1970; Evans *et al.*, 1969; Kinsman, 1969; Purser 1973, Schneider, 1975; Kinsman & Park, 1976; Park, 1977; Patterson & Kinsman, 1977, 1981, 1982; McKenzie *et al.*, 1980; McKenzie, 1981; Butler *et al.*, 1982; Shearman, 1983; Shinn, 1983; Warren & Kendall 1985; Kendall & Warren, 1988; Kenig *et al.*, 1990; Müller *et al.*, 1990; Kenig 1991; Alsharhan & Kendall, 1994, 2002; Baltzer *et al.*, 1994; Kendall *et al.*, 1994, 2002; Evans, 1995; Peebles *et al.*, 1995; Kirkham 1997, 1998; Alsharhan *et al.*, 1998; Sanford & Wood, 2001; Evans & Kirkham, 2002; Wood & Sanford, 2002; Wood *et al.*, 2002, 2005). The publication edited by Purser (1973) is still one of the most comprehensive compilations of studies carried out along the Abu Dhabi coastal region.

Abu Dhabi sabkha evaporites have often been used as analogues for ancient carbonate-evaporite dominated formations such as the Upper Jurassic Arab Formation of the Arabian Gulf region (Wood & Wolfe, 1969; Leeder & Zeidan, 1977; Alsharhan & Kendall, 1994, 2002). However, it is important to note that most of the thick ancient anhydrite deposits, like those of the Upper Permian (Khuff and Zechstein formations) and the Upper Jurassic (including the Arab Formation), mostly represent salina- or saltern-type (Warren, 1983, 1989, 1999; Warren & Kendall, 1985; Schreiber, 1988; Kendall, 1992) rather than sabkha-type deposits (Al-Silwadi et al., 1996; Strohmenger *et al.*, 1996; Steinhoff & Strohmenger, 1999; Kirkham, 2004).

As a complement to previous work, a suite of trenches and cores along a dip-transect (Fig. 1B) were chosen for investigation by petrographic thin-section microscopy, x-ray diffraction (XRD), and radiocarbon-dating techniques (Strohmenger *et al.*, 2004, 2007, 2010; Bontognali *et al.*, 2010). These trenches and cores afford an excellent overview of the major depositional and diagenetic facies associated with sabkha accumulations along the southern Arabian Gulf, thereby facilitating a better understanding of ancient sabkha deposits in the rock record. Additionally, they provide valuable constraints on the size, shape, distribution, and accumulation rates associated with sabkha deposits.

GENERAL SETTING

The Arabian Gulf is a semi-enclosed sea situated between 24°N and 30°N and covers approximately 226,000 km² (1000 km by 200–300 km). Average water depth is relatively shallow (35 m) attaining a maximum depth of approximately 100 m near its narrow entrance at the Straits of Hormuz (Purser & Seibold, 1973; Fig. 2A). The Arabian Gulf basin formed in the Tertiary in association with the uplift of the Zagros Mountains. Its elongate bathymetric axis separates two major geological provinces, the stable Arabian Foreland and the unstable Iranian Fold Belt. Basin bathymetry is asymmetric, being steeper and deeper along the eastern, Iranian side of the basin (Fig. 2A). Based on its bathymetry, the Arabian Gulf was divided into three sub-basins by Seibold & Vollbrecht (1969) and Purser & Seibold (1973): the Western and Central Basins on the Iranian side of the Gulf are separated by a

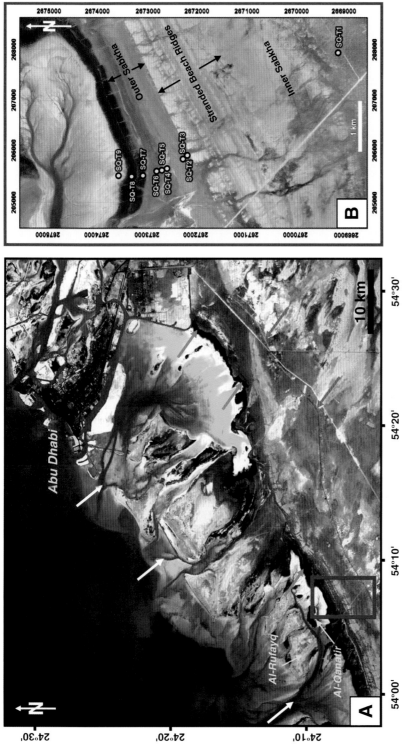

Fig. 1. Satellite images showing area studied and location of trenches and cores (SQ-T1 to SQ-T9). (A) Satellite image showing well-developed and wide sabkha, extending from the microbial mat (dark coloured band, green arrows) inland to Pleistocene-Holocene dunes (whitish colours, orange arrows). The offshore barrier complex is wide with diverse sub-environments. Ooid and coated-grain sand shoals are interspersed among islands and promontories and cut by tidal channels (white arrows). Flood-dominated tidal deltas are predominant. The slope of the Arabian Gulf dips gently, ramp-like from the coastline to the axis of the basin. Water depths towards the top of the image (upper left hand corner) do not exceed 40 m. The study area is shown by the red rectangle. Landsat-5 imagery, ca. 1990. Image processing provided by Earth Satellite Corporation. (B) Satellite image showing the position of trenches and cores (SQ-T1 to SQ-T9). Obvious are the stranded beach ridges (whitish stripes) and the palaeo-tidal-channels (darker, irregular bands cross-cutting the sabkha and the stranded beach ridges). Environments represented are: SQ-T1, inner, upper sabkha landward of the stranded beach ridges; SQ-T2, SQ-T3 and SQ-T4, outer, upper sabkha sub-environments seaward of the stranded beach ridges; SQ-T5, middle sabkha; SQ-T6, lower sabkha; SQ-T7, upper intertidal microbial mat; SQ-T8, lower intertidal microbial mat; SQ-T9, lowermost intertidal to shallow-subtidal (lagoonal and shallow tidal-channel/tidal-creek deposits), peloid-skeletal tidal-flat. DigitalGlobe QuickBird 61 cm resolution imagery, May 2004. Image processing provided by Maps Geosystems.

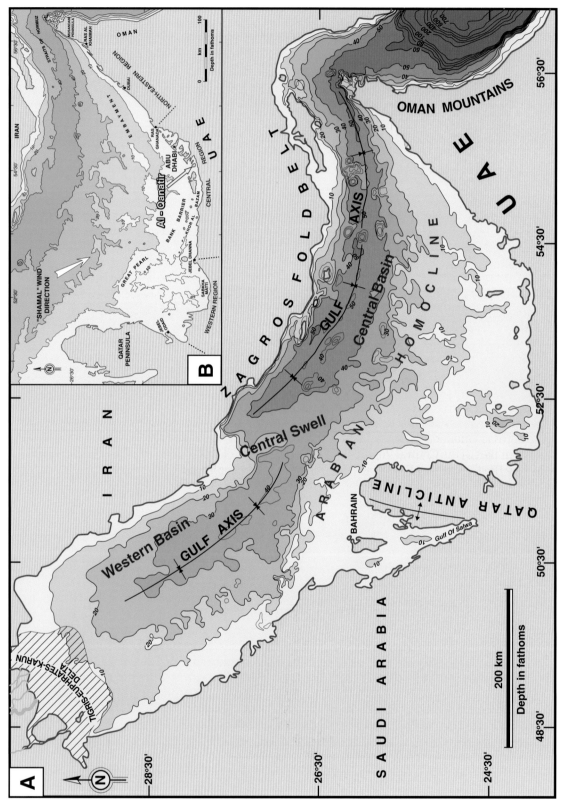

Fig. 2. Bathymetry and geological setting of the Arabian Gulf. Contour interval: 10 fathoms. 1 fathom = 6 feet (1.83 m). (A) Tectonic setting and bathymetry of the Arabian Gulf showing the position of the Gulf axis, the Western Basin and the Central Basin, separated by the Central Swell, as well as the Arabian Homocline. After Seibold & Vollbrecht (1969), Kassler (1973) and Purser & Seibold (1973). (B) Map of the UAE coastline showing the three major depositional provinces (regions): Western Region, Central Region, and North-Eastern Region. After Purser & Evans (1973). The Al-Qanatir study area is indicated (yellow arrow).

shallow ridge called the Central Swell (Fig. 2A). The Arabian side of the Gulf is called the Arabian Shallow Shelf or Arabian Homocline (Purser and Seibold, 1973; Kassler, 1973; Fig. 2A).

In contrast to carbonate platforms with well-developed shelf edges that are typical for the Caribbean, the Arabian platform exhibits a gentle bathymetric profile, thereby indicating a shallow ramp geometry (Burchette & Wright, 1992), ideal for depositing widespread and laterally continuous carbonate sediments. Whereas clastic influx to the Arabian Gulf is areally limited to the Tigris-Euphrates-Karun delta (Fig. 2A), the approximately 600 kilometre long coastline between the Qatar Peninsula near Jebal Odaid and the Musandam Peninsula is dominated by relatively pure carbonate sedimentation (Fig. 2B), and includes the following environments: (1) ooid beaches; (2) ooid shoals; (3) ooid tidal deltas; (4) coral reefs; (5) barrier islands; (6) lagoons; (7) mangrove swamps; (8) tidal flats; (9) microbial mats; and (10) aeolian carbonate sands. Many of these depositional environments correspond to those observed in cores from the subsurface of the Arabian Peninsula.

Three morphological provinces (regions) were distinguished by Purser & Evans (1973; Fig. 2B), the: Western Region (Jebal Odaid to Jebal Dhanna); Central Region (Jebal Dhanna to Ras Ghanada); and North-Eastern Region (Ras Ghanada to Musandam Peninsula). The eastern part of the Central Region is strongly influenced by the Great Pearl Bank Barrier system that separates the Khor Al Bazam Lagoon from the Arabian Gulf (Fig. 2B). In the eastern part of the Central Region the Great Pearl Bank Barrier and the associated Khor Al Bazam Lagoon are less defined and are replaced by a complex of islands and peninsulas (Fig. 2B). One of these islands is Al-Qanatir Island (close to Al-Rufayq Island, Al-Dabbiya area) in the vicinity of which the supratidal to the lowermost intertidal to shallow-subtidal coastal environments were studied (Fig. 2B).

Development of extensive sabkha deposits and associated carbonate and evaporite sediments is greatly facilitated by the arid, subtropical climate. The particular climatic regime and the fact that the Arabian Gulf is almost totally surrounded by land (absence of oceanic buffering) leads to marked seasonal fluctuations (Purser & Seibold, 1973). Thus, sedimentary packages are highly dynamic and spatially variable. Aridity, wind and high temperatures result in intricate intercalations of carbonate and evaporite minerals, and fine-grained, wind-blown siliciclastics. Extreme aridity plays a pronounced role in the dynamics of sabkha sedimentation. Annual rainfall is low, averaging less than 5 cm yr^{-1} (Evans et al., 1969). Humidity averages 40% during the day and 90% at night. Surface temperatures have an average annual maximum of 41 °C and an average annual minimum of 13 °C, but can reach more than 50 °C in summer and be as low as 0 °C in winter (Schneider, 1975). Surface water salinities in the Gulf range from 37 to 50‰, but can rise to more than 60‰ in remote lagoons and coastal embayments (Purser & Seibold, 1973).

Distribution of carbonate sediments is strongly influenced by the "Shamal" winds which blow mainly from the northwest in the northern parts of the Gulf, but tend to veer to the south as they approach the UAE coast in the southeast. Because of the "Shamal" winds, carbonate environments along the UAE coast are well swept by waves and surface currents facilitating the formation and dispersal of carbonate sediments. Tidal currents also affect the distribution of the carbonate sediments favoring the development of spectacular oolitic flood-dominated tidal deltas (Purser & Evans, 1973; Purser & Seibold, 1973; Fig. 1A). Tides are semi-diurnal with a spring tide range of 2.1 m and a neap tide range of 0.75 m. This tidal range can produce a high-tide inland reach of 3.5 kilometres. Strong seasonal "Shamal" winds influence wave heights and tidal level, and like spring tides, can move great volumes of water over the coastal sabkha and the coastal beach areas.

The environmental parameters along the Abu Dhabi coastline can be summarized as follows:

(1) Ramp carbonate setting;
(2) Moderate tidal range: microtidal, <1 m to 1.5 m tidal range; neap tide, 0.75 m; spring tide, 2.1 m.
(3) Quiescent tectonics and minimal subsidence.
(4) Very low onshore and offshore relief.
(5) Very low annual rainfall: 3–4 cm yr^{-1}.
(6) Average air temperatures: 13 °C (winter) to 41 °C (summer); range from 0 °C (winter) to >50 °C (summer).
(7) Average water temperature: 15 °C (winter) to 40 °C (summer).
(8) High evaporation rate: ca 124 cm yr^{-1}.
(9) High salinity: lagoons, >60‰; sabkha, >8× normal sea water.
(10) Abundant evaporites and dolomite.
(11) Dominantly onshore winds creating flood-dominated tidal deltas.
(12) Moderate to low terrigenous input.

The Arabian Gulf has experienced a number of desiccation and flooding events in its recent history. During the Pleistocene, sea-level periodically fell about 120 m below that of the present day (Kassler, 1973). The recent unconsolidated sediments are the product of the post-glacial Holocene transgression which began ca 18,000 years ago and reached its highest level during the Late Holocene, which is reported to have occurred between ca 4,000 (Evans *et al.*, 1969, 2002; Evans and Kirkham, 2002) and 6,000 years ago (Kassler, 1973; Lambeck, 1996; Lambeck *et al.*, 2002). Satellite images reveal prominent linear features several tens of kilometers long, sub-parallel to the present-day coastline (Fig. 1A and B). These features correspond to relict stranded beach ridges (storm beach ridges) that formed as the sea regressed to its present level (Evans *et al.*, 1969, 2002; Evans & Kirkham, 2002). The beach ridges are composed of gastropod-rich grainstones (cerithids) and indicate the progradation of the shoreline during falling sea-level.

Modern carbonate coastal sediments of the Arabian Gulf lie on Miocene and Pleistocene bedrock that outcrops along an arcuate, low-lying NE–SW oriented ridge (Fig. 1A). Undisturbed supratidal to intertidal and shallow-subtidal environments still exist along the Abu Dhabi coastline in the vicinity of Al-Qanatir (Al-Rufayq) Island (Fig. 1A). A transect from the supratidal to the lowermost intertidal to shallow subtidal was studied in detail (Fig. 1B), with emphasis on lateral and vertical facies successions, evaporite-mineral and non-evaporite-mineral distribution, age relationships of the different deposits, and the controls of sea-level variations (Strohmenger *et al.*, 2004, 2007).

METHODOLOGY

In the vicinity of the road to Al-Rufayq Island, close to Al-Qanatir Island, about 80 km southwest of Abu Dhabi City, a complete succession from the upper supratidal to the lowermost intertidal to shallow-subtidal environment (Kendall & Warren, 1988; Kendall *et al.*, 1994; Kirkham, 1997; Evans & Kirkham, 2002) has been examined (Strohmenger *et al.*, 2004, 2007; Bontognali *et al.*, 2010 Fig. 1A and B).

Surface morphology, the frequency of flooding, and the presence of a well-developed microbial mat were used as criteria to distinguish between supratidal, intertidal, and lowermost intertidal to shallow-subtidal environments. Furthermore, four supratidal and four intertidal environments, as well as three upper supratidal sub-environments were identified.

Nine trenches were dug within the upper supratidal to intertidal and lowermost intertidal to shallow-subtidal environments (Fig. 1B). The trenches have been described sedimentologically from the surface down to the water table or encountered hardgrounds. A total of 58 samples were collected for thin section and whole-rock and clay fraction x-ray diffraction (XRD) analyses. 15 samples have been taken for radiocarbon dating.

Subsequently, short (30 to 100 cm in length), four-inch (10.2 cm) diameter cores, called SQ-T1 to SQ-T9, were taken at each trench locality (Fig. 1B). The difference in thickness between trenches and corresponding cores due to compaction during coring was less than 10%. The core material was frozen, resinated, slabbed, and photographed by Core Laboratories (CoreLab) Abu Dhabi Branch. The trenches and the cores correspond to the inner sabkha: SQ-T1 (upper sabkha: landward, behind stranded beach ridges), the outer sabkha: SQ-T2 (high upper sabkha: seaward, close to stranded beach ridges), SQ-T3 (mid upper sabkha), SQ-T4 (low upper sabkha), SQ-T5 (middle sabkha), and SQ-T6 (lower sabkha), as well as the intertidal: SQ-T7 (upper intertidal) and SQ-T8: (lower intertidal), and lowermost intertidal to shallow-subtidal: SQ-T9 environments (Fig. 1B).

The following minerals were identified from XRD and thin-section analyses: quartz SiO_2; plagioclase $Na(Ca)AlSi_3O_8$; K-feldspar $KAlSi_3O_8$; aragonite $CaCO_3$; calcite $CaCO_3$ (<4 mol% $MgCO_3$); Mg-calcite $CaCO_3$ (4–22 mol% $MgCO_3$); dolomite $CaMg(CO_3)_2$; magnesite $MgCO_3$; gypsum $CaSO_4.2H_2O$; bassanite $CaSO_4.^1/_2H_2O$; anhydrite $CaSO_4$; halite $NaCl$. The values of the minerals (non-clay fraction) displayed in cumulative weight percentage along each photograph of cores SQ-T1 to SQ-T9 are considered only to be approximations. The amount of each mineral within a given sample may change considerably depending on the location of the sample selected, as well as on the timing of the sampling (summer: dry and extreme hot climate versus winter: temporarily wet and more moderate climate). However, the presence or absence of a mineral within each of the 58 analyzed samples and the relative ratio of the minerals within each sample is considered to be reliable. The XRD results were compared to petrographic thin-section analysis and adjusted where needed. XRD results also showed that the clay fraction of the samples is very low, mostly between below detection (the majority of the samples) and always <10%.

Salinities, determined from six water samples (SQ-T2: 268.64‰, SQ-T3: 275.62‰, SQ-T4: 274.54‰, SQ-T5: 262.92‰, SQ-T6: 257.24‰ and SQ-T7: 241.61‰) show a penesaline (gypsum/anhydrite/halite) brine stage (hypersaline). They also show a general trend of decreasing salinity seaward.

Radiocarbon dating was carried out on bulk samples of nine hardgrounds (SQ-T2-2, SQ-T3-2, SQ-T4-2, SQ-T5-2, SQ-T5-4, SQ-T6-2 SQ-T7-2, SQ-T8-2, and SQ-T9-2), three microbial mats (SQ-T3-5, SQ-T5-6, and SQ-T7-3) and two anhydrite layers (SQ-T1-4 and SQ-T4-7). The samples show a range from ca 3500 uncalibrated ^{14}C years before present (yr BP; outer, upper sabkha, seaward of stranded beach ridges: hardground) to ca 900 uncalibrated ^{14}C yr BP (upper intertidal: crinkly-laminated microbial mat), and thus substantiate the seaward progradation of the facies belts since the last Holocene sea-level highstand (formation of cerithid gastropod stranded beach ridges). Radiocarbon dating of an anhydrite sample of the inner, upper sabkha, landward of the stranded beach ridges (sample SQ-T1-2; Fig. 1B), yielded a radiocarbon age of ca 12,900 uncalibrated ^{14}C yr BP.

Conventional radiocarbon dating was used for carbonate samples representing the nine hardgrounds. However, the ^{14}C ages of three microbial mats and two anhydrite layers were determined

Fig. 3. Inner, upper sabkha sub-environment. (A) The red dot on the satellite image marks the location of trench and core SQ-T1. (B) Mid upper sabkha displaying buckled polygonal halite crust. (C) Close-up of buckled polygonal halite crust. (D) Trench showing from bottom to top: predominantly quartz- and carbonate-rich sand (Pleistocene aeolianite), thick (about 60 cm) dominantly distorted bedded mosaic anhydrite (lower part: incipiently cross-bedded carbonate-rich sand) to nodular to contorted mosaic (enterolithic or ptygmatic) anhydrite (upper part), and gypsum-rich, bioclastic packstone (lagoonal deposit). (E) Close-up of anhydrite layer, showing transition from distorted bedded mosaic anhydrite (lower part) to nodular to contorted mosaic (enterolithic or ptygmatic) anhydrite (upper part) most probably marking the Pleistocene-Holocene boundary. (F) Log of trench stratigraphy in (D).

using accelerator-mass-spectrometry (AMS), where only the organic matter fraction is used for dating after removal of the carbonate fraction. The radiocarbon dating results are given as the δ ^{13}C-corrected (conventional) ^{14}C age. The δ ^{13}C values are reported in ‰ V-PDB. Radiocarbon dating was carried out by Beta Analytic Inc., Miami. Radiocarbon dating results presented here are uncalibrated and as such are not directly comparable with 'calendar' ages.

Dating results of the hardgrounds actually represent a mixture of the age of the cement (diagenetic age) and the age of the sediment (depositional age). However, because the carbonate cement fraction of these samples is, with the exception of one hardground sample (sample SQ-T6-2), only minor, the radiocarbon dates obtained are interpreted to

mainly correspond to the depositional age. The influence of the cement is expected mostly to be within the error range of the radiocarbon dating method (±40–70 years). The various hardgrounds are interpreted to have been formed at different times due to rapid cementation at the vadose/phreatic interface of the groundwater table (favourable geochemical and hydrological conditions for carbonate precipitation), within the sediments, as discussed previously by Strasser *et al.* (1989). However, microbial activity may have further stimulated cementation (Strasser & Davaud, 1986; Strasser *et al.*, 1989).

AMS radiocarbon ages obtained from the anhydrite samples correspond to the age of the organic content of the host sediment and not to the age of the anhydrite formation. In the case of the

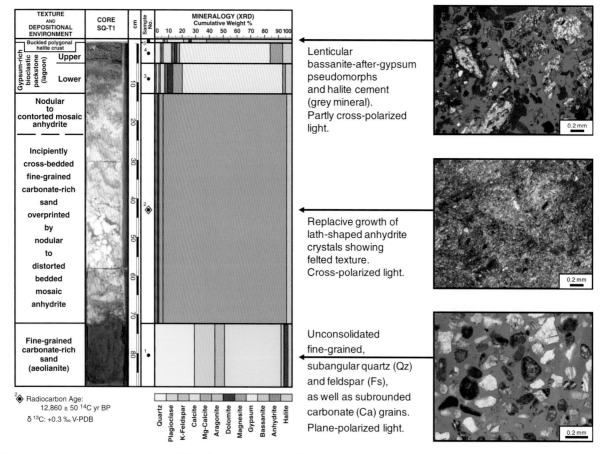

Fig. 4. Resinated core photograph (SQ-T1) and mineralogical composition of the trench shown in Fig. 3, and thin-section photomicrographs. The mineralogical composition was determined by X-ray diffraction (XRD) and reported as a cumulative weight percentage. The transition from distorted bedded mosaic anhydrite (lower part: the incipiently cross-bedded carbonate-rich sand) to nodular and contorted mosaic (enterolithic or ptygmatic) anhydrite (upper part) potentially corresponds to the boundary between the Pleistocene aeolian dune deposits and the Holocene gypsum-rich, bioclastic packstone facies (lagoonal deposit). Radiocarbon dating and carbon-isotope results of the anhydrite-dominated layer are also displayed. See Fig. 1 for location.

anhydrite sample of trench and core SQ-T1 (sample SQ-T1-2: anhydrite growing within aeolian deposits) the age of the host sediment and the age of the anhydrite are expected to be significantly different.

SUPRATIDAL, COASTAL SABKHA ENVIRONMENT: INNER SABKHA, LANDWARD OF STRANDED BEACH RIDGES

The inner sabkha (Fig. 1A and B) behind the stranded beach ridges can be subdivided into two sub-environments based on slightly varying surface morphologies: buckled halite crust and buckled polygonal halite crust. The width of the inner sabkha can exceed than 10 km. The upper supratidal inner sabkha (upper sabkha/upper salt-flat facies) characteristics are as follows:

(1) **High upper sabkha,** buckled halite crust displaying teepee structure:
   (a) Landward and seaward of stranded beach ridges;
   (b) Topographically higher areas surrounding topographically lower areas landward of stranded beach ridges;
   (c) Minor anhydrite but gypsum-rich upper layer below surface.

(2) **Mid upper sabkha,** buckled polygonal halite crust:
   (a) Landward and seaward of stranded beach ridges;
   (b) Topographically lower areas, surrounded by topographically higher areas landward of stranded beach ridges;
   (c) Thick, nodular to distorted bedded mosaic and contorted mosaic (enterolithic or

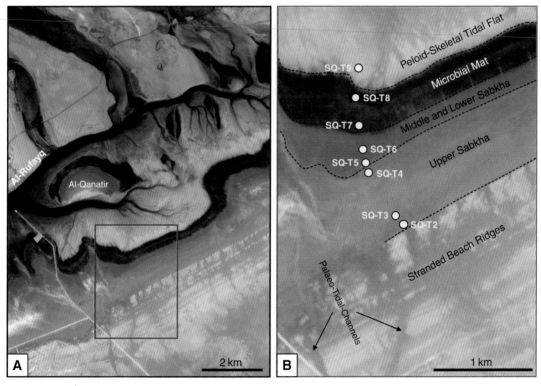

Fig. 5. Satellite image of the area SE of Al Rafeq Island (see Fig. 1). IKONOS remote sensing image, November 2002. Data copyright SpaceImaging.com, processed by Gene Rankey (RSMAS/University of Miami). (A) Clearly visible from landward (lower right hand corner) to seaward (upper left hand corner) are the stranded beach ridges (whitish stripes), the sabkha environment (supratidal, greyish area), the microbial mat (intertidal, blackish area), and the peloid-skeletal tidal flat (lowermost intertidal to shallow-subtidal, light blue to light-greyish area) environment. The peloid-skeletal tidal flat environment is cut by deep (dark blue) and shallow (light blue) tidal channels, and protected by barrier islands (brownish colour). Also visible are the palaeo-tidal-channels as darker, irregular bands cross-cutting the sabkha and the stranded beach ridges. (B) Close-up of Fig. 5A showing the facies belts as well as the position of trenches and cores SQ-T2 to SQ-T9. Also obvious are palaeo-tidal-channels, cross-cutting the stranded beach ridges (greyish, sabkha colour). Note that trench and core SQ-T2 most probably represents a facies succession corresponding to a palaeo-tidal-channel.

ptygmatic) anhydrite within Pleistocene aeolian dune and Holocene gypsum-rich, bioclastic lagoonal packstone deposits.

Trench and Core SQ-T1 (Figs 1B, 3 and 4)

Topographically lower areas of the inner, upper sabkha environment are recognized by a well-developed buckled polygonal halite crust. This part of the sabkha is often flooded during winter and early spring by ponded rain water (Kirkham, 1997, 1998). Obvious is the occurrence of a relatively thick (about 60 cm) intercalation of nodular to contorted mosaic (enterolithic or ptygmatic) and, dominantly, distorted bedded mosaic (Maiklem *et al.*, 1969) anhydrite, about 10 cm below the buckled polygonal halite crust (Figs 3 and 4). The lower, approximately 45 cm thick, distorted bedded mosaic anhydrite is interpreted to replace/displace incipiently cross-bedded aeolian carbonate-rich sands. These correspond to the unconsolidated brownish, quartz- and carbonate-rich Pleistocene aeolian dune deposits that underlie the anhydrite (Fig. 4). The upper approximately 15 cm thick nodular to contorted anhydrite is interpreted to have replaced a bioclastic lagoonal packstone, similar to the gypsum-rich lagoonal carbonates that overlie the anhydrite (Fig. 4). The boundary between the dominantly distorted bedded mosaic and the nodular to contorted mosaic (enterolithic or ptygmatic) anhydrite most probably marks the Pleistocene-Holocene boundary (Figs 3 and 4).

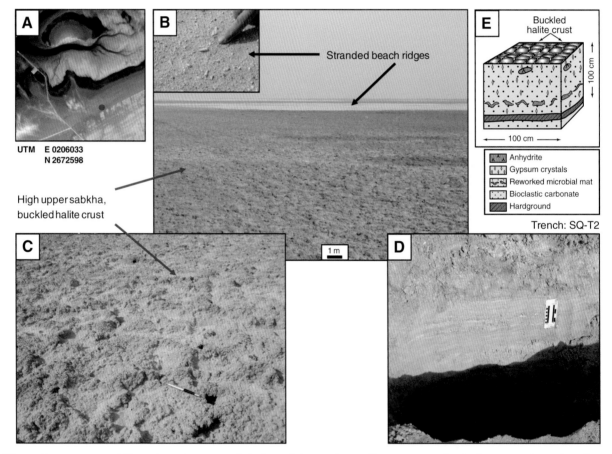

Fig. 6. Outer, upper sabkha sub-environment. (A) The red dot on the satellite image marks the location of trench and core SQ-T2. (B) Stranded beach ridges (whitish area, black arrows), composed of cerithid gastropods, and high upper sabkha displaying buckled halite crust (red arrows). (C) Close-up of buckled halite crust displaying teepee desiccation structures. (D) Trench showing from bottom to top: sand-rich, bioclastic packstone (lagoonal and shallow tidal-channel/tidal-creek deposits) with intercalated reddish (oxidation) horizon; sand-rich bioclastic grainstone (shallow tidal-channel/tidal-creek deposit); and gypsum-rich bioclastic packstone (lagoonal and shallow tidal-channel/tidal-creek deposits). (E) Log of trench stratigraphy in (D).

AMS radiocarbon dating of the anhydrite interval shows an age of 12,860 ± 50 ^{14}C yr BP (sample SQ-T1-2; Fig. 4). This sample demonstrates the importance to distinguish between salina- or saltern-type (subaqueous) and sabkha-type (subaerial) anhydrite (Strohmenger *et al.*, 1996; Steinhoff & Strohmenger, 1999; Kirkham, 2004). In the case of a salina-type anhydrite, the radiocarbon age would correspond to the organic matter deposited penecontemporaneously with anhydrite precipitation. The analyzed radiocarbon age would therefore approximately represent the age of the anhydrite. In the case of a sabkha-type anhydrite, the dated organic matter would correspond to the age of the host sediment and not to the age of the anhydrite. The age of the anhydrite might be considerably (many thousand years?) younger than the sediment that it was replacing.

The felted fabric of the anhydrite (Maiklem *et al.*, 1969), seen in petrographic thin-section photomicrographs from sample SQ-T1-2 (Fig. 4), supports the interpretation of a sabkha-type anhydrite (Steinhoff & Strohmenger, 1999). The Pleistocene age of the sample most probably corresponds to the age of the aeolian dune deposit that the anhydrite was replacing during Holocene time. Besides pervasive halite cementation, pseudomorphic replacement of bassanite-after-gypsum can be seen in thin section from the surface-buckled polygonal halite crust (sample SQ-T1-5; Fig. 4).

SUPRATIDAL, COASTAL SABKHA ENVIRONMENTS: STRANDED BEACH RIDGES AND OUTER SABKHA, SEAWARD OF STRANDED BEACH RIDGES

Based on surface morphology and vertical evaporite mineral distribution, the seaward portion of

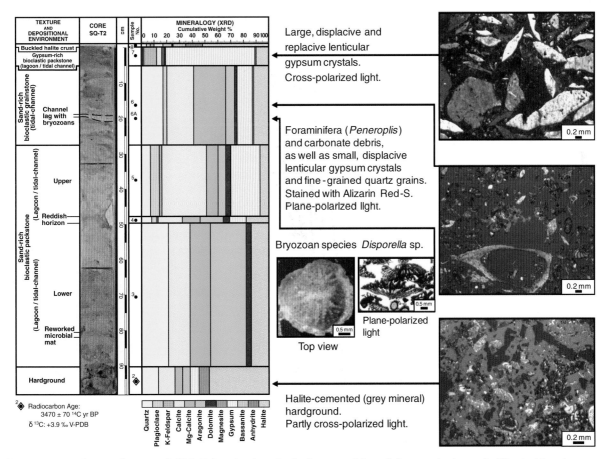

Fig. 7. Resinated core photograph (SQ-T2) and mineralogical composition of the trench shown in Fig. 6. Also shown are thin-section photomicrographs, as well as a photograph of the bryozoan species *Disporella* sp. The core photograph displays a thin channel-lag deposit dominated by the stenohaline bryozoan species *Disporella* sp. (sample SQ-T2-6A). Radiocarbon dating and carbon-isotope results of the microbial mat and the hardground are also displayed. See Fig. 5 for location.

the coastal sabkha (outer sabkha; Fig. 1A and B) can be subdivided into four facies belts (Butler *et al.*, 1982; Alsharhan & Kendall, 1994, 2002; Kendall &., 1994; Alsharhan *et al.*, 1998; Fig. 5). Furthermore, seaward of the stranded beach ridges, the morphology of the surface halite crust allows a subdivision of the outer, upper sabkha (upper supratidal) environment into three sub-environments (Strohmenger *et al.*, 2004, 2007). The distance between the seaward extent of the stranded beach ridges and the lowermost intertidal to shallow-subtidal, peloid-skeletal tidal flat environment (SQ-T9) is approximately 1500 m (Fig. 5B). The relief change over

this distance is less than 3 m. This translates into a very gentle slope angle of approximately 0.1 degree.

The seaward portion facies belts can be characterized as follows:

(1) **Upper Supratidal, Stranded Beach Ridges;** topographic highs, a few centimetres to decimetres above the adjacent upper sabkha environment:
   (a) identified as "whitish stripes" on satellite images (Figs 2B and 5);
   (b) approximately 1500 m wide;
   (c) composed mainly of cerithid gastropods.

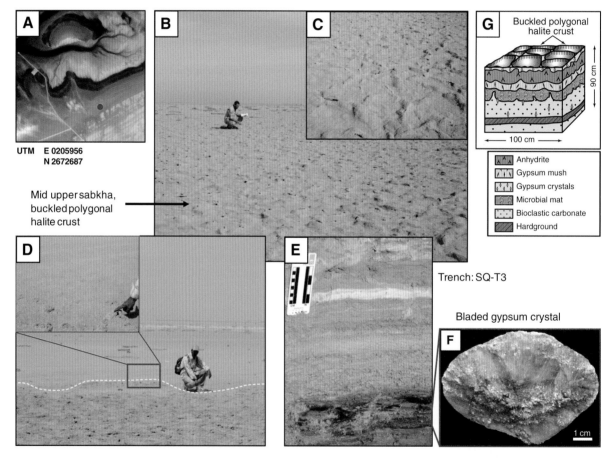

Fig. 8. Outer, upper sabkha sub-environment. (A) The red dot on the satellite image marks the location of trench and core SQ-T3. (B) Mid upper sabkha displaying buckled polygonal halite crust. (C) Close-up of buckled polygonal halite crust displaying desiccation polygons and teepee structures. (D) Very rare, exceptional marine flooding due to spring tides and/or strong seasonal "Shamal" winds during winter and spring months can lead to the dissolution of the mid upper sabkha buckled polygonal halite crust. The dashed line marks the boundary between high upper sabkha (unaffected by marine flooding) and the mid upper sabkha, showing remnants of dissolved buckled polygonal halite crust. (E) Trench showing, from bottom to top: sand-rich, bioclastic packstone (lagoonal and shallow tidal-channel/tidal-creek deposits), crinkly-laminated microbial mat (displaying erosive/burrowed upper surface and reworking), gypsum mush facies, anhydrite layer (whitish band), and anhydrite-rich, bioclastic packstone (lagoonal and shallow tidal-channel/tidal-creek deposits). (F) Large bladed (lensoid), poikilotopic gypsum crystal with intercalated microbial mat (dark layers within gypsum crystal). (G) Log of trench stratigraphy in (E).

(2) **Upper Supratidal** (upper sabkha/upper salt-flat facies):

(a) light gray area on satellite image (Fig. 5),

(b) approximately 600 m wide.

(c) **High upper sabkha**, buckled halite crust displaying teepee structures:

    (i) seaward and landward (topographically higher areas) of stranded beach ridges;

    (ii) minor anhydrite and gypsum-rich upper layer below surface.

(d) **Mid upper sabkha**, buckled polygonal halite crust displaying teepee and polygonal structures:

    (i) seaward and landward (topographically lower areas) of stranded beach ridges;

    (ii) anhydrite nodules and layers within the upper 20 cm below surface;

    (iii) buckled polygonal halite crust can very rarely be dissolved due to exceptional marine flooding during winter and spring months.

(e) **Low upper sabkha**, smooth polygonal halite crust:

    (i) transition between middle sabkha and upper sabkha;

    (ii) seaward limit corresponds to the high-water mark;

    (iii) anhydrite directly underlying surface;

    (iv) smooth polygonal halite crust is commonly dissolved by marine flooding during winter and spring months.

(3) **Middle Supratidal** (middle sabkha/middle salt-flat facies):

(a) dark grey area on satellite image (Fig. 5);

(b) approximately 100–200 m wide;

(c) finely-crystalline anhydrite: white, "cottage-cheese-like" appearance;

(d) oxidized anhydrite polygons on surface.

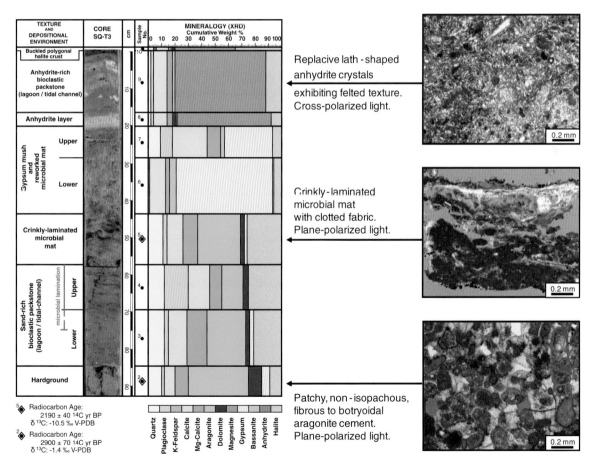

Fig. 9. Resinated core photograph (SQ-T3) and mineralogical composition of the trench shown in Fig. 8, and thin-section photomicrographs. Radiocarbon dating and carbon-isotope results of the microbial mat and the hardground are also displayed. See Fig. 5 for location.

(4) **Lower Supratidal** (lower sabkha/lower salt-flat facies; Fig. 5B):
   (a) dark grey area on satellite image (Fig. 5).
   (b) approximately 200–300 m wide;
   (c) gypsum mush facies (thickness, 5–25 cm);
   (d) shiny, sparkling (gypsum crystals) surface.

Trench and Core SQ-T2 (Figs 5, 6 and 7)

Topographically higher areas represent the high upper sabkha sub-environment landward and seaward of the stranded beach ridges (Figs 5 and 6B). This part of the sabkha is identified morphologically by a buckled halite crust (teepee structures) with faintly developed polygonal structures (Fig. 6A and B). It is represented by trench and core SQ-T2 (Figs 6C and 7). No continuous microbial mat was identified within the trench or core. The interval between a well-developed subsurface

hardground and the surface shows from base to top (Figs 6D and 7): sand-rich bioclastic packstone (lagoonal and shallow tidal-channel/tidal-creek deposits); sand-rich bioclastic grainstone with channel-lag deposits (shallow tidal-channel/tidal-creek deposit); and gypsum-rich bioclastic packstone (lagoonal and shallow tidal-channel/tidal-creek deposits). Only a minor amount of anhydrite and bassanite (bassanite-after-gypsum pseudomorphs) could be identified (Fig. 7). This trench and core most probably represent deposits of a palaeo-tidal-channel, cross-cutting the sabkha and the stranded beach ridges (Figs 1B and 5).

A thin channel lag deposit (sample SQ-T2-6A) is dominated by the bryozoan species *Disporella* sp. *Disporella* belongs to the Order Cyclostomata, Suborder Rectangulata, Family Lichenoporidae, and has not been recorded previously from the United Arab Emirates (Paul D. Taylor, 2006,

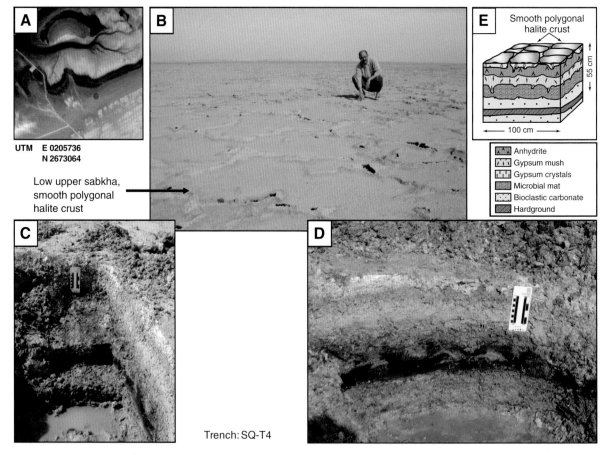

Fig. 10. Outer, upper sabkha sub-environment. (A) The red dot on the satellite image marks the location of trench and core SQ-T4. (B) Low upper sabkha smooth polygonal halite crust exhibiting desiccation polygons. (C) Trench showing from bottom to top: hardground, sand-rich bioclastic grainstone (shallow tidal-channel/tidal-creek deposit); crinkly-laminated microbial mat; gypsum mush facies; and discontinuous (whitish) anhydrite layer. (D) Close-up of crinkly-laminated microbial mat displaying erosive/burrowed upper surface and reworking. (E) Log of trench stratraphy in (D).

pers. comm.). Cyclostomes such as *Disporella* are benthic, epifaunal, and colonial. Colonies need hard (e.g. shells or rocks) or firm (e.g. aquatic plants) substrates for larval attachment. In the case of the UAE specimens, the substrate is not obvious as the colonies have apparently become detached. However, some of the specimens in thin section are broadly cup-shaped (Fig. 7), probably with a limited area of attachment to the substrate. This suggests that they were attached to flexible substrates (e.g. seagrass; Paul D. Taylor, 2006, pers. comm.). Like other suspension-feeding bryozoans, *Disporella* requires good water circulation to provision colonies with a phytoplanktonic food resource. All known cyclostomes are stenohaline and do not occur in environments with salinities appreciably above or below normal seawater (Paul D. Taylor, 2006, pers. comm.). The latter fact is interesting, as today the Arabian Gulf clearly shows elevated salinities, especially in lagoonal areas behind islands (as is the case at the studied location).

The occurrence of the *Disporella* in sediments presumably deposited ca 2000 years before present might indicate somewhat more normal marine conditions during times when sea level was slightly higher (2 to 3 m) and the islands offshore Abu Dhabi (e.g. Al-Rufayq Island and Al-Qanatir Island) were inundated. The absence of stenohaline bryozoans like *Disporella* from recent sediments indicates the salinity change of the Arabian Gulf over time, consistent with the global eustatic trend. Conventional radiocarbon dating carried out on one hardground sample (sample SQ-T2-2; Fig. 7) shows an age of 3470 ± 70 uncalibrated ^{14}C yr BP.

Trench and Core SQ-T3 (Figs 5, 8 and 9)

Topographically slightly lower areas representing the mid upper sabkha sub-environment. This part

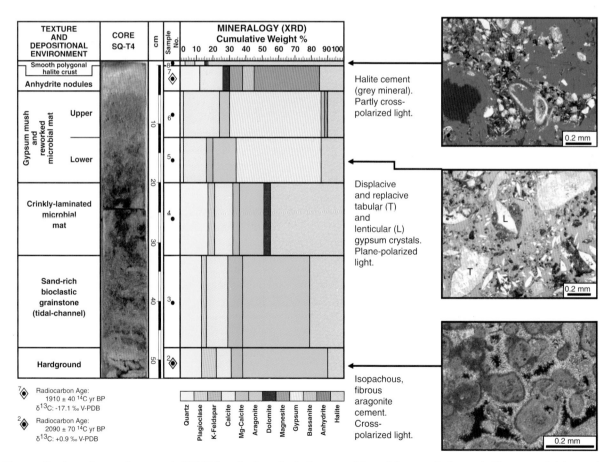

Fig. 11. Resinated core photograph (SQ-T4) and mineralogical composition of the trench shown in Fig. 10, and thin-section photomicrographs. Radiocarbon dating and carbon-isotope results of the anhydrite-dominated layer and the hardground are displayed. The analyzed radiocarbon age of the anhydrite sample SQ-T4-7 is interpreted to correspond to the age of the crinkly-laminated microbial mat (sample SQ-T4-4; see text for explanation). See Fig. 5 for location.

of the sabkha is identified morphologically by a buckled polygonal halite crust with well-developed polygonal and teepee structures (Fig. 8A and B). It is represented by trench and core SQ-T3 (Figs 8E and 9). Very rare, exceptional marine flooding due to spring tides and/or strong seasonal "Shamal" winds during winter and spring months can temporarily dissolve the buckled polygonal halite crust (Fig. 8D). The buckled polygonal halite crust that was dissolved in late spring (May) of 2005 (Fig. 8D) reformed during the summer months (June to September) of the same year. A well-developed microbial mat was identified about 40 cm below the surface (Fig. 8E). The interval between a subsurface hardground and the surface

shows, from base to top (Figs 8E and 9): sand-rich bioclastic packstone (lagoonal and shallow tidal-channel/tidal-creek deposits); crinkly-laminated microbial mat (Fig. 8E) containing replacive, bladed (lensoid) gypsum crystals (Fig. 8F); gypsum mush facies; anhydrite layer; and anhydrite-rich bioclastic packstone (lagoonal and shallow tidal-channel/tidal-creek deposits). The surface of the microbial mat shows erosion and/or burrowing, and remnants of the reworked microbial mat are intercalated within the overlying gypsum mush facies (Figs 8D and 9).

Radiocarbon dating (Fig. 9) carried out on one microbial mat (AMS) and one hardground sample (conventional) show an age of 2190 ± 40

Fig. 12. Transition of low upper sabkha to middle sabkha. (A) Satellite image showing the transition from low upper sabkha (black arrow) to middle sabkha (red arrow); clearly visible by change in colour. Light grey area (black arrow): low upper sabkha. Darker grey area (red arrow): middle sabkha. This transition also corresponds to the high-water mark (HWM, dotted white line). Very rare, exceptional marine flooding due to spring tides and/or strong seasonal "Shamal" winds during winter and spring months can move the high-water mark close to the stranded beach ridges (dashed white line, also marking the transition of high upper sabkha to mid upper sabkha; see Fig. 8D). (B) Boundary between low upper sabkha (background: darker area, black arrows) and middle sabkha (foreground: lighter area, red arrows). (C) Close-up of low upper sabkha smooth polygonal halite crust (black arrow). (D) Close-up of middle sabkha ("cottage-cheese-like") oxidized anhydrite polygons (red arrow). (E) Common marine flooding during winter and spring months causes dissolution of smooth polygonal halite crust and exposure of underlying anhydrite layer.

uncalibrated ^{14}C yr BP for the microbial mat (sample SQ-T3-5) and 2900 $\pm$ 70 uncalibrated ^{14}C yr BP for the hardground (sample SQ-T3-2; Fig. 9). The microbial mat also shows a distinctively more negative δ^{13}C value (Fig. 9). An accumulation rate of approximately 1 cm per 18 ^{14}C years (0.6 mm yr^{-1}) was calculated for the interval between the radiocarbon dated hardground (sample SQ-T3-2) and the microbial mat (sample SQ-T3-5; Fig. 9).

Trench and Core SQ-T4 (Figs 5, 10, 11 and 12)

Trench and core SQ-T4 (Figs 10 and 11) represent the low upper sabkha sub-environment, identified morphologically by a smooth polygonal halite crust with well-developed desiccation polygons (Fig. 10B). The seaward limit of the low upper sabkha corresponds to the high-water mark (Fig. 12A). It marks the transition between the upper (Fig. 12A, B and C) and middle sabkha (Fig. 12A, B and D). The surfacial smooth polygonal halite crust can temporarily get dissolved by marine flooding during winter and spring months (Fig. 12E). The interval between a subsurface hardground and the surface shows, from base to top (Figs 10B, C and 11): sand-rich bioclastic grainstone (shallow tidal-channel/tidal-creek deposit); crinkly-laminated microbial mat; gypsum mush facies; and a discontinuous (whitish) anhydrite layer close to the surface. The surface of

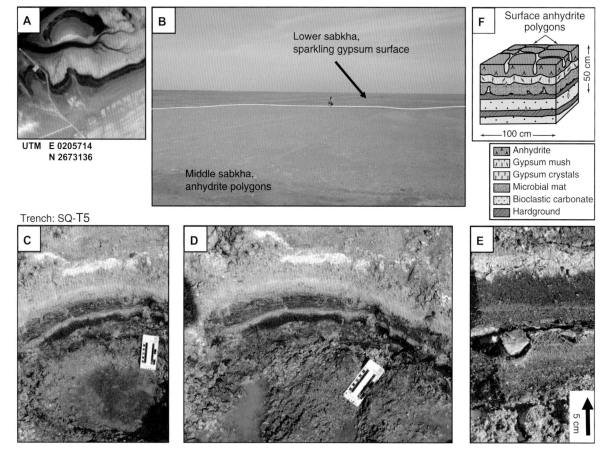

Fig. 13. Transition of middle sabkha to lower sabkha. (A) The red dot on the satellite image marks the location of trench and core SQ-T5. (B) Boundary between middle sabkha (foreground: lighter area) and lower sabkha (background: darker area, black arrow). (C) Trench, showing from bottom to top: lower hardground; uncemented sand-rich, bioclastic grainstone (shallow tidal-channel/tidal-creek deposit); irregular upper incipient hardground (cemented, reddish, sand-rich, bioclastic grainstone: shallow tidal-channel/tidal-creek deposit); crinkly-laminated microbial mat (displaying erosive/burrowed upper surface and reworking); gypsum mush facies; and discontinuous (whitish) anhydrite layer. (D) Close-up of uncemented and cemented, sand-rich bioclastic grainstone. (E) Close-up of thick (about 20 cm) crinkly-laminated microbial mat overlying upper, irregular incipient hardground (cemented, sand-rich bioclastic grainstone). (F) Log of trench stratigraphy shown in (C).

the microbial mat shows erosion and/or burrowing, and remnants of the reworked microbial mat are intercalated within the overlying gypsum mush facies (Figs 10D and 11).

Radiocarbon dating (Fig. 11) carried out on one anhydrite (AMS) and one hardground sample (conventional) show an age of 1910 ± 40 uncalibrated ^{14}C yr BP for the organic matter within the anhydrite (sample SQ-T4-7) and 2090 ± 70 uncalibrated ^{14}C yr BP for the hardground (sample SQ-T4-2; Fig. 11). The relative high amount of fibrous aragonite cement, identified in thin section representing the hardground sample (Fig. 11) appears to have influenced the dating result. In contrast to the other analyzed hardground samples, where the influence of the diagenetic age (age of the carbonate cement) is considered to be negligible, the radiocarbon age of sample SQ-T4-2 appears to be slightly too young (younger than the

depositional age) due to post-depositional cementation. The obtained radiocarbon age is therefore interpreted to represent a mixture of the diagenetic age (age of the fibrous aragonite cement, postdating the deposition of the carbonate grains) and the depositional age (age of the carbonate grains; Fig. 11). The depositional age of the sediment that formed the hardground might be slightly older (200 to 300 years?) than the analyzed 2090 uncalibrated ^{14}C yr BP.

The organic matter of the anhydrite sample shows a distinctively more negative $\delta^{13}C$ value (Fig. 11). An accumulation rate of approximately 1 cm per 3.6 ^{14}C years (3 mm yr^{-1}) was calculated for the interval between the radiocarbon dated hardground (sample SQ-T4-2) and the organic content of the anhydrite layer (sample SQ-T4-7, Fig. 11). This rate is anomalously high compared with accumulation rates of around 0.5 mm yr^{-1}

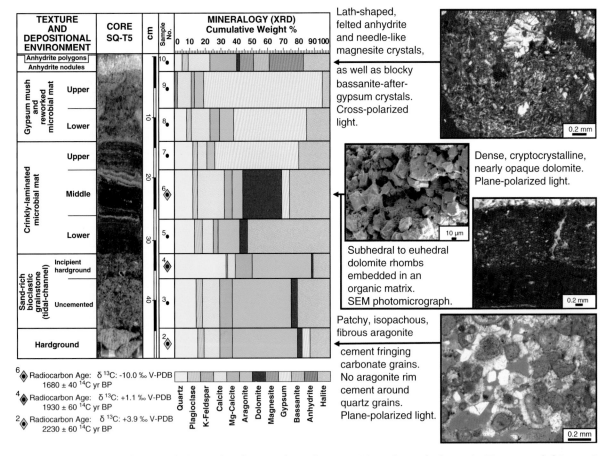

Fig. 14. Resinated core photograph (SQ-T5) and mineralogical composition of trench shown in Fig. 13, and thin-section photomicrographs, as well as scanning electron microscope (SEM) photomicrograph (SEM photomicrograph courtesy of the Geomicrobiology Group, ETH-Zurich). Radiocarbon dating and carbon-isotope results of the middle laminated microbial mat, the upper irregular incipient hardground (cemented, sand-rich bioclastic grainstone), and the lower hardground are also displayed. See Fig. 5 for location.

calculated for trenches and cores SQ-T3 (1 cm per 18 ^{14}C years), SQ-T5 (1 cm per 21 ^{14}C years) and SQ-T7 (1 cm per 22 ^{14}C years). It might partly be explained by the inferred anomalously young age of the hardground (see above) and, most probably, by erosion and reworking of the microbial mat.

As noted above, the radiocarbon age of the anhydrite layer corresponds to the radiocarbon age of the organic content within the sediment. Erosion is clearly visible on top of the microbial mat (Figs 10C and 11). During subsequent flooding, parts of the microbial mat have been reworked and incorporated into the overlying sediments. The negative δ^{13}C value ($-17.1‰$ V-PDB) also indicates an organic (microbial) origin of the dated matter. Provided the age given for the anhydrite is correct, the age of the anhydrite layer might actually represent the age of the microbial mat deposited approximately 25 cm above the radiocarbon dated hardground (Fig. 11). Assuming a ^{14}C age

approximately 300 years older for the hardground, the calculated accumulation rate for the interval between the hardground and the microbial mat would be approximately 1 cm per 19 ^{14}C years (0.5 mm yr^{-1}).

Trench and Core SQ-T5 (Figs 5, 12, 13 and 14)

Trench and core SQ-T5 (Figs 13 and 14) represent the middle sabkha environment (Fig. 5B), identified morphologically by "cottage-cheese-like" and oxidized anhydrite polygons at the surface (Figs 12D and 13A). The interval between a lower subsurface hardground and the surface shows from base to top (Figs 13C, D and 14): uncemented, sand-rich, bioclastic grainstone (shallow tidal-channel/tidal-creek deposit), cemented (aragonite cement), reddish, sand-rich, bioclastic grainstone (incipient hardground: shallow tidal-channel/tidal-creek deposit), crinkly-laminated microbial

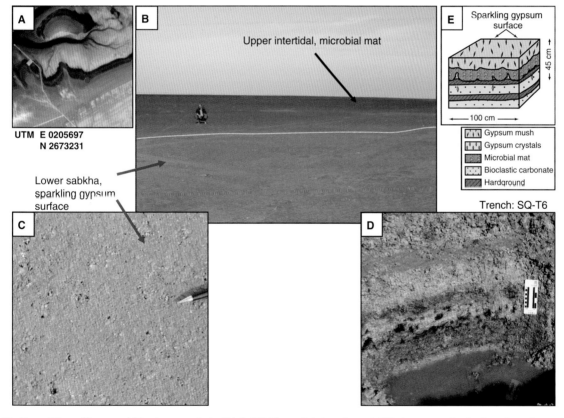

Fig. 15. Transition of lower sabkha to upper intertidal. (A) The red dot on the satellite image marks the location of trench and core SQ-T6. (B) Boundary between lower sabkha (foreground: lighter area, red arrow) and upper intertidal crinkled microbial mat (background: darker area, black arrow). (C) Close-up of lower sabkha, sparkling gypsum mush facies surface (red arrow). (D) Trench showing from bottom to top: hardground, uncemented and cemented (upper incipient hardground), sand-rich bioclastic grainstone (shallow tidal-channel/tidal-creek deposit); crinkly-laminated microbial mat (displaying erosive/burrowed upper surface and reworking); and gypsum mush facies. (E) Log of trench stratigraphy shown in (D).

mat, gypsum mush facies, and discontinuous (whitish) anhydrite layer. The crinkly-laminated microbial mat can be subdivided into lower, middle and upper microbial mat (Figs 13E and 14). The middle microbial mat shows a relatively high amount of microcrystalline dolomite (Fig. 14). The surface of the microbial mat shows minor erosion and/or burrowing, and remnants of the reworked microbial mat are intercalated within the overlying gypsum mush facies (Figs 13C, D and 14). Radiocarbon dating carried out on one microbial mat (AMS) and two hardground samples (conventional) show an age of 1680 ± 40 uncalibrated ^{14}C yr BP for the microbial mat (sample SQ-T5-6), 1930 ± 60 uncalibrated ^{14}C yr BP for the upper cemented incipient hardground layer (sample SQ-T5-4), and 2230 ± 60 uncalibrated ^{14}C yr BP for the lower hardground (sample SQ-T5-2; Fig. 14).

Using a scanning electron microscope equipped with a cryogenic preparation system, the fine-crystalline dolomite displays subhedral to euhedral dolomite rhombs embedded in an organic matrix of extrapolymeric substances (EPS; Bontognali et al., 2010; Fig. 14). Formation of dolomite therefore is interpreted to have been induced by Sulphate-reducing bacteria (SRB), as suggested by McKenzie (1981), Baltzer *et al.* (1994), Vasconcelos *et al.* (1995, 2005), Vasconcelos & McKenzie (1997), Warthmann *et al.* (2000) and van Lith *et al.* (2003a, 2003b). The organic matter of the microbial mat shows a distinctively more negative δ ^{13}C value compared with the hardground samples (Fig. 14). An accumulation rate of approximately 1 cm per 20 ^{14}C years (0.5 mm yr^{-1}) was calculated for the interval between the two radiocarbon dated hardgrounds (sample SQ-T5-2 and sample SQ-T5-4), as well as for the interval between the incipient hardground (sample SQ-T5-4) and the microbial mat (sample SQ-T5-6; Fig. 14).

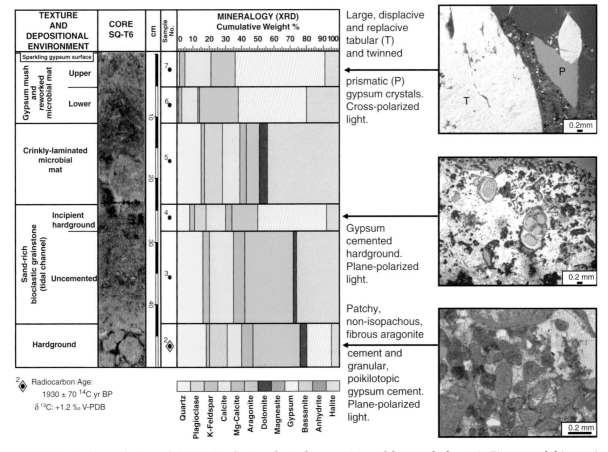

Fig. 16. Resinated core photograph (SQ-T6) and mineralogical composition of the trench shown in Fig. 15, and thin-section photomicrographs. Radiocarbon dating and carbon-isotope results of the hardground are also displayed. See Fig. 5 for location.

A thin-section photomicrograph of the lower hardground (sample SQ-5-2; Fig. 14) shows the substrate-dependency of the aragonite rim-cement. Only carbonate (aragonite) grains show isopachous fibrous rim cement. No cement overgrowth is observed on quartz grains.

Trench and Core SQ-T6 (Figs 5, 15 and 16)

Trench and core SQ-T6 (Figs 15 and 16) represent the lower sabkha environment (Fig. 5B). This, most seaward, part of the sabkha (Fig. 15B) is identified by a soft surface displaying shiny, sparkling gypsum crystals (Fig. 15C). The lower sabkha grades into the upper intertidal crinkled microbial mat environment (Fig. 15B). The interval between a subsurface hardground and the surface shows from base to top (Figs 15D and 16): uncemented sand-rich, bioclastic grainstone (shallow tidal-channel/tidal-creek deposit); cemented (gypsum cement), sand-rich bioclastic grainstone (incipient hard-ground: shallow tidal-channel/tidal-creek deposit); crinkly-laminated microbial mat; and gypsum mush facies. The surface of the microbial mat shows erosion and/or burrowing, and remnants of the reworked microbial mat are intercalated within the overlying gypsum mush facies (Figs 15D and 16). Radiocarbon dating carried out on hardground sample SQ-T6-2 (conventional) shows an age of 1930 ± 70 uncalibrated ^{14}C yr BP (Fig. 16).

INTERTIDAL, MICROBIAL MAT ENVIRONMENTS

Depending on the relative amount of subaerial exposure the morphology of the microbial mats

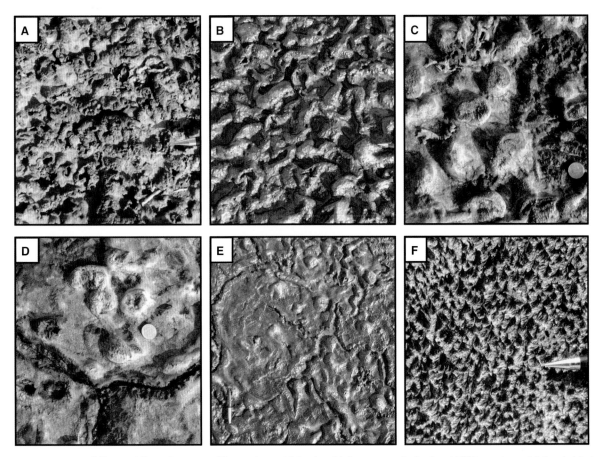

Fig. 17. Upper, middle, middle to lower, and lower intertidal microbial mat morphologies. (A) Upper intertidal: crinkled or crenulated microbial mat. (B) Middle intertidal: blistered microbial mat, often replacing pinnacle mat. (C) Middle intertidal: pinnacle or domed microbial mat. (D) Middle to lower intertidal: transition of pinnacle to polygonal microbial mat showing desiccation cracks with infill of reddish silty material. (E) Lower intertidal: smooth polygonal microbial mat. (F) Lower intertidal: tufted or cinder-like microbial mat.

varies. Four intertidal environments can be distinguished (Kendall & Skipwith, 1968, 1969; Park, 1977; Alsharhan & Kendall, 1994, 2002; Kendall *et al.*, 1994; Peebles *et al.*, 1995; Alsharhan *et al.*, 1998; Strohmenger *et al.*, 2004, 2007). The intertidal microbial mat is between approximately 300 to 500 m wide. The intertidal environments can be characterized as follows:

(1) **Upper Intertidal** (Fig. 17A), crinkled or crenulated microbial mat above gypsum mush facies:
  (a) thin, leathery, wrinkled microbial mat;
  (b) low preservation potential.
(2) **Middle Intertidal** (Fig. 17B–D):
  (a) blistered microbial mat (Fig. 17B):
    (i) variety of crinkled microbial mat;
    (ii) grading seaward and laterally into pinnacle and/or polygonal microbial mat;

(iii) may replace pinnacle microbial mat.
  (b) pinnacle, dome-like (domed) microbial mat (Fig. 17C):
    (i) grading seaward into polygonal microbial mat (Fig. 17D);
    (ii) may be replaced by blistered microbial mat (Fig. 17B).
  (c) low to moderate preservation potential.
(3) **Middle to Lower Intertidal** (Fig. 17D and E):
  (a) polygonal microbial mat with low-relief pinnacles (Fig. 17D);
  (b) smooth polygonal microbial mat (Fig. 17E);
  (c) grading seaward into tufted or cinder-like microbial mat;
  (d) characteristic upturned edges;
  (e) alternation of organic and sedimentary laminae;
  (f) high preservation potential.

Fig. 18. Upper intertidal microbial mat environment. (A) The red dot on the satellite image marks the location of trench and core SQ-T7. (B) Upper intertidal, crinkled microbial mat. (C) Close-up of upper intertidal, thin, leathery, crinkled microbial mat. (D) Trench showing from bottom to top: hardground; crinkly-laminated microbial mat (displaying minor erosive/burrowed upper surface and reworking); gypsum mush facies; and surface crinkled microbial mat. (E) Log of trench stratigraphy shown in (D).

(4) **Lower Intertidal** (Fig. 17F):
  (a) tufted or cinder-like microbial mat (Fig. 17F);
  (b) pustular microbial mat;
    (i) variety of tufted or cinder-like microbial mat;
  (d) low preservation potential.

Trench and Core SQ-T7 (Figs 5, 18 and 19)

Trench and core SQ-T7 (Figs 18 and 19) represent the upper intertidal environment (Fig. 5B), identified by a thin, leathery, crinkled microbial mat (Fig. 18B and C). The interval between a subsurface hardground and the surface crinkled microbial mat shows from base to top (Figs 18D and 19): subsurface irregular, crinkly-laminated microbial mat overlain by gypsum mush facies. The surface of the microbial mat shows erosion and/or burrowing, and remnants of the reworked microbial mat are intercalated within the overlying gypsum mush facies (Figs 18C and 19).

Radiocarbon dating (Fig. 18) carried out on one microbial mat (AMS) and one hardground sample (conventional) show an age of 880 ± 40 uncalibrated ^{14}C yr BP for the microbial mat (sample SQ-T7-3) and 1280 ± 60 uncalibrated ^{14}C yr BP for the hardground (sample SQ-T3-2; Fig. 19). The analyzed age for the hardground as well as its position directly beneath the crinkly-laminated microbial mat makes it very likely that these carbonates correspond to the upper cemented incipient hardground found in trench and core SQ-T6 (sample SQ-T6-4; Fig. 17). The microbial mat also shows a distinctively more negative δ^{13}C value (Fig. 19). An accumulation rate of approximately 1 cm per 22 ^{14}C years (0.5 mm yr^{-1}) was calculated for the interval between the radiocarbon age-dated hardground (sample SQ-T7-2) and the microbial mat (sample SQ-T7-3; Fig. 19).

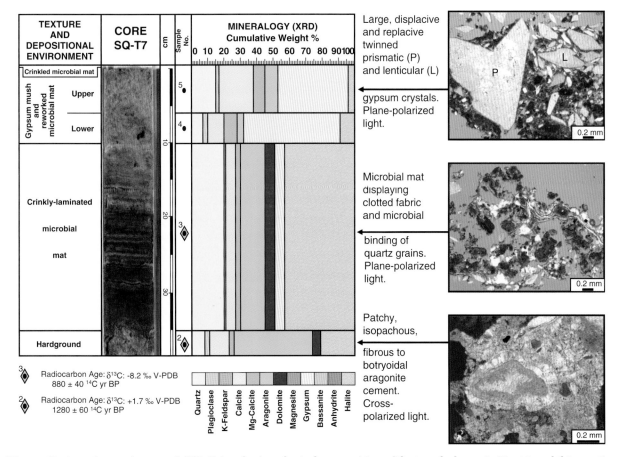

Fig. 19. Resinated core photograph (SQ-T7) and mineralogical composition of the trench shown in Fig. 18, and thin-section photomicrographs. Radiocarbon dating and carbon-isotope results of the hardground and the crinkly-laminated microbial mat are also displayed. See Fig. 5 for location.

Trench and Core SQ-T8 (Figs 5, 20 and 21)

Trench and core SQ-T8 (Figs 20 and 21) represent the lower intertidal environment (Fig. 5B), identified by a relatively thick, surface microbial mat displaying well-developed polygonal structures (Fig. 20B–E). The interval between a subsurface hardground and the surface smooth polygonal microbial mat shows from base to top (Figs 20F and 21): sand-rich, bioclastic grainstone (shallow tidal-channel/tidal-creek deposit), crinkly-laminated microbial mat, and sand-rich, bioclastic packstone (lagoonal and shallow tidal-channel/tidal-creek deposits: peloid-skeletal tidal-flat facies). The surface of the microbial mat shows erosion and/or burrowing, and remnants of the reworked microbial mat are intercalated within the overlying peloid-skeletal tidal-flat facies (Figs 20F and 21). The facies stacking pattern (more seaward facies: lowermost intertidal to shallow-subtidal, peloid-skeletal tidal-flat facies overlying more landward facies: lower to middle intertidal microbial mat) indicate a minor sea-level rise on top of the microbial mat.

Conventional radiocarbon dating (Fig. 21) carried out on two samples of the hardground show an age of 3460 ± 70 uncalibrated ^{14}C yr BP for hardground sample SQ-T8-2A and 1550 ± 60 uncalibrated ^{14}C yr BP for hardground sample SQ-T8-2B (Fig. 21). The unexpected "old" age of hardground sample SQ-T8-2A corresponds to dark grey rock fragments (cemented, organic-rich mudstone exhibiting evidence of relict microbial activities) that differ in colour and composition from the

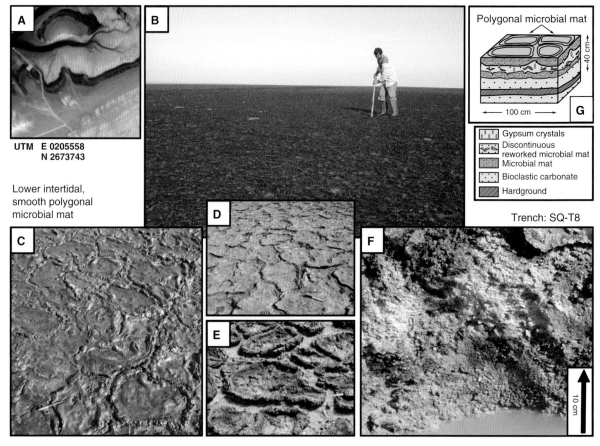

Fig. 20. Lower intertidal microbial mat environment. (A) The red dot on the satellite image marks the location of trench and core SQ-T8. (B) Lower intertidal, smooth polygonal microbial mat. (C) Close-up of wet lower intertidal, smooth polygonal microbial mat showing characteristic upturned edges. (D) Close-up of dry lower intertidal, polygonal microbial mat (same location as (B) showing buckling of polygon edges. (E) Transition of lower intertidal, polygonal microbial mat to tufted or cinder-like microbial mat. (F) Trench showing from bottom to top: sand-rich bioclastic grainstone (shallow tidal-channel/tidal-creek deposit); crinkly-laminated microbial mat (displaying pronounced erosive/burrowed upper surface and reworking); sand-rich bioclastic packstone with reworked microbial mat (lagoon and shallow tidal-channel/tidal-creek deposits); and surface smooth polygonal microbial mat. Log of trench stratigraphy shown in (F).

surrounding light grey cemented sediments (sample SQ-T8-2B; Fig. 21). Tidal channels/tidal creeks might have eroded down into and subsequently re-deposited "older" indurate sediments (hardground sample SQ-T8-2A: dark grey rock fragments). The actual age of the *in situ* carbonates is shown by hardground sample SQ-T8-2B (light grey rock fragments).

LOWERMOST INTERTIDAL TO SHALLOW-SUBTIDAL, PELOID-SKELETAL TIDAL FLAT ENVIRONMENT

The lowermost intertidal to shallow-subtidal environment is represented by the peloid-skeletal packstone facies (Strohmenger *et al.*, 2004, 2007; Fig. 22A). The sediment surface shows wave ripples (partially eroded during subaerial exposure; Fig. 22B), gastropod (cerithids and littorinids) grazing-traces within shallow-subtidal tidal channels/tidal creeks and tidal ponds (Fig. 22C), and *Skolithos*-type crab burrows (Fig. 22D). The peloid-skeletal packstone facies belt (lagoonal and shallow tidal-channel/tidal-creek deposits) is between approximately 2500 to 3500 m wide and displays the following characteristics:

Lowermost Intertidal to Shallow-Subtidal (Fig. 22):

(a) peloid-skeletal tidal flat;
(b) shallow lagoon and shallow tidal channels/tidal creeks;
(c) wave ripples;
(d) cerithid and littorinid gastropod grazing-traces;
(e) *Skolithos*-type burrows.

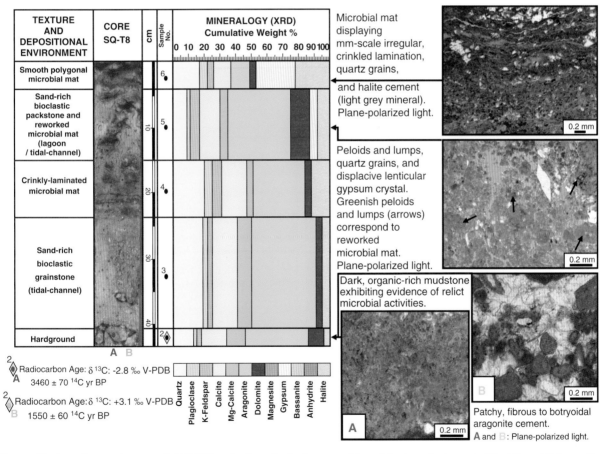

Fig. 21. Resinated core photograph (SQ-T8) and mineralogical composition of the trench shown in Fig. 20, and thin-section photomicrographs. Radiocarbon dating and carbon-isotope results of the hardground are also displayed. The core photograph of the hardground shows dark grey rock fragments (left half of hardground, sample SQ-T8-2A: interpreted to be older, reworked carbonates) and light grey rock fragments (right half of hardground, sample SQ-T8-2B: interpreted to be younger, *in situ* carbonates). The thin-section photomicrograph of hardground sample SQ-T8-2A corresponds to dark grey rock fragments. The thin-section photomicrograph of hardground sample SQ-T8-2B corresponds to light grey rock fragments. See Fig. 5 for location.

Trench and Core SQ-T9 (Figs 5, 23 and 24)

Trench and core SQ-T9 (Figs 23 and 24) represent the lowermost intertidal to shallow-subtidal environment (Fig. 5B). The peloid-skeletal tidal flat grades landward into the lower intertidal tufted or cinder-like microbial mat environment (Fig. 23B and C). The interval between a subsurface hardground and the surface is dominated by sand-rich, bioclastic packstone (lagoonal deposits; Fig. 23D). Conventional radiocarbon dating (Fig. 24) carried out on hardground sample SQ-T9-2 shows an age of 1320 ± 70 uncalibrated ^{14}C yr BP (Fig. 20). This age approximately corresponds to the age of the carbonates that form the hardground directly underlying the crinkly-laminated microbial mat of trench and core SQ-T7 (sample SQ-T7-2; Fig. 19).

CONCLUSIONS

Genuine supratidal sabkha to intertidal microbial mat and lowermost intertidal to shallow-subtidal, peloid-skeletal tidal-flat environments were studied in the vicinity of Al-Qanatir Island (close to Al-Rufayq Island, Al-Dabbiya area) about 80 km southwest of Abu Dhabi City (UAE; Figs 1, 2 and 5). Surface morphology and vertical evaporite mineral distribution allows the seaward portion of the coastal sabkha (outer sabkha) to be subdivided into four facies belts (Butler *et al.*, 1982; Alsharhan & Kendall, 1994, 2002; Kendall *et al.*, 1994; Alsharhan *et al.*, 1998, 2002; Fig. 5): (1) stranded beach ridges; (2) upper sabkha; (3) middle sabkha, and (4) lower sabkha.

The morphology of the surface halite crust allows a subdivision of the upper sabkha (upper supratidal) environment into three sub-environments:

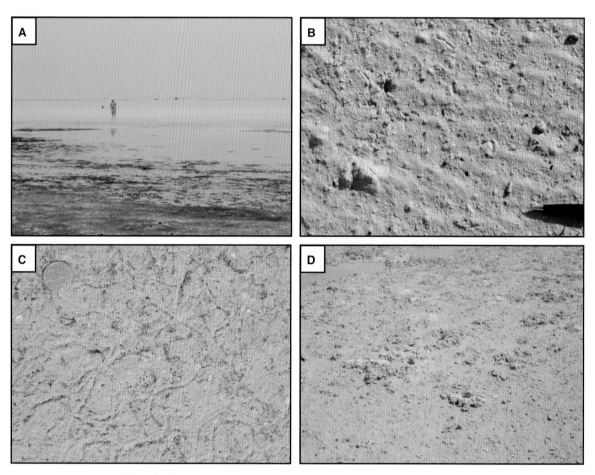

Fig. 22. Lowermost intertidal to shallow-subtidal, peloid-skeletal tidal flat: features and sedimentary structures. (A) Flooded peloid-skeletal tidal-flat environment during onset of high-tide. The red flag marks the position of core and trench SQ-T9 (Figs 23 and 24). (B) Rounded wave ripples showing reworking of ripple-crests during subaerial exposure. (C) Cerithid gastropod grazing-traces displayed in subtidal ponds and shallow tidal channels/tidal creeks. (D) *Skolithos*-type crab burrows.

high, mid, and low upper sabkha (Fig. 25). Very rare, exceptional marine flooding due to spring tides and/or strong seasonal "Shamal" winds during winter and spring months can lead to the dissolution of the low and mid upper sabkha surface halite crusts. The low upper sabkha smooth polygonal halite crust and the mid upper sabkha buckled polygonal halite crust that were dissolved in late spring (May) of 2005 reformed during the summer months (June to September) of the same year.

In comparison to humid carbonate settings (e.g. the Bahamas) where the microbial mat dominates the supratidal zone, under arid conditions (e.g. the Arabian Gulf) the microbial mat is restricted to the intertidal environment (Strohmenger *et al.*, 2008). Within the studied intertidal environment, the

frequency of flooding or, respectively, the relative length of exposure, influences the morphology of the intertidal microbial mat. Four microbial mat intertidal environments can be distinguished (Fig. 25): (1) upper intertidal, crinkled or crenulated microbial mat; (2) middle intertidal, blistered microbial mat that may grade seaward and laterally into pinnacle or domed microbial mat; (3) middle to lower intertidal, pinnacle to smooth polygonal microbial mat; and (4) lower intertidal, smooth polygonal microbial mat and tufted or cinder-like microbial mat (Figs 15B, 17, 18B and C, 20B–E, 23B and C, and 25).

Only the relatively thick polygonal microbial mat has preservation potential and also can be identified in cores and trenches representing the

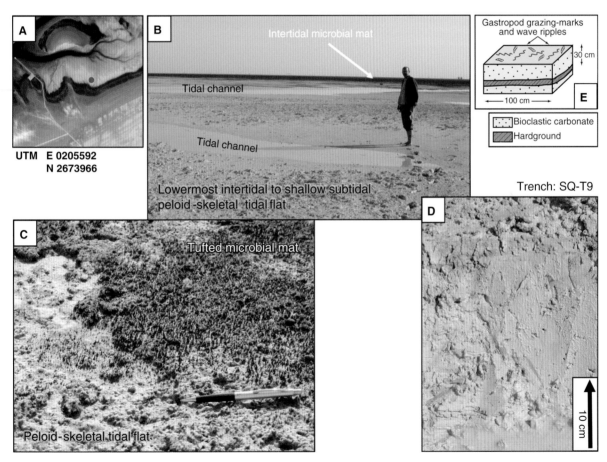

Fig. 23. Lowermost intertidal to shallow-subtidal, peloid-skeletal tidal-flat environment. (A) The red dot on the satellite image marks the location of trench and core SQ-T9. (B) Boundary between lower intertidal, microbial mat (background: dark area, white arrow) to lowermost intertidal to shallow-subtidal, peloid-skeletal tidal flat (foreground: lighter area). Also shown are shallow-subtidal tidal channels/tidal creeks cross-cutting the peloid-skeletal tidal-flat environment. (C) Close-up of the transition between lower intertidal, tufted or cinder-like microbial mat and lowermost intertidal to shallow-subtidal, peloid-skeletal tidal-flat environment. (D) Trench showing from bottom to top: hardground (at groundwater level); and sand-rich bioclastic packstone (lagoonal and tidal channels/tidal-creek deposits: peloid-skeletal tidal flat). (E) Log of trench stratigraphy shown in (D).

upper sabkha to lower intertidal microbial mat environments (Figs 8E, 9, 10C and D, 11, 13C–E, 14, 15D, 16, 18D, 19, 20F and 21). Because of the relatively high frequency of flooding and the resulting grazing-activity of small cerithid and littorinid gastropods (Fig. 22C), no subsurface microbial mat was found within trench and core of the lowermost intertidal to shallow-subtidal, peloid-skeletal tidal flat environment (Figs 23D, 24, and 25).

The distance between the seaward extent of the stranded beach ridges and the lowermost intertidal to shallow-subtidal, peloid-skeletal tidal-flat environment (SQ-T9) is approximately 1500 m, with a vertical topographic change of approximately 3 m. This translates into a very gentle slope angle of approximately 0.1 degree.

Analyses of satellite images, in combination with the observed surface morphologies of the studied coastal sabkha in the vicinity of Al-Qanatir

Island, allows an estimation of the approximate widths of the facies belts seaward of the stranded beach ridges to be made:

(1) stranded beach ridges: 1500 m
(2) upper sabkha: 600 m
(3) middle sabkha: 100–200 m
(4) lower sabkha: 200–300 m
(5) intertidal microbial mat: 300–500 m
(6) lowermost intertidal to shallow-subtidal (peloid-skeletal tidal flat): 2500–3500 m

A thin palaeo-channel-lag deposit (sample SQ-T2-6A; Fig. 7) of the outer, upper sabkha environment (high upper sabkha sub-environment) is dominated by the stenohaline bryozoan species *Disporella* sp., which has not been recorded previously from the United Arab Emirates (Paul D. Taylor, 2006, pers. comm.). The occurrence of the stenohaline bryozoan *Disporella* in older

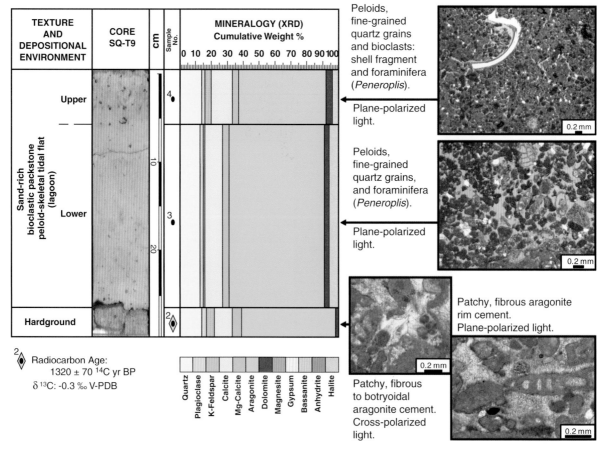

Fig. 24. Resinated core photograph (SQ-T9) and mineralogical composition of trench shown in Fig. 23, and thin-section photomicrographs. Radiocarbon dating and carbon-isotope results of the hardground are also displayed. See Fig. 5 for location.

sediments combined with its absence from recent sediments indicates a salinity increase in the Arabian Gulf over time, which is consistent with the global eustatic trend.

Significant amounts of dolomite were found within a sample from a palaeo-microbial mat within trench and core SQ-T5 (sample SQ-T5-6; Fig. 14). Scanning electron microscope studies show that the fine-crystalline dolomite rhombs are embedded in organic matter (EPS, Bontognali *et al.*, 2010; Strohmenger *et al.*, 2010). Sulphate-reducing bacteria (SRB) of the microbial mat are interpreted to be responsible for the precipitation of dolomite within the described peritidal environment (McKenzie, 1981; Baltzer *et al.*, 1994; Vasconcelos *et al.*, 1995, 2005; Vasconcelos & McKenzie, 1997; Warthmann *et al.*, 2000; van Lith *et al*, 2003a,b).

Radiocarbon dating results (uncalibrated) show an age range between ca 3500 and ca 1300 ^{14}C yr BP for the analyzed subsurface hardgrounds and between ca 2,200 and ca 900 yr BP for the analyzed

subsurface microbial mats; thereby supporting the seaward progradation of the facies belts since the last Holocene sea-level highstand (formation of cerithid gastropod stranded beach ridges). An anhydrite layer (sample SQ-T4-7) within core and trench SQ-T4, representing the low upper sabkha environment showed an uncalibrated radiocarbon age of ca 1900 ^{14}C yr BP; interpreted to correspond to the age of the reworked, underlying microbial mat (sample SQ-T4-4; Fig. 11). Radiocarbon dating of an anhydrite layer found within aeolian deposits of trench and core SQ-T1 (inner, upper sabkha subenvironment: landward of stranded beach ridges) showed an age of the host sediment of ca 12,900 uncalibrated ^{14}C yr BP, corresponding to Pleistocene dune deposits, pre-dating the Holocene flooding event.

Radiocarbon dating results were used to calculate the accumulation rates at four locations (trenches and cores SQ-T3: mid upper sabkha, SQ-T4: low upper sabkha, SQ-T5: middle sabkha,

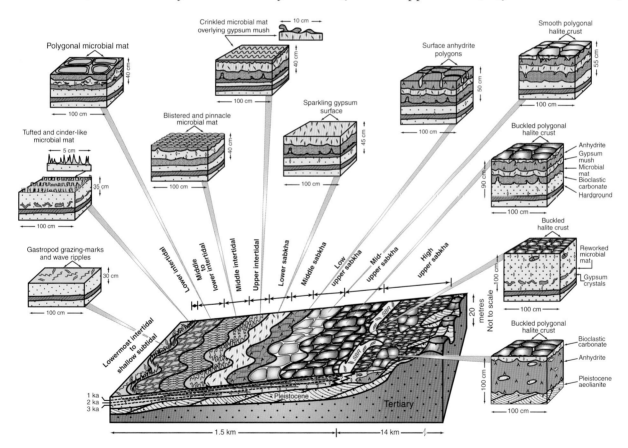

Fig. 25. Idealized schematic block diagram showing sediment and evaporite distribution, as well as the different supratidal to shallow-subtidal environments along the Abu Dhabi coastline (SBR: Stranded Beach Ridges). After Warren & Kendall (1985) and Kendall *et al.* (1994). Also shown are the vertical facies stacking patterns as identified from nine trenches and cores studied in the vicinity of Al-Qanatir (Al-Rufayq) Island. Note that time lines (stippled lines) cross-cut facies and diagenetic (hardgrounds) boundaries.

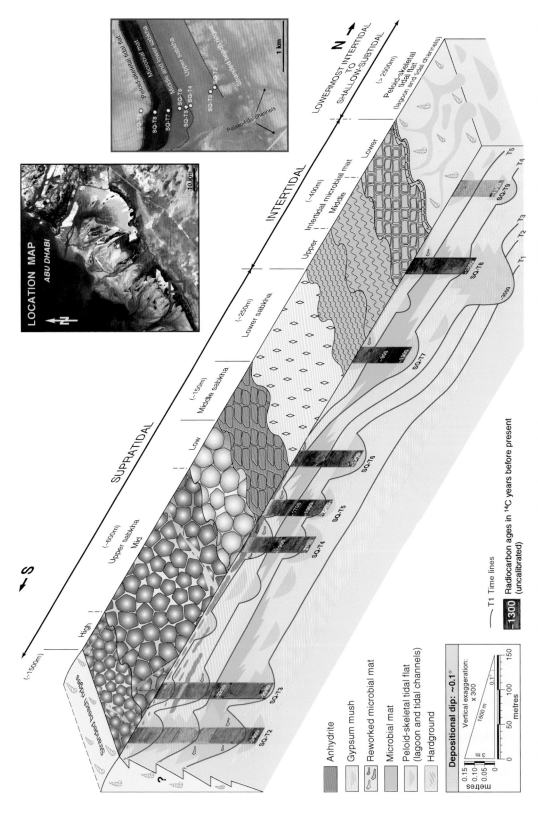

Fig. 26. Sketch showing the interpreted subsurface chronostratigraphic correlation between trenches and cores SQ-T2 (supratidal environment: high upper sabkha) and SQ-T9 (lowermost intertidal to shallow-subtidal environment: peloid-skeletal tidal flat). Time lines (in red) cross-cut facies (e.g. microbial mats) and diagenetic (e.g. hardgrounds) boundaries. Radiocarbon dating results of the microbial mats encountered in trenches and cores SQ-T3 to SQ-T8 show that the mats do not represent isochronous events (e.g. the microbial mat of SQ-T3 is approximately 500 ¹⁴C years older than the microbial mat of SQ-T5 and approximately 1300 ¹⁴C years older than the microbial mat of SQ-T7). Using the modern day microbial mat widths (approximately 300–500 m) as an analogue, it is evident that time-equivalent microbial mats cannot have a dip width significantly larger than 500 m. The geometry of modern analogues can therefore help guide subsurface correlations and interpretations.

and SQ-T7: upper intertidal; Figs 9, 11, 14, and 19). With the exception of core and trench SQ-T4 (see above), the calculated accumulation rate shows a narrow range of 1 cm per 18 to 22 ^{14}C yr (approximately 0.5 mm ^{14}C yr^{-1}).

The radiocarbon dating results underline the importance of chronostratigraphic correlation. Time lines clearly cross-cut facies boundaries (e.g. bioclastic carbonates and microbial mats) and diagenetic boundaries (hardgrounds; Fig. 25). Without the results of the radiocarbon dating, miscorrelation of the identified hardgrounds would be likely, as well as the microbial mats between trenches and cores representing the upper (SQ-T2, SQ-T3, SQ-T4), middle (SQ-T5), and lower sabkha (SQ-T6), with those representing the upper (SQ-T7) and lower (SQ-T8) intertidal microbial mat, and the lowermost intertidal to shallow-subtidal (lagoonal and shallow tidal-channel/tidal-creek deposits) peloid-skeletal tidal flat (SQ-T9) environment (Figs 25 and 26).

The overall facies successions and evaporite distribution show the expected regressive trend. However, all the studied subsurface microbial mats show erosion and/or burrowing on top, as well as re-deposition of remnants of reworked microbial mats within the overlying sediments. In addition, all subsurface microbial mats are clearly under- and overlain by somewhat more seaward facies, indicating minor sea-level rises on top of the microbial mats. The identified subsurface microbial mats therefore represent the tops of shallowing-upward cycles, with flooding surfaces on top (Fig. 26). This interpretation is in accordance with published data by Fairbridge (1961) and Al-Asfour (1982) who postulated several sea-level fluctuations above and below the present-day sea level during the last 6000 years. The sediments were subsequently strongly overprinted by evaporites (gypsum mush and anhydrite; Fig. 26).

The trenches and cores studied here only represent a snap shot in terms of geological time-scales. However, along the studied dip-transect, time lines clearly cross-cut depositional lithofacies and diagenetic boundaries (Fig. 26), demonstrating the advantage of sequence stratigraphy (chronostratigraphy) over lithostratigraphy. Thus, the coastal sabkha environment in the vicinity of Al-Qanatir (Al-Rufayq) Island represents a perfect training ground for teaching geologists, reservoir engineers, and geophysicists the importance of facies analysis (facies stacking patterns: "Walther's Law"; Walther, 1893–1894), diagenesis (dolomite and

hardground formations), radiometric dating, and sequence stratigraphic correlation. Furthermore, the analysis of modern analogues is one of the few means by which the high-resolution spatial complexity of stratigraphic systems can be described (Fig. 26). If the horizontal dimensions of facies belts are less than the typical well spacing, modern analogues, together with seismic and production data help to construct realistic geological and simulation models of subsurface reservoirs.

ACKNOWLEDGEMENTS

The authors gratefully acknowledge the management of Abu Dhabi Company for Onshore Oil Operations (ADCO) and Abu Dhabi National Oil Company (ADNOC), Abu Dhabi for permission to publish this paper. Prasad Patannakara and his team, especially Yousef A. Hamade, John Danial, and Issam Al-Haj (ADCO Drafting Department) are thanked for drafting the figures. Ozair A. Khan (ADCO Geodetics Department) is thanked for providing the satellite image shown on Fig. 1B and for positioning the locations of the studied trenches and cores on the ICONOS imagery. Appreciation is extended to ExxonMobil Exploration Company (EMEC), Houston for releasing the two satellite images shown on Figs 1A and 5. The IKONOS remote sensing image (Fig. 5) was processed by Gene Rankey (RSMAS/University of Miami).

We thank Judith A. McKenzie and the Geomicrobiology Group, ETH-Zurich for providing the dolomite SEM photomicrograph. Special thanks to Erkan Atalik, Hany El-Sahn, George Mani, and Fariz Srouji (Core Laboratories, Abu Dhabi Branch) for carrying out the core treatment and the core photography, as well as for their help in performing the x-ray diffraction and radiocarbon dating. Chris Patrick (Beta Analytic Inc., Miami) is thanked for carrying out the radiocarbon dating. We extend our thanks to Chegjie Liu (EMEC, Houston) for photographing the bryozoan species *Disporella* sp. and to Paul D. Taylor (Natural History Museum, London) for providing bryozoan taxonomy and ecology. We express our sincere appreciation to the management of ADCO for granting us the time and for financing this project.

For very valuable discussions we would like to thank J.A. McKenzie (Zurich, Switzerland), B.C. Schreiber (Seattle, USA), C.G.St.C. Kendall (Columbia, USA), G. Evans (Jersey, UK), J.K. Warren (Darussalam, Brunei), F. Kenig (Chicago,

USA), L.J. Weber (Houston, USA), and M.G. Kozar (Houston, USA).

The paper benefited from careful and constructive initial reviews of William L. Soroka (Abu Dhabi, UAE) and Sean A. Guidry (Houston, USA). The authors greatly appreciate the thorough and thoughtful reviews of Anthony Kirkham (Safron Walden, UK) and Ian West (Romsey, UK).

REFERENCES

Al-Asfour, T.A. (1982) *Changing Sea-Level along the North Coast of Kuwait Bay.* Kegan Paul International., London, 186 pp.

Al Silwadi, M.S., Kirkham, A., Simmons, M.D. and **Twombley, B.N.** (1996) New insights into regional correlation and sedimentology, Arab Formation (Upper Jurassic), offshore Abu Dhabi. *GeoArabia*, **1**, 6–27.

Alsharhan, A.S. and **Kendall, C.G.St.C.** (1994) Depositional setting of the Upper Jurassic Hith anhydrite of the Arabian Gulf: an analog to Holocene evaporites of the United Arab Emirates and Lake MacLeod of Western Australia. *AAPG Bull.*, **78**, 1075–1096.

Alsharhan, A.S. and **Kendall, C.G.St.C.** (2002) Holocene carbonate/evaporites of Abu Dhabi, and their Jurassic ancient analogs. In: *Sabkha Ecosystems. Volume I: The Arabian Peninsula and Adjacent Countries* (Eds H.-J. Barth and B. Böer), pp. 187–202. Kluwer Academic Publishers, Dordrecht.

Alsharhan, A.S., Kendall, C.G.St.C. and **Whittle, G.L.** (1998) Field trip guide to examine the Holocene carbonate/evaporites of Abu Dhabi, United Arab Emirates. *UAE University Publications Department, Al Ain*, 46 pp.

Baltzer, F., Kenig, F., Boichard, R., Plaziat, J.-C. and **Purser, B.H.** (1994) Organic matter distribution, water circulation and dolomitization beneath the Abu Dhabi sabkha (United Arab Emirates). In: *Dolomites. A Volume in Honour of Dolomieu,* (Eds B. Purser, M. Tucker and D. Zenger), *Int. Assoc. Sedimentol. Spec. Publ.*, **21**, 409–427.

Bontognali T.R.R., Vasconcelos C.G., Warthmann R.J., Bernasconi S.M., Dupraz C., Strohmenger C.J. and **Mckenzie J.A.** (2010) Dolomite formation within microbial mats in the coastal sabkha of Abu Dhabi (United Arab Emirates). *Sedimentology*, **57**, p. 824–844.

Burchette, T.P. and **Wright, V.P.** (1992) Carbonate ramp depositional systems. *Sed. Geol.*, **79**, 3–57.

Butler, G.P. (1969) Modern evaporite deposition and geochemistry of co-existing brines, the sabkha, Trucial Coast, Arabian Gulf. *J. Sed. Petrol.*, **39**, 70–89.

Butler, G.P. (1970) Holocene gypsum and anhydrite of the Abu Dhabi sabkha, Trucial Coast: an alternative explanation of origin. *Third Symposium on Salt, N. Ohio Geol. Soc.*, **1**, 120–152.

Butler, G.P., Harris, P.M. and **Kendall, C.G.St.C.** (1982) Recent evaporites from the Abu Dhabi coastal flats. In: *Depositional and Diagenetic Spectra of Evaporites – A Core workshop* (Eds C.R. Handford, R.G. Loucks and G.R. Davies), *SEPM Core Workshop*, **3**, 33–64.

Curtis R., Evans, G., Kinsman, D.J.J. and **Shearman, D.J.** (1963) Association of dolomite and anhydrite in the recent sediments of the Persian Gulf. *Nature*, **197**, 679–680.

Evans, G. (1995) The Arabian Gulf: a modern carbonate-evaporite factory; a review. *Cuad. Geol. Ibérica*, **19**, 61–96.

Evans, G. and **Kirkham, A.** (2002) The Abu Dhabi sabkha. In: *Sabkha Ecosystems. I: The Arabian Peninsula and Adjacent Countries* (Eds H.-J. Barth and B. Böer), pp. 7–20. Kluwer Academic Publishers, Dordrecht.

Evans, G., Schmidt, V., Bush, P. and **Nelson, H.** (1969) Stratigraphy and geologic history of the Sabkha, Abu Dhabi, Persian Gulf. *Sedimentology*, **12**, 145–159.

Evans, G., Kirkham, A. and **Carter, R.A.** (2002) Quaternary development of the United Arab Emirates Coast: new evidence from Marawah Island, Abu Dhabi. *GeoArabia*, **7**, 441–458.

Fairbridge, R.W. (1961) Eustatic changes in sea level. In: *Physics and Chemistry of the Earth* **4** (Eds L.H. Ahrens, F. Press, K. Rankama and S.K. Runcorn), pp. 99–185. Pergamon Press, London.

Kassler, P. (1973) The structural and geomorphic evolution of the Persian Gulf. In: *The Persian Gulf: Holocene Carbonate Sedimentation and Diagenesis in a Shallow Epicontinental Sea* (Ed. B.H. Purser), pp. 11–32. Springer-Verlag, Berlin.

Kendall, A.C. (1992) Evaporites. In: *Facies Models. Response to Sea Level Change* (Eds R.G. Walker and N.P. James), 4th edn, pp. 375–409. Geological Association of Canada, St. John's.

Kendall, C.G.St.C. and **Skipwith, P.A.d'E.** (1968) Recent algal mats of a Persian Gulf lagoon. *J. Sed. Petrol.*, **38**, 1040–1058.

Kendall, C.G.St.C. and **Skipwith, P.A. d'E.** (1969) Geomorphology of a recent shallow-water carbonate province: Khor Al Bazam, Trucial Coast, Southwest Persian Gulf. *Geol. Soc. Am. Bull.*, **80**, 865–892.

Kendall, C.G.St.C. and **Warren, J.K.** (1988) Peritidal evaporites and their sedimentary assemblages. In: *Evaporites and Hydrocarbons* (Ed. B.C. Schreiber), pp. 66–138. Columbia University Press, New York.

Kendall, C.G.St.C., Sadd, J.L. and **Alsharhan, A.** (1994) Holocene marine cement coatings on beach-rocks of the Abu Dhabi coastline (UAE): analogs for cement fabrics in ancient limestones. *Carbonates Evaporites*, **9**, 119–131.

Kendall, C.G.St.C., Alsharhan, A.S. and **Cohen, A.** (2002) The Holocene tidal flat complex of the Arabian Gulf coast of Abu Dhabi. In: *Sabkha Ecosystems. I: The Arabian Peninsula and Adjacent Countries* (Eds H.-J. Barth and B. Böer), pp. 21–35. Kluwer Academic Publishers, Dordrecht.

Kenig, F. (1991) *Sédimentation, Distribution et Diagenèse de la Matière Organique dans un Environnement Carbonaté Hypersalin: le Système Lagune-sabkha d'Abu Dhabi.* Unpubl. PhD thesis, Université d'Orléans, Orleans, 327 pp.

Kenig, F., Huc, A.Y., Purser, B.H. and **Oudin, J.L.** (1990) Sedimentation, distribution and diagenesis of organic-matter in a recent carbonate environment, Abu-Dhabi, UAE. *Org. Geochem.*, **16**, 735–747.

Kinsman, D.J.J. (1969) Modes of the formation, sedimentary associations, and diagnostic features of shallow-water and supratidal evaporites. *AAPG Bull.*, **53**, 830–840.

Kinsman, D.J.J. and Park, R.K. (1976) Algal belt and coastal sabkha evolution, Trucial Coast, Persian Gulf. In: *Stromatolites* (Ed. M.R. Walter). *Dev. Sedimentol.*, **20**, 421–433.

Kirkham, A. (1997) Shoreline evolution, aeolian deflation and anhydrite distribution of the Holocene, Abu Dhabi. *GeoArabia*, **2**, 403–416.

Kirkham, A. (1998) A Quaternary proximal foreland ramp and its continental fringe, Arabian Gulf, UAE. In: *Carbonates Ramps* (Eds V.P. Wright and T.P. Burchette), *Geol. Soc. London Spec. Publ.*, **149**, 15–41.

Kirkham, A. (2004) Patterned dolomites: microbial origins and clues to vanished evaporites in the Arab Formation, Upper Jurassic, Arabian Gulf. In: *The Geometry and Petrogenesis of Dolomite Hydrocarbon Reservoirs* (Eds C.J.R. Braithwaite, G. Rizzi and G. Darke), *Geol. Soc. London Spec. Publ.*, **235**, 301–308.

Lambeck, K. (1996) Shoreline reconstructions for the Persian Gulf since the last glacial maximum. *Earth Planet. Sci. Lett.*, **142**, 43–57.

Lambeck, K., Esat, T.M. and Potter, E.-K. (2002) Links between climate and sea levels for the past three million years. *Nature*, **419**, 199–206.

Leeder, M.R. and Zeidan, R. (1977) Giant late Jurassic sabkhas of Arabian Tethys. *Nature*, **268**, 42–44.

Maiklem, W.R., Bebout, D.G. and Glaister, R.P. (1969) Classification of anhydrite – a practical approach. *Bull. Can. Petrol. Geol.*, **17**, 194–233.

McKenzie, J.A. (1981) Holocene dolomitization of calcium carbonate sediments from the coastal sabkhas of Abu Dhabi, U.A.E.: a stable isotope study. *J. Geol.*, **89**, 185–198.

McKenzie, J.A., Hsü, K.J. and Schneider, J.F. (1980) Movement of subsurface waters under the sabkha, Abu Dhabi, UAE, and its relation to evaporative dolomite genesis. In: *Concepts and Models of Dolomitization* (Eds D.H. Zenger, J.B. Dunham and R.L. Ethington). *SEPM Spec. Publ.*, **28**, 11–30.

Müller, D.W., McKenzie, J.A. and Mueller, P.A. (1990) Abu Dhabi sabkha, Persian Gulf, revisited: application of strontium isotopes to test an early dolomitization model. *Geology*, **18**, 618–621.

Park, P.K. (1977) The preservation potential of some recent stromatolites. *Sedimentology*, **24**, 485–506.

Patterson, R.J. and Kinsman, D.J.J. (1977) Marine and continental groundwater sources in a Persian Gulf coastal sabkha. In: *Reefs and Related Carbonates – Ecology and Sedimentology* (Eds S.H. Frost, M.P. Weiss and J.B. Saunders). *AAPG Stud. Geol.*, **4**, 381–397.

Patterson, R.J. and Kinsman, D.J.J. (1981) Hydrologic framework of a sabkha along Arabian Gulf. *AAPG Bull.*, **65**, 1457–1475.

Patterson, R.J. and Kinsman, D.J.J. (1982) Formation of diagenetic dolomite in coastal sabkha along Arabian (Persian) Gulf. *AAPG Bull.*, **66**, 28–43.

Peebles, R.G., Shaner, M. and Kirkham, A. (1995) Arid coastline depositional environments: an AAPG International Field Seminar. *A Fieldguide to the Abu Dhabi Coast, United Arab Emirates*, 45 pp.

Purser, B.H. (Ed.) (1973) *The Persian Gulf: Holocene Carbonate Sedimentation and Diagenesis in a Shallow Epicontinental Sea.* Springer-Verlag, Berlin, 471 pp.

Purser, B.H. and Evans, G. (1973) Regional sedimentation along the Trucial Coast, SE Persian Gulf. In: *The Persian Gulf: Holocene Carbonate Sedimentation and Diagenesis in a Shallow Epicontinental Sea* (Ed. B.H. Purser), pp. 211–231. Springer-Verlag, Berlin.

Purser, B.H. and Seibold, E. (1973) The principle environmental factors influencing Holocene sedimentation and diagenesis in the Persian Gulf. In: *The Persian Gulf: Holocene Carbonate Sedimentation and Diagenesis in a Shallow Epicontinental Sea* (Ed. B.H. Purser), pp. 1–9. Springer-Verlag, Berlin.

Sanford, W.E. and Wood, W.W. (2001) Hydrology of the coastal sabkhas of Abu Dhabi, United Arab Emirates. *Hydrogeol. J.*, **9**, 358–366.

Schneider, F.J. (1975) Recent tidal deposits, Abu Dhabi, UAE, Arabian Gulf. In: *Tidal Deposits. A Casebook of Recent Examples and Fossil Counterparts* (Ed. R.N. Ginsburg), pp. 209–214. Springer-Verlag, Berlin.

Schreiber, B.C. (1988) Sabaqueous evaporite deposition. In: *Evaporites and Hydrocarbons* (Ed. B.C. Schreiber), pp. 182–255. Columbia University Press, New York.

Seibold, E. and Vollbrecht, K. (1969) Die Bodengestalt des Persischen Golfs. *Meteor Forschungsergeb.*, **C2** 29–56.

Shearman, D.J. (1983) Syndepositional and late diagenetic alteration of primary gypsum to anhydrite. *Sixth International Symposium on Salt, Toronto*, **1**, 41–50.

Shinn, E.A. (1983) Tidal flat environment. In: *Carbonate Depositional Environments* (Eds P.A. Scholle, D.G. Bebout and C.H. Moore), *AAPG Mem.*, **33**, 171–210.

Steinhoff, I. and Strohmenger, C. (1999) Facies differentiation and sequence stratigraphy in ancient evaporite basins – an example from the Basal Zechstein (Upper Permian of Germany). *Carbonates Evaporites*, **14**, 146–181.

Strasser, A. and Davaud, E. (1986) Formation of Holocene limestone sequences by progradation, cementation, and erosion: two examples from the Bahamas. *J. Sed. Petrol.*, **56**, 422–428.

Strasser, A., Davaud, E. and Jedoui, Y. (1989) Carbonate cements in Holocene beachrock: example from Bahiret el Biban, southeastern Tunisia. *Sed. Geol.*, **62**, 89–100.

Strohmenger, C., Antonini, M., Jäger, G., Rockenbauch, K. and Strauss, C. (1996) Zechstein 2 Carbonate reservoir facies prediction in relation to Zechstein sequence stratigraphy (Upper Permian, Northwest Germany): an integrated approach. *Bull. Centres Rech. Explor.-Prod. Elf-Aquitaine*, **20**, 1–35.

Strohmenger, C.J., Al-Mansoori, A., Shebl, H., Al-Jeelani, O., Al-Hosani, I., Al-Mehsin, K. and Al-Shamry, A. (2004) Modern carbonate-evaporite depositional environments of Abu Dhabi, (U.A.E.). *Unpubl. Field Trip Guide Book, 11th ADIPEC Meeting, First International Evaporite Conference,* Abu Dhabi.

Strohmenger, C.J., Shebl, H., Al-Mansoori, A., Al-Mehsin, K., Al-Jeelani, O., Al-Hosani, I., Al-Shamry, A. and Al-Baker, S. (2007) Depositional environment and sea-level history of the Abu Dhabi sabkha in the vicinity of Al-Qanatir Island (U.A.E.). *GeoArabia*, **12**, 219.

Strohmenger, C.J., Al-Mansoori, A., Al-Jeelani, O., Al-Hosani, I., Shebl, H., Al-Mehsin, K. and Al-Baker, S. (2008) The arid shallow subtidal to supratidal environment: a case study from the Abu Dhabi sabka (United Arab Emirates). *AAPG Ann. Conv. Exhib. Abstr.* Vol., **17**, 197.

Strohmenger C.J., Al-Mansoori A., Al-Jeelani O., Al-Shamry A., Al-Hosani I., Al-Mehsin K. and Shebl H. (2010) The sabkha sequence at Mussafah Channel (Abu Dhabi, United Arab Emirates). facies stacking patterns, microbial-mediated dolomite and evaporite overprint. *GeoArabia*, **15**, 49–90.

van Lith, Y., Warthmann, R., Vasconcelos, C. and **McKenzie J.A.** (2003a) Microbial fossilization in carbonate sediments: a result of the bacterial surface involvement in dolomite precipitation. *Sedimentology*, **50**, 237–245.

van Lith, Y, Warthmann, R., Vasconcelos, C. and **McKenzie, J.A.** (2003b) Sulphate-reducing bacteria induce low-temperature Ca-dolomite and high Mg-calcite formation. *Geobiology*, **1**, 71–79.

Vasconcelos, C. and **McKenzie, J.A.** (1997) Microbial mediation of modern dolomite precipitation and diagenesis under anoxic conditions (Lagoa Vermelha, Rio de Janeiro, Brazil). *J. Sed. Res.*, **67**, 378–390.

Vasconcelos, C., McKenzie, J.A., Bernasconi, S, Grujic, D. and **Tien, A.J.** (1995) Microbial mediation as a possible mechanism for natural dolomite formation at low temperatures. *Nature*, **377**, 220–222.

Vasconcelos, C., McKenzie, J.A., Warthmann, R. and **Bernasconi, S.M.** (2005) Calibration of the $\delta^{18}O$ paleothermometer for dolomite precipitated in microbial cultures and natural environments. *Geology*, **33**, 317–320.

Walther, J. (1893–1894) *Einleitung in die Geologie als historische Wissenschaft.* Fischer-Verlag, Jena, **1–3**, 1055 pp.

Warren, J.K. (1983) On the significance of evaporite lamination. *Sixth International Symposium on Salt, Toronto*, **1**, 161–170.

Warren, J.K. (1989) *Evaporite Sedimentology.* Prentice Hall Advanced Reference Series, Englewood Cliffs, 285 pp.

Warren, J.K. (1999) *Evaporites: Their Evolution and Economics.* Blackwell Science, Oxford, **438** pp.

Warren, J.K. and **Kendall, C.G.St.C.** (1985) Comparison of sequences formed in marine sabkha (subaerial) and salina (subaqueous) settings – modern and ancient. *AAPG Bull.*, **69**, 1013–1023.

Warthmann, R., van Lith, Y., Vasconcelos, C., McKenzie, J.A. and **Karpoff, A.M.** (2000) Bacterially induced dolomite precipitation in anoxic culture experiments. *Geology*, **28**, 1091–1094.

Wood, W.W. and **Sanford, W.E.** (2002) Hydrology and solute chemistry of the coastal-sabkha aquifer in the Emirate of Abu Dhabi. In: *Sabkha Ecosystems. I: The Arabian Peninsula and Adjacent Countries* (Eds H.-J. Barth and B. Böer) pp. 173–185. Kluwer Academic Publishers, Dordrecht.

Wood, G.V. and **Wolfe, M.J.** (1969) Sabkha cycles in the Arab/Darb Formation off the Trucial Coast of Arabia. *Sedimentology*, **12**, 165–191.

Wood, W.W., Sanford, W.E. and **Al Habshi, A.R.S.** (2002) Source of solutes to the coastal sabkha of Abu Dhabi. *Geol. Soc. Am. Bull.*, **114**, 259–268.

Wood, W.W., Sanford, W.E. and **Frape, S.K.** (2005) Chemical openness and potential for misinterpretation of the solute environment of coastal sabkhat. *Chem. Geol.*, **215**, 361–372.

Int. Assoc. Sedimentol. Spec. Publ. (2011) **43**, 183–204

Coastal and continental sabkhas of Barr Al Hikman, Sultanate of Oman

MONIQUE METTRAUX[*], PETER W. HOMEWOOD[*,†], ANDY Y. KWARTENG[‡] and JOERG MATTNER[§]

[*]GEOSOLUTIONS, 99bis rue d'Ossau, 64290 Gan, France (E-mail: moho1959@yahoo.fr)
[†]Oman Geo-Consultants, PO Box 194, PC 134, Shatti Al-Qurum, Oman
[‡]Remote Sensing and GIS Center, Sultan Qaboos University, PO Box 33, Al Khod, Oman
[§]GeoTech Consulting, PO Box 20393, Manama, Bahrain

ABSTRACT

The Barr Al Hikman sabkhas, on the east coast of the Sultanate of Oman, cover about $1400\,km^2$ of low-lying topography (the peninsula is about 30 km wide). They form within a thin irregular soft-sediment veneer over a complex geological substrate where recent sea-level fall, together with structural movement, has provided emergent, marine to lagoonal carbonate host sediments of varying thickness, as well as ponded depressions, suited to the development of evaporite deposits. The sabkhas develop under an extremely arid climate with high evaporation in spite of frequent early-morning high relative humidity. Evaporite and thermalite minerals form either as halite-dominated deposits in coastal low-lying to ponded environments, or as gypsum layers in soft grainy host sediments. Geomorphological analysis has been made on satellite images and has been tested by ground truthing. Sediment sampling and sedimentological analyses have also been carried out. The sabkhas may be grouped into two types: coastal sabkhas and continental sabkhas, based on their geomorphology and likely hydrogeology. Lower lying coastal sabkhas (1–5 m altitude) are occasionally flooded, sometimes by runoff after short heavy rainfall and sometimes by unusually high-water marine influx, to feed extensive perennial halite deposits. The slightly higher continental sabkhas (lying between 5 and 15 m altitude), may be fed exclusively by continental groundwaters, and are depleted in halite by surface runoff after occasional rains. A dual permeability system is proposed that might govern flow in the continental aquifer (moderate to low permeability in the non-lithified Quaternary to older lithified deposits; conjectural high permeability along fault and fracture conduits). Dolomite is forming in high-salinity coastal lagoons; a microbially mediated origin is preferred, as opposed to hypersaline or mixing-zone chemical precipitation. An older generation of sabkha deposits, preserved as terraces and patches of exposed evaporite, may possibly be linked to higher sea-level six thousand years ago. A summary comparison with the Abu Dhabi sabkhas reveals some commonalities, but many contrasting features.

Keywords: Sabkhas, evaporites, remote sensing, geomorphology, Holocene, Oman.

INTRODUCTION

The Barr Al Hikman peninsula is located on the eastern coast of the Sultanate of Oman on the northern coast of Masirah Bay, and has developed over the Pleistocene and Holocene as an *in situ* shallow-water carbonate/evaporite system. The system comprises a subordinate amount of siliciclastic sand and gravel fed from the alluvial systems of the Hajar Mountains, and a very minor input of siliciclastic sands from adjacent areas of Masirah

Island and Huqf. The peninsula is approximately 40 by 35 km with the longer axis trending NNE and northern edge merging with the southern part of the Wahaybah sand dunes (Fig. 1; Glennie 2005). Sabkha settings of the Barr Al Hikman peninsula have two main end members: coastal sabkhas fed in part by seawater, and continental sabkhas possibly fed by the continental aquifer. Lower lying topographic areas, one running in from the east, and three along the west coast of the peninsula, host sabkhas that develop with a pronounced horizontal

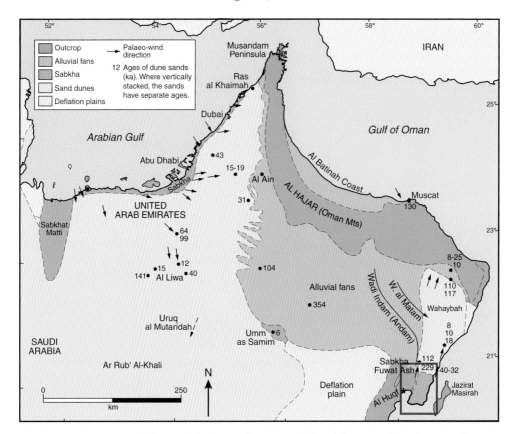

Fig. 1. Location of eastern Arabian sabkhas. Black dots are locations of dated dune sands with ages in kyr. Rectangle in SE corner locates Barr Al Hikman study area. Star in the western sector of inset indicates the location of Filim village. Reproduced from Glennie (2005) by kind permission of GeoArabia.

evaporite gradient. A general zonation may be observed in them that goes from marine lagoons and more saline supratidal flats at the seaside, to terminate inland with extensive zones of thick halite polygon growth. In contrast, unlithified Quaternary marine carbonate sediments, lying a little higher than the coastal sabkhas or farther inland on the peninsula, host a vertical evaporite gradient within the upper 50 cm of sediments, below the deflated sediment surface. In these continental sabkhas, a well-developed layer of bladed or "swallowtail" gypsum crystals (sometimes a genuine cemented gypcrete) develops at the top of the damp capillary fringe, and this is overlain by a shallower zone of fibrous to acicular gypsum nests, and finally by a dusting or wafer-thin crust of halite that occurs on the deflated surface.

The Barr Al Hikman peninsula, and its south and eastern coasts in particular, have been noted mostly for the abundant birdlife (Eriksen & Eriksen, 1999; Eriksen *et al.*, 2001), for turtle habitat, the passage of whales and dolphins in the

Masirah channel between the peninsula and Masirah Island, and for the unique monospecific *Montipora* coral carpets off the southern coast (Claereboudt 2006). Due to its uniqueness and importance, the area has been proposed as a World Heritage Site (Pilcher 2002).

The sedimentological study reported here is based essentially on satellite image geomorphological analysis, onsite ground truthing, reconnaissance surveying and limited salinity and petrographic analyses. The modern and recent carbonate systems of Barr Al Hikman, surrounding the peninsula and providing the host sediment to the sabkhas on the peninsula, have been summarily presented by Homewood *et al.*, (2007). From this preliminary study, a more complex picture of the Barr Al Hikman sabkha is established than the simple differentiation of recent sabkhas followed by modern sabkhas, given by previous authors (Dubreuil *et al.*, 1992, Le Metour *et al.*, 1995, Glennie 2005). Nevertheless, the succession of an older 1 m higher-standing abandoned sabkha, succeeded by today's lower-lying sabkha development is well documented by terracing of the coastal plain (Dubreuil *et al.*, 1992). Broadly speaking, the sabkhas of the Barr Al Hikman peninsula form an irregular, differentiated and polyphased carpet, over a complex geological substrate in which faults and fractures may possibly act as localised major high-permeability conduits for groundwater. Together with the sediment matrix, faults and fractures might create a dual-permeability system. Thus there could be a four-fold hydrological input to the aquifer, from: (1) precipitation, both very slight (on average) rainfall and from heavy morning dew; (2) lower permeability Quaternary, Tertiary and older sediment matrix aquifer groundwaters; (3) high-permeability subsurface aquifer groundwaters flowing along structural heterogeneities; and (4) seawater.

Further studies should provide answers to three major ongoing avenues of research: first the relationships between stages of Barr Al Hikman sabkha development, Quaternary sea-level variation and structural movement; second, the relationship between faults and fractures that might act as major fluid conduits, compared with the more general aquifer trends in the subsurface; and third, the diagenetic effect of the high permeability streaks hypothetically leading to lithification of the Pliocene to Holocene carbonate sediments.

METHODS

Satellite images

Successive generations of satellites with resolutions ranging from 80 m to 0.61 m per pixel, and multi-temporal imaging of the peninsula, provide an evolving regional low- to high-resolution database for the analysis of the geomorphology of the Barr Al Hikman sabkhas (Table 1). In this study, the following were used: Landsat Multi Spectral Scanner (MSS); Landsat Thematic Mapper (TM); Landsat Enhanced Thematic Mapper Plus (ETM +); Advanced Spaceborne Thermal Emission and Reflection Radiometer (ASTER); Russian TK-350; and QuickBird datasets. The characteristics of the datasets are presented in Table 1. Each satellite image was orthorectified using a digital elevation model (DEM) image of Oman and rectified to the UTM Zone 40 and WGS84 to facilitate their comparisons. Several digital enhancement techniques were used to map and differentiate the different lithologies and geomorphology, and structural features. A DEM image of the area (Fig. 2) was generated from TK-350 stereo images.

Geomorphological analysis

The main geomorphological analysis was initiated on a Landsat TM colour composite image of principal components (PC), PC1, PC2 and PC3 using TM's 6 optical bands (Fig. 3), on which the false colour combinations were helpful in distinguishing some of the main features, including four main coastal sabkha areas (1–4 on Fig. 4), and higher- or lower-lying areas (+ and − on Fig. 4) delimited by linear features thought to be fault trends (double lines on Fig. 4). Some features, such as the tidal channel system of the now raised central lagoon, were better observed on the ASTER image (Fig. 5).

Ground truthing

On-site, field verification of geomorphological features identified and interpreted on the satellite images, was made during a number of fieldtrips between 2002 and 2006, using a Garmin e-trek^tm GPS unit linked to a laptop computer. Ozie Explorer^tm software was used to track calibration positions directly on the satellite image during traverses, and several iterations of interpretation and calibration were carried out. Sabkha conditions have so far precluded making regularly spaced traverses with a grid format, and most calibration has been carried out using well-worn tracks made by light vehicles of the local fishing industry.

Sampling and thin-section preparation

Sediment sampling of numerous surface sites, in general included three separate samples from each locality: (a) loose surface sediment; (b) firm

Table 1. Types of satellite data used in this study.

Satellite	Acquisition date	Resolution pixel (m^2)
Landsat MSS	21 October, 1972	80
Landsat TM	27 January 1987	30
Landsat ETM +	23 January 2000	15, 30
ASTER	22 November 2000	15, 30, 90
TK-350	2002	10
QuickBird	January 2004	0.61, 2.4

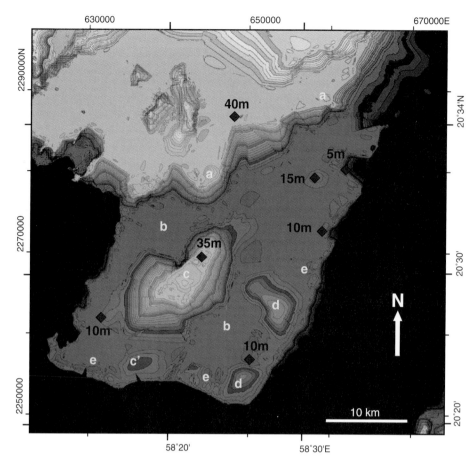

Fig. 2. TK-350-derived DEM of the Barr Al Hikman peninsula. a = Quaternary marine transgressive surface; b = raised marine embayments and lagoons; c, c′ and d = Hikman Formation highs; e = ancient and present-day beach ridges (see text for details). Spot heights (red diamonds) are from the 1501 series topographic base map, Joint Operations Graphic, UK Ministry of Defence, approved by the National Survey Authority, Sultanate of Oman. Blue figures are latitude and longitude; black numbers on grid are Petroleum Development Oman (PDO) map coordinates.

sediment immediately below the loose surface, and (c) sediment at 20–30 cm below the surface. Sediments were dried, consolidated in epoxy resin and then thin-sectioned. Sediment consolidation and thin section preparation was undertaken at Geosciences Rennes, Rennes University, France. Thin-sections were stained with potassium ferricyanide and Alizarin Red-S, following the method of Dickson (1966).

Near-coastal free-water level measurement

Field-based observations of free-water level height, in salt ponds near southern coast of the peninsula, were made over several years, recording a rise of some 20 cm from early 2002 to late 2004, but mostly accomplished over a six-month period in 2002.

Salinity measurements

Water salinities were measured by a hand-held ocular refractometer (type MC1011; salinity range 1–100 ‰, error +/− 1 ‰). Measurements were made for seawater off the southern coast, open and restricted lagoon waters, and at gushing spring sources that are exposed along the southern beach at low spring tides.

XRD, light microscope and SEM analyses

X-ray diffraction (XRD) analyses on bulk sediments were carried out at the Department of Earth Sciences, Sultan Qaboos University, Oman. Specific crystal forms of evaporite minerals from the sabkhas were hand separated, crushed and measured as bulk samples. Samples of the fine grain-size fractions of lagoon sediments were also

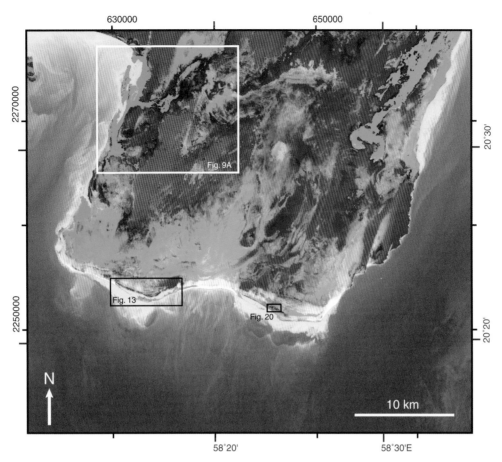

Fig. 3. Landsat TM principal component (PC) bands PC1, PC2 and PC3 colour image recorded on 20 January 2000 (courtesy of Rashid Al Hinai, Petroleum Development Oman). The locations of Figs 9A, 13 and 20 are indicated. Latitude, longitude and PDO grid coordinates are indicated.

analysed. Quantitative XRD analyses on bulk lagoon sediment and oriented fine grain-size fractions (<2 μm) of lagoon sediments were carried out by The Mineral Lab. Lakewood, Colorado, USA. Light microscope analyses were carried out with a standard polarizing microscope. SEM analyses were performed at Sultan Qaboos University and ETH Zurich, Switzerland.

Hydrological and meteorological data

Statistical data on rainfall, temperature humidity and wind strength were obtained from the Ministry of Transport and Communications (MTC, 2003) and the Ministry of Regional Municipality Environment and Water Resources (MRMEWR, 2005), from a report on the ecology of the Masirah Island and Barr Al Hikman areas for conservation planning purposes (Weidle, 1991; Anon, 1992), from numerous individual references, as cited in the

text, and through personal communication from Steven Fryberger in 2006.

SABKHAS OF BARR AL HIKMAN

Topography

The first impression of an extremely flat terrain, gained when driving over the peninsula, is borne out by regional maps, satellite images, and the digital elevation model of the peninsula generated from TK-350 datasets (DEM, Fig. 2). All of these emphasize the low-lying topography of the peninsula, as well as the remarkably linear eastern, southern and western coastlines. Spot heights from the UK Ministry of Defence topographical sheet, noted from the geological map (Dubreuil *et al.*, 1992), are broadly in agreement with the satellite DEM. The Landsat TM, ASTER and TK-350 images also reveal several less prominent topographic features: a

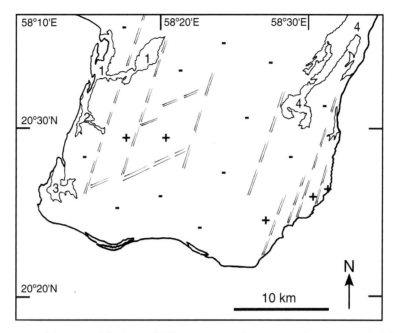

Fig. 4. Diagram of the principal features of the Barr Al Hikman peninsula. 1, 2, 3 and 4 are coastal sabkhas; + and − indicate higher and lower ground; double lines indicate possible fault zones.

northern scarp running WSW-ENE that terminates the low-lying peninsula to the north (and that was cut by Quaternary marine transgression: a on Fig. 2); a central to west-central higher area, mapped as outcropping Miocene–Pliocene Hikman Formation (Dubreuil *et al.*, 1992; c on Fig. 2); two SSW–NNE trending slightly higher areas, one on the east of the peninsula, one a little west of the centre running SSW from the higher Hikman Formation outcrops (c–c′ and d on Fig. 2); a slightly higher rim on the eastern and southern coastlines, mostly formed by ancient and present-day beach ridges (e on Fig. 2). Together with the barred southern coastline, these features surround lower lying areas that were ancient (probably Holocene) now raised marine embayments or lagoons (b on Fig. 2).

Climate, meteorology and wadi flow

The essential features of the climate at Barr Al Hikman are the extreme aridity and the summer monsoon wind regime, blowing strongly from the south. Rainfall is scarce, and the peninsula lies closely to the regional yearly 25 mm isohyet (as reconstructed by S. Fryberger, pers. comm. 2006, from MTC, 2003). Masirah Island rainfall records, certainly higher than for Barr Al Hikman, give a 1980–2003 yearly average of 58 mm yr^{-1}, with a range from 1 mm yr^{-1} to 279 mm yr^{-1} (MEMWR, 2005). The greatest rainfall ever recorded at

Masirah was in 1977, with 431 mm in 24 hr on 13th June. Between 1956 and 2003, highest average monthly rainfall ranged from 4.2 mm in September to 431 mm in June. Long-term monthly rainfall average figures between 1956 and 2003 are much lower, ranging from 0.1 mm (September) to 12.3 mm (June).

A visit to Barr Al Hikman in February 2006, just after a heavy rain shower, allowed observation of local gullying caused by run-off from higher ground, and flooding of the salt ponds on the southern coast. Such amounts of rain were only observed once between 2001 and 2006, although erosional features such as gullying may be observed around lower-lying sabkha areas on high-resolution QuickBird, Google Earth images and aerial photos. Ross (2005) gave a vivid account of the unnamed cyclone that struck Masirah in 1977, with a record deluge of 431 mm of rain. This hurricane, with sustained wind speeds of 160 km h^{-1} and gusts up to 222 km h^{-1}, must have caused a major gullying event on the neighbouring peninsula. This northwest sector of the Indian Ocean lies somewhat at the limit of the hurricane zone, with only one such event recorded over the 1970–1989 period (JISAO, 2006). Hurricanes must affect this region up to several times in a century, but with varying consequences for the peninsula. The 2002 cyclone that struck Salalah for example, did not result in any recorded anomaly at Barr Al Hikman.

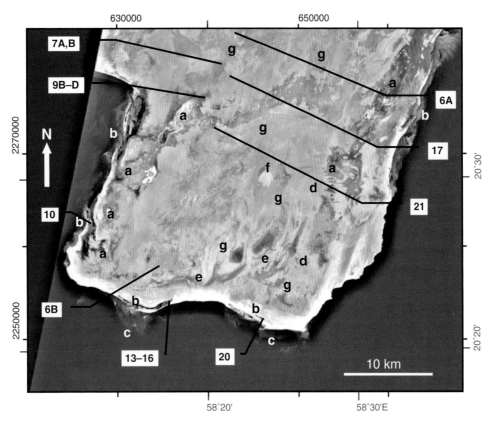

Fig. 5. ASTER visible and near-infrared bands 3, 2, and 1 composite image of the Barr Al Hikman peninsula recorded on 22 November 2000 (grey-scale reproduction). a = coastal sabkhas; b = spits and lagoons; c = coral carpets; d = drainage divide; e = ancient tidal channels; f = ancient lagoon (see text for details). Numbers in white boxes are Figure numbers of illustrated material.

A report on the Barr Al Hikman site, prepared for conservation planning of the Masirah Island and Barr Al Hikman area (Weidle, 1991; Anon, 1992), noted yearly evaporation of 1000 mm and maximum temperatures of 42 °C for the peninsula. Between 1956 and 2003 (MTC, 2003), the highest mean monthly temperature was 30.4 °C (May) going down to 22.3 °C (January). Fog and morning dew can be very heavy however, with records of relative humidity, again from Masirah Island meteorological station between 1994–2003, giving an average relative humidity maximum of 99.4 %, and minimum of 11.8 %. Monthly figures ranged between 100 % and 9 % (MEMWR, 2005).

According to Scott (1995), the wind direction between April and September is predominantly southwest, influenced by the southwest monsoon in the Indian Ocean; the mean July wind speed, when the influence of the monsoon is usually at its peak, is in excess of 30 km h^{-1}; outside the monsoon period, winds are lighter and more variable, often with an easterly component. Masirah meteorological station wind records, for 2003, give a monthly average of 28 km h^{-1} for August, and a yearly average of 18.5 km h^{-1}, both coming from 210° SE (MTC, 2003). The very coarse coral boulder debris piled up as successive beach ridges on the southern coast of Barr Al Hikman, with several ridges per year accumulated over the past 15 years (Homewood *et al.*, 2007), attests to the strength of these summer storm winds. With the lack of any accompanying rainfall, the strong winds from the summer monsoon must play a major role in the evaporation on the sabkhas.

Wadi Halfayn is a tributary to the lower reaches of Wadi Andam, a major ephemeral stream coming from the Oman Mountains, hundreds of kilometres to the north (Fig. 1). The streams run to Filim, a village at the head of the embayment that delimits the western coast of the Barr Al Hikman peninsula. The wadi was flowing in February 2006, after a heavy rainfall over the whole area between Barr Al Hikman and the mountains.

Geomorphological analysis

Both the Landsat TM and the ASTER image show numerous, both modern and more ancient geomorphological features. On the ASTER image (Fig. 5) modern features include: (a) coastal sabkhas; (b) spits and lagoons on the western and southern coastline; (c) dark patches close to the southern shore that are thriving coral carpets; (d) a divide between eastward and westward drainage. Older geomorphological features include: (e) tidal run-off channels from an ancient, now raised, central to southern lagoon; and (f) an ancient lagoon. However, the broadly developed continental sabkhas (g), are not conspicuous on the ASTER image.

From the foot of the scarp that provides the northern limit to the low-lying topography, and apart from the slightly elevated area of Hikman Formation outcrops, the extremely flat surface is the result of ongoing deflation over what was initially a marine terrace. Surface lags with abundant molluscs, indicating marine embayment to lagoonal environments, cover most of the lower, flat topography of the peninsula. A deflated lag deposit

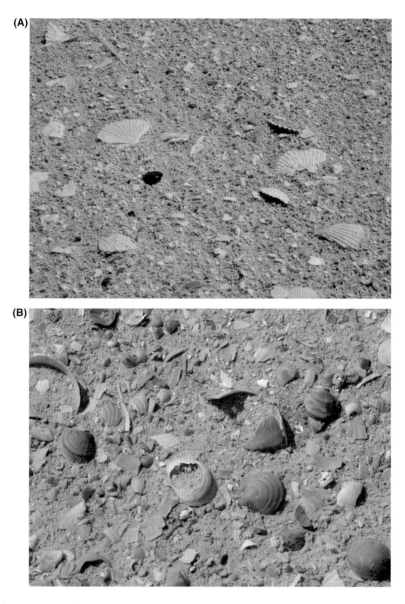

Fig. 6. Sabkha gravels. (A) Pectenid shells and gravel compose lag deposits strewn over the continental sabkha deflation surface. Larger shells are 8 cm across. (B) Bivalve-dominated assemblage of the deflated ancient lagoon. Shells are 2–3 cm. Photographs by H. Droste.

Fig. 7. Gypsum deposits. (A) Patches, several hundreds of metres in width, of exhumed large gypsum crystals formed in an older sabkha. (B) Detail of A: exfoliated and corroded large gypsum crystals from older sabkha deposits exhumed by deflation. Larger crystals in (A) and (B) are 12 cm long. Photographs by C. van Eden.

thus litters the surface, composed of residues of a Quaternary marine transgression and the subsequent phases of sabkha development, which occurred prior to more recent, Holocene relative sea-level fall. In this lag deposit, locally monospecific pectenid bivalves, give way more generally to diverse molluscan assemblages ranging from lower energy restricted environment fauna to higher energy marine community gastropods and bivalves (Fig. 6). These faunal elements are mixed with finer gravels and sands, with fine acicular gypsum

crystals, and with large corroded gypsum crystals, eroded from sabkha deposits that were formed earlier than those developing today (Fig. 7). The slightly higher standing rims around the eastern and southern coasts of the peninsula are formed by ancient beach ridges, present-day dune fields, and to the east, by several rocky outcrops of upraised and lithified, undated but presumably Pliocene to Pleistocene, reefal and littoral deposits.

The Landsat ETM + principal component false colour image emphasizes many of these features

(Fig. 3). Coastal sabkhas (probably flooded when this image was captured) stand out as bright green, whereas their inner drier portions are dark purple to blue; Hikman Formation outcrops and other rocky formations are pale pink to red; older, raised lagoon and tidal channel deposits are light blue; beach ridges are light blue; windblown sands are white; thicker continental sabkhas are red to dark purple to blue with green tinges in places; shallow marine sands, rubble and wave cut platforms are lighter yellow; coral carpets or reefs are dark patches limiting lighter yellow zones. Faint but clear east–west features to the north of the ancient central lagoon were confirmed by QuickBird images, and by ground-truthing, to be shelly beach ridges on the northern shore of the old lagoon.

The NNE-trending linear eastern and western shorelines run parallel to several features on the Landsat ETM + PC image (Fig. 3), suggesting a strong structural control on the morphology of the peninsula. Structural analysis of the image suggests that the modern coral carpets or reefs off the southern coast lie on structurally higher "noses" (Fig. 8). The lower trend between these highs, hosting a sandy marine shoreface and lacking coral development off the southern coast, would explain the now on-land abandoned lagoonal feature. The

rocky headlands of the south-eastern corner, alternating with non-lithified zones of carbonate sand, are interpreted to result from intersection of the slightly more NE-trending coastline with the NNE-trending lithified fault lines. The fault trends are closely similar to those of the nearby Huqf as extracted from the 1:250,000 geological maps (Fig. 8B).

Geology

There are three essential geological features of the substrate to the sabkhas of Barr Al Hikman (Dubreuil *et al.*, 1992; Le Metour *et al.*, 1992). These are the: (a) NNE-trending regional major normal and strike-slip fault pattern; (b) allochthonous units linked to the westward overthrust Masirah Ophiolite, which build the eastern flank of the peninsula (Hamrat Duru Group) and reach westwards to the lower lying central area; (c) downthrown Cretaceous and Tertiary authochthonous formations that compose the outcrops of the central portion, and the western flank of the peninsula (Aruma Group and Hadramaut Group).

Two boreholes for near-surface calibration of a vibro-seismic survey of the south of the peninsula were made for Petroleum Development Oman

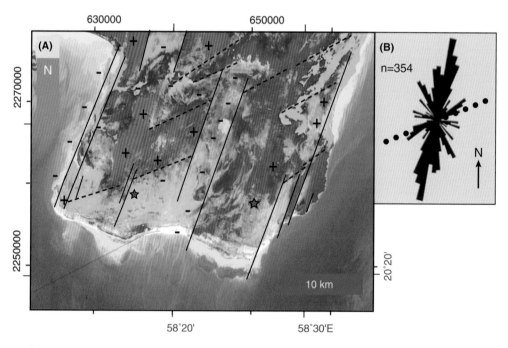

Fig. 8. Structural interpretation of the Barr Al Hikman peninsula. (A) Landsat TM principal component composite image with interpretations of faults and relative vertical component of movement. Red stars indicate approximate sites of boreholes (see text). (B) Fault patterns of the Huqf, extracted from the 1:250,000 geological maps. The dotted line shows the subordinate Barr Al Hikman trend.

(PDO) in 1970–1972 (stars on Fig. 8), and regional exploration marine seismic surveys of Masirah Bay were carried out by Sun Oil and Amoco between 1973 and 1983 (Henk Droste, pers. comm., 2006). These data give evidence for a general low-angle southward dip under the peninsula, with uplifted or downthrown SSW–NNE faulted compartments stepping progressively down to the east to be over-ridden by the Masirah Ophiolite complex. The NNE-trending faults run from the bay to the peninsula and those interpreted from marine seismic coincide well with the positions of those interpreted from on-land geomorphological analysis (unpublished internal PDO reports, H. Droste pers. comm., 2006). It is interesting to note that the structural geomorphological analysis given here was made before obtaining the seismic survey reports.

Bands of NNE-striking outcrops near the eastern coast of the peninsula are composed of littoral deposits, with faunal assemblages that are strikingly similar to the modern reefs. They possibly date between the Pliocene and the Pleistocene, and form headlands along the coast. These lithified shallow-marine to littoral sediments were described as beachrock by Dubreuil *et al.* (1992) and Le Metour *et al.* (1992), but in fact are composed of a succession of uplifted, lithified littoral sandwaves of carbonate debris, patch reefs on the platform built by the sandwaves, and beach and intertidal facies. These outcrops either lie on breccias of allochthonous Jurassic Hamrat Duru Group turbidites (themselves directly covering the Hamrat Duru rocks), or they abut autochthonous carbonates which, covering the thrust sheets, contain benthic foraminiferal assemblages possibly indicative of a Middle Eocene age (S. Packer pers. comm., 2006).

Although the general geological framework may be described in these fairly simple terms, the sabkhas of the peninsula in fact seal the thrust tip of the Masirah Ophiolite complex, as well as draping the concomitant and also later strike-slip to normal fault system, down-stepping eastwards from the Huqf high. In some respects this narrow zone is a diminutive foreland basin. Some faults or fractures are possibly active fluid conduits today (see below), and this complex geological framework is under re-evaluation as one part of ongoing studies.

Coastal sabkhas

The false-colour Landsat TM principal component image (Fig. 3) shows the flooded, lower-standing reaches of the major coastal sabkhas as bright green areas, connected to the sea at both the eastern and western shorelines of the peninsula. When this image was recorded in 2000, the sabkhas were flooded to a greater extent than usual. Rainfall records for Masirah do show higher amounts between 1995 and 1997 (yearly amounts of 140 mm, 95 mm, 117 mm) possibly explaining the higher water level, although a heavy shower that does not show up on records may have just preceded the recording of these images in 2000. However, very high spring tides may also have been responsible for the flooding.

The large eastern coastal sabkha (Figs 3 and 4), which resembles a scorpion in shape, is not easily accessed so no ground-truthing of the feature had been carried out at the time of writing. The comparison of images of this sabkha from 1972 (Landsat MSS), from 1987 (Landsat TM) and from 2000 (Landsat TM) with a 2004 QuickBird image, suggests that the full flooding of these areas, such as imaged in 2000, seems to be fairly rare or short lived. However, such high spring tide flooding of the eastern low-lying sabkha area was observed in March 2009 (by S. Dirner and J. Mattner) to take place during several successive days. The coastal sabkhas are thus shown to be only irregularly inundated. Figure 9A shows the large coastal sabkha linked to the lagoons on the western coast (1 on Fig. 4). This sabkha is crossed by several tracks, allowing easy access and observation. The brighter white area and the dark surrounding rim on Fig. 9A is a vast expanse of halite growth polygons with metre-scale diameters (Fig. 9B and C). Halite polygons are interbedded with and covered by wind-blown sand (Fig. 9D).

Closer to the coast, supratidal flats form the intermediate zone of the coastal sabkhas (Fig. 10B). Here, under an irregular and ephemeral veneer or crust of halite, alternate layers of tidal and windblown host sediments show a banding that is deformed by growth of gypsum nodules (Fig. 10A). Patchy red staining by iron oxides and hydroxides is concentrated at the limit of the damp capillary fringe zone: darker, damp sediments below compared to pale dry sediment above. Gypsum crystals and nodules deform the banding in the upper 10–20 cm, and the surface sediment is loose and vuggy (Fig. 10C).

The features that are common to all these coastal sabkhas are summarized in plan view on Fig. 11 and in cross section on Fig. 12. Seawater at 35–40 ‰ salinity (A on Figs 11 and 12) may be

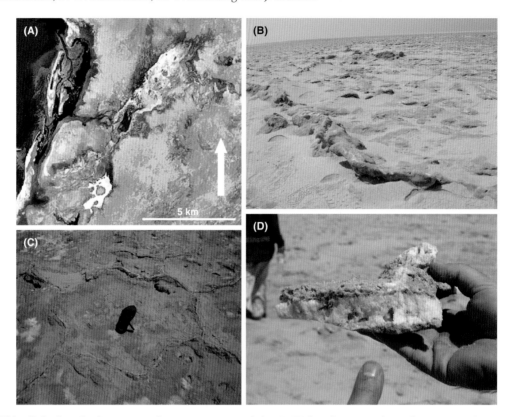

Fig. 9. Sabkha linked to the lagoons on the western coast. (A) ASTER bands 3, 2 and 1 colour composite image, central western coastal sabkha; see Fig. 3 for location. (B) Details of halite polygons on the sabkha. Sandal is 30 cm long. (C) Halite growth ridges or tepees. (D) Inter-bedded halite and windblown sand. Field photographs by H. Droste.

concentrated up to 70 ‰ salinity or more in barred intertidal lagoons (B on Figs 11 and 12) with the development of a variety of algal mats (1 on Fig. 12, described further below). Supratidal flats or slightly higher ground (2 on Figs 11 and 12; Fig. 10B) commonly show a halite surface crust or a vuggy layer (Fig. 10C) above laminated grainy sediments with growth of displacive gypsum (Fig. 10A). Lower ground, beyond the lagoon (Fig. 9A; 3 on Figs 11 and 12) runs inland but is linked to the lagoon and may be flooded from the lagoon (a on Fig. 12), or from runoff from surrounding continental sabkhas after short heavy rainfall (b on Figs 11 and 12), whereas the continental aquifer must also feed fluids and solutes to the sabkha (c on Fig. 12). These inner zones of the coastal sabkha system are therefore characterized by halite tepees and polygons.

The outer zone of the coastal sabkhas merges with coastal lagoons (Fig. 13 inset). Salinity measurements show the progressive concentration of brines from normal 36 ‰ seawater, to above 40 ‰ in protected lagoon heads, and even higher (>70 ‰)

in restricted ponds (respectively 4, 2 and 1 on Fig. 13). Algal and microbial mats are well developed in these environments (Fig. 14), showing rapid variation in thickness, texture, and layering. They evolve from thinner tough leathery textures to thicker and rubbery, or soft but reasonably cohesive mats, even during single neap-spring tidal cycles. Dolomite is found as cement rims on carbonate sand grains (Fig. 15), as well as forming small clearly zoned isolated euhedral dolomite crystals (Figs 15 and 16) where the mats are anoxic, giving off abundant H_2S. The sediment below the anoxic mat is essentially composed of carbonate grains (50 % aragonite, dolomite, calcite), with minor amounts of quartz sand, halite, traces of polyhalite and K-feldspar, but with surprisingly little *in-situ* mud, be it carbonate or wind-blown siliciclastic sediment. Dolomite forms some 15 % of the inorganic bulk sediment. The fine fraction, below 2 μm, forms between 4.6–10.0 wt % of the total sample. This comprises carbonates (0.5–6.4 %), and wind-blown clays (1.6 % mica/illite, 1.4–1.7 % chlorite).

Fig. 10. Supratidal flats forming the intermediate zone of the coastal sabkha. (A) Pit (60 cm deep) in the supratidal flats with fine layers of tidal and windblown sand disturbed by gypsum displacive growth. (B) Supratidal playa flats with an occasional irregular halite veneer (e.g. on track where car is parked). (C) Vuggy top layer of supratidal sand flats. Visible portion of knife is 14 cm long.

Continental sabkhas

The non-lithified Quaternary marine to lagoonal sediments of the peninsula host the major part of the sabkha in terms of surface area. The upper 10–20 cm of sabkha host sediment tends to be cohesionless but vuggy, allowing vehicles to sink quite deeply into the substrate. These sediments are generally sandy (coarse to fine mixed siliciclastic-carbonate to carbonate material, illustrated on the sediment surface of Fig. 17A and C, indicated as 5 on Fig. 18). Host sediments are richer in lithic grains and gravels (from the Oman Mountains to the north) closer to the northern erosional scarp limiting the Barr Al Hikman sabkhas (Fig. 2), and richer in carbonate grains in the older abandoned lagoon area towards the south of the peninsula. A coarser fraction of quartz grains is added to the host sediment in the south and west of the peninsula.

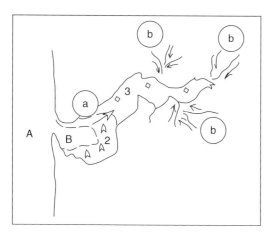

Fig. 11. Plan view of the coastal sabkha zonation. A = seawater; B = barred intertidal lagoon; 2 = supratidal flats; 3 = low ground; a = flooding from the lagoon; b = continental sabkha. Not to scale. See text for commentary.

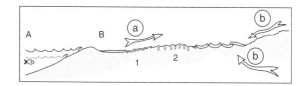

Fig. 12. Schematic cross section of coastal sabkhas. A = seawater; B = barred intertidal lagoon; a = low ground flooded from the lagoon; b = runoff from continental sabkha; 1 = algal mats; 2 = supratidal flats on higher ground. Not to scale. See text for commentary.

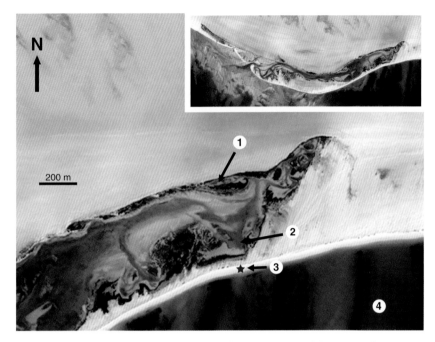

Fig. 13. QuickBird bands 3, 2 and 1 colour composite image showing a coastal lagoon with progressive concentration of brines. Insert shows whole lagoon; see Fig. 3 for regional location. Boxed numbers indicate locations of salinity measurements: 1, lagoonal pond >70‰; 2, protected intertidal lagoon 40–50‰; 3, beach springs 34–40‰, the location of the springs is marked by the blue star; 4, seawater 35–42 ‰. See text for commentary.

Fig. 14. Algal-microbial mats. Alternating bands of algal-microbial mat and carbonate sands in protected lower energy lagoon environments. (A) and (B) Anoxic mat, salinity > 70‰. Field of view 30 cm. (C) and (D) Aerated mat, salinity < 50‰. (C) Soft sediment slab is 20 cm high. (D) Sediment layer is 5 cm thick (knife blade tip for scale). Photographs by H. Droste.

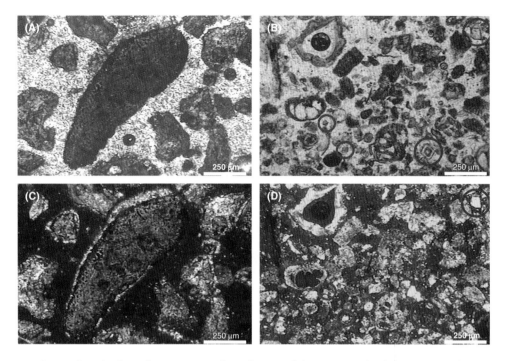

Fig. 15. Lagoon sediment from higher salinity restricted pond. Stained thin section of sediment set in colourless epoxy resin. (A) and (B) Plain polarized light views. (C) and (D) Crossed polarized light views. Note dolomite rim-cements and small euhedral dolomite crystals (circled).

Silicified Eocene limestones (Dammam Formation; Dubreuil *et al.*, 1992) outcrop around the southwest corner of the peninsula, where the abandoned tidal channels once flowed to flood and drain the now abandoned lagoon. Modern beach sands in this area are composed of well-rounded coarse to fine quartz grains, and this grain population is mixed into the older lagoon carbonate-sand

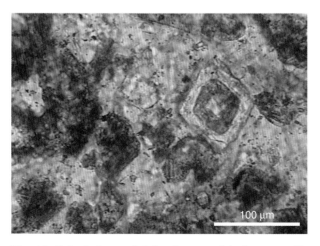

Fig. 16. Euhedral zoned dolomite crystal in lagoon sediment. Stained thin-section under plane polarized light. Sediment is set in colourless epoxy resin.

deposits, gradually decreasing in proportion to the east, away from the outcrops.

The shelly faunas of the old marine embayments and lagoon tend to be locally dominated by one or the other assemblage or dominant species of bivalve or gastropod, revealing a clear palaeoecological zonation across the area. Thus, although the shelly fauna are observed as a component of the coarse composite lag on a deflation surface (4 on Fig. 18), little to no movement of the larger clasts and fossils has taken place (Fig. 6). Close to the southern and eastern shores of the old lagoon (to the northern side of the ancient barrier barr protecting the lagoon, or to the west of the slightly higher-lying drainage divide, see Fig. 5), thick-shelled bivalves and gastropods indicate higher-energy environments, whereas farther north, away from the previous coastlines into the abandoned lagoon, localized patches dominated by cerithid gastropods give way to patches dominated by larger thin-shelled bivalves or smaller thicker-shelled bivalve assemblages. One or several irregular but thin layers of green silty sediment or clay occur in places, within the upper metre of sediment (Dubreuil *et al.*, 1992; 6 on Fig. 18).

Numerous pits dug into the sabkha (Fig. 17) show a common profile illustrated schematically

Fig. 17. Subsurface sabkha sediments. (A) Pit dug in loose sabkha sediment. Lens cap is 2.5 cm across. (B) Nest of fibrous gypsum crystals 10–20 cm below the sabkha surface. (C) Acicular gypsum crystals from near-surface nests in (A), but reworked in wind-blown sediment on the sabkha. Shovel blade is 20 cm across. (D) Large bladed gypsum crystal from 30 cm below the sabkha surface.

on Fig. 18. Evaporites form conspicuously in these host sediments (5 on Fig. 18). From the bottom up, they start with a layer of large, 5–15 cm bladed gypsum crystals 20–40 cm below the surface lag (4 on Fig. 18), and these crystals may locally coalesce into a 5–10 cm thick gypcrete crust (Fig. 17C and D; 3 on Fig. 18). These larger crystals grow towards the top of the permanently damp sediment of the capillary fringe. Above these large crystals,

clusters or nests of mm-wide by cm-long fibrous or acicular crystals of gypsum are formed in the day-time dried-out zone, between the large gypsum crystal layer and the sediment surface (Fig. 17A and B; 2 on Fig. 18). A dusting or extremely thin layer of halite (1 on Fig. 18), overlying both gypsum evaporites described above, is found at the sediment surface. After rain showers, the sediment surface is covered by a snow-white, fairly coherent layer of halite. This layer of halite is progressively broken up to the more common dusty material after several days of normal sunny weather.

The continental sabkhas may be summarized as illustrated on Fig. 19. The sediment surface cuts into Pleistocene gravel fans (B′) at the northern end and is a flat deflation surface (A) except where bedrock lies a little higher (possibly a Pleistocene to Holocene wave-cut platform). The abundant marine or lagoonal molluscs in the lag show that the deflation surface has cut down below an earlier embayment or lagoon floor. Near the coast, older beach ridges and storm spits (C) stand proud and

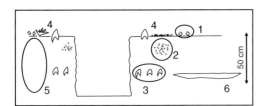

Fig. 18. Schematic cross section to show zonation in pits in continental sabkha. 1 = halite-dusted surface; 2 = clusters of fibrous or acicular gypsum; 3 = bladed gypsum crystals; 4 = deflation surface; 5 = sandy host sediment; 6 = thin silty/clay layers. Not to scale. See text for commentary.

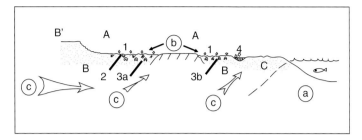

Fig. 19. Schematic cross section showing zonation of continental sabkhas. A = deflation surface; B' = Pleistocene gravel fans; B = sabkha host sediment; 1 = halite-dusted surface; 2 = nests of fibrous or acicular gypsum; 3a = large bladed or swallowtail gypsum; 3b = gypsum crust; 4 = halite crust on supersaturated brines in salt ponds. Fluid inputs: a = marine aquifer; b = rainfall runoff; c = continental aquifer. Not to scale. See text for commentary.

have more or less preserved their original morphology. From the sediment surface downwards, the evaporites develop in the host sediment (B) and are composed of: (1) a surface crust or dusting of halite; (2) nests of fibrous or acicular gypsum crystals; (3a) large bladed or swallowtail gypsum crystals; (3b) a gypsum crust or gypcrete; (4) a halite crust on supersaturated brines in salt ponds. The fluid inputs are ringed: (c) the continental aquifer and (b) rainfall runoff, give way at the coast to (a) the marine aquifer. The input from high relative humidity is not shown, but this may significantly contribute to remobilize the surface halite.

Coastal salt ponds

A group of salt ponds, regularly used by local fisherman to salt their catch, is located a few hundreds of metres northeast of the eastern lagoon on the southern coastline (Fig. 20). These salt ponds lie in topographic depressions close to the height of present-day sea level. They were formed by

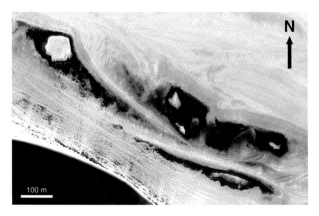

Fig. 20. QuickBird satellite bands 3, 2, and 1 composite image showing salinas or salt ponds on the south coast of the Barr Al Hikman peninsula. See Fig. 3 for location. The image was recorded on 26 October 2003.

abandonment of lagoonal lows, following the accretion of successive beach ridges that terminated in arcuate storm spits. The free-water level in the westernmost pond was observed from late 2001 until late 2006. During 2002, the level slowly rose by about 20–30 cm over a six to eight month period, to stabilize near the level maintained between 2004 and 2006, just below the sediment surface. In February 2006, the pond was flooded by a heavy rainfall, but the water level dropped again within weeks to slightly below the sediment surface, as before.

In spite of the proximity to the seashore and the coarse supposedly permeable nature of much of the sediment composing the beach ridges, the water level shows little influence from tidal variations, either from semi-diurnal (about 2 m range at spring tides) or from neap-spring variations. However, exceptionally high equinox spring tides, as observed in March 2007, led to a rise of the brine level and flooding of the salt pond, but not for more than a couple of days, after which the free-water level sank back to the longer term mark. Salinity here reaches supersaturation, with halite precipitating at the surface and forming growth polygons. The coarse carbonate gravel to sandy sediment contains a mush of mm- to cm-size gypsum crystals and a well-developed, layered microbial mat that can measure up to 20 cm thick. This mat passes laterally into buckled tepee structures at the edge of the salt pond.

Lower salinity beach springs

Low spring tides revealed a cluster of four to five spring sources on the present-day beach near the western lagoon on the south coast of Barr Al Hikman (3 on Fig. 13), and these were observed over a number of years between 2002 and 2007. The sources can flow at rates up to several tens of litres per minute, and did not dry up during low tide, as do the runnels and rivulets simply draining seawater from the beach. Repeated salinity measurements over a number of months in 2005–2006 showed a systematic 1–2 ‰ lower salinity than the seawater for these springs in spite of evaporation on the beach after high tide. Eleven measurements of seawater averaged 35.8 ‰ (minimum 34 ‰, maximum 37 ‰); 14 measurements of the beach spring water averaged 34.6 ‰ (minimum 33 ‰, maximum 36 ‰).

During February–March of 2007, considerable erosion of the beach (between 50 cm and 1 m), followed by reworking and rebuilding of the beach profile over the springs was observed. This was caused by winter storms and by very high equinox spring tides. Salinities of both the seawater and gushing springs varied considerably during this period, but with no clear pattern. A several hundred metre broad long-shore coastal plume of seawater developed after the rebuilding of the beach profile, with seawater salinity unexpectedly rising to 42 ‰. No regular difference between spring and seawater salinity was evident during this period, and it is clear that further work is necessary to clarify possible relationships between groundwater, beach spring water and seawater.

Older sabkhas

Older sabkha deposits comprise a variety of features that may or may not be directly related within a single phase of sabkha formation, comprising both coastal and continental sabkhas. Thinly bedded laminated and rippled wind-blown sands and coastal sabkha deposits lie above the modern coastal sabkhas on the north to western area of the peninsula, forming a metre-high terrace around their landward terminations (Dubreuil *et al.*, 1992; Le Metour *et al.*, 1992; Fig. 21). The sediments were deposited on an alternately shallow-flooded or emergent surface, similar to the upper intertidal to supratidal flats today. These sabkha deposits appear to have been eroded into, either by seawater encroaching from the western coast, or by gullying and headward erosion. They appear to represent older coastal sabkhas.

Large exfoliated and corroded gypsum crystals form patches several hundreds of metres in width on the present deflation surface (Fig. 7). These patches are to be found scattered within the surface layer of the soft sediment deposits, all over the peninsula. They correspond to an exhumed large

Fig. 21. Older sabkha deposits. (A) One-metre high terrace formed by recent erosion of older sabkha. (B) Older coastal sabkha deposits consisting of thinly bedded and laminated and rippled alternations of wind-blown and carbonate playa sands. Hammer is 35 cm long.

gypsum crystal/gypcrete layer, comparable to the layer forming today, and therefore are interpreted to have formed during an older phase of continental sabkha formation. They are now exposed by deflation that has cut down through the sabkha deposits.

These patches lie at various heights above the thinly bedded older coastal sabkha deposits described above. They may represent a single same phase of sabkha growth as these, or they may come from one or several independent phases of continental sabkha development.

DISCUSSION

Comparative setting

The Barr Al Hikman sabkhas are characterised by their rather coarse and grainy, predominantly marine or lagoonal carbonate host sediment, their predominantly gypsum-halite mineralogy, and in that their distribution is controlled by an irregular surface topography and falling relative sea-level. The general survey of the Barr Al Hikman sabkhas shows a number of features that are in contrast to the better-known Arabian sabkhas, particularly the Abu Dhabi sabkhas of the Arabian Gulf (Fig. 1; Glennie, 2005) that develop on a fairly regular low-angle ramp over a simpler geological substrate.

Alsharhan & Kendall (2003) provided a recent major review of the Arabian Gulf sabkhas, and Warren (2006) gave a milestone encyclopaedic review of evaporites in general, with many references to the Abu Dhabi sabkhas in particular. To summarize, the studies on the sabkhas of the Arabian Gulf established the classic models for sabkha development, and the Abu Dhabi sabkhas subsequently provided the basis for the notions first of seawater flooding, then of evaporative pumping, and very recently of upward aquifer leakage, as the dominant factor of the hydrological cycle, all moderated by strong evaporation. A study of the detailed hydrological budget on the Abu Dhabi sabkha (Sanford & Wood, 2001), demonstrated that rainfall and upward leakage from Tertiary formations are the main sources of replenishment of the aquifer there, contrary to the previous model assumptions. Strohmenger *et al.* (2004, 2007, this volume pp. 149–183,) have recently conducted detailed surveys of the Abu Dhabi sabkha, providing the latest generation of detailed descriptions of both sediment and evaporite mineral zonation there, as well as the ages and types of microbial mat growth, and a model for sabkha development that has been continuous, if cyclic, over the past few thousand years.

There are no precise dates for the older phase of sabkha formation at Barr Al Hikman. Circumstantial evidence for both sea-level fall and for differential sabkha-basement uplift come from observations on presumed Pliocene, Pleistocene and Holocene deposits There are several Pleistocene or older wave-cut terraces, at heights ranging from spring tide low water level to 1.5 m above high tide level. Pliocene or Pleistocene beach deposits lie about 1 m above high sea-level on the east coast of the peninsula. The Pleistocene to Holocene sabkha marine host sediment lies at heights of 1–10 m above sea-level. All of these observations suggest recent differential uplift of the peninsula, while absolute sea level here is taken to have fallen about 1 m over the last six thousand years (Lambeck *et al.*, 2002). A possible age for the very fresh, unlithified marine embayment and lagoonal sabkha host sediments could therefore be about six thousand years, during the latest high stand of sea-level given by Lambeck *et al.*, (2002).

The major differences of the Barr Al Hikman sabkhas compared to those of the Arabian Gulf lie in their contrasting climatic and oceanographic settings. The variety of environments at Barr Al Hikman are grouped into two basic assemblages, the coastal versus continental sabkhas, and these may also be distinguished by their hydrology and by the mineral development that they host. In general, the parts of Barr Al Hikman termed coastal sabkha in this paper arguably could be called halite salinas (cf. Warren, 2006), in that they are low-lying areas in temporary connection with the sea, in that they are occasionally flooded and that some of the solutes therefore must come from marine influx. However, they are dry most of the time, and much of the evaporite does not appear to be sub-aqueous. A certain amount of halite solute must also be provided by rainwater run-off, draining brines from higher ground after rain storms. The coastal salt ponds might also be termed halite salinas, but in their case there is no surface connection to the sea, and the link to the continental aquifer charge suggests upward leakage from the continental aquifer, as opposed to brine influx from the sea. The detailed hydrological and chemical demonstration of the role played by the continental aquifer in Abu Dhabi (Sanford & Wood 2001; Wood *et al.*, 2002) is taken to support the rather less well-established, more circumstantial hypothesis made here for the role of the continental aquifer in the Barr Al Hikman case.

Sabkha hydrology

Although no detailed hydrological budget is available yet for Barr Al Hikman, the conditions do appear quite different from those of Abu Dhabi. The two precipitation inputs, slight rainfall (Masirah station records between 1980–2003: average 58.3 mm rainfall; range 1–279 mm yr^{-1}) and the very high relative humidity with heavy morning dew, do not appear sufficient to compensate the high evaporation (1000 mm yr^{-1}).

The gradual rise in free-water level in the coastal salinas, six to eight months after a period of increasing rainfall in the Oman Mountains, corresponds to the time expected for the pressure build-up to be conveyed to this area. It is likely that groundwater from the mountains feeds the aquifer at Barr Al Hikman. Calculations using Darcy's law, with reasonable figures for permeability in Quaternary wadi sediments (B. Wassing pers. comm., 2006) do not contradict this hypothesis. Study of the major aquifers west of Wadi Halfayn has shown that groundwater flows south from the Oman Mountains to the Umm as Samim sabkha (Al Lamki & Terken, 1996). The aquifer fed by the Oman Mountains must also flow southwards to the east of Wadi Halfayn, under the Wahaybah sands towards Barr Al Hikman. The figures given by Al Lamki & Terken (1996) for flow rates in the Tertiary aquifer ($10 \, m \, yr^{-1}$) and the Palaeozoic aquifer ($3 \, m \, yr^{-1}$), based on radiocarbon dating, are much higher than those suggested for the Abu Dhabi sabkha by Sanford & Wood (2001). Whatever the rates of flow involved here, this evidence does support the hypothesis that in the case of Barr Al Hikman the aquifers may feed the continental sabkha from upward leakage, similarly to those of Abu Dhabi.

However, in the Barr Al Hikman case, the major NNE regional faults might serve as high permeability conduits. Circumstantial evidence comes from the freshwater spring that is located on such a fault line located between the nearby village of Filim and the peninsula. The slightly fresher-than-seawater springs, described above from the beach of the south Barr Al Hikman coast, are aligned with a trend of older, lower-lying beach ridges onshore at the eastern termination of the lagoon. These beach ridges, highly oblique to the present beach and dune ridge, are composed of very coarse gravel to boulder carbonate sediment, and may constitute a reservoir of seawater that flushes at very low tide. However, the lower salinities recorded during the 2005–2006 cool season, require that fresher waters rather than seawater provide part of the charge. In this context, it is relevant to note that the springs are located on the fault line bounding the eastern limit of the westerly structural nose. The non-negligible aquifer recharge on the peninsula, therefore, may possibly be made through the fault and fracture system, and there might be a major dual-permeability groundwater system at Barr Al Hikman. Halite is observed to cement surface sediment along a linear feature interpreted to be a recent fracture that cuts across several Holocene sedimentary geomorphological features near the southern coast of the peninsula. The halite cement is interpreted to be an evaporite "host-rock alteration" of the grainy carbonate, at the surface intersection of a higher permeability fracture. Further study should help to elucidate the possible fault-related brine upward leakage here.

Evaporites, thermalites and microbialites

The evaporite mineral composition of the Barr Al Hikman sabkhas, both coastal and continental, has proven to be limited essentially to gypsum and halite. The absence of anhydrite is thought to result from the only moderately high temperatures, rarely reaching above $40 \, °C$, with an absolute maximum of $45.3 \, °C$ recorded between 1943 and 2003 (Masirah weather station; MTC, 2003).

The large-bladed gypsum crystals forming 20–40 cm below the surface of the continental sabkhas at Barr Al Hikman closely resemble the similarly large, bladed gypsum crystals of the Abu Dhabi sabkha, both in the Mussafah Channel and in pits at Al Qanatir. These would then be more properly accounted for as thermalite minerals (Wood *et al.*, 2002), precipitated by the effect of increasing temperatures of the rising brines below the sabkha surface. The cm or longer, mm-thick acicular and fibrous gypsum crystals occurring in nests are different though, and these may be true evaporite minerals, deposited from evaporation of brines in the upper, unsaturated zone of the sabkha host sediment. However, they might alternatively form from dissolution of the older large euhedral but corroded crystals that litter the surface, followed by re-precipitation in the upper layers of sediment.

Authigenic dolomite has been found in sheltered ponds at the restricted terminus of the western lagoon on the south coast. A wind-blown origin of this dolomite (from the Huqf for instance) can be excluded since dolomite is observed both as crusts and rim cements around carbonate grains, and as zoned euhedral crystals of a range of sizes (Fig. 15). The euhedral crystal nuclei measure about 20 μm and the zoned crystals grow to about 100 μm. XRD analyses determine the mineral as a well-ordered dolomite and also a Mn-Mg-Fe variety of dolomite. The water salinity where the dolomite is forming is above 70 ‰, and the sediment also contains a little polyhalite. Given the anoxic environment and complex microbial mat development, dolomite formation is interpreted as being mediated by microbial processes (cf. Vasconcelos *et al.*, 1995; Vaconcelos & McKenzie, 1997) rather than being driven by mixing zone or hypersaline brine chemistry.

CONCLUSION

The Barr Al Hikman sabkhas, covering about 1400 km^2 of low-lying topography (the peninsula is about 30 km wide), form a thin irregular soft-sediment veneer over a complex geological substrate. Recent sea-level fall, together with structural movement, have provided emergent marine to lagoonal carbonate host sediments of varying thickness, as well as ponded depressions, for the development of evaporite deposits.

The sabkhas may be grouped into two types, coastal and continental, both from their geomorphology and from their likely hydrogeology. Episodic flooding of lower-lying coastal sabkhas (0–5 m elevation), produces extensive perennial halite deposits, both by marine influx and by remobilization of surface-crust halite during short, heavy, rainfall. A hypothetical dual permeability system is proposed to govern flow in the continental aquifer (moderate to low permeability in the non-lithified Quaternary to older lithified deposits; high permeability along fault and fracture conduits). The continental groundwaters are thought to feed sabkha development in the slightly higher sabkhas (5–15 m). In these continental sabkhas, gypsum forms layers of large bladed crystals and crusts at shallow depths ("thermalites"), with shallower nests of fine gypsum needles and fibres (possibly true evaporites) in the grainy carbonate host sediments. Authigenic dolomite, forming in highly saline coastal lagoon environments, is thought to be mediated by microbial activity rather than to result from purely chemical constraints.

Older, Holocene sabkha deposits lie a metre or more above the present sabkhas. These possibly 6 kyr old sabkhas form marginal terraces or deflated lags. Ongoing studies are investigating the structural evolution of the peninsula, the links between fault systems, evaporite deposition and lithification of Quaternary carbonates, as well as the link between microbial systems and the precipitation of dolomite.

ACKNOWLEDGEMENTS

Interest in Barr Al Hikman, for MM and PH, was sparked off by John Grotzinger, Volker Vahrenkamp and Mark Newall. Christian Strohmenger has helped us to bring Barr Al Hikman to the notice of the evaporite community. Hopefully this paper is some reward for their friendship and input. A review by Chris Kendall helped to clarify ambiguities and made us improve our presentation of the sabkhas. Many friends and colleagues have helped at different stages of this study, in particular Philippe Razin, Sabine Vahrenkamp, Carine Grelaud and Mathieu Rousseau who participated in sample collection; Luc van Schijndel who tracked the Landsat images while ground-truthing both on and off the sabkha tracks; Francois Guillocheau and Cecile Robin who ensured thin-section making after preparing a startling variety of fish dishes on the beach. Rashid Al Hinai made the first false colour image for us to work on, and Steve Fryberger helped to find meteorological data. Jean Paul Breton kindly contributed several analyses on clay minerals. Bart Wassing gave us a more solid foundation to gauge the possible time-lag between rainfall on the Oman Mountains and pressure response in the aquifer on the Peninsula. Together with Henk Droste and Volker Vahrenkamp, we have conducted numerous trips to Barr Al Hikman. Their staunch contribution, as well as the interest of trip participants has helped to mature our investigations. Moral support came from numerous family members and friends who accompanied us on field trips to this beautiful location. The Executive Committee of the JVRCCS Carbonate Centre, a Shell SQU joint venture at Sultan Qaboos University, were most supportive of the Barr Al Hikman project in general, and have also ensured some financial support.

REFERENCES

Al Lamki, M.S.S. and Terken, J.J.M. (1996) The role of hydrogeology in Petroleum Development Oman. *GeoArabia*, **1**, 495–510.

Alsharhan, A.S. and Kendall C.G.St.C. (2003) Holocene coastal carbonates and evaporites of the southern Arabian Gulf and their ancient analogues. *Earth Sci. Rev.*, **61**, 191–243.

Anon (1992) *Study for Wildlife and Conservation Areas – Master Plan for the Coastal Areas of Barr Al Hikman and Masirah Island.* Weidleplan Consulting GmbH.

Claereboudt, M.R. (2006) *Reef Corals and Coral Reefs of the Gulf of Oman.* The Historical Association of Oman, *Al Roya Press & Publishing House*, 344 pp.

Dickson J.A.D. (1966) Carbonate identification and genesis as revealed by staining. *J. Sed. Petrol.*, **36**, 491–505.

Dubreuil, J., Bechennec, F., Berthiaux, A., Le Metour, J., Platel, J.P., Roger, J. and Wyns, R. (1992) *Geological map of Khaluf, sheet NE 40-15 1:250000.* Directorate General of Minerals, Oman Ministry of Petroleum and Minerals.

Eriksen, H. and Eriksen, J. (1999) *Bird Life in Oman.* Al Roya Publishing, Muscat.

Eriksen, H., Eriksen, J., Sargeant, P. and Sargeant D.E. (2001) *Bird Watching Guide to Oman. Al Roya Publishing*, Muscat.

Glennie, K.W. (2005) *The Desert of Southeast Arabia. GeoArabia, Gulf PetroLink*, Bahrain, 215 pp.

Homewood, P.W., Vahrenkamp, V., Mettraux, M., Mattner, J., Vlaswinkel, B., Droste, H. and Kwarteng, A.Y. (2007) Barr al Hikman: a Modern Carbonate and Outcrop Analogue in Oman for Middle East Cretaceous Fields. *First Break*, **25**, 55–61.

JISAO (2006) *Gridded Data Sets.* Joint Institute for the Study of the Atmosphere and the Ocean, University of Washington Seattle. http://jisao.washington.edu/data_-sets/.

Lambeck K., Esat, T.M. and Potter E.K. (2002) Links between climate and sea levels for the past three million years. *Nature*, **419**, 199–206.

Le Metour, J., Bechennec, F.J., Roger, J., Platel, J.P. and Wyns, R. (1992) *Geological map of Al Masirah, sheet NE 40-16, 1:250000.* Directorate General of Minerals, Oman Ministry of Petroleum and Minerals.

Le Metour, J., Michel, J.C., Bechennec, F., Platel, J.P. and Roger, J. (1995) *Geology and Mineral Wealth of the Sultanate of Oman Directorate General of Minerals, Oman.* Ministry of Petroleum and Minerals, 285 pp.

MRMEWR (2005) *National Action Programme to Combat Desertification in the Sultanate of Oman.* Ministry of Regional Municipalities, Environment, and Water Resources in Collaboration with United Nation Environment Program/Regional Office of West Asia (UNEP/ROWA) and National Committee to follow up the Implementation of United Nations Convention to Combat Desertification (UNCCD).

MTC (2003) *Annual Climate Summary 2003.* Department of Meteorology, Climate section, Directorate General of Civil Aviation and Meteorology, Ministry of Transport and Communications, Sultanate of Oman, 154 pp.

Pilcher, N.J., (2002) *Potential Tropical Coastal, Marine and Small Island World Heritage Sites in the Middle East Region.* Report to IUCN/UNESCO/NOAA. World Heritage Biodiversity: Filling Critical Gaps and Promoting Multi-Site Approaches to New Nominations of Tropical Coastal, Marine and Small Island Ecosystems. Hanoi, Vietnam, 18 pp.

Ross, J.P., (2005) *Hurricane Effects on Nesting Caretta caretta.* Marine Turtle Newsletter, **108**, pp. 13–14. ISSN 0839-7708.

Sanford, W.E.,and Wood, W.W., (2001) Hydrology of the coastal sabkhas of Abu Dhabi, United Arab Emirates. *J. Hydrogeol.*, **9**, 358–366.

Scott, D.A., (1995) *A Directory of Wetlands in the Middle East: The Sultanate of Oman.* World Conservation Union.

Strohmenger, C.J., Al-Mansoori, A., Shebl, H., Al-Jeelani, O., Al-Hoseni, I., Al-Mehsin, K. and Al-Shamry, A. (2004) Modern Carbonate-Evaporite Depositional Environments of Abu Dhabi, (U.A.E.). Unpubl. Field trip guide book, 11th ADIPEC Meeting, 1st International Evaporite Conference.

Strohmenger, C.J., Shebl, H., Al-Mansoori, A., Al-Mehsin, K., Al-Jeelani, O., Al-Hoseni, I., Al-Shamry, A. and Al-Baker, S. (2007) Depositional environment and sea-level history of the Abu Dhabi sabkha in the vicinity of Al-Qanatir Island (U.A.E.). *GeoArabia*, **12**, 219.

Vasconcelos, C.O., McKenzie, J.A., Bernasconi, S., Grujic, D. and Tien, J. (1995) Microbial mediation as a possible mechanism for natural dolomite formation at low temperature. *Nature*, **337**, 220–222.

Vasconcelos, C.O. and McKenzie, J.A. (1997) Microbial mediation of modern dolomite precipitation and diagenesis under anoxic conditions (Lagoa Vermelha, Rio de Janeiro, Brazil). *J. Sed. Res.*, **67**, 378–390.

Warren, J.K. (2006) *Evaporites: Sediments, Resources and Hydrocarbons.* Springer-Verlag, Berlin Heidelberg, 1035 pp.

Weidle, A. (1991) *Study for Wildlife and Conservation Areas Master Plan for the Coastal Areas of the Barr Al Hikman and Masirah Island.* Ministry of Regional Municipalities and Environment, Muscat, Oman.

Wood, W.W., Sanford, W.E. and Frape, S.K. (2002) Source of solutes to the Coastal Sabkhas of Abu Dhabi. *Geol. Soc. Am. Bull.*, **114**, 259–268.

Int. Assoc. Sedimentol. Spec. Publ. (2011) **43**, 205–220

Coastal Holocene carbonates of Abu Dhabi, UAE: depositional setting, sediment distribution, and role of cyanobacteria in micritization

CHRISTOPHER G. St.C. KENDALL* and ABDULRAHMAN S. ALSHARHAN[†]

*Department of Geological Sciences, University of South Carolina, Colombia, South Carolina 29208, USA
(E-mail: kendall@geol.sc.edu)*
[†]*Faculty of Science, UAE University, P.O. Box 17551, Al-Ain, United Arab Emirates
(E-mail: sharhana@emirates.net.ae)*

ABSTRACT

The low-latitude coastal sediments of the United Arab Emirates accumulate adjacent to the gently sloping ramp-like bathymetry of the southern Arabian Gulf. Here, an extensive arid coastline system is expressed by a variety of different shallow carbonate and evaporite depositional settings. Off the western Abu Dhabi mainland coast, flanking the north and seaward side of the Khor Al Bazam lagoon, is an offshore bank. This bank is progressively accumulating carbonate sand and mud, extending an area of shoals and channels, coral banks, tidal deltas, and nearshore coastal terraces that prograde outwards from the offshore bank. South of the bank, where wave energy is minimal, cyanobacterial mats colonize protected intertidal sediments, building these seaward, and binding sediment washed onto them.

Ooids, grapestones, pellets, lime mud and bioclastic grains accumulate along the western Abu Dhabi coast. The distribution and characteristics of individual grains types is controlled largely by their hydrodynamic setting, with interactions between wave and current action, sediment mobility, burial rate and cementation. Aragonite mud is confined mainly to the sheltered areas; the source of lime mud is the abundant blooms or clouds of "whiting" formed by carbonate precipitation directly from seawater. Soft and cemented aragonitic pellets accumulate in sheltered lagoons and shallow shelves with moderate exposure to wind and waves. High salinities in evaporative water bodies likely promote cementation in lagoons and ooid formation on tidal deltas. The surfaces of many carbonate grains are altered by cyanobacteria. This micritization forms envelopes enclosing grains that matches those found in ancient carbonates, and represents a useful tool that can be used as evidence of lower rates of accumulation and longer periods of exposure on the shallow seafloor.

Keywords: Ooids, grapestones, pellets, bioclastic sediments, lime muds, cyanobacterial mats, micrite envelopes.

INTRODUCTION

The Holocene sediment province of the Arabian (Persian) Gulf coast contrasts with the isolated platforms of the Bahamas and is the arid equivalent of the shelf sediments of the Yucatan, British Honduras and Florida (Fig. 1). The depositional systems of the Gulf province consist of seaward reefs, islands and tidal deltas protecting inner sheltered saline lagoons and supratidal evaporite/carbonate flats or sabkhas extending along the United Arab Emirates (UAE) coast of the southern Arabian Gulf (Fig. 2). The climate along the coastline is arid and humid, with an average humidity that varies between 40 % to 90 % for the day and night respectively. Summer temperatures range from 28° to 45 °C and winter temperatures are around 12 °C. The ramp bathymetry, arid climate and the water circulating in the Arabian Gulf driven by the NW "Shamal" wind, collectively produce the distinctive depositional settings of the UAE coast described below.

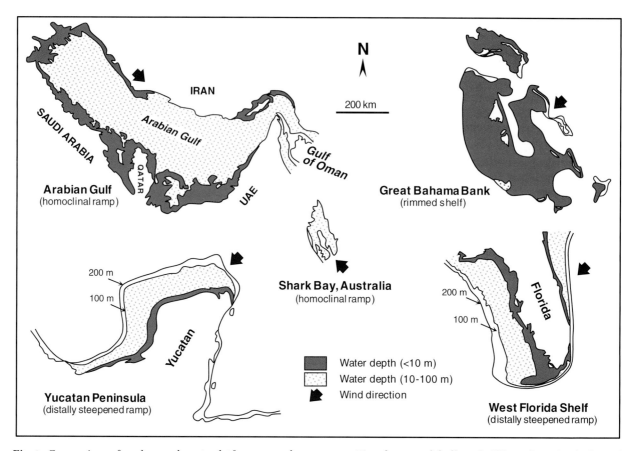

Fig. 1. Comparison of modern carbonate platform-ramp deposystems. Note the area of shallow shelf in each setting (coloured black in the figure). A sea-level fall of more than 10 m would expose the inner and mid-shelf in a ramp setting, while the mid- to outer shelf would become a favourable setting for shallow-water carbonate deposition. On a rimmed margin (e.g. the Great Bahama Bank), the entire shelf would be exposed, essentially causing a hiatus in carbonate deposition. Note 100 m and 200 m contours showing deeper areas. Modified from Burchette & Wright (1992).

Sediments of the region are dominated by shallow-water carbonates that include ooids, grapestones, pellets, muds and bioclastic debris (Fig. 2A). The surfaces of many of the grains of these facies are enclosed by cyanobacterially altered layers that are being micritized (Kendall & Skipwith 1969a). This alteration of grain carbonates matches that of the geological record where grains that have rested on the sea floor for some time have developed similar envelopes (Bathurst, 1966; Reid & Macintyre, 2000). In mixed sediment successions micritic envelopes are less common. One of the purposes of this paper is to describe the modern analogues of this occurrence of micritization and micritic envelopes so common in ancient carbonates.

Many studies of the Holocene marine sediments in the Arabian Gulf have been published since the first detailed study of the area by Houbolt (1957). These include: Sugden (1963); Curtis *et al.* (1963);

Evans *et al.* (1964a,b); Kinsman (1964a,b,c); Murray (1966); Kendall & Skipwith (1968, 1969a, b); Butler (1969); Shinn (1969); Wood & Wolfe (1969); Purser (1973); Hughes-Clarke & Keij (1973); Wagner & Van der Togt (1973); McKenzie *et al.* (1980); Patterson & Kinsman (1981); Butler *et al.* (1982); Warren & Kendall (1985); Crumbie (1987); Swart *et al.* (1987); Kenig *et al.* (1990, 1995); Muller *et al.* (1991); Burchette & Wright (1992); Baltzer *et al.* (1994); Hughes (1997); Kirkham (1997); Kendall *et al.* (1998); and more recently, Wood *et al.* (2002); Alsharhan & Kendall (2003); John (2005) and Kendall *et al.* (2007).

MORPHOLOGY OF THE HOLOCENE CARBONATES DEPOSITIONAL SETTING

Five major geomorphological units underlie the Holocene carbonate sediments currently

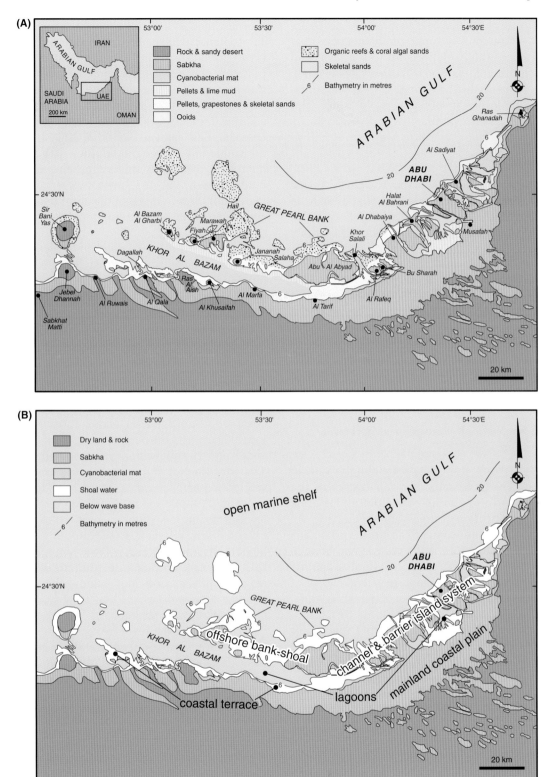

Fig. 2. Sediment distribution along the southern Arabian Gulf coast (Abu Dhabi, UAE). (A) Sediment types and facies belts. (B) Geomorphological units. Modified from Kendall & Skipwith (1969a, b).

accumulating in the coastal complex of Abu Dhabi (Fig. 2B). These may be separated from each other on the basis of their relationship to sea level and their geographical position: (1) open marine shelf; (2) offshore bank-shoal, channel and barrier island system; (3) lagoons; (4) coastal terrace; and (5) mainland coastal plain.

Open marine shelf

Just offshore from the coast of the Abu Dhabi a shelf platform forms numerous shoals that rise from water depths of 40 m and extend east–west at least 170 km. A thin veneer of sediment, largely composed of mollusc shells and their debris, covers these shoals and the shelf floor (Fig. 2; Houbolt, 1957; Kendall & Skipwith, 1968; Purser, 1973; Purser & Evans, 1973). The shoals act as the nucleus for coral reef growth. Some of the shoals have accretionary tails of coral debris that are aligned southeast, parallel with the direction of the Shamal winds. With time, these shoals will accrete and merge with the barrier complex to the south.

Offshore bank-shoal, channel and barrier island system

An offshore bank that is approximately 120 km long and 8 km wide separates the waters of the Khor Al Bazam and a series of lagoons and islands from the open waters of the Arabian Gulf (Fig. 2). This barrier passes eastwards into a shoal area dissected by channels with numerous islands between them. The province terminates against the N–S-oriented Al Dhabaiya peninsula. East of this promontory is a further area of barrier islands, lagoons, tidal channels and deltas.

The offshore bank is surrounded by water between 5 and 20 m deep and is seldom covered by water deeper than 2 m. It is dissected by numerous tidal channels and is surmounted by shoals, intertidal flats and islands. Its seaward edge is the site of coral reef growth, while its lee is marked by tidal deltas whose channels debouch into the Khor Al Bazam lagoon. The bank is mainly covered by coral debris, although locally, Pleistocene limestone, or "miliolite", forms the surface of wide stretches of the shoals. The term miliolite comes from the chambered benthic foraminifera *Miliola* that these rocks contain (Pilgrim, 1908).

The adjacent eastern shoal and channel setting consists of a large area of sand shoals fed and drained by a series of subparallel tidal channels. These usually have ebb deltas at their mouths and levees that flank some of their length. A "miliolite" platform of Pleistocene sediment underlies most of the area, and small erosional remnants of this protrude above the general surface. Carbonate sand accumulates around these remnants to form islands. The seaward edge of the shoal area is dominated by a large tidal delta with a single axial channel. East from this delta, the shoal banks continue without any break in slope to a north-facing coral reef. South of the coral bank is a back-reef lagoon in which small coral communities occur.

To the east of the shoals, islands and channels, lies the subdividing peninsula of Al Dhabaiya (an earlier barrier island now tied to the mainland by an extended tail of sediments carried to their landward location by the wind and storm wash-over) and the barrier islands of Halat Al Bahrani, Abu Dhabi and Al Sadiyat.

Lagoons

Khor Al Bazam, the largest Khor or lagoon of the coast of Abu Dhabi, is about 130 km long. It is located between the barrier complex of the offshore shoal and the mainland supratidal flats (Fig. 2). The western end of this lagoon is some 20 m deep, while the eastern end is intertidal. This swath of intertidal waters makes definition of the lagoon's eastern boundary fuzzy, particularly since it extends east between a complex of islands. The lagoon has a largely flat bottom, though westward this is locally eroded into numerous flat-topped ridges. This erosion may have either have been produced by westward tidal flow from the large channels, or more probably by fluvial erosion during Pleistocene sea-level lows.

The Khor is presently being filled by sediment from three sources: the offshore bank to the north carried by tidal delta accretion; (2) from the shoals as current-deposited sediment; and (3) from the southern coastal strip by coastal terrace accretion northwards. The Khor should eventually fill. In this case, the coastline will probably take on a N–S configuration that matches the area east of the Al Dhabaiya peninsula described by Kinsman (1964b, c). The present configuration of the coast and the way it has accreted supports this prediction.

In contrast, the lagoonal area to the east of the Khor Al Bazam and the Al Dhabaiya peninsula is flanked to the north by the barrier islands of Halat

Al Bahrani, Abu Dhabi and Al Sadiyat (Fig. 2). South of these are a series of shallow protected lagoons, seldom deeper than 2–3 m. These lagoons fill with carbonate sand to the north and carbonate mud and cyanobacterial mats to the south.

Coastal terrace

The Khor Al Bazam is flanked on its south side by an intertidal platform that ranges from 0.5 to 8 km in width and is covered by a thin veneer of sediment. To its north, or seaward, Pleistocene to lower Holocene miliolite often underlies this platform, but landward it is covered by Holocene beachrock. The intertidal/supratidal boundary of the coastal terrace is demarcated by a mix of Tertiary peninsulas with cliffs, berms and beach face, and cyanobacterial flats. Its seaward edge is marked by a distinct break in slope with a depth of between 2 and 4 m. In some areas this edge is the site of coral and coralline algal growth.

The terrace narrows around the headlands formed by Tertiary and Quaternary rocks, but is wider across embayments. At the Tertiary headlands, it is likely that the terrace represents an eroded step cut into the rock. Opposite the bays, are accreting reef flats growing on a foundation of cross-bedded miliolite. To the east, the lagoon becomes shallower and the terrace becomes progressively more indeterminate. To the extreme west, the coastal terrace is a complex consisting of a small offshore bank, lagoons and terrace. The small offshore bank is delimited by a poorly developed coral reef. To the south and behind the bank, the lagoons are progressively filling up and their southern edges are accreting seawards. The narrow coastal terrace is similar to the terrace to the east, but has no recognizable coral development along its northern edge.

Mainland coastal plain

Supratidal flats of "sabkha" (coastal salt flats) are composed of carbonates and evaporites. These characterize the mainland coastal plain just landward of the Holocene tidal flats. The sabkha stretches from the high-water mark, inland to the alluvial fans skirting a low escarpment of upper Tertiary rocks. These hills usually trend in the same direction as the shoreline, but in some areas of western Abu Dhabi, as at Al Marfa (Fig. 2), and along the western Khor Al Bazam lagoon, they reach the coast in a series of NW–SE-trending elongate spurs. Fans of outwash sediment surround the bases of the hills and stand slightly above the general surface of the coastal plain. Where the fans occur near the coast, they merge with the beach ridges and sometimes become cliffs.

DEPOSITIONAL SETTING AND SEDIMENT DISTRIBUTION

The shallow waters of the lagoons and shelves of the Abu Dhabi coast are sites of the accumulation of a variety of Holocene carbonates. The primary components of these carbonate sediments include calcium carbonate muds, skeletal grains and a variety of accretionary (ooids), pellets and aggregating (grapestones and botryoidal) grains of physiochemical origin (Fig. 3; Kendall & Skipwith 1969a; Purser & Evans, 1973). Based on the hydrodynamic activity within the area, these settings have been broadly classed as protected to moderately exposed. As seen in Figs 2 and 3, each of the local sediment types now accumulating within restricted areas has small aerial extent. However, the rapid sediment accretion along the Abu Dhabi coast means there is a good chance that these local sediment accumulations will be preserved. Vertical sections in the supratidal zone show lagoonal sediments at the base capped by intertidal sediments and overlain by windblown and storm-washover sediments. The addition of quartz sand and evaporites distinguishes the sabkha from the intertidal sediment. These sabkha facies are modified further by diagenetic alteration of aragonite to dolomite.

Sediments of protected settings

The protected settings of the region include enclosed lagoons and areas to the lee of banks and islands, and the deep water of the axial trough of the Arabian Gulf. The sediments range from aragonite mud to soft and cemented aragonite pellets. The aragonite mud is confined mainly to the sheltered lagoonal areas of eastern Abu Dhabi, and to the lee of banks and islands in the area (Figs 2 and 3). Soft and cemented aragonitic pellets accumulate in sheltered lagoons and shallow shelves with moderate exposure to wind and waves. The source of the lime mud is the abundant blooms or clouds of "whiting" formed by carbonate precipitation directly from the sea (Fig. 4; Kendall *et al.*, 2007).

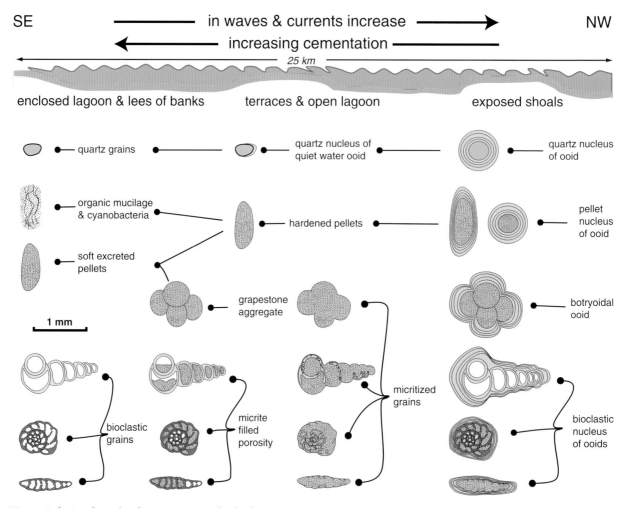

Fig. 3. Relationship of carbonate grains to hydrodynamic setting. Cyanobacteria often modify initial texture of all carbonate grains through the process of micritization.

The aragonite mud is sequestered in areas of low hydrodynamic energy, and once settled is seldom disturbed. It is often uncemented, probably because the static porewaters remain in equilibrium with the sediment. Associated with the mud are blebs of flocculent organic mucilage containing blue-green algae (cyanobacteria). The blebs are common to most of the shallow protected settings along the coast. It is suggested that in this protected setting, the gentlest movement of water causes the mucilagenous blebs, and excreted faecal pellets of low density (Fig. 3), to saltate. The surfaces of the mucilaginous blebs and pellets are sticky, and most likely accrete detrital aragonite. During saltation, the surfaces of accreting grains are exposed continuously to sunlight and to fresh solutions of seawater, possibly supersaturated with respect to calcium carbonate.

Kendall & Skipwith (1969a) hypothesized that as a result of the processes of photosynthesis and respiration, aragonite is precipitated within the mucilage and the grains become cemented and harden to form peloids whose various origins are indistinguishable (Fig. 3). The longer the pellets saltate the better cemented they become. The Kendall & Skipwith (1969a) model suggests that the balance between soft and hard pellets is probably determined by the relative intensity of the turbulence of the water and the length of saltation. Should one pellet adhere to another, they can no longer be lifted from the sea floor. Water movement in pellet areas is too gentle to lift them, and cyanobacterial growth binds them to other grains. Grains lying on the sea floor become buried beneath other grains. Cementation ceases when the cyanobacterial cells within the grains are no

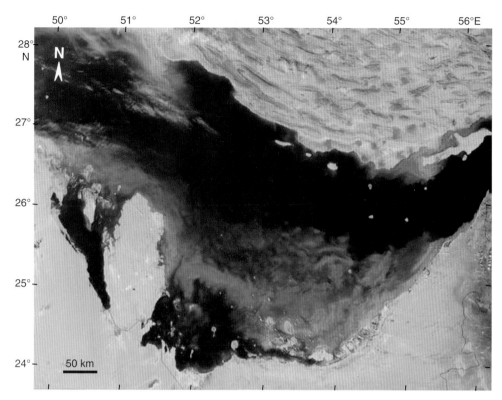

Fig. 4. NASA satellite image of cyanobacteria and phytoplankton blooms offshore from the southern Arabian Gulf coast (Abu Dhabi, UAE). Note the dust blowing from Iran into the Gulf in response to winds over 25 km hr^{-1} and the white clouds of material precipitating in the Gulf. This occurs following the settling of the land-derived dust (Kendall et al., 2007).

longer exposed to sunlight and, in addition, the sediment develops a chemical equilibrium with the surrounding porewaters.

Sediments of settings moderately exposed to wave and current energy

Moderately exposed settings are found along lagoon shores and offshore banks that are protected only partially from heavy seas and intense hydrodynamic activity. The sediments that accumulate here are characterized by aggregates of cemented aragonite pellets, grapestones, and skeletal fragments (Figs 3 and 5). Kendall & Skipwith (1969a) proposed that the aggregates probably form when pellets are at rest long enough to adhere to each other to form a thin crust, the adhesion being induced either by the process of photosynthesis of the algae mucilaginous material or beachrock cementation. In this setting, heavy wave action is common enough and intense enough to break the thin surface crust but not strong enough to split the grains into their original individual forms. Aggregates develop and are joined to other grains

and acquire cement both externally and internally. If washed into an exposed setting, they develop polished ooid coats (Figs 3 and 5).

Sediments of exposed settings

Exposed environments mark the seaward edge of the complex of banks and lagoons of the Abu Dhabi coastline and in the Khor Al Bazam. Sediments include ooid shoals and vast lateral accumulations of coral (Fig. 2). This setting is characterized by considerable turbulence, where the adhesion of one carbonate grain to another seldom occurs. The grains are never at rest long enough for a sufficiently strong cement to develop which will not be subsequently broken by wave action. As in the other settings, the grains grow by the dual process of mechanical and chemical accretion. However, instead of being ellipsoidal, they develop as spheres in response to a combination of carbonate precipitation and abrasion in this high-energy environment (Figs 3 and 5). Here it appears, elongate aragonite crystals are flattened tangentially to the surface of the grain, where the grains

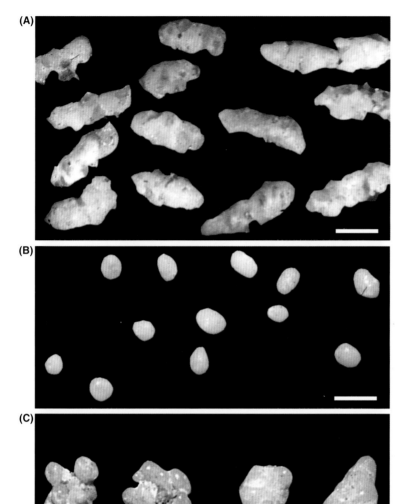

Fig. 5. Non-skeletal grains from the Abu Dhabi coast. (A) Irregular elongate partially cemented pellets probably excreted by worms. (B) Spherical hardened cemented and often polished peloids, whose origin may be excretion (faecal pellets), or alternatively, micritized bioclastic fragments or saltating mucilaginous blebs rich in cyanobacteria. (C) Hardened cemented aggregates of pellets and bioclastic material forming grapestones. Note that aggregates to the right have an ooid-like polish and form botryoidal grains.

impact with each other. The greater the momentum acquired by the grains the greater the impact.

DISCUSSION

The mosaic of carbonate mud to oolite facies in Abu Dhabi is distributed parallel a progressive increase in the turbulence of the sea water in which they form. This spread is similar to that described by Illing (1954), Purdy (1963a,b) and Bathurst (1976) for carbonate grains of physiochemical origin in the Bahamas. This apparent order to the development of carbonate grains may

also be related to variations of salinity. Though this variability in salinity probably does not change the chemical processes by which the aragonite of particles is acquired, it likely affects the speed of the processes. For instance, if the water body in which the particles are being formed is enclosed, its area controls the concentration of salts by evaporation. This may provide the chemical reason for the absence of ooids at the mouths of the most easterly and smallest lagoons of Abu Dhabi, their area perhaps being too small to elevate salinities sufficiently for carbonate to precipitate. Similarly, the presence of living coral, large patches of living marine weed, and living calcareous red algae, also may modify the calcium carbonate balance, inhibiting chemical means by which precipitation or crystallization of aragonite occurs.

The sedimentary facies that accumulate within the eastern barrier system are distinctly different from each other. There is a variation in their content and the character of the surface texture of the constituent components. These facies are found in distinct geomorphological settings and also can be related to variations in the hydrodynamic conditions of the depositional setting. Thus, in deeper waters, <10 m, below the action of tidal currents, skeletal carbonate grains accumulate with contents of interstitial carbonate mud that can locally be over 30%. The original surface texture of the skeletal components is commonly destroyed and most grains have a saccharoidal texture.

Where tidal currents sweep the bottom sediments, the interstitial carbonate mud is generally carried away, but the component grains become joined together to form grapestone "lumps". Within the influence of the wave base, carbonate sand grains may become rounded and coated, generating a sequence of sediment types related to increasing turbulence of the water at the sediment-water interface within the lagoon (Fig. 3). North of the barrier complex in the open Gulf, waves and currents follow a different pattern. There is an abrupt change from the skeletal carbonate sand of the open Gulf floor, to the coral and the calcareous red algae of the reef banks. This change is also expressed in the bottom topography, with an abrupt break in slope between the Gulf floor and the adjacent banks.

Cyanobacterially-induced alteration of carbonate sediments

As with other Holocene shallow-water carbonate settings, some of the sediments of the Khor Al Bazam of western Abu Dhabi show a distinctive type of alteration that affects all carbonate grains, irrespective of their origin (Kendall & Skipwith, 1969a). The texture of the grains is altered to a homogeneous microcrystalline fabric of aragonite. This alteration is termed "micritization" (Bathurst, 1966). In the final stage of alteration, the grains are virtually indistinguishable from each other. In many instances the external shape may be the only indication of the grain's origin. This process of micritization has been observed in the Bahamas (Illing, 1954; Purdy, 1963a,b), Portuguese Timor (Wolf & Conolly, 1965), the Arabian Gulf off Qatar (Houbolt, 1957), and in British Honduras

(Purdy, 1965). Purdy (1963a, b) concluded that organic matter trapped within the grains promotes the process. Bathurst (1966) and Reid & Macintyre (1998, 2000) ascribed the alteration to cyanobacteria.

Foraminifera and calcareous red algae seem particularly susceptible to micritization in the Holocene sediments of the Khor Al Bazam of western Abu Dhabi (Kendall & Skipwith, 1969a). In contrast, thick-walled shells and shells with a more coarsely crystalline fabric are less susceptible. Alteration in these tests generally is confined to their surface, but a few lobes of fine-grained carbonate often extend deep into the shell. Alteration also affects aggregates and pellets of the microcrystalline carbonate, including grapestones. These aggregates are initially loosely bound, open-textured grains that rapidly harden and cement as they alter (Fig. 5). Ooids are similarly affected. Crusts of radial acicular aragonite crystals have been precipitated in the small internal cavities of some mollusc shells, foraminifera and grapestone lumps. These infillings generally alter into disorganized microcrystalline mosaics of aragonite.

In ancient limestones, many of the detrital carbonate components have a fabric similar to that of the Holocene altered material, but are preserved as calcite rather than mixtures of aragonite and calcite. Banner & Wood (1964) attributed this to late diagenetic grain growth. Orme & Brown (1963) and Folk (1965), apparently describing the same phenomenon, postulated a late diagenetic process of grain diminution, confirming some of the findings of Bathurst (1964) and of Wolf & Conolly (1965). It was demonstrated that these changes are not exclusively late diagenetic, but also result from the alteration of the depositional fabric.

The various stages in the process of alteration can be studied in peneroplid foraminifera, which are particularly susceptible to alteration (Murray, 1966). Detailed study of specimens collected from the Khor Al Bazam of western Abu Dhabi show all stages of alteration (Fig. 6). The tests display a series of surface changes; these range from translucent and porcelaneous in fresh tests to opaque and saccharoidal in altered specimens. Rounding and pitting of tests accompany these changes. Many of the pits are seen to contain cyanobacteria. Alteration may be so pronounced that the tests are almost indistinguishable from aragonite ovoids of faecal and accretionary origin. Illing (1954), Purdy (1963a, b), Bathurst (1976) and Reid & Macintyre (1998, 2000) have noted similar changes in the foraminifera of the Bahamas.

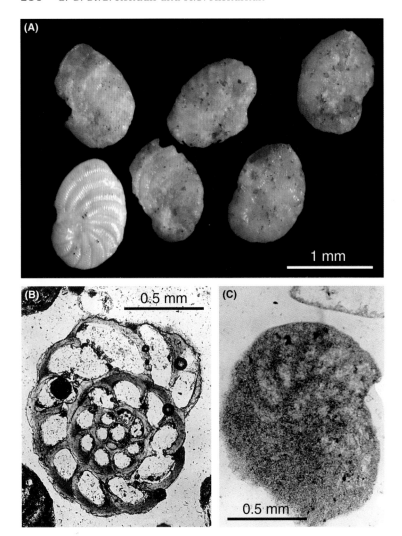

Fig. 6. Alteration and micritization of peneroplid benthic foraminiferal tests. (A) Progression from pristine highly-reflective polished tests (left), to matt micritized grains (right). (B) Thin section plain-polarized light view of an unaltered peneroplid test. (C) A heavily micritized test (plain polarized light).

Under plane-polarized light thin-sections, fresh peneroplid tests have a brown body colour. The tests display no crystal boundaries apart from a faint structure parallel with the walls (Fig. 6B) and a very faint granulation under high power. In reflected light they are milky white. When viewed through crossed polars the tests are seen to be composed of parallel sheaths of crystals that show low polarization colours, and lie parallel to the curvature of the walls. The sheaths are embedded in a fine, granular matrix of crystals, which show pinpoint polarization in greys and yellows of the first order. Thin sections of fresh tests stained with Feigel's solution suggest that the tests were not aragonite; x-ray analysis shows the tests to be of high-Mg calcite with a small amount of aragonite.

This aragonite was either part of the tests, or more probably, an infilling of the chambers, as proposed by Sollas (1921) and Wood (1948).

The brown body colour and the low-order polarization colours of the fresh tests are not optical properties of high-Mg calcite. Several theories have been proposed to explain these anomalies. For example, Sollas (1921) attributed the brown body colour in plane-polarized light, and the milky colour in reflected light, to the reflection of light through the felted crystals of the test. The low-order polarization colours were not explained. In contrast, Wood (1948) attributed the brown body colour to small amounts of lead and iron, although Frondel *et al.* (1942) were unable to trace any connection between the distribution of Fe, Cu,

Mn, Al, Sr, and Mg and the morphology and colour of the calcite crystals. Wood (1948) proposed that small crystals are optically compensating each other in the thin section producing the low polarization colours. However, this is unlikely, because thin sections of micrite (Folk, 1962) do not show low polarization colours. Chalcedony and turbid feldspar also are brown in transmitted light and milky under reflected light. These optical properties have been interpreted as being caused by minute water-filled vacuoles (Folk, 1955; Folk & Weaver, 1952). The same properties in the peneroplids may have a similar origin (Kendall & Skipwith, 1969a; Reid & Macintyre, 1998, 2000).

Electron photomicrographs of peneroplid tests show them to be composed of a felt of crystals about 0.5 μm long and 0.1 μm in diameter (Hay *et al.*, 1963). The disorganized arrangement of some of the crystals should leave voids at least 0.1 μm in diameter. The voids are probably filled with water, which produces a brown body colour, as in chalcedony and turbid feldspars. The presence of the water may be related to the low refractive index of the calcite, perpendicular to the c-axis (1.6), the low polarization colours, and the low specific gravity of 2.724 (Sollas, 1921). However, it does not explain why the test shows low polarization colours, and does not exclude the possibility that the low specific gravity may be caused by the presence of organic matter. If the voids were filled with air, the test would be black.

In thin sections, tests show that alteration proceeds from discrete patches that become enlarged and gradually obliterate the original texture (Fig. 9C). This process may be accompanied by the infilling of the chambers by microcrystalline aragonite. The altered areas are characteristically microgranular and lack the brown body colour shown by fresh tests. Under crossed polars the altered areas show high polarization colours, but do not extinguish. The size of the crystals is of the order of 1.5 mm. Feigel's solution has a patchy reaction on completely altered foraminifers, but x-ray analysis shows that those tests are almost entirely aragonite. The failure of Feigel's test may be the result of films of organic matter covering the crystals.

Cyanobacteria (*Entophysalis deusta*) grow on some of the peneroplids and on the ooids from Abu Dhabi. Newell *et al.* (1960) found this genus to be common in the Bahamas. Cyanobacteria collected in Abu Dhabi survived for more than

2–3 years in storage jars. On exposure to sunlight, they started to grow vigorously. The cyanobacteria of several samples were separated from the associated carbonate sand by a solution of weak hydrochloric acid. They consisted of several different types of globular cells, and tangles of filaments in a transparent mucilaginous matrix, believed to be a secretion of the cyanobacteria (Fritsch, 1952). Cyanobacteria cells and filaments lie just beneath the surface of altered tests, as shown by progressively dissolving the carbonate of the tests in 5% acetic acid or 0.2% HCl, and rendering the surface transparent with 40% hydrofluoric acid (Kendall & Skipwith, 1969a). Cyanobacteria could be made even more apparent by staining with malachite green, which stains cyanobacteria cells and the mucilage with different intensities. Recently, Reid & Macintyre (2000), examining Bahamian carbonates, have described a coccoid cyanobacterium, probably *Solentia* sp., that bores tunnels, penetrating grain surfaces, and extending through entire grains. They have demonstrated that this multicyclic boring and concurrent filling of bore holes lead to the eventual micritization of grains.

When unaltered foraminiferal tests are completely decalcified, they leave behind a thin, diaphanous, soft elastic membrane that retains the original form of the test (Kendall & Skipwith, 1969a). This is the tectin of Hyman (1940), and is the original organic material of the test that in life occurs in conjunction with the calcite. Such "microforaminiferal test linings" are a common constituent of acid-insoluble residues produced from modern and ancient shallow-marine sediments (e.g. Mathison & Chmura, 1995). Decalcified altered foraminifera, however, leave behind a much more translucent material, which also moulds the test and fills its chambers, but is markedly different in appearance from the tectin. In all ways it resembles the mucilage that is present in association with the blue-green algae. Supporting Reid & Macintyre's (2000) observations in the Bahamas, this mucilage has been reported as containing cyanobacterial cells and small quantities of minute mineral grains that probably adhered to the mucilage as the test rolled on the sea floor (Kendall & Skipwith, 1969a).

If foraminifera tests are treated with 40% hydrofluoric acid for more than 10 minutes, all the calcium carbonate is replaced, molecule-by-molecule, by transparent calcium fluoride (Grayson, 1956). The envelope of mucilage and algae resist solution and consequently stand out as translucent and green in altered areas. Treatment with 40 % hydrofluoric acid for 5 minutes makes the surface of the foraminifera transparent and reveals the network of radiating filaments and globular algae. Fresh foraminifera become completely transparent. The best optical results with the calcium fluoride specimens are obtained by viewing them under a film of water.

As Reid & Macintyre (2000) later found in the Bahamas, partially altered foraminifera examined in thin section from the Khor al Bazam have cyanobacteria filaments penetrating both the unaltered and altered areas (Kendall & Skipwith, 1969a). The filament bores generally are filled with microcrystalline aragonite, which preserves the filament shapes. Thus tubes of altered material may penetrate the unaltered test. The boundaries separating the unaffected high-Mg calcite areas of the test from the aragonitic altered areas commonly resemble the borings of cyanobacterial filaments (Bathurst, 1966; Kendall & Skipwith, 1969a; Reid & Macintyre, 2000). Under plane-polarized light, parts of the test adjacent to the bore shapes may show faint granulation, which is believed to be the first sign of alteration.

Thin sections of altered foraminiferal tests stained with aqueous malachite green show patchy colours (Kendall & Skipwith, 1969a). Deeper colours are present at the outside edges and in chambers devoid of carbonate. Etching thin sections with weak acid solutions of malachite green dissolves the calcium carbonate, leaving stained insoluble algae and mucilage. The deeper stain is usually at the edge of the former test and in some chambers, showing cyanobacterial cells and filaments occur both at the surface and penetrate the interior. Generally, the parts of the chambers not filled with cyanobacterial cells are stained less intensely, an indication that they are filled with mucilage. Dissolution and staining of the altered walls of the foraminifera leave only a trace of mucilage that stains lightly. This most likely results from the displacement of mucilage by the formation of microcrystalline aragonite crystals. The presence of this finely dispersed mucilage is confirmed by gently dissolving the whole tests. If the dissolution is too vigorous, the mucilage is ruptured and removed.

It can be demonstrated that there is an intimate relation of blue-green algae with the alteration of carbonate. All plants photosynthesize and respire. Dalrymple (1965) observed the ability of blue-green algae to precipitate calcium carbonate. Parks & Curl (1965) showed in the laboratory that a culture of blue-green algae in seawater produces measurable change in the electric conductivity of the water between day and night. They inferred that, in light, photosynthesis promotes the transformation of bicarbonate ions to carbonate, and in the dark, respiration causes the opposite effect. In the peneroplids of the Arabian Gulf, it is proposed that mucilaginous envelope generally creates a microenvironment within each test. Carbon dioxide given off during respiration promotes the dissolution of calcium carbonate in this restricted environment. Conversely, the carbon dioxide utilized during photosynthesis would cause precipitation. Although high-Mg calcite is dissolved, it is aragonite that is precipitated because it is apparently the more stable form of calcium carbonate under such conditions (Kinsman, 1964b). In this way, the high-Mg calcite of the foraminifera test is progressively dissolved and replaced by aragonite on a piecemeal basis. Once a foraminifera is altered, dissolution and reprecipitation of aragonite continues, ultimately destroying all evidence of the original mineralogy and fabric of the test. This process alone could account for the observed alteration.

Bathurst (1976) suggested that algal boring and reboring of a calcite test causes the alteration and solution of the test where it is in direct contact with the cyanobacterial filaments. Similarly, Reid & Macintyre's (2000) observations support multicyclic boring and concurrent filling of bore holes as the origin of the micritization of grains. However, the fact that the algae are restricted to the area near the surface of the test, and do not necessarily extend into all the affected parts, argues against alteration as the result of boring alone. In addition, the mucilaginous envelope secreted by the algae confines the carbon dioxide. This carbon dioxide can react at any point within the envelope.

The degree of alteration of carbonate grains could be controlled by several factors: the length and frequency of exposure of the grains to direct sunlight, the dimensions of the grains, and the size of their component crystals. For example, if the grain is buried quickly, it may not have time to alter, or will only alter if it has a thin, delicate shell and small component crystals. The shoal areas of the Khor Al Bazam are rippled in the areas where the altered foraminifera were collected by Kendall & Skipwith (1969a). If the ripples migrate slowly, the foraminifera being altered will be exposed only occasionally to direct sunlight. Consequently, the alteration process may be slow. Alteration is almost effective on the weed-covered shoals of the coastal terrace and parts of the offshore bank. Skeletal grains from those areas have abundant cyanobacteria-filled pits. The prolific plant growth must upset the carbon dioxide balance of the seawater and alteration therefore may be more rapid.

Wolf & Conolly (1965) pointed out that algae are generally good indicators of shallow water, because photosynthesis is depth controlled. Thus, alteration would not be expected to take place below the photic zone. Off Qatar, bioclastic material is rounded and the surface character obliterated in water less than 20 m deep (Houbolt, 1957). In rocks from similar ancient carbonate settings, the microcrystalline aragonite contained within the altered material generally is replaced by a mosaic of microcrystalline calcite in which cyanobacterial alteration is still recognizable (Shearman, 1966). The altered material commonly forms Bathurst's (1966) "micrite envelope", which Alexanderson ((1972)) went on to define as the alteration of a pre-existing fabric to micrite with no genetic implications. Nonetheless, Reid & Macintyre (1998, 2000) have shown that for the pervasive syndepositional recrystallization of the miliolid foraminifera *Archaias* and the green alga *Halimeda* that occurs on the sea floor off Florida, the Bahamas, and Belize, cyanobacteria play a major role in the micritization process.

Significance of ancient micrite envelopes

Any ancient carbonate grain can be expected to have sat on the sea floor exposed to potential of infestation by blue-green algae and to develop a consequent alteration of their surface to micrite (Fürsich *et al.*, 1992; Scasso *et al.*, 2005; Caron *et al.*, 2005; Boyd, 2007). Surprisingly, many shallow-water carbonate grains preserved in the geological record do not have such envelopes. This implies that those depositional settings did not favour the cyanobacterial activity needed for micritization, or that the grains only remained a short time at the sediment surface before they were buried. Trangressive surfaces and hardgrounds in successions commonly have grains resting on them that do have micrite envelopes (Fürsich *et al.*, 1992; Pomar & Kendall, 2008); this supports an inverse relationship between burial rate and the degree of micritization.

CONCLUSIONS

The coast of western Abu Dhabi is divided into: (1) offshore shelf and shoals with reefs and bioclastic sediments; (2) offshore bank, shoals, channel and barrier islands with coral reefs, bioclastic sediments and local oolite accumulations on ebb deltas; (3) protected lagoons with pelleted muds and cyanobacterial flats; (4) coastal terrace with reefs; and (5) mainland coastal plain or sabkha with evaporites and stranded beach ridges.

In this eclectic shallow-water carbonate setting of lagoons and shelves, carbonate mud and a variety of accretionary grains are presently accumulating. Cyanobacterial coatings on grains cause the micritization of the carbonate grains they enclose. The presence of micrite envelopes in ancient carbonate deposits suggests exposure to light in a setting rich in cyanobacteria. An absence of micritization, on the contrary, indicates either a lack of cyanobacteria in that setting or rapid burial before surface alteration could take place.

REFERENCES

Alsharhan, A.S. and **Kendall, C.G.St.C.** (2003) Holocene coastal carbonates and evaporites of the southern Arabian Gulf and their ancient analogues. *Earth-Sci. Rev.*, **61**, 191–243.

Alexanderson, E.T., (1972). Micritization of carbonate particles: process of precipitation and dissolution in modern shallow-marine sediments. *Universitet Uppsala, Geologiska Institut Bulletin*, **7**, 201–236.

Baltzer F., Kenig F., Boichard R., Plaziat J.-C. and **B. H. Purser** (1994) Organic matter distribution, water circulation and dolomitization beneath the Abu Dhabi sabkha (United Arab Emirates). In: *Dolomites. A Volume in Honour of Dolomieu* (Eds B. Purser, M. Tucker and D. Zenger). *Spec. Publ. Int. Assoc. Sedimentol.*, **21**, 409–427.

Banner, P.T. and **Wood, G.V.** (1964) Recrystallization in microfossiliferous limestones. *J. Geol.*, **4**, 21–34.

Bathurst, R.G.C. (1964) The replacement of aragonite by calcite in the molluscan shell wall. In: *Approaches to Paleoecology* (Eds J. Imbrie and N. D. Newell), pp. 357–376. John Wiley and Sons, New York.

Bathurst, R.G.C. (1966) *Boring Algae, Micrite Envelopes and Lithification of Molluscan Biosparites*. Elsevier, Amsterdam, 658 p.

Bathurst, R.G.C. (1976) *Carbonate Sediments and their Diagenesis* (2nd edn). Developments in Sedimentology, Elsevier, Amsterdam, 658 p.

Boyd, D. W. (2007) Morphology and diagenesis of *Dimorphosiphon talbotorum* n. sp., an Ordovician skeleton-building alga (Chlorophyta: Dimorphosiphonaceae). *J. Paleontol.*, **81**, 1–8.

Burchette, T.P. and **Wright, V.P.** (1992) Carbonate ramp depositional systems. *Sed. Geol.*, **79**, 3–57.

Butler, G.P. (1969) Modern evaporite deposition and geochemistry of co-existing brines, the sabkha, Trucial Coast, Arabian Gulf. *J. Sed. Petrol.*, **39**, 70–89.

Butler, G. P., Harris, P. M. and Kendall, C.G.St.C. (1982) Recent evaporites from Abu Dhabi coastal flats. In: *Depositional and Diagenetic Spectra of Evaporites* (Eds C.R. Handford, G.L. Robert and R.D. Graham), pp. 33-64. *SEPM Core Workshop*, **3**, Tulsa.

Caron, V., Nelson, C. S. and Kamp, P.J.J. (2005) Sequence stratigraphic context of syndepositional diagenesis in cool-water shelf carbonates: Pliocene limestones, New Zealand. *J. Sed. Res.*, **75**, 231–250.

Crumbie, M.C. (1987) *Avicennia marina,* the grey mangrove: general notes and observations. *Bull. Emirates Nat. Hist. Gr. (Abu Dhabi)*, **32**, 2–13.

Curtis, R., Evans, G., Kinsman, D.J.J. and Shearman, D.J. (1963) Association of dolomite and anhydrite in the recent sediments of the Persian Gulf. *Nature*, **197**, 679–680.

Dalrymple, D.W. (1965) Calcium carbonate deposition associated with blue-green algal mats, Baffin Bay, Texas. *Inst. Mar. Sci. Univ. Texas Publ.*, **10**, 187–200.

Evans, G., Kendall, C.G.St.C. and Skipwith, P.A. (1964b) Origin of the coastal flats, the sabkha, of the Trucial Coast. Persian Gulf. *Nature*, **202**, 759–761.

Evans, G., Kinsman, D.J.J. and Shearman, D.J. (1964a) A reconnaissance survey of the environment of Recent carbonate sedimentation along the Trucial Coast, Persian Gulf. In: *Deltaic and Shallow Marine Deposits* (Ed. L.M.J. U. van Straaten), **1**, 129–135. Elsevier, Amsterdam.

Folk, R.L. (1955) Note on the significance of "turbid" feldspars. *Am. Mineral.*, **40**, 356–357.

Folk, R.L. (1962) Spectral subdivision of limestone type. In: *Classification of Carbonate Rocks - A Symposium* (Ed. W. E. Ham). *AAPG Mem.*, **1**, 62-84.

Folk, R.L. (1965) Some aspects of recrystallization in ancient limestones. In: *Dolomitization and Limestone Diagenesis, a Symposium.* (Eds L.C. Pray and R.C. Murray). *SEPM Spec. Publ.*, **13**, 14–48.

Folk, R.L. and Weaver, C.E. (1952) A study of the texture and composition of chert. *Am. J. Sci.*, **250**, 498–510.

Fritsch, F.E. (1952). *Structure and Reproduction of the Algae, Volume 2.* Cambridge University Press, Cambridge, 939 p.

Frondel, C., Newhouse, W.H. and Jarrell, R.F. (1942) Spatial distribution of minor elements in single crystals. *Am. Mineral.*, **27**, 726–745.

Fürsich, F. T., Oschmann, W., Singh, I. B. and Jaitly, A. K. (1992) Hardgrounds, reworked concretion levels and condensed horizons in the Jurassic of western India: their significance for basin analysis. *J. Geol. Soc. London*, **149**, 313–331.

Grayson, J.F. (1956) The conversion of calcite to fluorite. *Micropaleontol.*, **2**, 71–78.

Hay, W.W., Towe, K.M. and Wright, R.C. (1963) Ultramicrostructure of some selected foraminiferal tests. *Micropaleontol.*, **9**, 171–195.

Houbolt, J.J.H.C. (1957) *Surface sediments of the Persian Gulf near the Qatar Peninsula.* PhD thesis Utrecht. University, Moutan and Co., Den Haag, 113 p.

Hughes G.W. (1997) The Great Pearl Bank barrier of the Arabian Gulf as a possible Shu'aiba analogue. *GeoArabia*, **2**, 279–304.

Hughes-Clarke, M.W. and Keij, A.J. (1973) Organisms as producers of carbonate sediment and indicators of environment in the southern Persian Gulf. In: *The Persian Gulf: Holocene Carbonate Sedimentation and Diagenesis in a Shallow Epicontinental Sea* (Ed. B. H. Purser), pp. 33–56. Springer Verlag, Berlin, New York.

Hyman, L.H. (Ed.) (1940) *The Invertebrates, 1. Proterozoa through Ctenophora.* McGraw Hill, New York.

Illing, L.V. (1954) Bahamian calcareous sands. *AAPG Bull.*, **38**, 1–95.

John, D. (2005) Marine plants. In: *Emirates Natural History*, pp. 161–167. Trident Press Ltd, London.

Kendall, C.G.St.C. and Skipwith, P.A. (1968) Recent algal mats of a Persian Gulf lagoon. *J. Sed. Petrol.*, **38**, 1040–1058.

Kendall, C.G.St.C. and Skipwith, P.A. (1969a) Holocene shallow water carbonate and evaporite sediments of Khor Al Bazam, Abu Dhabi, southwest Persian Gulf. *AAPG Bull.*, **53**, 841–869.

Kendall, C.G.St.C. and Skipwith, P.A. (1969b) Geomorphology of a Recent shallow water carbonate province: Khor Al Bazam, Trucial Coast, southwest Persian Gulf. *Geol. Soc. Am. Bull.*, **80**, 865–891.

Kendall, C.G.St.C., Alsharhan, A.S. and Whittle, G.L. (1998) The flood re-charge sabkha model supported by recent inversions of anhydrite to gypsum in the UAE sabkhas. In: *Quaternary Deserts and Climatic Change* (Eds A.S. Alsharhan, K.W. Glennie, G. L. Whittle and C. G. St. C. Kendall), pp. 29–42. A.A. Balkema, Rotterdam.

Kendall, C.G.St.C., Shinn E.A. and Janson, X. (2007) Holocene cyanobacterial mats and lime muds: links to Middle East carbonate source rock potential. *Proc. AAPG, SEPM (Soc. Sed. Geol.) Annual Meeting Long Beach*.

Kenig F., Huc A.Y., Purser B. H. and Oudin J. L. (1990) Sedimentation, distribution and diagenesis of organic matter, in a carbonate hypersaline environment, Abu Dhabi (U.A.E.). *Org. Geochem.*, **16**, 735–747.

Kenig F., Sinninghe Damste J., de Leeuw J. W. and A. Y. Huc (1995) Occurrence and origin of mono-, di- and trimethylalkanes in modern and Holocene microbial mats from Abu Dhabi (U.A.E.). *Geochim. Cosmochim. Acta*, **59**, 2999–3015.

Kinsman, D.J.J. (1964a) Reef coral tolerance of high temperatures and salinities. *Nature*, **202**, 1280–1282.

Kinsman, D.J.J. (1964b) *Recent Carbonate Sedimentation near Abu Dhabi, Trucial Coast, Persian Gulf.* Unpublished PhD Thesis, University of London.

Kinsman, D.J.J. (1964c) The recent carbonate sediments near Halat Al Bahraini, Trucial Coast, Persian Gulf. In: *Deltaic and Shallow Marine Deposits* (Ed. L.M.J.U. van Straaten), **1**, 189–192. Elsevier, Amsterdam.

Kirkham, A. (1997) Shoreline evolution, aeolian deflation and anhydrite distribution of the Holocene, Abu Dhabi. *GeoArabia*, **4**, 403–416.

Mathison, S.W. and Chmura, G.L. (1995) Utility of microforaminiferal test linings in palynological preparations. *Palynology*, **19**, 77–84.

McKenzie, J.A., Hsu, K.J. and Schneider, J.F. (1980) Movement of subsurface waters under the sabkha, Abu Dhabi, UAE, and its relation to evaporative dolomite genesis. In: *Concepts and Models of Dolomitization* (Eds D.H. Zenger, J.B. Dunham and R.L. Ethington), *SEPM Spec. Publ.*, **28**, 11–30.

Muller, D.W., McKenzie, J.A., Kendall, C.G.St.C. and Alsharhan, A.S. (1991) Application of strontium isotopes to the study of brine genesis during sabkha dolomitization (abstract). In: *Dolomieu Conference on Carbonate Platforms and Dolomitization*, (Eds A. Bosellini, R. Brander, E. Flugel, B. Purser, W. Schlager, M. Tucker and D. Zenger), p. 181.

Murray, J.W. (1966) The foraminiferids of the Persian Gulf, Khor al Bazam. *Palaeogeogr. Palaeoclimatol. Palaeoecol.*, **2**, 153–169.

Newell, N.D., Purdy, E.G. and Imbrie, J. (1960) Bahamian oolitic sand. *J. Geol.*, **68**, 481–497.

Orme, G.R. and Brown, W.W.M. (1963) Diagenetic fabrics in the Avonian limestones of Derbyshire and North Wales. *Proc. Yorks. Geol. Soc.*, **34**, 51–66.

Parks, R.K. and Curl, H.C. (1965) Effects of photosynthesis and respiration on electrical conductivity of seawater. *Nature*, **205**, 274–275.

Patterson, R.J. and Kinsman, D.J.J. (1981) Hydrologic framework of a sabkha along the Arabian Gulf. *AAPG Bull.*, **65**, 1457–1475.

Pilgrim, G.E. (1908) Geology of Persian Gulf and adjoining portions of the Persia and Arabia. *Geol. Surv. India Mem.* **34**, 1–179.

Pomar, L. and Kendall, C.G.St.C. (2008) Carbonate platform architecture; a response to hydrodynamics and evolving ecology. In: *Controls on Carbonate Platform and Reef Development* (Eds J. Lukasik and T. Simo) *SEPM Spec. Publ.*, **89**, 187–216.

Purdy, E.G. (1963a) Holocene calcium carbonate facies of the Great Bahama Bank, pt. 1. Petrography and reaction groups. *J. Geol.*, **71**, 334–355.

Purdy, E.G. (1963b) Holocene calcium carbonate facies of the Great Bahama Bank, pt. 2. Sedimentary facies. *J. Geol.*, **71**, 472–497.

Purdy, E.G. (1965) Diagenesis of recent marine carbonate sediments. In: *Dolomitization and Limestone Diagenesis, a Symposium* (Eds L.C. Pray and R.C. Murray). *SEPM Spec. Publ.*, **13**, 14–48.

Purser, B.H. (1973) Sedimentation around bathymetric highs in the southern Persian Gulf. In: *The Persian Gulf: Holocene Carbonate Sedimentation and Diagenesis in a Shallow Epicontinental Sea* (Ed. B.H. Purser), pp. 157–178. *Springer Verlag*, Berlin.

Purser, B.H. and Evans, G. (1973) Regional sedimentation along the Trucial Coast, SE Persian Gulf. In: *The Persian Gulf: Holocene Carbonate Sedimentation and Diagenesis in a Shallow Epicontinental Sea* (Ed. B. H. Purser), pp. 211–232. *Springer-Verlag*, Berlin, New York.

Reid, R.P. and Macintyre, I.G. (1998) Carbonate recrystallization in shallow marine environments: a widespread diagenetic process forming micritized grains. *J. Sed. Res.*, **68**, 928–946.

Reid, R.P. and Macintyre, I.G. (2000) Current ripples; microboring versus recrystallization; further insight into the micritization process. *J. Sed. Res.*, **70**, 24–28.

Scasso, R.A., Alonso, M.S., Lanés, S., Villar, H.J. and Laffitte G. (2005) Geochemistry and petrology of a Middle Tithonian limestone-marl rhythmite in the Neuquén Basin, Argentina. In: *The Neuquen Basin, Argentina: a Case Study in Sequence Stratigraphy and Basin Dynamics* (Eds G.D.Veiga, L.A. Spalletti, J.A. Howell, and E. Aschwarz). *Geol. Soc. London Spec. Publ.*, **252**, 207–229.

Shearman, D.J. (1966) Origin of marine evaporites by diagenesis. *Trans. Inst. Min. Metall. Section B.*, **75**, 208–215.

Shinn, E.A. (1969) Submarine lithification of Holocene carbonate sediments in the Persian Gulf. *Sedimentology*, **12**, 109–144.

Sollas, W.J. (1921) On *Saccamina carteri* Brady and the minute structure of the Foraminifera shell. *Q. J. Geol. Soc. London*, **77**, 193–212.

Sugden, W. (1963) Some aspects of sedimentation in the Persian Gulf. *J. Sed. Petrol.*, **33**, 355–64.

Swart, P.K., Shinn, E.A., Mckenzie, J.A., Kendall, C.G.St.C. and Hajari, S.E. (1987) Spontaneous dolomite precipitation in brines from Umm Said sabkha, Qatar. *Geol. Soc. Am. Ann. Meet. Abstract*, p. 862.

Wagner, C.W. and Van der Togt, C. (1973) Holocene sediment types and their distribution in the southern Persian Gulf. In: *The Persian Gulf: Holocene Carbonate Sedimentation and Diagenesis in a Shallow Epicontinental Sea* (Ed. B. H. Purser), pp. 123–156. Springer-Verlag, Berlin, New York.

Warren, J.K. and Kendall, C.G.St.C. (1985) Comparison of marine (subaerial) and salina (subaqueous) evaporites, modern and ancient. *AAPG Bull.*, **69**, 1013–1023.

Wolf, K.H. and Conolly, J.R. (1965) Petrogenesis and palaeoenvironment of limestone lenses in Upper Devonian red beds of New South Wales. *Palaeogeogr. Palaeoclimatol. Paleoecol.*, **1**, 69–111.

Wood, A. (1948) The structure of the wall of the test in the Foraminifera – its value in classification. *Q. J. Geol. Soc. London*, **104**, 229–256.

Wood, G.V. and Wolfe, M.J. (1969) Sabkha cycles in the Arab/Darb Formation off Trucial Coast of Arabia. *Sedimentology*, **12**, 165–192.

Wood, W.W., Sanford, W.E. and Al Habschi, A.R.S. (2002) The source of solutes in the coastal sabkha of Abu Dhabi. *Geol. Soc. Am. Bull.*, **114**, 259–268.

Int. Assoc. Sedimentol. Spec. Publ. (2011) **43**, 221–242

Engineering properties of the carbonate sediments along the Abu Dhabi coast, United Arab Emirates

ROGER J. EPPS

Consultant, 'Chy Fawwyth', 4 Goodyers, Ashdell Park, Alton, Hampshire GU34 2SH, UK Formerly Director, Fugro Engineering Services Limited (E-mail: rjepps@nildram.co.uk)

ABSTRACT

The sedimentology of the lagoons and island of Abu Dhabi was studied in detail by Douglas Shearman and his colleagues during the 1960s. Since that time, the island has undergone extensive development and a considerable database on the ground conditions in the area exists from site investigations. Despite this, there are few published data on the engineering characteristics of these sediments; this paper addresses this shortfall. A brief description of the environments of deposition and the sediments of the foreshore, tidal delta and the lagoons of the Abu Dhabi coast is presented, and the effects of both the origin and the diagenesis of these sediments on their engineering properties is examined. Results from a selection of site investigations are summarised to present characteristic values for the various sediments. Physical property data reported include: lithology and grain-size profiles; moisture contents; bulk and dry density; porosity; estimated degrees of saturation; unconfined compressive strength; standard penetration test (SPT) values; soil and groundwater chemistry (chloride and sulphate contents); and plasticity charts (plasticity index versus liquid limit). The ground conditions are also reviewed to identify significant issues that need to be addressed for the design and construction of engineering projects in Abu Dhabi.

The data show that there is a contrast between the strength and compressibility and salt content of the granular, often cemented deposits of the higher energy beach and nearshore environments, and the soft or very loose silts and fine sands of the lagoons and sabkha. A decrease in the strength of the underlying calcarenites with depth to about 10 m below ground level is thought to reflect the effect of diagenetic alteration of the carbonates from aragonite to high-magnesium calcite.

Keywords: Carbonate sediments, geotechnical properties, site investigations, engineering geology.

INTRODUCTION

During the 1960s, Douglas Shearman led a series of expeditions to the Trucial Coast (now the United Arab Emirates, UAE) to study the sedimentology and oceanography of the lagoons and foreshore surrounding Abu Dhabi Island, and the sabkha plain to the rear. These studies provided a detailed understanding of the processes of carbonate sedimentation and the sedimentary environments of this section of coast.

Since the early 1970s there has been continuing development and construction on the island of Abu Dhabi and there is a substantial database on the ground conditions of this area arising from

site investigations relating to such development. However, there are few publications concerning the engineering geology of Abu Dhabi, or indeed on the engineering properties of carbonate sediments more generally. Epps (1980) and Maurenbrecher & Van der Harst (1989) reviewed ground conditions and geotechnical practice in coastal regions of the UAE, and in May 1984 a symposium on ground conditions in the country (Symposium on the Nature and Characteristics of Soils in the UAE and Foundation Engineering) was held in Dubai, but no proceedings were published. The objective of this paper is to describe briefly the engineering properties of sediments from the differing sedimentary environments of the UAE coast using the results of a number of site investigations across Abu Dhabi Island, together with available published data.

GEOLOGICAL SETTING

Abu Dhabi is part of a complex of barrier islands located on the southeastern shoreline of the Arabian Gulf, in the UAE. The Arabian Gulf is an area of extensive carbonate sedimentation, and the deposits on the floor and of the coastline of the UAE are mostly Pleistocene or Holocene in age. The geological structure and geomorphological features of the Arabian Gulf have been described by Kässler (1973) and the general geographical setting is illustrated in Fig. 1.

The nature and distribution of the sediments is governed by the recent geological history and structural setting of the Gulf, the orientation of the coastline and the prevailing Shamal winds, which are from the northwest. Abu Dhabi Island has been formed by a process of both seaward and tail accretion on a core of "Miliolite Sandstone" of Pleistocene age, composed of quartzose calcarenitic limestone of aeolian origin. This in-turn overlies terrestrial clastic sediments of Miocene age,

designated variously the "Miocene Clastic" or the Baynunah Formation. With the rise in sea level during the late Pleistocene and early Holocene, dune sands were deposited across an earlier landscape formed in the Pleistocene and cut into the Miocene strata, until about 4000 BP. Since then, sea level has remained relatively constant and the sediments of the lagoons and islands have prograded seawards.

Within the lagoons and tidal flats behind the barrier islands, carbonate muds, silts and sands have accumulated and onshore deflation of wind-blown sand to near the water table, and subsequent capillary action and evaporation, have led to extensive development of sabkha deposits. Figure 2 shows the sub-environments of the Abu Dhabi–Sa'diyat Island Barrier Complex present in the early 1970s (Evans *et al.*, 1973). The following paragraphs give a brief description of the main sub-environments and their sediments, based on Evans *et al.* (1973). This does not take account of any of the considerable man-made changes to the physiography and morphology of the island that have been made in recent years. The reef and back-reef lagoons will not be discussed, as no data on their properties will be presented in the later

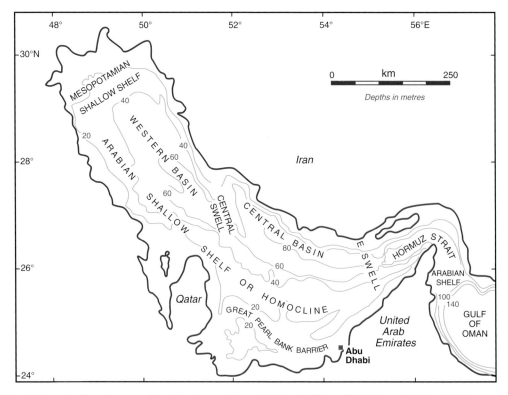

Fig. 1. Principal structural and geographical features of the Arabian Gulf (modified from Kässler, 1973).

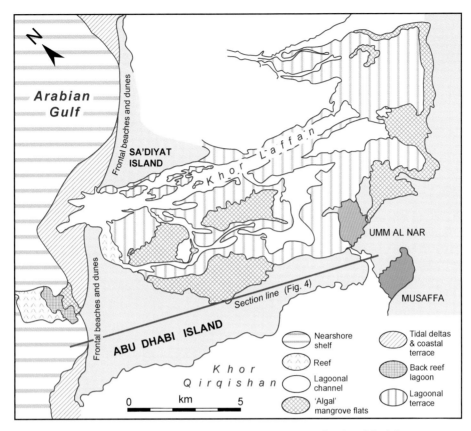

Fig. 2. Principal environments of the Abu Dhabi-Sadiyat Island barrier complex (modified from Evans *et al.*, 1973).

section of this paper on the geotechnical and chemical properties of the sediments.

Nearshore shelf

The nearshore shelf is a reasonably flat surface, which is covered with grey speckled sands with minor amounts of silt and clay. Typically the sands are composed of skeletal debris with a variable content of faecal pellets and composite grains.

Frontal beaches and dunes

The island is mostly fronted by beaches with well-developed beach faces and berms. The beach sediments are composed mainly of bioclastic and oolitic sands, containing large accumulations of coarse transported shell debris, concentrated either on the berm or the toe of the beach face. Prior to development of the area, transverse ridges of dunes with steep landward faces were found onshore from the beaches. The dunes comprised a mixture of bioclastic and oolitic sands, with oolitic sands predominant adjacent to the

tidal delta between Abu Dhabi Island and Sa'diyat Island.

Tidal delta

The sediments of the tidal delta are composed mainly of white oolitic sand with skeletal debris, pellets and composite grains. They are typically uniformly graded, being mainly of fine to medium sand size. Coarser material may be present as shell or cemented rock fragments.

Back-Reef lagoon

The sediments accumulating between the reef and the beach on Abu Dhabi Island are finer grained than found on the nearshore shelf, comprising carbonate silts, usually richer in organic carbon and supporting a dense growth of seagrass.

Lagoonal channels

The sediments of the lagoonal channels are described as mainly grey speckled sands with variable amounts of coarse gravel-sized debris, lithoclasts of Miliolite Sandstone and some fragments of weakly cemented younger rock or sediment. On the inner edge of the lagoons, the channel sediments comprise a mixture of carbonate sand and carbonate mud. The sand fraction is composed of skeletal debris, ooids (which are more abundant towards the tidal deltas), faecal pellets and composite grains.

Lagoonal terraces

The lagoonal terraces form a large part of the lagoons, particularly in the landward parts. Much of the lagoonal terraces comprises bare rock at the surface, which may be eroded "Miliolite Sandstone", but in other places represents *in-situ* cementation (submarine hardgrounds and beachrock), as it is underlain by soft sediment. Such layers are commonly labelled "Caprock" and have a significant impact on various aspects of engineering design and construction, as discussed later, in the section on engineering considerations. The sediments of the terrace banks are mainly oolitic sands, although they may sometimes be grey or speckled with local concentrations of coarser material (usually composed of shell fragments).

Intertidal flats ("algal" mangrove flats)

In many places the lagoonal terraces grade landwards into a zone of intertidal flats which are composed of skeletal and pelletal carbonate sands with minor amounts of coarser debris and finer carbonate mud. Large areas of the surface may be covered with a microbial mat ("blue-green algae") on a cemented crust that is often brecciated. In other areas the flats are vegetated with mangrove, floored with bioturbated pelletal carbonate mud.

Sabkha

The sabkha lies to south and east of the lagoons and is not shown in Fig. 2. It comprises a flat plain of halite-encrusted sediments formed by the denudation of aeolian deposits and the progradation of marine sediments. High rates of evaporation have resulted in high concentrations of anhydrite and/ or gypsum which are often present in nodular form.

Summary

The island has been formed by a process of progradation, and the sediments accumulating to seaward of the island have different characteristics from those in the lagoons. Essentially high-energy bioclastic and oolitic sands prograde seaward, and intertidal and sabkha sediments consisting of a mixture of carbonate sands composed of skeletal debris, composite grains, ooids and pellets and carbonate muds prograde into the lagoons, particularly on their inner (landward) edge. Figure 3 illustrates a schematic cross section through the Abu Dhabi coast with representative successions encountered in the prograding beach (Fig. 3, succession A) and prograding intertidal flat (Fig. 3, succession B). Where possible, the sediments and environments described above are identified in the sites discussed below and their physical and mechanical characteristics are summarised. Figure 4 shows the correlation of boreholes from three sites, one in the nearshore zone to the northwest of the island, one in the centre of the island, and one on the mainland to the southeast of the island. This shows the principal stratigraphic units encountered in the boreholes.

ENGINEERING DESCRIPTION AND CLASSIFICATION

When unconsolidated, the mechanical behaviour of any soil or sediment, whether of detrital or chemical or biochemical origin, is a reflection of its constituent grain size and grading, its particle shape, degree of compaction and mineralogy. Hence, the systematic description of soils and sediments in engineering terms must take account of such factors. Figure 5 sets out in tabular form the terminology used for the field description of soils as defined in BS 5930 (1999) and a glossary of the terms used in the discussions that follow is given in the Appendix.

Several schemes have been developed for the classification of carbonates in engineering terms, for example, Fookes & Higginbottom (1975), Clark & Walker (1977), Burnett & Epps (1979) and latterly, Fookes (1988). Each uses the basis of grain size, texture and fabric, mineral composition, degree of induration and, to a lesser extent, genetic origin, as the basis of classification. These schemes are referred to briefly below, while a short description of the Fookes & Higginbottom (1975) classification is given, and summarised in Figs 6 and 7.

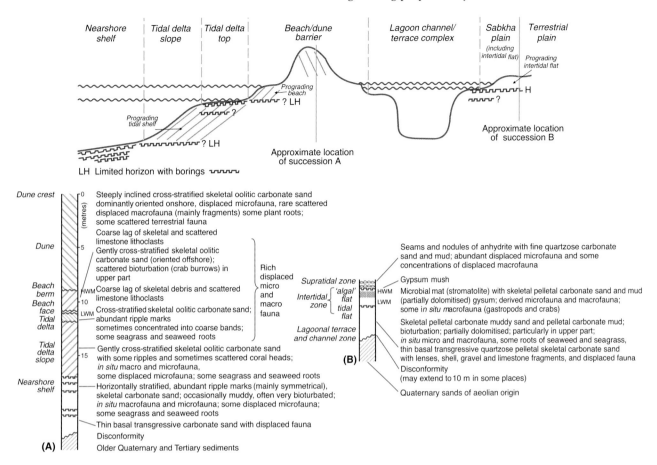

Fig. 3. Typical stratigraphic sections on Abu Dhabi Island, with associated sedimentary environments (modified from Evans *et al.* (1973).

Unconsolidated sediments are described as carbonate mud, carbonate silt, carbonate sand and carbonate gravel and their cemented equivalents as carbonate mudstone (or calcilutite), carbonate siltstone (or calcisiltite), carbonate sandstone (or calcarenite) and carbonate conglomerate (or calcirudite) (Fig. 6). The boundary between unconsolidated and cemented sediments is defined by

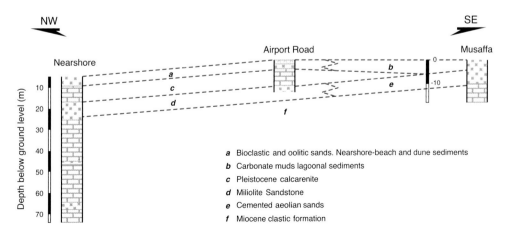

Fig. 4. Correlation of representative sections across Abu Dhabi Island. See Fig. 2 for location.

Category	Basic soil type	Particle size (mm)	Visual identification	Particle nature and plasticity
VERY COARSE SOILS	BOULDERS	— 200	Only seen complete in pits or exposures	**Particle shape:** Angular, Subangular, Subrounded, Rounded, Flat, Elongate
	COBBLES	— 60	Often difficult to recover from boreholes	
COARSE SOILS (typically > 65% sand and gravel)	GRAVEL	coarse — 20 / medium — 6 / fine — 2	Easily visible to naked eye; particle shape can be described; grading can be described. Well graded: wide range of grain sizes, well distributed. Poorly graded: may be uniform, or gap graded;	
	SAND	coarse — 0.6 / medium — 0.2 / fine	Visible to naked eye; no cohesion when dry; grading can be described. Well graded: wide range of grain sizes, well distributed. Poorly graded: may be uniform, or gap graded;	**Texture:** Rough, Smooth, Polished
FINE SOILS (typically >35% silt and clay)	SILT	coarse — 0.06 / medium — 0.02 / fine — 0.006	Only coarse silt visible with hand lens; exhibits little plasticity and marked dilatancy; slightly granular or silky to the touch. Disintegrates in water, lumps dry quickly; possesses cohesion but can be powdered easily between fingers.	
	CLAY/SILT	— 0.002	Intermediate behaviour between clay and silt: slightly dilatant.	
	CLAY		Drylumps can be broken but powdered between the fingers; they also disintegrate under water but more slowly than silt smooth to the touch; exhibits plasticity but no dilatancy; sticks to the fingers and dries slowly; shrinks appreciably on drying usually showing cracks. Intermediate and high plasticity clays show these properties to a moderate and high degree, respectively.	
ORGANIC SOIL	ORGANIC CLAY, SILT or SAND	Varies	Contains varying amounts of organic vegetable matter - defined by colour grey - slightly organic; dark grey - organic; black - very organic.	
	PEAT	Varies	Predominantly plant remains usually dark brown or black in colour, often with distinctive smell; low bulk density. May contain disseminated discrete mineral soils.	

Composite soil types (mixture of basic types)

Scale of secondary constituents with coarse and very coarse soils. Term either before or after constituent.

Term before	Principal	Term after	Approx % secondary
Slightly (sandy)		Used to describe components of secondary constituents e.g. Gravel is fine and medium subangular sandstone and mudstone.	< 5
--(Sandy)	SAND, GRAVEL, COBBLES or BOULDERS		5 to 20 +
Very (sandy)			20+ to 40+
--		and (sand) or and (cobbles+)	50+

* Fine or coarse soil type as appropriate;
+ very coarse soil type as appropriate or
+ described as fine soil depending on mass behaviour.

Scale of secondary constituents with fine soils. Terms either before or after principal constituent.

Term before	Principal	Term after	Approx % secondary
Slightly (sandy)		Used to describe components of secondary constituents e.g. Gravelly sandy CLAY. Gravel is coarse rounded quartzite fragments.	< 35
--(Sandy)	CLAY or SILT		35 to 65
Very (sandy)			> 65+

* Coarse soil type as appropriate or described as coarse soil on mass behaviour.
+

EXAMPLES OF COMPOSITE TYPES (indicating preferred order for description)

Loose brown very sandy subangular coarse GRAVEL with many pockets (<5mm across)of soft grey clay
Firm brown thinly interminated SILT and CLAY
Dense light brown clay and medium SAND.

(A)

Fig. 5. Identification and description of soils to BS 5930. (A) Terms for composition and grading. (B) Terms for strength and structure. See Appendix for further details.

(B)

Compactness/Strength

Term	Field test	
Loose	By inspection of voids and particle packing	
Dense		
Standard penetration test in borehole		
No of blows	Relative density	
< 4	Very loose	
4 - 10	Loose	
10 - 30	Medium dense	
30 - 50	Dense	
> 50	Very dense	
Slightly cemented	Visual Examination: pick removes soil in lumps which can be abraded.	
Uncompact	Easily moulded or crushed in the fingers.	
Compact	Can be moulded or crushed by strong pressure in the fingers.	
Very soft	Finger easily pushed in up to 25 mm	cu (kPa) <20
Soft	Finger pushed in up to 10 mm	20 to 40
Firm	Thumb makes impression easily	40 to 75
Stiff	Can be indented slightly by thumb	75 to 150
Very Stiff	Can be indented by thumbnail	150 to 300
Hard	Can be scratchod by thumbnail	>300
Firm	Fibres already compressed together	
Spongy	Very compressible and open structure	
Plastic	Can be moulded in hand and smears fingers	

Stucture

Terms	Field identification	Interval scales	
Homogeneous	Deposits consists essentially of one type.	Scale of bedding spacing	
		Term	Mean spacing (mm)
Interbedded or interlaminated	Alternating layers of varying types pre-qualified by thickness term if in equal proportions. Otherwise thickness of and spacing between subordinate layers defined	Very thickly bedded	Over 2000
		Thickly bedded	2000 - 600
		Medium bedded	600 - 200
		Thinly bedded	200 - 60
		Very thinly bedded	60 - 20
		Thickly laminated	20 - 6
		Thinly laminated	Under 6
Heterogeneous	A mixture of types	Spacing terms may also be used for distance between partings, isolated beds or laminae, desiccation cracks, rootlets etc.	
Weathered	Particles may be weakened and may show concentric layering		
Fissured	Breaks into blocks along unpolished discontinuties.		
Sheared	Breaks into blocks along polished discontinuties.		
Intact	No fissures.		
Homogeneous	Deposit consists essentially of one type.	Scale of spacing of other discontinuities.	
		Term	Mean spacing (mm)
Interbedded or inter-laminated	As described for coarse soils above.	Very widely spaced	Over 2000
		Widely spaced	2000 - 600
		Medium spaced	600 - 200
		Closely spaced	200 - 60
Weathered	Usually has crumb or columnar structure.	Very closely spaced	60 - 20
Fibrous	Plant remains recognisable and retain some strength	Extremely closely spaced	Under 20
Pseudo-fibrous	Plant remains recognisable strength lost		
Amorphous	Recognisable plant remains absent		

Fig. 5. (*Continued*)

a uniaxial compressive strength of between 0.6 MPa and 1.25 MPa, and the gradation from moderately weak to moderately strong (i.e. equivalent to a uniaxial compressive strength of 12.5 MPa) represents the transition to the term limestone. The description of impure sediments may be addressed by means of the ternary diagram shown in Fig. 7, using terms such as "sandy carbonate mud" or 'sandy carbonate silt' for unindurated equivalents of limestone, suggesting the term "siliceous carbonate sand" to avoid descriptions such as "sandy carbonate sand".

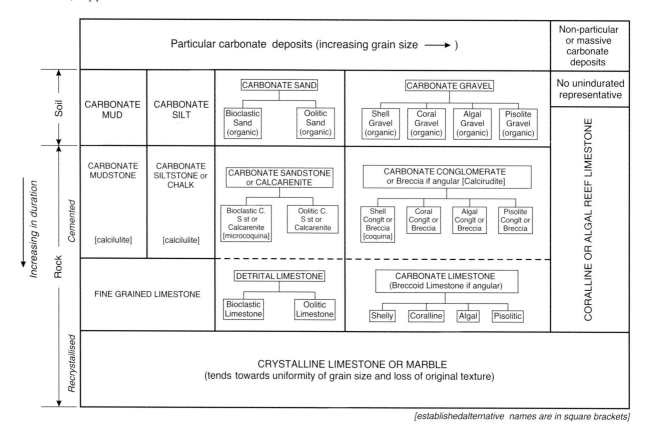

Fig. 6. Classification scheme for the engineering description of carbonate soils and rocks (after Fookes & Higginbottom, 1975).

The above classification was extended and refined by Clark & Walker (1977) to provide a range of terms for each of the categories proposed by Higginbottom & Fookes (1975), according to the content of silica or silicate minerals in the sediment or rock. Burnett & Epps (1979) presented a review of engineering geological and sedimentological carbonate classification schemes published at the time and synthesised these into their own single scheme, that also takes account of the dolomite content.

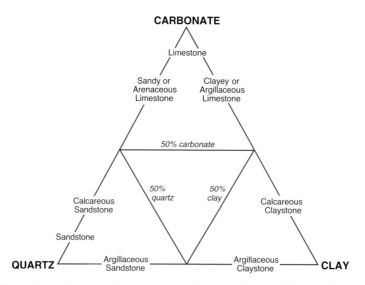

Fig. 7. Ternary classification of mixed indurated carbonate-sand-clay sediments (after Fookes & Higginbottom, 1975).

Fookes (1988) presented a later review and proposed a basic classification scheme derived from the original Fookes & Higginbottom (1975) classification. For engineering purposes, the use of a rock material classification based on the modulus ratio (tangent modulus of elasticity to unconfined compressive strength) was proposed, with rocks of a modulus ratio of >500 being of high modulus ratio and designated category H, and those with a modulus ratio of <200 being designated category L, or low. However, the system proposed by Fookes & Higginbottom (1975) is preferred by the author, as the boundaries of grain size and strength fall along accepted boundaries in terms of strength, and it has the merit of simplicity.

GEOTECHNICAL AND CHEMICAL PROPERTIES OF SEDIMENTS

General comments

Given the influence of particle size and particle shape on the mechanical behaviour of engineering soils (i.e. 'soft' sediments), the origin of nearshore or coastal carbonate sediments will have a significant effect on their properties. For example, the internal shearing resistance of the oolitic sands of the tidal deltas, which are essentially uniformly graded, generally well rounded, fine to medium sands, may be expected to differ from the bioclastic and composite sands found in the lagoon channels. The latter are less uniformly graded, with clasts of variable composition and varying in shape from subangular to subrounded. Valent (1979) reported the results of direct shear tests for coralline sand and oolitic sand, which showed average angles of internal shearing resistance of 34° and 39° respectively. These compare with direct shear test results from sites at Umm Al Nar (see Fig. 2 for location) and Sa'diyat Island, on sand containing silt bands and silty fine sand respectively, which indicate friction angles of 36° and 34°.

Finer grained carbonate muds and silts may be expected to exhibit some plasticity and cohesion, dependent on the clay mineral content and the degree of consolidation of the sediments and the composition of the porewater. It is pertinent to review some characteristics of the carbonate muds in the context of diagenesis and lithification. Table 1 presents some physical properties for carbonate muds from Abu Dhabi. The feature of interest is the apparent degree of "supersaturation" (a term coined here to indicate saturation with moisture of more than 100%) of these sediments and the high porosity (55–56%) and high voids ratio (1.2–1.3).

Bathurst (1975) stated that micritic (fine-grained) limestones, i.e. indurated carbonate muds, with very low porosity or void space often contain intact delicate skeletal debris, which would normally be crushed with burial, suggesting that the lithification process is one of early cementation with low-magnesium calcite. Whilst there is an increase in solid volume of 8% associated with the cementation, there remains a considerable shortfall from the 55–56% porosities reported in Table 1.

There are a number of experiments on the compaction of carbonate sediments reported in the literature. Terzaghi (1940) recorded the results of two tests in which carbonate mud from Andros Island in the Bahamas was compressed in a one dimensional consolidation cell. The first sample was incrementally loaded to 3200 kPa where the void ratio fell from 1.6 to 0.9. The second was loaded to 8500 kPa and showed a decrease in voids ratio from 2.1 to 1.2. Fruth et al. (1966) used a triaxial apparatus to consolidate a variety of carbonate sediments and recorded the compaction as the ratio of the change in volume to the original bulk volume. The mud facies sediments showed a compaction of 50% at a maximum confining pressure of 1000 bars (100 MPa). They also plotted porosity against confining pressure (Fig. 8) and

Table 1. Physical properties of some carbonate muds from Abu Dhabi

Moisture content (%)	Bulk density ($Mg\,m^{-3}$)	Dry Density ($Mg\,m^{-3}$)	Porosity (%)	Voids Ratio	Estimated degree of saturation (%)	Undrained cohesion (kPa)
52.6	1.85	1.21	55	1.227	116	7
53.4	1.88	1.23	55	1.203	120	6
52.4	1.80	1.18	56	1.286	110	6

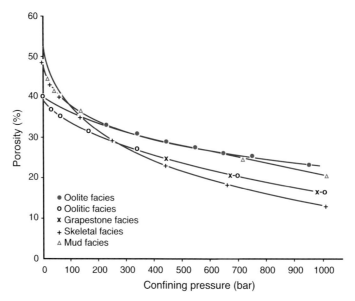

Fig. 8. Plot of porosity against confining pressure for selected carbonate facies (from Fruth *et al.*, 1966).

the results showed that the sediments maintain a porosity of 30% or greater up to 250 bars (25 MPa) of pressure, and porosities of the order of 20% at 1000 bars (100 MPa). A confining pressure of 250 bars represents a burial depth of the order of 2500 m of saturated sediment and 1000 bars about 10,000 m depth of burial. Fruth *et al.* (1966) concluded that their tests produced artificial limestones containing features found in natural limestones, but this was achieved only by applying pressures equivalent to at least 250 m depth of burial in order to obtain porosities of a similar order of magnitude to those recorded at much shallower depth in Abu Dhabi.

Ebhardt (1968) used a pressure cell with an oil hydraulic press to compact a variety of carbonate sediments. These included material described as silty clay and clayey silt from the Arabian Gulf, the clayey silt containing 65% carbonate and the silty clay 36% carbonate. The applied normal pressure ranged from 1000 kPa to 100 MPa and the voids ratio decreased from about 1.05 to 0.25 for the clayey silt. Whilst these experiments show that one possible mechanism of lithification may be through burial, it would seem from the history of the Abu Dhabi sediments that the more active process in these shallow-marine environments is early diagenetic alteration by cementation. Such processes make a significant impact on the "soil" properties. In general terms, the process of cementation and lithification will result in increasing

intact strength. In any entirely monomineralic rock, the intact strength is related to the dry density. With the reduction in pore space as the degree of lithification increases, the dry density of the rock should approach the specific gravity of the constituent mineral.

The intact properties of the rock would also reflect the properties of the constituent mineral. Aragonite has a Specific Gravity of 2.94 and a hardness of 3.5–4 on the Moh's Scale, whereas calcite has a Specific Gravity of 2.71 and a hardness of 3. Thus sediments composed of and cemented with aragonite would be expected to be stronger than calcite cemented sediments. Typically, the sediments of the Arabian Gulf are deposited as and cemented with aragonite which progressively converts to high-magnesium calcite. Cook (1997) reported a trend of decreasing intact strength with depth for weak calcarenites from Abu Dhabi, which was suggested to reflect the dominance of aragonite at shallower depth (less than 10 metres) and that of calcite at greater depth. However, other factors may also affect the strength of cemented carbonates, such as the volume changes associated with the change from aragonite to calcite, which could result in a decrease in intergranular pore space.

An alternative measure of the degree of cementation, particularly if there is a mixture of minerals present, is the voids ratio and porosity. Cook's (1997) studies of weak calcareous rocks in the UAE

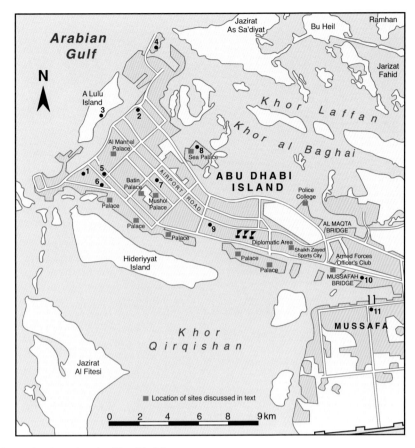

Fig. 9. Location of investigation sites in Abu Dhabi.

showed a number of relationships for various rock types which confirm the broad trend of increasing strength with increasing dry density. A significant increase in strength with decreasing moisture content was also demonstrated, and through a series of "drying out" tests, confirms the importance in the hot climates of the Middle East, of careful sample preservation to prevent any overestimation of the rock strength (Cook, 1999).

Site data

Figure 9 shows the approximate locations from where site investigation data have been provided. In order to evaluate these in relation to their sedimentary environments, these have been grouped as follows:

1. nearshore and beach sediments, Sites 2 and 3;
2. tidal delta, Site 4;
3. tidal flats and terraces, Site 8;
4. inner lagoon and sabkha, Sites 10 and 11.

In addition, a collection of data from sites on the island itself is presented (Sites 1, 5–7 and 9). The data are presented in summary form in Figs 9–16 and are described briefly below.

A number of summary strata descriptions are given in the presentation of the site data which mostly use the terms calcarenite and sandstone. It is the author's view that in some instances the term "sandstone", which tends to suggest the presence of "Miliolite Sandstone", may in fact reflect calcarenites, in particular on the tidal delta where the cemented sands are oolitic.

Nearshore and beach sediments

The profile presented in Fig. 10 may be regarded as reasonably representative of ground conditions beneath Sites 2 and 3, although the natural sediments underlie a substantial thickness of dredged hydraulic fill under Site 3. Typically, the natural strata comprise a variable thickness of sand overlying calcarenite with lenses/bands of sandstone,

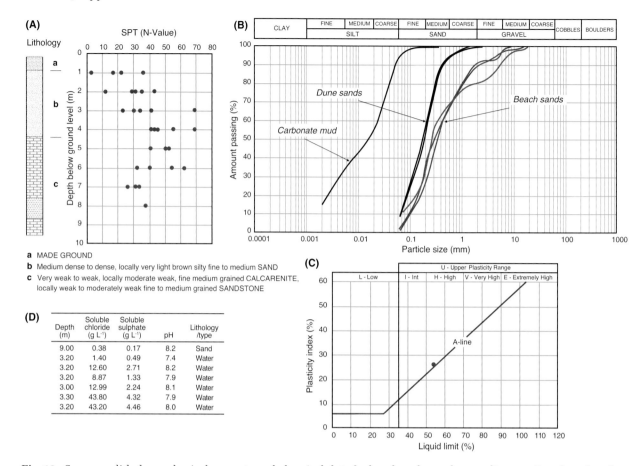

Fig. 10. Summary lithology, physical property and chemical data for beach and nearshore sediments. Based on data from Sites 2 and 3 (Fig. 9). (A) Lithology (schematic) and standard penetration test (SPT) profile. (B) Cumulative grain-size curves. (C) Plasticity chart. (D) Sediment and groundwater chemical test data.

although fine sediments were also reported at Site 3.

The sand varies in relative density from medium dense to very dense, with values of penetration resistance typically ranging from 18 to 68 (Fig. 10A; relative density terms are listed in Fig. 5 and are described in the Appendix). Grading curves for the sands show them to be generally well sorted, with three distinct groups of curves being apparent (Fig. 10B). The coarsest sediments, which are thought to represent beach deposits, are typically fine to coarse sands with a minor component of gravel-size material. The second group, composed of fine and medium sand with a little coarse sand, is thought to represent wind-blown dune sand. A single sample of carbonate mud encountered beneath the dredged fill shows approximately 15% clay-size particles with the remainder comprising fine to coarse silt. The Atterberg Limits of this sample plot as a clay of high plasticity on a Casagrande Plasticity Chart (Fig. 10C).

Chemical analyses of groundwater samples show chloride contents varying from 1.40 to 43.8 g L^{-1}, sulphate (as SO_4) contents varying from 0.49 to 4.46 g L^{-1}, and pH values ranging from 7.4 to 8.2 (Fig. 10D).

The measured bulk densities for the underlying rocks range from 1.73 to 2.16 Mg m^{-3}, with corresponding dry densities ranging from 1.52 to 1.82 Mg m^{-3} (Fig. 11). The porosity values, calculated assuming a particle density of 2.70, range from 26 to 44%. When the measured uniaxial compressive strengths are plotted against depth (Fig. 11B), there is an apparent trend of decreasing strength with depth between 9 and 14 m. Below 14 m, the results are scattered, with values ranging from 1.40 to 8.23 MPa.

Tidal delta

The data from the tidal delta are limited to one borehole from Site 4 and these are summarised in

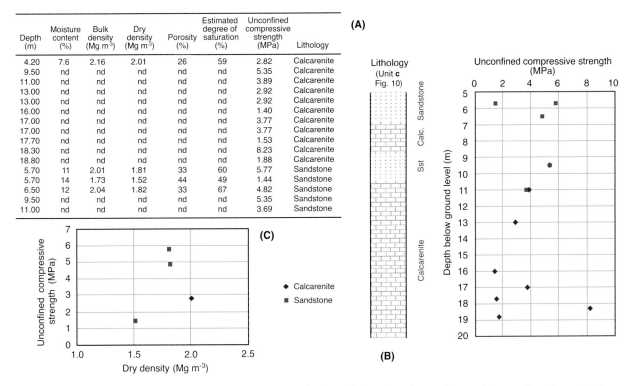

Depth (m)	Moisture content (%)	Bulk density (Mg m⁻³)	Dry density (Mg m⁻³)	Porosity (%)	Estimated degree of saturation (%)	Unconfined compressive strength (MPa)	Lithology
4.20	7.6	2.16	2.01	26	59	2.82	Calcarenite
9.50	nd	nd	nd	nd	nd	5.35	Calcarenite
11.00	nd	nd	nd	nd	nd	3.89	Calcarenite
13.00	nd	nd	nd	nd	nd	2.92	Calcarenite
13.00	nd	nd	nd	nd	nd	2.92	Calcarenite
16.00	nd	nd	nd	nd	nd	1.40	Calcarenite
17.00	nd	nd	nd	nd	nd	3.77	Calcarenite
17.00	nd	nd	nd	nd	nd	3.77	Calcarenite
17.70	nd	nd	nd	nd	nd	1.53	Calcarenite
18.30	nd	nd	nd	nd	nd	8.23	Calcarenite
18.80	nd	nd	nd	nd	nd	1.88	Calcarenite
5.70	11	2.01	1.81	33	60	5.77	Sandstone
5.70	14	1.73	1.52	44	49	1.44	Sandstone
6.50	12	2.04	1.82	33	67	4.82	Sandstone
9.50	nd	nd	nd	nd	nd	5.35	Sandstone
11.00	nd	nd	nd	nd	nd	3.69	Sandstone

Fig. 11. Summary lithology and physical properties of rocks beneath beach and nearshore sediments. Based on data from Sites 2 and 3 (Fig. 9). (A) Numerical data. (B) Lithology, and unconfined compressive strength versus depth. (C) Strength versus dry density plot.

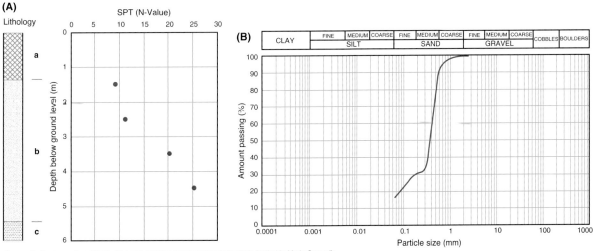

a Brown silty fine to medium SAND with some cobble to boulder sized rock debris (probable Made Ground)

b Medium dense pale brown silty fine SAND

c Slightly weathered light brown slightly shelly fine to medium very weak to locally weak SANDSTONE

(C)

Depth (m)	Soluble chloride (g L⁻¹)	Soluble sulphate (g L⁻¹)	pH	Lithology /type
1.50	2.04	0.59	7.2	Made Ground
2.50	3.42	0.67	7.7	Water

(D)

Depth (m)	Moisture content (%)	Bulk density (Mg m⁻³)	Dry density (Mg m⁻³)	Porosity (%)	Estimated degree of saturation (%)	Unconfined compressive strength (MPa)	Lithology
5.50	11	2.16	1.95	28	77	4.43	Sandstone

Fig. 12. Summary lithology, physical property and chemical data for tidal delta sediments. Data from Site 4 (Fig. 9). (A) Lithology (schematic) and standard penetration test (SPT) profile. (B) Cumulative grain-size curve. (C) Sediment and groundwater chemical test data. (D) Physical properties of sandstone.

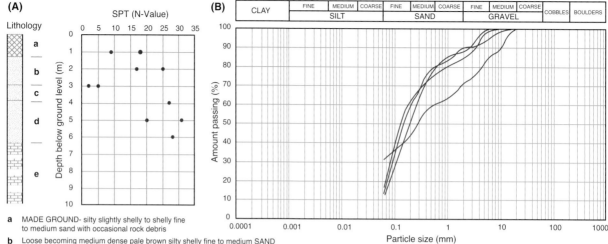

a   MADE GROUND- silty slightly shelly to shelly fine
to medium sand with occasional rock debris

b   Loose becoming medium dense pale brown silty shelly fine to medium SAND

c   Very loose grey silty shelly fine to medium SAND

d   Medium dense to dense grey silty fine to medium SAND. Many shells below 6.00 m

e   Slightly weathered greyish white fine grained very weak CALCARENITE

Note: *SANDSTONE recorded at 12.5 m and 13.0 m depth.*

(C)

Depth (m)	Soluble chloride (g L⁻¹)	Soluble sulphate (g L⁻¹)	pH	Lithology /type
1.00	0.43	1.55	8.4	Made Ground
1.60	0.43	0.91	8.0	Water
4.00	0.06	0.49	8.6	Sand

(D)

Depth (m)	Moisture content (%)	Bulk density (Mg m⁻³)	Dry density (Mg m⁻³)	Porosity (%)	Estimated degree of saturation (%)	Unconfined compressive strength (MPa)	Lithology
7.50	17	1.54	1.32	51	44	1.45	Calcarenite
6.70	17	1.54	1.32	51	44	0.55	Calcarenite
9.20	22	1.70	1.39	48	63	1.15	Calcarenite

Fig. 13. Summary lithology, physical property and chemical data for tidal flat and terrace sediments. Data from Site 8 (Fig. 9). (A) Lithology (schematic) and standard penetration test (SPT) profile. (B) Cumulative grain-size curves. (C) Sediment and groundwater chemical test data. (D) Physical properties of calcarenites.

Fig. 12. The natural strata are described as medium dense pale brown silty fine sand overlying a light brown shelly fine to medium, very weak to locally weak sandstone. It is a matter of speculation from the borehole descriptions whether the sand and sandstone are oolitic, although from the site location this is likely as the tidal delta sediments consist predominantly of oolitic sand. Standard penetration tests show an increase with depth, with N-values varying from 9 to 25 (Fig. 12A). One particle-size analysis is presented for the sand, which is primarily of medium grade, with up to 30% fine sand and silt (Fig. 12B).

A single analysis of groundwater is presented with measured concentrations of chloride of 3.42 g L⁻¹, of sulphate of 0.67 g L⁻¹, and a pH value of 7.7 (Fig. 12C).

The bulk and dry density of a single sample of sandstone measured 2.16 Mg m⁻³ and 1.95 Mg m⁻³ respectively, with an estimated porosity of 28%. The recorded uniaxial compressive strength was 4.43 MPa (Fig. 12D).

Tidal flats and terraces

The data for the tidal flats or terraces are limited to two boreholes from Site 8 and these are summarised in Fig. 13. The natural strata are shown to comprise a succession of shelly sands overlying calcarenites containing interbedded sandstones. The sands are shown to be medium dense, with a very loose layer between 3 and 4 m depth. N-values in this layer vary from 2 to 5, and in the layers above and below N-values vary from 9 to 31, with a trend of increasing relative density with depth (Fig. 13A). The grading curves for the sands show them to be poorly sorted, comprising mainly fine to coarse sand with between 10 and 30% gravel (Fig. 13B).

A single groundwater analysis recorded a chloride concentration of 0.43 g L⁻¹, a sulphate concentration of 0.91 g L⁻¹, and a pH value of 8.0 (Fig. 13C).

Three sets of tests on rock show bulk densities of between 1.54 and 1.70 Mg m⁻³, dry densities of 1.32 to 1.39 Mg m⁻³ and estimated porosity values

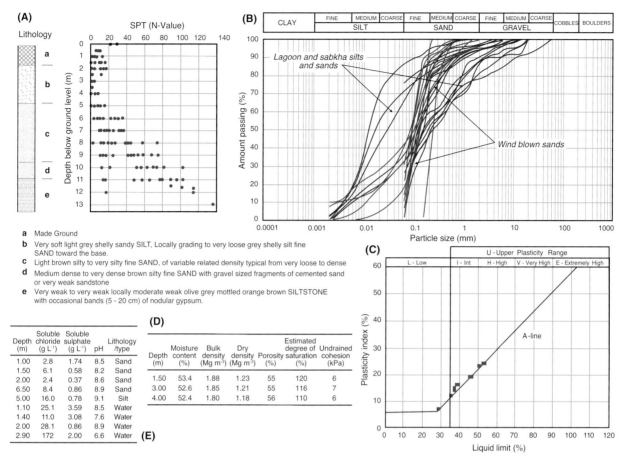

a Made Ground

b Very soft light grey shelly sandy SILT, Locally grading to very loose grey shelly silt fine SAND toward the base.

c Light brown silty to very silty fine SAND, of variable related density typical from very loose to dense

d Medium dense to very dense brown silty fine SAND with gravel sized fragments of cemented sand or very weak sandstone

e Very weak to very weak locally moderate weak olive grey mottled orange brown SILTSTONE with occasional bands (5 - 20 cm) of nodular gypsum.

Depth (m)	Soluble chloride (g L⁻¹)	Soluble sulphate (g L⁻¹)	pH	Lithology /type
1.00	2.8	1.74	8.5	Sand
1.50	6.1	0.58	8.2	Sand
2.00	2.4	0.37	8.6	Sand
6.50	8.4	0.86	8.9	Sand
5.00	16.0	0.78	9.1	Silt
1.10	25.1	3.59	8.5	Water
1.40	11.0	3.08	7.6	Water
2.00	28.1	0.86	8.9	Water
2.90	172	2.00	6.6	Water

Depth (m)	Moisture content (%)	Bulk density (Mg m⁻³)	Dry density (Mg m⁻³)	Porosity (%)	Estimated degree of saturation (%)	Undrained cohesion (kPa)
1.50	53.4	1.88	1.23	55	120	6
3.00	52.6	1.85	1.21	55	116	7
4.00	52.4	1.80	1.18	56	110	6

Fig. 14. Summary lithology, physical property and chemical data for inner lagoon and sabkha sediments. Data from Sites 10 and 11 (Fig. 9). (A) Lithology (schematic) and standard penetration test (SPT) profile. (B) Cumulative grain-size curves. (C) Sediment and groundwater chemical test data. (D) Physical properties of carbonate silts. (E) Sediment and groundwater chemical test data.

of 48 to 51%. The measured uniaxial compressive strengths vary from 0.55 to 1.45 MPa (Fig. 13D).

Inner lagoon and sabkha

The data for the inner lagoon and sabkha are derived from two sites on the mainland at Mussafa (Fig. 9, Sites 10 and 11), and show similar conditions; these are summarised in Figs. 14 and 15. Beneath the Made Ground, the boreholes recorded a very soft light grey to white sandy silt, grading to very loose shelly silty fine sand near the base. This overlies light brown silty fine sand, which grades from very loose to dense and a dense to very dense sand successively. This is thought to represent the older wind-blown sands described above (Geological Setting). The underlying bedrock consists of siltstone with bands of nodular gypsum, with a layer of calcarenite between 18 and 19 m depth,

and is thought to represent the Miocene clastic formation shown in Fig. 4.

The standard penetration tests in the sediments show N-values which are generally less than 10 in the silt and show a wide scatter of results in the underlying silty fine sands, suggesting that they are variably cemented (Fig. 14A). The Atterberg Limits of the silts plot in the categories from low to high plasticity (Fig. 14C) and they are shown by three unconsolidated undrained triaxial compression tests to be very soft with cohesion values of 6 to 7 kPa (Fig. 14D).

The particle size analyses for the sediments show two distinct groupings (Fig. 14B), the silts being generally poorly sorted, varying from medium to coarse silt to silty fine sand with a variable amount of gravel. The brown silty fine sand, by contrast is well-sorted fine sand with silt contents ranging up to 30%.

(A)

Depth (m)	Moisture content (%)	Dry density (Mg m^{-3})	Porosity (%)	Estimated degree of saturation (%)	Unconfined compressive strength (MPa)	Lithology
17.65	23	1.41	48	68	0.88	Calcarenite
18.00	15	1.50	44	51	0.70	Calcarenite
18.15	24	1.47	46	77	0.68	Calcarenite
18.50	13	1.89	30	82	1.49	Calcarenite
19.30	22	1.54	43	79	0.85	Calcarenite
19.60	17	1.54	43	61	1.11	Calcarenite
19.70	22	1.52	44	77	0.58	Calcarenite
10.00	4.9	2.01	26	39	1.79	Siltstone
10.90	15	1.65	39	64	1.44	Siltstone
11.00	8.1	1.85	31	48	1.31	Siltstone
11.50	24	1.52	44	84	0.39	Siltstone
11.70	31	1.42	47	93	0.67	Siltstone
12.10	37	1.28	53	88	1.20	Siltstone
12.50	18	1.68	38	78	3.23	Siltstone
12.50	19	1.71	37	88	1.94	Siltstone
12.50	31	1.44	47	85	0.54	Siltstone
12.70	21	1.47	48	68	1.47	Siltstone
12.80	22	1.81	40	88	0.74	Siltstone
12.80	20	1.65	39	85	3.68	Siltstone
13.00	24	1.51	44	82	0.49	Siltstone
13.00	11	1.75	35	85	0.82	Siltstone
13.20	17	1.75	35	85	4.26	Siltstone
13.40	24	1.52	44	84	3.07	Siltstone
14.00	21	1.49	45	70	4.65	Siltstone
14.00	15	1.72	36	71	3.52	Siltstone
14.00	13	1.81	33	71	1.83	Siltstone
14.50	16	1.64	39	67	4.33	Siltstone
14.55	20	1.73	38	97	1.60	Siltstone
14.60	17	1.68	38	75	4.32	Siltstone
15.10	19	1.71	37	89	1.66	Siltstone
15.10	15	1.46	46	48	1.68	Siltstone
15.50	20	1.73	36	96	1.53	Siltstone
15.50	17	1.74	35	84	4.18	Siltstone
15.55	25	1.60	41	95	4.42	Siltstone
15.60	19	1.68	37	88	3.43	Siltstone
15.70	20	1.59	41	78	3.32	Siltstone
16.40	18	1.73	36	87	3.11	Siltstone
16.95	23	1.60	41	91	2.09	Siltstone
17.00	18	1.71	37	74	3.45	Siltstone
17.20	21	1.63	40	86	0.79	Siltstone
17.50	19	1.69	37	88	3.37	Siltstone
18.55	18	1.64	39	75	0.90	Siltstone
18.60	13	1.53	43	46	1.73	Siltstone
19.00	18	1.69	38	81	3.57	Siltstone
19.50	17	1.61	40	87	4.73	Siltstone
19.75	19	1.50	44	65	0.62	Siltstone
21.00	17	1.48	46	54	0.91	Siltstone
21.40	12	1.71	37	56	2.02	Siltstone
23.50	12	2.00	26	93	6.11	Siltstone

Fig. 15. Summary lithology and physical properties of rocks beneath inner lagoon and sabkha deposits. Data from Sites 10 and 11 (Fig. 9). (A) Physical property values. (B) Lithology (schematic) and strength versus depth plot, with inferred trend lines for calcarenites and siltstones. (C) Strength versus dry density plot.

The groundwater chemistry shows chloride contents varying from 11 to $172\,g\,L^{-1}$, sulphate contents from 0.86 to $3.59\,g\,L^{-1}$ and pH values from 6.6 to 8.5 (Fig. 14E).

The test data for the rocks are summarised in Fig. 15. The profile of uniaxial compressive strength against depth, whilst exhibiting a degree of scatter, shows a trend of increasing strength with depth (Fig. 15B). This is illustrated in Fig. 15B by two inferred trend lines, one for the siltstones and the other for the calcarenites. A plot of unconfined compressive strength against dry density shows two distinct groups of data (Fig. 15C). The first, which relates to very weak to weak rocks with strengths varying from 0.5 to 2 MPa shows a broad trend of increasing strength with dry density. The second group, with strengths from 3 to 5 MPa shows no distinct pattern.

Sediments of Abu Dhabi Island

Site investigations on the island (Sites 1, 5–7, 9) show natural deposits comprising medium dense to dense sand overlying calcarenite, data for which are summarised in Figs 16 and 17. N-values from the standard penetration test vary between 4 and 43 (Fig. 16A). The sands again fall into two groups (Fig. 16B): a well-sorted fine to medium sand with a trace of gravel fraction; and a well-graded fine to coarse sand, locally sand and gravel. This variation in gravel content is a reflection of varying amounts of shell debris and cemented material.

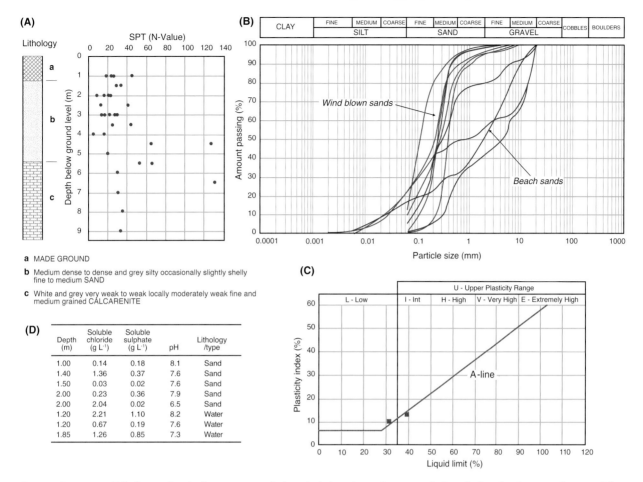

(A) Lithology

a MADE GROUND

b Medium dense to dense and grey silty occasionally slightly shelly fine to medium SAND

c White and grey very weak to weak locally moderately weak fine and medium grained CALCARENITE

(D)

Depth (m)	Soluble chloride (g L⁻¹)	Soluble sulphate (g L⁻¹)	pH	Lithology /type
1.00	0.14	0.18	8.1	Sand
1.40	1.36	0.37	7.6	Sand
1.50	0.03	0.02	7.6	Sand
2.00	0.23	0.36	7.9	Sand
2.00	2.04	0.02	6.5	Sand
1.20	2.21	1.10	8.2	Water
1.20	0.67	0.19	7.6	Water
1.85	1.26	0.85	7.3	Water

Fig. 16. Summary lithology, physical property and chemical data for sediments of Abu Dhabi Island. Compilation of data from Sites 1, 5–7 and 9 (Fig. 9). (A) Lithology (schematic) and standard penetration test (SPT) profile. (B) Cumulative grain-size curves. (C) Plasticity chart. (D) Sediment and groundwater chemical test data.

Two samples of fine sediment plot as clay of low to intermediate plasticity on a Casagrande plasticity chart (Fig. 16C).

Groundwater chemistry for the sites on the island shows chloride concentrations from 0.67 to 2.21 g L⁻¹, sulphate concentrations from 0.19 to 1.10 g L⁻¹ and pH values from 7.3 to 8.2 (Fig. 16D).

The calcarenite is in one location very weak and shows N-values of 20 to 35. Elsewhere, it has been recovered as cores and the profile of strength against depth shows a trend of decreasing strength from 4 to 7 m (Fig. 17B). Below this the rocks are generally uniformly weak or very weak with depth. The strength versus dry density plot (Fig. 17C) show two groups. Results of less than 2 MPa in strength are associated with dry densities ranging from 1.0 to 1.7 Mg m⁻³, which broadly increase with unconfined compressive strength. In the second group, unconfined compressive strengths measuring between 5 and 6 MPa are associated with dry densities of between 1.6 and 1.8 Mg m⁻³.

Summary

To summarise the data presented above, it is possible to recognise differences in the particle-size distributions of the granular sediments according to their depositional environment. However, their engineering properties mostly reflect the degree of cementation of the sands, and the finer grained sediments show generally similar characteristics. This would indicate that the investigation and testing techniques adopted, in which tests were made at discrete depths, are not sufficiently sensitive to identify the subtle changes in properties within an individual sedimentary environment. This point was made previously by Maurenbrecher & Van der Harst (1989), who advocated greater use of the static (Dutch) cone penetration test, as the

(A)

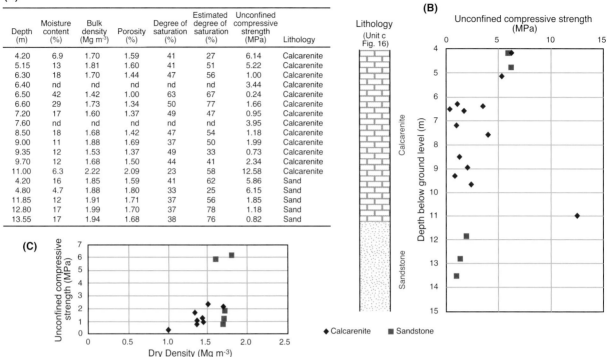

Depth (m)	Moisture content (%)	Bulk density (Mg m⁻³)	Porosity (%)	Degree of saturation (%)	Estimated degree of saturation (%)	Unconfined compressive strength (MPa)	Lithology
4.20	6.9	1.70	1.59	41	27	6.14	Calcarenite
5.15	13	1.81	1.60	41	51	5.22	Calcarenite
6.30	18	1.70	1.44	47	56	1.00	Calcarenite
6.40	nd	nd	nd	nd	nd	3.44	Calcarenite
6.50	42	1.42	1.00	63	67	0.24	Calcarenite
6.60	29	1.73	1.34	50	77	1.66	Calcarenite
7.20	17	1.60	1.37	49	47	0.95	Calcarenite
7.60	nd	nd	nd	nd	nd	3.95	Calcarenite
8.50	18	1.68	1.42	47	54	1.18	Calcarenite
9.00	11	1.88	1.69	37	50	1.99	Calcarenite
9.35	12	1.53	1.37	49	33	0.73	Calcarenite
9.70	12	1.68	1.50	44	41	2.34	Calcarenite
11.00	6.3	2.22	2.09	23	58	12.58	Calcarenite
4.20	16	1.85	1.59	41	62	5.86	Sand
4.80	4.7	1.88	1.80	33	25	6.15	Sand
11.85	12	1.91	1.71	37	56	1.85	Sand
12.80	17	1.99	1.70	37	78	1.18	Sand
13.55	17	1.94	1.68	38	76	0.82	Sand

Fig. 17. Summary lithology and physical properties of rocks beneath the sediments of Abu Dhabi Island. Compilation of data from Sites 1, 5–7 and 9 (Fig. 9). (A) Physical property values. (B) Lithology (schematic) and strength versus depth plot. (C) Strength versus dry density plot.

continuous recording of data is more sensitive to variations in strength and soil type. However, on a wider scale, the contrasting facies of the deposits of the beach, tidal delta and outer lagoon, and the deposits of the inner lagoon are apparent from the conditions described above. The higher energy deposits show varying degrees of cementation, but are generally of medium dense or greater relative density. By contrast, the sediments of the inner lagoon are very soft or very loose and appreciably weaker, with mostly silt and fine-sand grade particles.

The results of the chemical analyses show sulphate and chloride concentrations that range up to $3.59\,\text{g}\,\text{L}^{-1}$ and $172\,\text{g}\,\text{L}^{-1}$ respectively. These are significantly elevated when compared to soils from temperate climates, which typically display SO_4 values of only $1.2–1.8\,\text{g}\,\text{L}^{-1}$.

When the test data for the sandstones and calcarenites below the outer island and lagoons are viewed collectively, the trends identified for individual sites can still be recognised. Figure 18 shows a summary plot for all sites of strength against depth. Between 5 and 10 m there is a trend

of decreasing strength with depth and below 10 m there is a trend of increasing strength with depth.

The plots of strength against dry density (Fig. 19) show two distinct groups, one (Group A in Figure 19) in the 0–2 MPa strength range, exhibits a trend of increasing intact strength with dry density, as indicated by the dotted line in Fig. 19. Within the second (Group B), in the 4–6 MPa range, this trend is not apparent. The differentiation of two strength groups may reflect the presence of two mineral types. Although the higher strengths are not exclusively associated with the shallowest samples, this tends to support the suggestion by Cook (1997) that the shallowest rocks are cemented with aragonite that has not converted to calcite. To verify this, some chemical analyses of bulk sediments from a nearshore site investigation in the early 1980s and these data are plotted against depth in Fig. 20, which shows the variation in content of calcium, magnesium, carbonate and sulphate. The magnesium content is very low between 5 and 10 m depth, but shows a marked increase below about 14 m depth, and below 20 m the Mg content is about 10%. This suggests that

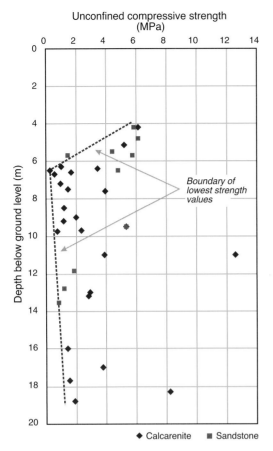

Fig. 18. Summary strength versus depth plot for calcarenites and sandstones.

there may be a progressive alteration to high-magnesium calcite with depth. Beneath the inner lagoon and sabkha deposits, the general trend within the fine-grained siltstones is of increasing strength with depth (Fig. 15B).

ENGINEERING CONSIDERATIONS

There are a number of conditions that can be identified from the foregoing discussion which have a significant impact on engineering design and construction in and around Abu Dhabi. Groundwater generally occurs at shallow depth (usually between 1 and 2 m and less than 1 m on the sabkha plain). Thus it is necessary to control groundwater ingress into excavations, the construction of which might otherwise be straightforward. Typically, the grading of the sands is suited to the use of dewatering by well points. Installation is mostly by jetting techniques, but where cemented layers are present, rotary or rotary percussive drilling methods of installation may be necessary. For the finer grained sediments, the use of a sheet piled or other cut-off wall provides support to the excavation and permits dewatering using sump pumps in the underlying bedrock. Where the permeability of the bedrock is too great for pumping from sumps, the use of deep wells would be necessary.

The high salt content of both the soils and the groundwater produces a chemical regime that is

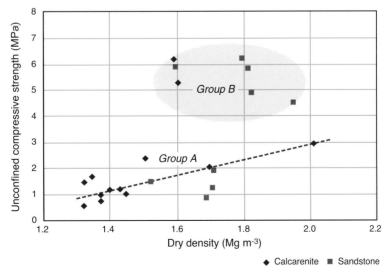

Fig. 19. Summary strength versus dry density plot for calcarenites and sandstones. Note two distinct groups, A and B (see text for discussion).

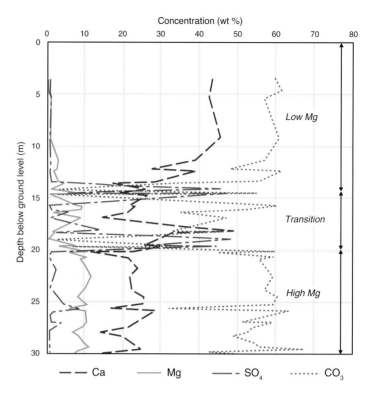

Fig. 20. Profile of bulk sediment chemical composition with depth for a nearshore site. The location approximates to the nearshore site indicated in Fig. 4.

highly aggressive to construction materials, particularly by sulphate attack on concrete and the corrosion of steel reinforcement and steelwork by chlorides. This is addressed by the choice of an appropriate cement, by quality control in the production of concrete, and by the selection of appropriate aggregates. In severe conditions, the use of bitumen or other inert coatings on preformed concrete or steel, or the provision of plastic linings or sleeves is required. Typically, a moderate sulphate-resistant cement such as a blend of Ordinary Portland Cement and blast furnace slag or ASTM Type II cement is used as standard sulphate-resistant cement offers less resistance to chloride penetration.

The variable composition of the sediments requires careful ground investigation and consideration of the proposed engineering design. Choice of foundation is important, as very often there may be strongly cemented layers overlying very soft or loose sediments. Whilst the stronger layers may have sufficient immediate strength to support the required loadings in the short term, the underlying weaker sediments are prone to longer term consolidation. Settlement of structures resulting from such consolidation can induce distress and visible damage (cracking) to the structure. Strongly cemented layers can also obstruct the penetration of rotary, driven or sheet piles. With modern plant there are usually means of removing or penetrating such obstructions, but nonetheless their presence must be recognised.

CONCLUSIONS

Review of site investigation data from Abu Dhabi coastal areas has shown a contrast in the strength and properties of the sediments of the outer lagoon tidal delta and beach environments compared to those of the inner lagoon and sabkha. The higher energy sediments are typically granular and show more evidence of cementation at shallow depths (<4–5 m), whereas the inner lagoon and sabkha sediments consist principally of silts and fine sand of lower permeability that are typically very soft or very loose, with little evidence of cementation above a depth of 10 m. Typically, the granular cemented sediments can support higher foundation loadings than the very soft low-energy sediments of the lagoons.

A review of the rock-test data confirms the observation made by Cook (1997) that the uniaxial compressive strength decreases with depth, down to a depth of about 10 m (Fig. 18). There are, for both the calcarenites and sandstones and the siltstones, two distinct groupings in terms of strength and dry density. It is tentatively suggested that this may reflect different mineral composition and states of diagenetic alteration; this needs to be clarified by further mineralogical and petrographic studies. For the design of deeper foundations, such as piles and deep basements, sufficiently detailed investigation is required to identify such trends on a site specific basis.

ACKNOWLEDGEMENTS

I would like to record my thanks and appreciation for the support provided by Fugro Middle East. Mr Maarten Van der Harst gave permission for me to use their data. In particular, I am deeply grateful to Mr Jeff Williams, who provided me with the site data and a number of additional references and has also reviewed this paper on my behalf. I would also like to thank Mr Jim Cook for the loan of his dissertation and my wife Jackie, for proofreading the typescripts.

APPENDIX: BRIEF EXPLANATION OF ENGINEERING TERMS USED

The strength of a soil or rock is characterised by its resistance to shearing. In granular or **cohesionless** soils this is derived from the friction between soil particles and is measured by determining the **angle of internal shearing resistance** (sometimes also known as the friction angle). In fine grained or **cohesive** soils, particularly clays, the soil strength is derived from both friction and bonding between and within the soil particles.

The strength of cohesive soils is measured by determining both its **cohesion** or shear strength and its angle of internal shearing resistance. In undrained conditions, such as initial foundation loading, the applied stress will generate a porewater pressure equal to the *in situ* stress plus the applied loading pressure and the **undrained cohesion** of the soil is the primary measure of strength. In drained conditions, the porewater pressure dissipates through drainage from the soil and the internal friction is the dominant measure of strength, although fine grained soils may typically also exhibit a small amount of cohesion.

The internal cohesion of clay is related to its mineralogy and also its moisture content. An empirical index of its likely behaviour is the **Atterberg Limits**, the **liquid limit** and **plastic limit**. This test determines the response of the soil to increasing moisture content. The plastic limit represents the transition from brittle behaviour to plastic, and the liquid limit represents the transition from the plastic phase to behaving effectively as a liquid. The **plasticity index** is the difference between the liquid limit and the plastic limit. Soils with moisture content at or near the plastic limit are relatively stiff and become softer with increasing moisture content. Montmorillonite clays, which retain a significant content of water within their molecular structure, will have very high liquid limits and plasticity indices. By contrast, carbonates are typically non-plastic, but carbonate mud may show some plasticity according to its content of clay minerals.

Measurement of the internal shearing resistance of granular or cohesionless soil is hindered by the difficulty in recovering representative undisturbed samples, as disturbance of the soil results in a loss of strength. The internal friction of an undisturbed sand or gravel will increase with compaction, and the relationship between relative density and angle of shearing resistance has been established as shown in Table A1.

An empirical relationship has been established between the **penetration resistance** or **N-Value** from the **Standard Penetration Test (SPT)** and relative density, which is shown in Fig. 5B under the heading compactness/strength. The standard penetration test is a simple test in which a sampling tube of standard dimensions is driven into the soil by a drop weight of 65 kg mass with a free fall of 760 mm. The N-Value is the number of blows required to drive the tube 300 mm after an initial penetration of 150 mm. An alternative measure of

Table A1. Relationship between relative density and angle of shearing resistance

Relative density (%)	Angle of shearing resistance	Descriptive term
0–15	<29°	Very Loose
15–35	29°–30°	Loose
35–65	30°–35°	Medium Dense
65–85	35°–40°	Dense
85–100	>40°	Very Dense

soil strength and density can be obtained through the static (Dutch) cone penetration test whereby a cone is pushed continuously into soil and the load required to push the cone is related to the soil strength. There are also laboratory tests that can be used to determine the strength of re-compacted sands or gravels.

The strength of intact rocks is derived from internal bonding of crystals or cementation of sediments and the standard test is to compress a core in unconfined conditions to determine its uniaxial compressive strength.

DOCUMENTATION

British Standards Institution (1990a). BS 1377. Part 2: Tests 4 & 5, Determination of the Liquid Limit and the Plastic Limit. In: *Methods of Test for Soils for Civil Engineering Purposes. Classification Tests.* BS Inst., London.

British Standards Institution (1990b). BS 1377. Part 9: 1990 Test 3.1, Determination of the penetration resistance using the fixed 60° cone and friction sleeve (static cone penetration test CPT). In: *Methods of Test for Soils for Civil Engineering Purposes. Classification Tests.* BS Inst., London.

British Standards Institution (1990c). BS 1377: Part 9: 1990 Test 3.3, Determination of the penetration resistance using the split barrel sampler (the standard penetration test SPT). In: *Methods of Test for Soils for Civil Engineering Purposes. Classification Tests.* BS Inst., London.

REFERENCES

Bathurst, R.G.C. (1975) Carbonate sediments and their diagenesis, 2nd Edn. *Dev. Sedimentol.*, **12**, Elsevier, Amsterdam, 658 pp.

British Standards Institution (1999) BS 5930. *Code of Practice for Site Investigations.*

Burnett, A.D. and Epps, R.J. (1979) The engineering geological description of carbonate suite rocks and soils. *Ground Eng.* (March), 41–48.

Clark, A.R. and Walker, B.F. (1977) A proposed scheme for the classification and nomenclature in the engineering description of Middle Eastern sedimentary rocks. *Geotechnique*, **27**, 93–99.

Cook, A.W. (1997) *Geotechnical engineering concerns with the weak calcareous rocks of the lower Arabian Gulf.* Unpubl. MSc Dissertation, Dept. of Civil Engineering, University of Surrey.

Cook, A.W. (1999) The rate of drying out of calcarenite. In: *Engineering for Calcareous Sediments* (Ed. K.A. Al-Shafei), pp. 229–232. *Balkema*, Rotterdam.

Ebhardt, G. (1968) Experimental compaction of carbonate sediments. In: *Recent Developments in Carbonate Sedimentology in Central Europe* (Eds G. Muller and G.M. Friedman), pp. 58–65. Springer Verlag, Berlin.

Epps, R.J. (1980) Geotechnical practice and ground conditions in coastal regions of the United Arab Emirates. *Ground Eng.*, **13**, 19–25.

Evans, G., Murray, J.W., Biggs, H.E.J., Bates, R. and Bush, P. R. (1973) The oceanography, ecology, sedimentology and geomorphology of parts of the Trucial Coast barrier island complex, Persian Gulf. In: *The Persian Gulf: Holocene Carbonate Sedimentation and Diagenesis in a Shallow Epicontinental Sea* (Ed. B.H. Purser), pp. 223–278. Springer Verlag, New York.

Fookes, P.G. (1988) The geology of carbonate rocks and their engineering characterisation and description. In: *Engineering for Calcareous Sediments* (Eds R.J. Jewell and M.S. Khorshid), pp. 787–806. Balkema, Rotterdam.

Fookes, P.G. and Higginbottom, I.E. (1975) The classification and description of nearshore carbonate sediments for engineering purposes. *Geotechnique*, **25**, 406–11.

Fruth, L.S., Jr., Orme, G.R. and Donath, F.A. (1966) Experimental compaction effects in carbonate sediments. *J. Sed. Petrol.*, **36**, 747–754.

Kässler, P. (1973) The structural and geomorphic evolution of the Persian Gulf. In: *The Persian Gulf: Holocene Carbonate Sedimentation and Diagenesis in a Shallow Epicontinental Sea* (Ed. B.H. Purser), pp. 11–32. Springer Verlag, New York.

Maurenbrecher, P.M. and Van der Harst, M. (1989) The geotechnics of the coastal lowlands of the United Arab Emirates. *Proc. KNGMG Symposium 'Coastal Lowlands Geology and Geotechnology'*, 321–335.

Terzaghi, R.A.D. (1940) Compaction of lime mud as a cause of secondary structure. *J. Sed. Petrol.*, **10**, 78–90.

Valent, P.J. (1979) Coefficients of friction of calcareous sands and some building materials and their significance. *Technical Note No N-1542, Naval Construction Battalion Centre, Port Hueneme, California.*

Part 2

Geochemistry of recent carbonate and evaporite sediments

Int. Assoc. Sedimentol. Spec. Publ. (2011) **43**, 243–254

An historical odyssey: the origin of solutes in the coastal sabkha of Abu Dhabi, United Arab Emirates

WARREN W. WOOD

206 Natural Science building, Michigan State University, East Lansing, MI 48824, USA (E-mail: wwwood@msu.edu)

ABSTRACT

The quest for a model to provide an explanation of dolomite has been one of the major challenges of the geological community during the last half of the 20th century. Over the last 40 years the coastal sabkha of Abu Dhabi has been an important area of research to resolve the dolomite problem. Essentially the sabkha conceptual models dealt with several different mechanisms for transporting seawater through permeable media. As a result of recent detailed investigations of the hydrogeology of the coastal sabkha system, it is now possible to put these earlier models in perspective. The new data indicate that seawater is not transported through this system but rather the solutes come from ascending deep-basin brines while the water is derived from local rainfall. The amount of dolomite forming in the system is small; the Abu Dhabi sabkha does not offer a modern analogue for the massive dolomitization observed in the geological record.

Keywords: Sabkha, solutes, groundwater, mass balance, Abu Dhabi, United Arab Emirates.

INTRODUCTION

Massive dolomites are ubiquitous in the geological record since the Precambrian, yet only minor amounts are forming today and then only in unusual or restricted environments. As the operational paradigm of geology is "the present is the key to the past" and massive dolomitization has not been observed to be occurring today, this lack of association became defined as the "dolomite problem" (Fairbridge, 1957). Workers in the late 1950s and early 1960s recognized the widespread occurrence of sabkha sediments in the geological record and, after Wells (1962) discovered Holocene-age dolomite in the coastal sabkha of Qatar, it was hypothesized that this might be the elusive environment in which current dolomitization was occurring. It is generally recognized that most massive dolomites in the geological record are derived from altered limestones rather than direct precipitation of dolomite, thus any model of dolomitization must include a mechanism of transporting large masses of magnesium through a pre-existing limestone. It was also generally assumed that seawater was the most likely source of magnesium for dolomitization, and many different models have been

suggested for transporting seawater through the sedimentary framework.

The first model proposed for the coastal sabkha of Abu Dhabi was the concept of "seawater flooding" (Butler 1969; Kinsman 1969; Butler *et al.*, 1973; Patterson & Kinsman, 1977, 1981, 1982). In this model, seawater occasionally floods the low-lying supratidal zone of the coastal sabkha, and because of the low topographic gradient and rough polygonal surface texture, some of the floodwater remains on the surface. It was postulated that the retained seawater evaporated, anhydrite, calcite, and gypsum precipitated and the density of the remaining water increased. The dense water with a large Mg/Ca ratio, resulting from precipitation of calcium minerals, was postulated to sink through the sabkha converting calcite to dolomite and then discharge as groundwater to the sea. Explicit in this model was that both water and solutes were derived from seawater and that the process was confined to the distal-supratidal zone of the coastal sabkha.

As an increased number of solute samples became available from both the supratidal and the entire coastal sabkha area, it became apparent that solutes from continental sources were present in significant quantities throughout much of the sabkha, making it necessary to modify this model. The modified model, referred to as the "evaporative-pumping" model (Hsü & Siegenthaler, 1969; Hsü & Schneider, 1973; McKenzie *et al.,* 1980; Müller *et al.,* 1990), postulated that water was removed by evaporation from the sabkha surface leaving the concentrated solutes. Both seawater from the distal edge and continental water from the proximal edge of the coastal sabkha replaced the evaporated water. These two waters mixed giving the observed solute distribution. Implicit in this model was that both water and solutes were from these two sources and there was some mechanism to mix solutes from the two sources.

It became necessary to abandon both of these conceptual models when measurements indicated that the hydraulic conductivity and hydraulic gradients were so low that there was little horizontal transport through the coastal sabkha. A third model referred to as "ascending-brine" model (Sanford & Wood, 2001; Wood *et al.,* 2002; Wood & Sanford, 2002; Wood *et al.,* 2005) incorporated the observations that water in the coastal sabkha is derived largely from local rainfall on the sabkha surface, while the solutes are derived largely from

underlying continental brines discharging to the base of the sabkha. The seawater component of the solute load is derived from seawater trapped by the seaward progressing sabkha and not seawater transported through the sediments. Capillary forces bring water and solutes to the surface where water evaporates and the solutes precipitate on the surface as highly soluble Ca, Mg, Na, and K chloride and nitrate minerals antarcticite, carnallite, halite, sylvite and nitre (saltpetre). Rain dissolves these minerals and transports their solutes into the sabkha through recharge thus mixing the solutes. Retrograde minerals of relatively low solubility (gypsum, anhydrite, calcite, and dolomite) are precipitated in the capillary zone as the water moves from the cooler groundwater to the warmer surface. This model treated the entire UAE coastal sabkha as a single hydrological system.

HYDROGEOLOGICAL FRAMEWORK

The coastal sabkha in the Emirates of Abu Dhabi is exposed as a strip of sediments approximately 300 km long and 15 km wide (Fig. 1). Beneath

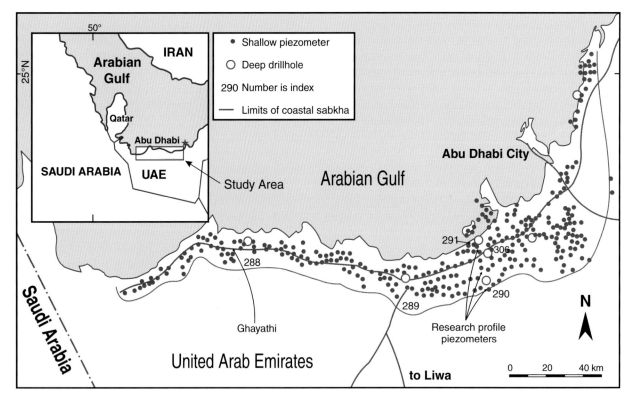

Fig. 1. Map showing the location of piezometers and wells on the coastal sabkha of the Emirates of Abu Dhabi (modified from Wood *et al.,* 2002).

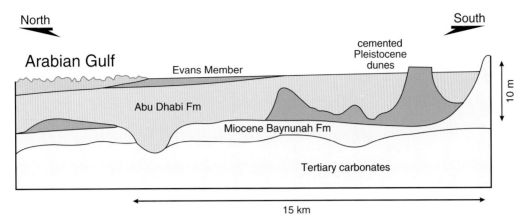

Fig. 2. Generalized geological cross-section of the coastal-sabkha aquifer system of Abu Dhabi (modified from Wood *et al.*, 2002).

the coastal sabkha lies a thick (6000–8000 m), Palaeozoic–Cenozoic-age, sedimentary sequence of carbonates and evaporates that accumulated on the stable Arabian Platform (Alsharhan & Nairn, 1997) (Fig. 2). Above this sequence is a siliciclastic deposit of later Miocene age derived from the rising mountains in western Saudi Arabia, Yemen and Oman, known by the descriptive name "Miocene Clastics Formation" (Alsharhan & Nairn, 1997), Fars, or the Baynunah Formation (Whybrow, 1989). In many areas of the coastal sabkha, the siliciclastics are approximately 30 m thick, but post-Miocene erosion makes the thickness highly variable, and the former has been completely removed in some areas at the western extent of the coastal sabkha.

Infrequent outcrops of the siliciclastic deposits exist near the proximal edges of the sabkha and are generally composed of fine quartz sand with carbonate cement, small numbers of quartzite pebbles, and occasional thin, discontinuous layers of gypsum. These sediments exhibit a reddish-pink Munsell colour (2.5YR-7/4). The siliciclastic deposits are overlain in some areas by carbonate-cemented sand dunes of Pleistocene age. The uppermost horizontal surfaces of several of these features have been dated at 26,000 yr BP by OSL (Optical Stimulated Luminescence; Wood *et al.*, 2002). Near the coast these remnants consist of 60–70% detrital carbonate grains with the reminder largely quartz indurated by carbonate cement. Their composition, however, becomes more quartz-rich inland from the coast (Hadley *et al.*, 1998). The dune remnants are yellowish-orange Munsell colour (10YR-6/4), exhibit well-developed foreset bedding, and do not contain beds of gypsum. Thus,

they are readily distinguishable in the field from outcrops of the siliciclastic deposits.

The Abu Dhabi Formation (Wood *et al.*, 2002), upon which the coastal sabkha deposits are formed, consists of Holocene age sediments deposited on the underlying Pleistocene and Tertiary sediments. The rise of sea level starting 18,000 yr BP reworked the Pleistocene dune sediments to create the Abu Dhabi Formation. The profile geometry of the Abu Dhabi Formation is variable, but it is generally wedge-shaped, starting from zero thickness at the proximal edge to approximately 15 m near the coast. The presence of palaeo-channels or erosional remnants of cemented siliciclastic deposits or Pleistocene dunes increases or decreases the thickness locally.

Gulf sea-level rose to approximately its present level by about 7000 yr BP (Evans, 1995). An OSL date of a quartz sample 1.5 m below the surface and 4.5 km inland from the present high tide line suggests that this sample was deposited 8000 yr BP (Wood *et al.*, 2002). Since the time of formation of the Abu Dhabi Formation, a 0–2 m thick, wedge-shaped sand deposit (Evans Member; Wood *et al.*, 2002) has been added to the distal portion of the coastal sabkha by progradation of the sediments approximately 6 km into the Gulf. Due to the low topographic gradient and abundant sand source, the progradation has been rapid at a rate approximately 1 km every thousand years (Evans, 1995). The coastal sabkha is topographically flat with a surface gradient of between 1 in 4000 and 1 in 5000 toward the Gulf.

The framework of the Abu Dhabi Formation below the water table consists of extremely uniform, fine sand (0.16–0.22 mm) composed of detrital carbonates (60%) and quartz (35%), with minor amounts of feldspars, anhydrite, and heavy minerals. There are no detectable authigenic evaporite or carbonate minerals below the water table. That is, the authigenic minerals have formed in the capillary zone and on the surface. The sand grains are covered with iron and manganese oxy-hydroxides, giving the sediment a light yellowish-orange Munsell colour (10YR-6/4). X-ray diffraction analysis of the grain coatings shows trace amounts of the clay minerals smectite, palygorskite, and illite. Several laboratory analyses of repacked sand yield a consistent porosity of 0.38 (38%). Because of the uniformity of the sand and its uncemented nature below the water table, it is believed that this value is also representative of the *in situ* porosity.

Retrograde authigenic minerals gypsum, anhydrite, calcite, and dolomite are confined largely to the capillary and unsaturated zone. Most of the authigenic anhydrite is massive and consists of clay-sized grains that give it a "toothpaste" consistency. "Chicken-wire" anhydrite nodules typically underlie this structureless anhydrite in many areas. Minor amounts of authigenic calcite and dolomite are intermixed with the anhydrite. Equidimensional crystals of gypsum up to 0.5 cm in length exist close to the water table. Rubbery mats of cyanobacteria (blue-green algae), typically 10 mm thick, cover the surface of the intertidal zone in many areas. Algal remnants are observed in the prograded sediments of the Evans Member, but not in the Abu Dhabi Formation. The polygonal surface features and thickness of the salt crust tend to be more strongly developed in the western area of the coastal sabkha that receives less rainfall.

Water fluxes

The general topographic slope of the sabkha is toward the coast and is controlled by the dynamic equilibrium among the aeolian process, groundwater elevation, and sediment supply. The water table typically lies 0.5–1 m below the coastal sabkha. Lateral groundwater flux through the sabkha-aquifer is calculated using Darcy's Law (Equation 1):

$$q = -k(dh/dl) \tag{1}$$

where q is the water flux in dimensions of L/T, k is the hydraulic conductivity of the porous medium in consistent units of L/T, and dh/dl is the gradient of the hydraulic head and is dimensionless (L is

length, and T is time). Using a hydraulic conductivity of $1 \, m \, day^{-1}$, and a gradient of 0.0002 (Sanford & Wood, 2001) in Equation 1, results in a lateral Darcy flux between 0.07 and $0.08 \, m \, yr^{-1}$. Dividing the flux by porosity (0.38) results in a seepage velocity of 0.18 to $0.21 \, m \, yr^{-1}$. That is, water and solutes moves laterally approximately 1 metre in 5 years.

The low topographic elevation means that the coastal sabkha acts as a regional groundwater discharge area for deeper Palaeozoic through Cenozoic formations, and hydraulic heads in the six deep piezometers finished in the Tertiary age carbonate and evaporite formations average about 10 m above land surface in the coastal areas. The hydraulic head for this regional flow system is assumed to be the Asir Mountains in western Saudi Arabia and Yemen, and the Hajar Mountains in Oman, where elevations exceed 3000 m. Upward vertical flux from the underlying formations is approximately $4 \, mm \, yr^{-1}$, determined from Equation 1 with a measured vertical gradient of 0.1 and vertical hydraulic conductivity of approximately $1 \times 10^{-4} \, m \, day^{-1}$ (Sanford & Wood, 2001). An almost identical flux has been calculated from solute concentration data (Wood *et al.*, 2002)

Potential evaporation from a free freshwater surface in the area is $3800 \, mm \, yr^{-1}$ potential (Bottomley, 1996). Actual evaporation, however, from the surface is known to be considerably less than this value. Evaporation measurements from the coastal-sabkha surface were made using a humidity chamber and a technique described by Stannard (1988). Evaporative flux (there is no vegetation and thus, no transpiration) at a site with $70 \, mm \, yr^{-1}$ rainfall was found to be $0.24 \, mm \, day^{-1}$ ($68 \, mm \, yr^{-1}$), averaged over four seasons. A second site where rainfall is $50 \, mm \, yr^{-1}$ was found to average $0.14 \, mm \, day^{-1}$ ($51 \, mm \, yr^{-1}$; Sanford & Wood, 2001). Thus, the evaporation rates were less than 3% of freshwater pan evaporation rates and surprisingly close to the precipitation values. That is, to maintain a steady-state system, recharge must equal discharge, so it can be concluded that on average the amount of rainfall is approximately equal to recharge. Typically for an arid area recharge is less than 2–3% of rainfall, thus special conditions exist in the sabkha environment, specifically a: (1) decrease in the thermodynamic activity of water by the large concentration of solutes, preventing evaporation; (2) low topographic gradient restricting runoff; and (3) wetted

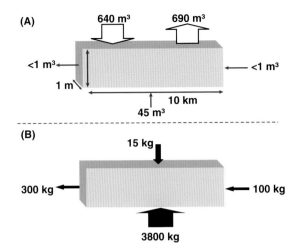

Fig. 3. Illustration of annual water and solute fluxes. (A) Water and (B) Solute fluxes in a control volume 1 m × 10 m × 10 km, representative of the coastal-sabkha aquifer (modified from Wood *et al.*, 2002).

capillary surface near the surface allowing rapid recharge (Wood *et al.*, 2002).

Inserting these annual water fluxes together into a representative rectilinear volume 1 m wide, 10 m deep, and 10 km long results in the following values: $<1\,m^3\,yr^{-1}$ enters and exits by lateral groundwater flow; $45\,m^3\,yr^{-1}$ enters by upward vertical leakage; $640\,m^3\,yr^{-1}$ enter by recharge of rainfall; and $690\,m^3\,yr^{-1}$ is lost to evaporation (Fig. 3A).

The average steady-state residence time for water in the same rectilinear volume is approximately 50 years, calculated using the following expression:

$$R_s - M_t/M_f, \qquad (2)$$

where, R_s is the average residence time of the water (T) in the sabkha-aquifer, M_t is the total mass of water in a representative rectilinear volume, and M_f is net mass flux to the representative rectilinear volume (M/T; M is mass, and T is time). This residence time is consistent with the observed tritium present throughout the vertical profile (McKenzie *et al.*, 1980; table 1 in Wood *et al.*, 2002).

Solute fluxes and chemistry

Four hundred and fifty shallow (less than 5 m) piezometers (details of construction are given in Wood *et al.*, 2002) were installed to collect samples for solute analyses, make water-level measurements and hydrological slug tests. Utilizing the water

fluxes above and multiplying them by average solute concentration for each source (table 2 in Wood *et al.*, 2002) results in a solute-mass flux for the same representative rectilinear volume. Approximately $100\,kg\,yr^{-1}$ enter from lateral flow; approximately $15\,kg\,yr^{-1}$ enter by rainfall; approximately $3700\,kg\,yr^{-1}$ enter by ascending vertical leakage; and approximately $200\,kg\,yr^{-1}$ are discharged to the Gulf by lateral flow (Fig. 3B).

Using the seepage velocity of 0.18 to $0.21\,m\,yr^{-1}$, it can be calculated that it requires between 47,600 and 55,500 years for a molecule of a conservative solute to traverse the 10-km width of the coastal sabkha. The average steady-state residence time of a conservative solute in this rectilinear volume is approximately 26,000 years, assuming that the average residence time is one-half of the total time required for a conservative solute to traverse the full length (10 km) of the rectilinear volume by horizontal flow. The calculated residence time is therefore greater than the age of the coastal sabkha, so it is clear that steady state for the solutes has not been achieved and less than 0.1 pore volume (Equation 3) has been discharge since the aquifer formed 7000 years ago. A pore volume is defined as:

$$PV = v\theta \qquad (3)$$

where PV is the volume of fluid in the interstitial pores of a volume of saturated aquifer, v is the volume of saturated aquifer (L^3) and θ is the porosity (dimensionless).

The flux data and residence time are compelling arguments for the majority of the solutes being derived from the underlying brines, but it must be shown that the observed solute and isotope chemistry in the sabkha-aquifer is consistent with this model. One approach to this requirement is to compare the concentrations of conservative solutes calculated from the hydrological data with the analogous concentration ratio (CR) calculated from the observed solute data. Vertical gradients and hydraulic conductivity suggests that approximately 7 pore volumes of water from the underlying Tertiary formations have discharged to the original coastal sabkha-aquifer in the 7000 years since its deposition (Sanford & Wood, 2001). If there are no other sources or sinks of solutes, then the concentration of a conservative solute should be approximately seven times the concentration of the input Tertiary brine, as there is little or no solute discharged to the Gulf. The concentration ratio for an individual ion for a closed system originally filled with seawater is calculated by use of the following relation:

$$CR = (C_o - C_s)/C_i \qquad (4)$$

where CR is the concentration ratio of the ion from solute data (dimensionless), C_o is the average observed concentration of the ion in the sabkha-aquifer ($M L^{-3}$), C_s is the concentration of the ion in seawater ($M L^{-3}$), and C_i is the concentration of the ion in the input solution ($M L^{-3}$).

Assuming that magnesium is conservative, it would be expected to have a CR value close to the seven predicted by the hydrological analysis, and this realization is achieved (Fig. 4). The loss of halite from the solution is clearly shown by the low CR values of sodium and chloride. Sulphate and calcium concentrations are less than seven because of the precipitation of calcite, anhydrite, and gypsum in the capillary zone. Magnesium is believed to be generally conservative in this system because there is no significant mass of clay that could act as an ion exchanger or structural sink for the mineral and the amount of dolomite is insignificant relative to the total magnesium mass. The thermodynamic saturation index, calculated using the code PHRQPITZ (Plummer et al., 1988), for the least soluble magnesium chloride mineral, carnallite, is <1, so it is unlikely that magnesium is removed by mineral precipitation of chloride minerals within the aquifer. The chemical conditions do not favour dolomitization (low concentrations of bicarbonate, low pH, and magnesium/calcium ratio of approximately 1:1 (table 2 in Wood et al., 2002). Thus, it does not appear that

a significant mass of magnesium is lost or gained from sources other than the ascending brine, and the CR is approximately seven (Fig. 4).

The relative concentrations of calcium and sulphate can also be used to define the source of the solutes. It is known from the chemical divide concept (Hardie & Eugster, 1970) that mineral precipitation in a closed system will cause the concentration of one element of a precipitating mineral to approach zero, while the other element or elements will increase in concentration. The initial input concentration of the elements determines which element will be limiting. In the case of seawater, there is a greater equivalent of sulphate than calcium; thus, with precipitation of calcite, anhydrite, and gypsum, the thermodynamic activity of calcium will approach zero. In the case of the Tertiary brines, the average equivalents of calcium are greater than the average equivalents of carbonate and sulphate; thus, precipitation of anhydrite and calcite will result in sulphate and carbonate approaching zero, while calcium increases in concentration. In this system, the calcium is observed to be increasing, while sulphate approaches zero in most cases (Fig. 6 in Wood et al., 2002), consistent with a Tertiary brine source. The few analyses of sulphate that are greater than calcium are from the supratidal zone. The concentration of sulphate does not go to zero because of complexes formed with magnesium and calcium. However, the thermodynamic activity does approach zero. This analysis assumes that no other process is removing sulphate from solution. The groundwater system is oxidizing, with dissolved oxygen values between 4 and 5 mg L^{-1}, except at the distal edge of the coastal sabkha in the presence of algal mats. Therefore, it is not likely that sulphate reduction is removing a substantial mass of sulphate to significantly alter the Ca/SO$_4$ ratio.

There appears to be a significant total loss of halite from the system, as indicated by the chemical evolution of sodium and chloride (Fig. 4). Mass-balance calculations from the 7 pore volumes of ascending brine quantify this loss and indicate that to account for the mass of chloride and sodium input to this system, the surface halite crust should be nearly 30 cm thick. The crust is generally less than 1 cm. The mechanism or combination of mechanisms by which this halite is lost is not known. Halite may be removed from the surface by aeolian processes or transported to the Gulf by rain or flooding of the Gulf waters. The salt tends to be firmly cemented, forming "ice-like" masses

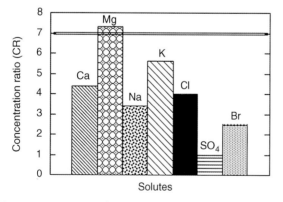

Fig. 4. Concentration factor (ratio of existing concentration in the sabkha-aquifer to input concentration of Tertiary brines) of the coastal-sabkhas aquifer. Line at a concentration factor of 7 is that calculated from physical hydrological information (modified from Wood et al., 2002).

rather than occurring as loose grains, and so does not appear to be sufficiently mobile for aeolian transport. Aeolian samples collected at two locations on the coastal sabkha, however, were not conclusive.

It is speculated that occasional major rainfall events dissolve the surface salts and transport them to the Gulf. It is known that approximately 200 mm of rain occurred in a 24 hour period at least once in the 30 years for which records exist (Bottomley, 1996). In the rainfall scenario, recharge preferentially selects the most soluble salts and returns them to the aquifer before they are transported to the Gulf. This selective process is aided by the geometry of the polygonal surface features that tend to trap the first rainfall, forming millions of disconnected pools. This trapped water becomes thermodynamically saturated with the most soluble minerals. Transport of this water into the small volume of the unsaturated zone quickly fills the available space, preventing less soluble salts from being recharged until later in the event. Only when the rain exceeds a certain flux do the pools coalesce and flow toward the Gulf. Thus, it is possible to selectively lose significant amounts of halite without the loss of associated, more soluble salts on the crusts. It seems unlikely that the salt is removed by flooding of Gulf water, where the flood water dissolves the salts and returns them to the Gulf, as halite is missing over the entire area, not just seaward of the strand line representing maximum flooding conditions.

In a closed system, precipitated anhydrite will have a sulphur stable-isotope ($\delta^{34}S$) signature that is isotopically heavier by about 1.8 ‰ than the initial solution, and the remaining sulphate in solution will be isotopically lighter by 1.8 ‰ (Thode & Monster, 1965). Seawater has a $\delta^{34}S$ value of approximately 20 ‰ relative to VCDT (Hoefs, 1973), while that of the Tertiary brines averages 18.5 ‰ (table 2 in Wood *et al.*, 2002). If the sulphate were derived from seawater, one would expect an anhydrite value of approximately 21.8 ‰ and the remaining dissolved sulfate to be 18.2 ‰. Conversely, anhydrite derived from Tertiary brines would be approximately 20 ‰ and the resulting dissolved sulphate would be approximately 16.7 ‰. The observed value for the anhydrite is 19.4 ‰, while that of the sulphate ion averages 15.8 ‰ (table 2 in Wood *et al.*, 2002). Thus, sulphur isotope values, while slightly lighter than predicted, probably due to input of some isotopically lighter groundwater (15.9 ‰, table 2 Wood *et al.*, 2002), are

consistent with an origin of sulphate from the ascending brines. Butler *et al.* (1973) sampled the supratidal zone, and their isotopically heavier values reflected seawater trapped by prograding sediments rather than continental brines.

Mixing within the sabkha and aquifer

The density of 10 samples of surface water ponded on the coastal sabkha averaged 1.2675 g cm^{-3} (table 4 in Wood *et al.*, 2002). The high density and solute concentration of this surface water is largely a result of dissolution of soluble effluorescent surface crusts (table 5 in Wood *et al.*, 2002). The average density of 161 samples from shallow (~2 m below the water table) sabkha-aquifer averaged 1.1858 g cm^{-3} (table 2 in Wood *et al.*, 2002). This contrast in density suggests that when surface water is recharged, it will descend to the bottom of the aquifer. The observation that tritium exists throughout the vertical profile (McKenzie *et al.*, 1980; table 1 in Wood *et al.*, 2002) is consistent with vertical circulation resulting from this density instability.

Analysis of 12 sets of paired piezometers provided verification of this process (table 6 in Wood *et al.*, 2002). In these pairs, one piezometer is placed within 1 m of the water table, while the other is placed within 1 m of the base of the aquifer. The ratios (deep to shallow) of calcium, magnesium, potassium, chloride and bromide average greater than one, consistent with a density convection model. Sulphate and bicarbonate are the only ions significantly below the value of one; this is readily explained by their loss in anhydrite, gypsum, and calcite in the unsaturated zone before reaching the surface. The sodium and pH ratios are within the margin of error.

Discussion of seawater models

Considering the information on groundwater hydrology and geochemistry presented above, it is possible to evaluate the proposed transport models. The seawater-flooding model suggests that the water and solutes are derived from the Gulf. The water balance calculated above, however, can account for the mass of water transported through the system without input from the Gulf. If ocean flooding transports water to the sabkha-aquifer, the water flux must be small and within the error of the difference between rainfall, recharge, and evaporation, all of which agree reasonably well.

It is also not clear that the density proposed by evaporating seawater would be sufficient to have the water sink through the brine in the aquifer. The above analysis also suggests that the combination of the hydraulic conductivity and the hydraulic gradient is so low that less than 0.1 pore volume has moved laterally through the system since it was formed 7000 yr BP. Thus, it is not possible to transport significant amounts of water or solute through this system in the time available. The solutes and isotopes are also consistent with a continental brine source rather than a seawater source.

The evaporative-pumping model was based on laboratory experiments (Hsü & Siegenthaler, 1969) that violated consideration of dimensional and dynamic similitude. That is, it is necessary to scale the hydraulic conductivity to correspond to the evaporative stress applied. Thus, the results of these experiments are incorrect. Simple mass-balance calculations clearly show that this model is physically inconsistent with the observations. If the volume of water required to replace the quoted evaporation were composed of seawater or local groundwater, as proposed, the entire aquifer would have become a block of salt within a few hundred years! Additionally, and perhaps more importantly by failing to adjust the observed water levels for density, barometric pressure changes, and earth and ocean tides, the direction of the groundwater gradient (McKenzie *et al.*, 1980) was incorrect. Flow is toward the Gulf not toward the continent. McKenzie *et al.* (1980) also over-estimated by nearly an order of magnitude the amount of water lost annually by evaporation from the coastal sabkha. As in the case of seawater flooding, the hydraulic conductivity and gradient are so low that it is physically impossible to move the required mass of water and solutes in the time available.

In the ascending brine model, most of the solutes are derived from ascending continental brines not seawater, while the water is provided from local rainfall (Fig. 10c in Wood *et al.*, 2002). When the sabkha-aquifer was formed from sand dunes 7000–8000 yr BP, it contained seawater in the interstitial pores and was in dynamic equilibrium with the water table, aeolian processes, and sediment supply. Brines from the underlying Tertiary formations immediately began discharging water and solutes to the newly formed sabkha-aquifer. Water and solutes were brought to the surface, precipitating retrograde minerals such as anhydrite, calcite, and gypsum in the capillary zone, and soluble salts on the coastal sabkha. Annual rainfall selectively dissolved these soluble salts and returned them to the aquifer. This recharge solution is denser than the water in the sabkha-aquifer and sinks to the bottom of the aquifer. These returned solutes are added to the input of solutes from the Tertiary brine and result in an increase of solutes with time and the accumulation of less soluble anhydrite, calcite, and gypsum in the capillary zone. Over time, solute ratios and concentrations of the original seawater have been modified by the addition of ascending brines. This coastal sabkha-aquifer system has not yet reached chemical equilibrium and will not for many thousands of years under the present climatic, hydraulic, and sea-level conditions. That is, it is the lack of lateral transport of solutes out of the system that permits solutes to accumulate in the coastal-sabkha aquifer over time.

The "Chloride Plateau" (Patterson & Kinsman, 1982) is an area of approximately equal chloride concentration, declining at both the distal and proximal edges. Water of relatively low total dissolved solids enters the aquifer by lateral flow at the proximal boundary, and entrapment of relatively dilute seawater at the distal boundary by sabkha progradation, explains the lower solute concentrations at the end of the flow path through the sabkha-aquifer. The plateau represents the area of the original sabkha-aquifer modified by contributions from the underlying continental brine. The prograded area has not had time to experience the same increase in dissolved solids. The "mixing zone" described by previous researchers results from the addition of continental brine from below to the trapped interstitial seawater; it is a vertical rather than the implied horizontal mixing phenomena.

Müller *et al.* (1990) observed a seaward isotopic gradient of strontium perpendicular to the coast and implied that it was a result of lateral mixing of continental brines and seawater. The observed mixing is readily explained as a function of the number of pore volumes of continental brine contributed to a specific area. That is, the distance from the shoreline and the number of pore volumes of Tertiary brine recharged to the sabkha-aquifer are directly proportional in the distal portion of the coastal sabkha. Thus the solutes have a more continental appearance as one moves toward the proximal edge. It is not necessary to provide lateral mixing or transport to obtain the observed strontium isotope distribution.

The area of dolomitization closest to the shore identified by Patterson & Kinsman (1982) has the largest amount of seawater solutes because it is the area most recently prograded. The mass of dolomite formed is limited to the magnesium and carbonate concentration in the initial pore volume of seawater trapped by the prograding sediments. The actual evaporative geochemical mechanism for dolomite formation proposed by Patterson & Kinsman (1982) appears to be correct. There is no significant seawater transport occurring in this system.

SUMMARY AND CONCLUSIONS

Early researchers made critical observations on the distribution of solutes and provided challenging conceptual models. Current geochemical and hydrological evidence suggests, however, that a large fraction of solute mass in the sabkha-aquifer of Abu Dhabi is derived from ascending continental brines rather than transport of seawater through the sediments. The seawater solutes observed in the supratidal zone are the result of trapped seawater in the prograding sediments rather than transport through the system. There is relatively little dolomite formed in this system and the likelihood of this as a model for massive dolomitization is nil. The current findings do not preclude other seawater transport models or transport on a larger time-scale. The study does illustrate, however, the importance of quantitatively evaluating both the physical hydrology and solute and isotope chemistry in the context of a transient hydrogeological system.

Sabkha deposits in the geological record are used to interpret the environmental conditions of deposition. Implicit in this use is the assumption that the solute system is chemically closed, that is, the authigenic minerals represent the composition of the fluids in their environment of origin. Thermodynamic and mass-balance calculations based on measurements of water and solute flux of the contemporary Abu Dhabi coastal sabkha system, however, demonstrate that the system is open for sodium and chloride, where nearly half of the input is lost, but closed for sulphur, where nearly 100% is retained (Wood *et al.*, 2005). Sulphur and chloride isotopes were consistent with this observation. If these coastal-sabkha deposits were preserved in the geological record, they would suggest a solute environment rich in sulphate and poor in chloride; yet, the reverse is true. In most coastal-sabkha environments, capillary forces bring solutes and water to the surface, where the water evaporates and halite, carnallite, sylvite, and other soluble minerals are precipitated. Retrograde minerals such as anhydrite, calcite, dolomite, and gypsum, however, precipitate and accumulate in the capillary zone beneath the surface of the coastal sabkha. Because they possess relatively low solubility and are below the surface, these retrograde minerals are protected from dissolution and physical erosion occurring from infrequent but intense rainfall events. Thus, they are more likely to be preserved in the geological record than highly soluble minerals formed on the surface.

ACKNOWLEDGEMENTS

I thank Dr. Abdulrahman S. Alsharhan for organizing this symposium with financial support from Ahmed Sager Al Suwaidi and ADNOC. I wish to also thank Mohamed B. Al-Qubaisi, ADNOC, for his interest and support of the sabkha research; David Clark, John Czarnecki, and Dennis Woodward, US Geological Survey (USGS/NDC), for supervising the installation of wells and piezometers, and collecting the water samples; and Terry Councell, USGS, for chemical analyses of samples. Special thanks go to Ward Sanford of the US Geological Survey without whose insight and many discussions, the analysis would not have been possible. Work on this project was done while the author was with the US Geological Survey.

REFERENCES

Alsharhan, A.S. and Nairn, A.E.M. (1997) *Sedimentary Basins and Petroleum Geology of the Middle East.* Elsevier, Amsterdam, 843 pp.

Bottomley, N. (1996) Recent climate in Abu Dhabi, In: *Desert Ecology of Abu Dhabi* (Ed. P.E. Osborne), pp. 36–49. Pisces Publications, Newbury, UK.

Butler, G.P. (1969) Modern evaporite deposition and geochemistry of coexisting brines, the sabkha, Trucial Coast, Arabian Gulf. *J. Sed. Petrol.*, **39**, 70–81.

Butler, G.P., Krouse, R.H. and Mitchell, R. (1973) Sulphur-isotope geochemistry of an arid, supratidal evaporite environment, Trucial Coast. In: *The Persian Gulf: Holocene Carbonate Sedimentation and Diagenesis in a Shallow Epicontinental Sea* (Ed. B.H. Purser), pp 453–467. Springer-Verlag, New York.

Evans, G. (1995) The Arabian Gulf: a modern carbonate-evaporite factory; a review. *Cuad. Geol. Ibérica*, **19**, 61–95.

Fairbridge, R.W. (1957) The dolomite question, In: *Regional Aspects of Carbonate Deposition* (Eds R. J. Le Blanc and J. G. Breeding). *SEPM Spec. Publ.*, **5**, 125–178.

Hadley, D.G., Brouwers, E.M. and **Bown, T.M.** (1998) Quaternary paleodunes, Arabian Gulf coast, Abu Dhabi Emirates: age and paleoenvironment evolution. In: *Quaternary Deserts and Climate Change* (Eds A.S. Alsharhan, K.W. Glennie, G.L. Whittle and C.G.St.C. Kendall), pp. 123–140. A.A. Balkema, Rotterdam.

Hardie, L.A. and **Eugster, H.P.** (1970) The evolution of closed-basin brines. *Min. Soc. Am. Spec. Publ.*, **3**, 273–290.

Hoefs, J. (1973). *Stable Isotope Geochemistry*. Springer-Verlag, New York, 140 pp.

Hsü, K.J. and **Siegenthaler, C.** (1969) Preliminary experiments on hydrodynamic movement induced by evaporation and their bearing on the dolomite problem. *Sedimentology*, **12**, 11–25.

Hsü, K.J. and **Schneider, J.** (1973) Progress report on dolomitization – hydrology of Abu Dhabi sabkhas, Arabian Gulf. In: *The Persian Gulf: Holocene Carbonate Sedimentation and Diagenesis in a Shallow Epicontinental Sea* (Ed. B.H. Purser), pp. 409–422. Springer-Verlag, New York.

Kinsman, D.J.J. (1969) Modes of formation, sedimentary associations, and diagnostic features of shallow-water and supratidal evaporites. *AAPG Bull.*, **53**, 830–840.

McKenzie, J.A., Hsü, K.J. and **Schneider, J.F.** (1980) Movement of subsurface waters under the sabkha, Abu Dhabi, UAE, and its relation to evaporative dolomite genesis. *SEPM Spec. Publ.*, **28**, 11–30.

Müller, D.W., McKenzie, J.A. and **Mueller, P.A.** (1990) Abu Dhabi sabkha, Persian Gulf, revisited: application of strontium isotopes to test an early dolomitization model. *Geology*, **18**, 618–621.

Patterson, R.J. and **Kinsman, D.J.J.** (1977) Marine and continental groundwater sources in a Persian Gulf coastal sabkha. *Stud. Geol.*, **4**, 381–397.

Patterson, R.J. and **Kinsman, D.J.J.** (1981) Hydrologic framework of a sabkha along Arabian Gulf. *AAPG Bull.*, **65**, 1457–1475.

Patterson, R.J. and **Kinsman, D.J.J.** (1982) Formation of diagenetic dolomite in coastal sabkha along Arabian (Persian) Gulf. *AAPG Bull.*, **66**, 28–43.

Plummer, L.N., Parkhurst, D.L., Fleming, G.W. and **Dunkle, S.A.** (1988) A computer program incorporating Pitzer's equations for calculation of geochemical reactions in brines. *US Geol. Surv. Water-Resour. Invest. Rep.*, **884153**, 310 pp.

Sanford, W.E. and **Wood, W.W.** (2001) Hydrology of the coastal sabkhas of Abu Dhabi, *United Arab Emirates*. *Hydrogeol. J.*, **9**, 358–366.

Stannard, D.I. (1988) Use of a hemispherical chamber for measurement of evapotranspiration. *US Geol. Surv. Open-File Rep.*, **88452**, 18 pp.

Thode, H.G. and **Monster, J.** (1965) Sulfur-isotope geochemistry of petroleum, evaporites, and ancient seas. *AAPG Mem.*, **4**, 367–377.

Wells, A.J. (1962) Recent dolomite in the Persian Gulf. *Nature*, **194**, 274–275.

Whybrow, P.J. (1989) New stratotype; the Baynunah Formation (Late Miocene), United Arab Emirates: lithology and paleontology. *Newsl. Stratigr.*, **21**, 1–9.

Wood, W.W. and **Sanford, W.E.** (2002) Hydrogeology and solute chemistry of the coastal-sabkha aquifer in the Emirate of Abu Dhabi. In: *Sabkha Ecosystem, 1: The Sabkhas of the Arabian Peninsula and Adjacent Countries* (Eds H.-J. Barth and B. Boer), Kluwer Academic Publishers, Dordrecht, 354 pp.

Wood, W.W., Sanford, W.E. and **Al Habschi, A.R.S.** (2002) The source of solutes in the coastal sabkha of Abu Dhabi. *Geol. Soc. Am. Bull.*, **114**, 259–268.

Wood, W.W., Sanford, W.E. and **Frape, S.** (2005) Chemical openness and potential for misinterpretation of the solute environment of coastal sabkhat. *Chem. Geol.*, **215**, 361–372.

Int. Assoc. Sedimentol. Spec. Publ. (2011) **43**, 255–264

Geochemistry and nature of organic matter of the Pleistocene–Holocene carbonate-evaporite sediments of Al-Khiran, Southeastern Kuwait

SUAD QABAZARD, FOWZIA H. ABDULLAH and ALI AL-TEMEEMI

Department of Earth and Environmental Sciences, Faculty of Sciences, Kuwait University, Kuwait,
(E-mail: fozabd20008@gmail.com)

ABSTRACT

The complex of Pleistocene–Holocene sediments in the Al-Khiran area of Kuwait represents a petroleum basin analogue where organic-rich sediments in lagoons are closely associated with porous potential reservoir facies of oolitic beach ridges. The petrography, mineralogy, elemental and organic geochemistry of the carbonate-evaporite successions exposed along the southern coast of Kuwait have been analyzed. The physical and chemical conditions of the depositional environments and the diagenetic history of the sediments have been evaluated.

Analysis of sediment grain-size distributions and parameters indicates that the coastal areas near their connection to the open Gulf and at the mouths of tidal creeks are dominated by coarse-grained, moderately-sorted carbonate beach sands, with low amounts of total organic matter (TOC = 0.3 wt%). Landward, at the ends of the creeks, the concentration of ooids decreases, and silt-clayey sediments composed mainly of pellets and calcareous mud interbedded with algal mats are found. This low-energy setting reveals a higher TOC content (0.9 wt%) than the sediments deposited at the mouths of the tidal creeks, indicating petroleum source-rock potential for the tidal creek facies.

Pleistocene oolitic limestones consist of elongate, cross-bedded, thinly-laminated ridges of carbonate sand, oriented parallel to the strong tidal currents. The sediments are composed entirely of well-sorted, coarse ooids (0.5 mm diameter). Study of the diagenetic processes and porosity in these rocks shows that they exhibit excellent reservoir potential as a result of prolonged exposure to freshwater leaching.

Low amounts of trace metals, low organic-matter contents, and relatively high O/C atomic ratios of the organic matter in the bioturbated beach sediments and ridges, indicate that they were laid down under highly oxygenated open-marine conditions, where current and wave action contributed to the destruction of the organic matter. However, sediments deposited at the ends of the creeks show relatively higher trace-metal and higher TOC contents. The physical and chemical conditions in the creeks allow the preservation of amorphous algal marine-type organic matter, enhancing the source-rock potential of the sediments.

Keywords: Organic-rich source rocks, oolite ridges, tidal creeks, beach sands, algal mats, tidal currents, trace metals.

INTRODUCTION

The role of carbonate sediments, especially oolites, in petroleum generation has provoked much discussion in the last decade. It has been emphasized that carbonate sediments may yield valuable amounts of organic matter and, thus serve as petroleum source beds (Hunt, 1967). Oolitic sediments, in particular, have been considered to be the source of oil accumulated in some limestone reservoirs (Ferguson, 1988). This became generally accepted, given that ooids are internally composed of layers of organic mucilage alternating with concentric carbonate-rich laminae (Shearman, 1965; Shearman *et al.*, 1970). A number of studies related to the identification of the amount and character of organic matter in ooids have been made in the ancient carbonate sequences of the Arabian Gulf Basin (Ibe, 1984, 1985; Abdullah & Kinghorn, 1996). Other geochemical investigations have concentrated on Holocene carbonate sediments as modern analogues for understanding the nature and preservation of organic matter in their ancient equivalents (Ferguson & Ibe, 1982; Ibe, 1983; Ferguson, 1987; Kenig *et al.*, 1990).

The Al-Khiran area, on the southern coast of Kuwait, is a site of extensive carbonate sedimentation and evaporite precipitation. The area has been the subject of detailed sedimentological and geomorphological investigations (Saleh, 1975; Picha & Saleh, 1977; Khalaf *et al.*, 1984; Al-Sarawi *et al.*, 1993; Cherif *et al.*, 1994; Duane & Al-Zamel, 1999). The present work is the first attempt to evaluate and identify the organic-matter type and richness in these Holocene carbonate sediments. The aim is to provide geochemical data useful for investigating carbonate settings and to better understand the processes of accumulation and preservation of organic matter in the carbonate sediments of the area.

Area of study and stratigraphy

Kuwait is located at the northwestern corner of the Arabian Gulf, where it forms an important carbonate-depositional setting (Fig. 1). The carbonate surface sediments of Kuwait constitute part of the upper member of the Pleistocene Dibdibha Formation. This consists of fluviatile sands and gravels, siltstones, calcareous sandstones, oolitic limestones and evaporites. Overlying Holocene sediments include coastal and inland sabkha deposits, and alluvial and aeolian sands (Picha & Saleh, 1977; Alsharhan & Nairn, 1997). The Quaternary sediments are underlain by Miocene–Pliocene siliciclastic sediments of the Kuwait Group and carbonate–evaporite sediments of the Hasa Group (Eocene–Palaeocene; Fig. 2).

The Quaternary carbonate-evaporite surface sediments are well represented in the Al-Khiran area where, on the southern coast of Kuwait (Fig. 3), the contribution of terrigenous materials from the Tigris-Euphrates delta has significantly decreased. The sediments are here mainly represented by calcareous and oolitic sands intercalated with sediments of the coastal sabkha (Khalaf *et al.*, 1984). The climate of the area is arid, with an average rainfall of 112 mm yr^{-1}. Holocene–Pleistocene oolitic sands are found along the coastline and are represented by three elongated, subaerially cemented coastal ridges that have elevations that range between 5 to 15 m above sea level (Al-Sarawi *et al.*, 1993). These include: (1) ridges of younger oolitic limestones; (2) ridges of older oolitic limestones; and (3) bodies of oolitic-quartzose sandstone that constitute the oldest ridges in the area (Fig. 3).

The low areas between the coastal ridges form sabkhas that have accumulated from organic-rich marine and wind-blown sandy and argillaceous sediment, in addition to gypsum and evaporite minerals (Saleh, 1975). The processes that formed the ridges were partly marine and partly aeolian, and these have been controlled by local climate change and the effects of tectonic and eustatic sea-level changes (glacial–interglacial cycles) that have prevailed during the Quaternary (Picha & Saleh, 1977; Al-Sarawi *et al.*, 1993). Traced from north to south, the Quaternary complex is cut by relatively large and wide tidal creeks, the Khor Al-A'ama, Khor Al-Mufateh and Khor Al-Mamlaha. The Khor Al-A'ama forms a broad tidal creek with a typical width of about 1 km, but having a narrower channel where it connects to the open sea. Both the Khor Al-Mufateh and Khor Al-Mamlaha have average widths of about 0.75 km and extend 5.5 km inland. The amplitude of the tidal movements in the area is about 1.5 m near the mouth of the creeks, and decreases gradually inland (Cherif *et al.*, 1994).

METHODS OF STUDY

Surface sediment samples (1–3 cm depth) of the Holocene oolitic deposits were collected along the Al-Khiran coast. Both surface-exposure and bore-hole samples were obtained from the Pleistocene older oolitic limestone ridge and the oldest oolitic-quartzose ridges (Fig. 3). The grain-size distributions of the Holocene sediments were determined using standard sieving techniques (Folk & Ward, 1957). The Quaternary sediments were examined under binocular and transmitted light microscope. A JSM-6300 scanning electron microscope (SEM) was used for more detailed petrographical analysis.

Inorganic geochemical analyses were employed to determine the mineralogy and chemistry of the sediments. Mineral identification and quantification were verified using a Siemens D500 X-ray diffractometer (XRD) and an energy-dispersive spectrometer (EDS) attached to a JSM-6300 SEM. The chemical composition of the sediments was investigated by determining selected minor-element (Mg, Sr, Fe, Mn) contents to complement the

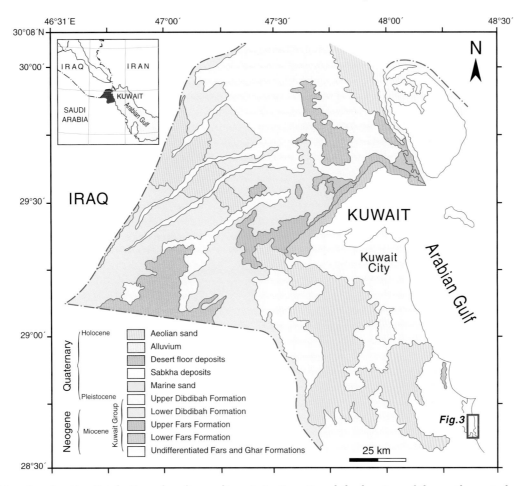

Fig. 1. Map showing the distribution of surface sediments in Kuwait and the location of the southern study area (after Al-Sulaimi & El-Rabaa, 1994).

mineralogical data, as well as certain trace metals (Cd, Co, Cr, Cu, Mo, Ni, V, Zn) using Inductively Coupled Plasma – Optical Emission Spectrometry (ICP-OES).

Standard organic geochemical analysis techniques that use acid maceration (HCl + HF) were followed, using the procedure of Ferguson & Ibe (1982), to establish the character of the organic matter type and richness in the samples. The quantity of organic matter in sediment and rock samples was determined by measuring their total organic carbon (TOC) content (wt%). The sediments were then prepared for kerogen elemental analysis and petrography, except for those with TOC contents below 0.3%. The acid maceration residues were centrifuged in heavy liquid (ZnBr₂) solution with a specific gravity of $2.2\,g\,cm^{-3}$ to isolate any possible inorganic and acid-insoluble heavy minerals. The organic matter that floated to the top of the solution was decanted into a separate

container, washed with 0.1M HCl and centrifuged. This was repeated twice, followed by three washes with deionized water to remove any trace of acid. After drying in an oven and homogenization, the kerogen concentrates were ready for analysis; smear slides of the kerogen concentrates were prepared and examined by optical microscope under transmitted light to determine kerogen structure and character.

RESULTS AND DISCUSSION

Petrographical study

Depositional microfacies

The main lithofacies recognized in the Quaternary carbonate sediments, following the classification of Dunham (1962) and the standard microfacies of Wilson (1975), may be summarized as follows:

Age			Lithostratigraphy			Lithology	
Era	Period	Epoch	Group	Formation	Member	Northern Kuwait	Southern Kuwait
Cenozoic	Quaternary	Holocene				Terrestrial deposits: sand and gravel	Coastal deposits: sand, mud, calcareous sandstone
		Pleistocene					
	Tertiary – Neogene	Pliocene	Kuwait	Dibdibba	Upper Lower	Coarse-grained pebbly sands with thin intercalations of clayey sand and clay; pebble gravel and conglomerate. Coarse-grained poorly sorted sandstone with carbonate cement and scattered pebbles.	
	Tertiary – Neogene	Miocene	Kuwait	Fars and Ghar	Upper Fars Fm	Interbedded well-sorted sand and sandstone, silty sand and sandstone with clay, and clayey sand; minor thin-bedded fossiliferous limestone in the east, prominent soft white calcareous sandstone in the west	Undifferentiated interbedded sand and clayey sand with subordinate clay, sandstone and soft white nodular limestone
	Tertiary – Paleogene	Oligocene	Kuwait	Fars and Ghar	Lower Fars Fm and Ghar Fm	Interbedded well-sorted sand and clayey sand with subordinate clay; prominent fawn cross-bedded sandstone layers with gypsum and carbonate cement	
						Unconformity	
	Tertiary – Paleogene	Eocene	Hasa	Dammam		Silicified limestone at top, nummulitic shale at base	
			Hasa	Rus		Anhydrite evaporites, limestone and some marl	
	Tertiary – Paleogene	Paleocene	Hasa	Umm Er-Radhuma		Marly limestone, dolostone and some anhydrite	

Fig. 2. Stratigraphy of Kuwait Cenozoic strata (after Al-Sulaimi & El-Rabaa, 1994).

Recent carbonate sands include beach sediments and tidal creek (Khor) deposits.

(1) Beach sediments are composed of a mixture of skeletal debris and non-skeletal particles (Fig. 4A). The non-skeletal components mainly consist of creamy and metallic, spherical to ovoid, well-sorted ooids, mixed with aggregates, intraclasts and quartz grains. The ooids are highly micritized by boring. Large molluscan shells, calcareous algal fragments and foraminifera are the main skeletal components. These sediments correspond to a skeletal ooid-grainstone microfacies.

(2) Khor sediments are mainly composed of faecal pellets, shell fragments, calcareous mud and superficial ooids. The distribution of these components in the sediments of the tidal creeks is governed by sediment locality with respect to the open sea, tidal movements and wind-generated currents (Cherif *et al.*, 1994). In areas close to where the Khors debouch into the sea, coarse-grained particles of skeletal debris, ooids and reworked ooids are the main components. However, the concentration of ooids decreases landward, where silt-clayey sediments composed of pellets and calcareous mud are found. The sediments correspond to a grainstone-packstone microfacies.

The petrographical results are consistent with those of the grain-size analysis of the samples. The study of the grain-size distribution and parameters of the Holocene oolitic sediments indicates that the coastal areas and the mouths of the tidal creeks near their connection to the open Gulf are dominated by coarse-grained, moderately-sorted beach sands in which the ooids are formed by wave agitation. The landward ends of the creeks are characterized by relatively low-energy sediment types that accumulated fines washed landward by tidal- and wind-generated currents, away from the seaward ends of the tidal creeks.

Pleistocene oolitic grainstones include fossil oolitic limestones and the oldest ridge deposits, composed of oolitic-quartzose sandstones.

(1) The older oolitic limestone ridges show well-preserved thin laminations and low-angle cross-beds of large scale (Fig. 5A). The sediments are composed almost entirely of ooids, which comprise more than 95% of the grains (Fig. 5B). The ooids are well-sorted and spherical or ovoid in

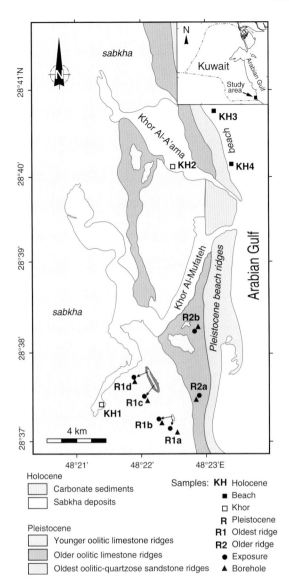

Fig. 3. Simplified surface lithofacies map of the Al-Khiran area showing the sample locations (after Picha & Saleh, 1977).

diameter from 0.2 to 0.7 mm. The ooids are superficial, often with nuclei of quartz and rarely feldspar. Skeletal fragments of bivalves and other molluscs are associated with the ooids.

The sedimentary structures present in both ridges, along with the dominance of ooids and macrofossils, are diagnostic of high-energy, marine-shoal settings (Boardman *et al.*, 1993).

Diagenetic processes

The main diagenetic processes observed through the petrographic analysis of the Quaternary sediments are cementation, dissolution and micritization. In the Pleistocene deposits, subaerial exposure of the oolitic limestones has caused lithification by two types of cement: (1) meniscus cements are commonly localized at grain contacts and bind grains together, reflecting an early diagenetic near-surface meteoric environment (Fig. 5B); (2) isopachous granular rims with subhedral to euhedral calcite crystals and drusy mosaics are randomly distributed at grain contacts and do not significantly bind them together (Fig. 6B), indicating high rates of pore-fluid influx (Longman, 1980; Moore, 1989).

The degree of dissolution of the Pleistocene oolites ranges from the development of secondary intraparticle pores to the complete removal of the interior of ooids and skeletal grains, leading to oomouldic and biomouldic porosity (Figs 5B, 6B and C). Micritization occurs less frequently than other diagenetic processes, mainly during mineralogical stabilization from aragonitic to calcitic ooids and by the boring activity of microbes (Swirydczuk, 1988). However, the original concentric cortices of ooids can still be recognized in the Quaternary sediments (Figs 5B and 6B).

Mineralogy and inorganic geochemistry

The results of X-ray diffraction analysis of the Quaternary oolitic sediments indicate that the Holocene ooids are mainly composed of aragonite. The Holocene oolitic samples contain insoluble residues of quartz and clay minerals. The dominant mineral constituting the Pleistocene oolitic limestones is calcite. Aragonite is a minor component in the older oolitic limestone ridge and quartz is abundant in the rocks of the oldest oolitic-quartzose deposits. The results of XRD and EDS analyses show that the sediments of the landward older

shape. The average diameter of the ooids is 0.5 mm. The kinds of ooids present are normal, micritized and superficial, coating nuclei of quartz grains or peloids. Some ooids have suffered partial or complete dissolution.

(2) The oldest oolitic-quartzose sandstone ridge sediments are cross-bedded, thinly laminated and burrowed by marine organisms (Fig. 6A). The sediments are mainly composed of rounded to subrounded quartz grains, in addition to skeletal fragments and ooids (Fig. 6B). The ooids comprise no more than 20 % of the rock volume and range in

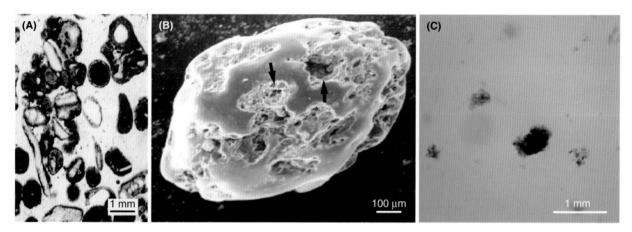

Fig. 4. Photomicrographs showing the organic and inorganic components of Holocene carbonate beach sediments in the Al-Khiran area. (A) Beach sediments composed of a mixture of ooids, peloids, aggregates, quartz grains and skeletal fragments (plain polarized light). (B) SEM photomicrograph of a Holocene ooid showing a heavily bored cortex (arrows). (C) Amorphous marine organic matter macerated from the Holocene beach sediments.

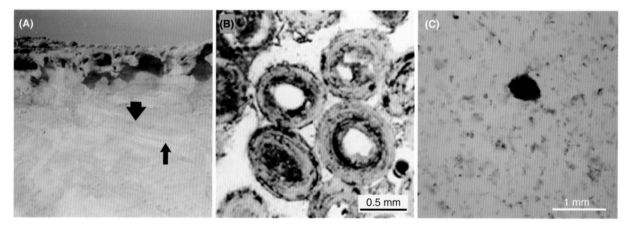

Fig. 5. Pleistocene older oolitic limestone ridge deposits of Al-Khiran. (A) Well-preserved large-scale, low-angle cross bedding (small arrow) and thin laminations (large arrow) displayed in a ridge outcrop. (B) Photomicrograph (plain polarized light) showing well-sorted, coarse ooids showing concentric cortex fabrics. Some nuclei have been removed by dissolution. (C) Amorphous organic matter of algal marine origin extracted from the older oolitic limestones.

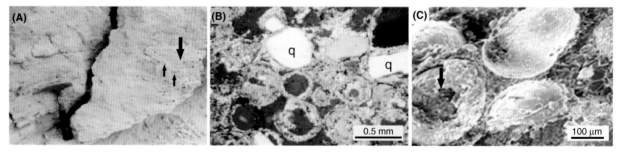

Fig. 6. Pleistocene oldest oolitic-quartzose sandstone ridge deposits of Al-Khiran. (A) Ridge outcrop (section is 1 m high) showing cross-bedded (large arrow), thinly laminated and burrowed marine sediments (small arrows). (B) Photomicrograph (crossed-polarized light) showing calcitized ooids and abundant quartz grains (q) with thin cortices. (C) SEM photomicrograph showing partial dissolution of the concentric laminae of an ooid (arrow).

oolitic limestone ridges are composed almost entirely (98%) of carbonates, but the carbonate percentage is lower (79%) in the very oldest oolitic-quartzose ridge sediments, where silicates form some 21% of the rock on average.

The average values of the minor- and trace-elements for the Quaternary samples are shown in Table 1, and the organic geochemical data are presented in Table 2. Mineralogical analyses show that aragonite is the main component of the Holocene sediments and the older oolitic limestones, and decreases in abundance in the oldest oolitic-quartzose sandstones. The decrease in aragonite and the increase in calcite contents are verified by a general decrease in Sr contents and an increase in Mg, Fe and Mn concentrations from the Holocene sediments and the older oolitic limestone ridge deposits, to those of the oldest oolitic quartzose ridges. Strontium varies inversely with Mg, Fe and Mn, whereas Mg, Fe and Mn are positively correlated (Table 3). This trend in the minor-element distribution is due to the mineralogical stabilization of aragonite to calcite during diagenesis (Kinsman, 1969; Heydari *et al.*, 1993; Morse, 2004). Petrographic observations support the interpretations made by the mineralogical and elemental results. Figures 4–6 show that the cortices of the Holocene and Pleistocene ooids exhibit tangential concentric fabric, indicating an original aragonitic composition (Loreau & Purser, 1973; Heydari *et al.*, 1993).

The contents of Cd, Co and Cu in the Quaternary sediments were generally below the limit of detection. Overall, the Quaternary samples exhibit low average trace-metal contents (Table 1), ranging between 0.2 and 35 $\mu g\,g^{-1}$ (parts per million) except for the Khor sediments which show higher metal contents on average. The high positive correlation between Ni, Zn, Cr, V, Mo and Mg, Fe Mn, and the weak correlation between the metals and organic carbon content of the Quaternary sediments (Table 3) indicate that metal concentrations are not influenced by changes in the amount of organic matter, but are mainly affected by the detrital components of the sediments. On the basis of structural similarities, Ni, Zn and V can substitute for Fe and Mg, and Cr can replace Mg (Francois, 1988). The significant relationship between the metals and carbonate phases indicate that metals are present in the carbonate structure and adsorbed on the surfaces of carbonate particles (Breit & Wanty, 1991; Rifaat *et al.*, 1992). However, Zn shows a moderate positive correlation with TOC, which suggests that Zn is partially associated with organic compounds (Francois, 1988; Rifaat *et al.*, 1992; Burland & Lohan, 2004). In general, the higher metal contents in the sediments containing higher amounts of fine-grained material indicates that the concentration and distribution of the trace metals in the Holocene coastal sediments is grain-size dependent (Francois, 1988; Rifaat *et al.*, 1992), and that the sediments accumulated under oxic conditions (Calvert & Pedersen, 1993).

Organic matter characterization and classification

The Quaternary sediments have TOC values between 0.08–1.27%, with an average value of about

Table 1. Geochemistry of Quaternary sediments, Al-Khiran area

Sample	Mg	Sr	Fe	Mn	Cd	Co	Cr	Cu	Mo	Ni	V	Zn
	($\mu g\,g^{-1}$)											
KH1	15200	1350	2950	120	bd	2.2	29	3.7	2.6	53	635	31
KH2	2980	4530	678	24	bd	bd	4.8	bd	0.7	6.5	112	22
KH3	2590	4890	277	18	bd	bd	3.3	bd	0.3	2.7	37	6.3
KH4	2490	4140	231	16	bd	bd	2.7	bd	bd	2.4	34	6.1
R1a	3380	5690	708	31	nd	nd	nd	nd	nd	nd	nd	nd
R1b	3610	1610	246	27	nd	nd	nd	nd	nd	nd	nd	nd
R1c	3680	2410	416	28	nd	nd	nd	nd	nd	nd	nd	nd
R1d	2860	1530	428	22	nd	nd	nd	nd	nd	nd	nd	nd
R1a*	3830	2990	1340	34	nd	nd	nd	nd	nd	nd	nd	nd
R1b*	3920	5440	757	24	nd	nd	nd	nd	nd	nd	nd	nd
R1c*	5330	1660	880	28	nd	nd	nd	nd	nd	nd	nd	nd
R1d*	5150	1490	947	37	nd	nd	nd	nd	nd	nd	nd	nd
R2a	2010	5590	11	10	nd	nd	nd	nd	nd	nd	nd	nd
R2b	1230	4300	bd	10	nd	nd	nd	nd	nd	nd	nd	nd
R2a*	2500	7130	243	14	bd	0.8	1.9	bd	bd	1.8	20	5.0
R2b*	2550	6070	260	14	0.1	0.9	2.3	bd	bd	1.9	23	2.6

*Denotes borehole sediments, bd = below detection, nd = not determined.

Table 2. Organic geochemical data for Quaternary sediments, Al-Khiran area.

Sample	TOC	C	H	N	S	O	C/N	H/C	O/C
	(wt%)	(%)							
KH1	0.52	43.5	1.96	1.07	1.13	nd	47.2	0.54	nd
KH2	1.27	33.7	5.12	2.48	1.78	20.1	15.8	1.82	0.45
KH3	0.33	47.8	6.54	4.94	2.73	27.2	11.3	1.64	0.43
KH4	0.28	28.2	4.97	2.86	0.38	24.3	11.5	2.11	0.64
R1a	0.21	nd	nd	nd	nd	nd	nd	nd	nd
R1b	0.20	nd	nd	nd	nd	nd	nd	nd	nd
R1c	0.14	nd	nd	nd	nd	nd	nd	nd	nd
R1d	0.12	nd	nd	nd	nd	nd	nd	nd	nd
R1a*	0.13	nd	nd	nd	nd	nd	nd	nd	nd
R1b*	0.08	nd	nd	nd	nd	nd	nd	nd	nd
R1c*	0.08	nd	nd	nd	nd	nd	nd	nd	nd
R1d*	0.08	nd	nd	nd	nd	nd	nd	nd	nd
R2a	0.25	nd	nd	nd	nd	nd	nd	nd	nd
R2b	0.21	nd	nd	nd	nd	nd	nd	nd	nd
R2a*	0.47	21.3	3.70	1.76	2.98	23.6	14.1	2.09	0.83
R2b*	0.31	25.7	4.25	2.32	1.54	22.3	13.0	1.98	0.65

Denotes borehole sediments, nd = not determined.

Table 3. Pearson correlation coefficients of elemental chemistry and TOC contents, Quaternary sediments, Al-Khiran area.

	Ca	Mg	Sr	Fe	Mn	Ni	Zn	Cr	V	Mo	TOC
Ca											
Mg	−0.910										
Sr	0.942	−0.837									
Fe	−0.913	0.992	−0.852								
Mn	−0.922	0.999	−0.855	0.997							
Ni	−0.919	0.999	−0.851	0.997	1.000						
Zn	−0.838	0.820	−0.818	0.883	0.847	0.848					
Cr	−0.922	0.998	−0.861	0.997	1.000	1.000	0.849				
V	−0.923	0.994	−0.862	0.999	0.998	0.998	0.875	0.998			
Mo	−0.898	0.950	−0.870	0.978	0.965	0.964	0.942	0.967	0.975		
TOC	−0.094	0.028	−0.124	0.149	0.071	0.075	0.575	0.073	0.127	0.296	

0.29% (Table 2). Holocene beach sediments exhibit lower TOC values than those reported for sediments from the Khors. This is mainly due to the high wave and current dynamics which prevent the preservation of organic matter (Kenig et al., 1990). In contrast, the stagnant conditions at the extensions of the Khors, as expressed by their higher amounts of fine-grained pelletal sediments, lead to higher quantities of organic matter accumulation (Tyson, 1987). TOC percentages for the Pleistocene ridges range between 0.08% and 0.47%. Rocks from the older oolitic limestone ridge contain higher amounts of organic carbon, with a range of 0.21–0.47%. The oldest oolitic-quartzose sandstones give lower organic matter contents. The petrographical observations (Figs 4–6) and mineralogical evidence show an increase in quartz content in the sediments of the oldest oolitic-quartzose sandstone ridges. Dilution of the organic matter

contained in these sediments with siliciclastic material lowers their TOC concentrations to about 0.08% (cf. Tyson, 1995).

The average C, H, N, S and O percentages for the organic matter of the Quaternary sediments are presented in Table 2. The organic matter isolated from the Holocene and Pleistocene carbonate sediments shows low carbon values and high oxygen contents, indicating an early diagenetic stage of kerogen thermal maturation, where the organic matter is immature (Vandenbroucke et al., 1993). The high O/C ratios indicate highly oxygenated depositional environments and early diagenetic aerobic degradation (Bordenave, 1993; Klemme & Ulmishek, 1991). The average values of C/N are indicative of sedimentary organic matter of marine origin (Stein, 1991; Tyson, 1995; Meyers, 1997). The elemental results are confirmed by petrographic examination which shows that the

Holocene oolitic deposits (Fig. 4C) and the Pleistocene oolitic limestones (Fig. 5C) both contain light-coloured, immature organic matter of algal, amorphous facies type.

SUMMARY AND CONCLUSIONS

The Quaternary oolitic sediments in southern Kuwait are high-energy carbonate facies deposited under highly oxygenated conditions in bioturbated marine environments. This is clearly reflected in the low content of trace metals in the sediments. The amount of organic matter in the Quaternary sediments is low (TOC < 0.5%) and it is characterized by high O/C ratios. The low amount of organic matter in the Holocene sediments is related to the high oxygen concentrations and strong wave and tidal action in their depositional environment. Such physical and chemical marine conditions favour the oxidation of organic matter before being absorbed onto the surfaces of particles or incorporated in the sediments.

The Pleistocene carbonate ridges were formed during periods of glaciation, when sea level dropped and the sedimentary environment was characterized by well-oxygenated, high-energy conditions. Thus, the Pleistocene sediments contain low concentrations of trace metals and oxidized organic matter. Moreover, the overall low TOC values and high O/C ratios of the degraded, amorphous organic matter present in the Quaternary sediments reflect the effect of early diagenetic aerobic degradation under increasing availability of oxygen. Consequently, although ooids initially may contain significant amounts of algal marine organic matter, the oolitic coastal ridges are not a favourable setting for the preservation of this organic matter. However, further investigations of the lagoonal and tidal creek settings, which include fine-grained low-energy pelletal facies, may reveal high organic-matter richness and preservation in the sediments.

ACKNOWLEDGEMENTS

The paper is part of an MSc thesis at Kuwait University. The authors are grateful to the University of Kuwait for financial support of this research. The assistance of Dr Abbas Saleh with the petrographic analysis is gratefully acknowledged. Many thanks are given to Mr Yousef Abdullah for preparing the thin sections.

REFERENCES

Abdullah, F.H. and **Kinghorn, R.R.F.** (1996) A preliminary evaluation of Lower and Middle Cretaceous source rocks in Kuwait. *J. Petrol. Geol.*, **19**, 461–480.

Al-Sarawi, M.A., Al-Zamel, A. and **Al-Refaiy, I.A.** (1993) Late Pleistocene and Holocene sediments of the Khiran area (South Kuwait). *J. Univ. Kuwait (Sci.)* **20**, 145–156.

Alsharhan, A.S. and **Nairn, A.E.M.** (1997) *Sedimentary Basins and Petroleum Geology of the Middle East.* Elsevier, New York, *843* pp.

Al-Sulaimi, J.S. and **El-Rabaa, S.M.** (1994) Morphological and morphostructural features of Kuwait. *Geomorphology*, **11**, 151–167.

Boardman, M.R., Carney, C. and **Bergstrand, P.M.** (1993) A Quaternary analog for interpretation of Mississippian oolites. In: *Mississippian Oolites and Modern Analogs* (Eds B.D. Keith and C.W. Zuppann). *AAPG Stud. Geol*, **35**, 227–240.

Bordenave, M.L. (1993) The sedimentation of organic matter. In: *Applied Petroleum Geochemistry* (Ed. M.L. Bordenave), pp. 17–76. Editions Technip, Paris.

Breit, G.N. and **Wanty, R.B.** (1991) Vanadium accumulation in carbonaceous rocks: A review of geochemical controls during deposition and diagenesis. *Chem. Geol.*, **91**, 83–97.

Burland, K.W. and **Lohan, M.C.** (2004) Controls of trace metals in seawater. In: *Treatise on Geochemistry: the Oceans and Marine Geochemistry* (Ed. H. Elderfield), pp. 23–47. Elsevier, UK.

Calvert, S.E. and **Pedersen, T.F.** (1993) Geochemistry of Recent oxic and anoxic marine sediments: implications for the geological record. *Mar. Geol.*, **113**, 67–88.

Cherif, O.H., Al-Rifaiy, I.A. and **Al-Zamel, A.** (1994) Sedimentary facies of the tidal creeks of Khor Al-Mufateh and Khor Al-Mamlaha, Khiran area, Kuwait. *J. Univ. Kuwait (Sci.)*, **21**, 87–105.

Dunham, R.J. (1962) Classification of carbonate rocks according to depositional texture. In: *Classification of Carbonate Rocks* (Ed. W.E. Ham). *AAPG Mem.*, **1**, 108–121.

Duane, M.J. and **Al-Zamel, A.Z.** (1999) Syngenetic textural evolution of modern sabkha stromatolite (Kuwait). *Sed. Geol.*, **127**, 237–245.

Ferguson, J. (1987) The significance of carbonate ooids in petroleum source-rock studies. In: *Marine Petroleum Source Rocks* (Eds J. Brooks and A.J. Fleet), pp. 207–215. Blackwell Scientific Publications, UK.

Ferguson, J. (1988) Oil generation and migration within marine carbonate sequences – A review. *J. Petrol. Geol.*, **11**, 389–401.

Ferguson, J. and **Ibe, A.C.** (1982) Some aspects of the occurrence of proto-kerogen in Recent ooids. *J. Petrol. Geol.*, **4**, 267–285.

Folk, R. L. and **Ward, W.C.** (1957) Brazos River bar: a study in the significance of grain size parameters. *J. Sed. Petrol.*, **27**, 3–26.

Francois, R. (1988) A study of the concentrations of some trace metals (Rb, Sr, Zn, Pb, Cu, V, Cr, Ni, Mn and Mo) in Saanich inlet sediments, British Columbia, *Canada. Mar. Geol.*, **83**, 285–308.

Heydari, E., Snelling, R.D., Dawson, W.C. and **Machain, M. L.** (1993) Ooid mineralogy and diagenesis of the Pitkin Formation, North-Central Arkansas. In: *Mississippian*

Oolites and Modern Analogs (Eds B.D. Keith and C.W. Zuppann). *AAPG Stud. Geol.*, **35**, pp. 175–184.

Hunt, J.M. (1967) The origin of petroleum in carbonate rocks. In: Carbonate Rocks. (Eds G.V. Chillingar, H.J. Bissel and R.W. Fairbridge), pp. 225–251. Elsevier, New York.

Ibe, A.C. (1983) Organic geochemistry of Recent marine ooids as a key to origin of petroleum in oolite reservoirs. *AAPG Bull.*, **67**, 486–487.

Ibe, A.C. (1984) In-situ formation of petroleum in oolites, Part I: Scheme of hydrocarbon generation and accumulation. *J. Petrol. Geol.*, **7**, 267–276.

Ibe, A.C. (1985) In-situ formation of petroleum in oolites, Part II: A case study of the Arab Formation oolites reservoirs. *J. Petrol. Geol.*, **8**, 331–341.

Kenig, F., Huc, A.Y., Purser, B.H. and **Oudin, J.** (1990) Sedimentation, distribution and diagenesis of organic matter in a Recent carbonate environment, Abu Dhabi, *U. A. E. Org. Geochem.*, **16**, 735–747.

Khalaf, F.I., Gharib, I.M. and **Al-Hashash, M.Z.** (1984) Types and characteristics of the Recent surface deposits of Kuwait, Arabian Gulf. *J. Arid Environ.*, **7**, 9–33.

Kinsman, D.J.J. (1969) Interpretation of Sr^{2+} concentrations in carbonate minerals and rocks. *J. Sed. Petrol.*, **39**, 486–508.

Klemme, H.D. and **Ulmishek, G.F.** (1991) Effective petroleum source rocks of the world: Stratigraphic distribution and controlling depositional factors. *AAPG Bull.*, **75**, 1809–1851.

Longman, M.W. (1980) Carbonate diagenesis textures from near surface diagenetic environments. *AAPG Bull*, **64**, 461–487.

Loreau, J.P. and **Purser, B.H.** (1973) Distribution and ultra-stucture of Holocene ooids in the Persian Gulf. In: *The Persian Gulf: Holocene Carbonate Sedimentation and Diagenesis in a Shallow Epicontinental Sea* (Ed. B.H. Purser), pp. 279–328. Springer-Verlag, New York.

Meyers, P.A. (1997) Organic geochemical proxies of paleoceanographic, paleolimnologic, and paleoclimatic processes. *Org. Geochem.*, **27**, 213–250.

Moore, C.H. (1989) *Carbonate Diagenesis and Porosity. Dev. Sedimentol.*, **46**, 338 pp. Elsevier, New York.

Morse, J.W. (2004) Formation and diagenesis of carbonate sediments. In: *Treatise on Geochemistry: Formation and Diagenesis of Carbonate Sediments* (Ed. F.T. Mackenzie), pp. 67–86. Elsevier, UK.

Picha, F. and **Saleh, A.A.M.** (1977) Quaternary sediments in Kuwait. *J. Univ. Kuwait (Sci.)*, **4**, 169–184.

Rifaat, A.E., El-Sayed, M.Kh., Beltagy, A., Morsy, M.A. and **Nawar, A.** (1992) Geochemical predictive models of manganese, zinc, nickel, copper and cadmium in Nile shelf sediments. *Mar. Geol.*, **108**, 59–71.

Saleh, A.A.M. (1975) *Pleistocene and Holocene Oolitic Sediments in Al-Khiran Area, Kuwait.* Unpubl. MSc Thesis, Kuwait University, Kuwait, 125 pp.

Shearman, D.J. (1965) Organic matter in Recent and ancient limestones and its role in their diagenesis. *Nature*, **208**, 1310–1311.

Shearman, D.J., Twyman, J. and **Zand Karimi, M.** (1970) The genesis and diagenesis of oolites. *Proc. Geol. Assoc.*, **81**, 561–575.

Stein, R. (1991) *Accumulation of Organic Carbon in Marine Sediments.* Lecture Notes in Earth Science. Springer-Verlag, New York, 214 pp.

Swirydczuk, K. (1988) Mineralogical control on porosity type in Upper Jurassic Smackover ooid grainstones, southern Arkansas and northern Louisiana. *J. Sed. Petrol.*, **58**, 339–347.

Tyson, R.V. (1987) The genesis and palynofacies characteristics of marine petroleum source rocks. In: *Marine Petroleum Source Rocks* (Eds J. Brooks and A. J. Fleet). *Geol. Soc. London Spec. Publ.*, **26**, 47–67.

Tyson, R.V. (1995) *Sedimentary Organic Matter, Organic Facies and Palynofacies.* Chapman & Hall, London, 615 pp.

Vandenbroucke, M., Bordenave, M.L. and **Durand, B.** (1993) Transformation of organic matter with increasing burial of sediments and the formation of petroleum source rocks. In: *Applied Petroleum Geochemistry* (Ed. M.L. Bordenave), pp. 103–121. Editions Technip, Paris.

Wilson, J.L. (1975) *Carbonate Facies in Geologic History.* Springer-Verlag, Berlin, 471 pp.

Int. Assoc. Sedimentol. Spec. Publ. (2011) **43**, 265–276

Halite, sulphates, sabkhat and salinas of the coastal regions and Sabkha Matti of Abu Dhabi, United Arab Emirates

ANTHONY KIRKHAM

Pen-yr-Allt, Village Road, Nannerch, Mold, Flintshire, Wales, UK. CH7 5RD
(E-mail: kirkhama@compuserve.com)

ABSTRACT

The gross distributions and morphologies of Holocene halite, gypsum and anhydrite deposits on and within Abu Dhabi coastal sabkhat and the mainly continental Sabkha Matti are discussed. Areas of present-day rainwater ponding landward of the Holocene marine sediment-evaporite wedge give rise to halite salinas. Tracts of early and late Holocene anhydrite are illustrated, and the approximate Flandrian transgressive limit on the coastal sabkhat is carbon dated at 5750–5900 yr BP. Holocene gypsum and anhydrite morphologies are discussed and a salina origin for Holocene banded anhydrites and their interbedded microbial mats is advocated. Attention is drawn to the common vuggy nature of these coastal sabkha anhydrites. Salina gypsum deposits from the Pleistocene carbonate succession and from continental Miocene deposits on the coastal sabkhat are briefly described. The occurrence of gypsum volcanoes occurring in continental Sabkha Matti is noted.

Keywords: Abu Dhabi, sabkhat, salina, evaporites, Sabkha Matti.

INTRODUCTION

This paper focuses on the coastal regions of Abu Dhabi Emirate which have been world famous since the 1960s for the occurrences of arid, coastal, Holocene evaporites as a result of a series of landmark papers, including Curtis *et al.* (1963), Shearman (1966), Butler (1969, 1970), Butler *et al.* (1982), Evans *et al.* (1969) and Kinsman (1969). It also touches on evaporites within the continental Sabkha Matti which extends inland from the coast for about 100 kilometres in a N–S direction (Fig. 1). The sabkhat (plural of sabkha) of the region are amongst the largest in the world and the coastal sabkhat are most famous for the recent occurrences of anhydrite. Halite, gypsum and (to a certain extent) dolomite also occur in abundance on the coastal sabkhat, but Holocene evaporites are not alone in this coastal region because Pleistocene and Miocene evaporites are also exposed.

Largely due to difficult and often treacherous access across the soft sabkha surfaces, the areal distribution of the Holocene evaporites was not well-established during the early research and so this paper partly illustrates their gross distributions. Even today, many geologists visit the sabkhat in the hope of examining the anhydrite, and expect to find it extensively developed (perhaps sheet-like over vast areas) but have difficulty in finding any. In fact, there are two distinct anhydrite zones which are regarded as "early" and "late" Holocene (Kirkham, 1997; Fig. 2). The Holocene gypsum and anhydrite were interpreted by the early workers as having formed within the subsurface of the sabkha sediments, but this paper presents evidence to indicate that significant amounts of former gypsum, which dehydrated to present-day anhydrite, formed subaqueously within salinas upon the sabkha surfaces. Furthermore, salina evaporites also formed during the Pleistocene and the Miocene.

HOLOCENE COASTAL SABKHAT

Individual coastal sabkhat may reach over 15 km in width and several tens of km in length as they extend along the mainland shoreline. To a lesser extent, they also occur on the offshore barrier islands. From them, earlier workers such as

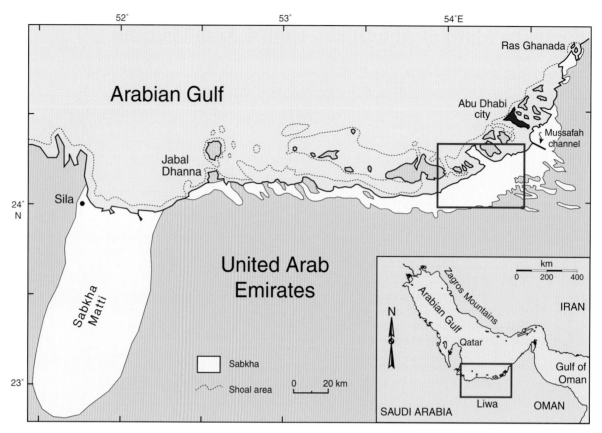

Fig. 1. General location map of western Abu Dhabi showing the distribution of coastal sabkhat and Sabkha Matti. Large red rectangle area indicates location of Fig. 2.

Evans *et al.* (1969) and Kenig *et al.* (1989) established the "sabkha sequence" which includes a basal, rapid Holocene transgression that climaxed about six thousand years ago when relative sea level was 1–2 metres higher than it is today. The most landward evidence of marine influence occurs in the form of relict beach ridges several kilometres inland, beyond the Holocene sulphates (Fig. 2), and one such relict has been carbon-dated as 5750–5900 years old. This climax was followed by a forced regression as sea-level fell to its approximate present level and gave rise to a progradational, sabkha sequence representing an upward transition from subtidal through intertidal to supratidal sediments with their associated evaporites.

Although the coastal sabkhat have expanded seawards by progradation over the last several thousand years, creating a Holocene marine wedge of carbonates and evaporites, their inner parts comprise Pleistocene aeolian quartzose sands, and to a lesser extent carbonate aeolianites. Prior to the availability of satellite imagery, the distinction between the marine and aeolian sediments was

not obvious, due to the sabkha surfaces being deflated to elevations that are barely above present-day sea-level. In fact, the coastal sabkhat are also expanding landwards due to wind deflation of the Pleistocene and Miocene strata on their inner margins (Kirkham, 1998).

Numerous, parallel storm beaches traversed by the remnants of narrow tidal channels are obvious from satellite imagery as the beaches extend along the coast behind the modern intertidal and supratidal zones. Their parallelism is interrupted occasionally where they formed "winged spits" at the seaward sides of sites where once stood probable Miocene mesas that have since disappeared from the topography as a result of advanced deflation (Kirkham 1997; Fig. 2). Today remnants of the parallel storm beaches are barely perceptible as elevated features, due to the deflation. However, they are recognisable partly by their surficial deflationary lag of bioclastic material dominated by cerithid gastropods, and partly from their sparse developments of halophytic plants which are limited to the areas of the deflated beach-ridge system. In the main, this storm beach system separates the

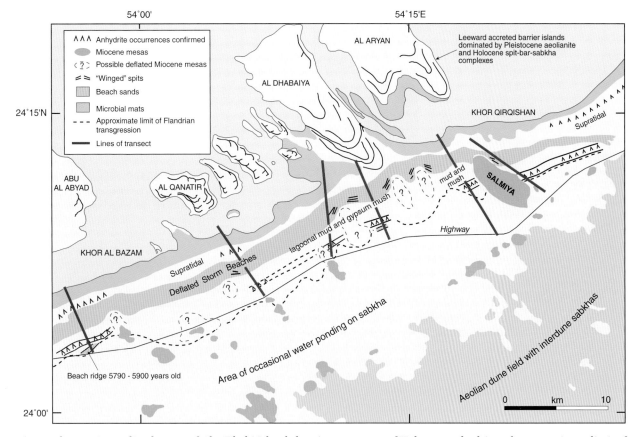

Fig. 2. The area immediately west of Abu Dhabi Island showing two tracts of Holocene anhydrites, the approximate limit of the Flandrian transgression and the area of periodic rainwater ponding on the sabkha. The transects were used to define the facies distributions with the aid of satellite imagery (modified after Kirkham, 1997, courtesy of GeoArabia).

early and late Holocene anhydrites. Immediately landward of these parallel storm beaches are often encountered Holocene sediments comprising former subtidal lagoonal and intertidal sediments which probably accumulated in the lee of this beach barrier system.

Holocene coastal halite accumulations/salinas

The approximate southern limit of the Holocene marine wedge is marked by the northern limit of periodically impounded rainwater after heavy precipitation. By processing satellite images to enhance the ponded areas and dampened sabkhat, the areas that are prone to such flooding are fairly easily identified, as is the trend of the early Holocene anhydrite. Such ponds are regular occurrences and some years may extend over many square kilometres. They may survive for many months during which they increase in salinity before desiccating completely with precipitation of halite. In other words, they evolve into halite salinas.

The ponds develop due to natural damning effects caused by the combination of: (1) jacking-up of the sediment surface as early Holocene anhydrite developed within mainly aeolian sediments; (2) deflation of the aeolian quartz sands to elevations that are slightly lower than those of the adjacent early Holocene anhydrite zone; and (3) the occasional occurrences of the relict beach ridges created at about the time of the climax of the Flandrian transgression. The rainwater ponding areas give rise to ephemeral halite layers which precipitate as the salinas evaporate. Each rainwater flood quickly dissolves the previous halite layer, immediately increases in salinity, and then slowly evaporates to precipitate a new halite layer. Due to its high solubility, this halite is never preserved as part of the sedimentary sequence. The fresh halite precipitates as a white sheet (Fig. 3) which is blinding to the human eye in bright sunlight, but gradually turns brown over the following few months as iron oxides also precipitate on the halite accumulation. Iron is relatively abundant within the subsurface waters as indicated, for instance,

Fig. 3. An extensive, white halite crust precipitated from a dessicated salina of ponded rainwater origin on a sabkha immediately landward of the Holocene marine sediment wedge.

by the repeated occurrences of subsurface bands of rusty colouration. These record different levels of the water tables and associated capillary fringes as they subsided under the effect of relative sea-level fall during the late Holocene. Halite also precipitates within the sabkha sediments to a depth of about one metre (Butler, 1970).

Holocene coastal sabkha gypsum and anhydrite

Gypsum and anhydrite are easily the dominant sulphate minerals within the coastal sabkhat. Gypsum develops either as a mush of sub-centimetre-sized crystals within the present-day upper intertidal zone, or as often decimetre-sized, (lozenge- or) discus-shaped crystals beneath the water table, usually within the buried lagoonal and intertidal deposits. These discus-shaped crystals may be transparent, or may be opaque due to poikilitically enclosing carbonate sediment or organic matter from the host microbial mats. Particularly in the inner parts of the marine wedge, the mush and the discus-shaped crystals may be exposed by extreme deflation.

The older gypsum mush behind the storm beaches tends to be more coarsely crystalline, probably due to the effects of repeated dehydration to anhydrite and re-hydration back to gypsum. Such re-crystallization may have been induced largely by repeated wetting by rainwater ponding or marine floods alternating with annual high-temperature desiccation. However, marine flooding of the lagoonal areas behind the old storm beach ridges very rarely occurs today because relative sea-level is so low.

Although some of the anhydrite is thought to have originated by dehydration of gypsum (Butler, 1970), much of it precipitates directly (Kinsman, 1969). It tends to form as anhydrite crystal laths within the capillary zone of the contemporaneous water table of the supratidal subsurface environment, where it may concentrate as nodules or mush. It also occurs as bands which are discussed below. The nodules are typically several centimetres in diameter, display irregular outlines, and tend to displace the host sediment by continued precipitation of anhydrite laths within the nodules. As individual nodules expand they begin to impinge on each other and tend towards the "chicken-wire" appearance typical of many ancient anhydrites. Collections of nodules are often vaguely aligned horizontally as if they have developed at the same still-stand of the parent capillary zone (Fig. 4). Impingement of adjacent nodules often creates enterolithic morphologies.

Less commonly, individual anhydrite nodules reach about 30 cm in size horizontally (Fig. 5). The reason for this unusual type of nodule is unknown but they sometimes display vertical, carbonate-filled features a few millimetres in diameter. Butler (1970) believed these nodules originated as gypsum mush which was later dehydrated to form anhydrite and considered the vertical features as possible sand-filled worm tubes. The present author has found vacant tubes in such anhydrite nodules that had formed in aeolian deposits. Douglas J. Shearman (pers. comm.) suspected that such tubes were insect borings. One is often plagued with swarms of mosquito-like insects on the sabkhat even during the height of summer and so such an origin is not unlikely.

Fig. 4. A pit dug into Pleistocene brown aeolian quartz sand to expose Holocene nodular anhydrite along the early Holocene anhydrite tract. A relatively thick layer of surficial brown-coloured gypsum mush occurs near the surface, about 15 cm above the anhydrite, at the approximate level of the spade handle (after Evans & Kirkham, 2005, courtesy of Trident Press).

Banded anhydrites: sabkha versus salina origin

Banded anhydrite occurs as the thin anhydrite seams (Fig. 6) described previously by Butler (1970). It generally occurs above the nodular forms, tends to be ptygmatically folded, and its "anticlinal" contortions are frequently exposed at the sabkha surface, where they may be eroded by deflation (Fig. 7). Anhydrite "mush" often forms within central cores of the "anticlinal" folds, which give the impression of being diapiric.

The precise origin of the banded anhydrites is debatable. Were they formed, as were the nodular anhydrites, within a sabkha subsurface, or did they form by dehydration of subaqueous gypsum precipitates within former salinas? Kirkham (1998) argued for the salina origin, and the following reasons are offered in support of that interpretation:

1. Despite often severe buckling or ptygmatic folding, the anhydrite bands tend to retain a fairly constant thickness with relatively smooth lower and upper boundaries. It is difficult to accept such constancy if they formed by coalescing anhydrite nodules within a capillary zone where merging nodules tend towards enterolithic structure. Internal expansion due to continued sulphate precipitation within the bands has undoubtedly given rise to the ptygmatic folding, but there was no need to firstly coalescence pre-existing nodules if the original sulphates constituted a continuous subaqueous sheet.

2. Notwithstanding the above comments regarding fairly constant thicknesses, the lower boundaries are generally more uniform than the upper. The upper boundary is often distinctly botryoidal (Fig. 8). This is probably due to

Fig. 5. A shallow (∼30 cm deep) trench exposing large, Holocene anhydrite nodules within Pleistocene cross-bedded aeolian quartz sand. This type of nodule seems to be prone to presumed insect borings (after Evans & Kirkham, 2005, courtesy of Trident Press).

Fig. 6. Several stacked anhydrite bands each 3–4 cm thick, in the bank of the Mussafah channel. Each band tends to have a planar base and sharp, slightly undulating top. Tool handle (top right) is 7 cm thick.

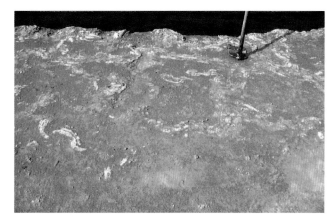

Fig. 7. Sabkha surface adjacent to the Mussafah channel. Note how the anhydrite bands have been contorted upwards to become exposed and truncated by deflation. Tool handle (top right) is 7 cm thick.

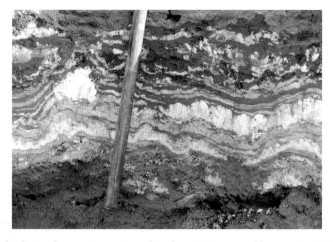

Fig. 8. Contorted bands of anhydrite. The most prominent band contains several bouquets of what appears to be a collection of merged, sub-vertical anhydrite nodules, each of which probably representing dehydrated former gypsum crystals. Note that the shallowest anhydrite band is less contorted than the ones below and is interpreted as being younger. Tool handle is 7 cm thick. Bank of the Mussafah channel (after Evans & Kirkham, 2005; courtesy of Trident Press).

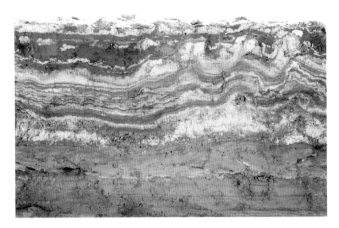

Fig. 9. Several, stacked, banded anhydrites contorted upwards in unison except for the topmost band which is essentially flat. Some layers have planar bases but irregular tops due to bouquets of dehydrated vertical gypsum crystals. On the right, the bands have anhydrite mush at the core of a small diapiric structure. Thickness of illustrated section is approximately 60 cm. Bank of the Mussafah channel.

bouquets of crystals that grew vertically upwards out of the original layer of subaqueous gypsum. Such features are typical of modern gypsum salinas. The author has never seen evidence of the original gypsum within these botryoids, presumably because of complete dehydration of the gypsum to pasty anhydrite.

3. Where several stacked bands of anhydrite occur at the same location, the upper bands tend to be significantly less deformed, such as in the banks of the Mussafah channel (Fig. 9). The upper bands are therefore presumed to be younger than the lower bands. This would be difficult to explain if the anhydrite bands were forming within the subsurface sediments of a prograding, regressive sequence under falling sea-level with reduced elevation of the water table and its associated capillary zone, unless they were related to perched water tables that formed above the impervious, deeper anhydrite bands. There is no reason to believe that they are equivalent to the transgressive, supratidal nodular anhydrites seen elsewhere on the coastal sabkhat by Kirkham (1997). A more likely explanation is that the shallower, less deformed anhydrite bands (in fact, all the banded anhydrites) originated as younger salina gypsum deposits.

4. Hardie (1986) described what appear to be excellent analogues to these banded anhydrites forming in the modern, gypsiferous Salina Omotepec, Baja California.

5. The banded anhydrites of the inner Mussafah channel are interbedded with microbial carbonate laminites (Fig. 10). Such microbialites

could not have formed in the subsurface, and so strengthen the case for their interbedded, banded anhydrites having formed as surficial evaporites, i.e. within salinas.

Apart from an arid climate, gypsum salina formation requires one of two basic types of situations: (a) a ponding area or lagoon of impounded storm surge or neap high-tide seawater linked to more open-marine conditions by a narrow channel(s); (b) a ponding area of lower elevation than mean sea level behind a barrier through which there is continual seepage of seawater. A regular supply of marine water is required in order to build up a significant thickness of subaqueous gypsum precipitate.

Gypsum salinas are not known to exist anywhere on the present-day coastal sabkhat of Abu Dhabi, although Butler (1970) observed flooding recharge of an Abu Dhabi sabkha leaving a thin prismatic surficial layer of gypsum crystals which, if repeated and combined with dehydration, could explain the formation of the banded anhydrites. Loreau & Purser (1973) described subaqueous gypsum in Khor Odaid, SE Qatar, which would be analogous to situation (a) above. Gunatilaka & Shearman (1988) described modern gypsum laminites in a supratidal sabkha "pool" of southern Kuwait which would also be analogous to situation (a) above.

Barriers are known to have existed on the Abu Dhabi sabkhat during the earlier Holocene when the relative sea level was 1–2 m higher than today, when the modern barrier island complex was less well-developed, and when the mainland coast was

Fig. 10. Microbially laminated carbonate sediment (arrowed) sandwiched between two anhydrite bands. Tool handle (right) is 7 cm thick. Bank of the Mussafah channel.

exposed to higher energy wave action and slightly greater tidal range that enabled easier marine flooding of the lagoons behind the parallel storm beaches (Kirkham, 1998). Such a system could explain the occurrence of the banded anhydrite originating from salinas within the early Holocene anhydrite zone. It is interesting to note that the banded anhydrite featured on the AAPG video entitled *"Arid Carbonate Coastlines"* (Scholle *et al.,* 1986) was located on this early Holocene anhydrite zone (Fig. 2).

The early Holocene barrier system, however, does not explain the occurrence of the even better developed and more numerous banded anhydrites in the inner reaches of the Mussafah channel, about 7 km inland from the present-day lagoon. These formed within a location that was seaward of the parallel storm beaches, but they were located deep into what appears to have been an embayment formed between two distinct winged-spit systems (Kirkham, 1997). Their general setting bears loose comparison to the Khor Odaid gypsum salina. The restriction may have been enhanced by, for instance, temporary and periodic oolitic bars formed seaward of the banded anhydrite of the Mussafah channel. Superficial ooids dominate the part of the vertical section separating the transgressive and later regressive sequences exposed in the banks of the Mussafah channel (Kirkham, 1998).

Indistinct layers of discrete anhydrite nodules within the present-day supratidal zones seaward of the parallel, deflated storm beach ridges also often exhibit folding and erosional truncation by either sea or wind. This is not unreasonable under conditions where enlarging, layered anhydrite nodules require space to expand into. The natural inclination is for them to expand in the direction of least confinement, i.e. upwards.

The author has not directly correlated banded anhydrite over appreciable distances, but it is not unreasonable to assume that banded anhydrite of subaqueous salina origin may pass laterally into layers of nodular anhydrite formed in the contemporaneous sabkhat bordering the salinas (Warren & Kendall, 1985).

Vuggy porosity

Vugs are not uncommon within the sabkha anhydrites and reach up to 4 cm across. Their origins may lie partly in the volume-loss associated with the dehydration of gypsum to anhydrite, but leaching is believed to be the governing factor. Such vugs are often lined by a film of iron oxide which must have precipitated from groundwaters migrating through connected vug systems. Sometimes, vugs and sub-vertical pipes (up to a centimetre in diameter) within the anhydrites appear to have been rounded as if to mimic micro-speleothem galaries. Such effects have presumably been caused by the downward percolation of water previously ponded on the sabkha surfaces. Gaping holes up to 30 cm across have been observed by the author on the sabkha surface in close proximity to some of the vuggy anhydrites at Mussafah channel. They were interpreted as karstic effects, although the possibility that they were created by lizards could not be completely eliminated. Butler (1970) recorded uniform porosities of 20% in a seam of "chicken-wire" anhydrite, although the implication was that it was intercrystalline and similar to the porosity recorded by Peebles *et al.* (1995).

HOLOCENE SABKHA MATTI EVAPORITES

Sabkha Matti, near the western extremity of the Emirate, represents by far the largest sabkha in the UAE, but most of it is continental sabkha (Goodall, 1997). Its marine-influenced coastal strip, about 8 km wide on average, is similar in many ways to the Abu Dhabi coastal sabkhat farther east, except that anhydrites are poorly developed. The surface of the southern margin of the coastal sabkha is covered by a deflationary marine shelly lag which passes into the northern part of the continental sabkha. This is covered by a veneer of coarse sand and gravel representing the deflation lag from vanished aeolian dunes and former fluvial channels.

The Holocene, soft-sediment terrain behind the coastal sabkha commonly has a slightly reddened, "leopard-skin" appearance (Evans & Kirkham, 2005). This term is thought to have been informally introduced by Ken Glennie. It constitutes a hummock-and-hollow relief up to a metre vertical scale. The hummocks tend to be darker in colour than the hollows, hence the derivation of the "leopard skin" descriptive term. Within half a metre of the surface, the hummocks in particular are generally underlain by concentrations of fibrous halite, gypsum and palygorskite (R. Peebles, pers. comm.; Fig. 11). Locally, selenite gypsum veins protrude through the surface of the "leopard skin" terrain. Although the timings of the

protrusions are unknown, the gypsum is probably of Miocene origin and the veins are similar to selenite veins observed in the Miocene outcrops immediately south of Sila, at the NW corner of Sabkha Matti.

Several modern gypsum volcanoes have also been observed by both the present author and Warren Wood (pers. comm.). They may reach a metre in height and create a gypsum cone up to 50 m in diameter around a central vent. They seem to occur near the centres of large depressions on Sabkha Matti and are probably an artesian effect. The gypsum is likely of Miocene or Quaternary origin, but at least some of the volcanoes are known to be modern, as they spread across recent vehicle tracks.

PLEISTOCENE GYPSUM

Apart from possible Pleistocene gypcretes in the continental sabkhat of the Liwa, only one other occurrence of Pleistocene evaporites is known onshore, east of Abu Dhabi Island. This constitutes a massive gypsum bed up to 30 cm in thickness and sandwiched between underlying marine carbonates and overlying aeolian carbonates (miliolite; Fig. 12). It is interpreted here to represent a salina deposit.

MIOCENE GYPSUM

In addition to the Miocene selenite veins referred to above, bedded selenite comprising large, thin glassy crystals is a feature of the upper Miocene Shuweihat Formation and are particularly well

Fig. 11. A shallow trench revealing a subsurface mixture of palygorskite and gypsum concentrated within soft aeolian quartz sand, Sabkha Matti. Handle of knife is 11 cm long.

Fig. 12. A massive gypsum bed (arrowed) underlain by marine carbonate and overlain by carbonate aeolianite. All three strata are of Pleistocene age. Mainland east of Abu Dhabi Island.

exposed on Shuweihat Island, about 25 km west of Jebel Dhanna, western Abu Dhabi. Massive gypsum beds sometimes up to about a metre thick and comprising intergrown selenite crystals are also a feature of the fluvial-dominated Miocene Baynunah Formation, especially in the region immediately west of Abu Dhabi Island, and they are exposed in the Miocene mesas that form inliers on the coastal sabkhat (Fig. 13). Again they are interpreted as salina deposits.

Massive gypsum layers occur buried beneath the inner coastal sabkhat. Recent excavation of a deep trench along the front of a Miocene scarp west of Mussafah, for the purpose of pipe-laying, exposed a gypsum layer well over one metre thick. It was interpreted as a former salina gypsum deposit because sabkha anhydrites are generally

accepted as being unable to exceed one metre in thickness due to limitations imposed by capillary zone heights. The white, intergrown gypsum crystals were very coarsely crystalline and contained anhydrite inclusions that suggested at least one cycle of dehydration and rehydration back to gypsum. Its intercrystalline porosity and permeability was so high that groundwater literally streamed through and held up the trenching operation for many days. Such Miocene bedded sulphates are known to have been mis-identified as Holocene by Godfrey Butler during the 1960s (P. Bush, pers. comm.). Slightly deeper Miocene anhydrite layers of salina origin (hypersaline lagoon/ephemeral lake) were cored during shallow drilling along the length of the Mussafah channel (Peebles *et al.*, 1995).

Fig. 13. A massive bed of coarse, salina gypsum within the Miocene Baynunah Formation. It is now exposed at an inlier on a mainland coastal sabkha west of Abu Dhabi Island. The yellow tube is 7 cm long.

CONCLUSIONS

Whilst the Abu Dhabi coastal sabkhat are renowned as the basis for understanding sabkha evaporite sequences, it is clear that the Holocene anhydrites developed in two main stages, each forming discrete tracts separated by well-defined storm beaches. It is, however, debatable whether all the so-called sabkha sulphates formed within the sabkha subsurface. The evidence suggests that salinas have also played a significant role in forming the coastal evaporites.

Extensive, ephemeral halite deposits precipitate almost annually from ponded rainwater on deflated Pleistocene aeolian sediments situated immediately landward of the limit of the Holocene marine sediment wedge. In places, that limit is defined by relict beach ridges, one of which has been carbon-dated as 5750–5900 years old, and is taken to approximate to the Flandrian transgressive limit. Each successive rainwater ponding episode dissolves the previous halite deposit and therefore rapidly increases the water salinity to create salinas which then evaporate and reprecipitate a new halite layer. These repetitive events render the preservation potentials of such halite layers essentially almost non-existent.

There are no known present-day gypsum salinas of marine origin in Abu Dhabi, but the characteristics of banded anhydrites within the upper parts of some Holocene coastal sabkha sequences suggest their previous existence. The banded anhydrites replaced subaqueous gypsum that precipitated on the floors of salinas flooded by seawater. They occur stratigraphically above the nodular, enterolithic, often vuggy anhydrites that precipitated within the sabkha sediment from a progressively lowering capillary zone as relative sea-level fell during the late Holocene regression. Occasionally, microbial mats developed on the floors of the gypsum salinas and are now interbedded with the banded anhydrites. Salina gypsum deposits also occur within the Pleistocene and Miocene strata of the coastal regions, and care must be taken not to confuse these with Holocene gypsum.

Subsurface gypsum and halite, together with palygorskite, is forming extensively within the uppermost sediments of the Holocene 'leopard skin' terrain of continental Sabkha Matti. A limited number of present-day gypsum volcanoes are also forming there under artesian influence, although the source of the sulphate is uncertain.

ACKNOWLEDGEMENTS

As a former student of both Doug Shearman and Graham Evans in the early 1970s, the author would like to express enormous thanks to them for instilling in him such interest and enthusiasm in the Abu Dhabi evaporites. Additional thanks are also due to the following experts who, although some may not entirely agree with the interpretations herein, have provided such stimulating discussions in the field, especially with regard to the origins of the banded anhydrites: Peter Homewood, Alan Kendall, John Warren, Warren Wood, Chris Kendall, Peter Bush, Charlotte Schreiber and Graham Evans. Thanks also go Mohammed Al Dabal for guiding the author to the AAPG video pits, and to Peter Hellyer who indirectly provided the funding for much of the fieldwork via the Environmental Agency Abu Dhabi (formerly known as ERWDA) and the Abu Dhabi International Archaeological Society (ADIAS). Kate Davis of Southampton University kindly assisted with the drafting as also did Nestor Buhay II of GeoArabia. The author is grateful to Trident Press for allowing reproduction of some illustrations.

REFERENCES

Butler, G.P. (1969) Modern evaporite deposition and geochemistry of co-existing brines, the sabkha, Trucial Coast, Arabian Gulf. *J. Sed. Petrol.*, **39**, 70–89.

Butler, G.P. (1970) Holocene gypsum and anhydrite of the Abu Dhabi sabkha, Trucial Coast: an alternative explanation of origin. In: *Third Symposium on Salt* (Eds J.L. Rau and L.F. Dellwig), **1**, pp. 120–152. Northern Ohio Geological Society.

Butler, G.P., Harris, P.M. and **Kendall, C.G.St.C.** (1982) Recent evaporites from the Abu Dhabi coastal flats. In: *Deposition and Diagenetic Spectra of Evaporites* (Eds G.R. Hanford, R.G. Loucks and G.R. Davies). SEPM Core Workshop, **3**, 33–64.

Curtis, R., Evans, G., Kinsman, D.J.J. and **Shearman, D.J.** (1963) Association of dolomite and anhydrite in the Recent sediments of the Persian Gulf. *Nature*, **197**, 679–680.

Evans, G. and **Kirkham, A.** (2005) The Quaternary deposits. In: *The Emirates: A Natural History* (Eds P. Hellyer and S. Aspinall), pp. 65–78. Trident Press, London.

Evans, G., Schmidt, V., Bush, P. and **Nelson, H.** (1969) Stratigraphy and geological history of the sabkha, Abu Dhabi, Persian Gulf. *Sedimentology*, **12**, 145–159.

Goodall, T.M. (1997) The Sabkhat Matti – a forgotten wadi system. *Tribulus*, **4.2** 10–13.

Gunatilaka, H.A. and **Shearman, D.J.** (1988) Gypsum-carbonate laminites in a recent sabkha, Kuwait. *Carbonates Evaporites*, **3**, 67–73.

Hardie, L.A. (1986) Ancient carbonate tidal-flat deposits. *Q.J. Colorado Sch. Min.*, **81**, pp. 37–57.

Kenig, F., Huc, A.Y., Purser, B.H. and **Oudin, J.-L.** (1989) Sedimentation, distribution and diagenesis of organic matter in a recent carbonate environment, Abu Dhabi, U.A.E. *Adv. Org. Geochem., Org. Chem.*, **16**, 735–747.

Kinsman, D.J.J. (1969) Modes of formation, sedimentary associations, and diagenetic features of shallow water and supratidal evaporites. *AAPG Bull.*, **52**, 830–840.

Kirkham, A. (1997) Shoreline evolution, aeolian deflation and anhydrite distribution of the Holocene, *Abu Dhabi. GeoArabia*, **2**, 403–416.

Kirkham, A. (1998) A Quaternary proximal foreland ramp and its continental fringe, Arabian Gulf, UAE. In: *Carbonate Ramps* (Eds **V.P. Wright** and **T.P. Burchette**). *Geol. Soc. London Spec. Publ*, **149**, 15–41.

Loreau, J.-P. and **Purser, B.H.** (1973) Distribution and ultrastructure of Holocene ooids in the Persian Gulf. In: *The Persian Gulf: Holocene Carbonate Sedimentation in a Shallow Epicontinental Sea* (Ed. B.H. Purser), pp. 279–328. Springer-Verlag, New York.

Peebles, R.G., Suzuki, M. and **Shaner, M.** (1995) The effects of long-term shallow-burial diagenesis on carbonate-evaporite successions. In: *Selected Middle East Papers from the Middle East Geoscience Conference 1994, Bahrain* (Ed. M.I. Al-Husseini), pp. 761–769.

Scholle, P.A., Shinn, E.A., Halley, M. and **Harris, P.M.** (1986) *Arid Carbonate Coastlines*. AAPG Video, Tulsa OK.

Shearman, D.J. (1966) Origin of marine evaporites by diagenesis. *Trans. Inst. Min. Metall., Sect. B*, **75**, 208–215.

Warren, J.K. and **Kendall, C.G.St.C.** (1985) Comparison of sequences formed in marine sabkha (subaerial) and salina (subaqueous) settings – modern and ancient. *AAPG Bull.*, **69**, 1013–1023.

Int. Assoc. Sedimentol. Spec. Publ. (2011) **43**, 277–298

Distribution of organic matter in the transgressive and regressive Holocene sabkha sediments of Abu Dhabi, United Arab Emirates

FABIEN KENIG

Department of Earth and Environmental Sciences, University of Illinois at Chicago, M/C 186, 845 West Taylor Street, Chicago, IL60607-7059, USA (E-mail: fkenig@uic.edu)

ABSTRACT

Two navigation channels dredged in the 1980s across the hypersaline coastal plain of Abu Dhabi (UAE) offered a unique opportunity to study the distribution of organic matter in transgressive and regressive, subtidal, intertidal and supratidal sediments that accumulated during the Holocene. In modern Abu Dhabi lagoons, intertidal microbial mats and *Avicennia marina* mangals (mangroves), as well as subtidal seagrass fields dominated by *Halodule uninervis*, provide significant amounts of organic matter to modern sediments. Each of these primary producers is associated to a distinct sedimentary facies and biocoenosis recognizable in Holocene sediments. The distribution of these biocoenoses obeys a number of rules deriving from the ability of each of the primary producers to adapt to salinity changes in response to lagoon closure, rate of sea-level change, and associated rate of displacement of the intertidal zone, as well as changing progradation and aggradation rates. The subtidal seagrass biocoenosis is mostly part of the transgressive sequence, as the salinity in the Abu Dhabi lagoon was lower prior to barrier island development. In the transgressive sequence, *Avicennia marina* mangal palaeosols are sparse and formed only where the slope of the substrate was steep enough to allow minimal lateral displacement of the mangal community during transgression. The thin intertidal transgressive sequence is mostly characterized by burrowed microbial mat.

The regressive sequence is dominated by the microbial mat biocoenosis in areas of fast progradation of the sabkha ($> 1\,\mathrm{m\,yr^{-1}}$) as this community is well adapted to the hypersaline conditions of the intertidal zone. A regressive *A. marina* mangal palaeosol was observed only where progradation rates were between 0.35–$0.75\,\mathrm{m\,yr^{-1}}$. The mangal community in Abu Dhabi lagoons is stressed by salinity and fast progradation rates. This contrasts with the dominance of mangal soils in humid tropical environments, where highly productive mangals produce thick and organic matter-rich soils irrespective of the rate of sea-level change and rates of progradation.

Keywords: Sabkha, organic matter, palaeosol, mangal, microbial mat, seagrass.

INTRODUCTION

The Abu Dhabi lagoon carbonate sedimentary system is dominated by aragonite, which constitutes ~90 % of the sediments (e.g. Loreau, 1982). This depositional setting is a product of the Holocene transgression which reached Abu Dhabi ~8000 yr BP and the relative regression that started ~4000 years ago (e.g. Evans *et al.*, 1969). Subject to arid climate conditions, these carbonates are associated with secondary evaporites (gypsum and anhydrite) during their fossilization via lateral accretion of a wide hypersaline coastal plain, the sabkha (Kendall & Skipwith, 1969a; Purser & Evans, 1973).

Organic matter is a significant component in modern Abu Dhabi subtidal and intertidal sediments. Total organic carbon (TOC) content may locally reach 11 wt % (Kenig *et al.*, 1990; Kenig, 1991) and organic matter is playing an important role in diagenetic processes that include dolomitization (e.g. Baltzer *et al.*, 1994). Intertidal microbial mats and *Avicennia marina* mangals (mangroves), as well as subtidal seagrass fields dominated by *Halodule uninervis*, provide significant amounts of organic matter to modern sediments, each forming a distinct sedimentary facies or biocoenoses (Kenig, 1991; Kenig *et al.*, 1990). These facies were identified in Holocene lagoon/sabkha sediments as part of a transgressive and a regressive sequence and have been integrated in a transgressive and regressive virtual sabkha sequence (Kenig *et al.*, 1990; Kenig, 1991).

The Holocene sediments forming the subtidal coastal plain of Abu Dhabi have been the subject of many studies (e.g. Evans *et al.*, 1964, 1969; Kendall & Skipwith, 1968b, 1969b; Butler, 1969; Evans & Bush, 1969; Kinsman & Park, 1976; Patterson & Kinsman, 1977; McKenzie *et al.*, 1980; Butler *et al.*, 1983; Hadley *et al.*, 1998), but few of these were concerned with organic matter distribution, and prior to Kenig *et al.* (1990), they mostly focused on microbial mats (e.g. Kendall & Skipwith, 1968a). These microbial mats have fascinated the pioneers of sedimentological and geochemical exploration of the Abu Dhabi lagoon system as they are prominent features of the vast upper intertidal zone of Abu Dhabi lagoons.

During the 1970s, the urban and industrial development of Abu Dhabi was initially limited to Abu Dhabi Island (Fig. 1). Since the 1980s, the remarkable expansion of Abu Dhabi led to the sprawl of habitation, commercial, industrial, and recreational complexes on the sabkha. This anthropogenic evolution of the sabkha is well illustrated by comparing the aerial photos of the Abu Dhabi lagoon and sabkha shown in Purser (1973) to recent satellite images. The development of the sabkha was associated with dredging of large channels such as the Mussafah and Umm al Nar channels (Fig. 1). In 1987 and 1989, the banks of these channels were fully accessible and allowed the continuous observation of Holocene sediments across the sabkha over large distances.

In this paper, cross-sections of the sabkha based on observations made along the banks of the Mussafah and Umm al Nar channel are described. These cross-sections are used to determine the factors affecting the distribution of organic matter in these Holocene lagoon/sabkha sediments.

METHODS

In the field, samples were kept on ice and deep frozen on the day of collection. Samples remained frozen until lyophilization. Total organic carbon (wt % TOC) was measured by Rock-Eval analysis (Espitalié *et al.*, 1985) and confirmed with a LECO carbon analyser (Kenig, 1991). Mineralogical analyses by X-ray diffractometry were performed at the Institut Français du Pétrole (Rueil Malmaison, France).

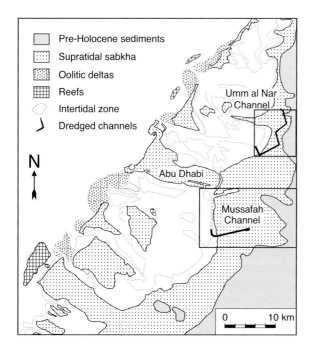

Fig. 1. Map of the lagoon and sabkha sedimentary systems of Abu Dhabi showing the location of the Umm al Nar and Mussafah channels. This map is based on a SPOT satellite image collected in 1987.

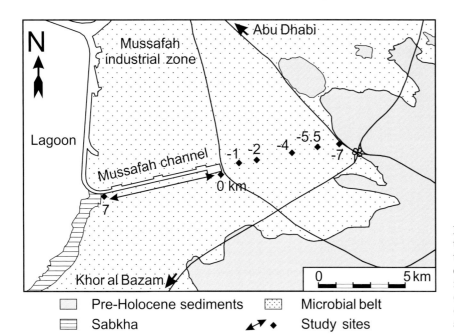

Fig. 2. Map showing the location of the Mussafah channel and the relationship between the Mussafah channel and the living microbial belt in the west and the Pleistocene sand dunes in the east. This map was drawn using SPOT satellite multispectral data obtained in 1987.

RESULTS

Mussafah channel

The Mussafah channel is located to the southwest of the city of Abu Dhabi and south of the industrial zone of Mussafah (Fig. 2). This channel was dredged in 1985 and 1986 perpendicular to the progradation line of the sabkha on the western flank of the main lagoon of Abu Dhabi. In this area, the microbial belt of the mid to upper intertidal zone of Abu Dhabi lagoons (e.g. Kendall & Skipwith, 1968c; Park, 1977; Kenig *et al.*, 1990) is relatively narrow (~300 m) and the mangrove, often present in Abu Dhabi lagoon lower intertidal zone, is absent. The Mussafah channel cuts the sabkha from west to east over 7.5 km in an area where the sabkha is 12 km wide (Fig. 2). The Mussafah channel crosses the so-called "Evans line," a site of previous studies of the Abu Dhabi sabkha (Evans *et al.*, 1969).

Along the southern bank of the channel, vertical profiles were measured every 200 m from the end of the channel to its mouth, in the main lagoon of Abu Dhabi. These 36 vertical profiles were numbered according to their kilometric distances from the landward end of the channel (e.g. km 4.2). The relative elevation of each profile was obtained by measuring the distance between the top of the profile and sea level during a low-tide slack water.

All elevation measurements were completed within 45 minutes by three motorized teams. It is estimated that sea level did not vary more than 5 cm during these measurements. As a result, a synthetic cross-section of the southern bank of the Mussafah channel could be drawn (Kenig, 1991; Fig. 3). The continuity between profiles of each of the facies was verified by following the channel bank. The depth of the layers located below average low tide was measured on cores. The sedimentary facies observed on the bank of the Mussafah channel are described in the following paragraphs with a particular emphasis on their organic matter content.

Pleistocene sediments

Pre-Holocene sediments were observed between km 0 and 2.2 (Fig. 3), as well as along the face forming the bottom end of the Mussafah channel. These sediments are Pleistocene cross-bedded aeolian carbonate sands with abundant quartz. The top of this Pleistocene substrate is located 25 cm below the low-tide reference level at km 1.8 and 1 m above the low-tide reference level at km 0. This steep slope is probably a reflection of the aeolian dune morphology of the Pleistocene substrate of Abu Dhabi (e.g. Evans *et al.*, 1969; Hadley *et al.*, 1998). As observed by Evans *et al.* (1969),

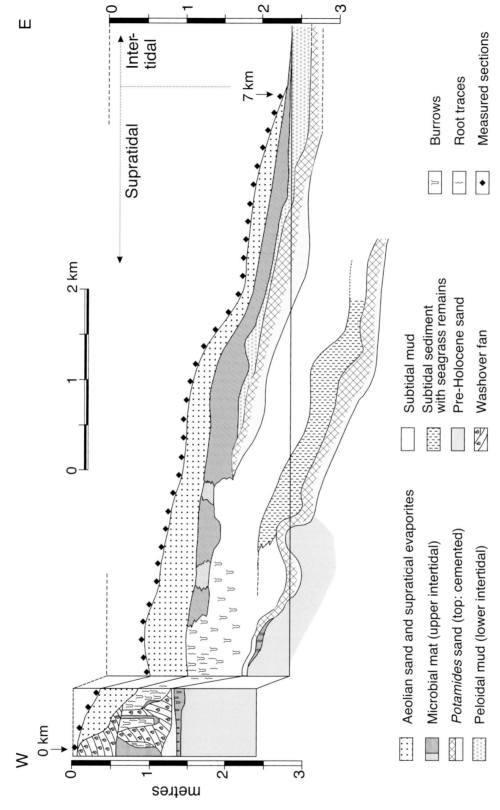

Fig. 3. Cross-section of the Holocene sediments along the bank of the Mussafah channel (adapted from Kenig 1991).

this aeolian sand was locally reworked during the Holocene transgression to form an individual layer. This sand does not contain measurable amounts of organic carbon.

Microbial mats

Two distinct levels of microbial mats can be observed on the banks of the Mussafah channel. The lowest one lies directly on pre-Holocene sediments and is overlain by lower intertidal and subtidal sediments. Thus, this mat represents upper intertidal sediments deposition during the Holocene transgression. Aragonitic shells of gastropods collected just above this mat at km 0.2 were dated 5110 ± 167 yr BP. The second microbial mat, located higher in the section, is laterally connected to the microbial mat of the modern upper intertidal zone, and thus represents regressive upper intertidal sediments.

Transgressive microbial mat. The transgressive microbial mat, 15–35 cm thick, directly overlies the pre-Holocene substrate and forms the first member of a transgressive sequence. It was first reported by Patterson & Kinsman (1977). Its extension seems to be limited to the most elevated part of the pre-Holocene substrate (between km 0 and 1.0; Fig. 3). It is also observed over the whole length of the channel eastern extremity. Petrographic observation (Kenig *et al.*, 1990; Kenig, 1991) indicates that the transgressive mat is genetically similar to modern microbial mats: alternance of thin organic laminae with mineral laminae, the former being mostly formed of the glycocalix of cyanobacteria.

The transgressive mat can be subdivided in two parts. The lower part, ~10 cm thick, is divided by desiccation polygons (Fig. 4A), suggesting its deposition in the upper intertidal zone and progressive increase in the frequency of mat inundation. The lower mat is moderately burrowed. In contrast, the upper part of the transgressive microbial mat seems to be less affected by desiccation, as organic laminae are more continuous than in the lower part of the mat. The thickness of carbonate laminae increases upward reflecting the increased proximity of the lower intertidal sediment source. The upper mat is disrupted by burrows (Fig. 4B and C), which increase in density upwards, suggesting a progressive decrease in the salinity during deposition. The burrows are filled by the overlying bioclastic sand (including numerous *Potamides* gastropods) and, as a result, the surface of the mat

is very irregular and sometimes difficult to define. Burrowing seems to have resulted in a loss of approximately 50% of the mat volume. Thus, the morphology of the transgressive microbial mat displays an internal sequence characteristic of its deposition during a transgression, with upward: (1) increasing frequency of flooding; (2) increasing

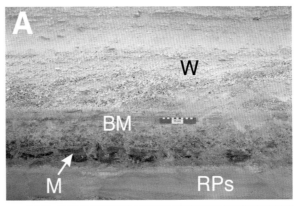

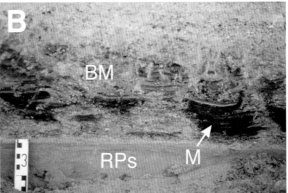

Fig. 4. Transgressive sequence at km −0.2 along the southern bank of the Mussafah channel. (A) General view of section. (B) Detail of A showing the microbial mat in the lower part of the succession. (C) Detail of B showing burrows penetrating the microbial mat. Scales in centimetres. M, microbial mat; BM, burrowed microbial mat; B, burrows; RPs, reworked Pleistocene sand; W, washover fan deposits.

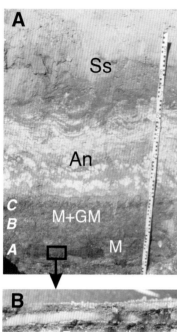

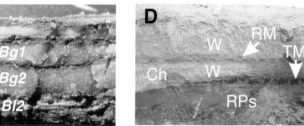

Fig. 5. Mussafah channel section. (A) Pit at −2 km, east of km 0 of Mussafah channel. Ss, sabkha aeolian sand; An, anhydrite; M + GM, microbial mat and gypsum mush. (B) Detail of (A) showing alternation of fine gypsum layers and microbial laminae. Letters in italics indicate the positions of samples listed in Table 1. (C) Subtidal sediments with vertical seagrass root remains (Sr) and lucine bivalves (L) in life position (Mussafah km 4.2). (D) Channel fill (Ch) cross-cutting the transgressive microbial mat (TM), reworked Pleistocene sand (RPs), and regressive microbial mat (RM) at the east end of the Mussafah channel (km 0). W, washover fan deposits.

proximity to sediment source; and (3) decreasing salinity of the depositional environment.

Changes in the mat morphology from base to top are associated with variations in its total organic carbon content (TOC, wt %). The lower mat, which is more massive, has TOC values varying between 0.7 and 2.4 % with an average of 1.5% (n = 11). In the upper, intensely burrowed part of the transgressive microbial mat, TOC never exceeds 1%.

Regressive microbial mat.   The regressive microbial mat is continuous over 6 km from km 1.2 to the modern microbial belt at km 7.2 (Fig. 3). Moreover, the regressive microbial mat was observed at km 0 and −2 (Fig. 5A and B), 7.2 and 9.2 km respectively from its modern equivalent. The total thickness of the microbial mat remains constant (∼40 cm) between km 1.2 and 6, but decreases to ∼20 cm between km 6 and the modern microbial belt.

At the base of the regressive microbial mat, the carbonate laminae are thick (> 1 cm), indicating proximity to the lower intertidal sediment source, and the TOC content of this part of the mat never exceeds 0.5 %. The thickness of the carbonate laminae decrease progressively upwards and the organic laminae become preponderant. The TOC

content of the more massive part of the microbial mat reaches 2.0 %, with an average TOC content of 1.4 % (n = 8). This increased concentration of organic laminae, reflecting the growing distance to the lower intertidal zone, persists until gypsum becomes an important component of the mineral laminae and the microbial mat loses its laminated morphology.

This form of microbial mat is typical of the limit of the upper intertidal and supratidal domains in the modern microbial belt (Kendall & Skipwith, 1968c; Kenig *et al.*, 1990). Though the mat gypsum is green coloured, as a result of the incorporation of microbial pigments during growth, the TOC content of the top part of the mat does not exceed 0.5 % (average 0.3 %, n = 5). Well-defined mud cracks are often present in the organic matter-rich part and in the gypsum-rich part of the regressive mat. The regressive microbial mat is thus characterized by an internal regressive succession from: (1) a lower part with continuous but sparse organic laminae, formed in the mid-intertidal zone; (2) a central part with thin mineral laminae and presence of mud cracks; and (3) an upper part, gypsum-rich, deposited in the upper intertidal zone.

Potamides conicus *sand and grainstone*

These facies are characterized by abundant *Potamides conicus* (Blainville). This taxon is a gastropod of the Cerithioidea superfamily that is able to withstand high salinity, and *P. conicus* sand and grainstone represent lower intertidal deposits. Two layers of sediments containing abundant *P. conicus* were observed along the Mussafah channel (Fig. 3). The higher level overlays subtidal sediments and is overlain by intertidal sediments, and thus belongs to the regressive sequence where it separates subtidal and intertidal sediments. The lower level of sediments containing abundant *P. conicus* generally overlies pre-Holocene aeolian sediments, except where it rests on the transgressive microbial mat. Thus, the lower level of sediments containing *P. conicus*, which is overlain by subtidal lagoonal sediments and/or seagrass-bearing sediments, belongs to the transgressive sequence.

The *P. conicus* sands (in both transgressive and regressive sequences) are mostly formed of gastropod fragments (dominated by *P. conicus*), various bivalves, *Acetabularia* (green alga) stipes, and foraminifera, mostly *Peneroplis* and *Ammonia* (Granier, 1988). The TOC content is very low, always less than 0.1%. Biogenic constituents of the *P. conicus* grainstone are similar to those of the sand, but numerous bioclasts of the grainstone are micritized and numerous grains without structure cannot be identified. Early lithification of the sand is induced by fibrous aragonitic cement in radial position around grains. The summit of the lower grainstone (transgressive sequence) is usually a flat erosion surface cross-cutting cemented gastropods and bivalves.

Subtidal sediments

Subtidal mud with seagrass roots. Subtidal sediments containing seagrass remains were observed over 2.4 km (between km 2.2 and 4.6) along the bank of the Mussafah channel (Fig. 3). This facies is easily recognizable from the vertical traces of seagrass roots (diameter < 1 mm) of *Halodule uninervis*, preserved in life position (Fig. 5C). This sediment is generally made of muddy peloid sand and contains *Cerithium*, *Trochides* and *Mitrella*. It is mostly characterized by the common presence of a deep infaunal bivalve, *Anodontia edentula* (lucine), preserved in life position (Fig. 5C), ~40 cm under its contemporary water–sediment interface (Kenig, 1991). The mineral composition of three samples of subtidal sediments containing seagrass roots collected on the bank of the Mussafah channel is variable. Two samples are similar (km 3.2 and 4.2) with ~40% aragonite, ~30% calcite, ~10% Mg-calcite and ~10% dolomite, but the third sample, sampled the farthest from the lagoon (km 2.6), contains a high percentage of dolomite (57% of the carbonate fraction; Table 1). The TOC contents of the three samples of subtidal sediments containing seagrass roots, collected at km 4.2, 3.2 and 2.6 km, are 0.9%, 0.6% and 0.6%, respectively. This homogeneity in TOC content suggests that dolomitization in the sample collected at km 2.6 did not occur at the expense of the TOC content.

Lucine bivalve shells were sampled for ^{14}C dating at km 4.2 in the lagoonal sediment containing seagrass roots. These lucine shells were collected ~45 cm below the top of the lagoonal sediment containing seagrass. The age obtained, 4280 ± 186 yr BP, corresponds to that proposed by Evans *et al.* (1969) for the end of the Holocene transgression in the Abu Dhabi region. As lucines live 40 cm below the water–sediment interface, the lagoonal sediment containing seagrass roots belong, for the most part, to the transgressive sequence.

Other lagoonal muds. Lagoonal muds without seagrass remains form the largest part of the subtidal sediments observed along the bank of the Mussafah channel. These muds are made of pelletized aragonite needles with abundant foraminifera. At the base of this facies, *Cerithium scabridum* are more abundant than *Potamides*. Toward the top, this ratio is reversed and *Potamides* are associated with *Mitrella* and small-size *Trochides* in an assemblage suggesting increased confinement of the lagoon. This trend, in turn, suggests that the lagoonal muds without seagrass roots belong mostly to the regressive sequence.

The carbonate fraction of the lagoonal sediments without seagrass roots collected at km 4.2 is dominated (Table 1) by aragonite (45%) and calcite (26%). This contrasts with the composition observed at km 1.6 (three samples) where dolomite, which represents 51% to 85% of the carbonate fraction, appears to have developed at the expense of aragonite and Mg-calcite. The TOC content of lagoonal sediments without seagrass roots is very low, always below 0.3% (n = 6, Table 1).

Table 1. Mineral composition and total organic matter contents of Holocene sediments collected along the Mussafah channel.

Sample location	IFP number	Facies	TOC (wt %)	Aragonite	Calcite	Mg-calcite	Dolomite	Σ Carbonate	Quartz	Albite	Gypsum	Aragonite	Calcite	Mg-calcite	Dolomite
				(% of total fraction)								(% of carbonate fraction)			
−2 km A	73055	Regressive microbial mat	1.9	nd	nd	nd	nd	nd	nd	nd	nd	nd	nd	nd	nd
−2 km B	73056	Regressive microbial mat	1.0	nd	nd	nd	nd	nd	nd	nd	nd	nd	nd	nd	nd
−2 km C	73057	Green gypsum mush	0.1	nd	nd	nd	nd	nd	nd	nd	nd	nd	nd	nd	nd
−2 km BL2	73058	Regressive microbial mat	1.5	nd	nd	nd	nd	nd	nd	nd	nd	nd	nd	nd	nd
−2 km BG 1	73059	Gypsum lamina	0.0	nd	nd	nd	nd	nd	nd	nd	nd	nd	nd	nd	nd
−2 km BG 2	73060	Gypsum lamina	0.1	nd	nd	nd	nd	nd	nd	nd	nd	nd	nd	nd	nd
0 km B	73051	Laminated peloid sand	0.6	nd	nd	nd	nd	nd	nd	nd	nd	nd	nd	nd	nd
0 km C	73052	Transgressive microbial mat	0.4	nd	nd	nd	nd	nd	nd	nd	nd	nd	nd	nd	nd
0 km D	73053	Transgressive microbial mat	1.0	nd	nd	nd	nd	nd	nd	nd	nd	nd	nd	nd	nd
0 km F	73054	Transgressive microbial mat	1.2	nd	nd	nd	nd	nd	nd	nd	nd	nd	nd	nd	nd
RE 0.2km	73061	Transgressive microbial mat	2.4	15	17	bd	13	45	45	10	bd	34	38	bd	28
0.6 km E	73049	Regressive microbial mat	1.2	nd	nd	nd	nd	nd	nd	nd	nd	nd	nd	nd	nd
1.6 km B	73045	Lagoonal sediment with seagrass roots	0.6	22	11	7	54	94	5	bd	bd	23	12	7	57
1.6 km C	73046	Lagoonal sediment	0.2	3	10	bd	79	92	6	2	bd	3	11	bd	86
1.6 km D	73047	Lagoonal sediment	0.2	25	8	4	49	86	13	bd	bd	29	9	4	58
1.6 km E	73048	Lagoonal sediment	0.2	31	6	5	45	87	13	bd	bd	35	6	6	52
2.6 km A	73043	Regressive microbial mat	1.0	nd	nd	nd	nd	nd	nd	nd	nd	nd	nd	nd	nd

			TOC												
2.6 km B	73044	Lagoonal sediment with seagrass roots	0.6	nd	nd	nd	nd	nd	nd	nd	nd	nd	nd	nd	nd
3.2 km 0.4	89683	Regressive microbial mat	1.7	22	20	bd	27	69	26	4	bd	32	28	bd	39
3.2 km 1.4	89684	Lagoonal sediment with seagrass roots	0.6	38	22	14	10	84	12	4	bd	45	26	17	12
3.6 km A	73041	Regressive microbial mat	1.6	nd	nd	nd	nd	nd	nd	nd	nd	nd	nd	nd	nd
4.2 km E1	73036	Gypsum mush	0.4	nd	nd	nd	nd	nd	nd	nd	nd	nd	nd	nd	nd
4.2 km E2	73037	Regressive microbial mat (top)	1.0	4	34	bd	14	52	18	10	20	7	65	bd	27
4.2 km K	73038	Lagoonal sediment	0.3	40	7	6	44	97	3	bd	bd	41	8	6	45
4.2 km L	73039	Lagoonal sediment with seagrass roots	0.9	31	20	20	12	83	16	bd	bd	37	25	24	14
4.2 km F	73040	Regressive microbial mat (bottom)	2.1	19	33	bd	9	61	33	6	bd	31	54	bd	14
6 km A	73033	Regressive microbial mat	1.8	nd	nd	nd	nd	nd	nd	nd	nd	nd	nd	nd	nd
6 km B	73034	Regressive microbial mat	1.0	nd	nd	nd	nd	nd	nd	nd	nd	nd	nd	nd	nd
6 km C	73035	Lower intertidal mud	0.5	nd	nd	nd	nd	nd	nd	nd	nd	nd	nd	nd	nd

TOC = total organic carbon content, nd = not determined, bd = below detection.

Supratidal sediments

All of the Holocene intertidal sediments of the Mussafah channel are covered by a layer of fine brown aeolian sand. This sand is comparable to the pre-Holocene sand but includes many bioclasts. The carbonate fraction (Table 1) is dominated by calcite (80 %) and a smaller amount of dolomite (15 %). Aragonite is usually absent, but quartz represents more than 20 % of the total fraction.

The thickness of supratidal sediments increases landward, from a few centimetres to 50 cm to 65 cm, at distances of 440 m, 5.6 km and 6 km, respectively, from the modern intertidal zone. In the lower supratidal zone, below the microbial film and above the intertidal microbial mat, sandy sediments made of millimetric gypsum crystals occur. This is the so-called "gypsum mush" of Butler *et al.* (1983). As observed by these authors, this gypsum sand can be divided in two distinct parts on the basis of the substrate in which it grows. The lower part, with a green colour, grows at the expense of the upper intertidal microbial mat from the sub-millimetric gypsum crystals that precipitated between the mat laminae in the upper intertidal zone. The secondary growth of these gypsum crystals causes the disruption of the top of the intertidal microbial mat. The second part of the "gypsum mush," beige in colour, grows in the aeolian sabkha sediment deposited above the intertidal mat. Together, these two layers may reach 30 cm, their thickness depending on the age of the observed profile.

Toward the eastern end of the channel, 2.4 km from the modern intertidal zone, anhydrite becomes dominant in supratidal sediments. Anhydrite can be observed as dispersed or packed nodules, the so-called "chicken wire" morphology, as well as enterolithic and banded anhydrite. These morphologies were previously described in the Abu Dhabi sabkha by Butler (1969) and Butler *et al.* (1983). Evaporite zonation in the sabkha, dominated by an evolution of gypsum to anhydrite in aging sediments, was studied by Butler *et al.* (1983) who suggested that it is controlled by flooding frequency. In aging sabkha sediments, anhydrite replaces the "gypsum mush," and secondary gypsum crystals form deeper in the sediments, notably, but not exclusively in the microbial mat.

Washover fans

Two cross-stratified accumulations of shells and bioclasts were observed between 0 and 0.5 km of the Mussafah channel, as well as on the bank forming the end of the channel. The fraction above 250 μm represent more than 70 % by weight of these sediments for two samples and more than 80 % and 90 % for two others. Bioclasts represent ~60 % of this fraction, the remainder being formed of aggregates of bioclasts. This sand is made of 30 % bivalve and gastropod shells, 10–15 % foraminifera (*Peneroplis* being dominant), 10 % *Acetabularia*, 10 % *Spirorbis* (polychaete worm), and up to 5 % bryozoans (Granier, 1988). The bioclasts correspond to different biocoenosis, with *P. conicus*, *Cerithium*, *Mitrelles*, *Trochoides* and *Brachydontes*. Bioclastic beds are erosive and have a landward dip of 5° to 10°. Certain groups of beds crosscut others. This description matches well the definition of storm-washover fan deposits provided by Reineck & Singh (1980), though other high-energy processes cannot be fully excluded at this point. Aragonitic shells of the lowest shell and bioclasts layer were dated and gave ages of 5400 ± 126 yr and 5110 ± 167 yr BP. The lower unit thus belongs to the transgressive cycle.

Between km 0 and 0.2 and on the bank forming the extremity of the Mussafah channel, a second package of washover fan sediments covers the regressive intertidal sediments (Fig. 5D). This washover fan, which is covered by aeolian sabkha sediments, covers a microbial mat belonging to the regressive sequence and is considered part of the regressive sequence. The topographic high formed by the pre-Holocene sediments was the site of events of high hydrodynamic energy necessary for the formation of these washover fans.

Relative elevation of sedimentary facies

Between km 0 and 0.2, the top of the regressive microbial mat is 1.8 m above the reference level (low-tide slack water) but the top of the modern intertidal microbial mat is approximately 25 cm above the reference level. As the top of the microbial mat, including the green gypsum mush, represents the top of the intertidal zone, this difference in relative elevation reflects a draw down of the upper intertidal zone elevation during progradation of the sabkha. The slope of the regressive microbial mat ($\sim$0.2 m km^{-1}) is relatively constant between km 1.2 and the modern microbial belt, though local variations are observed (Fig. 3). The average slope of the sabkha surface is steeper ($\sim$0.3 m km^{-1}) because of the landward increase in thickness of the supratidal sediments. The slope of the regressive *Potamides* grainstone is equal to that of the regressive microbial mat between km 3.2 and the modern microbial belt. Thus, there is little variability between the base of the *Potamides* facies and the top of the microbial mat (62–74 cm). As the *Potamides* grainstone forms in the lower intertidal domain (see above), the constant vertical distance between the top of microbial mat and the *Potamides* grainstone suggests that the tidal range did not vary in the Mussafah area during the Holocene. Evans *et al.* (1969), by contrast, suggested a progressive reduction of the tidal range in response to the growth of barrier islands during the Holocene. Thus, the change in elevation of the regressive microbial mat during the Holocene reflects a relative sea-level elevation change.

Progradation and aggradation at Mussafah

Evans *et al.* (1969) determined that the accumulation of the regressive sequence started ~4000 yr BP. A regressive microbial mat was observed at the highest point of the Mussafah channel section (km 0), 7 km away from it modern equivalent. Based on these data, the minimum average rate of lateral accretion for regressive intertidal sediments is 1.75 m yr^{-1}. This calculation does not use the maximum width of the sabkha at Mussafah (12 km) because the microbial mats observed at km −2, landward of the relative high of pre-Holocene sediments around km 0, were formed in a lagoon individualized by this relative high and are possibly more recent than the regressive mat observed at the highest point of the Mussafah section. This suggests that the sabkha around Mussafah did not prograde from the modern continental limit of the sabkha but from relative highs of pre-Holocene aeolian sediments.

Potamides shells collected just below the regressive mat at km 4.2 of the Mussafah section were dated by ^{14}C and gave an age of 1580 ± 186 yr BP. These lower intertidal *Potamides* were sampled ~3.3 km away from the modern lower intertidal. Thus, the lateral accretion rate of the intertidal zone between these two points is approximately 2.1 m yr^{-1}. The lateral accumulation rate of the intertidal sediments seems to have been higher during the last 1500 years (~2.1 m yr^{-1}) than during the previous 2500 years (~1.3 m yr^{-1}).

The top of the lagoonal sediments containing seagrass roots at km 4.2 was dated 4280 ± 186 yr BP (see above). The two dated sediments at km 4.2, which are vertically separated by 95 cm of subtidal and lower intertidal sediments, permit to estimate an average aggradation rate during the regressive phase of ~0.35 mm yr^{-1}.

Umm al Nar channel

The Umm al Nar channel, located northeast of the city of Abu Dhabi, was dredged in an area where the sabkha is 4 km wide (Figs 1 and 6). The two extremities of the channel reach a lagoon now profoundly transformed by the development of Abu Dhabi harbour and the dredging of navigation channels between the barrier islands. Fortunately, Evans & Bush (1969) and Evans *et al.* (1973) extensively studied this lagoon prior to major anthropogenic influence. The Umm al Nar channel banks were still accessible in 1989, but could not be accessed in 2004, following urbanization. Dredging of the channel and the related urbanization changed the hydrology of this area but the hydrological framework of the Umm al Nar area was studied by Patterson & Kinsman (1981, 1982) prior to disruption of the sabkha.

The Umm al Nar channel, as studied in 1987 and 1989, was formed of four straight sections (designated A, B, C, D, from north to south; Fig. 6). Section D was not studied because of the poor preservation of the channel banks. The section from A–C, dredged shortly before the 1987 survey, offered well-preserved and easily accessible flanks. Additionally, three pits were dug along an east–west line joining the angle formed by sections B and C of the channel and the intertidal zone. Relative elevations of the vertical profiles were obtained following the method used for the Mussafah channel. Vertical elevations of the pits were obtained by leveling from a reference point on the channel bank.

All the measured vertical sections were assembled to form a fence diagram of the sabkha showing the distribution of organo-sedimentary facies and their spatial relationship to the Pleistocene substrate (Fig. 7). Sedimentary and organic facies along the Um al Nar channel are generally similar to those observed in the Mussafah channel. Thus, the following description will concern facies that were not observed along the Mussafah channel and local particularities of facies already described for Mussafah. Particular attention will be given to the spatial distribution of organo-sedimentary facies (mangal palaeosol, microbial mats, and lagoonal sediments containing seagrass roots).

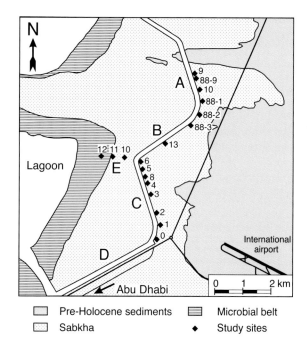

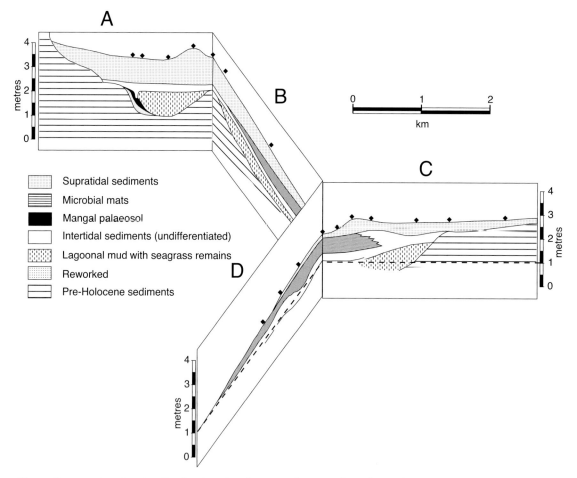

Fig. 6. Map of the Umm al Nar channel. This map was drawn using SPOT satellite multispectral data obtained in 1987.

Fig. 7. Fence diagram showing the distribution of Holocene sediments along the bank of the Umm al Nar channel (adapted from Kenig 1991).

Pre-Holocene sediments

The Holocene transgression deposited carbonate sediments on a landscape of Pleistocene aeolian sand dunes. These slightly cemented aeolian sediments are partly covered by the sediments of the sabkha and disappear westward of the channel under sabkha and intertidal sediments (Fig. 6). The east–west orientation of the dunes of their cross bedding with a southern dip indicates the dominance of the northerly winds during pre-Holocene aeolian sediment accumulation.

Section A cuts across a slightly lithified, cross-bedded, aeolian dune (Fig. 7). This dune, which stands up to 4 m above the sabkha surface, is part of the pre-Holocene sedimentary system limiting the sabkha to the east (Fig. 6). Westward of the channel, the flank of the palaeodune is onlapped by sabkha sediments. Along branch A, to the south, Holocene sediments rest disconformably on cross-bedded aeolian sediments.

The pre-Holocene highs outcropping in the sabkha or along the bank of the Umm al Nar channel are extensions of the pre-Holocene sediments observed on the continental side of the sabkha. These pre-Holocene highs restricted the area of deposition of Holocene carbonates. During the Holocene transgression and regression, these aeolian sediments formed relative highs separating the two distinct zones of sediment accumulation along the Umm al Nar channel: Sections A, B and C (Fig. 7).

Mangal palaeosol

Along Section A, a mangal palaeosol outcrops over 70 m in an area where the pre-Holocene sediment surface has a relatively steep slope (Figs 7 and 8). This palaeosol rests on a lens of grey sand, which can be interpreted as reworked aeolian Pleistocene sand. The mangal palaeosol is overlain by lagoonal sediments with seagrass roots (Fig. 8) deposited in a subtidal environment. The mangrove soil, which was deposited in a mid-intertidal environment, belongs to the transgressive sequence. A tree stump and aragonite shells of *Potamides* collected in the organic matter-rich part of the mangal palaeosol were ^{14}C dated at 5950 ± 80 yr BP and 5470 ± 1670 yr BP, respectively (note that the very large error around the age of the *Potamides* shells is the result of the small sample size).

The thickness of the mangrove palaeosol reaches 50 cm, but the palaeosol can be divided in two horizons (Fig. 8). The lower horizon is made of reworked Pleistocene aeolian sand colonized by a sparse vertical and horizontal root system (Fig. 9A and B). The top horizon of the palaeosol is made of 20 cm of dark, muddy, peloid sand where the organic-carbon concentration can be relatively high (see below) and where tree stumps are commonly found. Roots in the palaeosol form a horizontal network from which depart vertical roots, as in the modern *Avicennia marina* root system, in which aeriferous roots (pneumatophores) form an upward extension of a radiating underground root-system (e.g. Plaziat, 1995). The distribution of the fossil pneumatophores is in all points similar to that observed in the modern *Avicennia* mangals. Observation of the attachment between vertical pneumatophores and the horizontal root network is diagnostic of *Avicennia* (Plaziat, 1995). Tree stumps with radiating horizontal root system were observed in life position. As mentioned by Plaziat (1995) and Kenig (1991), these tree stumps

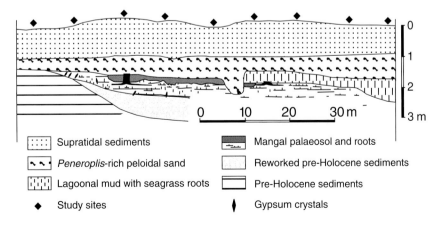

Fig. 8. Profile of the bank of the Umm al Nar channel at Site 88-9 (see Fig. 6 for location) showing the transgressive *Avicennia marina* mangal palaeosol.

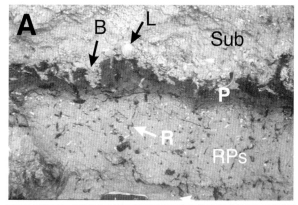

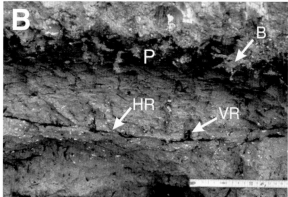

Fig. 9. Mangal palaeosol. (A and B) Transgressive *Avicennia marina* mangal palaeosol on the bank of the Umm al Nar channel at Site 9 (see Fig. 6 for location). (C) Isolated *Avivennia marina* trunk. Sub, subtidal sediments; P, palaeosol; R, roots; RPs, reworked Pleistocene sand; B, burrows; L, lucine bivalve; HR, horizontal roots; VR, vertical roots (pneumatophores).

were identified as *Avicennia* by J.C. Koeniguer on the basis of the wood structure of stumps and roots.

The upper, organic-rich part of the mangal palaeosol is burrowed. These burrows, which penetrate more than 15 cm in the organic-rich palaeosol, are filled by peloid sand including abundant foraminifera, notably *Peneroplis*. The

palaeosol was eroded in the southern part of the section prior to the deposition of the overlying lagoonal sediments containing vertical seagrass rootlets.

The TOC content of the mangrove palaeosol varies as a function of the root network density (Table 2). In the lower part of the palaeosol, apart from the roots, the sediment is organic matter depleted (TOC < 0.1 %; n = 2). In the dark, upper part of the palaeosol, the TOC content varies between 0.3 and 4.7 % (average 1.6 %, n = 5). The TOC content of the burrows fill is ≤0.1 % (n = 1). Microscopic observation of thin sections of resin-impregnated palaeosol samples suggest that the organic matter is almost exclusively made of elements of the root system (pneumatophores, horizontal roots and feeding rootlets). A very small amount of leaf material was observed exclusively in the top part of the palaeosol, but the majority of the organic matter of the TOC-rich part of the soil is made of small horizontal roots and feeding rootlets.

Lagoonal sediment with seagrass roots

This facies was described previously, as it outcrops along the bank of the Mussafah channel. However, the density of seagrass roots seems to be less at Umm al Nar than at Mussafah, and at Umm al Nar the seagrass facies was mostly recognized by its typical faunal assemblage and by the preservation of seagrass remains on the shells of *Anodontia edentula* (lucine). The TOC content of the seagrass facies at Umm al Nar varies between 0.1 and 0.3 % (average 0.2 %; n = 3) and is generally lower than at Mussafah.

The lagoonal sediments containing seagrass roots were observed along sections A–C. This facies is dominant along Section A, reaching 1 m of thickness, but thins on the palaeo-slopes formed by the pre-Holocene sediments and the mangal palaeosol. Along Section B, the lagoonal sediments containing seagrass roots onlap pre-Holocene sediments and increase in thickness toward the southwest until Site 13; beyond this, the lagoonal sediments containing seagrass roots can no longer be observed as they pass under the reference low-tide slack water level. It is likely that this facies extends farther south and west as it appears again on Section C of the channel, were it overlays subtidal sediments with large *Cardium* and *Cerithes* shells that suggest the existence of a more open lagoonal

Table 2. Mineral composition and total organic matter contents of Holocene sediments collected along the Umm al Nar channel.

Sample location	IFP number	Facies	TOC (wt %)	Aragonite	Calcite	Mg-calcite	Dolomite	Σ Carbonate	Quartz	Albite	Gypsum	Aragonite	Calcite	Mg-calcite	Dolomite
				(% of total fraction)								(% of carbonate fraction)			
2.0 km	73062	Lagoonal sediment with seagrass roots	0.1	nd	nd	nd	nd	nd	nd	nd	nd	nd	nd	nd	nd
2.7 km 8A	73063	Regressive microbial film and mud	bd	nd	nd	nd	nd	nd	nd	nd	nd	nd	nd	nd	nd
2.7 km 8B	73064	Regressive microbial mat	0.6	nd	nd	nd	nd	nd	nd	nd	nd	nd	nd	nd	nd
2.7 km 8C	73065	Intertidal peloid sand	0.9	nd	nd	nd	nd	nd	nd	nd	nd	nd	nd	nd	nd
3.0 km 5A	73066	Regressive microbial mat	1.8	nd	nd	nd	nd	nd	nd	nd	nd	nd	nd	nd	nd
3.0 km 5B	73067	Regressive microbial mat	1.2	nd	nd	nd	nd	nd	nd	nd	nd	nd	nd	nd	nd
3.0 km 5C	73068	Intertidal peloid sand	0.8	nd	nd	nd	nd	nd	nd	nd	nd	nd	nd	nd	nd
3.2 km 6W	73069	Regressive microbial mat (top)	0.4	19	12	bd	bd	21	14	2	nd	61	39	bd	bd
3.2 km 6X	73070	Regressive microbial mat	2.1	59	27	bd	4	90	10	bd	bd	66	30	bd	4
3.2 km 6Y	73071	Regressive microbial mat	1.5	68	11	3	4	86	14	bd	bd	79	13	3	5
3.2 km 6Z	73072	Lower intertidal	1.0	51	22	8	6	88	10	3	bd	58	25	9	8
9	73073	Transgressive mangal palaeosol	4.7	3	17	3	21	44	44	12	bd	6	39	6	48
9	73074	Transgressive mangal palaeosol-roots	1.0	nd	nd	nd	nd	nd	nd	nd	nd	nd	nd	nd	nd
9	73075	Transgressive mangal roots and sediment	0.1	nd	nd	nd	nd	nd	nd	nd	nd	nd	nd	nd	nd
9	73076	Wood stump	13.6	nd	nd	nd	nd	nd	nd	nd	nd	nd	nd	nd	nd
9-4	89686	Transgressive mangal palaeosol (top)	0.6	16	40	bd	20	76	20	4	bd	21	52	bd	27
9-4	89687	Transgressive mangal palaeosol (mid-)	0.3	12	41	bd	21	74	26	bd	bd	16	55	bd	29

Table 2 (*Continued*)

Sample location	IFP number	Facies	TOC (wt %)	Aragonite	Calcite	Mg-calcite	Dolomite	Σ Carbonate	Quartz	Albite	Gypsum	Aragonite	Calcite	Mg-calcite	Dolomite
				(% of total fraction)								(% of carbonate fraction)			
9-4	89688	Lagoonal sediment with seagrass roots	0.2	12	20	9	23	64	29	6	bd	19	31	14	36
9-4	89689	Pre-Holocene aeolian sand	0.9	26	17	4	27	74	18	5	3	35	23	5	37
10V	73077	Regressive microbial mat	2.0	61	12	5	3	81	19	bd	bd	76	15	6	3
10W	73078	Regressive microbial mat	2.0	56	15	6	5	82	13	5	bd	68	18	8	6
10X	73079	Regressive microbial mat	1.1	64	17	7	5	93	7	bd	bd	69	19	7	5
10Y	73080	Regressive microbial mat	1.9	54	15	8	8	85	14	bd	bd	62	18	10	10
10Z	73081	Regressive microbial mat	1.1	59	16	7	8	90	10	bd	bd	65	18	8	9
11X	73082	Regressive microbial mat	1.0	10	7	bd	bd	17	9	5	68	60	40	bd	bd
11Y	73083	Regressive microbial mat	1.3	nd	nd	nd	nd	nd	nd	nd	nd	nd	nd	nd	nd
11Z	73084	Lagoonal mud	0.3	nd	nd	nd	nd	nd	nd	nd	nd	nd	nd	nd	nd
12 A	73085	Regressive microbial mat	1.8	nd	nd	nd	nd	nd	nd	nd	nd	nd	nd	nd	nd
12 B	73086	Regressive mat with gypsum mush	0.3	nd	nd	nd	nd	nd	nd	nd	nd	nd	nd	nd	nd
88-1 (13-19)	89707	Sabkha aeolian sediment	bd	bd	32	bd	bd	32	2	bd	65	bd	100	bd	bd
88-1 (24-31)	89706	Regressive microbial mat (top)	bd	11	18	bd	bd	29	bd	bd	71	37	62	bd	bd
88-1 (25-35)	89704	Regressive microbial mat	1.8	66	17	4	2	89	11	bd	bd	74	19	5	3
88-1 (35-40)	89703	Regressive microbial mat	1.4	58	21	5	3	87	13	bd	bd	66	24	6	4
88-1 (37-47)	89705	Regressive microbial mat	2.2	55	18	bd	bd	73	20	bd	7	76	24	bd	bd
88-1 (48-54)	89702	Regressive microbial mat	1.2	55	21	5	3	84	15	bd	bd	65	25	5	4
88-1 (66-77)	89701	Regressive microbial mat	0.4	50	27	4	8	89	11	bd	bd	56	30	5	8
88-2	89694	*Peneroplis*-rich sand	0.1	9	33	3	19	64	34	2	bd	14	51	5	29

TOC = total organic carbon content, nd = not determined, bd = below detection.

environment before the development of the sea-grass community.

The lagoonal sediments containing seagrass roots directly overlie eroded pre-Holocene sediments, a mangal palaeosol dated 5950 ± 160 yr BP, or more open lagoon sediments (Site 8). Thus, these subtidal sediments were deposited, in part, during the transgression. The lagoonal sediments containing seagrass roots are also locally covered by subtidal sediments without seagrass roots and by the lower intertidal *Potamides*-rich peloid sand, similar to that observed at Mussafah. Thus, the top of the lagoonal sediment containing seagrass roots may have been locally deposited during the regressive phase. This hypothesis is concordant with the age of lucine aragonitic shells (3020 ± 160 yr BP) contemporaneous to the accumulation of the top of the lagoonal sediments containing seagrass roots.

Microbial mats

The transgressive microbial mat does not outcrop on the bank of the Umm al Nar channel, though its presence was detected by Patterson & Kinsman (1981) just south of the area where the Umm al Nar channel was dredged (Fig. 1). In contrast, the regressive microbial mat outcrops spectacularly along branches B and C (Fig. 10). The regressive microbial mat appears a few tens of metres from the intersection of branches A and B and progressively increases in thickness to reach 60 cm at intersections of branches B and C (Site 6, Fig. 10). The regressive microbial mat is continuous for 2.8 km on the banks of branches B and C of the channel, and was observed to extend to the modern micro-bial belt along Section E. On the banks of branch C, stratified sand (Site 3) passes laterally to muddy sediments containing microbial films (Site 4) with an increasing density of microbial laminae toward Site 6. This facies change reflects a change in hydrodynamics, probably resulting from the progressive development of the coastal barrier island system.

The elevation of the base of the regressive microbial mat at its highest point along branch B of the channel is less than that of the regressive lower intertidal sediments of Section A. Thus, the microbial mat outcropping on the bank of Umm al Nar channel is not contemporaneous with the onset of the regression.

Vertical change in morphology of the microbial mats on the banks of Umm al Nar channel is

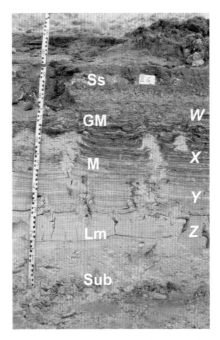

Fig. 10. Regressive microbial mat along the bank of the Umm al Nar channel (Site 6; see Fig. 6 for location). Letters in italics indicate the positions of samples listed in Table 2. Sub, subtidal sediments; Lm, microbial lamination in lower intertidal mud; M, microbial mat; GM, gypsum mush; Ss, sabkha aeolian sand.

identical to that observed at Mussafah (Fig. 10). However, at Umm al Nar, microbial mats are thicker than at Mussafah, at the expense of the mid-intertidal sediments. This suggests that conditions for the preservation of microbial lamination were found lower in the intertidal zone at Umm al Nar than at Mussafah. This difference is perhaps a reflection of higher salinity at Umm al Nar, limiting the activity of heterotrophs lower in the intertidal zone than at Mussafah.

The TOC content of the microbial mat is dependent on the density of organic laminae. The lower part of the mat, where organic laminae are sparse, has a TOC content of 0.8 % (n = 5). The TOC content increases with organic laminae density to reach 2.2 %. The average TOC in the organic lamina-rich part of the mat is 1.8 % (n = 16). The upper part of the mat altered by growth of "gypsum mush" is TOC depleted, with an average of 0.2 % (n = 5).

Hypothesis on the progradation and aggradation rates at Umm al Nar

The outcrop of lagoonal sediments with seagrass roots on Section C of the Umm al Nar channel is found at the same elevation as its homologue on Section A. These outcrops are found relatively high in the profiles and no outcrops of their contemporaneous intertidal sediments are visible on the bank of the Umm al Nar channel. The transgressive intertidal zone is not expressed with lagoonal sediments directly overlying pre-Holocene aeolian sediments. The intertidal sediments overlying the lagoonal sediments with seagrass roots belong to the regressive sequence and must be younger than the age of the topmost seagrass sediments (3020 ± 160 yr BP). It can thus be supposed that the regressive microbial mat colonized the upper intertidal zone relatively late in the regressive phase. Although the microbial mat was not dated, its age can be approximated using the average aggradation rate calculated for the intertidal sediments of the Mussafah channel (0.37 mm yr^{-1}). The base of the microbial mat on Profile 8, located 50 cm above the lagoonal sediments containing seagrass roots, would thus be $\sim$1350 years younger, with an age of $\sim$1670 yr BP. Using that approximation, the progradation between the base of the fossil mat at Site 8 and its modern equivalent, exactly 2 km away, would be $\sim$1.2 m yr^{-1}. This progradation rate is less, but of the same order of magnitude than that calculated for the Mussafah area (1.75–2.1 m yr^{-1}). A lower rate was to be expected as the sabkha at Umm la Nar is narrower than at Mussafah. Moreover, this value is compatible with the average progradation rate at Umm al Nar ($\sim$1 m yr^{-1}) calculated from the age of maximum transgression (4000 yr BP; Evans *et al.*,1969) and the total width of the sabkha (4000 m).

DISCUSSION

Distribution of organic matter in Abu Dhabi sabkha sediments

Observations made along the channel banks of the Mussafah and Umm al Nar channels indicate that the various organic matter-bearing facies of the sabkha are not distributed evenly. For example, the regressive mangal palaeosol that was observed by Kenig (1991) and Baltzer *et al.* (1994) at Ras Ghanadah was not observed at all on the channel banks. The distribution of the organic facies in both the transgressive and regressive sequences seems to be dictated by a number of parameters: rate of sea-level change; aggradation and progradation rate; slope of the substrate; and formation of the barrier islands.

Transgressive sequence

The transgressive sequence is generally represented by a layer of reworked aeolian Pleistocene sand with marine carbonate overlain sometimes by a thin layer of intertidal sediments, and by subtidal sediments containing seagrass remains. Kinsman & Park (1976) and Patterson & Kinsman (1981) noted the irregular presence of thin, upper intertidal, 2–5 cm microbial lamination, sometimes burrowed, overlain by lower intertidal skeletal and peloid sand with abundant cerithids (probably *Potamides*). Transgressive microbial mats up to 10 cm thick were also noted by Schneider (1975), although no indication of the location of these mats was provided. A much thicker transgressive microbial mat ($>$ 20 cm) was observed along the Mussafah channel (Fig. 3), but the transgressive microbial mats are always much thinner than those observed in Holocene regressive sequences. A transgressive mangal palaeosol was observed only at Umm al Nar, and both the transgressive microbial mats and mangal palaeosol were observed on the slopes of Pleistocene highs (Figs 3 and 7).

The limited thickness (and sometimes absence) of intertidal transgressive sediments is interpreted to be a consequence of the relatively rapid Holocene transgression (Kinsman & Park, 1976) as the rate of intertidal sediment accumulation was too slow to keep pace with the transgression rate. The rapid landward displacement of the intertidal zone was accompanied by the rapid displacement of the ecological zones favourable to the growth of intertidal communities of primary producers (microbial mats and *Avicennia* mangals). Thus, the presence of these communities at a given location was not long enough to allow accumulation of significant amounts of organic matter. In extreme cases, if the rate of lateral displacement of the ecological zones favoured by primary producers is too rapid, the renewal of the intertidal communities cannot occur, resulting in community disappearance. In contrast, the steep slopes of the Pleistocene highs allowed a slower lateral movement of the intertidal zone during transgression. The communities of intertidal primary producers, mangal and microbial mats, remained in place for longer periods of time, resulting in significant accumulation of organic matter.

Other factors may also help explain the local distribution of the transgressive microbial mat and mangal palaeosol. Mangals can only develop on soft substrates, favourable to the development of their root system; at Mussafah, the palaeosol lies on a soft lens of reworked Pleistocene aeolian sand mixed with marine carbonates. The indurated Pleistocene substrate would not have allowed mangal development. It is also probable that a significant portion of intertidal deposits were eroded during transgression. This was observed for part of the transgressive mangal palaeosol at Umm al Nar (Fig. 8) and for the microbial mat of Mussafah (Fig. 5D).

In contrast to the thin intertidal transgressive sediments, the lowest intertidal and subtidal facies of the transgressive sequence can reach a significant thickness (>1 m). The subtidal sediments containing seagrass roots were observed along the banks of both the Mussafah (~50 cm) and Umm al Nar channels (>1 m thick). The stratigraphic position of this facies and ^{14}C dates obtained on its lucine bivalve shells suggest that this facies was deposited mostly during the transgressive phase. At Mussafah, the subtidal sediments containing seagrass roots directly overlie the eroded surface of the lower intertidal *Potamides* grainstone (Fig. 3), suggesting that part of the lower intertidal sediments was eroded before subtidal sediments accumulated.

In the modern environment, *Halodule uninervis* patches are located in the centre of lagoons where salinity is lower than 50‰ (Evans *et al.*, 1973). In the main Abu Dhabi lagoon, southwest of the city, the seagrass patches are located mostly near the tidal channels, at least 3–6 km away from the modern intertidal microbial belt, as the salinity of the lagoon waters increase from 40‰ in near the open marine system up to 65‰ in the lagoon west of the Umm al Nar channel (Evans *et al.*, 1973), and up to 75‰ in the most restricted part of the main lagoon (Kenig, 1991). In areas with salinity > 50‰, subtidal sediments are organic-matter poor. The presence of abundant subtidal sediments containing seagrass roots along both the channel banks suggest that during transgression, the salinity of the lagoon waters remained mostly well below 50‰. This is concordant with the progressive formation of the barrier island after the transgression ended (Purser & Evans, 1973).

The TOC contents of the transgressive mangrove palaeosol and microbial mats are higher than those measured for the subtidal sediments containing seagrass roots (Tables 1 and 2). However, the very localized nature of the transgressive intertidal organic facies contrasts with the widespread presence of the subtidal sediments containing seagrass roots, which were observed for ~2.8 km at both Mussafah and Umm al Nar. Thus, the seagrass remains form the bulk of the sedimentary organic matter in the transgressive sequence.

Regressive sequence

The distribution of sedimentary organic matter in the regressive sequence is also heterogeneous. A regressive mangal palaeosol was not observed at Mussafah or Umm al Nar. The past presence of mangals is only indicated by rare *Avicennia* horizontal roots and pneumatophores below the microbial mat. A 25 cm thick mangal palaeosol with a TOC content of up to 8.2% was observed at Ras Ganadah, where the narrow sabkha had a progradation rate of 0.35–0.75 m yr^{-1} (Kenig *et al.*, 1990; Kenig, 1991; Baltzer *et al.*, 1994). This low rate, associated with a slow lateral movement of the intertidal zone, allowed several generations of *Avicennia* to grow at the same place, providing enough sedimentary organic carbon to create a significant accumulation.

At Mussafah and Umm al Nar, the lateral accretion rates were estimated to be 1.75–2.1 m yr^{-1} and 1–1.5 m yr^{-1}, respectively. These fast rates, at least double of those estimated for Ras Ganadah, probably did not enable the persistence of the mangal. In the modern lagoons, the mangal community does not produce significant accumulation of organic matter (Kenig, 1991). The mangrove community appears stressed by both the increase in salinity associated with the progressive closure of the barrier island system and the associated increase in lateral accretion rate of the intertidal zone.

The abundance of regressive microbial mats seems to be affected less than that of the mangal by lagoon closure and by changes in lateral accretion rates. The microbial mat is more adapted to the high salinity, arid conditions of the modern Abu Dhabi lagoons. However, the continuity of the regressive microbial mats along the bank of the Mussafah channel (> 5 km) does not match the heterogeneous distribution of the microbial mats in the modern microbial belt of the main Abu Dhabi lagoon. This difference seems to result from a recent increase in the lateral accretion rate of the lower intertidal zone. This evolution is visible at Mussafah, where the thickness of lower intertidal aragonite mud and peloid sand increases at the expense of the microbial mat (km 6.5–7.5; Fig. 3). In the modern microbial belt of the main Abu Dhabi lagoon, the massive microbial mats are mostly found in tidal pools and tidal channels. This increase in lateral accretion rate led to the dominance of a thin microbial mat or microbial film with low preservation potential.

Subtidal sediments containing seagrass roots are only a minor part of the regressive sequence, notably at Mussafah. The disappearance of this facies during the regressive phase may be associated with an increase in the salinity of the lagoon.

Major controls on organic matter accumulation in intertidal sediments

A theoretical model including the main factors influencing the distribution of organic matter-rich facies in intertidal sediments may be proposed on the basis of the observations made along the Mussafah and Umm al al Nar channels, as well as the Ras Ganadah section described by Kenig (1991) and Baltzer *et al.* (1994) and other published reports on intertidal organic matter-rich sediments in tropical environments (e.g. Baltzer, 1970b; Risk & Rhodes, 1985). This model takes into account rate of sea-level change, expressed as lateral displacement of the intertidal zone, sediment accumulation rate (including lateral accretion), morphology of the substrate, and climate.

The optimal development of organic matter-rich intertidal facies occurs when the intertidal zone that includes the primary producing communities is stable or moves sufficiently slowly to allow the long-term presence of these communities. During regression or transgression, if relative sea level changes rapidly, the intertidal zone will be displaced laterally at a fast rate, and the lateral displacement of intertidal biocoenosis will result in their poor development. The survival of these biological communities will depend on their ability to colonize new substrate. In Abu Dhabi, in a climate setting stressful to the mangal, a rapid displacement of the mangal's ecological zone leads to its disappearance.

During sea-level rise or fall, the lateral displacement of the intertidal zone can be slowed if the substrate has relatively steep slopes. Several generations of a biological community can produce organic matter at the same site, resulting in accumulation of sedimentary organic matter. Fast aggradation and progradation of intertidal sediments can have the same effects as rapid sea-level fall. If sea level is stable, a fast lateral accretion rate in response to fast sediment accumulation in the intertidal environment results in a rapid lateral displacement of the intertidal zone. This displacement could lead to the disappearance of the biocoenosis at the origin of sedimentary organic matter, or at least prevent significant accumulation of organic matter.

If the climatic conditions favour a highly productive mangal, such as in humid tropical environments, a significant palaeosol can form even during rapid transgression. Such palaeosols were observed by Baltzer (1970b) in the siliciclastic Mara wetlands of New Caledonia and by Risk & Rhodes (1985) in carbonate deposits of Missionary Bay (Australia). These transgressive palaeosols, formed during the Flandrian transgression, are prolonged by a sub-horizontal palaeosol fossilized by lateral accretion of intertidal sediments. The transgressive palaeosol (50–60 % TOC) and organic matter-rich intertidal mud (up to 10 % TOC) described by Risk & Rhodes (1985) have a ~10 km extension, a thickness reaching more than 2 m, and were observed up to 16 m below average sea-level.

In the Mara wetlands, the mangal palaeosol has a km-scale extension, a thickness reaching 1 m, and was observed down to 5 m below sea-level where it was ^{14}C dated to be 7300 ± 170 yr BP (Baltzer, 1970a,b). On the other hand, a humid climate does not favour the development of microbial mats.

CONCLUSIONS

The channels dug across Abu Dhabi sabkha for the development of new harbour, industrial, and recreational facilities offered a unique opportunity to observe the distribution of organic matter in the Holocene subtidal, intertidal, and supratidal sediments of the coastal plain, the sabkha. The three major biological communities of the modern Abu Dhabi lagoon providing sedimentary organic matter, subtidal seagrass patches of *Halodule uninervis*, intertidal *Avicennia marina* mangal, and intertidal microbial mats, are represented in the both the transgressive and regressive sequences.

The distribution of organic biocoenoses is not homogenous and seems to obey a number of rules deriving from the ability of the primary producers to adapt to salinity changes, rate of sea-level change and associated rate of displacement of the intertidal zone as well as progradation and aggradation rates.

The subtidal seagrass biocoenosis is mostly represented in the transgressive sequence as the salinity in the Abu Dhabi lagoon was lower prior to barrier-island development. The distribution of *Avicennia marina* mangal palaeosol is very limited as it was observed only in the transgressive sequence where the slope of the substrate was steep enough to allow minimal lateral displacement of the mangal community during transgression. The thin intertidal transgressive sequence is mostly characterized by burrowed microbial mat.

The regressive sequence is dominated by the microbial mat biocoenosis in areas of fast progradation of the sabkha ($> 1 \, \mathrm{m \, yr^{-1}}$) as this community is well adapted to the hypersaline conditions of the intertidal zone. The *Avicennia marina* mangal palaeosol was only observed in an area where the progradation was between $0.35–0.75 \, \mathrm{m \, yr^{-1}}$. The limited distribution of the mangal palaeosol in Abu Dhabi sediments, constrained by salinity and progradation rate, contrasts with the dominance of this biocenosis in humid tropical environments where highly productive mangals produce thick and organic matter-rich soils, preserved in both regressive and transgressive sequences, irrespective of the rate of sea-level change and rates of progradation.

ACKNOWLEDGEMENTS

I would like to thank Dr. Alain Huc (IFP), Professor Bruce Purser, Dr. Jean-Claude Plaziat, and Dr. Fréderic Baltzer (Université d'Orsay, France), as well as Dr. Robert Boichard (TOTAL) for their great help and advice during the measurement of the Mussafah and Umm al Nar channel sections. I would like to thank TOTAL (France) and TOTAL ABK (Abu Dhabi) for financial and infrastructure support during my Doctoral thesis field work in 1987 and 1989. We thank Dr. Christian Strohmenger for his review of the manuscript.

REFERENCES

Baltzer, F. (1970a) Datation absolue de la transgression holocene sur la cote ouest de Nouvelle-Caledonie sur des echantillons de tourbes a paletuviers. Interpretation neotectonique. *CR Acad. Sci. Paris, Sér. D*, **271**, 2251–2254.

Baltzer, F. (1970b) *Etude Sédimentologique du Marais de Mara (Cote Ouest de la Nouvelle Calédonie) et de Formations Qualernaires Voisines.* Expédition française sur les récifs coraliens de la Nouvelle-Calédonie, **4**. Edition de la Fondation Singer-Polignac, Paris, 146 pp.

Baltzer, F., Kenig, F., Boichard, R., Plaziat, J.C. and Purser, B.H. (1994) Organic matter distribution, water circulation and dolomitization beneath the Abu Dhabi sabkha (United Arab Emirates). *Int. Assoc. Sedimentol. Spec. Publ.*, **21**, 409–427.

Blainville, H.M. (1829) *Dictionnaire des Sciences Naturelles, Zoologie, Conchyliologie et Malacologie.* Levrault, Strasbourg, 113 pp.

Butler, G.P. (1969) Modern evaporite deposition and geochemistry of coexisting brines, the Sabkha, Trucial coast, Arabian gulf. *J. Sed. Petrol.*, **39**, 70–89.

Butler, G.P., Harris, P.M. and Kendall, C.G.St.C. (1983) Recent evaporites from the Abu Dhabi coastal flats, Persian Gulf. *Shale Shaker*, **33**, 44–58.

Espitalié, J., Deroo, G. and Marquis, F. (1985) La pyrolyse Rock-Eval et ses applications. *Rev. Inst. Fr. Pétrol.*, **40**, 563–579.

Evans, G. and Bush, P. (1969) Some oceanographical and sedimentological observations on a Persian Gulf lagoon. In: *Lugunas Costeras, Un Simposio.* Memorias de Simposio Internacional, Laagunas Costeras, pp. 155–170. UNAM-UNESC.O.

Evans, G., Kinsman, D.J.J. and Shearman, D.J. (1964) A reconnaissance survey of the environment of Recent carbonate sedimentation along the Trucial Coast, *Persian Gulf. Dev. Sedimentol.*, **1**, 129–135.

Evans, G., Schmidt, V., Bush, P. and Nelson, H. (1969) Stratigraphy and geologic history of the Sabkha, Abu Dhabi, Persian Gulf. *Sedimentology*, **12**, 145–159.

Evans, G., Murray, J.W., Biggs, H.E.J., Bate, R. and Bush, P.R. (1973) The Oceanography, Ecology, Sedimentology and Geomorphology of Parts of the Trucial Coast Barrier Island Complex, Persian Gulf. In: *The Persian Gulf: Holocene Carbonate Sedimentation and Diagenesis in a Shallow Epicontinental sea* (Ed. B.H. Purser), pp. 234–269. Springer, New York.

Granier, B. (1988) Etude pétro-sédimentologique du canal de Mussafah (Abu Dhabi, Emirats Arabe Unis). Rapport Interne TOTAL CFP, RL4324.

Hadley, D.G., Brouwers, E.M. and Bown, T.M. (1998) Quaternary paleodunes, Arabian Gulf coast, Abu Dhabi Emirate: Age and paleoenvironmental evolution. In: *Quaternary Deserts and Climatic Change* (Eds A.S. Alsharhan, K.W. Glennie, G.L. Whittle and C.G.St.C. Kendall), pp. 123–139. Balkema, Rotterdam.

Kendall, C.G.St.C. and Skipwith, P. (1968a) Recent algal stromatolites of the Khor al Bazam, Abu Dhabi, the southwest Persian Gulf. *Geol. Soc. Am. Spec. Pap.*, **101**, 108.

Kendall, C.G.St.C. and Skipwith, P. (1968b) Recent shallow-water carbonate and evaporite sediments of the Khor

298   *F. Kenig*

al Bazam, Abu Dhabi, southwest Persian Gulf. *Geol. Soc. Am. Spec. Pap.*, **115**, 371–372.

Kendall, C.G.St.C. and Skipwith, P.A. (1968c) Recent algal mats of a Persian Gulf lagoon. *J. Sed. Petrol.*, **38**, 1040–1058.

Kendall, C.G.St.C. and Skipwith, P.A. (1969a) Geomorphology of a recent shallow-water carbonate province; Khor Al Bazam, Trucial Coast, southwest Persian Gulf. *Geol. Soc. Am. Bull.*, **80**, 865–891.

Kendall, C.G.St.C. and Skipwith, P.A. (1969b) Holocene shallow-water carbonate and evaporite sediments of Khor al Bazam, Abu Dhabi, southwest Persian Gulf. *AAPG Bull.*, **53**, 841–869.

Kenig, F. (1991) *Sédimentation, Distribution et Diagénèse de la Matiere Organique dans un Environnement Carbonaté Hypersalin: le Système Lagune-sabkha d'Abu Dhabi.* Unpubl. PhD thesis, Université d'Orléans, Orléans, France, 327 pp.

Kenig, F., Huc, A.Y., Purser, B.H. and Oudin, J.-L. (1990) Sedimentation, distribution and diagenesis of organic matter in a Recent carbonate environment, Abu Dhabi, U. A. E. *Org. Geochem.*, **16**, 735–747.

Kinsman, D.J.J. and Park, R.K. (1976) Algal belt and coastal sabkha evolution, Trucial Coast, Persian Gulf. In: *Stromatolites* (Ed. M.R. Walter), pp. 421–433. Elsevier, Amsterdam.

Loreau, J.P. (1982) *Sédiments Aragonitiques et leurs Génèse.* **Mém. Mus. Hist. Nat. Sér. C, XLVII**, 311 pp.

McKenzie, J.A., Hsü, K.J. and Schneider, J. (1980) Movement of subsurface waters under sabkha, Abu Dhabi, United Arab Emirates, and its relation to evaporite dolomite genesis. In: *Concepts and Models of Dolomitization* (Eds D.H. Zenger, J.B. Dunham and R.L. Ethington), *SEPM Spec. Publ.*, **28**, 11–30.

Park, R.K. (1977) The preservation potential of some Recent stromatolites. *Sedimentology*, **24**, 485–506.

Patterson, R.J. and Kinsman, D.J.J. (1977) Marine and continental groundwater sources in a Persian Gulf coastal sabkha. *Stud. Geol.*, **4**, 381–397.

Patterson, R.J. and Kinsman, D.J.J. (1981) Hydrologic framework of a sabkha along Arabian Gulf. *AAPG Bull.*, **65**, 1457–1475.

Patterson, R.J. and Kinsman, J.J. (1982) Formation of diagenetic dolomite in coastal sabkha along Arabian (Persian) Gulf. *AAPG Bull.*, **66**, 28–43.

Plaziat, J.-C. (1995) Modern and fossil mangroves and mangals; their climatic and biogeographic variability. *Geol. Soc. London Spec. Publ.*, **83**, 73–96.

Purser, B.H. (Ed.) (1973) *The Persian Gulf: Holocene Carbonate Sedimentation and Diagenesis in a Shallow Epicontinental Sea.* Springer, New York, 471 pp.

Purser, B.H. and Evans, G. (1973) Regional Sedimentation along the Trucial Coast, SE Persian Gulf. In: *The Persian Gulf: Holocene Carbonate Sedimentation and Diagenesis in a Shallow Epicontinental Sea* (Ed. B.H. Purser), pp. 211–232. Springer, New York.

Reineck, H.-E. and Singh, I.B. (1980) *Depositional Sedimentary Environments.* Springer-Verlag, Berlin, 549 pp.

Risk, M.J. and Rhodes, E.J. (1985) From mangrove to petroleum precursors: An example from tropical northeast Australia. *AAPG Bull.*, **69**, 1230–1240.

Schneider, J.F. (1975) Recent tidal deposits, Abu Dhabi, UAE, Arabian Gulf. In: *Tidal Deposits* (Ed. R.N. Ginsburg), pp. 209–214. Springer, New York.

Int. Assoc. Sedimentol. Spec. Publ. (2011) **43**, 299–314

The role of bacterial sulphate reduction in carbonate replacement of vanished evaporites: examples from the Holocene, Jurassic and Neoarchaean

DAVID T. WRIGHT* and ANTHONY KIRKHAM[†]

Department of Geology, University of Leicester, Leicester LE1 7RH, UK (E-mail: davywright2@ntlworld.com)
[†]*5 Greys Hollow, Rickling Green, Saffron Walden, Essex CB11 3YB, UK (E-mail: kirkhama@compuserve.com)*

ABSTRACT

Vanished evaporites are well documented in the geological record, but many may go unrecognized because of a lack of obvious field or petrographic evidence, particularly in carbonates deposited in restricted settings where bacterial sulphate reduction (BSR) was active. An early stage in evaporite replacement by carbonates may be observed in modern salterns in Eilat, where gypsum crusts are colonized by stratified, endoevaporitic, microbial communities including sulphate-reducing bacteria (SRB). Here, BSR can be associated with gypsum dissolution and replacive carbonate formation. Rapid consumption of aqueous sulphate by SRB is illustrated in the ephemeral Coorong lakes of South Australia, where extremely high concentrations are completely removed by intense BSR during evaporation, so that no solid sulphate precipitates. Here, carbonates form as the consequence of mediation of ambient waters by BSR, with sedimentary pyrite providing the only indication of the former presence of dissolved sulphate in the lake waters.

Pyrite patches in the patterned dolomite facies of the largely evaporitic Jurassic Arab Formation have been interpreted as 'birds eyes' or fenestrae, but show no signs of visible porosity. A hostile environment mitigates against their interpretation as burrows, although some form of root pseudomorph is not ruled out. However, it is possible that the pyrite is related to BSR operating in a hypersaline, microbially-dominated environment in which solid sulphate removal and sulphide formation is followed by dolomite formation.

Field, petrographic, and stable-isotopic evidence from bedded strata and thin sections of the Neoarchaean Gamohaan and Kogelbeen carbonate formations of South Africa argue for the former existence of evaporites, and suggest deposition in mainly shallow subtidal and marine sabkha environments. On the macroscale, cross-stratified grainstones showing multiple directions of sediment transport, together with abundant microbialites and rare, well-preserved cyanobacterial fossils, indicate shallow-water deposition and microbial growth in the photic zone. Large tepee-like fold structures, intrastratal karst, pseudomorphs after evaporites, replacive calcite after selenite, solution collapse breccias, autobrecciation, corrosion surfaces, nucleation cones, irregular bedding contacts and flowage structures are all indicative of evaporites, their dissolution and their replacement by carbonate.

Evaporite replacement in the Neoarchaean carbonates was largely fabric destructive, leaving few obvious clues at the microscale to the former presence of vanished evaporites. Nonetheless, diagenetic recrystallization, deformation, disruption, and cross-cutting relationships point to their previous existence, and the occurrence of length-slow chalcedony in silicified stromatolite heads may suggest the former presence of sulphates. Fold shapes indicative of enterolithic gypsum, picked out by trails of degraded organic matter and dolomitized clusters of folded filaments preserved within replacive calcite, are evident in thin sections of contorted microbialites. Sulphur isotopes of pyrite samples from the Cambellrand carbonates show a range of $\delta^{34}S$ values from $-6.2‰$ to $+1.4‰$ V-CDT, indicating biogenic fractionation of sulphate.

Keywords: Sulphate-reducing bacteria, evaporites, cyanobacteria, Holocene, Neoarchaean, Jurassic

INTRODUCTION

The dissolution and/or replacement of evaporites may be a more frequent occurrence in the geological past than is commonly recognized. The loss of bedded evaporites can involve several mechanisms that operate on a spectrum of scales, giving rise to completely different textures, ranging from replacement by silica or authigenic carbonates preserving original crystal fabrics as pseudomorphs, through boxwork, nodular and cauliform structures, to plastic flow, dissolution surfaces, autobreccias, multi-stage breccias with filled voids, and slumped horizons where complete dissolution has resulted in a collapsed chaotic aggregation of broken beds and insoluble residue (e.g. Anadón *et al.*, 1992; Taberner *et al.*, 1996; Russell *et al.*, 1998; Kendall, 2001; Rouchy *et al.*, 2001; Sanz-Rubio *et al.*, 2001). Pierre & Rouchy (1988) and Ulmer-Scholle & Scholle (1994) reported replacement of evaporites linked to early burial prior to hydrocarbon migration, and concluded that multistage evaporite dissolution and replacement may well be the norm rather than the exception in the geological record. The absence of relicts of original evaporitic minerals may limit field observations and interpretations of unusual fabrics and lead to consequent misinterpretation of breccias as originating from re-sedimentation or fracturing. Many acknowledged calcitized evaporites are associated with meteoric dissolution of halite, anhydrite and gypsum, which are vulnerable to attack by undersaturated groundwater at various times during their history, both during deposition and after evaporiteforming conditions have ended.

However, large-scale carbonate replacement of solid sulphate through the actions of sulphatereducing bacteria (SRB) is also possible (e.g. Taberner *et al.*, 1999; Kirkham, 2004; Gandin *et al.*, 2005), especially in the case of endoevaporitic microbial sediments. Although SRB can exploit solid as well as dissolved sulphate (e.g. Kowalski *et al.*, 2003; Sørensen *et al.*, 2004), little is known about the process of microbially-driven dissolution of evaporites and the nature and morphology of the replacement carbonate. In this paper, it is argued that relatively small-scale, modern examples of carbonate precipitation driven by bacterial sulphate reduction of solid sulphate are representative of a process that operated on a much larger scale at times in the past, and that evidence for ancient, vanished evaporites can be preserved in sedimentary structures and stable-isotope compositions of the replacive carbonate rocks. The implications of large-scale, bacterial sulphate reduction for understanding the origin of thick, Neoarchaean microbial carbonates in South Africa (dated at 2516 ± 4 Ma by Altermann & Nelson, 1998) will be discussed.

Holocene examples of bacterial sulphate reduction of gypsum associated with calcite

An early stage in evaporite replacement by carbonates may be observed in modern salterns in Eilat, where hard crystalline gypsum forms on the bottom of ponds in solar salterns used for the industrial production of NaCl from seawater. Gypsum crusts are colonized at elevated salinities by endoevaporitic, stratified microbial mat communities about 5 cm thick, including cyanobacteria and sulphate-reducing bacteria (Oren *et al.*, 1995; Sørensen *et al.*, 2004). The photic zone extends 2–3 cm into the crusts, through brown and green cyanobacterial layers Fig. 1). Below these is a white zone, underlain by anoxic purple laminae, and then a zone that is black in response to precipitation of metal sulphides. Sulphate reduction rates peak at depths between 1 and 2 cm, above the permanently anoxic zone. Sulphate reduction is an important process in the crust, and reoxidation of formed sulphide accounts for a major part of the oxygen budget. Methanogens are well adapted to the *in situ* salinity but contribute little to the

Fig. 1. Cross section through a gypsum crust from saltern in Eilat crystallizer pond. The gypsum exhibits a range of structures, from crystal clusters and crusts to undulatory beds and fanned domes (Schreiber & El Tabakh, 2000; Sørensen *et al.*, 2004), and supports a prolific stratified, endoevaporitic, benthic microbial community. Photosynthetic (brown and green) filamentous cyanobacteria overlie purple, photosynthetic, sulphate-oxidizing coccoid bacteria, with a black basal layer dominated by sulphate-reducing bacteria (SRB). Carbonate is associated with the filaments and SRB-dominated layers, and becomes incorporated into the gypsum. Thickness of crust = 5 cm.

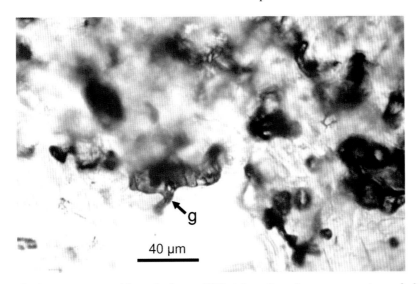

Fig. 2. Photomicrograph of a gypsum crystal from the lower, SRB-rich section of a gypsum crust sample from the Eilat saltern, showing uneven, corroded crystal margin (arrowed, g) adjacent to degraded organic material, calcite and fragmented gypsum. Bacterial sulphate reduction leaves ambient waters undersaturated in sulphate, and these then dissolve the gypsum. Liberated calcium ions become available for precipitation as calcite, as carbonate alkalinity is raised as a by-product of organic degradation.

anaerobic mineralization in the crust, being largely inhibited by the activity of SRB (Sørensen et al., 2004). Gypsum dissolution is recognized by microscopic observation of leaching and disintegration of crystals accompanied by patches of microcrystalline carbonate (Fig. 2), in association with bacterial sulphate reduction (BSR).

Carbonate replacement after sulphate and endoevaporitic microbial communities have been documented from a number of locations (e.g. Taberner *et al.*, 1996; Russell *et al.*, 1998; Taberner, 1999; Douglas & Yang, 2002; Spear *et al.*, 2003; Sørensen *et al.*, 2004), and microbial metabolic rates in hypersaline conditions can be remarkably high. In coastal solar salterns where seawater is evaporated during salt production, microbial mats may be present at salinities of ~130‰, while photosynthetic subevaporitic or endoevaporitic stratified bacterial communities develop associated with gypsum crusts on the bottom of the salterns at salinities of ~200‰ (e.g. Caumette, 1993). Phototrophic prokaryote communities associated with heterotrophic bacteria and a black layer in which bacterial dissimilatory sulphate reduction occurs have been described from saltern evaporation ponds in France (Caumette, 1993; Caumette *et al.*, 1994; Cornée, 1984) and Israel (Oren *et al.*, 1995; Canfield *et al.*, 2004; Sørensen *et al.*, 2004), and in natural lagoons with evaporating seawater in Guerrero Negro, Baja California, Mexico (Rothschild *et al.*, 1994).

SRB also play an active role in the destruction of solid sulphate in saline aqueous environments, where the solubility of anhydrite and gypsum strongly increases with salinity (e.g. Douglas & Yang, 2002; Wagner *et al.*, 2005). Gavrieli *et al.* (2001) showed that gypsum dissolution by gypsum-saturated water can be induced through the removal of sulphate from the water by bacterial sulphate reduction, while Vinogradov (2003) showed that sulphate sediments in evaporite sequences may become as much as two times thinner as a result of bacterial sulphate reduction.

The ability of SRB to metabolize gypsum and precipitate carbonate has also been exploited in industrial bioremediation techniques (e.g. Atlas *et al.*, 1988; Lal Gauri & Chowdhury, 1988; Kowalski *et al.*, 2003), although there is little information on the nature of carbonate replacement fabrics. There is therefore abundant evidence from modern settings that calcitization of solid sulphate driven by BSR is likely to have been a major process in similar environments in the geological past, although reports in the literature are scarce.

Microbial mediation of saline waters

Recent microbial mats, which flourish only in environmental conditions that are hostile and stressful to higher forms of life, are considered to be analogues for some of the oldest biotic communities on Earth. The interior of subaqueous microbial mats is typically devoid of oxygen and provides a habitat for anaerobic metabolic guilds, dominated by sulphate-reducing and fermenting bacteria. However, SRB in the oxic zone of phototrophic mats can successfully compete with aerobic heterotrophic bacteria for organic substrates, and are associated with laminar carbonate precipitation (Reid *et al.*, 2000). Fixation of inorganic carbon into organic biomass is mediated by photoautotrophic and chemoautotrophic microorganisms and the cycling of major elements (C, N, S) is very tight; the abundance of labile organic carbon is a major regulator of sulphate reduction rates in marine environments. Anaerobic groups play a vital role in the cycling of carbon and sulphur compounds, and sulphur cycling is an important pathway for carbon degradation in stromatolites. SRB have been found to metabolize more than half of the carbon produced within mat structures (e.g. Visscher *et al.*, 1998).

A syntrophic interaction between cyanobacteria and SRB appears to exist in microbial mats where both types of microorganisms occur in close spatial proximity, if not intermixed with each other. In these ecosystems, cyanobacterial organic carbon substrates may provide the electron-donating substrates for SRB (Jørgensen & Cohen, 1977; Skyring & Bauld, 1990; Fründ & Cohen, 1992). Microbial reduction of sulphate is accompanied by carbonate precipitation and replacement in many modern environments (e.g. Berner, 1971; Hendry, 1993; Wright, 1997; Visscher *et al.*, 1998; Reid *et al.*, 2000; Raiswell & Fisher, 2004; Wright & Wacey, 2004, 2005), and is likely to have been common in the past, especially in the Precambrian when microbial ecosystems dominated the biosphere.

Sulphate-reducing bacteria and other microbes oxidize cyanobacterial and other organic matter to support their metabolism, producing ammonia from enzymatic breakdown of proteins, which is rapidly absorbed by ambient waters, thereby increasing pH and carbonate alkalinity to levels necessary for carbonate formation (Berner, 1980; Durand, 1980; Slaughter & Hill, 1991):

$$CH_2NH_2COOH + 1.5\,O_2 \rightarrow 2CO_2 + H_2O + NH_3 \tag{1}$$

$$NH_3 + H_2O \rightarrow NH_4^+ + OH^- \tag{2}$$

$$OH^- + HCO_3^- \rightarrow H_2O + CO_3^{2-} \tag{3}$$

Calcite, like aragonite and dolomite, is an ionic compound, and can only form from its component ions, so CO_3^{2-} ions must be available (Lippmann, 1973; Zuddas & Mucci, 1998). Despite the innumerable assumptions in the literature that calcium carbonate may precipitate by calcium ions reacting with bicarbonate ions, laboratory studies (Zhong & Mucci, 1993; Zuddas & Mucci, 1998) have determined that the principal reaction mechanism of calcium-carbonate precipitation in seawater is:

$$Ca^{2+} + CO_3^{2-} \rightarrow CaCO_3 \tag{4}$$

In normal seawater, where $\sim$90% of the carbonate occurs as HCO_3^- ions, the activity of the CO_3^{2-} ion is generally two orders of magnitude lower than that of Ca^{2+} ions, and this can only be increased by raising carbonate alkalinity; bacterial sulphate reduction is the only natural phenomenon that can do this on a large scale (Berner, 1980; Wright, 2000).

Vanished evaporites in the Jurassic Arab Formation of the Arabian Gulf

The largely evaporitic Jurassic Arab Formation has furnished much of the oil produced to date in Saudi Arabia, Qatar and Abu Dhabi. Subaqueous anhydrites comprising palisades of subvertically orientated nodules exist at all levels within the upper Arab Formation, and thicker, nodular anhydrites interbedded with stromatolites and crypto-microbial laminites, may also have formed subaqueously (Al-Silwadi *et al.*, 1996). The Arab Formation includes a common and very distinctive dolofacies known as "patterned dolomite" (Fig. 3) that may have been overlooked because of difficulties in understanding its origins. This facies is generally buff-coloured, partly due to oil staining, and is finely crystalline, with pyrite-rich patches and mottles (Fig. 4) that have been interpreted as "birds eyes" (Wood & Wolfe, 1969), but show no signs of visible porosity.

A hostile environment mitigates against interpretation of the pyrite-rich patches as burrows, although some form of root pseudomorph is not ruled out. The absence of dolomite fabrics crosscutting facies boundaries argues against a seepage reflux origin for these dolostones. It is more likely that the pyrite is related to sulphate reduction of sulphate dissolved in the hypersaline, ambient waters and on discrete gypsum and/or anhydrite crystals in association with abundant organic matter in the form of degraded cyanobacteria (Berner *et al.*, 1979; Kirkham, 2004). BSR operating in a hypersaline, microbially-dominated environment

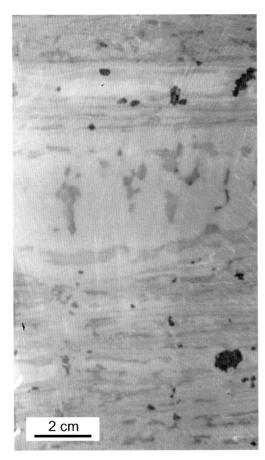

2 cm

Fig. 3. Core slab of patterned dolomicrite pervaded with irregular iron sulphide patches which are by-products of bacterial sulphate reduction. The upper and lower intervals display crypto-microbial laminations with sulphide patches that parallel the laminations. The sulphide patches are dominantly vertical in the central non-laminated interval, which may represent bacterial reduction of muds that formerly contained penecontemporaneous, sub-vertical, subaqueously formed gypsum crystals. The black spots are replacive and include poikilitic anhydrite crystals that are later diagenetic. Jurassic Arab Formation, Arabian Gulf.

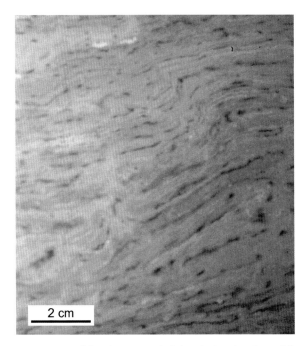

2 cm

Fig. 4. Core slab of patterned dolomicrite dominated by contorted crypto-microbial laminations and pervaded with dark patches of iron sulphide which are by-products of bacterial sulphate reduction. Arab Formation.

can drive dolomite formation (e.g. Gunatilaka *et al.*, 1984; Vasconcelos & McKenzie, 1997; Wright, 1999; Wright & Wacey, 2004, 2005), and the patterned dolomite may thus represent vanished evaporites interbedded with microbial carbonates (Kirkham, 2004).

Vanished evaporites in the Neoarchaean Campbellrand carbonates, South Africa

The Precambrian Campbellrand Subgroup carbonates of South Africa are exceptionally well-

preserved and largely unmetamorphosed, forming part of a vast succession of extensive carbonates some 1500 km thick and covering an estimated 600,000 km^2 of the Kaapvaal Craton (Fig. 5). The carbonates have been interpreted as a ramp to rimmed-platform succession formed in an open-marine environment that was ultimately drowned (Beukes, 1987; Sumner, 1997). However, Altermann & Siegfried (1997) considered the depositional environment to be generally shallow-water and identified several transgressive-regressive episodes. Altermann (1997) presented evidence for the former presence of evaporites in the Kogelbeen and Reivilo formations, which host large Palaeoproterozoic Pb-Zn Mississippi-Valley-type mineral deposits, and speculated that evaporites and organic matter at the shallow platform rim supplied the chloride, sulphur and organic complexes necessary for metal transport and precipitation.

Sumner & Grotzinger (2000) recorded halite pseudomorphs from supratidal and intertidal facies of the Cambellrand-Malmani carbonates, but considered sulphate concentrations to be low. Others have reported evidence for the former presence of evaporites in the contemporaneous Carawine Dolomite of the Hamersley Basin, Western Australia (Simonson *et al.*, 1993; Winhusen, 2001),

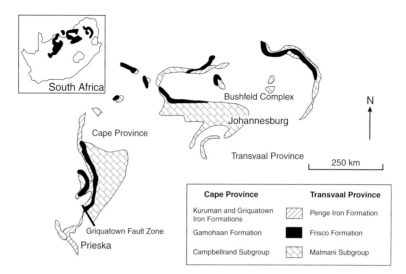

Fig. 5. Location map and simplified geological setting of the Neoarchaean Cambellrand and Malmani Subgroups of South Africa. The carbonate sediments are some 1.5 km thick, deposited over 600,000 km^2 of the Kaapvaal Craton, and are the best preserved carbonates of that age known. Dates of 2642.2 ± 2.3 Ma (Walraven & Martini, 1995) and 2669 ± 5 Ma (Gutzmer & Beukes, 1997) have been obtained using zircons from tuffs in the Vryburg Formation underlying the carbonates, giving a maximum age, while the uppermost carbonates of the Gamohaan Formation have been dated at 2516 ± 4 Ma (Altermann & Nelson, 1998).

while Hardie (2003) argued that gypsum, not aragonite, was the precursor mineral for radiating calcite fans in the Cambellrand-Malmani carbonates. The focus here will be on the sedimentological and petrographic features present in the upper Kogelbeen and lower to middle Gamohaan formations of the Neoarchaean Campbellrand Subgroup carbonates of South Africa that, taken

together with geochemical constraints, suggest deposition in shallow evaporitic subaqueous to sabkha environments. The Gamohaan Formation has been dated at 2516 ± 4 Ma (Altermann & Nelson, 1998).

Many features typical of evaporites are found as carbonate, including leaching and dissolution surfaces (Figs 6 and 8), nucleation cones (Fig. 7),

Fig. 6. Ductile deformation and flowage in a bed from the Neoarchaean upper Kogelbeen Formation, Campbellrand Subgroup, at Kuruman Kop, South Africa. Dissolution surfaces form when influxes of less saline water dissolve the uppermost part of the previously deposited evaporite; note the collapsed "bowl", which probably represents a response to dissolution or flowage of underlying evaporites. The formation is overlain conformably by the Gamohaan Formation, dated at ∼2.5 Ga (Altermann & Nelson, 1998).

Fig. 7. Nucleation cones replaced by carbonate in the Neoarchaean lower Gamohaan Formation of the Campbellrand Subgroup, dated at ~2.5 Ga (Altermann & Nelson, 1998). Nucleation cones are typical structures of evaporites in which underlying, pre-existing but unlithified beds and laminae are depressed, thinned and pierced (arrowed), while overlying laminae may curve downwards. Selenite crystals in nucleation cones form clusters oriented normal and oblique to bedding.

tepee-like, antiform structures of early diagenetic origin, and calcite pseudomorphs after selenite (Fig. 8). Diagenetic recrystallization and overprinting of original fabrics and textures is a feature of a majority of beds, with abundant evidence of roll-ups (Fig. 9), growth structures, early cementation, solution breccias, vertical piping, displacement, deformation, disruption and cross-cutting relationships (Gandin *et al.*, 2005). The uneven texture of much of the limestone resembles the "mammillated" texture of evaporites referred to by West (1965), or the fabrics produced by repeated evaporite precipitation, dissolution and sediment collapse that produce a chaotic mix of evaporite crystals within a fine-grained matrix, with sedimentary structures further masked by plastic flow (e.g. Dronkert, 1977). Other similarities include deformational and collapse structures interbedded with non-deformed fabrics, and

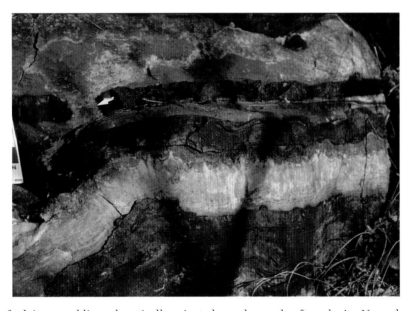

Fig. 8. A thick band of calcite resembling subvertically-oriented pseudomorphs after selenite. Neoarchaean upper Kogelbeen Formation, Campbellrand Subgroup. Bedding contacts are irregular, with vertical "piping" (yellow arrow) connecting dissolution surfaces, interpreted as forming when influxes of less saline water dissolved the uppermost part of a previously deposited evaporite. The calcite band passes laterally into a dissolution breccia, interpreted as semi-lithified sediment collapse into space created by the leaching of former evaporites.

Fig. 9. Contorted but largely uncompressed organic laminae partially set in a white calcitic matrix truncate, erode, penetrate and encompass earlier flat-lying dolomite, Neoarchaean Gamohaan Formation. These structures resemble microbial mat "roll-ups" reported from evaporite ponds (e.g. Schreiber & El Tabakh 2000), suggesting storm-generated transport of flexible, organic-rich mats formerly interlayered with evaporite "mush".

ductile and soft-sediment deformation of laminae (Kirkham, 2004).

Petrographic analysis of white, calcitic nodules (Fig. 10) previously interpreted as fenestrae fills in microbialites (Sumner, 1997) suggests that the "fenestrae" are not bounded by the margins of a former void, but that the calcite is traversed by strands and bundles of laminae (Fig. 11). Fenestrae are defined as irregular voids or cavities that form during the decay of organic material, gas expansion and/or dewatering of carbonate muds, and are preserved by early lithification and may be filled by cement. In the case of the "fenestrae", the deformation of the microbial laminae, now encased in calcite, can be clearly seen in thin section: the laminae do not bound voids that were

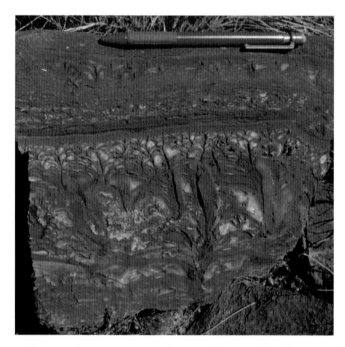

Fig. 10. "Cuspate" facies from the Gamohaan Formation, showing characteristic "fenestrae" of white calcite. Calcitized evaporites are often preserved as coarse white spar (Ulmer-Scholle & Scholle, 1994) and the "cusps", separated here by thin dolomitic laminae, resemble the net-like texture of macrocells of anhydrite, associated with selenite crystals (oblique to section). Pencil is 14 cm long.

Fig. 11. Thin section of "cuspate" facies sample from the Gamohaan Formation, in which the white calcite "fenestrae" are both convex- and concave-up. The filmy laminae are interpreted as representing strands of formerly flat-lying mat deformed during diapiric expansion of evaporites. The 'fenestrae' are interpreted here as solid calcite traversed by deformed organic laminae, with a net-like macrocell texture developing as impurities were pushed towards the margins of nodules during rehydration of anhydrite, before the evaporite was replaced by calcite during bacterial sulphate reduction. Width of section is 4 cm.

later filled by calcite, but form a net-like pattern reminiscent of macrocells of "chicken-wire" anhydrite. Subvertical clots of partially-degraded organic material represent organic material

trapped and compressed between coalescing evaporite nodules or the limbs of enterolithic folds.

Enterolithic gypsum consists typically of subvertical, isoclinal folds of almost constant amplitude, with the tightly-folded limbs separated only by a thin wafer of impurities (Fig. 12). Some veins are less tightly folded and can be irregular, while others show a clear transition with nodular texture. Enterolithic folds are thus not formed by lateral compression: the internal structure of the gypsum is deformed during expansion or diapiric growth, and any microbial matter within the gypsum will also be deformed, and may be preserved. It is proposed that continued progressive growth of nodular and enterolithic gypsum, and its possible conversion to anhydrite (cf. Kirkham, this volume, pp. 265–276), prior to calcitization led to the development of a "net texture"; with residual organic material increasingly pushed to the margins of crystals, where it concentrates in the old "hinge". A significant proportion of the organic material would have been consumed during BSR, which was presumably limited only by sulphate availability.

Stable isotope evidence

Pyrite is abundant in the Gamohaan Formation and is often associated with calcite in stromatolites and contorted microbial mat (e.g. Wright & Altermann, 2000). The intimate intergrowth of pyrite, calcite and partially-degraded microbial material strongly indicates that they are early diagenetic and genetically related (Kendall, 2001, p. 51). Laser-ablation analyses of sulphur isotopes from individual pyrite nodules from the Cambellrand carbonates show a range of $\delta^{34}S$ values from $-6.2‰$ to $+1.4‰$ V-CDT, indicating fractionation by sulphate-reducing bacteria (Gandin *et al.*, 2005). This

Fig. 12. Enterolithic gypsum of Messinian age from Faltona Quarry, Tuscany, Italy, showing subvertical, isoclinal folds and wavy, overturned, crest lines. Folding may be more open, and often develops into small thrusts; the mechanics of enterolithic folding are not fully understood, but West (1965) reported that a flat sheet of gypsum in sediment with little overburden (as in modern sabkhas), when hydrated, will expand by buckling in a direction normal to the sheet.

Table 1. Carbon- and oxygen-isotope data for Late Archaean carbonates, Gamohaan Formation, South Africa.

Sample	Sub-sample	$\delta^{13}C$ (V-PDB)	$\delta^{18}O$ (V-PDB)	Sample description
SA-1A	10	−1.1	−10.1	Micritic calcite
SA-1A	4	−1.3	−9.9	Micritic calcite
SA-1A	11	−1.4	−10.6	Sparry calcite cement (herringbone calcite)
SA-1A	12	−2.5	−12.1	Micritic calcite
SA-1A	2	−3.1	−12.8	Micritic calcite
SA-1A	7	0.8	−7.4	Dolomicrite forming bedding-parallel lenses
SA-1A	8	0.6	−6.6	Dolomicrite forming bedding-parallel lenses
SA-1A	9	−0.03	−7.7	Micritic calcite

association is supported by sulphur isotopes in pyrite from lower Campbellrand Subgroup samples, which indicate microbial fractionation associated with bacterial sulphate reduction (Strauss & Beukes, 1996).

In the Gamohaan Formation carbonates, $\delta^{13}C$ values of carbonates interpreted as replacive after sulphate due to BSR, range from −3.1‰ to +0.8‰ (Table 1) and therefore may not suggest an obvious organogenic origin. Kendall (2001) noted that in Mississippian pyrite samples, sulphur isotope values were consistent with bacterial sulphate reduction, but calcite replacive after evaporites did not carry the light carbon signature expected. However, such apparent anomalies can be explained: carbon isotopes alone are not a sufficient tool for determining the role of organic matter in carbonate formation (Hill, 1990). The degree of ^{13}C depletion through sulphate reduction is controlled by the carbon source, the extent of organic diagenesis, and the contribution from the ambient water reservoir (Mazzullo, 2000; Wright, 2000). Inorganic "marine" HCO_3^- is abundant in normal seawater (and therefore porewaters), but in response to increased alkalinity driven by sulphate reduction, HCO_3^- dissociates during buffering (Slaughter & Hill, 1991) according to:

$$OH^- + HCO_3^- \rightarrow H_2O + CO_3^{2-} \qquad (5)$$

thereby increasing in ambient waters the proportion of CO_3^{2-} derived from marine bicarbonate, and diluting the proportion of isotopically light carbon derived from organic diagenesis.

There are other features that support an evaporitic environment for both the Campbellrand-Malmani carbonates and the Carawine Dolomite. Both Sumner & Grotzinger (2000) and Simonson *et al.* (1993) reported "halite pseudomorphs," clear evidence for highly evaporative conditions. Indeed, Simonson *et al.* (1993) identified dolomite

pseudomorphs after gypsum associated with halite moulds. It must be pointed out that whether considering periods of "calcite seas" or "aragonite seas," gypsum will precipitate from evaporating seawater after the $CaCO_3$ stage but before halite precipitation, unless removed by SRB, as in the dolomitic Coorong distal lakes (Wright, 1999).

Hardie (2003) pointed to the problem of distinguishing between pseudomorphs after gypsum (monoclinic) and pseudomorphs after aragonite (orthorhombic). Both minerals typically form as twins that yield pseudohexagonal crystals with very similar interfacial angles. The large crystal fans described and illustrated by Sumner & Grotzinger (2000) are very similar in scale and morphology to the Solfifera radiating prismatic gypsum, while Simonson (1993) noted that the "distinctive square-tipped terminations" of radiating aragonite crystals were absent; the interested reader should compare Fig. 3A of Sumner & Grotzinger (2000) with Fig. 23 in Hardie & Eugster (1971).

Precambrian seawater sulphate concentrations and evaporites

The scarcity of preserved evaporites in Precambrian successions has been attributed to the high solubility and low preservation potential of gypsum and anhydrite, and is reinforced by the assumption that sulphate concentrations were absent or low in the Archaean and Early Proterozoic oceans. Seawater sulphate concentrations through time have been the subject of much debate, and estimates rely essentially on stable sulphur isotopic evidence from pyrite, supported by C/S ratios (Canfield & Raiswell, 1999; Grassineau *et al.*, 2001). Through most of the Archaean, sulphur isotope composition in sedimentary pyrite has a spread of ∼±5‰ around the mantle value of ∼0‰ V-CDT. Although such small fractionations have been attributed to BSR under sulphate-limiting conditions as far back as 3.4 Ga (Ohmoto *et al.*, 1993; Shen *et al.*, 2001), the similarity with magma-derived pyrite values makes discrimination between biogenic and non-biogenic pyrite difficult.

It is likely that some of the earliest organisms on Earth gained energy from the metabolism of sulphur compounds (Stetter, 1996), an ability which is widespread throughout the Bacteria and Archaea. Although anoxygenic photosynthesis may have been an important source of Archaean seawater sulphate (e.g. Woese, 1987), Canfield & Raiswell (1999) argued that the first evidence for sulphate reduction is found between 2.7 and 2.5 Ga, with levels of Archaean seawater sulphate below 1 mM. This did not increase until around 2.3 Ga, when an increase in sulphate levels may have been promoted by a rise in atmospheric oxygen concentration. Nevertheless, the earliest unequivocal evidence for oxygen production by oxygenic photosynthesis is found at ~2.8 Ga (e.g. Des Marais, 2000). The mass colonization of shallow submerged cratons (e.g. the Kaapvaal craton) by photosynthesizing microbes, especially cyanobacteria, not only produced a dramatic increase in the primary production of organic material (Canfield & Raiswell, 1999), but liberated oxygen into reducing seas on an unprecedented scale and stimulated further evolution by promoting a massive increase in ecosystem space.

The discovery of mass-independent fractionation (MIF) of sulphur isotopes in Archaean sulphide and sulphate minerals raised hopes that a quantitative, time-resolved history of the oxygenation of Earth's atmosphere during the Palaeoproterozoic had been found (Farquhar *et al.*, 2000). However, a more complex picture emerged when it was realized that sulphur from all sources became fully oxidized and part of the oceanic dissolved sulphate reservoir, and any MIF signature that might have been produced by photochemical reactions was erased when the different sulphur species became homogenized. While measurements support the existence of MIF in sulphur isotopes in Archaean rocks, Ono *et al.* (2003) concluded that pyrite precipitation was mediated by microbial enzymatic catalysis that superimposed mass-dependent fractionation on mass-independent atmospheric effects.

Sulphur isotope signatures from the Mt McRae Shale (~2.5 Ga) and the Jeerinah Formation (~2.7 Ga), both outcropping in the Hamersley Basin,

Western Australia, show a strong correlation between $\Delta^{33}S$ and $\delta^{34}S$, whereas values of $\Delta^{33}S$ and $\delta^{34}S$ do not correlate for sulphides from associated platform carbonates. Ono *et al.* (2003) concluded that bacterial mass-dependent fractionation signatures were superimposed on atmospheric MIF signatures, and that while microbial sulphate reduction was probably widespread in the Archaean ocean, SRB were particularly important in the carbonate platform environment in the late Archaean.

Farquhar *et al.* (2000) argued that in the Archaean, atmospheric deposition of sulphuric acid could have been a dominant source for the seawater sulphate, with subsequent precipitation of sedimentary pyrite as the main sink for seawater sulphate by the late Archaean. The negative $\Delta^{33}S$ ($-2.5‰$ to $-0.4‰$) of pyrite sulphur in the ~2.6 Ga Carawine Dolomite ($\delta^{34}S$ $+4.6‰$ to $+20.9‰$) is attributed to bacterial reduction of seawater sulphate because this process is expected to produce pyrite with variable $\delta^{34}S$ and negative $\Delta^{33}S$ inherited from seawater sulphate. The relatively high $\delta^{34}S$ may indicate sulphate reduction in a closed system with respect to seawater sulphate (Ohmoto, 2004).

Evidence for the presence of sulphate in Mesoproterozoic oceans was provided by Lyons *et al.* (2000), who reported trace amounts of carbonate-associated sulphate (CAS) trapped within abiotic limestones and dolostones of Precambrian successions. At one locality, this isotopic variability in CAS is corroborated by trends within coexisting gypsum, which helps to constrain the amount and isotopic composition of sulphate in ancient seawater. It was argued that the sulphur isotope response suggests a relatively small oceanic sulphate reservoir that helps explain the abundance of isotopically heavy bacteriogenic iron sulphides. However, a large demand for sulphate from SRB could also explain the data.

Grassineau *et al.* (2001) performed sulphur and carbon isotopic analyses on a large number (161 for S-isotopes) of small samples of kerogens and sulphide minerals from biogenic and non-biogenic sediments of the 2.7 Ga Belingwe Greenstone Belt (Zimbabwe). Sulphur isotope compositions displayed a wider range of biological fractionation than hitherto reported from the Archaean from small numbers of samples (e.g. Habicht & Canfield, 1996), and it was concluded that in shallow-water environments, microbial mats enabled photosynthetic oxygenation of sulphate and microbial sulphur-cycling capable of producing the observed wide range of isotopic values. Grassineau et al., (2001) proposed that by 2.7 Ga, a full microbial sulphur cycle was already in operation, though on a very fine scale compared with bioturbated Phanerozoic muds.

Other detailed investigations have also revealed a large variation in the $\delta^{34}S$ signal from Archaean sedimentary pyrite older than 2.2 Ga (e.g. Kakegawa 2000; Shen et al., 2001; Gandin et al., 2005). Evidence for significant sulphate concentrations in late Archaean seawater was provided by Kakegawa et al. (1998), who reported pyrite $\delta^{34}S$ values from the 2.5 Ga Mt McRae Shale of the Hamersley Basin, Western Australia of $-6.3‰$ to $+7.1‰$, interpreted to be formed by bacterial sulphate reduction of seawater sulphate. It was further proposed that the data indicate that sulphate concentrations in 2.5 Ga seawater were around one third of the present value in the modern ocean, and that the activity of sulphate reducing bacteria was generally higher.

Analyses of individual millimetre-size sulphide and pyrite grains from the 2.7 Ga Belingwe sediments show large depletions of light Fe isotopes, as well as a large variation in Fe-isotope composition (ranging from $-0.7‰$ to $-2.7‰$) correlating with depletions in light S isotopes measured from the same samples (ranging from $-3‰$ to $-18‰$; Archer & Vance, 2004). This correlated relationship showing coupled Rayleigh depletion in the light isotopes of both Fe and S can be explained in terms of a reducing sedimentary environment in a closed system in which sulphate is significantly depleted, a process that generally does not occur today because of high porewater sulphate levels. Ono et al. (2003) concluded from sulphur-isotope studies in the Hamersley Basin of Western Australia that the relatively high $\delta^{34}S$ indicates a high rate of BSR in a closed system.

DISCUSSION

Molecular evidence confirms the presence of cyanobacteria in rocks dating from 2.7 Ga (Brocks et al., 1999; Summons et al., 1999). It is thus highly probable that Neoarchaean benthic microbial communities would have been vertically-stratified, with cyanobacterial populations in consortial syntrophic association with SRB, as they are today. Microbialites flourished across the 600,000 km² Kaapvaal Craton between ~2640 Ma and ~2500

Ma (Eriksson & Altermann, 1998), and cyanobacteria in microbial mats and stromatolites likely formed the basis for a dominant ecosystem that extended across huge areas in shallow epeiric seas over ~150 million years. During this time, the output of vast volumes of cyanobacterial photosynthetic oxygen was available for rapid reaction with reduced Fe and S in the oceans, inevitably producing high concentrations of aqueous sulphate in surface seawater capable of sustaining enormous populations of SRB. Widespread sulphate reduction across the epeiric platforms would necessarily modify ambient waters by removing the kinetic inhibitors to carbonate precipitation, thus facilitating the bulk precipitation of limestone and dolomite.

Sulphate produced in surface seawater by oxidation from photosynthesizing cyanobacteria in shallow epeiric seas could also conceivably form widespread evaporites under appropriate conditions. Endoevaporitic microbial mat communities could then have developed over extensive areas, bringing together the necessary ingredients for large-scale, coupled oxidation and reduction processes driven by microbial metabolisms.

In the distal ephemeral lakes of the Coorong area, South Australia, solid sulphate appears not to form (Wright, 1999), though concentrations of sulphate and other solutes increase during late spring and summer during evaporation, when intense BSR in the uppermost, anoxic lake sediments exhausts the sulphate and, by modifying ambient waters, drives dolomite formation (Wright, 1999). In adjacent non-dolomitic lakes, BSR is less intense (Wacey, 2002), and other minerals, principally aragonite and hydromagnesite, are precipitated (Wright, 1999; Wright & Wacey, 2005).

Postulated low sulphate concentrations in the Neoarchaean (based on low sulphur isotope fractionation values) may have been due to a predominantly reducing ocean, or to intense and sustained regional microbial sulphate reduction in prolific SRB populations occupying an anoxic environment in close proximity to photosynthesizing cyanobacteria in the benthic microbial communities that colonized and oxidized the extensive, shallow epeiric seas. These microbial associations typically formed widespread stromatolites, laminites and other microbialites (Beukes, 1987; Altermann & Siegfried, 1997; Altermann & Nelson, 1998). Oxygenation of shallow seawater above regionally extensive microbialites is witnessed by abundant preserved biogenic pyrite both in the studied rocks and elsewhere in the Neoarchaean (e.g. microbially-precipitated pyrite constitutes up to 35% by volume of shales in the 2.7 Ga Belingwe Greenstone Belt; Grassineau, pers. comm., 2004).

Bacterial sulphate reduction may thus have been an important but underestimated process responsible for the large-scale transfer of carbon, sulphur, calcium and magnesium from the biosphere to the lithosphere throughout geological history. Mass microbial colonization of extensive shallow-water platforms in the Neoarchaean set the scene for the microbiogeochemical transformation of Earth's hydrosphere, lithosphere and (later) atmosphere, with the potential for bacterial sulphate reduction of both aqueous and solid sulphate to drive carbonate precipitation on a massive scale, leading to thick accumulations of carbonate rocks and vanished evaporites in restricted basins.

REFERENCES

Al-Silwadi, M.S., Kirkham, A., Simmons, M.D. and **Twombley, B.N.** (1996) New Insights into regional correlation and sedimentology, Arab Formation (Upper Jurassic), Offshore Abu Dhabi. *GeoArabia*, **1**, 6–27.

Altermann, W. (1997) Sedimentological evaluation of Pb-Zn exploration potential of the Precambrian Griquatown Fault Zone in the Northern Cape Province, South Africa. *Mineral. Deposita*, **32**, 382–391.

Altermann, W. and **Nelson, D.R.** (1998) Sedimentation rates, basin analysis and regional correlations of three Neoarchean and Palaeoproterozoic sub-basins of the Kaapvaal craton as inferred from precise U-Pb zircon ages from volcaniclastic sediments. *Sed. Geol.*, **120**, 225–256.

Altermann, W. and **Siegfried, H.P.** (1997) Sedimentology and facies development of an Archean shelf – carbonate platform transition in the Kaapvaal Craton, as deduced from a deep borehole at Kathu, South Africa. *J. Afr. Earth Sci.*, **24**, 391–410.

Anadón, P., Rosell, L. and **Talbot, M.R.** (1992) Carbonate replacement of lacustrine gypsum deposits in two Neogene continental basins, eastern Spain. *Sed. Geol.*, **78**, 201–216.

Archer, C. and **Vance, D.** (2004) Coupled Fe and S isotope evidence for Archaean microbial Fe(III) and sulphate reduction. *Eos Trans. Am. Geophys. Union*, **85**, Fall Meet. Suppl., Abstracts.

Atlas, R.M., Chowdhury, A.N. and **Lal Gauri, K.** (1988) Microbial calcification of gypsum-rock and sulphated marble. *Stud. Conserv.*, **33**, 149–153.

Berner, R.A. (1971) *Principles of Chemical Sedimentology*: New York, McGraw-Hill, 240 pp.

Berner, R.A. (1980) *Early Diagenesis: a Theoretical Approach*. Princeton University Press, Princeton, N.J., 241 pp.

Berner, R.A., Baldwin, T. and **HoldrenJr, G.A.** (1979) Authigenic iron sulphides as palaeosalinity indicators. *J. Sed. Petrol.*, **4**, 1345–1350.

Beukes, N.J. (1987) Facies relations, depositional environments and diagenesis in a major Early Proterozoic stromatolitic carbonate platform to basinal sequence, Campbellrand Subgroup, Transvaal Supergroup, Southern Africa. *Sed. Geol.*, **54**, 1–46.

Brocks, J.J., Logan G.A., Buick R. and **Summons R.E.** (1999) Archean molecular fossils and the early rise of eukaryotes. *Science*, **285**, 1033–1036.

Canfield, D.E. and **Raiswell, R.** (1999) The evolution of the sulphur cycle. *Am. J. Sci.*, **299**, 697–723.

Canfield, D.E. Sørensen, K.B. and **Oren, A.** (2004) Biogeochemistry of a gypsum-encrusted microbial system. *Geobiology*, **2**, 133–150.

Caumette, P. (1993) Ecology and physiology of phototrophic bacteria and sulphate-reducing bacteria in marine salterns. *Experientia*, **49**, 473–486.

Caumette, P., Matheron, R., Raymond, N. and **Relexans, J.-C.** (1994) Microbial mats in the hypersaline ponds of Mediterranean salterns (Salins-de-Giraud, France). *FEMS Microbial Ecol.*, **13**, 273–286.

Cornée, A. (1984) Etude préliminaire des bactéries des saumures et des sédiments des salins de Santa Pola (Espagne). Comparison avec les marais salants de Salin-de-Giraud (Sud de la France). *Rev. Inv. Geol.*, **38/39**, 109–122.

Des Marais, D.J. (2000) When did photosynthesis emerge on earth? *Science*, **289**, 1703–1705.

Douglas, S. and **Yang, H.** (2002) Mineral biosignatures in evaporites: presence of rosickyite in an endoevaporitic microbial community from Death Valley, California. *Geology*, **30**, 1075–1078.

Dronkert, H. (1977) A preliminary note on the geological setting of the gypsum in the province of Almeria (SE Spain). *Abstracts, Messinian Seminar III*, Malaga, Granada, 6 pp.

Durand, B. (1980) Sedimentary organic matter and kerogen. Definition and quantitative importance of kerogen. In: *Kerogen: Insoluble Organic Matter from Sedimentary Rocks* (Ed. B. Durand), pp. 13–34. Éditions Technip, Paris.

Eriksson, P.G. and **Altermann, W.** (1998) An overview of the geology of the Transvaal Supergroup dolomites (South Africa). *Environ. Geol.*, **36**, 179–188.

Farquhar, J., Bao, H. and **Thiemens, M.H.** (2000) Atmospheric influence of Earth's earliest sulphur cycle. *Science*, **289**, 756–759.

Fründ, C. and **Cohen, Y.** (1992) Diurnal cycles of sulphate reduction under oxic conditions in cyanobacterial mats. *Appl. Environ. Microbiol.*, **58**, 70–77.

Gandin, A., Wright, D.T. and **Melezhik, V.** (2005) Vanished evaporites and carbonate formation in the Neoarchaean Kogelbeen and Gamohaan Formations of South Africa. Presidential Review 8. *J. Afr. Earth Sci.*, **41**, 1–23.

Gavrieli, I., Stein, M., Yechieli, Y., Spiro, B. and **Bein, A.** (2001) Steady state sulfur isotopic composition during concurrent bacterial sulphate reduction and gypsum dissolution. *Geol. Soc. Am. Annual Meeting (Abstract 37)*, November 5–8.

Grassineau, N.V., Nisbet, E.G., Bickle, M.J., Fowler, C.M.R., Lowry, D., Mattey, D.P., Abell, P. and **Martin A.** (2001) Antiquity of the biological sulphur cycle:

evidence from sulphur and carbon isotopes in 2700 million-year-old rocks of the Belingwe Belt, Zimbabwe. *Biol. Sci.*, **268**, 1471–2954.

Gunatilaka, A., Saleh, A., Al-Temeemi, A. and Nassar, N. (1984) Occurrence of subtidal dolomite in a hypersaline lagoon, Kuwait. *Nature*, **311**, 450–452.

Gutzmer, J. and Beukes, N.J. (1995) Fault controlled metasomatic alteration of Early Proterozoic sedimentary manganese ores in the Kalahari manganese field, South Africa. *Econ. Geol.*, **90**, 823–844.

Habicht, K.S. and Canfield, D.E. (1996) Sulphur isotope fractionation in modern microbial mats and the evolution of the sulphur cycle. *Nature*, **382**, 342–343.

Hardie, L.A. (2003) Secular variations in Precambrian seawater chemistry and the timing of Precambrian aragonite seas and calcite seas. *Geology*, **31**, 785–788.

Hardie, L.A. and Eugster, H.P. (1971) The depositional environment of marine evaporites, a case for shallow, clastic accumulation. *Sedimentology*, **16**, 187–220.

Hendry, J.P. (1993) Calcite cementation during bacterial manganese, iron, and sulphate reduction in Jurassic shallow marine carbonates. *Sedimentology*, **40**, 87–106.

Hill, R.J. (1990) *Field Evidence for the Role of Organic Matter in Dolomitization.* Unpubl. MSc Thesis, Colorado School of Mines, Golden, CO.

Jørgensen, B.B. and Cohen, Y. (1977) Solar Lake (Sinai). The sulphur cycle of the benthic microbial mats. *Limnol. Oceanogr.*, **22**, 657–666.

Kakegawa, T. (2000) Biological activities and evolution of life during early stages of earth history: constraints from stable isotope records. *Viva Origino*, **28**, 191–208.

Kakegawa, T., Kawai, H. and Ohmoto, H. (1998) Origins of pyrites in the ∼2.5 Ga Mt. McRae Shale, the Hamersley District, Western Australia. *Geochim. Cosmochim. Acta*, **62**, 3205–3220.

Kendall, A.C. (2001) Late diagenetic calcitization of anhydrite from the Mississippian of Saskatchewan, western Canada. *Sedimentology*, **48**, 29–55.

Kirkham, A. (2004) Patterned dolomites: microbial origins and clues to vanished evaporites in the Arab Formation, Upper Jurassic. In: *The Geometry and Petrogenesis of Dolomite Hydrocarbon Reservoirs* (Eds C.J.R. Braithwaite, G. Rizzi and G. Darke). *Geol. Soc. London Spec. Publ.*, **235**, 301–310.

Kowalski, W., Holub, W., Wolicka, D., Przytocka-Jusiak, M. and Błaszczyk, M. (2003) Sulphur balance in anaerobic cultures of microorganisms in medium with phosphogypsum and sodium lactate. *Arch. Mineral.*, **2001–2**, 33–40.

Lal Gauri, K. and Chowdhury, A.N. (1988) Experimental studies on conversion of gypsum to calcite by microbes. *Proc. VIth Int. Congress on Deterioration and Conservation of Stone, Torun*, 545–550.

Lippmann, F. (1973). *Sedimentary Carbonate Minerals.* Springer-Verlag, Berlin, 228 pp.

Lyons, T.W., Luepke, J.J., Schreiber, M.E. and Zieg, G.A. (2000) Sulphur geochemical constraints on Mesoproterozoic restricted marine deposition: Lower Belt Supergroup, northwestern U.S. *Geochim. Cosmochim. Acta*, **64**, 427–437.

Mazzullo, S.J. (2000) Organogenic dolomitization in peritidal to deepsea sediments. *J. Sed. Res.*, **70**, 10–23.

Ohmoto, H. (2004) Archean atmosphere, hydrosphere, and biosphere. In: *The Precambrian Earth: Tempos and Events* (Eds P.G. Eriksson, W. Altermann, D.R. Nelson, W.U. Mueller and O. Catuneanu), pp 3–30. Elsevier, New York.

Ohmoto, H., Kakegawa, T. and Lowe, D. (1993) 3.4-billion-year-old biogenic pyrite from Barberton, South Africa: sulphur isotope evidence, *Science*, **262**, 555–557.

Ono, S., Eigenbrode, J.L., Pavlov, A.A., Kharecha, P., Rumble, D., Kasting, J.F. and Freeman K.H. (2003) New insights into Archaean sulphur cycle from mass-independent sulphur isotope records from the Hamersley Basin, Australia. *Earth Planet. Sci. Lett.*, **213**, 15–30.

Oren, A., Kühl, M. and Karsten, U. (1995) An endoevaporitic microbial mat within a gypsum crust: zonation of phototrophs, photopigments, and light penetration. *Mar. Ecol. Progr. Ser.*, **128**, 151–159.

Pierre, C. and Rouchy, J.M. (1988) Carbonate replacements after sulphate evaporites in the Middle Miocene of Egypt. *J. Sed. Petrol.*, **58**, 446 456.

Raiswell, R. and Fisher, Q.J. (2004) Rates of carbonate cementation associated with sulphate reduction in DSDP sediments: implications for the formation of carbonate concretions. *Chem. Geol.*, **211**, 71–85.

Reid, R.P., Bebout, B.M., Dupraz, C. and Macintyre, I.G. (2000) The role of microbes in accretion, lamination and early lithification of modern marine stromatolites. *Nature*, **406**, 989–992.

Rothschild, L.J., Giver, L.J., White, M.R. and Mancinelli, R.L. (1994) Metabolic activity of microorganisms in gypsum-halite crusts. *J. Phycol.*, **30**, 431–438.

Rouchy, J.M., Taberner, C. and Peryt, T.M. (2001) Sedimentary and diagenetic transitions between carbonates and evaporites. *Sed. Geol.*, **140**, 1–8.

Russell, M., Taberner, C., Rouchy, J.M. and Grimalt, J.O. (1998) Formation of diagenetic carbonates after bacterial sulphate reduction of evaporites associated to organic-rich deposits. 15th Int. Sedimentological Congress *Abstract*, Alicante 1998.

Schreiber, B.C. and El Tabakh, M. (2000) Deposition and early alteration of evaporites. *Sedimentology*, **47**, 215–238.

Shen, Y., Buick, R. and Canfield, D.E. (2001) Isotope evidence for microbial sulphate reduction in the early Archaean era. *Nature*, **410**, 77–81.

Simonson, B.M., Schubel, K.A. and Hassler, S.W. (1993) Carbonate sedimentology of the early Precambrian Hamersley Group of Western Australia. *Precambrian Res.*, **60**, 287–335.

Skyring, G.W. and Bauld, J. (1990) Microbial mats in coastal environments. *Adv. Microbial Ecol.*, **11**, 461–498.

Slaughter, M. and Hill, R.J. (1991) The influence of organic matter in organogenic dolomitization. *J. Sed. Petrol.*, **61**, 296–303.

Sørensen, K.B., Canfield, D.E. and Oren, A. (2004) Salinity responses of benthic microbial communities in a Solar Saltern (Eilat, Israel). *Appl. Environ. Microbiol.*, **70**, 1608–1616.

Spear, J.R., Ley, R.E., Berger, A.B. and Pace, N.R. (2003) Complexity in natural microbial ecosystems: the Guerrero Negro experience. *Biol. Bull.*, **204**, 168–73.

Stetter, K.O. (1996) Hyperthermophiles in the history of life. In: *Evolution of Hydrothermal Ecosystems on Earth (and Mars?)* (Eds G.R. Bock and J.A. Goode), pp. 1–10. Wiley & Sons, New York.

Strauss, H. and **Beukes, N.J.** (1996) Carbon and sulphur isotopic compositions of organic carbon and pyrite in sediments from the Transvaal Supergroup, South Africa. *Precambrian Res.*, **79**, 57–71.

Summons, R.E., Jahnke, LL., Logan, G.A. and **Hope, J.M.** (1999) 2-Methylhopanoids as biomarkers for cyanobacterial oxygenic photosynthesis. *Nature*, **398**, 554–57.

Sumner, D.Y. (1997). Carbonate precipitation and oxygen stratification in late Archean seawater as deduced from facies and stratigraphy of the Gamohaan and Frisco formations, Transvaal Supergroup, South Africa. *Am. J. Sci.*, **297**, 455–487.

Sumner, D.Y. and **Grotzinger, J.P.** (2000) Late Archaean aragonite precipitation: petrography, facies associations, and environmental significance. In: *Carbonate Sedimentation and Diagenesis in the Evolving Precambrian World* (Eds J. P. Grotzinger and N. P. James). *SEPM Spec. Publ.*, **67**, 123–144.

Taberner, C., Rouchy, J.M., Russell, M. and **Grimalt, J.O.** (1996) Timing of BSR and carbonate replacement after evaporites in organic-rich deposits. Lorca basin, Messinian, SE Spain. *AAPG Meeting, Abstract*, Dallas, TX.

Taberner, C., Rouchy, J.M., Pueyo, J.J., Marshall, J. and **Russell, M.** (1999) Carbonate replacement after sulphate in the Messinian of the Lorca Basin, SE Spain. Stages of development and relation to SO formation. *Bathurst Meeting, Abstract.*

Ulmer-Scholle, D.S. and **Scholle, P.A.** (1994) Replacement of evaporites within the Permian Park City Formation, Bighorn basin, Wyoming, U.S.A. *Sedimentology*, **41**, 1203–1222.

Vasconcelos, C. and **McKenzie, J.A.** (1997) Microbial mediation of modern dolomite precipitation and diagenesis under anoxic conditions (Lagoa Vermelha, Rio de Janeiro, Brazil). *J. Sed. Res.*, **67**, 378–390.

Vinogradov, V.I. (2003) Some features of epigenesis based on isotope geochemistry. *Lithol. Mineral. Resour.*, **38**, 332–349.

Visscher, P.T., Reid, P.R., Bebout, B.M., Hoeft, S.E., Macintyre, I.G. and **Thompson Jr, J.A.** (1998) Formation of lithified micritic laminae in modern marine stromatolites (Bahamas): the role of sulphur cycling. *Am. Mineral.*, **83**, 1482–1493.

Wacey, D. (2002) *The Origin of Dolomite in Distal Ephemeral Lakes of the Coorong Region of South Australia.* Unpubl. DPhil thesis, University of Oxford.

Wagner, R., Kuhn, M., Meyn, V., Pape, H., Vath, U. and **Clauser, C.** (2005) Numerical simulation of pore space clogging in geothermal reservoirs by precipitation of anhydrite. *Int. J. Rock Mech. Min. Sci.*, **42**, 1070–1081.

Walraven, F. and **Martini, J.** (1995) Zircon Pb-evaporation age determination of the Oaktree Formation, Chuniespoort Group, Transvaal Sequence: implications for Transvaal-Griqualand West basin correlations. *S. Afr. J. Geol.*, **98**, 58–67.

West, I.M. (1965) Macrocell structure and enterolithic veins in British Purbeck gypsum and anhydrite. *Proc. Yorks. Geol. Soc.*, **35**, 47–58.

Winhusen, E. (2001) *Precambrian Seawater Temperature Analysis Using Oxygen Isotopes from Hamersley Carbonates, Western Australia.* Unpubl. MS thesis, University of Cincinnati, Arts & Sciences, Geology, 146 pp.

Woese, C.R. (1987) Bacterial evolution. *Microbiol. Rev.*, **51**, 221–271.

Wood, G.V. and **Wolfe, M.J.** (1969) Sabkha cycles in the Arab/Darb Formation off the Trucial Coast of Arabia. *Sedimentology*, **12**, 165–191.

Wright, D.T. (1997) An organogenic origin for widespread dolomite in the Cambrian Eilean Dubh Formation, northwestern Scotland. *J. Sed. Res.*, **67**, 54–64.

Wright, D.T. (1999) The role of sulphate-reducing bacteria and cyanobacteria in dolomite formation in distal ephemeral lakes of the Coorong region, South Australia. *Sed. Geol.*, **126**, 147–157.

Wright, D.T. (2000) Benthic microbial communities and dolomite formation in marine and lacustrine environments – a new dolomite model: In: *Marine Authigenesis: from Global to Microbial* (Eds C.R. Glenn, J. Lucas and L. Prevot-Lucas), *SEPM Spec. Publ.*, **66**, 7–20.

Wright, D.T. and **Altermann, W.** (2000) Microfacies development in Late Archaean stromatolites and oolites of the Ghaap Group of South Africa. In: *Carbonate Platform Systems: Components and Interactions* (Eds E. Insalaco, P.W. Skelton and T.J. Palmer), *Geol. Soc. London Spec. Publ.*, **179**, 51–70.

Wright, D.T. and **Wacey, D.** (2004) Sedimentary dolomite – a reality check. In: *The Geometry and Petrogenesis of Dolomite Hydrocarbon Reservoirs* (Eds C.J.R. Braithwaite, G. Rizzi and G. Darke), *Geol. Soc. London Spec. Publ.*, **235**, 65–74.

Wright, D.T. and **Wacey, D.** (2005) Dolomite precipitation in experiments using sulphate reducing bacterial populations in simulated lake and pore waters from distal ephemeral lakes, Coorong region, South Australia. *Sedimentology*, **52**, 987–1008.

Zhong, S. and **Mucci, A.** (1993) Calcite precipitation in seawater using a constant addition technique: a new overall reaction kinetic expression. *Geochim. Cosmochim. Acta*, **57**, 1409–1417.

Zuddas, P. and **Mucci, A.** (1998) Kinetics of calcite precipitation from seawater, II: the influence of ionic strength. *Geochim. Cosmochim. Acta*, **62**: 757–766.

Part 3

Ancient carbonates and evaporites

Int. Assoc. Sedimentol. Spec. Publ. (2011) **43**, 315–392

Evaporitic source rocks: mesohaline responses to cycles of "famine or feast" in layered brines

JOHN K. WARREN

International Master Program in Petroleum Geoscience, Department of Geology, Faculty of Science, Chulalongkorn University, Bangkok, 10330, Thailand. Formerly: Shell Professor in Carbonate Studies, Oil and Gas Research Centre, Sultan Qaboos University, Muscat, Oman (E-mail: jkwarren@ozemail.com.au)

ABSTRACT

Organic matter in modern saline systems tends to accumulate in bottom sediments beneath a density-stratified mass of saline water where layered hydrologies are subject to oscillations in salinity and brine level. Organic matter is not produced at a constant rate in such systems; rather, it is generated in pulses by a halotolerant community in response to relatively short times of less stressful conditions (brackish to mesohaline) that occur in the upper part of the layered hydrology. Accumulations of organic matter can occur in any layered brine lake or epeiric seaway when an upper less-saline water mass forms on top of nutrient-rich brines, or in wet mudflats wherever waters freshen in and above the uppermost few millimetres of a microbial mat. A flourishing community of halotolerant algae, bacteria and archaeal photosynthesizers drives the resulting biomass bloom. Brine freshening is a time of "feast" characterized by very high levels of organic productivity. In a stratified brine column (oligotrophic and meromictic) the typical producers are planktonic algal or cyanobacterial communities inhabiting the upper mesohaline portion of the stratified water mass. In mesohaline holomictic waters, where light penetrates to the water bottom, the organic-producing layer is typically the upper algal and bacterial portion of a benthic laminated microbialite characterized by elevated numbers of cyanobacteria.

Short pulses of extremely high productivity in the upper freshened part of a stratified brine column create a high volume of organic detritus settling through the column and/or the enhanced construction of benthic microbial mats in regions where freshened waters reach the hypersaline base of the column. With the end of the freshening event, ongoing intensely arid conditions mean that salinity, temperature and osmotic stress increase rapidly in the previously freshened water mass. This leads to a time of mass die-off of the once flourishing mesohaline community ("famine"). First, these increasingly salty waters no longer support haloxene forms. Then, halotolerant life dies back, and finally by the halite precipitation stage, only a few halophilic archaea and bacteria remain in the brine column (typically acting as heterotrophs and fermenters). Repeated pulses of organic matter, created during a feasting event followed by a famine event, create laminated organic-enriched sediment on an anoxic bottom.

There are three, possibly four, major mesohaline density-stratified settings where organic-rich laminites (petroleum source rocks) accumulated in ancient "famine or feast" saline settings: (1) basin-centre lows in marine-fed evaporitic drawdown basins, associated with basinwide evaporites; (2) mesohaline intra-shelf lows on epeiric platforms, associated with platform evaporites; (3) saline-bottomed lows in under-filled perennial saline lacustrine basins; (4) closed seafloor depressions in halokinetic deepwater marine slope and rise terrains. An inherently restrictive hydrology means that the same mesohaline settings show a propensity to evolve into regions characterized by the accumulation of widespread evaporite salts. If this happens soon after deposition of a layer of organic-rich sediment there is an increased likelihood of evaporite plugging in the source-rock layer, this in-turn decreases expulsion efficiency and downgrades the laminite bed's ability to act as a prolific source rock.

Keywords: Oil and gas, source rocks, halobiota, hypersaline, salinity

INTRODUCTION

Back in 1988, Evans & Kirkland made the observation that some 50% of the world's oil may have been sourced in evaporitic carbonates. Heresy or not, the notion that much of the oil sealed by evaporite salts may also have also been sourced in sediments deposited in earlier less saline, but still related, evaporitic (mesohaline) conditions, is worthy of consideration. The association between saline waters, the accumulation of organic-rich sediments and the evolution of the resulting evaporitic carbonates into source rocks has been noted by many, including: Woolnough (1937), Sloss (1953), Moody (1959), Dembicki *et al.* (1976), Oehler *et al.* (1979), Malek-Aslani (1980), Kirkland and Evans (1981), Hite *et al.* (1984), Jones (1984), Eugster (1985), Sonnenfeld (1985), ten Haven *et al.* (1985), Warren 1986, 2006, Busson (1988), Evans & Kirkland (1988), Edgell (1991), Hite & Anders (1991), Beydoun (1993), Benali *et al.* (1995), Billo (1996), Carroll (1998) and Schreiber *et al.* (2001). The current paper draws on material and references discussed in greater detail in Warren (2006).

As long ago as the middle of the last century, Weeks 1958, 1961 emphasized the importance of evaporites as caprocks to major hydrocarbon accumulations. Weeks 1958, 1961 pointed out that many of the cycles of deposition that involve organic-rich carbonate marls or muds also end with evaporites. Many authors have noted the association of Type I–II hydrogen-prone kerogens in evaporitic source rocks and related their occurrence to the ability of[1] halotolerant photosynthetic algae and cyanobacteria to flourish in saline settings (Fig. 1A). Such kerogens tend to be oil-prone rather than gas prone (Fig. 1B). This paper examines chemical pathways and the conditions that favour various species that live in hypersaline environments. It compares modern saline settings with ancient systems and attempts to define those ancient saline settings that facilitated deposition of organic matter and subsequent evaporitic source rocks.

First, usage of the terms hypersaline and mesohaline in saline brines and the associated mineral precipitates is defined (Table 1). For seawater-derived and thalassic (seawater-like) brines the

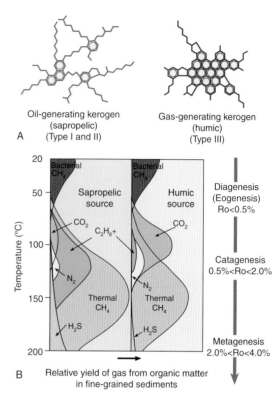

Fig. 1. Organic matter and source rocks. (A) Molecular structure of typical oil-generating (sapropelic) and gas-generating (humic) organic matter showing the greater preponderance of long-chain hydrocarbons in the sapropelic material (after Hunt, 1996). (B) Relative volumes of gases and liquids derived from a sapropelic and a humic source rock (after Hunt, 1996). Evaporitic source rocks tend to be sapropelic and contain less iron as a reductant so they are also likely to generate more H_2S.

use of the term "hypersaline" encompasses all waters more saline than seawater (mesohaline, penesaline and supersaline). There is an associated predictable suite of evaporite minerals, starting with carbonate (usually aragonite) in modern mesohaline waters, and evolving through gypsum and halite in penesaline waters, into bitterns from supersaline waters. However, ionic proportions in seawater have changed over time (Warren 2006). Accordingly, some ancient evaporite primary carbonates were dolomitic, especially in the Precambrian (although reflux dolomite has always been a widespread early diagenetic evaporite precipitate), and Archaean water probably precipitated a trona-halite-dominated set of salts rather then the halite-gypsum salts of the Phanerozoic.

Bitter salts from modern seawater concentrates show a carnallite-$MgSO_4$ association and a propensity to accumulate aragonite in the early stages of concentration, as they did also in the Permian

---

[1] (see Glossary at the end of this article for explanation of the specialist terminology used herein).

Table 1. Salinity-based classification of concentrated seawater and thalassic (seawater-like brines). Water density and dominant mineral precipitates are indicated (adapted from Warren, 2006)

Brine Stage		Mineral precipitate	Salinity (‰)	Degree of evaporation	Water loss (%)	Density (g cm^{-3})
Brackish		None	<35	<1	0	1.000–1.040
Normal marine seawater		Alkaline earth carbonates (aragonite, Mg-calcite)	35	1	0	1.040
Hypersaline	Mesohaline or vitahaline	Alkaline earth carbonates (aragonite, Mg-calcite, dolomite)	35–140	1–4	0–75	1.040–1.100
	Penesaline	CaSO$_4$ (gypsum/anhydrite)	140–250	4–7	75-85	1.100–1.126
		CaSO$_4$ ± halite	250–350	7–11	85–90	1.126–1.214
	Supersaline	Halite (NaCl)	>350	>11	>90	>1.214
		Bittern salts (K-Mg)	Extreme	>60	~99	>1.290

(icehouse climate mode). Bitterns at other times in the Phanerozoic (greenhouse climate mode) tend to lack the MgSO$_4$ minerals and have a propensity to form Mg-calcites in the mesohaline phase.

Use of the term hypersaline in modern continental (non-marine) waters is much less well defined and is variably applied to waters with total dissolved salt contents (TDS) that are more saline than 3 to 5‰. Nonmarine or continental waters can have ionic proportions considerably different to seawater (athalassic), and so a broader range of minerals can precipitate at all stages of brine concentration.

METABOLISM IN PRODUCERS AND CONSUMERS

Cyanobacteria and algae are the dominant primary producers in most mesohaline waters, along with lower and varying inputs from the higher plants and photosynthesizing bacteria. Algae and cyanobacteria typically make up the bulk of the planktonic biomass in the upper layer of a stratified mesohaline brine column and construct the upper parts of microbial mats in oxygenated settings where light penetrates to the water bottom. Cyanobacteria are prokaryotal (cell lacks a nucleus) bacteria and, like eukaryotic (cell with nucleus) algae and higher plants, are photoautotrophic and require oxygen to photosynthesize. The typical aerobic photosynthetic reaction for all three groups is:

$$CO_2 + 2H_2O = CH_2O + H_2O + O_2 \qquad (1)$$

That is, carbon dioxide and water react in the presence of chlorophyll and sunlight to form carbohydrate plus water plus oxygen.

All aerobic photosynthesizing organisms contain the pigments chlorophyll-α and phycobilin and use water as their electron source in reactions that generate oxygen. Chlorophyll-α is a green pigment that absorbs red and blue-violet light, hence the typical green colour of photosynthesizers (Fig. 2A, B). It is made up of molecules that contain a porphyrin 'head' and a phytol 'tail' (Fig. 2A). The polar (water-soluble) head is made up of a tetrapyrrole ring and a magnesium ion complexed to the nitrogen atoms of the ring, its phytol tail typically extends into the lipid layer of the thylakoid membrane. Phycobilisomes (phycocyanin, phycoerythrin) are proteins that absorb light of more energetic wavelengths than chlorophyll and are widespread pigments in the cyanobacteria and red algae (Fig. 2B). Their presence allows some members of the halotolerant community to photosynthesize into deeper brine and sediment depths, where light with longer wavelengths is less transmitted and therefore less available directly to chlorophyllic photosynthesizers (blue-green spectrum). Phycobilisome molecules are linear tetrapyrroles and are structurally related to chlorophyll-α, but lack the phytol side chain and the magnesium ion. In chlorophyll-rich oxygen-evolving reaction centres in a cell (chloroplasts) the phycobilisomes help transmit photic energy and in cyanobacteria they are located in chloroplasts in an extensive, intracellular system of flattened, membranous sacs, called thylakoids, the outer surfaces of which are studded with regular arrays of phycobilisome granules. Cyanobacteria are the only prokaryotes that contain chlorophyll-α in their chloroplast and so show strong affinity with the algae and the green plants. In the pre-genomic era of microbial studies this meant they were typically classified as

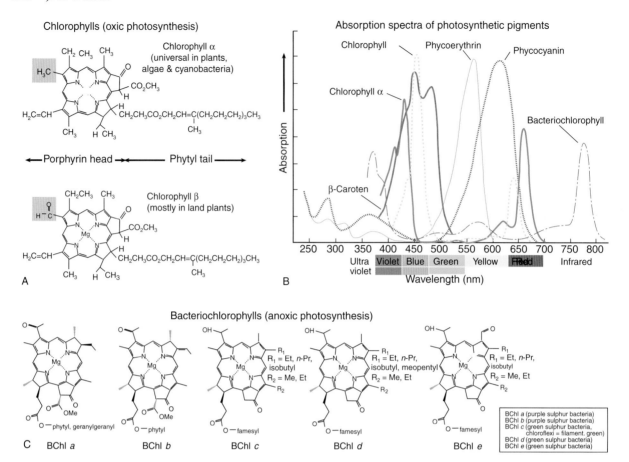

Fig. 2. Structures and features of the main photosynthetic pigments. (A) Chlorophyll-α and -β showing porphyrin head and phytyl tail. Pink box shows difference in structure that defines the two forms. (B) Absorption spectra of the chlorophylls, β-carotene and phycobilisomes (phycoerythrin in dominant in blue-coloured cyanobacteria and phycocyanin in red-coloured cyanobacteria. (C) Bacteriochlorophyll structures.

"blue-green algae." However, in terms of their genomic signature and the fact that the cell lacks a nucleus, they are better classified as prokaryotic bacteria.

In contrast to the cyanobacteria, all other photosynthetic bacteria in saline waters are anaerobes and utilize a single type of photosynthetic reaction centre with a different pigment (bacteriochlorophyll). When compared to chlorophyll and phycobilin, bacteriochlorophyll can absorb light with longer, less energy-rich wavelengths that extends into the infrared spectrum. Such light-requiring bacteria (e.g. purple bacteria) live in anoxic conditions and their ability to utilize the infrared spectrum means they can thrive deeper in a sediment or water column than aerobic photosynthesizers (Figs 2B and 13). The green sulphur bacteria (Chlorobiaceae) and purple sulphur bacteria (Chromatiaceae) use elemental sulphur, sulphide, thiosulphate, or hydrogen gas as the electron donor, whereas the purple and green nonsulphur bacteria use electrons from hydrogen or organic substrates (Table 2). As in aerobic photosynthesis, the resulting electrons (derived from the "light reaction" of photosynthesis) are stored in glucose and then used for CO_2 fixation (aka the "dark reaction" of photosynthesis).

The following simplified equation typifies anoxic photosynthesis of purple sulphur bacteria:

$$CO_2 + 2H_2S = CH_2O + H_2O + 2S \qquad (2)$$

That is, carbon dioxide and hydrogen sulphide react with bacteriochlorophyll and sunlight to form a carbohydrate (such as glucose) as well as water and sulphur. The light-absorbing pigments of the purple and green bacteria consist of bacterial chlorophylls and carotenoids. Phycobilins, characteristic of the cyanobacteria, are not found in these bacteria. All photoautotrophs (including

Table 2. Summary of Prokaryotes focusing on archaea and halophilic bacteria

Phylum or family			Informal group	Metabolic lifestyles (selected halophilic species)
Prokaryotes	Halo-adapted bacteria	Cyanobacteria	Cyanobacteria	Aerobic photoautotroph with chlorophyll-α plus phycobilicins, split water and generate oxygen (*Arthrospira platensis, Dactylococcopsis sauna, Aphanothece halophytica, Microcoleus chthonoplastes*)
		Proteobacteria	Purple sulphur bacteria	Anaerobic obligate photoautotroph – use H$_2$S or sulphur as electron donor (*Halorhodospira halophila* may be a significant primary producer in some settings, e.g. at a halocline; *Chromatium glycolicum, Ectothiorhodospira marismortui*)
			Purple non-sulphur bacteria	Anaerobic photoautotroph (tolerate low oxygen levels) – use hydrogen, but not from water, as the electron donor (e.g. *Rhodospirillum sp.*)
		Chlorobiaceae	Green sulphur bacteria	Anaerobic obligate photoautotroph – use H$_2$S or sulphur as electron donor (*Chlorobium limicola*)
		Flexibacteria	Green non-sulphur bacteria	Anaerobic (thermophilic) photoautotroph – use hydrogen, but not from water, as the electron donor (*Chloroflexus aurautiacus*)
		Spirochetes	Spirochetes	Free-living obligate anaerobe (*Spirochete halophila* oxidizes Fe and Mn). Chemolithotrophic, some species can be symbiotic or parasitic
		Bacteriacea	Flavobacteria	Anaerobic organotrophs and chemoautotrophs, some species can be parasitic (*Flavobacterium gondwanese* and *F. salegens* are psychrotolerant halophiles isolated from Antarctic Lakes) (*Salinibacter ruber* is phylogenetically affiliated with this group)
	Archaea	Euryarchaeota	Halophile	Aerobic photoautotroph, chemoautotroph (family Halobacteriales, includes: *Halobacterium* sp.; *Haloarcula* sp.; *Halobaculum* sp.; *Halococcus* sp.; *Haloferax* sp.; *Natroncoccus* sp.)
			Methanogen	Anaerobic chemoautotroph (*Methanocalculus* sp., *Methanohalophilus* sp., *Methanococcus* sp.)
		Crenarchaeota (sulpho-archaea)	Thermophile	Anaerobic chemoautotroph (Crenarchaeota-domain clone sequences documented in hypersaline sediments including Shark Bay stromatolites, anthropogenic salterns and sediments of Lake Qinghai, China)
			Acidophile	Aerobic- anaerobic chemoautotroph (as above), organotroph
		Nanoarchaeota	Symbiont on Ignoccus	Hosted on hot vent Archaea in Iceland. Its cell is only 400 nm long
		Korarchaeota	New group of extreme thermophile	Only described from RNA samples (if it is a real phylum then it is the closest to life's universal ancestor of the earliest Archaean ~3.7 Ga)

See Fig. 3 for additional information.

higher plants, cyanobacteria and bacteria) contain carotenoid pigments, which are red-orange and yellow in colour as they absorb the blue-violet and blue-green wavelengths missed by chlorophyll (Fig. 2B). The green sulphur bacteria are probably the most energetically efficient of all phototrophic organisms and can live in environments with long-term light intensities that are less than 0.01% of typical daylight. Many purple and green sulphur bacteria store elemental sulphur as a reserve material that, like glucose, can be further oxidized, in this case to SO_4, and so act as a photosynthetic electron donor.

Modern chemoautotroph communities in saline microbial communities generally form a metabolizing layer beneath the photosynthesizers (in both mats and stratified brine columns) and on top of the methanogens; some species in the chemoautotrophic layer are lithotrophs. Lithotrophic sulphur oxidizers active in saline environments include a number of species of both bacteria (e.g. *Thiobacillus*) and archaea (e.g. *Sulfolobus sp.*). *Thiobacillus* is widespread in marine and hypersaline microbialites and oxidizes thiosulphate and elemental sulphur to sulphate. Other lithotrophic bacteria use substances such as nitrate and reduce it to nitrite and include at least two species (*Nitrobacter akalicus* and *Nitrosomas halophila*) that flourish in the freshened water stages in the Kenyan soda lakes (Sorokin & Kuenen, 2005). *Sulfolobus sp.* tend to be hydrothermal vent dwellers that prefer settings characterised by hot, saline and acidic waters (see later).

Sulfolobus is a species within the archaeal domain (formerly known as the archaebacterial domain), which constitutes the "third domain of life" (see Glossary). While members of the archaea resemble bacteria in morphology and genomic organization, they resemble eukarya in their method of genomic replication. The archaeal domain is further subdivided into two kingdoms: the Crenarchaeota, of which all members presently isolated are extreme thermophiles, and the Euryarchaeota, a diverse group, that includes the halophilic order Halobacteriaceae, certain thermophiles and all members of the methanogen group (Table 2). All archaea are characterized by: (1) the presence of characteristic tRNAs and ribosomal RNAs; (2) the absence of peptidoglycan cell walls; (3) the presence of ether-linked lipids built from branched-chain subunits; and (4) their occurrence in unusual habitats (that seem extreme from an anthropocentric worldview).

Archaea are mostly extremophiles, that is they thrive in extreme conditions including: extremely saline (halophile), extremely hot (thermophile), extremely dry (xerophile), extremely acid (acidophile), extremely alkaline (alkaliphile), and extremely cold (psychrophile) environments. Some extremophiles are capable of surviving the multiplicity of extreme conditions that characterize arid environments and are known as polyextremophiles (Kates *et al.*, 1993).

Archaea and bacteria that require extreme salinities to metabolize are called halophiles, those that can tolerate salinities in excess of 25% are termed hyperhalophiles, and almost all documented modern hyperhalophiles are archaea. One of the more impressive halophilic archaea is *Halobacterium lacusprofundi*; it is both a halophile and a psychrophile, it thrives in bottom waters of the supersaline ($\sim$350‰) and cold ($<0°$–11.7 °C) Deep Lake, Antarctica (Franzmann *et al.*, 1988). The lake is too saline to freeze over, is thermally stratified and lies in an endoheic depression in the Vestfold Hills with a water surface that is some 50 m below sea level.

Throughout this paper, it will be discussed in detail why microbes (archaea and bacteria) in saline waters are halophilic, but it is notable that many halophiles are also thermophiles. Worldwide, aerobic halophilic archaea of the order Halobacteriales tend to have high growth optima at warm temperatures, typically between 35° and 50 °C and sometimes even higher (Oren, 2006). This may be an adaptation to the elevated temperatures that characterize stratified heliothermic waters in salt lakes worldwide. Likewise, within the anaerobic halophilic bacteria of the order Haloanaerobiales there are several moderately thermophilic representatives. *Halothermothrix orenii*, the first truly thermophilic halophile discovered, was isolated from Chott El Guettar, a warm saline lake in Tunisia. It grows optimally at 60 °C and up to 68 °C at salt concentrations as high as 200‰ (Cayol *et al.*, 1994). The halophilic *Acetohalobium arabaticum* strain Z7492 has a temperature optimum of 55 °C, can grow at salinities between 100‰ and 250‰ with an optimum around 150–180‰ and, like the halophilic archaea, has adapted to high salinities via the uptake of intracellular KCl (Zahran, 1997).

Even more extreme in its environmental preferences is the archaea *Sulfolobus tokodaii;* it is a hot spring dweller and intracellularly transforms H_2S to elemental sulphur. It flourishes at temperatures in excess of 90 °C and can live in saline environments where pH hovers around 1. Many such sulphur-dependant archaea are anaerobic chemoautotrophs that live in the vicinity of hydrothermal/volcanogenic vents in saline lacustrine rifts, or in marine hydrothermal brine springs, or in the basal waters of density-stratified brine lakes and seaways. They also inhabit deep seafloor brine pools on top of dissolving salt allochthons and they can survive and possibly metabolize hundreds of metres below the surface in bathyphreatic halite beds and sulphate caprocks (Grant *et al.*, 1998). Like these archaea, some chemoautotrophic bacteria can survive in extreme conditions, even living in saline subterranean settings, as evidenced by the recovery of viable halophilic chemolithotrophic bacteria (*Haloanaerobium sp.*) from highly saline oil fields brines in Oklahoma where pore-water salinities are ~150–200‰ (Bhupathiraju *et al.*, 1999).

Many halotolerant and halophilic bacteria and archaea are heterotrophs feeding on the remains of other organisms that had flourished in the same water mass at lower salinities. Heterotrophic metabolism shows variation in the terminal electron acceptor. In aerobic respiration it is oxygen and the breakdown product is CO_2, in anaerobic respiration it may be sulphate or nitrate and the breakdown product is H_2S or nitrite.

Halotolerant bacteria isolated from mesohaline settings tend to be anoxygenic photosynthesizers such as *Chromatium salexigens, Thiocapsa halophila,* and *Rhodospirillum salinarum,* with optimal growth in these species between 60 and 110‰ (Fig. 3; Caumette, 1993). In contrast to the cyanobacteria, which are oxygenic phototrophs, these anoxic phototrophs use H_2S, organic compounds or sulphur compounds as electron donors to produce various oxidized sulphur metabolites, the final product being sulphate. These anoxygenic halotolerant to halophilic photobacteria create dense biogenic laminae in a variety of anoxic, generally poorly lit environments. Phototrophic sulphur-oxidizing bacteria in many salt lakes grow in narrow niches defined by interfaces that are lit and also contain sulphide. In microbial mats they grow below the cyanobacterial and the phototrophic bacterial layers, while in stratified brine columns they inhabit waters of the halocline. Intense blooms of purple or green sulphur bacteria

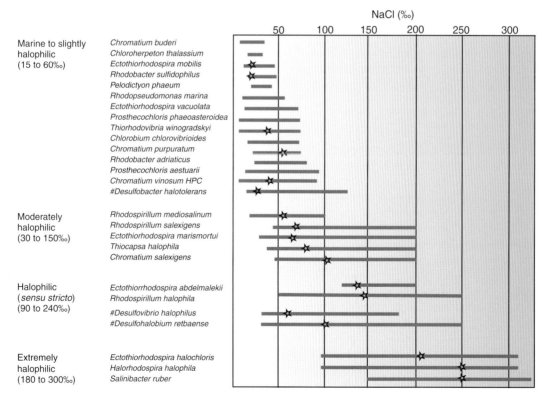

Fig. 3. Salinity tolerances of phototrophic halophilic bacteria, sulphate-reducers (#) and *Salinibacter ruber*. Star on bar indicates the salinity for optimal growth (after Ollivier *et al.*, 1994; Caumette, 1993; Imhoff, 1988; Ventosa, 2006).

can form brightly coloured water masses or sediment layers. Most of the H_2S they metabolize is biogenic rising from below and generated by microbial sulphate reduction, except around brine springs and hydrothermal vents where there can be high levels of abiogenic H_2S (Knauth, 1998). The biodiversity of bacterial photosynthesizers fall off as higher salinities are reached (Pedrós-Alió, 2005).

Purple and green bacteria growth is only possible in settings where the chemical gradient of sulphide is stabilized against vertical mixing. In layered microbial mats this occurs at the interface between the oxygenated layer created by the photosynthesizing cyanobacteria and the underlying anoxic layer created by the sulphate reducers. In stratified brine columns, chemical gradients are stabilized by density and temperature differences between the cooler less saline surface waters and the warmer more saline bottom layers and so they tend to inhabit the halocline. Examples of sulphur-oxidizing bacteria in hypersaline settings include *Atium volutans* from Solar Lake, *Beggiatoa alba* from Guerrero Negro, and *B. leptiformis* from Solar Lake (DasSarma & Arora, 2001). A unicellular halophilic, chemoautotrophic sulphur-oxidizing bacterium, *Thiobacillus halophilus*, has also been described in a hypersaline Western Australia lake where it grows in waters with salinities of up to 240‰ (Wood & Kelly, 1991).

All modern halophilic sulphur-oxidizing bacteria are filamentous, CO_2-fixers, typically motile and utilize sulphide as an electron donor for their photosynthesis. Purple sulphur bacteria seem to flourish across a broad range of continental brine salinities. For example *Ectothiorhodospira* halophilic species dominate in the microbial sediments of alkaline soda lakes in Egypt and Central Africa, while the moderate halophile *Ectothiorhodospira marismortui* is a strict anaerobe found in the waters of hypersaline waters (170‰) of sulphur springs in the Dead Sea, where spring waters had a pH around 5.2 and a temperature of 40 °C (Fig. 3). Isolates of this species grew poorly at a pH or 5.5 but thrived in the neutral pH range between 7 and 8 (Oren *et al.*, 1989). Another extremely halophilic sulphate-oxidizer *Ectothiorhodospira halochloris* was isolated from Wadi Natrun in 1977, in the laboratory it showed optimal growth in salinities in the range 140–270‰, a pH-range between 8.1 and 9.1 and temperatures between 47°C and 50°C (Fig. 3).

Dissimilatory sulphate reduction can occur up to quite high salt concentrations and black,

sulphide-containing sediments form on the bottom of modern coastal salinas and salt works well into the penesaline salinity range (Oren, 2001). Amongst the bacteria, most salt-tolerant halophilic sulphate-reducer thus far found is *Desulfohalobium retbaense*; it is a lactate oxidizer and was isolated from Lake Retba in Senegal. In the laboratory it remained viable at NaCl concentrations of up to 240‰, with a growth optima of 100‰ (Ollivier *et al.*, 1994). Two other sulphate-reducing halophilic isolates are *Desulfovibrio halophilus* and *Desulfovibrio oxyclinae*, which can tolerate periods of NaCl concentration of up to 225‰ (Fig. 3; Caumette, 1993; Krekeler *et al.*, 1997). Another isolate, similar to *Desulfovibrio halophilus* has been identified in brine pool samples collected in the allochthon-associated deep-water brine lakes on the floor of the Red Sea where it grows in salinities up to 240‰, but only slowly.

In sulphate reduction, the oxidation of acetate and CO_2 yields more energy per cell compared to the precursor oxidation of lactate. Recently the first halophilic acetate-oxidizing sulphate-reducing bacterium was isolated: *Desulfobacter halotolerans* (or *Desulfohalobium utahense*). It was collected from the hypersaline bottom sediments of the North Arm of Great Salt Lake, Utah, and was found to have a rather restricted salinity tolerance, it grows optimally at 10–20‰ NaCl, does not grow above 130‰ and is unable to survive above 240‰ NaCl (Fig. 3; Brandt & Ingvorsen, 1997; Jakobsen *et al.*, 2006). Even so, it possesses the highest NaCl-tolerance reported for any member of the genus *Desulfobacter*. In terms of metabolic pathways in the various salt-tolerant or halophilic sulphate reducers, it seems species that oxidize lactate, such as *Desulfovibrio halophilus*, can tolerate higher salinities of up to 250‰ than those species capable of oxidizing acetate, such as *Desulfobacter halotolerans*, which do best in salinities of less than 100–120‰ (Figs 3 and 4). In contrast to the sulphate-reducers, it seems no viable sulphur-oxidizing communities can survive at similar extreme salinities compared to the lactate oxidizers (Fig. 4).

Living below or among the sulphate-reducer communities in both microbial mats and stratified brine columns are the fermenters and the methanogens. One group of halophilic bacteria especially well adapted to the fermentative lifestyle is the Haloanaerobiales (Oren, 1992). Fermenters live in anaerobic saline conditions when there is no electron acceptor, and different sugars and in some cases amino acids are fermented to products

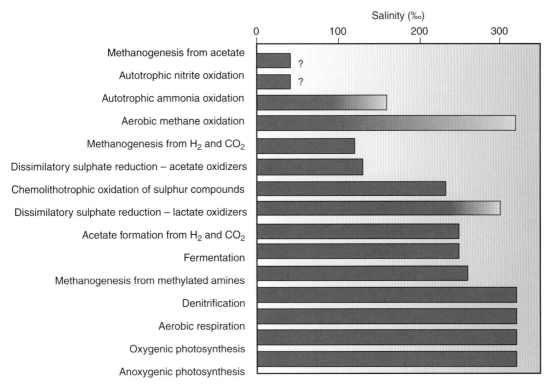

Fig. 4. Approximate upper salinity limits of selected microbial metabolic pathways (after Oren, 2001 and references therein). Values presented are based on laboratory studies of pure cultures (solid brown bars) and on activity measurements of natural communities in hypersaline environments (graded-colour extensions to bars).

such as acetate, ethanol, butyrate, hydrogen, and carbon dioxide. Fermentation breaks down an organic compound, such as a sugar or an amino acid, into smaller organic molecules that then accept the electrons released during the breakdown of the energy source. When glucose is broken down to lactic acid, each molecule of glucose yields only two molecules of ATP, and considerable quantities of glucose must be degraded to provide sufficient energy for microbial growth. Fermentation means only a relatively small output of energy per glucose molecule consumed as any organic molecule is only partially oxidized. Fermentation tends to be the sole metabolic mechanism in the well-adapted halotolerant and halophilic species of the Haloanaerobiales that inhabit the lowers parts of a microbial mat or brine column. It can be an emergency metabolic system in other microbes that find themselves in this same environment. The presence of biomarkers of fermentative microbes in saline and hypersaline Phanerozoic sediments or oils indicates depositional conditions that were anoxic and highly stressed with respect to higher life forms. Such brine- and pore-water conditions tend to facilitate

the preservation of hydrogen-prone organic end-products (proto-kerogens).

Methanogens are archaea that create methane gas as they remove excess hydrogen and fermentation products produced by other forms of anaerobic respiration. The methane then rises through the sediment or brine column to be utilized by the sulphate-reducers. Today some two-thirds to three quarters of the biogenic methane passed into the atmosphere worldwide is the work of a few dozen species of methanogenic archaea most of which live in relatively low salinity environments. The remainder of the world's methane comes from higher plants (Keppler *et al.*, 2006; Houweling *et al.*, 2006). All methanogenic archaea are obligate anaerobes thriving in three habitats: (a) bodies of anoxic fresh to hypersaline surface and subsurface water (e.g. the bottom brines of Solar Lake; (b) the vicinity of thermal brine springs; and (c) the digestive tracts of ruminants).

The main methanogenic processes in freshwater environments and in the guts of ruminants are the reduction of CO_2 with hydrogen and the aceticlastic split mechanism where organic matter is split by oxido-reduction into methane and carbon dioxide, but neither of these reactions has been shown to occur at high salt concentrations (Oren, 2006 and references therein). It seems that oxidation of organic matter is incomplete in settings with high levels of NaCl when compared with that in other ecosystems (temperate lacustrine and marine ecosystems) where acetate, $H_2 + CO_2$ are efficiently metabolized to produce CH_4, or are oxidized by sulphate-reducing bacteria whenever sulphate is available. At salinities higher than 150‰, the mineralization of organic compounds is limited by poor rates, or the complete absence, of microbial sulphate reduction of acetate and the inherently slow rates of methanogenesis utilizing H_2 (Fig. 4). The highest salt concentration at which such methanogenesis (from $H_2 + CO_2$) occurs in nature is in the 88‰ bottom waters of Mono Lake, California (Fig. 4: Oremland & King, 1989). The resulting accumulation of hydrogen and volatile fatty acids (VFA) in saline sediment beneath brine columns more saline than this, implies that catabolism via interspecies hydrogen transfer hardly occurs at all in more hypersaline environments (Ollivier *et al.*, 1994). Hence sediment from the floor of the Great Salt Lake contains up to 200 μM dissolved H_2. Similar results were obtained from Dead Sea sediments stimulated to reduce sulphate with H_2 and formate, although the same sediments did not respond with acetate, propionate, or lactate treatments. Energetic constraints may explain the apparent lack of truly halophilic methanogens capable of growing on $H_2 + CO_2$ or on acetate, as these methanogens use the energetically more expensive option of synthesizing organic osmotic solutes (see later, "low salt in" discussion).

Methanogenesis occurs at much higher salt concentrations but only via a less productive metabolic pathway that utilizes methylated amines, methanol, and dimethylsulphide substrates (Fig. 4; Oren, 2001, 2006). The most salt-tolerant methanogens known are *Methanohalobium evestigatum* and *M. portucalensis*, which grow in up to 250–260‰ NaCl. Additional moderately halophilic methanogens have been isolated, growing optimally at 40–120‰ salt (e.g. *Methanohalophilus mahii*, *M. halophilus*, *M. portocalensis*, *M. zhilinae*). Modern halophilic methanogens can form rich microbial communities in the density-stratified waters of deep-seafloor brine lakes, including suprasalt allochthon areas on the deep Mediterranean seafloor where some anoxic bottom brines have evolved to the bittern stage ($MgCl_2$ saturation; van der Wielen *et al.*, 2005). It may well be that

Earth-bound primordial life first evolved as methanogens in hypersaline waters (Dundas, 1998; Knauth, 1998).

Fermenters live in anaerobic saline conditions using a metabolic pathway where there is no electron acceptor, where different sugars, and in some cases amino acids, are fermented to products such as acetate, ethanol, butyrate, hydrogen and carbon dioxide (Hite & Anders, 1991). When evaporite salts precipitate, these liquid organics are often encased as brine inclusions in halite and other salt crystals and so are not picked up in standard TOC determinations. Their relative contribution to any source rock potential is poorly understood in subsurface evaporitic systems where cross-flowing basinal brines are dissolving buried halite beds. They may be locally significant contributors volumetrically to the hydrocarbons moving in carrier beds in contact with dissolving evaporites.

The last groupings to be considered in this discussion of the various metabolic pathways active in hypersaline settings are the haloviruses and bacteriophages that must infect the haloarchaea and bacteria. Compared to the literature-base for the halophilic archaea and bacteria little is known and most published studies to date deal with the haloviruses infecting the haloarchaea (Ventosa 2006; Dyall-Smith *et al.*, 2005). Nonetheless, given that in the natural world virus populations are greater than those of their prokaryotic hosts and that each host species is susceptible to infection by several different viruses, it may be deduced that a great diversity of such viruses must exist in hypersaline environments.

Electron microscopy of hypersaline waters shows that they maintain high levels of virus-like particles (about 10 times higher than the cell population), with recognizable morphotypes including head–tail and lemon-shaped particles (Ventosa, 2006 and references therein). Guixa-Boixereu *et al.* (1996) determined the abundance of viruses in two different salterns in Spain and observed that the number of viruses increased in parallel to that of prokaryotes, from $10^7\,mL^{-1}$ in the lowest salinity ponds to $10^9\,mL^{-1}$ in the most concentrated ponds (crystallizers), thus maintaining a proportion of 10 virions per prokaryotic cell throughout the salinity gradient. A lemon-shaped virus was found infecting square archaea; its abundance increased along the salinity gradient together with the abundance of the square archaea. Following a *Dunaliella sp.* bloom in freshened surface waters of the Dead Sea, Oren *et al.* (1997) observed the presence of large numbers of virus-like particles in water (up to 10^7 virus-like particles mL^{-1}), with a variety of morphologies, from spindle-shaped to polyhedral and tailed phages. Details of the halo-biotal bloom in the Dead Sea will be discussed later. It may be concluded that viruses and bacter-iophages are abundant in the archaea and bacteria that inhabit hypersaline waters, but the various metabolic pathways operating within this infest-ing community have yet to be fully established and await further study.

In summary, halophilic bacteria as a group are capable of using several metabolic styles and so are more biochemically divergent than the halophilic archaea. Some purple bacteria and gram-positive bacteria are chemoautotrophic halophiles, while halophilic cyanobacteria are oxygenic photoauto-trophs. Under suitably stressed conditions many halophilic bacteria become heterotrophic, utilizing either anaerobic or aerobic respiration pathways. By evolving oxygen respiration some 2.5 billion years ago, the cyanobacteria came to flourish and ultimately displace the archaea. Ancient cyano-bacteria lived in symbiosis with early eukaryotes; by the Mesoproterozoic this had allowed eukaryo-tic photosynthesis to develop and ultimately led to the evolution of the higher plants and animals. Once oxygen was plentiful in the atmosphere (~2.2 Ga), it was difficult for anaerobes (includ-ing photosynthesizing archaea and bacteria) to flourish in most surface waters. Some retreated to anaerobic environments, some acquired aerobic bacteria as endosymbionts, and some continued to thrive in aerobic/dysaerobic conditions that typify the hypersaline bottom waters of most density-stratified brine bodies.

BIOADAPTATIONS TO INCREASING SALINITY

In 1957, in his classic book on evaporites, Lotze made the statement: "Nicht das Leben, sondern der Tod beherrscht die Salzbildungsstätten" (Not life, but death rules the locales of salt deposition). Saline environments were considered biotal deserts, largely devoid of biological activity and hence were considered to be sedimentary systems with little source rock potential. The following discussion shows that this is not the case. Rather, saline environments and water bodies are sites of periodic but intense organic activity where numer-ous highly specialized algal, bacterial and archaeal species grow and die (cycles of "famine or feast"). They can leave behind substantial volumes of oil-prone organic residues in the bottom sediments, especially in laminated mesohaline carbonates, which on burial can evolve into prolific source rocks. Even in modern saline environments, where evaporite depositional settings are not as diverse or widespread as those of much of the past, conditions are such that mesohaline carbonates can preserve high levels of total organic carbon (TOC; Fig. 5).

Salinity ranges of the modern halobiota

Progressive brine concentration leads to sequential blooms of a macro and microbial biota adapted to different ranges of salinity in both marine and continental settings (Fig. 6A and B). As a modern surface brine is concentrated from 60‰ to around 200‰, dense eukaryotic algal and cyanobacterial populations appear, grazed by ostracods, brine shrimp and brine fly larvae. Halotolerant protozoa can be found feeding in this salinity range, as can yeasts and other fungi (Gunde-Cimerman *et al.*, 2005). Halotolerant microbial mats cover the bot-tom of many hypersaline ponds and shallow lakes in this salinity range. In anoxic waters in this salin-ity range there are a variety of sulphur-oxidizing, sulphate-reducing, homoacetogenic, methanogenic and heterotrophic bacteria and archaea, especially near the base of stratified brine columns or in the lower parts of mesohaline microbial mats.

From about 240‰ to more than 320‰, red-orange haloarchaea (halophilic and hyperhalophi-lic archaea) and halobacteria (e.g. *Salinibacter ruber*) come to dominate, mostly living on the products of decomposing organic matter left over from an earlier lower salinity time (Pedrós-Alió, 2005). At the elevated salinities where these halo-philes and hyperhalophiles flourish, only a few eukaryotes such as brine shrimp (*Artemia* and *Parartemia sp.*) can survive to graze on the remain-ing species of cyanobacteria and algae, especially *Dunaliella sp.* – a unicellular green alga (Fig. 6A). Above 300‰ no aerobic photosynthesis occurs, or at least is not known to occur based on the absence of chlorophyll-α in the brines. As ever more ele-vated salinities are attained, most other microbial activity first slows and then ceases.

A variety of obligate and facultative halophytic plants can survive in the moderate-to-high salin-ity soils that surround saline pans and lakes, e.g. *Atriplex halimus, Mesembryanthemum crystalli-num* and *Salicornia sp.* Only a few species of

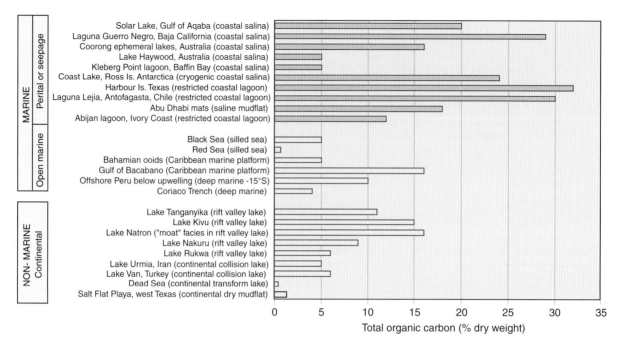

Fig. 5. Total organic carbon (TOC, dry wt%) in modern saline settings. Compiled from sources listed in Warren (1986, 2006).

animal can tolerate hypersaline conditions (Fig. 6A and B; DasSarma & Arora, 2001). The highest salinity at which viable vertebrates have been observed is around 60‰. For example, the white-lipped or alkaline tilapid fish (*Oreochromis alcalica*) flourishes in the mesohaline waters of the African rift lakes. This setting includes thermal spring waters at 36–40 °C along the moat-like edges of Lake Natron and Lake Magadi, where waters can have a pH as high as 10.5–11 and salinities in excess of 65–70‰.

Large numbers of invertebrates can survive in low to moderately mesohaline environments. Examples include salt-tolerant rotifers (aka wheel animalcules) such as *Brachionus angularis* and *Keratella quadrata*, and tubellarian worms (flatworms) such as *Macrostomum sp.* Insects from hypersaline environments include halotolerant brine flies, such as *Ephydra hians* and *E. gracillis*, which feed and lay their eggs in microbial mats and organic scum about the strandlines of modern saline lakes. Likewise, brine fly larva are adapted to saline conditions, *Ephydrella marshalli* larvae collected from commercial salt work lagoons on Port Phillip Bay, Victoria, have survived several days immersed in hypersaline sodium chloride media at an osmotic concentration of 5848 mOsm L^{-1} (175‰ NaCl; Marshall *et al.*, 1995).

Halotolerant crustaceans include ostracods such as *Cypridis torosa, Paracyprideinae sp., Diacypris*

compacta and *Reticypris herbsti*, copepods such as *Nitocra lacustris* and *Robertsonia salsa*, and brine shrimps such as *Artemia salina, Parartemia zietzianis* and other related species. Brine shrimp survive a wide range of salinities from 25 to 240‰ with their optimal range around 60–100‰; they are well adapted to the fluctuating oxygen and salinity levels that characterize many mesohaline ecosystems. Geddes 1975a–1975c argued that increased osmotic stress, not oxygen level, is the dominant limit on the upper salinity tolerance of Australian brine shrimp (*Parartemia zietzianis*), and this is probably also true for *Artemia* sp. (Geddes, 1981).

Halotolerant ostracods and brine shrimp feed on plankton living in the upper water mass of density-stratified brine columns and ingest benthic mats where this less-saline water mass intersects the lake margin or when bottom-water salinities are suitably lowered (de Deckker, 1981; De Deckker & Geddes, 1980). They produce pelleted carbonate detritus, which is a common component of many modern and ancient evaporitic laminites (Warren, 2006).

Diatoms (halotolerant bacillariophytal yellow-green algae with siliceous tests) are common planktonic forms in mesohaline and penesaline waters. They flourish at times of bloom in freshened salt-lake waters or in zones of perennial seepage and dissolution ponds (posas) about the edges of some salars. Their siliceous tests become a major detrital

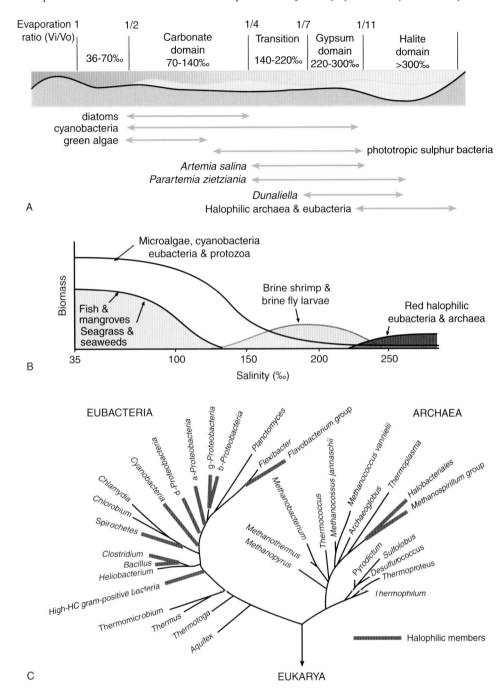

Fig. 6. Salinity tolerances of key biota. (A) Typical salinity ranges of the halotolerant biota where Vi is the volume of inflow to the basin and Vo is the volume of outflow (includes evaporation and reflux; after Barbé *et al.*, 1990). (B) Typical salinity ranges and biomass proportions of the biota in modern marine saltwork ponds. (C) Phylogenetic tree of the bacteria and the archaea, based on 16S rRNA sequencing comparisons. Thick red lines indicate branches containing representatives able to grow at or near optimal rates at NaCl concentrations exceeding 15% (after Ventosa *et al.*, 1998).

component in some lacustrine stromatolites (Winsborough *et al.*, 1994). Some halotolerant varieties of diatoms live in mesohaline lake brines with salinities around 120‰, while the upper limit for diatom growth is around 180‰ (Clavero

et al., 2000). The most halotolerant diatom taxa in the saltern ponds of Guerrero Negro, Baja California Sur, Mexico, are *Amphora subacutiuscula*, *Nitzschia fusiformis* (both *Amphora* taxa), and *Entomoneis sp.*; all grow well in salinities ranging from 5 to 150‰, Three strains of the diatom *Pleurosigma strigosum* are unable to grow in salinities of less than 50‰ and so are true halophilic alga.

In most moderately hypersaline (mesohaline to penesaline; 60–200‰) settings halotolerant cyanobacteria and green algae, rather than diatoms, are the dominant primary organic producers and green algal densities can be more than 10^5 mL^{-1}. These planktonic blooms can support many birds; one of the most spectacular examples is the pink flamingo population of African rift lakes. Mesohaline algae are mostly obligately aerobic, photosynthetic, unicellular microorganisms with some species producing large quantities of orange-coloured β-carotene pigment. Planktonic blooms in saline lakes can colour saline waters pink or red, especially at the higher end of this salinity range. Green algae of the genus *Dunaliella* (e.g. *Dunaliella salina, D. parva,* and *D. viridis*) are ubiquitous in modern Northern Hemisphere brine lakes, such as Great Salt Lake and the Dead Sea, and are the main source of food for brine shrimp and larvae of brine flies (DasSarma & Arora, 2001). Most species of green algae are moderate halophiles, with only a few extremely halophilic species (e.g. *Dunaliella salina* and *Asteromonas gracilis*) that survive in marginward refugia, even when the upper and low water layers in the Dead Sea are NaCl-saturated (see later).

Protozoa occur in most mesohaline waters, they are chemoheterotrophic protists that lack a cell wall and typically ingest algae and bacteria (Hauer & Rogerson, 2005). Identified species include the moderate halophile *Fabrea salina* from a west Australian salt lake and the extreme halophile *Porodon utahensis* from the Great Salt Lake. The number of known protozoan species identified in hypersaline waters (at salinities ranging from 90‰ to 220‰) has grown from <50 in the 1970s to more than 200 today, and include a range of amoebae, ciliates and flagellated protozoa. Fungi, mostly yeasts, are another heterotrophic component of a modern saline salt lake biomass (DasSarma & Arora, 2001; Gunde-Cimerman *et al.*, 2005). They are chemoheterotrophic eukaryotes, some of which are well adapted to tolerate markedly hypersaline environments. They grow best in aerobic conditions on carbohydrate substrates at moderate temperatures and in acidic to neutral pH brines. *Debaromyces hansenii* is a halotolerant yeast that, when isolated from seawater, can grow aerobically in salinities up to 250‰ and is capable of assimilating hydrocarbons. Another saprophytic hyphomycete (fungi), *Cladosporium glycolicum*, was

found at a salinity exceeding 260‰ growing on submerged wood panels in the Great Salt Lake. Halophilic fungi (e.g. *Polypaecilum pisce* and *Basipetospora halophila*) have also been isolated as a cause of spoilage in salted fish and bacon, but their distribution in hypersaline waters is not well known.

Above 200–240‰ most photosynthetic algae and halotolerant cyanobacteria cease to function and the halophilic archaea and bacteria come to dominate, mostly as decomposers (Fig. 6A and B; Pedrós-Alió, 2005). The transition into a more archaeal-rich biota means that while eukaryotic metabolic pathways and biochemistry dominates mesohaline waters, prokaryotic and archaeal metabolisms come to dominate at higher salinities (see later biomarker discussion).

"Halobacteria" flourishing in supersaline waters belong to the archaeal family Halobacteriaceae. All known extremely halophilic archaea stain gram negative, do not form resting stages or spores, and reproduce by binary fission (Oren, 2006). They are highly specialized micro-organisms, most of which will not grow at total salt concentrations below 2.5–3 M (145–175‰ NaCl) and are destroyed by lower salinities (see cellular adaptations). At present with the archaeal family Halobacteriacea there are 14 known genera and 35 validly described species. Six of these genera are monotypic: *Halobacterium; Halobaculum; Natrosomonas; Natrosobacterium; Halogeometricum; Haloterrigena.* The other genera (Table 3; Oren, 2006) are: *Natrialba* (2 species); *Natronococcus* (2 species); *Natrinema* (2 species); *Natrorubrum* (2 species); *Halococcus* (3 species); *Haloferax* (4 species); *Halorubrum* (7 species), *Haloarcula* (7 species). *Halococcus sp.* are strict aerobes, their name reflects their coccoid cell shapes and they form pairs, tetrads, sarcinae or irregular clusters. Cells produce several orange or red carotenoid and retinal pigments needed to better cope with high levels of ultraviolet light that characterize supersaline brine settings. *Haloferax* prefers high-Mg waters and is a major constituent in the halophilic biota of the Dead Sea. The various Halococcoid species, along with *Haloferax* and *Haloarcula*, flourish worldwide in hypersaline lakes, coastal salinas and brine lakes on the deep sea floor.

Most interesting, but most problematic, are the square-shaped archaea growing in many halite-precipitating brines. In 1980, A. E. Walsby described the abundant presence of "square bacteria" in

Table 3. Constituent groups of the archaeal family Halobacteriaceae (after Oren, 2006)

Family	Genus	Species
Halobacteriaceae	*Halobacterium*[T]	*Halobacterium salinarum*[T]
	Halobaculum	*Halobaculum gomorrense*[T]
	Halorubrum	*Halorubrum saccharovorum*[T]
		H. sodomense
		H. lacusprofundi
		H. coriense
		H. distributum
		H. vacuolatum
		H. trapanicum
	Haloarcula	*Haloarcula vallismortis*[T]
		H. marismortui
		H. hispanica
		H. japonica
		H. argentinensis
		H. mukohataei
		H. quadrata
	Natronomonas	*Natronomonas pharaonis*[T]
	Halococcus	*Halococcus morrhuae*[T]
		H. saccharolyticus
		H. salifodinae
	Natrialba	*Natrialba asiatica*[T]
		N. magadii
	Natronobacterium	*Natronobacterium gregoryi*[T]
	Halogeometricum	*Halogeometricum borinquense*[T]
	Natronococcus	*Natronococcus occultus*[T]
		N. amylolyticus
	Haloferax	*Haloferax volcanii*[T]
		H.x gibbonsii
		H. denitrificans
		H. mediterranei
	Natrinema	*Natrinema pellirubrum*[T]
		N. pallidum
	Haloterrigena	*Haloterrigena turkmenica*[T]
	Natronorubrum	*Natronorubrum bangense*[T]
		N. tibetense

[T] = type genus of the family or type species of the genus.

a small, halite-saturated brine pool, in the Sinai Peninsula (Walsby, 1980). These micro-organisms were collected from the surface of the pool, had a large number of gas vesicles, and presented unique cell morphologies never observed previously in the microbial world. The cells are squares and very thin, with sizes from 1.5 to 11 µm and a thickness of about 0.2 µm. Division planes were observed, with an arrangement indicating that division occurred in two planes alternating at right angles, so that each square grows to a rectangle which then divides into two equal squares, producing sheets divided like postage stamps (Walsby, 1980). In addition, different gas vesicles were observed, from spindle-shaped to cylindrical with conical ends and, in many cases, they were concentrated at the cell periphery. Ultrastructure studies confirmed that the square

cells observed by Walsby were micro-organisms with a typical prokaryote structure.

Direct observations of brines of salterns and other hypersaline environments showed that these square cells were very abundant in such habitats, especially in the most concentrated ponds with salinities higher than 3–4 M NaCl (at halite saturation, 175–233‰ NaCl), and they have been reported to occur in many geographical locations in both the Northern and Southern Hemispheres (Oren, 1999a). For more than 20 years these were considered uncultivable, until the work of Burns *et al.* (2004) and Bolhuis *et al.* (2004), who independently cultivated what Bolhius and coworkers informally named *"Haloquadratum walsbyi"*. The isolation of the square archaea by two independent research groups introduced a new methodology that now permits the isolation of new fresh isolates of square archaea from various areas. Future work will establish whether they are represented by a single or several novel haloarchaeal species capable of living amongst halite crystals. Work to date suggests that it is a new genus of the Halobacteriaceae, rather than a species within the existing genera of the order. The question arises as to whether the square cellular shapes of this archaeal group, and the cubic outline of pores and inclusions in and amongst the halite meshwork it inhabits, are more than fortuitous. Two haloarchaeal groups live only in high-pH high salinity waters of soda lakes such as Wadi Natrun and Lake Magadi; these are *Natronobacterium* and *Natronococcus*. They grow only in hypersaline waters with high pH (8.5–11.0) and very low levels of Mg (<0.024 g L^{-1}), and so are known as alkaliphiles. They have not been found in marine-fed salterns or less alkaline continental systems. However, the precise definition of genome systematics in halophiles living in soda lakes is still in a state of flux (Grant *et al.*, 1999).

Haloarchaeal species growing at salt concentrations above 150–200‰ characterize not only the Halobacteriaceae but also moderately halophilic methanogenic archaea, such as species of *Methanohalophilus* and *Methanococcus* (Table 2; Pflüger *et al.*, 2005). The existence of methanogenic archaea in hypersaline environments is related to the presence of noncompetitive organic substrates such as methylamines, which originated mainly from the breakdown of osmoregulator amines, rather than the presence of carbon dioxide, acetate and hydrogen (Fig. 4). The extremely halophilic methanogen, *Methanohalobium evestigatum*, with a NaCl optimum of 260‰, is also a thermophile with a temperature optimum of 50 °C. Halophilic methanogens are common contributors to the biota of methane hydrates associated with saline seeps and deep-sea brine pools above salt allochthons in the Gulf of Mexico (Lanoil *et al.*, 2001). Interestingly, unlike the Halobacticaea, these various methanogenic species seem to adapt to the increased salinity by a build up of osmolytes in the cell, and so are not obligate halophiles.

As well as halophilic archaea, there are halophilic bacteria flourishing in waters with salinities in excess of 240‰. One of the most interesting is *Salinibacter ruber*, which can constitute between 5 and 25% of the prokaryotic community in a number of halite-saturated settings worldwide (Antón *et al.*, 2005; Bardavid *et al.*, 2007).

Until the late 1990s, when *Salinibacter ruber* was first isolated and cultured from samples in anthropogenic salterns, the microbial population of halite-saturated waters was thought to be made up of a few species of archaea, including the distinctive archaeal phenotype with its square cell, which because of its unique four-sided morphology is widely known as the SHOW phylotype (Square Haloarchaea Of Walsby; Bolhuis, 2005; Dyall-Smith *et al.*, 2005).

It now seems that *Salinibacter ruber* is ubiquitous (some 5–25% of the biomass) in halite-saturated waters worldwide, and is typically commingled with the SHOW phylotype (Antón *et al.*, 2005). Most workers now agree that *S. ruber* and the square haloarchaea are the dominant microbial species at these elevated halite-saturated salinities (Ventosa, 2006). Once isolated in pure culture, it was evident that the similarity of *S. ruber* to many haloarchaea could have hampered the isolation of the bacterium. Both types of prokaryotes are extremely similar at the phenotypic level: both are extreme halophiles, aerobes, and heterotrophs and pigmented by carotenoids. *S. ruber*, like many haloarchaea, accumulates high concentrations of K+ to counterbalance the osmotic pressure of the medium, has a high proportion of acidic amino acids in its proteins, as well as enzymes functional at high salt concentrations and even has a high proportion of G+C in its genome (65–70% for *S. ruber*, in the range of 59–70% for haloarchaea; Antón *et al.*, 2005.). All these similarities, together with the fact that *S. ruber* and "*Haloquadratum walsbyi*" dominate the biota in halite-saturated waters whenever prokaryotic density is high, pose interesting questions as to the evolutionary process. For example, what environmental conditions in the Precambrian led to such a high degree of metabolic similarity between two groups that now occur in very genetically separate domains of life? It also shows that *S. ruber* is a bacterium that is as halophilic and as salt dependant as any member of the Halobacteriaceae.

Other than *Salinibacter ruber*, two groups of halophilic bacteria are well represented in salinity ranges in excess of 150‰; the fermentative bacteria belonging to the family Haloanaerobiaceae and the phototrophic sulphur-oxidizing bacteria of the family Ectothiorodhospiraceae (Figs 3 and 6C). Both groups are mostly made up of moderate halophiles. Phototrophic halophilic bacteria are further divisible into purple and green bacteria according to their respective bacteriochlorophylls and carotenoids (Fig. 2C). For example, the shallow parts of soda lakes in the Wadi Natrun may be intensively red coloured due to the development of *Halorhodospira halophila* (purple sulphur bacteria), while green non-sulphur bacteria dominate the photosynthesizing layer of the hypersaline microbial mats in brines of similar salinity in Guerrero Negro, Baja Mexico (see "red waters and brines" in Glossary). Small brine puddles in many hypersaline locations may show separate development of green- and red-coloured species of *Halorhodospira*, while top layers of the sediments of Wadi Natrun lakes (Egypt) show separate layers of the green-coloured and the red-coloured *Halorhodospira* species (Imhoff *et al.*, 1979). *Halorhodospira abdelmalekii, Halorhodospira halochloris*, and in particular *Halorhodospira halophila* are among the most halophilic phototrophic bacteria known (Imhoff, 1988). Isolates of *Halorhodospira halophila* from soda lakes in the Wadi Natrun have salt optima of 250‰ total salts and can survive in halite-saturated solutions (Fig. 3).

Oren (2001) has shown that in general in saline ecosystems the bioenergetic constraints of various metabolic pathways at the cellular level define the upper salinity limit at which the different dissimilatory processes can occur (Fig. 4). Dissimilatory sulphate reduction of a lactate substrate provides relatively little energy, and so the need to spend a substantial part of the cell's available energy for the production of organic osmotic solutes probably sets the relative low upper limit to the salt concentration at which sulphate-reducing bacteria can grow (Oren, 2001).

Our understanding of which microbial species dominate the halobiota at elevated salinities is not yet fully developed. As Ventosa (2006) emphasized, numerous ecological studies have demonstrate that haloarchaea may reach high cell densities ($>10^7$ cells mL^{-1}) in such hypersaline waters. Traditional microbial studies based on cultivation of viable cells from samples collected in such waters suggested that the predominant species found in most neutrophilic hypersaline environments were related to the genera *Halobacterium, Halorubrum, Haloferax* and *Haloarcula*. However, more recent molecular ecological studies based on cultivation-independent methods indicate that, in most natural hypersaline environments, members of these now well-documented halophilic genera constitute no more than a small proportion of the microbial biomass. They are seen more often

because they are more readily cultivated. Several more recent studies carried out in actual hypersaline environments and not the culture dish have allowed some general conclusions to be drawn (Ventosa, 2006): (a) square haloarchaea are very abundant, but they have not been isolated (cultivated) until recently; (b) many environmental clones within the haloarchaeal group are also obtained from natural hypersaline settings that are not phylogenetically closely related to previously described cultivated species of archaea and bacteria. In other words, the search to define the number of halophilic species and their relative contribution to the biomass made up haloadaptive microbes has only just begun.

Cellular biochemical adaptations to hypersalinity

Halotolerant and halophilic species have developed special biochemical mechanisms to cope with life's fatal propensity to desiccate in mesohaline and hypersaline waters, namely; motility, slime, osmolytes and halo-adapted cellular membranes and proteins. Many of the halotolerant cyanobacteria (still called "blue-green algae" by many geologists) form a significant component of photosynthesizing microbial mats in modern sabkhas and shallow brine lakes subject to both periodic desiccation and hypersalinity. Such microbial mats show a worldwide recurrence of a few cosmopolitan cyanobacterial species: *Microcoleus chthonoplastes; Lyngbya sp.; Entophysalis sp.;* and *Synecococcus sp.* All of these mat-forming taxa are embedded in matrices of extracellular polymeric mucous (slime), which hold large amounts of water to protect the photosynthesizing cell from osmotic stress and act as a buffer against extreme temperatures and ultraviolet rays (Gerdes *et al.*, 2000a, 2000b).

Aside from a greater likelihood of subaerial desiccation, one of the effects of increasing salinity is to increase osmotic pressure on all cellular life floating in a brine column. It may well be that the dominant biomass control in subaqueous hypersaline and supersaline environments is not lowered oxygen levels but increased osmotic pressure. At a concentration of 300‰, a sodium chloride brine exerts an osmotic pressure on a cell of more than 100 atmospheres (Bass-Becking, 1928; Reed *et al.*, 1984). Cells of haloxene species that typically inhabit marine or brackish waters quickly desiccate in such brines, leading to rapid death (e.g. Geddes, 1975b).

Accidental exposure of pelagic and bottom-living metazoans to osmotic stress and anoxia associated with hypersaline bottom waters has contributed to many ancient *lagerstatte* associations (fossil "death communities"), along with the inherent excellent preservation of fine details of the entrained largely undecomposed fossil forms. For example, there are the numerous "death-dance" trackways in the Cretaceous Solnhofen micritic plattenkalk of Bavaria. The tracks lead to the fossil remains of intimately preserved marine crustaceans that once wandered into the high osmotic pressures associated with hypersaline and anoxic bottom-waters of the lagoon that now preserves their remains (Viohl, 1996). Likewise, the feathers and skeleton of the bird-like dinosaur *Archeopteryx* were well preserved by the same anoxic hypersaline bottom waters.

Communities of haloxene chemosynthetic mussels, which inhabit nutrient-rich deep dark bottoms about the edges of modern anoxic brine lakes atop dissolving salt allochthons, can suffer mass extermination and immaculate preservation when sediment slumps occur and these mollusc biostromes are carried into lethal anoxic H_2S-rich hypersaline waters of the brine lake on the deep seafloor (MacDonald, 1992). Rich fossil-fish layers characterize subaqueous lacustrine laminated sediments of the Laney Shale Member of Eocene Green River Formation. Almost all the fish in the *lagerstatte* are freshwater species, and this death assemblage indicates the presence of inimical bottom waters tied to a salinity-stratified water column. This stratified column formed during the freshened highwater stage of the lake when dissolution of the underlying sodic evaporites of the Wilkins Peak Member released brines that created a condition of perennial anoxia in the deeper parts of lake water column (Boyer, 1982). A similar brine column stratification probably favoured the preservation of organic matter in the muddy lacustrine/lagoonal sediments that overlie the Mesozoic Purbeck evaporites that crop out along the Dorset coast (Schnyder et al., 2009).

Microbes that flourish in hypersaline waters have adapted at the cellular level to survive the increased osmotic stress that accompanies increased salinity. At the intracellular molecular level, *Dunaliella parva,* a halotolerant unicellular green alga that flourishes in mesohaline waters worldwide (optimum ~120‰, survival 20–340‰), distributes glycerol throughout its cytoplasm. This increases the total solute concentration in the cell and so the

algae can better cope with the high osmotic pressure of its high salinity surrounds. Glycerol is an osmotically active substance (osmolyte) that takes up space in the cellular fluids. It lowers the relative water content of the cell and so raises the internal osmotic pressure. Cytoplasmic concentrations of glycerol in *Dunaliella parva* can reach $7\,mol\,L^{-1}$ and can constitute over 50% of the dry weight of cells growing in salinities ~280‰. Glycerol is also used as an osmolyte by other halotolerant and halophilic algae, as well as yeasts, fungi and brine shrimp. Halotolerant and halophilic prokaryotes can use a variety of other sugars, sugar alcohols, amino acids and compounds derived from them (such as glycine, betaine and ecotoine) as osmolytes, leading to characteristic biomarker assemblages in ancient saline sediments.

In contrast to an osmolyte solution, halophilic archaea have two main defence mechanisms to cope with the extreme osmotic gradients induced by hypersaline to supersaline waters. One is a modification of the plasma (cell) membrane; the other is an adaptation the protein structure itself. Unlike most cell membranes, the plasma membrane of halophilic archaea is dominated by rhodopsins and the cell membrane does not actively exclude salts, so the overall internal salt concentration in halo-archaeal cytoplasm is high. This adaptive mechanism is sometimes called the "salt-in" strategy. However, the cell membrane is selective, it tends to exclude sodium, while actively pumping potassium into the cell against a thousand-fold concentration gradient. The total ionic strength remains the same on both sides of the plasma membrane, but the ratios of specific ions differ inside and outside the cell wall. Potassium is the dominant cation within the cell, many cell functions require potassium, but would be disrupted by high levels of sodium. Cells of halophilic archaea can contain up to $4\,mol\,L^{-1}$ of potassium chloride, which is eight times its concentration in seawater. Anaerobic bacteria of the order *Haloanaerobiales* use a similar "high potassium in the cell" strategy, as does *Salinibacter ruber* (Oren, 1999b).

Halobacterium halobium is a well-studied example of a well-adapted halophilic archaea; its cell carries a combination of orange-red carotenoid pigments, mostly β-carotene (C-40), and bacterioruberins (C-50). It is an aerobe-facultative anaerobe. When oxygen levels fall, but light penetrates the brine, photophosphorylation takes place in the absence of chlorophylls and redox carriers, so waters and sediments where it flourishes can take on a blue or purple tint. Bacteriorhodopsin's characteristic purple colour has led to many haloarchaea in their photo-active stage being described historically (pre-genomic) as purple bacteria, not archaea. The main cellular metabolism of *Halobacterium halobium* is heterotrophic in brines with higher oxygen levels, and it flourishes in such conditions in Great Salt Lake, Solar Lake and Lake Magadi.

Cells in the heterotrophic mode are mostly coloured by carotenoids, and waters where it is thriving appear red. This normal respiratory metabolic chain depends on b and c cytochromes as well as cytochrome oxidase to create ATP. This is not a particularly unusual metabolic pathway for most life forms, except that in this case the cell itself is halophilic. As a result, in brines with relatively high levels of oxygen, these highly salt-adapted halophilic cells respire and produce ATP using their red membrane (carotenoids). Under anaerobic conditions they can survive but produce lesser volumes of ATP, either by fermentation of arginine or by anaerobic respiration with fumarate as an electron receptor. It means that they can still live, although at lower metabolic rates, in well-lit dysaerobic to anoxic hypersaline conditions as they utilize their purple membrane patches to pump protons and so maintain their cellular metabolism (Oesterhelt & Marwan, 1993; Kates *et al.*, 1993).

Some halophilic archaeal strains, including *Halobacterium*, contain another rhodopsin pigment in their purple membrane called halorhodopsin. This acts as a Cl-pump, once again suitable for generating energy in highly saline conditions. It is usually present in the cell membrane in much smaller amounts than bacteriorhodopsin, but contains the same retinal pigment. It too absorbs light, which it converts to an ion gradient. However, instead of pumping protons out, like bacteriorhodopsin, it pumps chloride ions inwards, which helps maintain osmotic balance in highly saline conditions. Since chloride ions have a negative charge, moving chloride inwards is equivalent in terms of energy to moving a proton outwards. Thus halorhodopsin generates a chloride ion gradient, which also supplies energy to the cell.

Membrane adaptation shifts the osmotic balance problem from the cellular to the molecular level. Ions within the cell's cytoplasm still compete with proteins and other biomolecules for that universal solvent, water. In order to cope with high osmotic stress at the molecular level, halophilic archaeal proteins have evolved so as to maintain a high negative charge on the protein surface. This attracts water molecules and envelops the protein in a protective shroud of bound water.

Consider the halophilic archaea *Haloarcula marismortui*, which flourishes during times of freshened surface waters in the Dead Sea. Proteins of its metabolic enzyme (malate dehydrogenase) and its electron transfer protein (ferredoxin) are wrapped in coats of acidic amino acid sidechains so that both proteins carry a negative charge in neutral media. A predominance of charged amino acids on the surfaces of cellular enzymes and ribosomes better stabilizes the hydration shell of the various cytoplasm molecules, even when the extracellular brines are well into the halite saturation field. This high-density surface charge binds water molecules to the protein surface much more effectively than any mesophile protein. Hence, *Haloarcula* proteins do not dehydrate or unfold (denature) even when exposed to extremely high salinities. However, the high levels of shielding cations are lost by diffusion in low salinity environments and an excess of negatively charged ions then destabilizes the molecular structure by ionic repulsion and so intracellular proteins tend to lyse or disintegrate. This means any species using a "salt-in" cytoplasm strategy has an innate restriction to highly saline environments (obligate halophiles) and does not cope well with rapidly oscillating schizohaline waters. Ironically, intracellular acids that unfold normal proteins in non-halophiles counteract this tendency in the halophiles.

In summary, bioadaptations at the cellular level to highly saline conditions can be divided into two groups of responses (Oren, 2001):

(a) *The "high-salt-in" option*

Accumulation of salts in the cytoplasm are maintained at concentrations equal to or higher than those of the outside medium. Generally KCl is used as the main intracellular salt. Aerobic halophilic archaea (family Halobacteriaceae) utilize this strategy, as do fermenting bacteria, eubacterial acetogenic anaerobes (*Haloanaerobium, Halobacteroides, Sporohalobacter, Acetohalobium*), the extremely halophilic *Salinibacter ruber* and some sulphate-reducers. Energetically, this strategy is relatively inexpensive, but requires far-reaching adaptations of the intracellular enzymatic machinery to the presence of high salt concentrations. Cells utilizing this strategy show limited adaptability to changing salt concentrations (obligate halophiles) and do best in environments where long-term salinities remain elevated. That is, they do best in the perennial hypersaline portions of salt lakes such as the upper water mass of the Dead Sea and do not do as well in parts of the depositional setting characterised by rapid and frequent bouts of schizohalinity.

(b) *The "low-salt-in" option*
Exclusion of salts from the cytoplasm and the accumulation of organic osmotic "compatible" solutes (osmolytes) may be employed to provide osmotic balance. This strategy is used by halophilic and halotolerant eukaryotal microorganisms, by most salt-requiring and salt tolerant bacteria, and also by halophilic methanogenic archaea. This strategy is energetically expensive; the energetic cost depends on the type of organic solute synthesized. No major modification of the intracellular machinery is needed compared to non-halophiles, and in most cases cells can rapidly adapt to changes in salinity. As such, they do better than the "salt-in" biota in conditions of rapid and frequent schizohalinity.

As Oren (2001) clearly illustrated, organisms living in increasingly high salinity environments have to devote increasing large proportions of their bioenergetic budgets to osmoregulation, either to synthesize organic solutes or to activate and maintain ionic pumps. This creates a natural salinity-related control over various metabolic pathways and their energy efficiencies and hence to biomarker distributions in the resulting sediments. The following metabolic adaptations and groups function best at high salinities (Fig. 4):

(1) Processes that use light as energy source and are generally not energy-limited;
(2) Heterotrophs (bacteria as well as archaea) that perform aerobic respiration, denitrification, and other dissimilatory processes that yield large amounts of ATP;
(3) All types of metabolism performed by organisms that use the "high-salt-in" strategy, even if the efficiency and amount of ATP obtained in their dissimilatory metabolism is low.

Thus, the growth of phototrophic microorganisms in hypersaline surface waters is generally limited by the availability of inorganic nutrients and not by lack of light energy (Oren, 2001). Accordingly, both oxygenic (*Dunaliella*, cyanobacteria) and anoxygenic phototrophs (*Halorhodospira*, *Thiohalocapsa*, *Halochromatium*) produce organic solutes for osmotic stabilization. They may be found up to very high salt concentrations and in several cases survive well into the halite-saturation field. However, in the case of chemolithotrophs and anaerobic heterotrophs, the amount of energy generated in the course of their dissimilatory metabolism is often insufficient to supply the demands of cell growth, osmotic adaptation, and life maintenance. Long-term growth and survival in hypersaline environments is not possible in such species. The "high-salt-in" strategy is preferable for strongly energy-limited organisms (haloarchaea) as it is energetically much less costly than the production of organic osmotic solutes. Oren (2001) documented an excellent correlation between the amount of energy generated in the course of the different dissimilatory processes and the salinity-related occurrences of various metabolic styles (Fig. 4). Clearly the resulting biomarkers from the various metabolic pathways will likewise preserve salinity-related indicator markers in the rock record.

Biomarkers and microbial responses to salinity changes

The salinity-controlled transition between a halotolerant and halophilic biota can been seen in their organic residues and subsequent biomarkers preserved in evaporitic source rocks. Biomarkers can be thought of as molecular or chemical fossils whose basic carbon skeleton is derived from once-living organisms and are found in modern sediments, as well as in petroleum and rock extracts. They can provide information about species diversity, depositional environment, thermal maturity, migration pathways and hydrocarbon alteration (Peters *et al.*, 2005a, b).

Commonly accepted environmental generalizations based on biomarkers include the notion that variation in the ratio between the isoprenoids pristane and phytane indicates oxidizing versus reducing conditions (pristane and phytane are both breakdown products of chlorophyll). In an oxidizing environment the cleavage of the phytol side chain of chlorophyll is followed by decarboxylation to produce phytane. In a reducing environment the sidechain cleavage of chlorophyll is followed by its reduction to ultimately produce pristane. Low pristane/phytane ratios in marine and terrestrial settings are thought to indicate reducing conditions, while higher values (>1) indicate oxidizing conditions (Table 4). Philp & Lewis (1987) have shown that the chemistry of chlorophyll breakdown is much more complicated in many natural systems, and that variations in the ratio may also indicate varying inputs from archaeal membranes, which contain much higher levels of phytane chains than bacteria and can dominate the halobiota at higher salinities. Phytane can also come from bacteriochlorophyll-a and tocopherols, which can also be locally commonplace (Peters *et al.*, 2005a, b; Koopmans *et al.*, 1999).

Likewise, differences in the distribution of n-alkanes are thought to indicate depositional differences in the source contributors. Waxes in

Table 4. Depositional significance of selected biomarkers (see text for detail)

Setting	Biomarker/indicator	Detail
Hypersaline	Gammacerane	High relative to C_{31} hopanes in oils derived from sources deposited under hypersaline depositional conditions. High values indicate stratified water column during source deposition (Sinninghe Damste *et al.*, 1995).
	Pristane/phytane	Very low values (<0.5) in oils derived from source rocks deposited under hypersaline conditions, indicates contribution of phytane from halophilic bacteria/archaea (ten Haven *et al.*, 1987, 1988).
	Acyclic isoprenoids	Acyclic isoprenoid hydrocarbons are the predominant components in the organic matter extracted from sedimentary cores and oils of various hypersaline settings (Wang, 1998).
	Low diasteranes	Oils from the hypersaline sources typically have very low amounts of diasteranes, which commonly suggests a source rock that has low content of catalytic clays, consistent with carbonate or evaporite source rocks (Mello & Maxwell, 1991; Peters & Moldowan, 1993). The amount of diasteranes in oils also decreases in highly reducing noncarbonate environments with increasing salinity (Philp & Lewis, 1987).
	Other biomarkers	Even-over-odd dominance in the n-alkanes, presence of β-carotene and γ-carotene, dominance of the C_{27} steranes.
	Organic sulphur compounds	There is a high abundance of organic sulphur compounds (thiolanes, thianes, thiophenes and benzo-thiophenes) in many anoxic hypersaline settings (Wang, 1998). The distributions of the C-20 isoprenoid thiophenes in combination with those of the methylated 2-methyl-2-(4,8,12-trimethyltridecyl) chromans can be used to discriminate non-hypersaline from hypersaline paleoenvironments (Sinninghe Damste *et al.*, 1989)
	Carbon isotopes	Heavy ^{13}C values (−20 to −30‰) are consistent with typical saline lake environments and reflect depletion of CO_2 by photosynthetic organisms (Peters *et al.*, 1996).
Anoxic	C_{35} homohopanes	High relative to total hopanes in oils derived from source rocks deposited under anoxic conditions (Peters & Moldowan, 1993). Abundance of C_{35} homohopanes in oils (relative to C_{31}–C_{34} homopanes) is correlated with source rock hydrogen index (Dahl *et al.*, 1994).
	Pristane/phytane	>1.0 can indicate anoxic conditions, but the ratio is affected by many other factors.
	Isorenieratane & derivatives	Presence in oil indicates anoxic photic zone during source rock deposition, these compounds are biomarkers for green sulphur bacteria (Summons & Powell, 1987).
	V/(V + Ni) Porphyrins	High = reducing conditions (Lewan, 1984).
	28,30-bisnorhopane	High in certain reducing environments (Schoell *et al.*, 1992; Moldowan *et al.*, 1994).
Lacustrine	Botryococcane	Presence = lacustrine source. Absence = meaningless (e.g. Moldowan *et al.*, 1985).
	β-Carotane	Presence = lacustrine source. Absence = meaningless (Jiang & Fowler, 1986).
	Sterane/Hopanes	Low in oils derived from lacustrine source rocks. (Moldowan *et al.*, 1985)
	C_{26}/C_{25} tricyclic terpanes	>1 in many lacustrine-shale-sourced oils (Zumberge, 1987)
	Tetracyclic polyprenoids	High in oils from freshwater lacustrine sources (Holba *et al.*, 2000).

higher plants have significant concentrations of the long chain C_{22-36} alkanes with a pronounced odd/even distribution (Fig. 1A). Such organics tend to be relatively rare in evaporitic source rocks, unless surface waters were periodically freshened by run-off from the land. Much of the terrestrial organic material entering an evaporitic depression has been oxidized and biodegraded before it makes it to the final site of deposition on the brine pool floor. In contrast, a high algal input is thought to be characterized by n-alkanes in the C_{16-18} region.

Waples *et al.* (1974), along with ten Haven et al. (1986), noted that Tertiary sediments deposited in saline evaporitic lagoons possess high concentrations of regular C_{25} isoprenoids (Table 4). The activities of *Chlorobiaceae*, an anaerobic green sulphur bacteria, thought to flourish in mesohaline to hypersaline seaways, leads to the preservation of a series of 1-alkyl-2,3,6-trimethyl benzenes, thought to be derived from the breakdown of aromatic carotenoids in sulphate- and sulphide-rich brines (Summons & Powell, 1987). ten Haven *et al.* (1988) went on to propose a number of biomarker indicators of hypersalinity, including short-sidechain steranes and $5\alpha(H)$, $14\beta(H)$, $17\beta(H)$ pregnanes and homopregnanes, as well as high gammacerane indices (gammacerane/C_{30} hopane). Gammacerane is a pentacyclic C_{30} triterpane thought to be a diagenetic alteration product of tetrahymenol, a natural product produced by bactiverous ciliates, such as *Tetrahymena* (Table 4).

At its simplest, much of the utility of biomarkers in saline environments comes from salinity-related differences in contribution to organic matter of the general categories of primary producers (autotrophs), namely prokaryotes (cyanobacteria and bacteria) and eukaryotes (higher plants, algae) and the archaea. Most triterpanes are associated with prokaryotic sources, whereas eukaryotic organisms produce steranes. Thus, the triterpane/sterane ratio can be a rough measure of the prokaryote/eukaryote contribution to the organic material. As salinity increases, the less salinity-tolerant eukaryotic organisms (mostly sterane-producing green algae) give way to more halotolerant bacteria and cyanobacteria (tricyclic and hopane producers) with a corresponding increase in the triterpane/sterane ratio. Thus high alkalinity/salinity settings are characterized by tricyclics (C_{20}–C_{24}; m/z 191), β-carotene ($C_{40}H_{56}$ compound; m/z 125), gammacerane (C_{30} triterpane; m/z 191), all with prokaryote sources. Hence, high levels of gammacerane and β-carotene are typically associated with non-marine highly saline environments (Peters & Moldowan, 1993).

For archaea, the cell membrane itself, due to its inherent high thermal stability, is a very good candidate for a biomarker. Archaeal cytoplasmic membranes do not contain the same lipids that prokaryotes and eukaryotes do. Instead, their membranes are formed from isoprene chains (ether lipids) made up from C_5 isoprenoid units (as for the side chains of ubiquinone) rather than C_2 units (ester lipids) in the normal fatty acids of the non-archaea (Fig. 7A). Moreover, the isoprenoid chains are typically attached to glycerol by ether linkages instead of esters. Thus the cell wall of archaea is constructed of lipids, mainly of glycerol, connected to phytanyl chains 20 carbons in length by ether bonds to form phytanylic diether. Typically these are organized in bilayers. In the case of archaea living in extreme conditions, two glycerol molecules can be connected to a double chain of phytanol to create a tetraether structure of forty carbons (Fig. 7B). Isoprene derivatives indicative of ancient archaea have been found in Mesozoic, Palaeozoic, and Precambrian sediments (Hahn & Haug, 1986). Interestingly, their chemical traces have even been found in sediments from the Isua district of West Greenland, the oldest known sediments on Earth, some 3.8 billion years old.

Lipids entrained as organic residues in modern evaporitic carbonate and gypsum from modern saline pans (mesohaline and lower penesaline waters), are mostly derived from cyanobacteria and heterotrophic bacteria. Their n-alkane distributions show a high predominance of n-docosane. Thus, in the evaporitic carbonate domain (mesohaline waters), the presence of the C_{20} highly-branched isoprenoid olephines, tetrahymanol and large amounts of phytol constitute precursors to most lipids found in buried evaporitic sediments. In contrast, the main lipid contributors in organics accumulating in halites and bittern beds can be the extremely halophilic archaea and their organic signatures are enriched in the isoprenoids, especially phytane (Barbé *et al.*, 1990; Wang, 1998). However, both archaea and halophilic bacteria can flourish in upper penesaline and supersaline waters at salinities higher than 200–250‰, so the biomarker signatures are not mutually exclusive (Caumette, 1993; Ollivier *et al.*, 1994). Any high-salinity biota also includes the metabolic products of fermentators, homoacetogens, sulphate-reducers and methanogens.

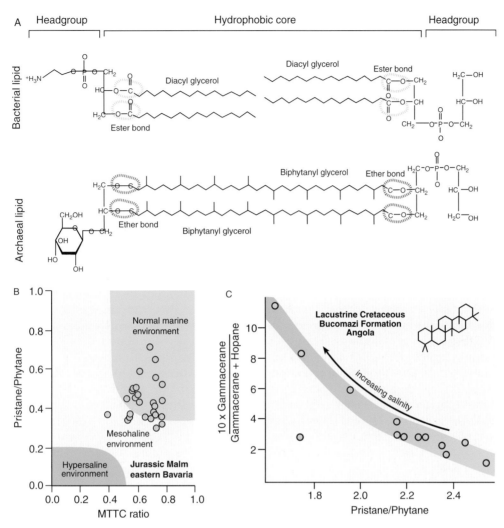

Fig. 7. Biomarkers: lipid structures and some rock characteristics. (A) Lipids from archaea and bacteria. Upper portion shows how phosphatidylanolamine forms a bi-layer lipid in bacterial cells. The acyl chain is usually, but not always straight. Some bacterial lipids have a methyl branch of a cyclohexal group at the end of the acyl chain, other lipids have one or more unsaturated bonds and the connection of glycerol to the acyl chain is via an ester linkage. The lower part shows how lipids are monolayers in archaea and the phytanyl chain contains isoprenoid-like branches. An ether linkage connects the phytanyl chain. Archaeal membranes also contain bilayer-forming diether lipids. (B) Relationship between pristane/phytane and the methyltrimethyltridecylchhroman (MTTC) ratio in Jurassic Malm carbonates of eastern Bavaria (after Schwark *et al.*, 1998). The MTTC ratio is defined as 5,7,8-trimethylchroman/total MTTCs. (C) Variations in pristane/phytane ratio and the gammacerane index for oils from lacustrine source rocks in Angola (inset shows gammacerane structure). Independent biomarker evidence suggests a marine source for the point that lies off the shaded trend (after Peters *et al.*, 2005a).

A predominance of degraded green algal and cyanobacterial biomarkers in mesohaline settings is indicated by biomarkers preserved in organically immature, evaporitic mesohaline laminates at the base of the Permian Zechstein evaporite succession in NW Europe (Bechtel & Puttmann, 1997; Pancost *et al.*, 2002). There the degree of methylation of 2-methyl-2-trimethyl-tridecylchromans (MTTCs) and the abundance of maleimides and bacteriochlorophylls in the organic-rich Kupferschiefer shales at the base of the Zechstein

evaporite succession implies a subaqueous euhaline to mesohaline ($\geq$30–40‰) transition into a hypersaline marine-fed drawdown basin. These biomarkers are thought to have been derived from green/purple sulphur bacteria (decomposers) within organic-rich laminites and suggest that the bottom waters were saturated with H_2S at the time of deposition. Maximum water depths were probably less than 100 metres (it was a stratified mesohaline drawdown basin at the onset of Zechstein salinity; Warren, 2006). Decomposers probably lived near the brine boundary between the photic zone and the anoxic (euxinic) bottom water at typical water depths of 10–30 metres below the water surface, as in modern marine-fed density-stratified systems, much like in Lake Mahoney today (Overmann *et al.*, 1996). Primary production in the upper water column was dominated by photosynthetic cyanobacteria or green algae, while sulphate reduction in the sediment was tied to the availability of abundant sulphate and organic detritus from the overlying water column.

Methanogenesis was also active in or at the base of the water column during Kupferschiefer deposition. This is reflected in the light carbon-isotopic composition of organic matter that had originated via recycling of CO_2 generated by methane-oxidizing bacteria in the water column (Bechtel & Puttmann, 1997). Saccate pollen is the only morphologically preserved body fossil in organic matter within the laminated Kupferschiefer sediment, all other traces of cellular morphologies are gone. Euxinic (anoxic) conditions were confirmed by Pancost *et al.* (2002) over a much larger area of Kupferschiefer deposition than studied by Bechtel and Puttmann (1997). They concluded that almost all of the Kupferschiefer seaway was subject to periods of photic zone stagnation and stratification during the early mesohaline history of the Zechstein Sea.

By themselves, high pristane/phytane ratios, high amounts of gammacerane, and high MTTC ratios are not totally reliable indicators of hypersalinity. More reliable determinations can be made when two or more of these biomarkers are cross-plotted and the resulting output shows a consistent trend. Figure 7B plots pristane/phytane ratio against the MTTC ratio in the Jurassic Malm Zeta laminites of eastern Bavaria in SW Germany (Schwark *et al.*, 1998). The plot shows a covariant decrease in the two measures, which is interpreted as indicating a trend from normal marine to meso-haline deposition.

A similar covariant trend can be seen in saline lacustrine source rocks when the pristane/phytane ratio is crossplotted against the gammacerane index in Mesozoic source rocks from offshore West Africa (Peters *et al.*, 2005a). Figure 7C plots variations in pristane/phytane (a redox and archaeal indicator) and gammacerane index for oils from a set of lacustrine source rocks. Increasing salinity in the depositional setting explains the covariant trend in the plot, whereby higher salinity was accompanied by density stratification and

associated reduced oxygen-levels in bottom waters. Independent biomarker evidence suggests a marine source for the point that plots off the shaded salinity trend in this figure.

Although the levels of entrained organics are very low, similar covariant trends in pristane/phytane and gammacerane occur in C_{21-25} isoprenoids, along with relatively heavy $\delta^{13}C$ signatures, in biomarkers extracted from the Miocene halites of the halokinetic Sedom Formation in Israel. They indicate a predominance of halophilic archaea when the salt was deposited in a CO_2 limited system (Grice *et al.*, 1998; Schouten *et al.*, 2001).

Light and sulphide in modern hypersaline environments typically occur in opposing gradients, so the growth of halotolerant and halophilic phototrophic sulphur bacteria is confined to narrow zones of overlap in stratified brine or water columns. In Precambrian sediments bacterial life forms were much more widespread in any water mass as the column was mostly anoxic. For example, in Neoproterozoic and early Cambrian hypersaline settings such as the Ara Salt Basin of Oman there were no higher life forms present in the depositional setting, and so this possibly normal-marine setting, and the subsequent encasement in salt, led to preservation of a unique set of bacterial biomarkers in blocks of salt-encased marine platform source rocks (Terken *et al.*, 2001; Schoenherr *et al.*, 2007).

The presence of mesohaline and marine carbonate muds, rather than aluminosilicate clays, in the matrix of many buried organic-rich evaporitic source rocks can change maturity and migration parameters compared with shale source rocks with non-carbonate matrices. For example, mesohaline carbonate source rocks are susceptible to early generation of bitumens and volatiles compared with terrigenous shales (ten Haven *et al.*, 1986; di Primio & Horsfield, 1996). The proportion of bitumens (extractables) in the organics of carbonate source rock is high compared with that of shales, as most mesohaline carbonates are largely isolated from high levels of terrigenous influx and the associated load of "spent" or oxidized terrestrial organic matter (Warren, 1986). For the same reason, much of the early asphaltic (hydrogen-prone) product in a mesohaline carbonate laminite begins to migrate early in the burial history, driven in-part by recrystallization of the carbonate matrix, whereas similar materials are held to greater burial depths in the lattice of aluminosilicate clays (Warren, 1986; Cordell, 1992).

Espitalie *et al.*, (1980) showed that carbonate source rocks are likely to provide larger volumes of hydrocarbons for pooling compared with analogous terrigenous shales for the same levels of TOC. Geochemical analyses show that heavy oils from carbonate source rocks can form at marginally mature stages and at lower temperatures than required for oil generation from terrigenous shales (Jones, 1984).

Lastly, some widely-used biomarker-derived maturity indicators formulated in marine sediments, are not as reliable in relatively young saline-lacustrine basins or in hypersaline oils with high sulphur contents, such as those sourced from hypersaline Oligocene laminites in the northern Qaidam Basin, China or the Miocene mesohaline carbonates of the Mediterranean (Hanson *et al.*, 2001; di Primio & Horsfield, 1996). The discrepancy between mature indications coming from vitrinite in these very young evaporitic source rocks and the immature indications derived from various sterane isomerization ratios or other standard source rock evaluation techniques, are in part explained by a lack of time for various sterane equilibria to develop and by the complicating effects of high sulphur levels on reaction kinetics in systems with high levels of early asphaltenes. Some of these hypersaline heavy oils in China come from hydrogen-rich type-I lacustrine sources that may be as little as 3 million years old in the Qaidam Basin.

LAYERED BRINES AND ORGANIC SIGNATURE

So far the biological factors and metabolic pathways that control the type and volume of organic matter accumulating in saline settings have been discussed, without much consideration of the physical conditions or hydrologies that control the condition or position of the brine and pore fluids in which the organics accumulated. The presence or absence of organics in an evaporite basin is predicated on suitable hydrologies to manufacture the requisite factors to create anoxia. Penecontemporaneously, the hydrological conditions of early burial control whether organics are to be preserved in its evaporitic matrix until any protokerogen is sufficiently buried to evolve into a source rock. Brine density increases with increasing salinity, so influxes of less dense brines into an evaporitic depression create a layered system where lower

salinity waters tend to float on top of higher salinity waters. Depositional hydrologies in all modern and ancient evaporite systems are accordingly layered and, as will be demonstrated below, the stability, temperature and salinity range of this layering largely controls whether and where organic matter accumulates in an evaporitic depression.

Almost all primary organic matter in an evaporitic source rock was first photosynthesized in surface brines outcropping at the top of the active phreatic zone or its capillary fringe. An active phreatic flow regime encompasses the zones of seepage outflow, brine/seaway ponding, brine reflux and meteoric flow. The brine surface of a perennial brine lake or seaway is the outcropping expression of the regional water table, and in marine-drawdown basins is by definition lower than sea level. In the surrounds, where the regional water table comes near the landsurface, the top of a capillary fringe may be the outcrop expression of the same regional water table and so is the depositional area where sabkhas dominate (Warren, 2006). Solar evaporation concentrates waters in both ponded and vadose settings to form matrices of bottom-nucleated and sabkha salts, respectively. Depressions with surface and near-surface brines in arid settings indicate groundwater discharge zones and their long-term outcropping creates discharge playas and ancient evaporite seaways.

Where areas of strong topographic relief define the margin of a modern evaporite basin, both unconfined and confined meteoric waters typically discharge into the edges of a brine-saturated depression, as in modern continental rift valleys of the East African rift, or in a transtensional basin-and-range setting of the American southwest, or in the Dead Sea of the Middle East. Farther out, beneath the more central and topographically lowest parts of the evaporite depression, density-induced brine reflux, and not meteoric throughput, is the dominant process driving shallow-subsurface brine flow. The saline water mass driving this reflux may be an ephemeral or perennial holomictic surface water. Which hydrological style of surface water dominates in an evaporite basin (ephemeral versus perennial) is in part related to the steepness of the surrounding topography and the presence or absence of a hydrological conduit for water outflow (Warren, 2006).

In a large evaporite drawdown basin, such as in the Mediterranean in the Late Miocene, the basin typically had no surface connection to the ocean, and the level of the discharge zone and the water surface of the perennial brine discharge lake was at times thousands of metres below sea level. Gravity-induced seepage and associated evaporation of surface waters cycled huge volumes of marine/meteoric waters through the depression's surrounds and into the drawdown depression. Reflux beneath a stratified perennial meromictic water mass is at its maximum when the brine column is holomictic (Warren, 2006).

Hydrologies in saline basins

Hydrological settings that allow most evaporitic carbonates and associated organics to accumulate can be divided into two end member situations:

(1) a perennial seaway or a brine lake that at times is density stratified; and (2) an evaporitic mudflat (sabkha) with a surface defined by the top of the capillary fringe and its most saline water lying at or just below the sediment surface. A common intermediary stage is an evaporitic mudflat occasionally covered by thin sluggish brine sheets (salt pan). The hydrology of an evaporitic mudflat and its landward surrounds is controlled by the position of the water table in the sediments (Fig. 8A). Pores below the water table are filled by brine and make up the phreatic or saturated zone where the water-volume saturation is total (equal to 1 or 100%). The phreatic zone is the region where gravity-driven groundwaters seep down the potentiometric slope

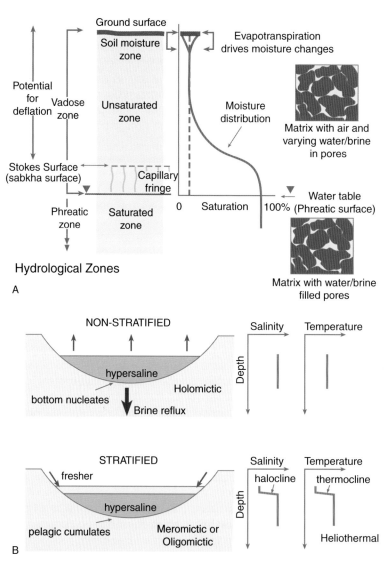

Fig. 8. Hydrological classification in saline settings. (A) Hydrological zonation in an evaporitic mudflat (sabkha) and its landward surrounds. (B) Water mass zonation in a perennial brine lake or seaway (after Warren, 2006).

toward discharge zones (forced-convection) and dense free-convecting brines sink into underlying strata. Above the water table is the vadose zone, where pores are filled by a combination of soil air and varying levels of water and brine and water-volume saturation is less than one.

Vadose hydrologies

The vadose zone in dry mudflats contains an uppermost interval where the water content and the oxygen/CO_2 content of the pore waters can vary considerably and is known as the soil moisture zone. Light rains on a dry mudflat typically penetrate the soil moisture zone but are not sufficient in volume to saturate the pore space of the whole vadose zone, and so are returned to the atmosphere via evapotranspiration without ever replenishing the water table (Wood *et al.*, 2002). Fluctuating evapotranspiration levels in this zone vary widely between storms, driven in part by the metabolic activities of the soil biota, by mean moisture contents, and by varying O_2 and CO_2 levels. Pedogenic carbonates and more soluble salts are precipitated as surface crusts and near-surface cements as a result of these fluctuations.

Ongoing and periodic oxidation of this vadose hydrology means any hydrogen-rich microbial material tends to be decomposed/oxidized prior to burial. Dry mudflats in saline depressions are not areas with any potential for long-term preservation of elevated amounts of hydrogen-prone organics. With the next heavy rain the less soluble carbonate precipitates remain, while the more soluble salts tend to be flushed through a now-saturated vadose profile to enter the water table. There they continue to flow as dissolved phreatic salts into the discharge depression where evaporative concentration once again reprecipitates them as wet sabkha and brine pool salts.

The lower part of the vadose zone beneath a dry mudflat is made up of the capillary fringe. In parts of an evaporitic mudflat (sabkha) depression that are actively accumulating displacive salts, the top of capillary fringe is at or above the land surface (Fig. 8A). This allows the concentration of pore brines to the attain hypersaline levels needed to precipitate the nodular and displacive salts that typify wet sabkhas and saline mudflats. Unlike the dry parts of a playa or mudflat, the inherently wet capillary surface facilitates long-term evaporative concentration of the near-surface pore brines and the ongoing crossflow of dense brine into

sediments below the water table (sabkha-driven brine reflux). As long as brine can be supplied from an updip source, closed cellular flow beneath the wet sabkha flat is the result. In a capillary zone-bound system that is also subsiding, such wet sabkha hydrologies can retain continual hypersaline brine saturation with associated dysaerobia or anoxia, and so sequences of hydrogen-prone microbial organic matter can accumulate as mats and so be preserved into the burial environment.

Given enough time, waters in a brine plume beneath a wet sabkha or a periodically holomictic brine lake will reach the bottom of the basin fill or be dispersed back into the regional cross flow on top of any aquiclude it intersects on its way down (Wood *et al.*, 2002; Warren, 2006). In either case, the refluxed waters start to spread laterally and back out toward the basin fringes, so mixing with fresher forced-convection waters beneath the lake, playa or seaway margin. Ultimately this brine returns to the playa surface, usually in a diluted form as a component of water-edge spring or seep waters. This convective flow is an effective way of moving salt load through large volumes of the basin sediments beneath an accumulating salt bed, and yet still maintain anoxic pore waters beneath a wet sabkha or salt lake (see discussion of brine curtains in Warren 2006). Ongoing recycling of anoxic brine beneath a salt lake or seaway in-turn aids in the preservation of organic material. Thus, organics deposited in a perennial wet sabkha or saline mudflats with a long-term reflux curtain have the potential to be preserved as hydrogen-prone organics until the sediment passes into the zone of catagenesis.

Even so, outcropping capillary-fringe areas where brine-saturated microbial mats reside about the edge of a saline water body, can be subject to episodes of water table lowering or to flushing by fresh oxygenated meteoric waters. Then the microbial organic matter may be intermittently stripped of its hydrogen by aerobic decomposers that will flourish on top of and within the mats during times of exposure or pore-water freshening.

Perennial phreatic hydrologies

Permanent brine lakes and ancient epeiric seaways with variably holomictic water columns driving stable long-term reflux curtains or plumes, are much more likely than wet sabkhas to preserve hydrogen-rich organics into the zone of catagenetic burial. Surface water-masses within perennial brine lakes or seaways routinely fluctuate between stratified and nonstratified conditions, while the bottom waters typically have a greater propensity to maintain long-term anoxia (Warren, 2006). An upper fresher (less dense) water layer sits on top of a more saline (denser) lower water mass in a stratified system in a perennially brine-filled depression (Fig. 8B). The narrow zone of transition in the brine column between the two waters is called a halocline or pycnocline. Often a stratified saline lacustrine/seaway system is also thermally stratified and heliothermic, with warmer waters constituting the lower water mass. The zone of transition separating the upper and lower water masses is the called the thermocline, or the chemocline, or the halocline, or the pycnocline. Thermal stratification of brine masses with a cool top and a hot base is opposite to that found seasonally in many temperate freshwater lakes and so is sometimes called reverse thermal stratification, but heliothermal is a more appropriate term.

Heliothermy is a result of the contrast in specific heat response between less saline and more saline waters. Specific heat of a brine decreases as the salinity increases (Kaufmann, 1960). Specific heat is the amount of heat needed to raise one gram of a substance by 1 °C. For a given amount of heat input, a unit volume of hypersaline water will show a greater increase in temperature than a less saline water. The resulting greater increase in temperature in a lower hypersaline water mass, compared to an upper less-saline water for the same degree of insolation, makes brine lakes and seaways heliothermic; independent of any salinity (osmotic) or oxygen stresses, heliothermy can induce pronounced heat stress in any benthic halobiota. Modern heliothermal lakes occur in appropriately stratified saline water bodies on all continents and include numerous saline lakes on the steppes of Russia, Siberia and Canada, Solar Lake in the Middle East (Fig. 11B), numerous coastal salinas in Australia, Lake Meggarine in Algeria, Hot Lake in Washington State and Red Pond in Arizona (Hammer, 1986). One of the more unusual examples is Lake Vanda in Antarctica, which has a maximum water temperature of 25 °C below the chemocline (at a depth of 64 m in the water column), yet the water surface remains ice-covered.

Dissolved gas levels and temperature changes associated with periodically heliothermal bottom waters create extreme levels of environmental stress in bottom communities attempting to cope with hypersalinity. For example, oxygen is supplied to any water body via exchange with the atmosphere and as a byproduct of photosynthesis of the biota living in the waters. In well-oxygenated water columns most of the organics produced by primary production the upper water mass are destroyed by microbial respiration and the conversion of organic matter back to CO_2. Much of the improved preservation potential associated with "higher than marine" bottom salinities in layered mesohaline to hypersaline settings reflects the effect on biotal diversity of increasingly low levels of dissolved oxygen and CO_2 in the bottom and pore waters on biotal diversity.

Oxygen levels in natural brines tend to decrease with increasing salinity, independent of biological respiration. If not enhanced by the activities of bottom-living photosynthesizers, natural bottom brines will be dysaerobic to anaerobic by the time salinities reach halite saturation (Fig. 9; Kinsman, 1973; Warren 1986). Dissolved CO_2 shows a similar depletion; dissolved CO_2 levels in marine-fed modern salt pans decrease to around 50% of the original value when the brine concentrations increase from 1.5 to 4 times that of seawater (Lazar & Erez, 1992). CO_2 concentration in brine is also pH dependent; at a pH of 9 (reached by concentrating seawater to around 70‰) the CO_2 content is near zero (Moberg *et al.*, 1932). Those algae and

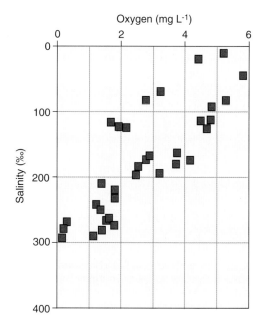

Fig. 9. Oxygen levels in present-day surface brines under increasing salinity. Oxygen content of modern seawater is ~5.4 mg L^{-1} (after Warren, 1986).

cyanobacteria, which are dependant on dissolved CO_2 for photosynthesis, will suffer or die at higher salinities.

Stability of brine stratification and anoxia in modern stratified brine lakes

Brine stratification occurs today in all perennial saline lakes, including many coastal salinas in Australia, rift valley lakes of the African Rift Valley and the Dead Sea, Israel and most are meromictic and oligomictic (Warren, 2006). Lake Hayward, a coastal salina in Western Australia, typifies such stratified brine system. It is a sub-sea-level water-filled marine seepage depression with bottom brines possessing ionic proportions similar to that of seawater (thalassic). The seawater seeps into the basin as groundwater and there is no hydrographic (at-surface) connection to the nearby ocean. Gypsiferous carbonate muds with organic contents of 1–5% accumulate on its microbially-bound floor beneath its small shallow perennial brine pool ($0.6\,km^2$ and less than 2–3 m deep). In the permanently inundated centre of the lake, where waters are 1–2 m deep even in late summer, laminated benthic microbial mat communities flourish and are dominated by photosynthesizing cyanobacteria (mostly *Cyanothece sp.*). Filamentous *Microcoleus sp.*, along with coccoids, dominate the laminated to pustular mats in the seasonally-desiccated strand zone. Sediments in the lake centre accumulate under a limnology that is density-stratified and heliothermal for much of the year (Rosen *et al.*, 1995). Meteoric winter inflow creates a well-defined long-term mixolimnion (Fig. 10A) where an upper, less dense and cooler water mass has salinities ranging from 50 to 210‰. It exists from late autumn to early summer

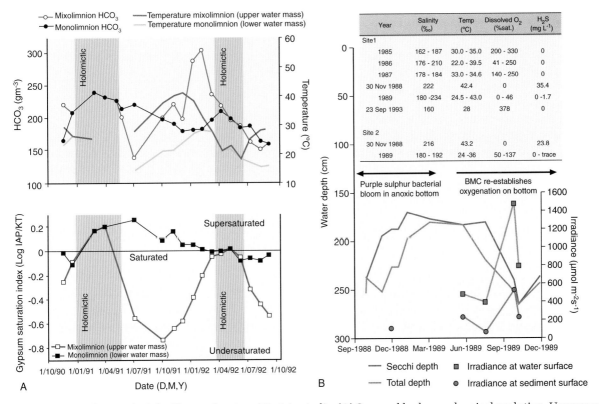

Fig. 10. Brine stratification in Lake Hayward waters, West Australia. (A) Seasonal hydrogeochemical evolution. Uppermost plot shows temperature regime in upper (mixolimnion) and lower (monolimnion) water masses and bicarbonate content. As the lake mixes, the thermal and density stratification disappears. Immediately prior to mixing the bicarbonate concentration decreases, suggesting precipitation of calcium carbonate (aragonite; after Rosen *et al.*, 1995). Lower plot shows saturation state of waters with respect to gypsum. The monimolimnion was saturated with respect to gypsum for summer 1991–1992 and was near saturation at the start of summer 1992–1993. The mixolimnion is only saturated when the lake water is homogeneous (after Rosen *et al.*, 1996). (B) Effect of gypsum precipitation on benthic microbial communities and associated oxygenation in Lake Hayward, West Australia (compiled from data in Burke & Knott, 1997).

(May to February) as it floats on top of a lower denser and warmer water mass (monolimnion) with salinities in the range 150 to 210‰.

Stratification disappears for a few months each year from mid-summer to mid-to-late-autumn. During summer the waters of the upper water mass evaporate and concentrate as their bicarbonate content steadily increases. From the onset of stratification in late autumn, across mid-winter and on into to early summer the temperature trends of the two water masses are parallel (Fig. 10A). They form a heliothermic system where the lower water mass is some 15–20 °C hotter than the upper water mass. By mid-summer (e.g. January 1992) lower water mass began to cool, while the temperature of the upper water mass continued to rise. Once the temperatures (and densities) of the two water masses equalize, they mix as the lake overturns (holomixis) and waters of the lower water mass (the monolimnion) came into contact with the atmosphere once more.

While the water masses are stratified, the chemocline and the thermocline are sharply defined across a 10 cm interface with a salinity contrast that may be as much as 135–140‰ and a temperature difference of up to 19 °C. The time of mixing is immediately preceded by a sharp fall in the level of bicarbonate in the mixolimnion, suggesting that from late summer to autumn the precipitation of pelagic calcium carbonate (mostly aragonite) then gypsum occurs in the upper water mass (Fig. 10A, upper plot; Rosen *et al.*, 1995). For example, the upper water mass in Lake Hayward was supersaturated with respect to gypsum and anhydrite from October 1991 to February 1992 (Fig. 10A, lower plot; Rosen *et al.*, 1996). When the lake mixed from March 1992 until early May 1992, the entire water body was at gypsum saturation, but slightly undersaturated with respect to anhydrite.

A 'whiting' (a cloudy white appearance to the water body) was observed in the lake in March, just at the time of first mixing of the lake waters. Analysis by scanning electron microscopy of the collected filtrate indicated that the "whiting" was composed of gypsum and silica from diatom tests. At that time, there was a thin (10–20 mm) crust of gypsum accumulating on the lake bottom. After the "whiting", when the water was unstratified (holomictic), both the monimolimnion and mixolimnion were near saturation with respect to gypsum. After mixing, with the next influx of meteoric waters building across the lake water surface, salinity stratification re-established itself. Prior to the

long-term stratification that defined the winter months, brine layering in the transition to stratification was not stable as the upper and lower water bodies remixed a number of times extending over periods of days before stable stratification set in by mid-winter (Fig. 10A, May–June 1992; M. Rosen pers. comm., 2006).

Lake Hayward bottom sediments also illustrate an aspect of pelagic evaporite precipitation that is probably true for many ancient "deep water" density-stratified saline settings. Namely, that while precipitation of pelagic evaporitic sediment is seasonal, and it occurs in the surface water mass even as the lake waters are density stratified, bottom salts can only accumulate when the whole brine column is saturated and unstratified (holomictic). This is so even if the chemistry of the lower water mass is at gypsum or halite saturation almost all year round. Only pelagic sediments (including carbonates and organic residues) and terrigenous detritus (including eolian dust) make it to the bottom when a saline lake is perennially stratified, as was the case in the Dead Sea for the 400 years prior to 1979 (Warren, 2006).

Intensity of light penetration to the subaqueous bottom of an evaporite-accumulating depression is a significant control on how much photosynthetic oxygen can be generated by the bottom-living microbial community. In Lake Hayward a healthy mat of photosynthesizing halotolerant cyanobacteria drastically alters the levels of anoxia in shallow bottom brines. Photosynthesis, via its benthic microbial mat community, is normally sufficient to maintain supersaturation of the Lake Hayward bottom waters with dissolved oxygen, even during periods of brine column stratification (Fig. 10B).

Burke & Knott (1997) found that after an unusually dry year in 1987, the salinity of Lake Hayward increased to 260‰ as gypsum precipitated throughout the lake waters. Brine turbidity generated by the markedly increased pelagic crystallization of gypsum in the surface waters in that year was sufficient to obscure the benthos completely, despite the shallowness of the lake. As a result the amount of oxygenic photosynthesis and aerobic decomposition was greatly reduced in the benthic mats and they degraded. During brine stratification in the following winter (1988), bottom waters became anoxic and sulphurous. During the next few years, after each subsequent brine column overturn, healthy microbial mats were re-established across the lake floor and they were able to again supersaturate the bottom waters with oxygen year round, as was seen in 1992, which was the time of the Rosen *et al.* (1995, 1996) study.

Development of a healthy mat community in the Lake Hayward salina prevents periodic bottom anoxia. Nearby lakes, that do not contain a healthy microbial benthos, tend to anoxia during stratification. Comparisons between Lake Hayward and nearby lakes clearly show that healthy microbial mats function as a strong homeostatic mechanism in the lake hydrology. Shallow meromictic but perennial salt lakes, with long-term stability of photosynthetically elevated oxygen levels in the bottom, tend to have lower levels of preserved TOC due to the higher efficiencies of halotolerant aerobic decomposers acting on the benthic organic matter. This can be seen when comparing TOC levels in the microbial laminites of Lake Hayward and Solar Lake (Fig. 5).

Solar Lake in the Middle East is another small meromictic hypersaline lake (coastal salina) with a well-documented heliothermal hydrology. It is located on the Sinai coast of the Gulf of Aqaba (Cytryn *et al.*, 2000) and is separated from the ocean by a narrow permeable sand barrier. The barrier allows seawater to seep into the sub-sea-level hydrographically-isolated lake depression and so replace waters lost to evaporation. Seepage-derived seawater can then pond on top of the lake floor, along with lesser volumes of waters supplied by occasional winter rains (Fig. 11A). Brine depth in the lake centre fluctuates between 4 and 6 m, driven by seasonal changes in the intensities of evaporation and the related levels of seawater seepage and brine reflux.

Thick cyanobacterial mats carpet much of the Solar Lake sediment surface and contain a highly diverse microbial community of cyanobacteria and haloarchaea. Unlike the microbial laminites of Lake Haywood, a structureless organic-rich gypsiferous mush constitutes the sediment across the m-deep bottom of the anoxic lake centre (Fig. 11A; see Warren, 2006 for geological detail). Fluctuating

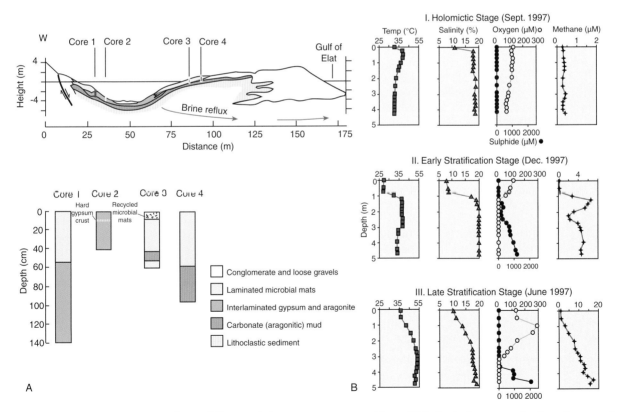

Fig. 11. Solar Lake, Sinai Peninsula, Middle East. (A) East–west topographic and lithological cross section based on cored sediments showing peripheral algal facies and a more central organic-rich gypsum zone (after Aharon *et al.*, 1977). (B) Effects of the seasonal development of heliothermic and oxic stratification in the water column of Solar Lake (after Cytryn *et al.*, 2000).

planktonic microbial activity mirrors seasonal changes in brine column layering by exporting varying volumes of oxygen, sulphide, and methane into the brine column as this mush accumulates (Fig. 11B).

High evaporation and aridity in the summer months raises Solar Lake salinity to 200‰. At that time the brine column is completely mixed (holomictic) and the entire water column is oxygenated (Fig. 11B, September, 1997). Temperature and salinity in the holomictic state vary little throughout the water column and sulphide and methane concentrations are very low. By autumn, the water column becomes stratified as a newly introduced seawater brine (50‰) layer overlies the residual, highly saline bottom water (180–200‰). A well-defined density gradient (halocline or pycnocline) separates an upper oxygenated cooler (16 °C) brine mass (epilimnion) from the lower hotter (45–55 °C) anoxic sulphide-rich brine mass (hypolimnion; Fig. 11b, December 1997). A high sulphide concentration develops in the anoxic waters below the halocline and comes from the activities of sulphate-reducing bacteria both in the water column and in the underlying degrading cyanobacterial mats. Crossing the halocline into the lower brine there is a rapid downward increase in sulphide concentration, which reaches a maximum value of 1240 mM near the sediment-brine contact (Cytryn *et al.*, 2000). At the same time a peak in methane concentration develops directly below the halocline, where a maximum value of 6.5 µM was measured in the December 1997 profile and was in-part a reflection of the decomposer activities of the methanogenic haloarchaea.

As winter passes into spring gradual heating, along with an increase in the salinity of the upper water mass, degrades the halocline (Fig. 11B). Thus the June 1997 data portray a transitional state between late stratification and the beginning of holomixis. Salinity and temperature gradients across the halocline are more gradual and deeper in the water column than those observed in the December 1997 profile. The deepening halocline of June 1997, when the chemocline was located 3.25 m below the water surface, allowed oxygen to penetrate much deeper into the brine column, and so extend farther out across the subaqueous lake bottom than it had in the earlier stages of stratification. Oxygen concentration in the upper water mass in June 1997 reached a maximum value of 300 µM at a depth of 1 m below the water surface. These supersaturated values of oxygen imply intense oxygenic photosynthetic activity at the brine surface and in underlying shallow salina waters. Deeper in the brine column, and well below the chemocline, a sharp rise in sulphide concentration occurs reaching a maximum value of 2150 µM at the water-sediment interface (Cytryn *et al.*, 2000). Methane concentration increases linearly below the brine surface and reaches a maximum value of 17 µM adjacent to the lake bottom, implying methane, like the bacterially derived H_2S, is escaping from the sediments. Increased evaporation rates in the mid- to later summer further increase the salinity of the surface layer and ultimately disrupt the stratification and so lead into summer holomixis (as seen in the September 1997 profile). Unlike Lake Hayward, conditions beneath the perennial lake floor of Solar Lake are anoxic for longer periods, oxygenic species, including cyanobacteria, tend not thrive in bottom waters for long periods, especially in the deep central parts of the lake, where levels of preserved organics tend to be higher (5–20% versus 1–5% TOC in Lake Hayward; Fig. 5).

Substantial density differences between the upper and lower masses in most density-stratified systems means diffusive mixing across any halocline is insignificant. When a hypersaline water mass is density stratified there is no mechanism to drive changes in saturation state and so there is no widespread bottom nucleation of salts. Sedimentation on the floor of the lower water layer (especially if it is below the euphotic zone) is mostly by a pelagic rain of organics and evaporite crystallites formed in the upper water mass at the halocline via brine mixing. There may also be local changes in brine chemistry and the associated mineral precipitation driven by local influxes of spring and seep waters. In contrast, when the upper and lower water masses equilibrate and homogenize, bottom nucleation of salts is possible even at the base of deep brine columns, as is occurring with the precipitation of coarsely crystalline halite at the base of the 350 m holomictic brine column in the Dead Sea today (Warren, 2006). In some perennial saline lakes in more temperate climates the holomictic stage is seasonal and so the sediment layers are varves (e.g. Lake Hayward, Solar Lake, Lake Van, Lake Urmia), in others, holomictic mixing of surface with bottom water masses occurs across a longer time frame (e.g. the 400 year stability of the lower water layer in the Dead Sea driving episodes of laminite versus coarse crystalline halite beds on the deep lake floor).

Whenever a homogenized brine mass restratifies, the lack of ongoing concentration in the lower water mass means the rate of brine reflux through bottom sediments beneath the sediment brine interface slows and ultimately stops. Hypersaline pore brines remain but in a stagnant condition as there is no ongoing mechanism to resupply brines denser than existing pore waters in the column substrate. High pore water salinities, a propensity for anoxia, anoxic pore brines and elevated bottom temperatures, all tend to exclude aerobic burrowing and grazing animals from evaporite bottom sediments beneath a stratified brine column. This minimizes the effects of oxygenic biota that would otherwise oxidize and destroy mesohaline microbial debris, as in sediments deposited in less saline settings.

Although the daily temperature/stratification records of Lake Hayward and Solar Lake are probably the best documented examples of the long-term hydrological cycling of a naturally stratified brine pool, they both have a number of limitations when attempting to use them as direct analogues for the hydrology of ancient stratified brine seaways. Unfortunately, neither lake can act as a same scale hydrological analogue for ancient evaporite seaways, intraplatform depressions or even some of the larger ancient evaporitic lakes. They suffer from the scaling limitations that enfeeble all modern marine-associated evaporite analogues (Warren, 2006). Although hydrographically isolated and marine-seepage fed, neither lake is situated in a hydrological setting where enough time or aerially extensive drawdown stability has transpired to allow substantial thicknesses of mesohaline carbonates, halite or gypsum to accumulate. Their small aerial extents ($\leq 0.6\,\mathrm{km}^2$) and their centripetal meteoric inflow hydrologies (driven by winter rains, a marked seasonal lowering of evaporation intensity, and a high-relief hinterland immediately adjacent to the water body) means they cannot be directly compared to ancient mesohaline epeiric brine seaways where inherently low-relief topographies meant strandlines moved up to hundreds of kilometres in a wet-dry cycle, and where unfractionated meteoric runoff from the hinterland was much less significant in the freshening process.

Biological responses to layered hydrologies

Brines in modern evaporitic depressions are typically layered in terms of the annual or longer-term hydrological cycle, with denser more saline waters supporting an ephemeral mass of less dense inflow water. The lower, denser part of the water mass tends to be more stable and its physical conditions change less. In contrast, the conditions in the floating less-dense layer are much more variable and its physical chemistry tends to oscillate. This upper layer periodically freshens to where it supports a widespread planktonic bloom (feast). Then, as ongoing evaporation concentrates the upper layer, conditions become increasingly less suitable for life (famine) and a mass die-off occurs, first of the haloxene species, then the halotolerants, then the halophiles.

The die-back creates a pulse of organic matter that can swamp the abilities of the decomposers in the water column to strip it of its hydrogen, and so it can reach the sediment interface with its long-chain alkanes still intact. Even in conditions where light penetrates to the bottom of the brine column and the bottom waters immediately above the sediment surface are oxygenated, the halotolerant primary producers can be sufficiently productive and the underlying anoxic decomposers sufficiently inefficient that substantial volumes of hydrogen-rich microbial mater can be retained into burial. Some modern examples of hydrologically induced community layering in modern evaporitic settings will now be considered.

Halotolerant plankton and water column layering

The significance of periodic freshening on biomass in stratified waters in creating conditions of "famine or feast" for halotolerant and halophilic species is clearly seen in the present productivity cycle of the Dead Sea (Fig. 12; Oren 2005; Oren & Gurevich, 1995; Oren *et al.*, 1995). The main component of the halobiota in the Dead Sea waters is a unicellular green alga, *Dunaliella parva* being the sole primary producer in the lake. Several types of halophilic archaea of the family Halobacteriaceae then metabolize the glycerol and other organic compounds produced by the alga. As in all ecosystems, producers and consumers in the brine-stratified water columns are codependent. Halophilic archaea (*Halobacterium and Halococcus*) survive into much higher salinities than *Dunaliella parva* as they decompose the glycerol-rich remains.

Three distinct periods of organic productivity were seen in thirteen years of quantitative studies

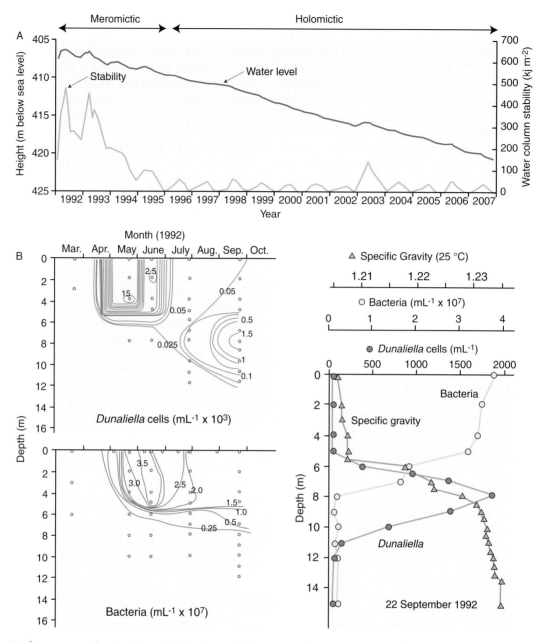

Fig. 12. Biodynamics in the Dead Sea, Middle East. (A) Long-term changes in quasi-salinity and temperature in the deep water body (below 100 m) of the Dead Sea in a meromictic to holomictic period (after Gertman & Hecht, 2002; with additional data from http://isramar.ocean.org.il/DeadSea, last accessed 29 September 2004). (B) Changes in the vertical distribution of *Dunaliella* and halobacteria during the rain flood-induced density stratification of the Dead Sea in 1992 (after Oren, 1993). Note the vertical distribution of *Dunaliella* and halobacteria relative to the halocline on 22 September 1992. See Warren (2006) for full discussion of long-term sedimentary dynamics of the Dead Sea and Lake Lisan depressions.

of Dead Sea microbiology starting in 1980 (Oren, 1993). Unlike more temperate lakes, organic blooms in the Dead Sea are not an annual event. Mass developments of *Dunaliella* (up to 8800 cells mL^{-1}) and red halophilic archaea (2×10^7 cells mL^{-1}) were observed in 1980, following a dilution of the saline upper water layers by rain floods and

an associated rise in lake level of 1.5 m. This bloom had disappeared at the end of 1982, following complete mixing of the water column and the associated salinity increase in the upper water mass (holomixis).

During the period 1983–1991 the Dead Sea was holomictic, and no *Dunaliella* cells were observed.

Viable halophilic and halotolerant archaea were present during this period, but in very low numbers. Then heavy rain floods during the winter of 1991–1992 raised the lake level by 2 m and caused new salinity stratification as the upper five metres of the water column was diluted to 70% of the normal surface salinity (Fig. 12A). The meromictic phase lasted until 1996 when the current holomictic conditions with seasonal stratification were established, clearly illustrating the oligotrophic hydrology of the Dead Sea brine column. *Dunaliella* reappeared in the upper less saline water layer (up to 3×10^4 cells mL^{-1} at the beginning of May 1992, rapidly declining to less than 40 cells mL^{-1} at the end of July), while the bloom of the archaea (3×10^7 cells mL^{-1}) continued and imparted a reddish colour to the Dead Sea brines (Fig. 12B). The reddish hue of the surface waters at this time was due to elevated levels of the carotenoid pigment bacterioruberin, which absorbs light at wavelengths between 420–550 nm.

Archaeal numbers in the Dead Sea decreased only a little after the July 1992 decline of the *Dunaliella* bloom, and most of the haloarchaeal community was still present at the end of 1993. However, the amount of carotenoid pigment per cell decreased 2–3-fold between June 1992 and August 1993 (Fig. 12B; Oren & Gurevich, 1995). No new algal and archaeal blooms developed after the winter floods of 1992–1993, in spite of the fact that salinity values in the surface layer were sufficiently low to support a new algal bloom. A remnant of the 1992 *Dunaliella* bloom maintained itself at the lower end of the pycnocline (halocline) at depths between 7 and 13 m (September 1992–August 1993). Its photosynthetic activity was small, and very little stimulation of archaeal growth and activity was associated with this algal community (Fig. 12B). This pattern suggests that the photosynthetic activity of the algae may have been limited by a lack of nutrients in the upper brine column.

Colouration of the Dead Sea waters from the initial algal bloom allowed Oren & Ben Yosef (1997) to use Landsat images, collected in May 1991 and in April and June 1992, to plot the development of the *Dunaliella parva* bloom. In contrast, the carotenoids of the subsequent archaea bloom did not produce a recognizable signal in the Landsat images. The April 1992 image obtained at the time of the onset of the algal bloom, prior to its lake-wide spread, suggested it originated in the shallow areas near the shore of the lake, where light penetrated to the bottom of the brine column and a newly established freshened water layer intersected the lake floor. The resulting algal blossoming was instigated by resting cells that had survived near the sediment surface around the lake margin or in species refugia around freshwater springs.

The decline of the lake-wide bloom of halophilic archaea in the Dead Sea appears to be related to viral infection (Oren *et al.*, 1997). For example, in October 1994, there were between 0.9 and 7.3×10^7 virus-like particles per mL of brine during declension of halophilic archaea in the upper 20 m of the Dead Sea water column and virus-like particles outnumbered archaea by a factor of 0.9–9.5 (averaging 4.4). By 1995 all water samples contained low numbers of both archaea and virus-like particles; $1.9–2.6 \times 10^6$ and $0.8–4.6 \times 10^7$ mL^{-1} respectively in April 1995. Viral numbers declined even further so that by November 1995–January 1996 all samples contained less than 10^4 particles mL^{-1}. Oren *et al.* (1997) further suggested that viruses play a major role in the decline of halophilic archaeal communities in many other hypersaline settings at salinities where protozoa and fungi are absent.

Community layering in mesohaline microbialites

Laminated organic-rich sediments that typify many evaporitic microbial carbonates (possible petroleum source rocks) and described by sedimentologists using the general, and perhaps dated, terms algal mats and cryptalgal-laminites, are in effect microbial "high rises," as are many living stromatolites. They, and their entrained biomarkers, indicate the activities of an ecologically layered microbial community that can thrive in both subaerial and subaqueous settings. Like microbial layering in the Dead Sea brine column, layering in a microbial mat is divisible into an upper aerobic community and a lower layer of anaerobic decomposers (Fig. 13). As a microbial mat accretes, its organic constituents are buried. The biochemical make-up of the initial halobiota, which flourished in the hypersaline conditions (typically oxygenic) at or above the mat surface, is altered by the metabolic activities of various decomposers flourishing below.

In order to investigate early or syndepositional changes in the dominant organic constituents and the main diagenetic processes influencing organic character, Grimalt *et al.* (1992) studied organic profiles in cross sections through two different

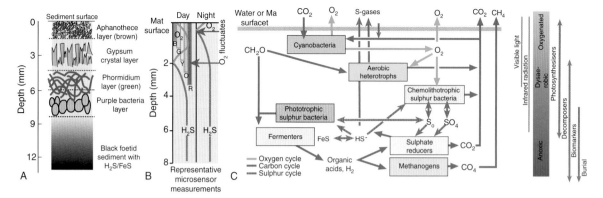

Fig. 13. Layering and metabolism in microbial mats. (A) Typical microbial community in an evaporitic laminite in a modern coastal salina with salinity in the range 130 to 200‰ (after Caumette, 1993). (B) Inter-relationships between microbes and the oxygen, carbon and sulphur cycles in the uppermost 5–10 mm of a microbial mat. The oxic-anoxic transition shifts lower in the mat in the daytime and higher in the mat at night (when photosynthesis has stopped). (C) Biogeochemical cycles operating within the mat versus depth.

subaqueous cyanobacterial mats in a modern coastal salina. Changes in lipid composition were compared with the vertical distributions of the various microbial populations. Vertical distributions in the lipid patterns were defined using enrichment cultures from typical species of cyanobacteria, diatoms, purple bacteria, sulphate-reducers and methanogens obtained from the layered mats. Cyanobacteria *Phormidium valderianum* and *Microcoleus chthonoplastes* were the dominant primary producers in the sampled mats, they occurred almost as monocultures in the upper 6 mm of the mats (Fig. 13A). Worldwide, these and other cosmopolitan filamentous cyanobacterial photosynthesizers are the most obvious organisms in most mesohaline algal mats. Their chlorophyll creates a green layer with a sticky or slimy surface and defines the upper few millimetres of the mat community. As layers of mucilaginous sheaths aggrade via periodic growth spurts, it creates biolamination.

Below the photosynthesizing cyanobacteria is a layer of aerobic heterotrophs (bacteria and archaea), which in the presence of oxygen break down organics produced by the cyanobacteria (mostly abandoned algal sheaths and slime). However, because oxygen is only produced by photosynthesis, these organisms can only perform this process during daylight. The next layer below is dominated by chemolithotrophic sulphur-oxidizing bacteria. These are life forms that use the redox energy contained in the biochemical gradient between reduced sulphur compounds produced by sulphate reducers (below) and oxygen produced by the cyanobacteria (above). They are anaerobic

photosynthesizers that fix sugars by utilizing the infrared portion of the light spectrum as they convert H_2S to SO_4. Infrared radiation penetrates deeper into microbial sediment than the visible light used by the cyanobacteria (Fig. 13B). Utilizing the redox gradient these purple sulphur-oxidizing bacteria synthesize organic matter and create the characteristic purple-orange layer seen below the green layer in an aggrading microbial mat. The sulphur-oxidizing bacteria can only metabolize in a suitable redox gradient, and they need to locate themselves exactly within this gradient, so they are typically the most motile forms in a microbial mat. The redox gradient lies deeper in the mat community during the day due to oxygen production by the cyanobacteria, but it rises and can even reach the mat surface in the dark of night (Fig. 13B).

Below the phototrophic sulphur-oxidizing bacteria are the sulphate reducers, the fermenters and the methanogens (Fig. 13C). In saline settings they are relatively inefficient decomposers and tend to be active at salinities that are less than 150‰ (Figs 3 and 4; Ollivier *et al.*, 1994). During the day the sulphate-reducers metabolize organics deep in the microbial mat via anaerobic respiration, utilizing a variety of chemical constituents including manganese, nitrate, sulphate and even CO_2. At night they rise somewhat with redox gradient and at times of prolonged bottom-brine anoxia and brine-column turbidity can even reach the mat surface or enter the lower parts of a density-stratified brine column. They are very important decomposers in most mesohaline microbial mats and give the lower parts of a mat its characteristic "rotten egg" smell. They generally consume up to one third of the organic carbon present in a typical mesohaline microbial mat and are most efficient at salinities that are less than 120–140‰.

Fermenters and reducers do not completely break down the organics but create alcohols, lactates and other organic compounds (Ollivier *et al.*, 1994). The other group of anaerobic decomposers deep in the mat is the methanogens. They are archaea and utilize simple compounds created by the fermenters (CO_2 and hydrogen, along with organic components such as methanol, acetic acid, formic acid and methylamines) to produce methane. This metabolic pathway produces little biochemical energy and is usually the "last resort" community that makes up the lowermost living layer in the microbial high-rise community that inhabits a microbial mat or a stratified brine column.

Stratification of the microbiota in response to, and in biofeedback with, oxygen levels has important implications in terms of the biomarkers found in preserved saline sediments (and source rocks). Grimalt *et al.* (1992) found that although the uppermost layers of saline microbial mats are dominated by cyanobacteria, they typically leave no more than minor traces in the solvent-extractable lipids (biomarkers) in the buried sediment. Rather, the predominant fatty acid distributions in the lower parts of mesohaline mat sediments parallel the compositions observed in the enrichment cultures of sulphur-oxidizing bacteria, and appear to be mixed with acids characteristic of heterotrophic bacteria, including purple bacteria and sulphate-reducers. In other words, the retained organic signatures, even as the mat lamina were accumulating a few millimetres above, were those of the decomposers, not the primary producers (Fig. 13C). This implies that many biomarkers used to typify organic production in modern and ancient hypersaline sediments typically do not come from organics of the primary producers but from the products of the decomposer community.

The overprinting effects of the decomposers in microbial laminites are clearly seen when molecular characteristics of recent hypersaline sediment from the Ejinur salt lake (northern China) are compared to Tertiary (Eocene) core samples from the Qianjiang Formation (hypersaline lacustrine) of the Jianghan Basin, central eastern China (Fig. 14; Wang *et al.*, 1998). N-alkanoic acids in sediments from both areas (Ejinur and Jianghan) show a pronounced even-over-odd predominance and a bimodal distribution. In the lower molecular

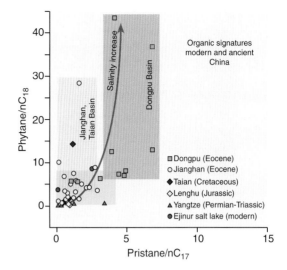

Fig. 14. Crossplot of pristane/nC_{17} versus phytane/nC_{18} ratios for various saline/hypersaline settings in China (after Wang, 1998).

weight range, the C_{16} and C_{18} components are prominent, with the former dominant. For higher homologues (greater than or equal to C_{20}), docosanoic (C_{22}) and tetracosanoic (C_{24}) acids dominate the n-alkanoic acid homologues for the Jianghan and Ejinur samples, respectively. Alkanoic acids with an isoprenoid skeleton are more abundant in Jianghan samples, including C_{20}, C_{21}, C_{24}, C_{25} and C_{30} homologues, with a C_{25} component (3,7,11,15,19-pentamethyleicosanoic acid) most pronounced in the lower part of the Qianjiang Fm. The carbon skeletons of these isoprenoid acids are attributed to archaeal decomposers. *Iso* and *anteiso* branched carboxylic acids are prevalent in both the Lake Ejinur samples and the upper portion of the Qianjiang Formation. (Wang, 1998). They derive from bacteria, probably sulphate-reducing bacteria, and their abundances clearly show again the importance of bacterial decomposers, along with haloarchaea as contributors to the biochemical signature of organic matter in both modern and ancient saline lake sediments. The presence of hopanoid acids and a 3-carboxy steroidal acid in both further attest to contributions from bacterial and eukaryotic sources, respectively. The occurrence of particular carboxylic acids in the Jianghan samples illustrate these compounds, indicators of halotolerant and halophilic decomposers, can survive as biomarkers in source rocks deposited in hypersaline settings.

Defining where ancient halotolerant and halophilic decomposers lived ("in brine column" or "in mat") is not possible from an analysis of sediment biomarkers. It requires sedimentological interpretation of the rock matrix, which typically is a laminite. As discussed in Warren (1986), any assumption of "lamination indicating deeper brines" is fraught with difficulty, as ancient evaporitic laminites and biolaminites were subaqueous sediments with near-identical mm-scale textures accumulating beneath stratified and unstratified perennial brines columns at depths ranging from less than one metre to hundreds of metres. Determination of water depth requires sedimentary analysis of evaporitic sediments that are intercalated with the laminites.

ORGANIC ENRICHMENT

At the broader scale of petroleum source rock creation, the amount of organic matter preserved in evaporitic sediment, which becomes a potential source rock, is a function of three factors that can be expressed in the equation (Bohacs *et al.*, 2000):

$$\text{Organic Enrichment} = (P\text{-De})/(Di) \qquad (3)$$

where P is production, De is destruction (oxidation and biodegradation and Di is dilution (detrital influx and matrix precipitation). Maximum enrichment occurs when production is maximized and destruction and dilution are minimized.

Production at mesohaline salinities is mostly via photosynthetic fixation of CO_2. It includes autochthonous organic matter, formed at or near the site of accumulation and derived from algae, bacteria and aquatic plants, as well as allochthonous organic matter carried in during floods (Bohacs *et al.*, 2000). Katz (1990) showed that varying levels of solar input and water chemistry/stratification exert the largest effects on overall primary production in mesohaline settings. Initial destruction of organic matter is mostly a function of the efficiency of the various scavengers and grazers, which is typically a function of the availability of oxygen (Demaison & Moore, 1980). In higher salinity settings the scavenger/decomposer population is dominated by bacteria and archaea, while the density of the metazoan grazers is largely inhibited by the same higher salinities. Dilution by an influx of mineral precipitates to possible areas of organic enrichment can limit levels of organic enrichment by decreasing the proportion of organic matter relative to inorganic matrix. The source of dilution can be detrital siliciclastics or precipitates of gypsum and halite.

Salinity controls on organic productivity: the flamingo connection

When biodiversity is low and salinity is high, a proliferation of well-adapted mesohaline species can generate organic productivity levels in surface waters that are much higher than those observed in most non-evaporitic settings. These saline systems are indicative of general principles seen in stressed ecosystems, namely, as the harshness or abiotic stress in a system increases, the biotic stress decreases on a few well-adapted species (as there are fewer predators and grazers). The total biodiversity (number of species) decreases, while at the same time the biomass (total mass of organisms in the system) initially increases (e.g. Garcia & Niell, 1993). Ultimately, as salinity rises further, biomass declines as conditions become unsuitable for any life. One of the most visually impressive indications of periodic very high levels of organic productivity in a modern salinity-stressed water body resides in the "flamingo connection", a connection between flamingos, mesohaline planktonic blooms and saline lakes, well documented in Lake Nakuru by Vareschi (1982) and first noted in the geological literature by Kirkland & Evans (1981).

Flamingos are filter feeders thriving on the dense cyanobacterial blooms in the shallows of saline lakes around the world. For example, two species of flamingo intermittently occur in huge numbers in the various East African rift lakes. These are the greater and the lesser flamingo (*Phoenicoptus ruber roseus* and *P. minor*); the lesser flamingo is the bird with the spectacular pink-red colouration. These bright pink birds feed and breed in rift lakes where cyanobacterial blooms can be so dense that a secchi disc disappears within a few centimetres of the lake's water surface (Warren, 1986). Lake Natron is a major breeding ground for flamingos in East Africa and is the only regular breeding site for the lesser flamingo in Africa (Simmons, 1995). In fact, Lake Natron has the highest concentration of breeding flamingos of any lake in East Africa. Both the greater and the lesser flamingo are found there, with the lesser flamingo outnumbering the greater by a hundred to one. Lesser flamingos bred at Lake Natron in 9 out of 14 years from 1954 to 1967. While Lake Natron (a trona lake) is an essential breeding site, it is not a focal feeding site for flamingos. Major feeding sites in the rift valley are Lakes Nakuru and Bogoria in Kenya, which are somewhat less saline than Lake Natron and so have a greater abundance of mesohaline plankton.

The main phytoplankton component in Lake Nakuru waters is the cyanobacterium, *Arthrospira platensis* (previously known as *Spirulina platensis*), it forms the main food component of the lesser flamingo. In Nakuru it is also consumed by one species of tilapid fish and one species of copepod and a crustacean. Rotifers, waterboatmen, and midge larvae also flourish in the waters of Lake Nakuru. The mouth-breeding tilapia fish *Sarotherodon alcalicum grahami* was introduced to the lake in the 1950s to control the mosquito problem and they have flourished ever since, at times displacing the flamingos as the primary consumers of planktonic algae. Their introduction also increased the number of fish-eating birds residing in the lake (Vareschi, 1978).

In a good year more than a million flamingos consume more than 200 tons of plankton per day from Lake Nakuru, a shallow saline soda lake with a pH ~10.5 and a typical annual salinity range of 10–120‰. The lake measures some 6.5 km by 10 km, with waters up to 4.5 m deep in the lower parts of the lake during a "lake-full" period. Lake levels are subject to change, both annually and on a longer time scale, with permanent eutrophic waters developed in the lake centre during times of higher water (Fig. 15A and B; Vareschi, 1978–1982).

The feeding style of the birds is to wade through the shallows, with heads upside down and beaks waving side to side across the water surface. Flamingo beaks have evolved to skim plankton from saline to brackish surface waters and are equipped with a filter-feeding system unlike any other bird on Earth and more akin to feeding apparatus of the krill-ingesting Great Whales. Flamingos siphon the lake water through the beak filters to trap *A. platensis* and other plankton by swinging their upside-down heads from side to side and using their fat tongues to swish water through their beaks. They can filter as many as 20 beak-fulls of plankton-rich water in a second. Spacings of the filters in the beak of the greater flamingo are wider than those in the beaks of the lesser flamingo. Greater flamingos feed mostly on zooplankton, while lesser flamingos feed almost solely on *Spirulina*, thus in any saline lake where the two species co-exist they do not compete for the same food source.

Arthrospira platensis provides high levels of the red pigment phycoerythrin to the food chain and it accumulates in flamingo feathers to give the birds their world famous colouration (hence the "flamingo connection"). This is the same pigment that turns waters red in Lake Natron and Lake Magadi, but in this case it comes mostly from the halophilic archaea. Lesser flamingos, with their narrower spacings of beak filters, survive solely on *Arthrospira* and so have a more intense pink colour than greater flamingos, a species that sits higher in the lake food chain and get its feather colour second-hand from the lake zooplankton.

As well as possessing very high levels of phycoerythrin in its cytoplasm, *Spirulina* is also unusual among the cyanobacteria in its unusually high protein content. In Lake Chad and in some saline lakes in Mexico it accumulates as a lake edge scum that is harvested by the local people and used to make nutritious biscuits. In the 16th century it was mentioned in the diary of one of Cortez's soldiers who described how the Aztecs sold hard flat cakes made from the dried remains of *Arthrospira*. The Aztec word for it was Tecuitlatl, which translates literally as "excrement of stones." The lack of cellulose in the *A. platensis* cell wall means it is a source of plant protein readily absorbed by the human gut, making it a potentially harvestable food source in water bodies in regions of desertification and it has become a popular alternative protein source in the developed world.

In 1972, Lake Nakuru waters held a surface biomass of 270 g m^{-3} and an average biomass of 194 g m^{-3} but, as in most hypersaline ecosystems, Nakuru's organic production rate varies drastically from year to year as water conditions fluctuate (Fig. 15C; Vareschi, 1978). *Arthrospira platensis* was in a long-lasting bloom in 1971–1973, and accounted for 80–100% of the large phytoplankton biomass in those years. In 1974, however, it almost disappeared from the lake and was replaced by coccoid cyanobacteria such as *Anabaenopsis sp.*, *Chroococcus sp.* and by diatoms, all species that dealt better with elevated salinities. This change in biota was tied to a serious reduction in biomass, which in 1974 was down to 71 g m^{-3} in surface waters, and averaged 137 g m^{-3} in the total water mass. As a result, the flamingo population in the lake declined from 1 million to several thousands (Vareschi, 1978).

The lower salinity limit for *Arthrospira platensis* growth is ~5‰ and in terms of water conductivity it does best in the range 10–50 mS cm^{-1}. The driving mechanism for the change in biomass in Lake Nakuru in the mid-1970s was not clearly documented. It was thought to be related to increased salinity and lowering of lake levels, along with

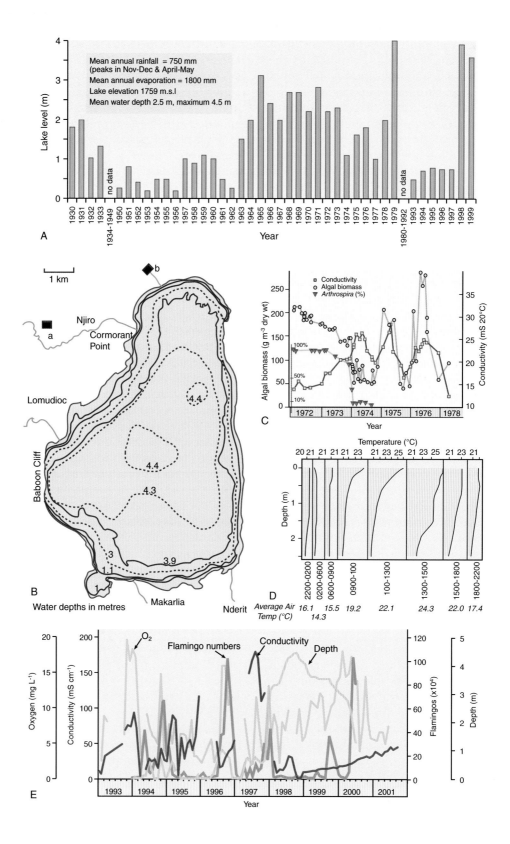

the growth and bottom shading by planktonic algal species that were better adapted to higher salinity. These better adapted were present in lesser volumes and with lower per cell protein cells (Fig. 15C; Vareschi, 1978).

During the mid- to late 1970s and in the 1980s Lake Nakuru returned from a relatively saline dry phase to more typical annual schizohaline oscillations in water level and salinity (Fig. 15A). Seasonally, in the 1980s, the *A. platensis* and flamingo populations returned to impressive numbers. From the 1990s more reliable long-term data on physical lake condition and flamingo numbers have being compiled (Fig. 15E). In 1995 and 1998 feeding flamingo populations in Lake Nakuru were once again at very low levels and the remaining population was stressed (Fig. 15E). In 1998, unlike 1974, the stress was related to freshening driving a decrease in *A. platensis* biomass, not increased salinity and desiccation. This freshening of the lake waters favoured a cyanobacterial assemblage that also produced toxins (see Cyanobacterial toxins in Glossary). In the preceding bountiful year (1996), the *A. platensis*-dominated biomass had bloomed at times when salinities were favourable, and died back at times of elevated salinities and lake desiccation, as in 1974. By 2000, the somewhat lower salinity cycles had once again increased, making surface waters suitable for another widespread *A. platensis* bloom and the associated return of high numbers of feeding flamingos.

It seems that breeding flamingos come to Lake Nakuru to feed in large numbers when there is water in the lake with appropriate salinity and nutrient levels to facilitate an *A. platensis* bloom. In some years when heavy rains occur, lake levels rise significantly and the lake waters, although perennial, stay in the lower salinity tolerance range for *A. platensis*, as in the El Niño period between October 1997 and April 1998. Once lake levels start to fall, salinities and rates of salinity change return to higher levels, then water conditions once again become appropriate for *A. platensis*. Environmental stress on the flamingos also comes with drier periods that elevate lake salinities into the upper end of *A. platensis* tolerance. Ongoing drought ultimately leads to lake desiccation and the complete loss of the food supply in the lake.

One of the reasons why Lake Nakuru is so suitable for cyanobacterial growth at times of *A. platensis* bloom, is the maintenance of suitable temperatures and associated oxygenation in the upper water mass. The lake develops a remarkable daily thermocline in the upper 1.5 m of the column that dissipates each day via wind mixing in the late afternoon, and so recycles nutrients back to the oxygenated surface waters to facilitate the ongoing plankton bloom of the next diurnal cycle (Fig. 15D).

Some environmentalists have argued in the popular press that lower numbers of flamingos in Lake Nakuru, such as occurred in 1995 when more than 50,000 birds died, are an indicator of: uncontrolled forest clearance; an uncontrolled increase in sewerage encouraging eutrophication; an increase in heavy metals from increasing industrial pollutants in the lake; and general stress on the bird population from tourists and the drastic increase in local human population centered on the town of Nakuru (now the third largest in Kenya). Numbers of people in the town, which is the main city in the rift valley, have grown by an average of 10% every decade for the past 30 years.

Like many environmental doomsday arguments, the above are more based on opinionated prediction than on scientific fact. Rise and fall of lake levels, drastic changes in salinity, a periodically stressed biota, and a lack of predictability in water characteristics are endemic to saline ecosystems (cycles of "famine or feast"). Oscillations in one or more of these factors are not necessarily

Fig. 15. Physical and chemical characteristics, algal biomass and flamingo numbers of Lake Nakuru, Kenya. (A) Lake levels from 1930 to 1999. Data are not sufficiently detailed to record times of complete lake drying or drought, and there are also substantial times when no data were collected/published (1934–1949 and 1978–1992). (B) Lake bathymetry (December 1979) and strand-zone changes. Solid isopleths are actual shorelines at different lake levels over a 20-year period; from margin to centre these are: December 1979; January 1969; January 1967; and January 1961. Dashed isopleths are based on soundings taken in December, 1971. Two sewage plants are located at a and b. (C) Trends of algal biomass and conductivity of lake water during 1972–1978. The percent contribution of *Arthrospira platensis* (formerly *Spirulina platensis*) to the total algal biomass is also shown. (D) Mean water temperature profiles at different times of day from 1972–1973. (A–D) after Vareschi (1982). (E) Monitored water depths, conductivities, oxygen levels and flamingo numbers at Lake Nakuru for the period 1993 to mid-2001. Water depth and flamingo data only available from 1994 to mid-2000 (data are digitized and replotted to a common time scale from a number of figures and pdf reports downloaded from http://www.worldlakes.org on 2 February 2004).

anthropogenically induced. For example, the amounts of most heavy metals (Cd, Cr, Cu, Hg, Ni, Pb, Zn) in the lake sediments were found to be in the typical range of metals in natural lakes worldwide (Svengren, 2002). The exception was cadmium, which is elevated in the lake sediment and could perhaps be assigned to anthropogenic pollution. All other metals are present at low levels, especially if one considers that Lake Nakuru lies within a labile catchment where bedrock is an active volcanogenic-magmatic terrane. At the present time, sufficient base a scientific data are not yet available in Lake Nakuru, and so scientifically accurate determinations cannot be made of the relative effects of humanity verses natural baseline environmental stresses on bird numbers in this schizohaline ecosystem.

Nearby lakes Magadi and Natron have extensive trona platforms covered with brine sheets that are characterized by short periods of high productivity and times of bright red waters. In this higher salinity ecosystem, the dominant producers of the brine colour are haloalkaliphilic bacteria and archaea. Archaeal species belonging to the genera *Natronococcus, Natronobacterium, Natrialba, Halorubrum, Natronorubrum* and *Natronomonas*, all occur in the lakes. Lake centre brines when this biota flourishes are at trona/halite saturation, with a pH ~12. Stratified moat waters around the trona platform edge are at times almost as chemically extreme, and the moat bottom sediments are made up of aragonite/dolomite/detrital laminites that can preserve elevated levels of organics (~6–8%). As well as algae and cyanobacteria, Lake Magadi and Natron moat waters also harbour a varied anaerobic bacterial community including cellulolytic, proteolytic, saccharolytic, and homoacetogenic bacteria (Shiba & Horikoshi, 1988; Zhilina & Zavarzin, 1994; Zhilina *et al.*, 1996). When the homoacetogen *Natroniella acetigena* was isolated from Lake Magadi its pH growth optimum was found to be 9.8–10.0, and it continued to grow in waters with pH up to 10.7, making it an exemplary alkaphile (Zhilina *et al.*, 1996).

Regionally, salinities in the East Africa rift-valley lakes range from around 5‰ total salts (w/v) in the more northerly lakes (Bogoria, Nakuru, Elmenteita, and Sonachi) to trona and halite saturation (>200‰) in lakes to the south (Lakes Magadi and Natron). Yet across this salinity range, a combination of high ambient temperature, high light intensity and a continuous resupply of CO_2, makes these soda lakes amongst the highest in the world in terms of their seasonal planktonic biomass (Grant *et al.*, 1999) and also places them among the world's most productive ecosystems (Melack & Kilham, 1974). As demonstrated above, the less saline soda lakes are dominated by periodic blooms of cyanobacteria, especially *Arthrospira platensis*, while the hypersaline lakes, such as Magadi, can on occasion support blooms of both cyanobacteria and alkaphilic phototrophic bacteria belonging to the genera *Ectothiorhodospira* and *Halorhodospira* (Jones *et al.*, 1998).

The halotolerant and halophilic biota living in the layered water columns of these soda lakes constitute small-scale "famine or feast" ecosystems, which at times of "feast" are far more productive than zones of marine upwelling (Fig. 16A). A general observation of short periods of enhanced organic productivity between somewhat less and somewhat more saline episodes in schizohaline lake waters worldwide, reflects the general ecological principle that increased environmental stress favours the survival of a few well-adapted halotolerant species, to the detriment of others better suited to life in a less saline environment (see inset in Fig. 16A). Ultimately, abiogenic stresses associated with more elevated salinity means all life, including the halophiles, ceases to function.

The same general principle that schizohaline ecosystems encourage dominance by a few better-adapted species can be clearly seen in the decrease in invertebrate species (grazers and predators) with increasing salinity in the carbonate lakes of the Coorong of Southern Australia, where only brine shrimp remain alive in waters with salinities in excess of 200‰ (Fig. 16B). It reflects a general trend of decreasing faunal biomass with increasing salinity (Fig. 16C). A decreased level of biomass in a mesohaline system does not necessarily mean less source rock potential in the resulting sediments. Preservation levels of organic matter and suitability of the protokerogen (type I to type III) depends on what biotal contributors constitute the biomass preserved in the sediment.

Examples of "famine or feast" productivity have so far focused on continental salt lakes. Similar enhancements of primary productivity can occur in modern coastal salinas during episodes of periodic freshening as blooms of haloxene and halotolerant microbes (oxygenic photosynthesizers) take place across formerly exposed mats. For example, a study of biolaminites collected from a modern salina (Storr's Lake in the Bahamas) showed that a reduction in brine salinity from 90 to 45‰

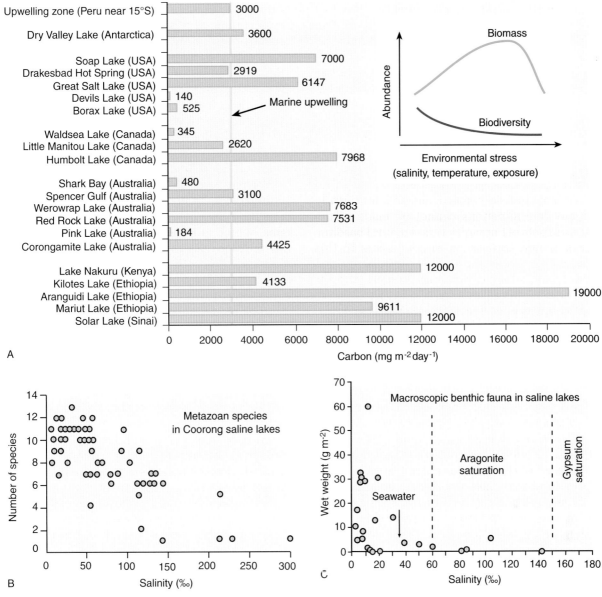

Fig. 16. Productivity and biodiversity in saline ecosystems. (A) Organic productivity in various saline ecosystems. Typical marine upwelling zone (offshore Peru) is ~2000–3000 mg C m^{-2} day^{-1} (yellow line), open-marine waters average 20 mg C m^{-2} day^{-1} and coastal waters 40–50 mg C m^{-2} day^{-1}. Preservation is more important than production rate in terms of source potential. Compiled from sources listed in Warren (1986, 2006). (B) The effect of increasing salinity on metazoan species diversity in the saline lakes of the Coorong region, South Australia (after de Decker & Geddes, 1980). (C) Standing crop (g m^{-2}) of the macroscopic benthic fauna in selected athalassic (non-marine) saline lakes of the world (after Hammer, 1986).

significantly enhanced CO$_2$ and N$_2$ fixation rates, but additions of inorganic nutrients and dissolved organic carbons did not show any significantly enhanced rates compared with the salinity controls (Pinckney *et al.*, 1995). That is, dissolved organic carbon/dissolved organic nitrogen uptake was not influenced over the entire range of salinities (45–90‰) seen in this study of mesohaline mat growth.

In other words, abiotic stress, induced by hypersaline conditions in the Bahamian lagoons, created lower productivity in the saline-tolerant microbial mats, even when nutrient levels were high. Once the osmotic stress on the biota was relieved by a lowering of salinity, mats underwent enhanced primary production and nitrogen fixation as haloxene producers could once again photosynthesize. Changing nutrient levels were far less significant than changing salinity levels as controls on mat productivity in this stressed setting. Salinity-induced stress in these halotolerant microbial mats outweighs typical limiting factors regulating primary production in phototrophic communities at the lower salinities that typify marine and brackish waters.

Organic production in saline waters

When surface water salinities are suitable, evaporite basins have some of the highest measured rates of organic productivity in the world (Fig. 16A; Warren, 1986). Times of elevated primary production ("feast") in density-stratified hypersaline brine lakes and evaporitic seaways appear to be controlled by two factors: (1) the ephemeral presence of a less saline surface water; and (2) a supply of nutrients (nitrate/ammonia and phosphate) from the lower hypersaline water mass. The biota contributing the organics that ultimately reach the bottom sediments to form organic-rich laminae (and so control their future biomarker signature) are largely dependent on the timing of photosynthetic activity. Perennial saline lakes subject to an annual freshening of surface waters will deposit varved carbonates (e.g. Lake Tanganyika with up to 7–11% TOC in deep bottom laminites; Cohen, 1989), while other saline lakes and seaways with less predictable stratification may form similar carbonate laminites, but the organic layering may not be annual, and the resulting organic biomarkers contain elevated levels of isoprenoids due to the greater contribution from the haloarchaea. Microbial mats in some shallower saline lakes may be subject to short periods of photosynthetic oxygenation and even subaerial exposure, yet still preserve elevated levels of hydrogen-prone organics (Warren, 1986).

Once brine moves into the gypsum precipitation field and beyond, the amount of organics accumulating on the bottom is diluted by inherently rapid rates of crystal precipitation and accumulation, especially if the bottom brine is photosynthetically oxygenated by its microbial community. Hypertrophication, in combination with shading by pelagic crystallites, can overwhelm the delivery of oxygen to the bottom, as in Solar Lake and in Lake Hayward in the late 1980s. Evaporitic systems subject to longer-term eutrophication can facilitate high levels of pelagic organic delivery to bottom sediments, even beneath shallow penesaline waters (Horsfield *et al.*, 1994; Warren 1986). In a study

of eutrophication in man-made saltfields, Javor (2000) found that in shallow gypsum and pre-gypsum ponds there was an interesting inverse relationship between the level of nutrients (nitrates and phosphates) in the brine column and the development of microbial mats on the water covered pan floor. In a high-nutrient brine system (eutrophic or "well-nourished pan") the algae and bacteria flourished as planktonic forms and the biomass buildup in the upper parts of the brine column tended to shade the sediment surface. There was little or no development of an oxygenated microbial mat and bottom waters were dysaerobic to anoxic. In a low nutrient system (oligotrophic or "poorly-nourished pan") there was little or no planktonic development of algae or bacteria and microbial mats tended to flourish on the perennial pond floor. Benthic mats formed from salinities of 50‰ to gypsum saturation in the poorly-nourished pans. In pans at even lower salinities the benthic mats were destroyed by the feeding activities of invertebrates and fish.

Preservation of microbial mats into early burial in hypersaline systems depends, in large part, on the efficiency of nutrient recycling by the bacterial and archaeal decomposers that making up the underside of the living part of a mat. Even if nitrogen and phosphorus levels are low in the overlying brine, there may be sufficient recycling of nutrients within the mat community to allow the benthic mat to continue to grow. Javor (1989) found that once a gypsum crust becomes well established over the floor of a crystallizer pond, there was little recycling of nutrients from beneath the crust to the uppermost photosynthesizers, and from this point downstream in the crystallizer pans, nitrate and phosphate nutrients were concentrated by evaporation and recycled into the biomass living in the brine column. In oligotrophic gypsum ponds the gypsum crust is stable and the bottom waters are oxygenated. In eutrophic gypsum ponds at the same salinities, the bottom waters become anoxic and sulphate-reducing bacteria flourish. In these systems, gypsum dissolves and a black slimy organic-rich carbonate/detrital mud (a potential source rock in ancient counterparts) covers the bottom and contains no more than a few embayed and dissolved gypsum remnants. If portions of this slimy bottom become suspended and are flushed into the next stage of crystallizer they create a problem in the final salt product known as "black spot".

The difference in organic levels between oligotrophic systems such as Lake Hayward and perennial eutrophic systems like Solar Lake probably explains why the latter retains much higher levels of hydrogen-prone organic material in its preserved bottom sediments (Fig. 5). Javor's work also demonstrated that as higher salinity brines tend to concentrate nutrients, they will supply these nutrients at times or at interfaces where supersaline waters in a natural schizohaline system are subject to periodic freshening or mixing.

Proximity of organic-rich mesohaline laminites to bedded evaporites in ancient successions has led some geologists to postulate that bedded salts are potentially also organic-rich sediments, with the same level of source potential as mesohaline carbonates. This is not borne out by TOC analysis of evaporitic salts; total organic matter levels in ancient $CaSO_4$ and $NaCl$ lithologies tend to be normal or depleted (Katz *et al.*, 1987), although levels of dissolved organic matter and volatile fatty acids may be high in the entrained brine inclusions (Hite & Anders, 1991). The typically low organic matter levels in bedded gypsum and halite may reflect the higher depositional rates of evaporitic salts when compared with evaporitic carbonates (dilution effect). Only in settings where bottom waters are anoxic in the gypsum stage can the build up of H_2S from sulphate reduction become high enough to remove $CaSO_4$ minerals and so allow preservation of elevated levels of partially degraded organics (as in Solar Lake today).

That organic enrichment mostly occurs in the mesohaline carbonate stage in an evaporite basin is clearly seen in organic contents of the evaporitic marls and salts of the Oligocene Mulhouse Basin (Fig. 17; Hofmann *et al.*, 1993a, b). This deposit is one of a few ancient evaporites where organic chemistry has been studied in a logical and detailed fashion. Organic matter occurs mostly in finely laminated carbonate marls (varves), which can contain up to 7% TOC, anhydrite beds contain 0.08–0.78% TOC, halite beds 0.01–0.25% TOC and the potash beds <0.1% TOC. The marls are characterized by thin, continuous bedding parallel laminae that are not disturbed by any bioturbation. They are interpreted as seasonal varves with the clay and carbonate forming alternate layers in the couplet (Fig. 17, inset). In spring and summer the increased temperature in surface waters favoured precipitation of micritic carbonate, in autumn and winter the phytoplankton died back and rainfall/runoff transported siliciclastic debris into the basin. On the basis that the laminar couplets are varves, Hofmann *et al.* (1993a)

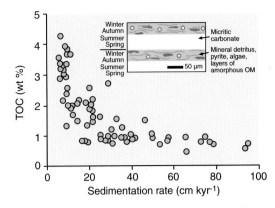

Fig. 17. Sedimentation rate versus percent of total organic carbon (TOC) in laminite marls from the Oligocene evaporites of the Mulhouse Basin, France. Inset shows the scale and mineral/organic distribution in the laminae (varve couplets) of the sampled intervals (after Hofmann *et al.*, 1993b).

were able to plot the relationship between sedimentation rate and TOC content in the marls (Fig. 17). Marls deposited during times of low sedimentation tend to be organic-rich, while marls deposited with higher sedimentation rates tend to contain lower amounts of organics. They extrapolated that the even lower levels of organics in the more saline salt beds (anhydrite and halite) indicate greater dilution because of the inherently higher sedimentation rates of halite and gypsum compared to mesohaline carbonate.

Salinity-induced environmental stress ("famine or feast") facilitates organic preservation

Organic matter in modern saline systems tend to accumulate in bottom sediments beneath density-stratified saline water and sediment columns in evaporitic settings where layered hydrologies are subject to oscillations in salinity and brine level. Organic matter is not produced at a constant rate in such systems, rather, it is produced as pulses by a halotolerant community in response to relatively short times when less stressful conditions occur in the upper part of the layered hydrology (Fig. 18A). This happens when an upper less-saline water mass forms on top of nutrient-rich brines or within wet mudflats when waters in and on top of the uppermost few millimetres of any microbial mat freshen. The resulting bloom (a time of "feasting") by halotolerant algae and cyanobacteria is a time characterized by very high levels of organic productivity. In mesohaline waters where light penetrates to the bottom, the organic-producing

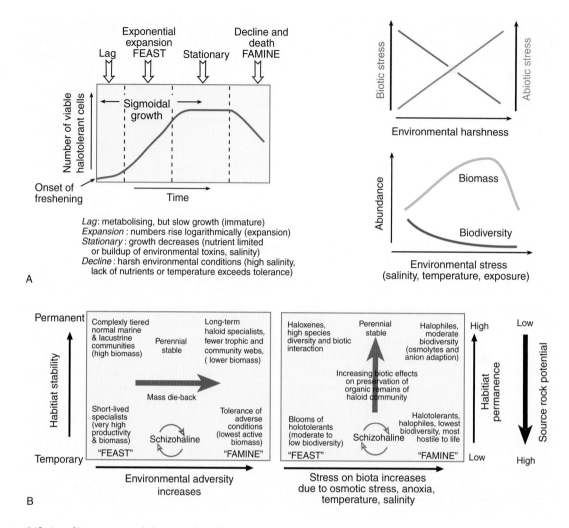

Fig. 18. Life in saline systems. (A) General ecological principles in stressed saline ecosystems. Sigmoidal growth curve of life moving into a newly created niche (as created by a freshening event setting up a layered brine). As environmental harshness increases due to abiogenic stress factors (such as increasing salinity, osmotic stress and temperature) the level of biogenic stress (predation or grazing decreases, allowing a few well-adapted species to flourish, leading to a decrease in biodiversity and a short-term increase in biomass. (B) Schematic summary of the inter-relationships between habitat stability, environmental stress or adversity, biodegradation, anoxia and source rock potential in an evaporite basin, showing why schizohaline environmentally stressed mesohaline systems tend to favour the accumulation of organic matter in bottom sediments.

layer is generally the upper algal and bacterial portion of benthic laminated microbialites (typically characterized by elevated numbers of cyanobacteria). In stratified brine columns, the typical producers are the planktonic algal or cyanobacterial community inhabiting the upper water mass. Halotolerant autotrophic microbial communities, living off the remains of this plankton, float midwater at the halocline of the density-stratified brine columns. Similar aerobic decomposers can also constitute bottom-layer communities in zones of oxygenated oligotrophic waters.

Short pulses of extremely high organic productivity during times of freshened mesohaline surface waters in-turn create a high volume of organic detritus settling through the water column and/or construction of benthic microbial mats. Then, with the end of the freshening event, the ongoing intensely arid climate that characterizes evaporitic depressions means that salinities, temperatures and osmotic stresses increase rapidly in the previously freshened water mass. This leads to a time of mass die-off, in the once flourishing mesohaline community ("famine;" Fig. 18B). First, these increasingly salty waters can no longer support haloxene forms. Then halotolerant life dies back and finally, by the halite precipitation stage, only a few halophilic archaea and bacteria remain in the brine column (typically acting as heterotrophs and fermenters). Repeated pulses of organic matter created during a freshening event in a stratified brine column thus create laminated bottom sediment.

Hydrogen-rich algal and eubacterial debris is best preserved within sediments where pore waters are anoxic and remain so until the resulting kerogen is sufficiently buried for it to generate liquid and gaseous hydrocarbons (Demaison & Moore, 1980). Thus the preservation of organics with long-chain hydrocarbons intact is most likely in settings characterized by long-term anoxia (less efficient decomposers). This tends to occur beneath stratified eutrophic saline brine columns with a permanent anoxic mesohaline bottom water mass and a transiently freshened upper water mass. Preservation is also possible in benthic mats in shallow oligotrophic systems with permanently hypersaline pore waters, but the propensity for periodic bottom oxygenation means preservation potential of oil-prone mats is less likely in this setting, as is also the case in mats of the strandzone.

Increasing rates of mineral precipitation at higher salinities means levels of hydrogen-prone organics are highest in sediments deposited in schizohaline waters that remain in the mesohaline field. Once ambient salinities pass into the penesaline and supersaline, the proportion of organics in the sediment becomes insignificant as it is rapidly diluted by large volumes of gypsum and halite (or trona and glauberite in some continental penesaline/supersaline waters).

Feast and famine cycles are a biotal response to environmental salinity conditions that at first are eminently suitable and then increasingly adverse to life. It reflects the typical biological response to rapid niche creation and expansion as the freshened water layer sets up in the evaporitic depression, and is known as the sigmoidal growth response (Fig. 18A). Freshening of surface waters creates a vacant ecological niche into which microbial life rapidly expands, flourishes and then once more dies back as drying or brine concentration shrinks suitable niche space.

Long-term stability in any ecosystem leads to domination by "specialist" species that are well adapted, diverse and efficient in terms of energy flow-nutrient utilization (Fig. 18B). This biodiverse community takes up and recycles carbon very efficiently through the system and the potential for large volumes of carbon passing into the burial realm is low and hence source potential is also low (a good example in the marine realm is a tropical coralgal reef).

In contrast, a schizohaline system is one defined by high habitat instability, its lack of habitat permanence encourages opportunistic lifestyles (cycles of "famine or feast"). Opportunist species expand rapidly into any suitable niche space, but then just as quickly die back or encyst as conditions become adverse. Such systems are characterised by high biomass pulses and low species diversity, leading to pulses of organics accumulating in the evaporitic depression, with the remains likely to be buried and "pickled" beneath anoxic bottom brines and pore waters, giving such systems a higher source rock potential (Fig. 18B).

Some hypersaline settings are intermediates in that their waters are constantly hypersaline, stressful to most life, but suitable to a few well-adapted species. These regions are usually populated by a few very specialized species (e.g. "high-salt-in" halophiles) that can survive and flourish in conditions where most life dies. They photosynthesize and metabolize successfully in these hypersaline conditions (often for example living near haloclines), but do so at rates that are much slower than the growth rates attained by halotolerant "opportunists" at times of "feast." The long-term cycling of organic materials means bottom sediment layers tend to have lower levels of entrained organics preserved compared to the schizohaline-mesohaline.

To take an understanding of community and controls from the modern into ancient saline systems, it is necessary to account for changes in both scale and diversity when comparing the style and setting of "famine or feast" in modern saline settings to those of the past. Put simply, there are no modern same-scale analogues for ancient marine-fed evaporite depressions, and so there are no modern same-scale marine-fed evaporitic source rocks (Warren 2006, 2010).

ANCIENT EVAPORITIC SOURCE ROCKS

The common theme to deposition of all prolific source rocks, both evaporitic and non-evaporitic, is that they require anoxic conditions to accumulate and preserve their organics (Demaison &

Moore, 1980). This in-turn requires some way of restricting the influx of oxygenated waters and is usually accomplished by thermal or density stratification of a water body, in a setting where planktonic-producers flourish periodically in the upper water mass. This is why so many ancient evaporitic settings are also places where organics tend to accumulate. By definition, evaporites require restricted inflow conditions to accumulate, they are also typically areas of density-stratified waters with long-term elevated salinities in the lower denser water mass. Levels of dissolved gases are negligible to non-existent in the lower brine body, as it does not easily exchange with the atmosphere, so encouraging long-term bottom anoxia.

This does not mean that all of the world's anoxia-associated source rocks were deposited in evaporitic settings, only that the restricted conditions that favour the accumulation of evaporites are in their early stages similar to conditions that favour the accumulation of organics. In arid climates any minor changes in inflow conditions can easily force a transition from one into the other. Hence the common association of source rocks with the freshened, typically early phases of the onset of evaporitic sedimentation in many modern and ancient lacustrine settings (Table 5). In the marine realm, the same need for tectonic or eustatic restriction to create basins suitable for source rock accumulations (intrashelf, circular or linear sags) also explains why, if further restriction and isolation from surface connections to the oceans ensue, the same regions are easily covered by a sequence of salts (Fig. 19).

Tectonic and eustatic styles in ancient evaporitic basins

Almost all of the information that has been gathered and analyzed so far in this paper comes from organic accumulations in saline Quaternary lacustrine and marine-margin settings. Likewise, all the largest and thickest examples of Quaternary sabkhas, saline pans and salinas are continental deposits (Warren, 2006). Quaternary saline systems only define same-scale analogs for ancient continental lacustrine deposition and are not same-scale analogues for ancient marine-fed epeiric and basin-centre salt basins, or their entrained source rocks (Fig. 19; Warren 2006, 2010). Yet compared to lacustrine sediments, these ancient marine-associated deposits entrain and seal far greater volumes of evaporitic source rocks (Table 5).

Most ancient marine-fed evaporite deposits are one or two orders of magnitude larger than today's marine-fed hypersaline systems and accumulated in huge marine-seepage fed sub-sea-level tectonic depressions with no surface connection to the ocean (Fig. 19). Well-documented source rock-entraining ancient basinwide evaporite deposits include the Jurassic Hith, Permian Zechstein and Miocene Messinian evaporites, but none have a same-scale modern counterpart. Likewise, no widespread marine-fed platform evaporites have accumulated since the Eocene. The Mediterranean Messinian salts (~5.5 Ma) are the youngest known basinwide evaporite. Well-documented marine-fed platform evaporites include the evaporite-capped cycles of the Jurassic Arab Formation in Saudi Arabia and the United Arab Emirates, the Permian Khuff Formation in the same region, and the evaporite-capped Permian backreef strata of Guadalupian age in West Texas and New Mexico.

Although platform and basinwide evaporites may have no Quaternary counterpart in terms of scale or thickness, they can be interpreted using process models that are combinations of the hydrologies of Quaternary sabkhas, saline pans and salinas. Accumulations of thick evaporite accumulations (continental or marine) and associated mesohaline stages require a stable long-term brine curtain (~100 kyr–1 Myr) to accumulate to substantial thicknesses and lateral extents (Warren, 2003, 2006). The hydrological position of the active top of an aggrading brine curtain, with respect to the evaporite depositional surface, in both platform and basinwide settings, defines the dominant textural signature of the resulting salt sequence (saline-pan, evaporitic mudflat/sabkha, saltern, deeper slope and basin).

High amplitude, high-frequency 4th-order sea-level oscillations of the current "icehouse" climate do not allow the set up of stable brine curtains behind laterally-continuous seepage shoals in present-day carbonate platforms, and so there are no Neogene examples of platform evaporites with stable intrashelf depressions. Platform evaporites require greenhouse eustacy to form. Nor are there suitable sub-sea-level rift-induced intracratonic sags or soft-collision belts in arid marine-fed settings where conditions are suitable for the creation of basinwide marine drawdown deposits (Warren, 2006).

The need for sub-sea-level tectonically-induced basins explains why times of worldwide continent-continent collision followed by continental

Table 5. Characteristics and settings of evaporitic source rocks. Basin styles: 1 = basin centre; 2 = intrashelf; 3 = lacustrine. See Warren (2006) for further detail on the significance of different basin styles

Formation/Group locality and age	Basin Style	Characteristics and depositional setting	Reference
Messinian carbonates Lorca Basin, Spain Upper Miocene	1a Basinwide evaporite marginal basin	Organic-rich laminated mudstones deposited in pre-evaporitic mesohaline marginal basins with density- and salinity-stratified brine columns. The basin underwent sporadic phases of circulatory restriction with marked production and preservation of organic matter, culminating in evaporative sedimentation (latest Miocene). As the water in this basin evolved toward evaporative conditions, a number of organic-rich depositional phases (>25 wt% TOC) occurred. During the early parts of these phases, the upper water was nutrient rich and comparatively normal marine, and the bottom water was anoxic and more saline. This was followed by rising salinities in the surface waters, holomixis and short phases of evaporite formation.	Benali *et al.*, 1995 Russell *et al.*, 1997 Rouchy *et al.*, 1998
Gessoso-Solfifera Formation, Sicily Upper Miocene (Messinian)	1a Basinwide evaporite marginal basin	Biomarker compositions from immature organic sulphur-rich marl samples from 10 of the 14 evaporite cycles in an Italian Messinian evaporitic basin (Vena del Gesso) indicate large variations in the composition of species in the water column (e.g. dinoflagellates, diatoms and other algae, cyanobacteria, methanogens, green sulphur bacteria and bacterivorous ciliates) within each marl bed and between marl beds. The marl beds were deposited within a stratified lagoon where anoxic conditions extended into the photic zone for much of the time.	Damste *et al.*, 1995 Keely *et al.*, 1995 Lugli *et al.*, 2007
Organic-rich marls, Rudeis, Kareem and Belayim Fms. Gulf of Suez Miocene	3 Rift basin continental to transitional marine	Inclusion studies show evaporitic environment of deposition (reducing conditions and high salinities), favoured oil-generating kerogen. Carbonate minerals and inclusions trapped in gypsum indicate possible mixing of marine water with a brine of restricted occurrence at the time source rocks were deposited.	Kholief & Barakat, 1986 Rouchy *et al.*, 1995 Barakat *et al.*, 1997 Alsharhan, 2003
Bresse and Valence salt basins, France Paleogene	3 Rift basin continental to transitional marine	Organic matter is mostly immature and occurs in intercalated non-halite beds. Type-III kerogen is tied to terrigenous deposition. Type-I kerogen is abundant in mesohaline evaporitic laminites of Valence basin. Type-II is more abundant in Bresse Basin. Organics accumulated beneath a perennially stratified brine column of up to tens of metres. Syndepositional dissolution of halite may have aided the accumulation of significant amounts of oil-prone organic matter in non-soluble brecciated residues.	Curial *et al.* 1990

(*continued*)

Table 5. (*Continued*)

Formation/Group locality and age	Basin Style	Characteristics and depositional setting	Reference
Salt IV Formation, Mulhouse Salt Basin, Europe Lower Oligocene	3 Rift basin continental to transitional marine	Biomarker assemblages are dominated by acyclic isoprenoids, especially phytane ($0.13 < Pr/Ph < 0.52$), n -alkanes with specific distribution patterns, and steranes. Large differences in distribution indicate changes in paleoenvironment. The largest changes are in desmethyl- and methylsterane distributions and are probably linked to occasional reconnection of the stratified evaporite basin to the sea, leading to a dinoflagellate bloom in the upper waters of a density-stratified brine column.	Damste *et al.*, 1993 Hofmann *et al.*, 1993a, b Hollander *et al.*, 1993 Keely *et al.*, 1993
Eocene Qianjiang and Lower Eocene-Paleocene Xingouzhui Formations of Jiangling-Dangyang area, Jianghan Basin, Northwest China Eocene-Paleocene	3 Foreland flexure basins, mountain-front fault depressions	Anoxic evaporitic lacustrine source rocks generated most of the crude oils. High salinity and low Eh enhanced preservation of oil-prone organic matter and facilitated incorporation of sulphur. Anoxia and the unusual presence of abundant sulphate as gypsum resulted in microbial reduction of sulphate to sulfide and incorporation of this sulphur into the kerogen. Biomarkers in Qianjiang Formation show that some source rocks (Sha 13 well -1322 m) was deposited under more saline, lower Eh conditions than others) Ling 80 well - 1808 m). Sha 13 sample is more organic-rich (6.62 vs 1.27 wt% TOC), has a higher hydrogen index (794 vs 501 mg HC g^{-1} TOC) and faster reaction kinetics. Kerogen from the Sha 13 sample is Type I.	Jiang & Fowler, 1986 Huang & Shao, 1993 Peters *et al.*, 1996 Ritts *et al.*, 1999 Hanson *et al.*, 2001
Green River Formation, Wyoming and Utah Eocene	3 Foreland flexure basin from Laramide orogeny	Laminated organic mudstones and oil shales deposited in deeper anoxic bottom waters of a perennial saline and density-stratified alkaline lake. Gilsonite and tabbyite bitumens are associated with Parachute Creek Member, deposited during a major expansion and freshening of ancient Lake Uinta. Compound specific isotopic analyses of β-carotene and phytane ($\delta^{13}C = -32.6$ to $-32.1‰$) from these bitumens reflect input from primary photosynthetic producers such as cyanobacteria. Sterane $\delta^{13}C$ values (-34.5 to $-29.2‰$) reflect contributions from lacustrine algae, while extremely depleted $\delta^{13}C$ values for methylhopanes (-58.1 to $-61.5‰$) suggest input from methanotrophic bacteria/archaea. Variations in the $\delta^{13}C$ values of the α−β-hopanes (-51.4 to $37.7‰$) imply additional input from other bacterial sources. The wurtzilite bitumen generated from the saline facies of the Green River Formation was deposited during a later regression of Lake Uinta. Compound specific isotopic analyses of phytane ($\delta^{13}C = -30.1‰$) and steranes ($\delta^{13}C = -29.6$ to $-26.7‰$) from this bitumen indicate continued input from primary producers and eukaryotes. The higher relative concentrations of gammacerane ($\delta^{13}C = -26.9‰$) indicate increasing input from aerobic species. Slight enrichment in $\delta^{13}C$ in the wurtzilite extract (and several biomarkers) suggests sulphate-reducing bacteria outcompeted methanogens, thereby, eliminating the influence of methanotrophs in this later saline stage of deposition.	Ruble *et al.*, 1994 Katz, 1995

Formation/location	Type	Description	References
Kongdian Formation (Ek) and Es4 member of the Shahejie Formation Zhaolanzhuang field, Jizhong depression, Bohai Bay Basin, China Eocene–Oligocene	3 Continental half-graben Palaeogene rift with Neogene subsidence	Natural gas contains the highest proportions of H$_2$S (40–92%) among the sour gases encountered in China. The sedimentary sequence consists of halite, anhydrite, carbonate, sandstone and shale interbeds deposited in the evaporative brackish water lacustrine- salt lake setting. In the deepest part of the Jinxian sag, the total thickness of evaporites is more than 1000 m, of which halite accounts for over 40%. Various organic-rich mudstones intercalated with the evaporites are currently within the conventional hydrocarbon window (with a depth of 2500–3500 m), and likely the source for the oil and sour gas in the Zhaolanzhuang field. The temperatures of the gas reservoirs range from 75 to 100 °C, too low for significant thermochemical sulphate reduction. The co-occurrence of abundant elemental sulphur with the sour gas and the δ^{34}S values of the various sulphur-containing compounds indicate that the H2S gases were most likely derived from much deeper source kitchens where significant thermochemical sulphate reduction has occurred.	Chen *et al.*, 1996 Zhang *et al.*, 2005
Bucomazi Formation, Cabinda, offshore Angola Cretaceous	3 Rift basin continental to transitional marine	The Lower Congo hydrocarbon habitat is dominated by the Pre-Salt Bucomazi petroleum system. These lacustrine, often super-rich, laminated evaporitic sediments reveal considerable organofacies variations between their early lacustrine basin fill and later sheet drape development with three major depositional regimes reflecting salinity and water depth control. ^{13}C depleted basal sediments, showing strong gammacerane/ 4-methyl sterane signatures, were segregated as seen by a major isotopic excursion. This event represents the transition from an early saline playa lake to a deeper water salinity-stratified mesohaline lake supporting a high level of bottom anoxia. For the intermediate sediments, sterane and tricyclic diterpane abundances, plus sterane/hopane ratios, had marine connotations and can be interpreted as the result of intermittent marine incursions. Foreshadowing irreversible oceanic ingression, a resurgence of gammacerane abundance in the uppermost sediments typified littoral-shallow marine depositional conditions.	Burwood *et al.*, 1995, 1992 Burwood & Mycke, 1996 Schoellkopf & Patterson, 2000
Lagoa Feia Formation, Campos Basin, Brazil Lower Cretaceous	3 Rift basin continental to transitional marine	Rift-stage lacustrine sediments are the source rock of all petroleum so far discovered in the Campos Basin, the most prolific oil province in Brazil. These organic-rich shales were deposited in anoxic brackish to saline lakes. Petroleum migration pathways involve direct contact between source and reservoir rocks within the rift sequence. Later migration to the marine sequence reservoirs is related to windows in the halokinetic salt layer connected to growth faults and unconformities.	Trindade *et al.*, 1995 Mello & Maxwell, 1991

(*continued*)

Table 5. (*Continued*)

Formation/Group locality and age	Basin Style	Characteristics and depositional setting	Reference
Muribeca and Maceió Formations, Sergipe Basin, Brazil Lower Cretaceous	3 Rift basin continental to transitional marine	Most hydrocarbon accumulations discovered in the Sergipe subbasin are sourced by the protomarine Aptian marls and calcareous shales of the Muribeca and Maceió formations both below and above the Aptian salt (Ibura Member). These source rocks average about 6% TOC and are composed mainly of hydrogen-rich type II kerogen.	Mello et al., 1988
Kimmeridgian Shale, North Sea Upper Jurassic	1b Sediment-starved deep marine basin centre generated by ongoing Permian-Mesozoic rifting and surrounded by Purbeck evaporite-carbonate platform	Seawater flowed southwards from Boreal Ocean into Tethyan Ocean but evaporation in the shallow waters of the archipelago of islands and shoals created waters of increased salinity and in some cases local strong brines and evaporites, which sank and flowed as saline bottom currents (from 34–42‰) into deeper water. Areas of ponded deeper mesohaline water were commonplace in the more rapidly subsiding grabens, separated by sills, denser waters accumulated until they were able to spill over the sills. In this way more saline waters migrated along the seabottom depressions, their paths being determined by local tectonic features, but mostly in a north to south direction. A stable water stratification ensued, the halocline being at about 50 m water depth and periods of widespread increased salinity are marked by "hot" shales.	Cooper et al., 1995 Miller, 1990
Purbeckian lacustrine beds, Dorset coast UK, Lower Cretaceous	3 Continental lagoons and lakes in sea-margin position at edge of carbonate platform	Within these beds, which immediately overlie the Purbeck evaporites, a large OM accumulation is recorded, with total organic carbon (TOC) of up to 8.5 wt%. High hydrogen index (HI) values (up to 956 mgHC g^{-1} TOC) point to a Type I OM, generally considered as derived from algal-bacterial biomass. This contrasts with the OM present in the underlying and overlying intervals, displaying in general lower TOC and HI values, and consisting of degraded algal-bacterial material with higher proportions of terrestrial OM. This organic-rich accumulation can be interpreted as a period of enhanced primary productivity within coastal lagoonal/lacustrine settings at times of low sea level and has strong affinities to modern Coorong Lakes in Australia.	Schnyder et al., 2009
Shelf margin laminites, Arabian Gulf from NW Iraq through the Arabian Gulf to central Oman (e.g. Diyab, Tuwaiq and Hanifa Formations) Upper Jurassic	2 Density-stratified intrashelf basins surrounded by evaporitic platform. Some lows are possibly tied to salt withdrawal	Organic-rich laminated mudstones and wackestones deposited in pre-salt intrashelf, mesohaline basins on a gently downwarped epeiric platform. Intrashelf basins that retain organics all have a density and salinity stratified water column with mesohaline bottom carbonates. These basins source more than 90% of the hydrocarbons in the Middle East. They are argillaceous dolomitic limestones that feed the most prolific petroleum system in the world and underlie the world's richest reservoir: the Arab D in the Arab Formation. One of the most prolific source units is the Hanifa Fm. in the Arabian Basin of Saudi Arabia. In the deeper parts of the intrashelf Arabian Basin (waters ~30 m deep) the Hanifa is composed of laminite muds and marls, while	Ayres et al., 1982 Palacas, 1984 Evans & Kirkland, 1988 Droste, 1990 Beydoun, 1993 Carrigan et al., 1995 Whittle & Alsharhan, 1996 Ibrahim et al., 2002

Location / Formation / Age	No.	Basin type	Description	References
(continued)			in the surrounding shelf it is dominated by grainstones and packstones. Within the Hanifa laminites there are discrete but minor primary laminae of anhydrite (originally gypsum), as well as early diagenetic anhydrite nodules, both indicators of occasional hypersalinity. There are also laminae of fibrous calcite, which some authors have interpreted as another indicator of hypersalinity. The Hanifa is a widespread and prolific source, it can entrain more than 30 metres of laminites; total organic carbon exceeds 1%, with some sections having more than 5%. The kerogen is predominantly hydrogen rich, the generated oils have a high sulphur content, a Pr/Py < 1, and a predominance of even-numbered normal alkanes.	Oehler, 1984
Smackover trend of Mississippi, Alabama and Florida, northeastern Gulf of Mexico — Upper Jurassic	2	Shallow restricted-circulation salinity stratified intrashelf basin on a broad carbonate platform	Oil and gas in this trend are generated from algal-rich light-brown to black laminated lime mudstones (argillaceous content <6%) of the Lower Smackover Formation. Deposition occurred in intraplatform depressions surrounded by a broad carbonate platform. Bottom waters were stratified with slightly saline waters supplied from the broad platform surrounds and the shoaling carbonate system was sealed by the sulfate evaporites of the Buckner Anhydrite.	Sassen, 1990 Mancini *et al.*, 2003
Todilto Formation — Todilto Basin, USA — Middle Jurassic	3	Foreland flexural basin (lacustrine?)	The Limestone Member of the Todilto Fm. is a widespread carbonate laminite up to 3 m thick, overlain in the subsurface by up to 30 m of $CaSO_4$ known as the Gypsum Member, although it is mostly anhydrite. The Limestone Member is the source rock for 6 known hydrocarbon accumulations in the sands of the underlying Entrada Fm, while it and the overlying $CaSO_4$ are the seal. The quantity of TOC is low (~1%) for such a thin source rock and it has some unusual characteristics compared to most evaporitic source rocks. Laminite couplets are 0.15 mm thick yet are laterally continuous for up to 3.7 km and form widespread cohesive "sheets", which invariably contain remnants of vascular plants. The Todilto is probably lacustrine with a marine seep hydrology, similar in hydrology to the coastal salinas of southern Australia. The organics reflect the growth of a healthy conifer cover in its surrounds and its waters were density-stratified lacustrine, probably fed via a marine seepage inflow and periodic seasonal runoff. This maintained a density-stratified brine column, first at carbonate saturation during deposition of the Limestone Member, then at gypsum saturation during the deposition of the Gypsum Member.	Vincelette & Chittum, 1981 Evans & Kirkland, 1988
Organic mudstones intercalated with evaporites, Mandawa Basin, Tanzania — Triassic	3	Rift basin continental to transitional marine	Total organic carbon (TOC) values range from 1.23 to 7.41 wt% of kerogen Type II/III. The evaporite (hypersaline) influence is indicated by the presence of C-24-tetracyclic terpanes, gammacerane and C-35-homohopanes. The relative abundance of tricyclic terpanes (C-19-C-28 +) with respect to pentacyclic terpanes (C-27-C-35,) seems to change with depth, whereby	Kagya, 1996

(continued)

Table 5. (*Continued*)

Formation/Group locality and age	Basin Style	Characteristics and depositional setting	Reference
		the tricyclic/hopanes ratio apparently increases with depth. The mixed organic input from marine and terrestrial precursors are indicated by mixed abundances of C-27-, C-28-, C-29- steranes.	Fowler *et al.*, 1993
Sub salt-allochthon oil seeps, Windsor Group, Nova Scotia, Canada Carboniferous	3 Rift basin continental to transitional marine	The oil seeps are either associated with upper Horton Group (Ainslie Formation) or basal Windsor Group (Macumber Formation) sediments. The biomarker distributions of the samples are similar to Stoney Creek oils and their lacustrine carbonate source rock (Albert Shale) of the Moncton Sub-basin, New Brunswick, as are oils from seeps in the Pugwash Salt mine.	
Black shales in the Paradox Member of the Hermosa Formation, Paradox Basin, USA Carboniferous (Pennsylvanian)	1a Foreland flexural basin	Organic-rich black shales (informally known as the Cane Creek, Chimney Rock, Gothic shales) form a carbonate laminite basal sequence to a number of salting-upward evaporite cycles in the Paradox Member. Mineralogically the shales are between 30–50% calcite or dolomite with clays and quartz sand forming the remainder of the matrix. TOC values of 5% of hydrogen-prone organics in the black shales are usual, with values of 10% not uncommon and the highest values more than 20%. The organics are mixtures of halotolerant debris and marine-style organics formed when surface waters were at marine salinities, as well as occasional terrestrial organics washed into the basin from the surrounding hinterland.	Evans & Kirkland, 1988 Hite *et al.*, 1984
Oil Shales, Junggar Basin, NW China Upper Permian	3 Controversial, with tectonic interpretations ranging from foreland flexure to regional transtensional basin	Junggar Basin is one of the largest oil-producing basins in China, its Upper Permian oil shales are among the thickest and richest lacustrine source rocks in the world. Together the Jingjingzigou, Lucaogou, and Hongyanchi Formations of the southern Junggar Basin comprise over 1000 m of organic-rich lacustrine facies. They record an evolution from relatively shallow, evaporative lakes to freshwater lakes with fluvial systems. Jingjingzigou Formation was deposited in a perennial saline lake characterized by low TOC and HI, and biomarker features (such as abundant β-carotane) consistent with a specialized saline or hypersaline biota. Biomarker distributions in Jingjingzigou Formation extracts most closely resemble oils from the giant Karamay oilfield. Overlying Lucaogou Formation represents one of the richest and thickest lacustrine source rock intervals in the world, yet it contradicts conventional lacustrine source rock models in at least two important aspects. First, deposition occurred at middle palaeolatitudes (39–43°N) rather than in the tropics. Second, limited nutrient supply in a drainage basin dominated by intermediate volcanic rocks appears to have caused low to moderate primary productivities. Stable salinity stratification and low inorganic sedimentation rates in a deep lake nonetheless resulted in deposits with up to 20% TOC and HI near 800.	Carroll, 1998 Carroll *et al.*, 1992 Tang *et al.*, 1997

Formation	Basin type	Description	References
		Overlying Hongyanchi Formation has 1–5% TOC but low HI, and was deposited in freshwater oxic to sub-oxic lakes.	
Ravnefjeld Formation, East Greenland Upper Permian	3 Rift basin continental to transitional marine	The Ravnefjeld Formation is subdivided into five units that can be traced throughout the Upper Permian depositional basin. Two of the units are laminated and organic rich, and were deposited under anoxic conditions. They are considered good to excellent source rocks for liquid hydrocarbons with initial average TOC (total organic carbon) values between 4 and 5% and HI (hydrogen index) between 300 and 400. The cumulative source rock thickness is between 15 and 20 m. The source rocks are separated and enclosed by three units of bioturbated siltstone with a TOC of less than 0.5% and an HI of less than 100. These siltstones were deposited under relatively oxic conditions. The organic geochemistry of the source rocks is typical for marine source rocks with some features normally associated with carbonate/evaporite rift environments [low Pr/Ph (pristane/phytane), low CPI (carbon preference index), distribution of tricyclic and pentacyclic terpanes]. The establishment of anoxic conditions and subsequent source rock deposition was controlled by eustatic sea level changes.	Christiansen et al., 1993
Phosphoria Formation, Little Sheep Creek, Montana Permian	1b Foreland flexure basin	Correlations between biomarker indicators of anoxia and salinity suggest that anoxia was in part the result of a chemocline separating normal marine waters above from more saline bottom waters. Anoxia and salinity in the bottom waters increased with time making conditions in the basin progressively more hostile to benthic organisms.	Dahl et al., 1993
Kupferschiefer Formation, NW Europe Permian	1a Rift basin, restricted marine	Thin, widespread organic-rich laminated mudstone deposited in shallow mesohaline marginal sea-lakes with waters which were <100m deep and mostly 10–30 m deep. Surface exchange with Zechstein ocean was restricted by palaeohighs.	Bechtel & Puttmann, 1997
Anhydritic lacustrine beds, East Shetland Platform, North Sea Middle Devonian	3 Rift basin, restricted marine	The UKCS well 9/16-3 drilled on the western flank of the Beryl Embayment indicates local development of hypersaline environments equivalent to the Achanarras/Sandwick fish bed. Shows close affinity with contemporaneous Middle Devonian source rocks of the Inner Moray Firth and crude oil in Beatrice oil field.	Duncan & Buxton, 1995
Muskeg Formation and Lower Keg River Member, Alberta, Canada Devonian	1a Rift basin, restricted marine	Laminated organic-rich bituminous mudstones deposited in pre-evaporitic density stratified water columns. Vertical migration from the Muskeg formation to the Muskeg Reservoirs may have occurred through local fracturing of anhydrites driven by the dissolution of Black Creek Halite.	Clark & Philp, 1989 Stasiuk, 1994; Chen et al., 2005
Laminated evaporitic A-1 mudstones, Michigan Basin Silurian	1a Rift basin, restricted marine	Biomarker characteristics indicate a carbonate/evaporite source rock deposited under hypersaline conditions in a strongly reducing environment. The source rocks occur in the basin centre in organic-rich laminites of the inter-reef A-1 Salina	Gardner & Bray, 1984 Obermajer et al., 1998, 2000

(continued)

Table 5. (*Continued*)

Formation/Group locality and age	Basin Style	Characteristics and depositional setting	Reference
		Formation, a mesohaline carbonate deposited beneath a density stratified brine column.	Bradshaw, 1988
Chandler Formation, Amadeus Basin, Australia Lower Cambrian	1a Foreland flexure, Restricted marine	Thin bituminous pre-evaporitic carbonate mudstone deposited subaqueously in a restricted basin immediately prior to the deposition of the thick halites of the Chandler Formation.	
Hormuz Series, Arabian Gulf and counterparts in Oman, India and Pakistan Neoproterozoic to Early Cambrian	1 Foreland flexure, restricted marine	Almost all the Persian Gulf and large areas of southern Iran and northeastern Arabia are underlain by a thick sequence of sediments, known as the Hormuz Series, or the Huqf Group in Oman. It is made up of interbedded salt, anhydrite, dolomite, shale and sandstone. It is not only the cause of many salt-dome-related oil and gas fields but is also considered to have been a major source rock for hydrocarbons in Ordovician, Devonian, Carboniferous, Permian and perhaps younger reservoirs. These oils have characteristic biomarkers and highly depleted carbon isotopes signatures indicative of their prokaryotic precursor.	Amthor, 2000 Edgell, 1991 Grantham *et al.*, 1988 Peters *et al.*, 1995 Terken & Frewin, 2000 Schoenherr *et al.*, 2007, 2009

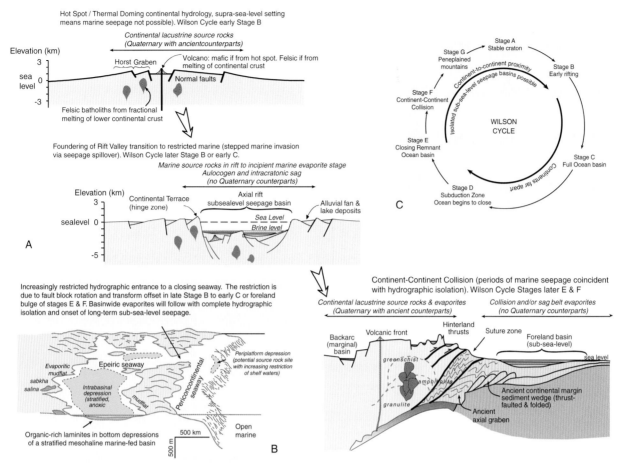

Fig. 19. Tectonic settings of ancient evaporate source rocks. (A) Ancient evaporite source rocks and regions with Quaternary counterparts (in part after Warren, 2006). (B) Styles of intrashelf and platform depressions where anoxic mesohaline bottom brines tended to accumulate in an ancient evaporitic platform. (C) Characteristics of the Wilson Cycle.

rifting ("zip then split") tectonics encourage the accumulation of substantial volumes of salts and organic-rich evaporitic sediments in sub-sea-level tectonic depressions (Fig. 19). In the past 800 Myr there have been two such tectonic cycles defined by supercontinent accretion and disaggregation (i.e. the accretion and disaggregation of Phanerozoic Pangea and Neoproterozoic Rodinia). Marine-fed evaporitic source rocks tend to form at times when the tectonic depression is undergoing hydrographic isolation from the open ocean. The basin still has a surface (hydrographic) connection to the ocean, but inflow is restricted and the basin is typically surrounded by widespread salty platforms with sheets of salterns and evaporitic mudflats (Fig. 19). When the surface connection to the open ocean is completely cut-off, the basin becomes a marine-fed sub-sea-level seepage depression and widespread thick salts precipitate

in the lower parts of the basin floor in waters ranging from hundreds of metres deep to ephemeral.

Ancient mesohaline source rocks

Worldwide, studies of ancient evaporitic basins have shown that organic-rich mesohaline sediments accumulate in ephemeral surface brines, saltern sediments, or basin and slope settings in both marine and continental regimes (Evans & Kirkland, 1981; Oehler, 1984; Warren, 1986; Busson, 1988; Kirkland & Evans, 1988; Rouchy 1988). The most prolific accumulations of organics in ancient evaporitic sediments tend to have been deposited as laminated micritic carbonates that accumulated beneath density-stratified moderately-saline (mesohaline) anoxic water columns of varying brine depth. There are three, possibly

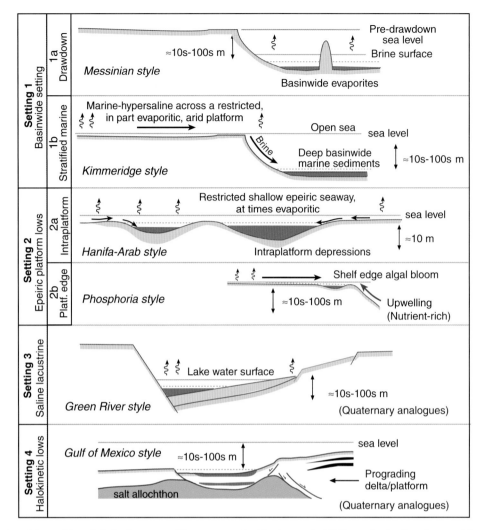

Fig. 20. Dominant depositional styles for evaporitic source rocks (dark green). All the water bodies show evidence of salinity-related density stratification (indicated by dashed line); this is also reflected in an associated heliothermic stratification (after Warren, 2006).

four, major mesohaline density-stratified settings where organic-rich laminites (source rocks) accumulate in saline settings that are also associated with, or evolve into, evaporite deposits (Fig. 20; Table 5). These are:

(1) Basin-centre lows in marine-fed evaporitic drawdown basins:
    (a) evaporite-plugged mesohaline basin centre;
    (b) evaporite platform rim surrounding restricted and stratified slightly-mesohaline carbonate basin centre.
(2) Mesohaline intrashelf lows on top of epeiric evaporitic platforms;
    (a) intraplatform depressions with layered mesohaline brines;

    (b) shelf-edge depressions with layered mesohaline brines.
(3) Saline-bottomed lows in perennial underfilled saline lacustrine basins.
(4) Closed seafloor depressions in halokinetic deep-water marine slope and rise terrains.

Basin Setting 1

There are two depositional styles for basin-centre or basinwide mesohaline source rocks: (a) basin-centre drawdown laminites intercalated and often sealed by basinwide salts; (b) basin-centre laminites where the basin margin is a shallow, restricted, at times evaporitic platform that surrounds a slightly mesohaline deep-water starved-basin centre. Bottom waters in this deeper part of the basin were never drawndown or concentrated to salinities where basinwide salts precipitated. Matrices in Setting 1b basin-centre laminites range from limestones through marls to siliciclastics (Fig. 20).

Basin Setting (1a) In this setting the basin is largely isolated from a surface marine connection and high levels of organics accumulate as "black shales" (typically carbonate laminites), which are intercalated with, and overlain by, basinwide evaporite salts (typically gypsum/halite). These laminites accumulate on the bottom of mesohaline density-stratified pools in topographically closed parts of the basin floor during the early stages of isolation and drawdown. Inactive, typically subaerial, shelf margin reefs and pinnacles typically lay updip or encircled the basin-centre lows (Warren, 2006). Subsequent encasement of these platform buildups and other potential reservoirs by "fill and spill" evaporites means these basins easily become highly focused hydrocarbon flow systems. Hydrocarbons accumulate in the reefs or shelf margin shoals beneath basinwide evaporite seals. Such systems typify the marine-fed Silurian reef/salina salt association in the Michigan Basin, USA (Gardner & Bray, 1984), the Devonian Rainbow Reef reservoirs of the Black Creek/Muskeg Basin of Alberta (Clark & Philp, 1989), and the Messinian marls of the Lorca Basin, Spain (Russell *et al.*, 1997).

Expulsion efficiencies suffer from encasement and widespread salt plugging of the potential source rocks in many evaporite-rich Setting 1a basins. This style of deposit tends to retain volatiles in the source beds until the whole system becomes overmature. For example, in the Gibson Dome No. 1 well in the Paradox Basin, almost all of the black shales in the Upper Carboniferous Paradox Member are mature enough to have generated oil (Hite *et al.*, 1984), but the hydrocarbons have been largely retained in this mature source rock, reflecting the highly efficient sealing effect of the intercalated evaporite beds and salt plugs (mostly halite). Only where the black shales are intercalated with porous carbonates have the hydrocarbons escaped. The source potential of these shales is very high; the "Gothic shale" member is capable of generating almost 5000 barrels per acre. But that potential is never more than partially realized. Paradox "shales" are responsible for an accumulative production of a little more than 400 million barrels of oil and about 1 TCF of gas in the Paradox Basin in fields such as Ismay and Aneth (Peterson & Hite, 1969).

The problem of evaporite plugging plagues many basinal source rocks where the basin centre evolved to accumulate basinwide evaporates, and is one of the main reasons for the dearth of large onshore oilfields beneath bedded salt or within intrasalt carbonates in the Zechstein Basin of NW Europe. Organic-rich source intervals such as the Stinkdolomit, Stink-kalk, and Stinkscheifer are widespread, but the associated oil accumulations are never more than localized and are small scale. This setting also created the salt-encased self-sourced low-permeability reservoirs in the Cambrian Ara Group platform carbonate stringers and basin centre silicilytes of the South Oman Salt Basin (Al-Siyabi, 2005; Amthor *et al.*, 2005; Schoenherr *et al.*, 2007, 2009). Very rich organic source beds can characterize Setting 1a sediments, but the intercalated seals prevent most of the volatiles from ever escaping into a reservoir. Sometimes, as in the silicilytes of Oman, the source beds can become self-reservoiring. In later burial diagenesis the trapped hydrocarbons can play an important role as a reductant and metal fixer as ongoing burial dissolution of salt interbeds focuses long-term contact with basinal waters carrying Cu, Pb and Zn (Warren, 2000).

Basin Setting 1b Basin-centre mesohaline accumulations in this setting are much more efficient petroleum producers than Setting 1a as they lack intercalated beds of basinwide gypsum or halite, which otherwise create salt-plugged and salt-cemented source intervals. Setting 1b source rocks develop where dense mesohaline bottom-waters pond on the bottom of a deep density-stratified restricted-marine basins. Slightly elevated bottom-brine salinities are maintained by a series of dense brine underflows, fed from nearby shoalwater evaporitic platforms (Fig. 20). Such basins are typified by a deep-water "marine" centre and the presence of widespread evaporitic and shoal water carbonate sediments in the surrounding platforms and shelves. Setting 1b basins are best thought of as stunted or underdeveloped basinwide evaporite systems. Development of the basinwide evaporite hydrology was suspended prior to the basin centre drawdown reaching the state of sea-surface (hydrographic) isolation, which is needed for widespread salt precipitation across the basin centre. Nonetheless, evaporitic conditions dominated in isolated platform depressions about the basin margin. Typical examples of this style of association are the Upper Jurassic Kimmeridge shales of the North Sea (Miller, 1990; Cooper *et al.*, 1995), the organic-rich laminites of the Cherry Canyon Formation in the Permian Delaware Basin of West Texas (Jones, 1984), and possibly the Jurassic Lower Smackover Formation in the Proto-Gulf of Mexico (Sassen, 1990).

In Setting 1b, the deep-water basin centre remains "sediment starved" and density-stratified, with the upper less-saline part of the water column typified by normal marine water, maintained through surface connections to the open ocean. The little sediment matrix that does accumulate in the basin centre settles into an anoxic thermally/density stratified bottom, so that the sediment matrix of the source laminites is dominated by a planktonic or nektonic biota and a more or less constant elevated organic content. Carbonate banks or reefs in the surrounding platform grade landward through lagoonal and evaporitic platform facies into terrestrial siliciclastics. At times of slightly lowered sea-level these epeiric platforms can go evaporitic and so deposit widespread platform salts (Warren, 2006). Such basins show a centrifugal salinity gradient where the highest salinity areas are the shallow evaporite platforms, which supply dense saline waters that trickle and seep basinward to pond in seafloor lows at the anoxic base of a density-stratified water column.

With Setting 1b, the formation of evaporites and saline waters about the basin edge is not typically followed by the deposition of basin-centre evaporites. At the time the deep-water organic laminites are accumulating, the deep basin-centre waters lie at the bottom of the thick density-stratified water column in a semi-enclosed seaway, and bottom brines may be no more than 3–5‰ more saline than the normal marine waters. For example, at the time the Upper Jurassic Purbeck evaporites were accumulating on the platforms of northern Europe, the anoxic bottom waters in the marine basin centre where the Kimmeridge Clay was accumulating had salinities around 42‰, temperatures ~30°C and densities ~1.027. The overlying near-normal marine waters had salinities of 34–39‰, surface temperatures of 26°C in the south and as low as 6°C farther north in the region of the Boreal Ocean inflow, and densities around 1.0267 (Miller, 1990). As the Kimmeridge Clay source rocks accumulated, halokinesis and salt solution of Zechstein mother salt probably also played a role in generating local briny anoxic bottoms on the deeper parts of the basin floor (Clark *et al.*, 1999).

Basin Setting 2

Upper Jurassic source rocks of the Middle East are well-studied examples of the epeiric shelf or marine evaporitic platform settings and typify Setting 2a source rocks, which are deposited within local somewhat deeper, density-stratified mesohaline intrashelf lows (Figs 19, 20 and 21; Table 5; Ayres *et al.*, 1982; Evans & Kirkland, 1988; Droste, 1990; Warren, 2006). In the Tethyan of the Middle East this style of source rock feeds a petroleum system that is probably the most efficient in the world. It evolved from Upper Jurassic source rock accumulators (typically with TOC's ~1%, occasionally up to 13%) into evaporitic mudflats sealing the mostly bioclastic carbonate reservoirs of the Arab Formation cycles, into the regional Hith Anhydrite seal, which was deposited on a saltern-covered platform.

The laminite source rocks in the various intrashelf depressions were deposited at times of widespread epeiric carbonate deposition and stagnant oceanic circulation. It was a time when a warm shallow greenhouse sea covered much of the Arabian platform and was precipitating carbonates and evaporites over much of its extent. Denser warmer waters, generated by evaporation of these epeiric waters, along with waters supplied by the surrounding evaporitic mudflats, were concentrated to where they were somewhat more saline than normal seawater. These dense brines then trickled and seeped over the shallow seafloor into the lower parts of restricted intrashelf basins to create density-stratified bottom waters that were not more than tens of metres deep and no more than 10–30‰ more saline than seawater (Ayres *et al.*, 1982). Ongoing mixing and dilution of the bottom-hugging brines during their passage to the lows meant that the basal waters on the intrashelf lows, although saline, rarely attained gypsum saturation (mostly mesohaline anoxic brines). The overlying column was typically at near-normal marine salinities. Salt withdrawal and dissolution of salt diapirs and allochthons, sourced in the Precambrian Hormuz salt, perhaps helped create the intrashelf depressions and may also have played a role in generating a portion of the bottom brines.

One of the most prolific intrashelf source rocks in the Middle East is the Upper Jurassic Hanifa

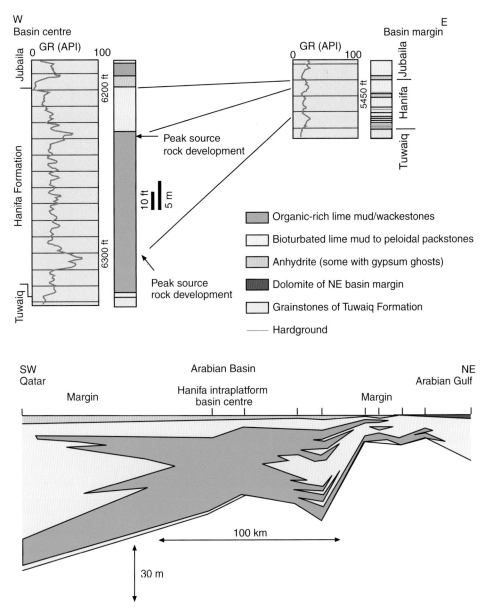

Fig. 21. Correlation and lithofacies of the Upper Jurassic Hanifa Formation, Saudi Arabia. This Setting 2 intraplatform basin source rock was deposited beneath relatively shallow and anoxic mesohaline bottom waters in a slight depression on the shelf and was surrounded by epeiric evaporitic sedimentation, as indicated by the presence of anhydrite in the Hanifa Formation (after Droste, 1990). Peaks in the organic content of these carbonate mudstones are indicated by peaks in the gamma curve. Cross section redrafted from Carrigan *et al.* (1995).

Formation deposited in the intrashelf Arabian Basin of Saudi Arabia (Fig. 21; Droste, 1990; Carrigan *et al.*, 1995). It sources most of the hydrocarbons reservoired in the Arab cycles of Saudi Arabia and is regionally sealed by the Hith Anhydrite. The organic-rich portions of the Hanifa and its equivalents elsewhere in the Middle East were deposited in the "deeper" parts of a number of Jurassic intrashelf basins (waters ~30 m deep). Sediments accumulating beneath the slightly

mesohaline layered bottom waters are composed of organic-rich laminite muds and marls, while sediments of the surrounding Arabian shelf are dominated by grainstones and packstones. As the various intrashelf depressions filled with laminites they shallowed to where the stratification was lost and the sediment became an open-marine packstone-wackestone equivalent to sediments deposited over the rest of the Arabian Platform.

Successive platform isolation, via buildups of a transgressive platform rim and a slight fall in sea level, then drove deposition of anhydritic salt-ern or mudflat caps of the various Arab cycles (Cycles A–D; Warren, 2006). Within the Hanifa laminites, there are discrete laminae and thin beds of primary anhydrite, as well as early diagenetic anhydrite nodules, both indicators of periodic hypersalinity even as the laminites were accumulating. Likewise, the evaporitic cap to each of the source cycles in the Hanifa Formation in the Arabian Basin preserves aligned gypsum ghosts indicating a subaqueous gypsum precursor. Similar, but volumetrically less prolific, intrashelf source rocks characterize the Lower Smackover Formation (Upper Jurassic) of the northeastern Gulf of Mexico (Oehler, 1984; Sassen, 1990) and the Cretaceous Sunniland Formation of Florida (Palacas *et al.*, 1984).

A variation on this density-stratified mesohaline platform-depression style of source rock accumulation is the phosphate and organic-rich Setting 2b source rocks sediment of the Phosphoria Formation in the USA (Stephens & Carroll, 1999). This unit sources much of the oil in Wyoming and was deposited in somewhat-saline shelf-edge depressions of the Phosphoria Sea adjacent to an area of marine upwelling. The nutrient-rich waters facilitated algal blooms near and above the shelf edge, while bottom-hugging brines, seeping seaward from the evaporitic hinterland, ponded in stagnant shelf-edge depressions. Overlap of the two systems created local anoxic bottom-waters in the shelf-edge depressions that preserved the organics fed by the detritus of the overlying algal blooms (Fig. 20).

Basin Setting 3

These source rocks accumulated in sediment-starved or underfilled saline lakes in arid climates. The laminites typically accumulated in the early continental stage of infill in an opening rift, prior to its opening into a marine-fed marginal evaporite basin (Figs 19 and 20), or in the restricted lacustrine stage of a foreland flexure in a collision belt. Modern underfilled saline rift lakes have some of the highest rates of organic productivity known and can accumulate high levels of organics within carbonate laminites. Some of these lacustrine laminites accumulate beneath deep density-stratified water columns, while others accumulate as "moat facies" or microbial mudflats around lake margins (Fig. 20; Table 5).

The likelihood of numerous water level changes during the life of a saline lake especially affects laminites deposited in shoal-water moats and margin mudflats and so can lower their preservation potential (Bohacs *et al.*, 2000). Most oil-prone Setting 3 lacustrine source rocks were deposited at the bottom of perennial salinity-stratified hydrologies, with water column depths measured in metres to tens of metres, not as subaerially exposed algal mudflats. Examples include the Wilkins Peak Member of the Eocene Green River Formation, USA, the Lower Cretaceous Lagoa Feia Formation of the Alagoas portion of the Campos Basin in Brazil, and the Permian Jingjingzigou Formation in the Junggar Basin, China (Table 5).

Desiccation needed to form an underfilled lacustrine succession typically means that saline lacustrine phases have lesser aerial extents compared to units deposited at times of fresher water in the lake depression. This does not necessarily mean the evaporitic lacustrine source rocks are less prolific than their freshwater counterparts. Within the Green River Formation there are two associations (fresh and saline) of lacustrine organic laminites exemplified by the Luman Tongue and the Laney Member (Fig. 22; Horsfield *et al.*, 1994). The Luman Tongue consists of organic matter-rich mudstones deposited as profundal sediments at a time when lake waters were relatively fresh (Fig. 22A). Proximal organic matter-poor sediments, as well as coals and thin sandstones, were deposited on the lake plain. The highest levels of TOC in the Luman Tongue occur in lake-margin deposits, which contain a mixture of alginite and vitrinite, but they have a low hydrogen index and a low petroleum generation potential (Fig. 22B, C). In fact, the petroleum generation potential of these freshwater deposits is no more than moderate.

In contrast, the organic percentages of the carbonate laminites of the evaporitic Laney Shale are much higher (6–22%) and these are hydrogen prone (Fig. 22B, C). The most organic-rich samples were deposited in the lowest parts of the lake and are composed mostly of alginite with relatively minor contributions from higher plants. The much higher petroleum potential of the Laney Shale reflects the marked density stratification of Lake Gosiute waters at the time the Laney was accumulating. At that time bottom anoxia was stabilized by the perennial highly saline bottom waters (Boyer, 1982; Fischer & Roberts, 1991). The dissolution of underlying evaporites to create density-stratified lacustine waters also characterized the

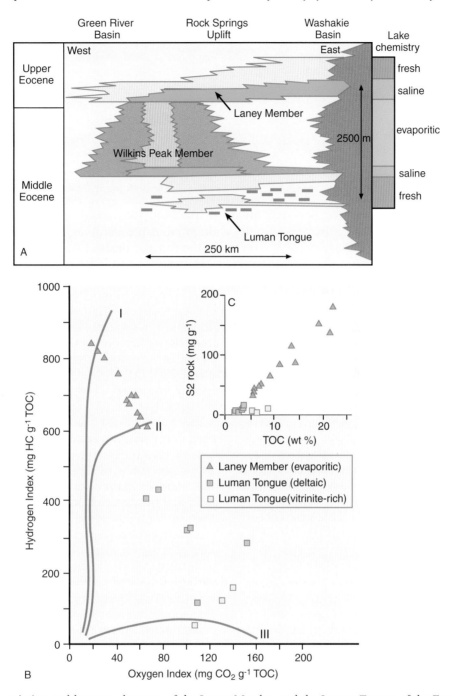

Fig. 22. Lateral variation and kerogen character of the Laney Member and the Luman Tongue of the Eocene Green River Formation, USA. (A) Schematic E–W cross section showing lateral extent and thickness of various lacustrine units (after Bohacs *et al.*, 2000). (B) Hydrogen and oxygen indices. (C) Pyrolysis yields of hydrocarbon-equivalents (mg g⁻¹ rock) plotted against organic carbon percentages. (B and C) after Horsfield *et al.* (1994).

Cretaceous Purbeck strata of Dorset, although in this case the underlying evaporites were marine and the depositional setting was akin to the sea-margin meteoric lakes of today's Coorong region of Australia (Schnyder *et al.*, 2009).

Basin Setting 4

This is an increasingly documented open-marine evaporite-associated source rock setting where density-stratified saline bottom brines and organic-enriched laminites accumulate in brine-bottomed depressions on top of and adjacent to shallow and dissolving salt allochthons (Fig. 20). With burial these laminites may act as source rocks, but unlike the restricted settings of the preceding three styles, deposition typically occurs in open-marine waters in continental slope and rise environments. The laminites accumulate on the floor of deep brine pools that form on or near the margins of actively flowing and dissolving shallow salt allochthons (Warren, 2006). This is also a significant setting for the formation of base metal-hosting laminites as in Tunisia and the Red Sea (Warren, 2000). As yet it is not a widely documented environment for hydrocarbon source rocks, but it may in-part explain the higher than expected source potential of many prodelta and slope muds in the passive margin fill of evaporite-floored halokinetic rifts.

Source rocks of settings 1a, 2 and 3 are typified by the accumulation of mm-scale layers of oil-prone organics typically in a matrix of fine carbonates. Setting 1b typically contains higher levels of siliciclastic fines and in many cases the host to the organics is a basinal marl or less often a laminated calcareous shale. The matrix of Setting 4 laminites tends to be composed of whatever detrital sediment is accumulating above the salt allochthon. In many situations, such as the Gulf of Mexico and the Mediterranean allochthon terranes, this matrix is a siliciclastic mud/clay laid down at the distal ends of prodelta wedges. In others, such as the Jurassic and Cretaceous of the Arabian Gulf and the Triassic of Tunisia, the matrix is carbonate.

CONCLUSION

It seems that there are no same-scale "famine or feast" analogues for modern and ancient saline lacustrine settings (Table 6, Setting 3). Organic-rich carbonate laminites forming in the African rift valley lakes have been documented and similar-scale counterparts exist in the Mesozoic source rocks found for example in the Cabinda Basin of Angola and the lacustrine basins of South America (Table 5). Likewise, Quaternary examples of supra-allochthon brine lakes with anoxic bottoms rich in organics are now being documented (Table 6, Setting 3).

However, there are no modern counterparts for most ancient large-scale marine evaporite depressions with mesohaline "famine or feast" laminites (Table 6, settings 1 and 2). These Pre-Quaternary evaporitic source rocks generated far greater hydrocarbon volumes than all lacustrine source rocks combined, yet many source rock models for evaporitic mesohaline carbonates still draw strong comparisons to underfilled Quaternary lacustrine settings. Such "strict" uniformitarian comparisons lead to the conclusion that evaporitic mesohaline source sources are probably not all that significant as contributors to the volume of exploitable hydrocarbons in the world.

It is the author's opinion that depositional settings in arid environments are still being considered in the same way that Lotze (1957) viewed the abundance of life in modern deserts. Organic accumulations in ancient evaporitic source rocks should be seen as biological responses to the "feast or famine" cycles that still characterize life in arid depositional settings. Formation of ancient marine-fed mesohaline depressions required varying combinations of greenhouse eustacy (epeiric seaways) and tectonics-climate, which are also needed to create huge sub-sea-level depressions filling with evaporites (epeiric platform evaporites, marine-seepage rifts, soft collision belts and far-field intracratonic sags). These systems do not have same-scale Quaternary counterparts (Warren, 2010). Yet, it was in salinity-layered intraplatform epeiric depressions and restricted tectonically induced mesohaline basin centres where the requisite schizohaline water columns and stable long-term brine bottoms occurred. They did so across time scales and across areas that not only allowed widespread periodic halotolerant blooms in the surface waters of whole mesohaline-floored seaways, but also formed in settings that facilitated long-term stable anoxic bottom brines. This anoxic brine-soaked bottom in-turn allowed widespread mesohaline source rocks to accumulate and to be preserved across the areally-extensive depressions (Figs 19 and 20; Table 5).

As for most ancient evaporite styles, the present is not a suitable time for studying scales of marine mesohaline source rock development. However, process and texture relevant to these ancient mesohaline source rocks can be seen in "famine or feast" cycles in small-scale Quaternary continental deposits. Rather than the words of Lotze (1957), a more apt aphorism to describe life in modern and ancient saline systems is found in the words of John Morley, a 19th Century Scottish statesman (1838–1923), agnostic, and the First

Table 6. Summary of evaporitic source rock associations

Setting	Other significant characteristics	Some ancient examples	Quaternary counterparts	Tectonic/eustatic associations
Setting 1. Basin-centre lows in marine-fed evaporitic drawdown basins, typically with evaporite platform rim	1a. Drawdown evaporite-sealed and plugged mesohaline basin centre carbonates	Cambrian Ara stringer and silicilytes, Oman; Silurian reef/Salina Salt, Michigan Basin, USA; Devonian Rainbow Fm of the Black Creek/Muskeg Basin, Canada; Permian Stinkdolomit Zechstein Basin, Europe; Carboniferous Gothic Shales, Paradox Basin, USA; Messinian marls, Lorca Basin, Spain	None	Rifts, collision belts and intracratonic sags with complete hydrographic isolation of marine-seepage basin centre
	1b. Density-stratified restricted-marine slightly-mesohaline basin centres with r.o immediately overlying evaporite seal	Permian Cherry Canyon Fm, Delaware Basin, USA: Jurassic Kimmeridge shales, North Sea; Jurassic Egret Member, Jeanne d'Arc Basin, offshore Canada; Cretaceous Sirte Shale, Sirte Basin, Libya	Partial	Often tied to early opening stages of rifts in semi arid climatic settings in greenhouse climate mode. Some aspects of Red Sea/Gulf of Suez are similar, but high aridity and icehouse eustacy do not allow a direct comparison
Setting 2. Mesohaline intrashelf lows within an epeiric evaporitic platform	2a. Intraplatform lows with adjacent epeiric mesohaline shoals	Hanifa Formation in the intrashelf Arabian Basin, Saudi Arabia and its equivalents throughout the Middle East; Lower Smackover Formation (Upper Jurassic) of the northeastern Gulf of Mexico; Cretaceous Sunniland Formation of Florida	None	Greenhouse eustacy creates continuous shelf- or ramp-edge rim, the dissolution of the crests of nearby active halokinetic structures may contribute brines to the seafloor lows
	2b. Platform edge low with epeiric mesohaline hinterland	1b: Phosphoria Fm, USA	None	Greenhouse eustacy creates continuous shelf- or ramp-edge rim
Setting 3. Mesohaline-upper column in perennial saline lacustrine basins.	Underfilled stages alternating with balanced fill stages in response to tectonic/climatic changes	Permian Jingjingzigou Formation in the Junggar Basin, China: Eocene Green River Formation, USA; Cretaceous Lagoa Feia Formation, Campos Basin in Brazil; Tertiary saline lacustrine basins, China	East African Rift Valley lakes; Lake Van, Turkey; Lake Urmia, Iran	Enhoheic basins in rift or continent-continent collision settings
Setting 4. Closed mesohaline-bottomed deep seafloor depressions in halokinetic belts	Brine seeps that create the stratified brine lakes are tied to ongoing salt dissolution and shallow subseafloor salt flows	Possible explanation for the enhanced source potential of marine deep-water shales in various circum-Atlantic halokinetic basins	Poorly documented stratified brine lakes in suprasalt allochthon provinces	Halokinetic provinces in the deep waters of the passive margin slope and rise of the Gulf of Mexico and the Mediterranean Ridge collision belt

See Fig. 20.

Viscount of Blackburn, who once said: "The great business of life is to be, to do, to do without, and to depart."

GLOSSARY

Aerobes are organisms that can survive and grow in an oxygenated environment.

Aliphatic hydrocarbons are any chemical compound belonging to the organic class in which the atoms are not linked together to form a ring. They are divided into three main groups according to the types of bonds they contain: alkanes, alkenes, and alkynes. Alkanes (n-alkanes) have only single bonds and a continuous chain structure, alkenes contain a carbon-carbon double bond, and alkynes contain a carbon-carbon triple bond. Aromatic hydrocarbons are classified as either arenes, which contain a benzene ring as a structural unit, or non-benzenoid aromatic hydrocarbons, which are characterized by stability in burial but lack a benzene ring as a structural unit.

Anaerobes are organisms that do not require oxygen for growth and may even die in its presence. Obligate (or strict) anaerobes are unable to live in even low oxygen concentrations. Facultative anaerobes are able to live in low or even normal oxygen concentration as well. All anaerobes are simple microorganisms such as bacteria, archaea and some fungi. Archaea are usually strict anaerobes.

Autotrophs (literally "self-feeders") are organisms capable of producing organic compounds from simple inorganic compounds.

Basinwide evaporites are one of the two major styles of ancient marine-fed salt accumulations known as: (1) marine-fed basinwides and (2) marine-fed platform evaporites. Neither setting is active anywhere on the world's current landsurface. Basinwide salt fills tend be thick (>100 m) and relatively pure with deposits accumulating rapidly across time frames of less than one million years. Basinwide evaporites are most likely to accumulate on the floors of isolated sub-sea-level depressions at times of close proximity of drifting landmasses (Warren 2006, 2008, 2010).

Catabolism is the breakdown in living organisms of more complex substances into simpler ones together with a release of energy.

Cellulolytic bacteria decompose cellulose; proteolytic bacteria breakdown proteins into simpler, soluble substances such as peptides and amino acids; saccharolytic bacteria breakdown sugars, while homoacetogens are obligately anaerobic bacteria that make acetate from either (H_2 + CO_2) or from the fermentation of sugars.

Chemoautotrophs use endogenous light-independent reactions to obtain energy, these reactions involve inorganic molecules and an electron donor other than water and do not release oxygen.

Cyanobacterial toxins. Intoxication of lesser flamingo flocks by cyanobacterial toxins, and sometimes even mass fatalities have occurred in nearby Lake Bogoria when the birds ingest detached cyanobacterial cells from cyanobacterial mats flourishing in the vicinity of hot springs (Krienitz *et al.*, 2003). This is because the flamingos feeding at night on plankton blooms in a saline lakes need to drink fresh or brackish water after feeding, and to wash their feathers daily. They tend to do this in waters in the vicinity of the hot springs, where salinities are lower than in the main waterbody of the lake where they have been feeding. Mycosystin heptatoxins can characterize the benthic microbial community in those hot spring mats and waters can be dominated by the potentially toxic cyanobacterial association of *Phormidium terebriformis, Oscillatoria willei, Spirulina subsalsa* and *Synechococcus bigranulatus.*

Denitrification is the microbial process by which nitrates are removed from an aqueous liquid.

Dissimilatory metabolic processes drive the conversion of food or other nutrients into products plus energy-containing compounds.

Dissimilatory sulphate reduction. Sulphate-reducing bacteria gain energy for cell synthesis and growth by coupling the oxidation of organic compounds or molecular hydrogen to the reduction of sulphate to sulphide (H_2S, HS^-). This process is called "dissimilatory sulphate reduction", to allow clear differentiation from assimilatory sulphate reduction. Assimilatory sulphate reduction generates reduced sulphur for biosynthesis (e.g. of cysteine). It is a widespread biochemical capacity in prokaryotes and plants and does not lead to the excretion of sulphide. Only upon decay (putrefaction) of the biomass is the assimilated reduced sulphur released as sulphide.

Endosymbionts are organisms that live within the body or cells of another organism without deleterious effects.

Extremophiles are organisms that thrive in extreme conditions including: extremely saline (halophile), extremely hot (thermophile), extremely dry (xerophile), extremely acid (acidophile), extremely alkaline (alkaliphile), and extremely cold (psychrophile) environments.

Fermentation is of a series of anaerobic transformation processes whereby organic matter is broken down into compounds of smaller size, which are more reduced or more oxidized and eventually more assimilatable by living matter. When fermentation leads to organic acids and a lowering of pH, it is called acidogenesis. The microorganisms responsible for this are called acidogens.

Gram-negative and *Gram-positive* are responses to the Gram staining technique whereby micro-organisms are first stained with crystal violet, then treated with an iodine solution, decolourized with alcohol, and counterstained with safranine. Gram-positive bacteria retain the violet stain; Gram-negative bacteria do not.

"Halobacteria". Prior to hybridization studies and genome mapping, classification was based on morphology and staining, the evolutionary relationships among various members of the bacterial and archaeal domains were contradictory and poorly defined. Today some of these contradictions and confusions remain in microbial taxonomy of the group of microbial organisms broadly known as halobacteria, as distinct from the archaeal group Halobacteriales.

Halophilic organisms thrive at very high salt concentrations and can be killed by lower salinities.

Halotolerant (aka halovariant) organisms can tolerate high salt concentrations but grow better at somewhat lower salinities.

Haloxene organisms cannot tolerate high concentrations of salts.

Heterotrophs (literally "feeders on others") use organic molecules synthesized outside their body as a source of energy and carbon.

Holomixis indicates the entire mixing of the water column. Meromixis indicates partial or incomplete mixing.

Hopanoids are group of compounds (triterpenoids) produced by prokaryotic organisms, and the diagenetic alteration products of these compounds (found in oils, rock extracts and sediment extracts). Just as steroids (steranes) are a useful group of biomarkers for identifying input from various eukaryotic organisms (e.g. plants and animals), an analogous group of compounds, hopanoids, are a useful group of biomarkers for identifying input from various bacteria. Hopanoids serve the same function in bacteria as sterols do in eukaryotes: they act as cell wall rigidifiers. In petroleum and their source rocks, hopanoid biomarkers exist as a subset of a group of compounds called triterpanes (isoprenoids).

Hypersaline encompasses all waters more saline than seawater (mesohaline, penesaline and supersaline).

Hypertrophication describes the presence of excessive nutrients in a water body.

Isoprenoids are a major class of nonsaponifiable lipids that occur in plants, animals, and bacteria and are characterized by chains of modular groups of five carbon atoms in which the typical pattern has four of the carbon atoms in a linear chain and a single carbon attached at the carbon one position removed from the end of the chain. The term isoprenoid is derived from the name of the five-carbon, doubly unsaturated branched hydrocarbon isoprene, which could in principle be the simplest monomeric chemical precursor for this class of compounds. Isoprenoids are also known as terpenes. Terpenes are usually grouped according to the number of isoprene (C_5H_8) units in the molecule· monoterpenes ($C_{10}H_{16}$) contain two such units; sesquiterpenes ($C_{15}H_{24}$), three; diterpenes ($C_{20}H_{32}$), four; triterpenes ($C_{30}H_{48}$), six; and tetraterpenes ($C_{40}H_{64}$), eight. The carotenoid pigments are the best known tetraterpenes.

Lithotrophs are organisms that use an inorganic substrate (usually of mineral origin) to obtain reducing equivalents for use in biosynthesis (e.g. carbon dioxide fixation) or energy conservation via aerobic or anaerobic respiration.

Meromixis indicates partial or incomplete mixing. The term *meromictic* is used to describe a permanently stratified water mass where surface layers may mix, but the bottom layer does not. Many modern brine lakes are meromictic, the upper water mass comes and goes.

Mesophiles are organisms that grow best in moderate temperatures, neither too hot nor too cold, typically between 15 and 40 °C.

Mixolimnion. An upper water mass that periodically mixes is the mixolimnion, it sits atop the monolimnion.

Monolimnion. A lower, permanent unmixed water mass.

Neutrophilic organisms are not stained strongly or definitely by either acid or basic dyes, but are stained readily by neutral dyes.

Oligomictic or *oligotrophic* is used to describe stratified water masses that mix or homogenize for short irregular periods every few years.

Photoautotrophs use light as a source of energy and CO_2 as a source of carbon. Term comes from *autotrophs* (literally "self-feeders"), which describes organisms capable of producing organic compounds from simple inorganic compounds.

Photophosphorylation is the production of ATP using the energy of sunlight.

Platform evaporites are the other major type of ancient marine-fed evaporite. They formed on continental margins throughout much more of Phanerozoic time than basinwides. Salt fills are 10–50 m thick, mostly $CaSO_4$ and typically interbedded with normal-restricted marine carbonates. Stacked platform sections characterised by this style of accumulation encompass time frames of ~1–10 Myr and are largely tied to times in earth history when climate was in greenhouse mode and the associated eustacy favoured widespread epicontinental seaways subject to periodic restriction (Warren, 2006, 2010).

Red waters and brines. The pink to purple colours that typify many hypersaline water bodies come mostly from concentrations of carotenoid pigments present in the cytoplasm of various halotolerant and halophilic microorganisms. Most haloarchaea are red due to a high content of C-50 carotenoids of the bacterioruberin series. Photosynthetic cyanobacteria and eukaryotes (e.g. unicellular green algae of the genus *Dunaliella*) contribute to the pigmentation of the hypersaline waters thanks to the presence of chlorophylls and C-40 carotenoids (mostly all-*trans*- and 9-*cis*-β-carotene). Chlorophylls absorb red and blue wavelengths much more strongly than they absorb green wavelengths, which is why chlorophyll-bearing cyanobacteria appear green (Fig. 2B). The carotenoids and phycobiliproteins, on the other hand, strongly absorb green wavelengths. Algae and microbes with large amounts of carotenoid appear yellow to brown (such as carotene-rich forms of *Dunaliella sp.*), those microbes with large amounts of phycocyanin appear blue, and those with large amounts of phycoerythrin appear red. Pigment levels can indicate the stratification of the microbial community in any photoresponsive biomass in a brine column. Red wavelengths (long wavelengths) are absorbed in the first few metres of a brine column or the uppermost millimetre or two of a microbial mat (where chlorophyll utilizers flourish). Blue and green wavelengths (shorter) reach deeper into the brine column or into the sediment.

Saprophytes obtain nourishment by absorbing dissolved organic material; especially from the products of organic breakdown and decay.

Schizohaline describes waters subject to substantial ongoing salinity changes and periodically pass from brackish to hypersaline.

Third domain of life. In the late 1970s, Professor Carl Woese proposed, on the basis of ribosomal RNA affiliations (gene mapping), that life be divided into three domains instead of two, namely; Eukaryota, Eubacteria, and Archaebacteria (Woese, 1993). He later decided that the term Archaebacteria was a misnomer, and shortened it to Archaea and Eubacteria to Bacteria. Since the 1970s, DNA base-pair studies (aka genomic studies or gene sequencing) have shown that Archaea are as different from Bacteria as from *Homo sapiens*. This new approach to taxonomy is still working its way through the scientific community and some books and articles still ocassionally refer to archaea as types of bacteria. Prior to genomic studies, and based on their morphology and staining response, the archaeal Halobacteriaceae were grouped with the bacterial Gram-negative rods (a gram positive or gram negative description indicates whether or not the bacterial cell wall reacts with Gram's stain).

REFERENCES

Aharon, P., Kolodny, Y. and **Sass, E.** (1977) Recent hot brine dolomitization in the "Solar Lake", Gulf of Elat; isotopic, chemical, and mineralogical study. *J. Geol.*, **85**, 27–48.

Al-Siyabi, H.A. (2005) Exploration history of the Ara intra-salt carbonate stringers in the South Oman Salt Basin. *GeoArabia*, **10**, 39–72.

Alsharhan, A.S. (2003) Petroleum geology and potential hydrocarbon plays in the Gulf of Suez rift basin, Egypt. *AAPG Bull.*, **87**, 143–180.

Amthor, J.E. (2000) Precambrian carbonates of Oman: A regional perspective (abstr.). *GeoArabia* **5**, 47.

Amthor, J.E., Ramseyer, K., Faulkner, T. and **Lucas, P.** (2005) Stratigraphy and sedimentology of a chert reservoir at the Precambrian-Cambrian boundary: the Al Shomou Silicilyte, South Oman Salt Basin. *GeoArabia*, **10**, 89–122.

Antón, J., Peña, A., Valens, M., Santos, F., Glöckner, F.-O., Bauer, M., Dopazo, J. Herrero, J., Rosselló-Mora, R. and **Amann, R.** (2005) Salinibacter Ruber: Genomics and Biogeography. In: *Adaptation to Life at High Salt Concentrations in Archaea, Bacteria, and Eukarya* (Eds N.Gunde-Cimerman, A.Oren, and A. Plemenitaš), pp 255–266. *Springer*, Dordrecht, Netherlands.

Ayres, M.G., Bilal, M., Jones, R.W., Slenz, L.W., Tartir, M. and **Wilson, A.O.** (1982) Hydrocarbon Habitat in Main Producing Areas, Saudi Arabia. *AAPG Bull.*, **66**, 1–9.

Barakat, A.O., Mostafa, A., El-Gayar, M.S. and **Rullkötter, J.** (1997) Source-dependent biomarker properties of five crude oils from the Gulf of Suez, Egypt. *Org. Geochem.*, **26**, 441–450.

Barbé, A., Grimalt, J.O., Pueyo, J.J. and **Albaiges, J.** (1990) Characterization of model vaporitic environments through the study of lipid components. *Org. Geochem.*, **16**, 815–828.

Bardavid, R.E., Hollen, B.J., Bagaley, D.R., Small, A.M., McKay, C., Ionescu, D., Oren, A. and **Rainey, F.A.** (2007) Selective enrichment, isolation and molecular detection of *Salinibacter* and related extremely halophilic Bacteria from hypersaline environments. *Hydrobiologia*, **576**, 3–13.

Bass-Becking, L.G.M. (1928) On organisms living in concentrated brines. *Tijdschr. Ned. Dierk. Ver. Ser.* **3**, 6–9.

Bechtel, A. and **Puttmann, W.** (1997) Palaeoceanography of the early Zechstein Sea during Kupferschiefer deposition in the Lower Rhine Basin (Germany) A reappraisal from stable isotope and organic geochemical investigations. *Palaeogeogr. Palaeoclimatol. Palaeoecol.*, **136**, 331–358.

Benali, S., Schreiber, B.C., Helman, M.L. and **Philp, R.P.** (1995) Characterisation of organic matter from a restricted/evaporative sedimentary environment – Late Miocene of Lorca Basin, southeastern Spain. *AAPG Bull.*, **79**, 816–830.

Beydoun, Z.R. (1993) Evolution of the northeastern Arabian plate margin and shelf – Hydrocarbon habitat and conceptual future potential. *Rev. Inst. Fr. Pétrol.*, **48**, 311–345.

Bhupathiraju, V.K., McInerney, M.J., Woese, C.R. and **Tanner, R.S.** (1999) *Haloanaerobium kushneri* sp. nov., an obligately halophilic, anaerobic bacterium from an oil brine. *Int. J. Syst. Bacteriol.*, **49**, 953–960.

Billo, S.M. (1996) Geology of marine evaporites favorable for oil, gas exploration. *Oil Gas J.*, **94**, 69–73.

Bohacs, K.M., Carroll, A.R., Neal, J.E. and **Mankiewicz, P.J.** (2000) Lake-basin type, source potential, and hydrocarbon character: An integrated sequence-stratigraphic – geochemical framework. In: *Lake Basins through Space and Time* (Eds E.H. Gierlowski-Kodesch and K. Kelts). *AAPG Stud. Geol.*, **46**, 3–34.

Bolhuis, H. (2005) Walsby's Square Archaeon. In: *Adaptation to Life at High Salt Concentrations in Archaea, Bacteria, and Eukarya* (Eds N. Gunde-Cimerman, A. Oren and A. Plemenitaš), pp. 185–199. *Springer*, Dordrecht, Netherlands.

Bolhuis, H., te Poele, E.M. and **Rodríguez-Valera, F.** (2004) Isolation and cultivation of Walsby's square archaeon. *Environ. Microbiol.*, **6**, 87–1291.

Boyer, B.W. (1982) Green River laminites: Does the playa-lake model really invalidate the stratified lake model? *Geology*, **10**, 321–324.

Bradshaw, J. (1988) *The Depositional, Diagenetic and Structural History of the Chandler Formation and Related Units, Amadeus Basin, Central Australia.* Unpubl. PhD thesis, University of New South Wales (Sydney, Australia).

Brandt, K.K. and **Ingvorsen, K.** (1997) *Desulfobacter halotolerans* sp nov, a halotolerant acetate-oxidizing sulfate-reducing bacterium isolated from sediments of Great Salt Lake, Utah. *Syst. Appl. Microbiol.*, **20**, 366–373.

Burke, C.M. and **Knott, B.** (1997) Homeostatic interactions between the benthic microbial communities and the waters of a saline lake. *Mar. Freshwat. Res.*, **48**, 623–631.

Burns, D.G., Camakaris, H.M., Janssen, P.H. and **Dyall-Smith, M.L.** (2004) Cultivation of Walsby's square haloarchaeon. *FEMS Microbiol. Lett.*, **238**, 469–473.

Burwood, R. and **Mycke, B.** (1996) Coastal Angola and Zaire; a geochemical contrast of the lower Congo and Kwanza Basin hydrocarbon habitats (abstr.). *AAPG Bull.*, **80**, 1277.

Burwood, R., de Witte, S.M., Mycke, B. and **Paulet, J.** (1995) Petroleum geochemical characterisation of the lower Congo Coastal basin Bucomazi Formation. In: *Petroleum Source Rocks* (Ed. B. Katz), pp. 235–263. *Springer Verlag*, Berlin.

Burwood, R., Leplat, P., Mycke, B. and **Paulet, J.** (1992) Rifted margin source rock deposition – a carbon isotope and biomarker study of a West African Lower Cretaceous lacustrine section. *Org. Geochem.*, **19** 41–52.

Busson, G. (1988) Relationship between different types of evaporitic deposits and the occurrence of organic-rich layers (potential source rocks) [French]. *Oil Gas Sci. Technol. Rev. Inst. Fr. Pétrol.*, **43**, 181–215.

Carrigan, W.J., Cole, G.A., Colling, E.L. and **Jones, P.J.** (1995) Geochemistry of the Upper Jurassic Tuwaiq Mountain and Hanifa Formation petroleum source rocks of eastern Saudi Arabia. In: *Petroluem Source Rocks* (Ed B. Katz), pp. 67–87. *Springer Verlag*, Berlin.

Carroll, A.R. (1998) Upper Permian lacustrine organic facies evolution, southern Junggar basin, NW China. *Org. Geochem.*, **28**, 649–667.

Carroll, A.R., Brasell, S.C. and **Graham, S.A.** (1992) Upper Permian lacustrine oil shales, southern Junggar Basin, Northwest China. *AAPG Bull.*, **76**, 1874–1902.

Caumette, P. (1993) Ecology and physiology of phototropic bacteria and sulfate-reducing bacteria in marine salterns. *Experientia*, **49**, 473–481.

Cayol, J.L., Ollivier, B., Patel, B.K.C., Prensier, G., Guezennec, J. and Garcia, J.-L. (1994) Isolation and characterization of *Halothermothrix orenii* gen. nov., sp. nov., a halophilic, thermophilic, fermentative strictly anaerobic bacterium. *Int. J. Syst. Bacteriol.*, **44**, 534–540.

Chen, J.Y., Bi, Y. P., Zhang, J.G. and Li, S.F. (1996) Oil-source correlation in the Fulin basin, Shengli petroleum province, East China. *Org. Geochem.*, **24**, 931–940.

Chen, Z., Osadetz, K.G. and Li, M. (2005) Spatial characteristics of Middle Devonian oils and non-associated gases in the Rainbow area, northwest Alberta. *Mar. Petrol. Geol.*, **22**, 391–401.

Christiansen, F.G., Piasecki, S., Stemmerik, L. and Telnaes, N. (1993) Depositional environment and organic geochemistry of the Upper Permian Ravnjeld Formation source rock in East Greenland. *AAPG Bull*, **77**, 1519–1537.

Clark, J.A., Cartwright, J.A. and Stewart, S.A. (1999) Mesozoic dissolution tectonics on the West Central Shelf, UK Central North Sea. *Mar. Petrol. Geol.*, **16**, 283–300.

Clark, J.P. and Philp, R.P. (1989) Geochemical characterization of evaporite and carbonate depositional environments and correlation of associated crude oils in the Black Creek basin, Alberta. *Bull. Can. Petrol. Geol.*, **37**, 401–416.

Clavero, E., Hernandez-Marine, M., Grimalt, J.O. and Garcia-Pichel, F. (2000) Salinity tolerance of diatoms from thalassic hypersaline environments. *J. Phycol.*, **36**, 1021–1034.

Cohen, A.S. (1989) Facies relationships and sedimentation in large rift lakes and implications for hydrocarbon exploration: Examples from Lake Turkana and Tanganyika. *Palaeogeogr. Palaeoclimatol. Palaeoecol.*, **70**, 65–80.

Cooper, B.S., Barnard, P.C. and Telnaes, N. (1995) The Kimmeridge Clay Formation of the North Sea. In: *Petroleum Source Rocks* (Ed. B. Katz), pp. 89–110. *Springer Verlag*, Berlin.

Cordell, R.J. (1992) Carbonates as hydrocarbon source rocks. In: Carbonate Reservoir Characterisation: a Geologic Engineering Analysis, Part 1 (Eds G.V. Chilingarian, S.J. Mazzullo and H.H. Rieke). *Dev. Petrol. Sci.*, **30**, 271–329.

Curial, A., Dumas, D. and Dromart, G. (1990) *Organic matter and evaporites in the Paleogene West European Rift: the Bresse and Valence salt basins (France)*. In: Deposition of Organic Facies (Ed. A.Y. Huc). *AAPG Stud. Geol.*, **30**, 119–132.

Cytryn, E., Minz, D., Oremland, R.S. and Cohen, Y. (2000) Distribution and Diversity of Archaea Corresponding to the Limnological Cycle of a Hypersaline Stratified Lake (Solar Lake, Sinai, Egypt). *Appl. Environ. Microbiol.*, **66**, 3269–3276.

Dahl, J., Moldowan, J.M. and Sundararaman, P. (1993) Relationship of biomarker distribution to depositional environment – Phosphoria Formation, Montana, USA. *Org. Geochem.*, **7**, 1001–1017.

Dahl, J.E., Moldowan, J.M., Teerman, S.C., McCaffrey, M.A., Sundararaman, P., Pena, M. and Stelting, C.E. (1994) Source rock quality determination from oil biomarkers I. - An example from the Aspen Shale, Scully's Gap, Wyoming. *AAPG Bull.*, **78**, 1507–1026.

Damste, J.S.S., De Leeuw, J.W., Betts, S., Ling, Yue and Hofmann, P.M. (1993) Hydrocarbon biomarkers of different lithofacies of the Salt IV Formation of the Mulhouse Basin, France. *Org. Geochem.*, **20**, 1187–1200.

Damste, J.S.S., Frewin, N.L., Kenig, F. and Deleeuw, J.W. (1995) Molecular indicators or palaeoenvironmental change in a Messinian evaporitic sequence (Vena del Gesso, Italy). 1. Variations in extractable organic matter of ten cyclically deposited marl beds. *Org. Geochem.*, **23**, 471–483.

DasSarma, S. and Arora, P. (2001) *Halophiles: Encyclopedia of Life Sciences* (*Nature Publishing Group* www.els. net), 1–9.

de Deckker, P. (1981) Ostracods of Athalassic Saline Lakes - a Review. *Hydrobiologia*, **81–2** 131–144.

de Deckker, P. and Geddes, M.C. (1980) Seasonal fauna of ephemeral saline lakes near the Coorong Lagoon, South Australia. *Aust. J. Mar. Freshwat. Res.*, **31**, 677–699.

Demaison, G.J. and Moore, G.T. (1980) Anoxic environments and oil source bed genesis. *AAPG Bull.*, **64**, 1179–1209.

Dembicki, H.J., Meinschein, W.G. and Hattin, D.E. (1976) Possible ecological and environmental significance of the predominance of even carbon number C20-C30 n-alkanes. *Geochim. Cosmochim. Acta*, **40**, 203–208.

di Primio, R., and Horsfield, B. (1996) Predicting the generation of heavy oils in carbonate/evaporitic environments using pyrolysis methods. *Org. Geochem.*, **24**, 999–1016.

Droste, H. (1990) Depositional cycles and source rock development in an epeiric intra-platform basin; the Hanifa Formation of the Arabian Peninsula. *Sed. Geol.*, **69**, 281–296.

Duncan, W.I. and Buxton, N.W.K. (1995) New evidence for evaporitic Middle Devonian Lacustrine sediments with hydrocarbon source potential on the East Shetland Platform, North Sea. *J. Geol. Soc. London*, **152**, 251–258.

Dundas, I. (1998) Was the environment for primordial life hypersaline? *Extremophiles*, **2**, 375–377.

Dyall-Smith, M., Burns, D., Camakaris, H., Janssen, P., Russ, B. and Porter, K. (2005) Haloviruses and Their Hosts. In: *Adaptation to Life at High Salt Concentrations in Archaea, Bacteria, and Eukarya* (Eds N. Gunde-Cimerman, A. Oren and A. Plemenitaš), pp. 553–563. Springer, Dordrecht, Netherlands.

Edgell, H.S. (1991) Proterozoic salt basins of the Persian Gulf area and their role in hydrocarbon generation. *Precambrian Res.*, **54**, 1–14.

Espitalie, J., Madee, M. and Tissot, B. (1980) Role of mineral matrix in kerogen pyrolysis: influence on petroleum generation and migration. *AAPG Bull.*, **64**, 59–66.

Eugster, H.P. (1985) Oil shales, evaporites and ore deposits. *Geochim. Cosmochimica. Acta*, **49**, 619–635.

Evans, R. and Kirkland, D.W. (1988) Evaporitic environments as a source of petroleum. In: *Evaporites and Hydrocarbons* (Ed. B.C. Schreiber), pp. 256–299. *Columbia Univ. Press*, New York.

Fischer, A.G. and Roberts, L.T. (1991) Cyclicity in the Green River Formation (lacustrine Eocene) of Wyoming. *J. Sed. Petrol.*, **61**, 1146–1154.

Fowler, M.G., Hamblin, A.P., MacDonald, D.J. and McMahon, P.G. (1993) Geological occurrence and geochemistry of some oil shows in Nova Scotia. *Bull. Can. Petrol. Geol.* **41**, 422–436.

Franzmann, P.D., Stackebrandt, E., Sanderson, K., Volkman, J.K., Cameron, D.E., Stevenson, P.L., McMeekin, T.A. and Burton, H.R. (1988) *Halobacterium lacusprofundi* sp. nov., a halophilic bacterium isolated from Deep Lake, Antarctica. *Syst. Appl. Microbiol.* **11**, 20–27.

Garcia, C.M. and Niell, F.X. (1993) Seasonal Change in a Saline Temporary Lake (Fuente-De-Piedra, Southern Spain). *Hydrobiologia*, **267**, 211–223.

Gardner, W.C. and Bray E.E. (1984) Oil and source rocks of the Niagaran Reefs (Silurian) in the Michigan Basin. In: *Petroleum Geochemistry and Source Rock Potential of Carbonate Rocks* (Ed. J.G. Palacas). *AAPG Stud. Geol.*, **18**, 33–44.

Geddes, M.C. (1975a) Studies on an Australian brine shrimp, *Parartemia zeitziana* Sayce (Crustacea: Anostraca). I. Salinity tolerance. *Comp. Biochem. Physiol.*, **51A**, 553–559.

Geddes, M.C. (1975b) Studies on an Australian brine shrimp, *Parartemia zeitziana* Sayce (Crustacea: Anostraca). II. Osmotic and ionic regulation. *Comp. Biochem. Physiol.*, **51A**, 561–571.

Geddes, M.C. (1975c) Studies on an Australian brine shrimp, *Parartemia zeitziana* Sayce (Crustacea: Anostraca). III. The mechanisms of osmotic and ionic regulation. *Comp. Biochem. Physiol.*, **51A**, 573–578.

Geddes, M.C. (1981) The Brine Shrimps Artemia and Parartemia – Comparative Physiology and Distribution in Australia. *Hydrobiologia*, **81–2** 169–179.

Gerdes, G., Klenke, T. and Noffke, N. (2000a) Microbial signatures in peritidal siliciclastic sediments: a catalogue. *Sedimentology*, **47**, 279–308.

Gerdes, G., Krumbein, W.E. and Noffke, N. (2000b) Evaporite Microbial Sediment. In: *Microbial Sediments* (Eds R. E. Riding and S.M. Awramik), pp. 196–208. *Springer-Verlag*, Berlin, Heidelberg.

Gertman, I. and Hecht, A. (2002) The Dead Sea hydrography from 1992 to 2000. *J. Mar. Syst.*, **35**, 169–181.

Grant, W.D., Gemmell, R.T. and McGenity, T.J. (1998) Halobacteria: the evidence for longevity. *Extremophiles*, **2**, 279–287.

Grant, S., Grant, W.D., Jones, B.E., Kato, C. and Li, L. (1999) Novel archaeal phylotypes from an East African alkaline saltern. *Extremophiles*, **3**, 139–145.

Grantham, P.J., Lijmbach, G.W.M., Posthuma, J., Hughes Clarke, M.W. and Willink, R.J. (1988) Origin of crude oils in Oman. *J. Petrol. Geol.*, **11**, 61–80.

Grice, K., Schouten, S., Nissenbaum, A., Charrach, J. and Damste, J.S.S. (1998) Isotopically heavy carbon in the C-21 to C-25 regular isoprenoids in halite-rich deposits from the Sedom Formation, Dead Sea basin, Israel. *Org. Geochem.*, **28**, 349–359.

Grimalt, J.O., De Wit, R., Teixidor, P. and Albaiges, J. (1992) Lipid biogeochemistry of Phormidium and Microcoleus mats. *Org. Geochem.*, **19**, 509–530.

Guixa-Boixereu, N., Calderón-Paz, J.I., Heldal, M., Bratbak, G. and Pedrós-Alió, C. (1996) Viral lysis and bacteriovory as prokaryotic loss factors along a salinity gradient. *Aquat. Microb. Ecol.*, **11**, 215–227.

Gunde-Cimerman, N., Butinar, L., Sonjak, S., Turk, M., Uršič, V., Zalar, P. and Plemenitaš, A. (2005) Halotolerant and Halophilic Fungi from Coastal Environments in the Arctics. In: *Adaptation to Life at High Salt Concentrations in Archaea, Bacteria, and Eukarya* (Eds N. Gunde-Cimerman, A. Oren and A. Plemenitaš), pp. 397–423. *Springer*, Dordrecht, Netherlands.

Hahn, J. and Haug, P. (1986) Traces of Archaebacteria in ancient sediments. *Syst. Appl. Microbiol.*, **7**, 178–183.

Hammer, U.T. (1986) *Saline Lake Ecosystems of the World* (Monographiae Biologicae, **59**). Dr. W. Junk, Dordrecht, Nederlands, 632 pp.

Hanson, A.D., Ritts, B.D., Zinniker, D., Moldowan, J.M. and Biffi, U. (2001) Upper Oligocene lacustrine source rocks and petroleum systems of the northern Qaidam basin, northwest China. *AAPG Bull.*, **85**, 601–619.

Hauer, G., and Rogerson, A. (2005) Heterotrophic Protozoa from Hypersaline Environments. In: *Adaptation to Life at High Salt Concentrations in Archaea, Bacteria, and Eukarya* (Eds N. Gunde-Cimerman, A. Oren and A. Plemenitaš), pp. 519–539. *Springer*, Dordrecht, Netherlands.

Hite, R.J. and Anders, D.E. (1991) Petroleum and evaporates. In: *Evaporites, Petroleum and Mineral Resources* (Ed. J.L. Melvin). *Dev. Sedimentol.*, **50**, 477–533.

Hite, R.J., Anders, D.E. and Jing, T. G. (1984) Organic-rich source rocks of Pennsylvanian age in the Paradox Basin of Utah and Colorado. In: *Hydrocarbon Source Rocks of the Greater Rocky Mountain Region* (Eds J. Woodward, F.F. Meissner and J.L. Clayton), pp. 255–274. Rocky Mountain Assoc. Geologists, Denver, CO.

Hofmann, P., Huc, A.Y., Carpentier, B., Schaeffer, P., Albrecht, P., Keely, B., Maxwell, J.R., Sinninghe, D.J.S., de Leeuw, J.W. and Leythaeuser, D. (1993a) Organic matter of the Mulhouse Basin, France: a synthesis. *Org. Geochem.*, **20**, 1105–1123.

Hofmann, P., Leythaeuser, D. and Carpentier, B. (1993b) Palaeoclimate controlled accumulation of organic matter in Oligocene evaporite sediments of the Mulhouse Basin. *Org. Geochem.*, **20**, 1125–1138.

Holba, A.G., Tegelaar, E., Ellis, L., Singletary, M.S. and Albrecht, P. (2000) Tetracyclic polyprenoids: Indicators of freshwater (lacustrine) algal input *Geology*, **28**, 251–254.

Hollander, D.J., Huc, A.Y., Damste, J.S.S., Hayes, J.M. and de Leeuw, J.W. (1993) Molecular and bulk isotopic analyses of organic matter in marls of the Mulhouse Basin (Tertiary, Alsace, France). *Org. Geochem.*, **20**, 1253–1263.

Horsfield, B., Curry, D.J., Bohacs, K.M., Carroll, A.R., Littke, R., Mann, U., Radke, M., Schaefer, R.G., Isaksen, G.H., Schenk, H.G., Witte, E.G. and Rulkotter, J. (1994) Organic geochemistry of freshwater and alkaline lacustrine environments, Green River Formation, *Wyoming*. *Org. Geochem.*, **22**, 415–450.

Houweling, S., Röckmann, T., Aben, I., Keppler, F., Krol, M., Meirink, J.F., Dlugokencky, E.J. and Frankenberg, C. (2006) Atmospheric constraints on global emissions of methane from plants. *Geophys. Res. Lett.*, **33**, L15821.

Huang, X.Z. and Shao, H.S. (1993) Sedimentary characteristics and types of hydrocarbon source rocks in the Tertiary semiarid to arid lake basins of Northwest China: *Palaeogeogr. Palaeoclimatol. Palaeoecol.*, **105**, 33–43.

Hunt, J.M. (1996) *Petroleum Geochemistry and Geology.* W. H. Freeman & Co., New York, 743 pp.

Ibrahim, M.I.A., Al-Saad, H. and **Kholeif, S.E.** (2002) Chronostratigraphy, palynofacies, source-rock potential, and organic thermal maturity of Jurassic rocks from Qatar. *GeoArabia*, **7**, 675–696.

Imhoff, J.F. (1988) Halophilic phototrophic bacteria. In: *Halophilic Bacteria*, **1** (Ed. F. Rodriguez-Valera), pp. 85–108. *CRC Press*, Boca Raton, FL.

Imhoff, J.F., Sahl, H.G., Soliman, G.S.H. and **Trüper, H.G.** (1979) The Wadi Natrun: Chemical composition and microbial mass developments in alkaline brines of eutrophic desert lakes. *Geomicrobiol. J.*, **1**, 219–234.

Jakobsen, T.F., Kjeldsen, K.U. and **Ingvorsen, K.** (2006) *Desulfohalobium utahense* sp. nov., a moderately halophilic, sulfate-reducing bacterium isolated from Great Salt Lake. *Int. J. Syst. Evol. Microbiol.*, **56**, 2063–2069.

Javor, B.J. (1989) *Hypersaline Environments. Springer Verlag*, Heidelberg, New York.

Javor, B.J. (2000) Biogeochemical models of solar salterns. In: *8th World Salt Symposium*, **2R** (Ed. M. Geertmann), pp. 877–882. *Elsevier*, Amsterdam.

Jiang, Z. and **Fowler, M.G.** (1986) Carentenoid-derived alkanes in oils derived from northwestern China. *Org. Geochem.*, **10**, 831–839.

Jones, B.E., Grant, W.D., Duckworth, A.W. and **Owenson, G.G.** (1998) Microbial diversity of soda lakes. *Extremophiles*, **2**, 191–200.

Jones, R.W. (1984) Comparison of carbonate and shale source rocks. In: Petroleum Geochemistry and Source Rock Potential of Carbonate Rocks (Ed. J.G. Palacas). *AAPG Stud. Geol.*, **18**, 163–180.

Kagya, M.L.N. (1996) Geochemical characterization of Triassic petroleum source rock in the Mandawa Basin, Tanzania. *J. Afr. Earth Sci. Middle East*, **23**, 73–88.

Kates, M., Kushner, D.J. and **Matheson, A.T.** (Eds) (1993) *The Biochemistry of the Archaea.* New Comprehensive Biochemistry, **26**, *Elsevier*, Amsterdam, 582 pp.

Katz, B.J. (1990) *Lacustrine Basin Exploration - Case Studies and Modern Analogues. AAPG Mem.*, **50**, 340 pp.

Katz, B. (1995) *Petroleum Source Rocks: Casebooks in Earth Sciences. Springer Verlag*, Berlin, 327 pp.

Katz, B.J., Bissada, K.K. and **Wood, J.W.** (1987) Factors limiting potential of evaporites as hydrocarbon source rocks (abst.). *AAPG Bull.*, **71**, 575.

Kaufmann, D.W. (1960) *Sodium Chloride.* Reinhold Publishing Corp., New York.

Keely, B.J., De Leeuw, J.W., Maxwell, J.R., Damste, J.S.S., Betts, S.E. and **Yue Ling** (1993) A molecular stratigraphic approach to palaeoenvironmental assessment and the recognition of changes in source inputs in marls of the Mulhouse Basin (Alsace, France). *Org. Geochem.*, **20**, 1165–1186.

Keely, B.J., Blake, S.R., Schaeffer, P. and **Maxwell, J.R.** (1995) Distribution of pigments in the organic matter of marls from the Vena del Gesso evaporitic sequence. *Org. Geochem.*, **23**, 527–593.

Keppler, F., Hamilton, J.T.G., Brav, M. and **Rockmann, T.** (2006) Methane emissions from terrestrial plants under aerobic conditions. *Nature*, **439**, 187–191.

Kholief, M.M. and **Barakat, M.A.** (1986) New evidence for a petroleum source rock in a Miocene evaporite sequence, Gulf of Suez, Egypt. *J. Petrol. Geol.*, **9**, 217–226.

Kinsman, D.J.J. (1973) Evaporite Basins and the Availability of Oxygen in Natural Brines. *Int. Symp. Salt, Tech. Program Abstr. Book.*

Kirkland, D.W. and **Evans, R.** (1981) Source-rock potential of evaporitic environment. *AAPG Bull.*, **65**, 181–190.

Knauth, L.P. (1998) Salinity history of the Earth's early ocean. *Nature*, **395**, 554–555.

Koopmans, M.P., Rijpstra, W.I.C., Klapwijk, M.M., de Leeuw, J.W., Lewan, M.D. and **Sinninghe Damste, J.S.** (1999) A thermal and chemical degradation approach to decipher pristane and phytane precursors in sedimentary organic matter. *Org. Geochem.*, **30**, 1089–1104.

Krekeler, D., Sigalevich, P., Teske, A., Cypionka, H. and **Cohen, Y.** (1997) A sulfate-reducing bacterium from the oxic layer of a microbial mat from Solar Lake (Sinai), *Desulfovibrio oxyclinae* sp. nov. *Arch. Microbiol.*, **167**, 369–375.

Krienitz, L., Ballot, A., Kotut, K., Wiegand, C., Putz, S., Metcalf, J.S., Codd, G.A. and **Pflugmacher, S.** (2003) Contribution of hot spring cyanobacteria to the mysterious deaths of Lesser Flamingos at Lake Bogoria, Kenya. *FEMS Microbiol. Ecol.*, **43**, 141–148.

Lanoil, B.D., Sassen, R., La Duc, M.T., Sweet, S.T. and **Nealson, K.H.** (2001) Bacteria and Archaea physically associated with Gulf of Mexico gas hydrates. *Appl. Environ. Microbiol.*, **67**, 5143–5153.

Lazar, B. and **Erez, J.** (1992) Carbon geochemisty of marine-derived brines. 1. C-13 depletions due to intense photosynthesis. *Geochim. Cosmochim. Acta*, **56**, 335–345.

Lewan, M.D. (1984) Factors controlling the proportionality of vanadium to nickel in crude oils. *Geochim. Cosmochim. Acta*, **48**, 2231–2238.

Lotze, F. (1957) *Steinsalz und Kalisalze. Gebruder Borntraeger*, Berlin.

Lugli, S., Bassetti, M.A., Manzi, V., Barbieri, M., Longinelli, A. and **Roveri, M.** (2007) The Messinian 'Vena del Gesso' evaporites revisited: Characterization of isotopic composition and organic matter. *Geol. Soc. London Spec. Publ.*, **285**, 179–190.

MacDonald, I.R. (1992) Sea-floor brine pools affect behavior, mortality, and preservation of fishes in the Gulf of Mexico: lagerstatten in the making? *Palaios*, **7**, 383–387.

Malek-Aslani, M. (1980) Environmental and diagenetic controls of carbonate and evaporite source rocks. *Trans. Gulf Coast Assoc. Geol. Soc.*, **30**, 445–456.

Mancini, E.A., Parcell, W.C., Puckett, T.M. and **Benson, D.J.** (2003) Upper Jurassic (Oxfordian) Smackover carbonate petroleum system characterization and modeling, Mississippi interior salt basin area, Northeastern Gulf of Mexico, USA. *Carbonates Evaporites*, **18**, 125–150.

Marshall, A.T., Kyriakou, P., Cooper, P.D., Coy, P. and **Wright, A.** (1995) Osmolality of rectal fluid from two species of osmoregulating brine fly larvae (Diptera: Ephyridae). *J. Insect Physiol.*, **41**, 413–418.

Melack, J.M. and **Kilham, P.** (1974) Photosynthetic rates of phytoplankton in East-African alkaline saline lakes. *Limnol. Oceanogr.*, **19**, 743–755.

Mello, M.R. and **Maxwell, J.R.** (1991) Organic geochemical and biological marker characterization of source rocks and oils derived from lacustrine environments in the Brazilian continental margin. In: *Lacustrine Basin*

Exploration Case Studies and Modern Analogs (Ed. B.J. Katz). *AAPG Mem.*, **50**, 77–97.

Mello, M.R., **Gaglianone, P.C.**, **Brassel, S.C.** and **Maxwell, J.R.** (1988) Geochemical and biological marker assessment of depositional environment using Brazilian "offshore" oils. *Mar. Petrol. Geol.*, **5**, 205–223.

Miller, R.G. (1990) A paleogeographic approach to Kimmeridge Shale formation In: *Deposition of Organic Facies* (Ed. A.Y. Huc). *AAPG Stud. Geol.*, **30**, 13–26.

Moberg, E.G., **Greenberg, D.M.**, **Revelle, R.** and **Allen, E.C.** (1932) The buffer mechanism of seawater. *Scripps Inst. Oceanogr. Tech. Serv. Bull.*, **3**, 231–278.

Moldowan, J.M., **Seifert, W.K.** and **Gallegos, E.J.** (1985) Relationship between petroleum composition and depositional environment of petroleum source rocks. *AAPG Bull.*, **69**, 1255–1268.

Moldowan, J.M., **Dahl, J.**, **Huizinga, B.J.**, **Fago, F.J.**, **Hickey, L.J.**, **Peakman, T.M.** and **Taylor, D.W.** (1994) The molecular fossil record of oleanane and its relation to angiosperms. *Science*, **265**, 768–771.

Moody, J.D. (1959) Relationship of primary evaporites to oil accumulation. *5th World Petroleum Congress, New York*, **1**, 134–138.

Obermajer, M., **Fowler, M.G.** and **Snowdon, L.R.** (1998) A geochemical characterization and a biomarker reappraisal of the oil families from southwestern Ontario. *Bull. Can. Petrol. Geol.*, **46**, 350–378.

Obermajer, M., **Fowler, M.G.**, **Snowdon, L.R.** and **Macqueen, R.W.** (2000) Compositional variability of crude oils and source kerogen in the Silurian carbonate-evaporite sequences of the eastern Michigan Basin, Ontario, Canada. *Bull. Can. Petrol. Geol.*, **48**, 307–322.

Oehler, D.Z., **Oehler, J.H.** and **Stewart, A.J.** (1979) Algal fossils from a Late Precambrian, hypersaline lagoon. *Science*, **205**, 338–340.

Oehler, J.H. (1984) Carbonate source rocks in the Jurassic Smackover trend of Mississippi, Alabama, and Florida. In: *Petroleum Geochemistry and Source Rock Potential of Carbonate Rocks* (Ed. J. G. Palacas). *AAPG Stud. Geol.*, **18**, 63–69.

Oesterhelt, D. and **Marwan, W.** (1993) Signal transduction in Halobacteria. In: *The Biochemistry of the Archaea*, **26** (Eds M. Kates, D. J. Kushner and A.T. Matheson), pp. 173 187. Elsevier, Amsterdam.

Ollivier, B., **Caumette, P.**, **Garcia, J.L.** and **Mah, R.A.** (1994) Anaerobic bacteria from hypersaline environments. *Microbiol. Rev.*, **58**, 27–38.

Oremland, R.S. and **King, G.M.** (1989) Methanogenesis in hypersaline environments. In: *Microbial Mats: Physiological Ecology of Benthic Microbial Communities* (Eds Y. Cohen and E. Rosenberg), pp. 180–190. Am. Soc. Microbiol., Washington, DC.

Oren, A. (1992) The genera Haloanaerobium, Halobacteroides, and Sporohalobacter. In: *The Prokaryotes: a Handbook on the Biology of Bacteria: Ecophysiology, Isolation, Identification, Applications* (Eds A. Balows, H.G. Trüper, M. Dworkin, W. Harder and K.-H. Schleifer), 2nd edn, pp. 1893–1900. Springer, New York.

Oren, A. (1993). The Dead Sea – alive again. *Experientia*, **49** 518–522.

Oren, A. (1999a) The enigma of square and triangular halophilic archaea. In: *Enigmatic Microorganisms and Life in Extreme Environments* (Ed. J. Seckbach), pp. 337–355. *Kluwer*, Dordrecht.

Oren, A. (1999b) Bioenergetic aspects of halophilism. *Microbiol. Molec. Biol. Rev.*, **63**, 334–348.

Oren, A. (2001) The bioenergetic basis for the decrease in metabolic diversity at increasing salt concentrations: implications for the functioning of salt lake ecosystems. *Hydrobiologia.*, **466**, 61–72.

Oren, A. (2005) A century of Dunaliella research: 1905–2005). In: *Adaption to Life at High Salt Concentrations in Archaea, Bacteria and Eukarya* (Eds N. Gunde-Cimerman, A. Oren and A. Plemenitaš), pp. 491–502. *Springer*, Dordrecht, Netherlands.

Oren, A. (2006) The Order Halobacteriales. In: *The Prokaryotes*, **3**: *Archaea. Bacteria: Firmicutes, Actinomycetes* (Eds M. Dworkin, S. Falkow, E. Rosenberg, K.-H. Schleifer and E. Stackebrandt), pp. 113–164. Springer-Verlag, New York.

Oren, A. and **Ben Yosef, N.** (1997) Development and spatial distribution of an algal bloom in the Dead Sea: A remote sensing study. *Aquat. Microb. Ecol.*, **13**, 219–223.

Oren, A. and **Gurevich, P.** (1995) Dynamics of a bloom of halophilic Archaea in the Dead Sea. *Hydrobiologia*, **315**, 149–158.

Oren, A., **Kessel, M.** and **Stackebrandt, E.** (1989) *Ectothiorhodospira marismortui* sp-nov, an obligately anaerobic, moderately halophilic purple sulphur bacterium from a hypersaline sulfur spring on the shore of the Dead-Sea. *Arch. Microbiol.*, **151**, 524–529.

Oren, A., **Gurevich, P.**, **Anati, D.A.**, **Barkan, E.** and **Luz, B.** (1995) A bloom of *Dunaliella* parva in the Dead Sea in 1992: biological and biogeochemical aspects. *Hydrobiologia*, **297**, 173–185.

Oren, A., **Bratbak, G.** and **Heldal, M.** (1997) Occurrence of virus-like particles in the Dead Sea. *Extremophiles*, **1**, 143–149.

Overmann, J., **Beatty, J.T.** and **Hall, K.J.** (1996) Purple sulfur bacteria control the growth of aerobic heterotrophic bacterioplankton in a meromictic salt lake. *Appl. Environ. Microbiol.*, **62**, 3251–3258.

Palacas, J.G. (1984) Petroleum geochemistry and source rock potential of carbonate rocks. *AAPG Stud. Geol.*, **18**, 208 pp.

Palacas, J.G., **Anders, D.E.** and **King, J.D.** (1984) South Florida Basin – A prime example of carbonate source rocks of petroleum. In: *Petroleum Geochemistry and Source Rock Potential of Carbonate Rocks* (Ed. J.G. Palacas). *AAPG Stud. Geol.*, **18**, 71–96.

Pancost, R.D., **Crawford, N.** and **Maxwell, J.R.** (2002) Molecular evidence for basin-scale photic zone euxinia in the Permian Zechstein Sea. *Chem. Geol.*, **188**, 217–227.

Pedrós-Alió, C. (2005) Diversity of Microbial Communities: The Case of Solar Salterns. In: *Adaptation to Life at High Salt Concentrations in Archaea, Bacteria, and Eukarya* (Eds N. Gunde-Cimerman, A. Oren and A. Plemenitaš), pp. 71–90. *Springer*, Dordrecht, Netherlands.

Peters, K.E. and **Moldowan, J.M.** (1993) *The Biomarker Guide: Interpreting Molecular Fossils in Petroleum and Ancient Sediments*. Prentice-Hall, Englewood Cliffs, N.J., 363 pp.

Peters, K.E., **Clark, M.E.**, **Dasgupta, U.**, **Mccaffrey, M.A.** and **Lee, C.Y.** (1995) Recognition of an Infracambrian source rock based on biomarkers in the Bahewala-1 oil, India. *AAPG Bull.*, **79**, 1481–1494.

Peters, K.E., Cunningham, A.E., Walters, C.C., Jiang, J.G. and Fan, Z.A. (1996) Petroleum systems in the Jiangling-Dangyang area, Jianghan basin, China. *Org. Geochem.*, **24**, 1035–1060.

Peters, K.E., Moldowan, J.M. and Walters, C.C. (2005a) *The Biomarker Guide*: **1**, *Biomarkers and Isotopes in the Environment and Human History.* Cambridge University Press, 490 pp.

Peters, K.E., Walters, C.C. and Moldowan, J.M. (2005b) *The Biomarker Guide*: **2**, *Biomarkers and Isotopes in Petroleum Systems and Earth History.* Cambridge University Press, 700 pp.

Peterson, J.A. and Hite R.J. (1969) Pennsylvanian evaporite-carbonate cycles and their relation to petroleum occurrence, southern Rocky Mountains. *AAPG Bull.*, **53**, 884–908.

Pflüger, K., Wieland, H. and Müller, V. (2005) Osmoadaptation in Methanogenic Archaea: Recent Insights from a Genomic Perspective. In: *Adaptation to Life at High Salt Concentrations in Archaea, Bacteria, and Eukarya* (Eds N. Gunde-Cimerman, A. Oren, and A. Plemenitaš), pp. 239–251. Springer, Dordrecht, Netherlands.

Philp, R.P. and Lewis, C.A. (1987) Organic Geochemistry of Biomarkers. *Ann. Rev. Earth Planet. Sci.*, **15**, 363–395.

Pinckney, J., Paerl, H.W. and Bebout, B.M. (1995) Salinity control of benthic microbial mat community production in a Bahamian hypersaline lagoon. *J. Exp. Mar. Biol. Ecol.*, **187**, 223–237.

Reed, R.H., Chudek, J.A., Foster, R. and Stewart, W.D.P. (1984) Osmotic adjustment in cyanobacteria from hypersaline environments. *Arch. Microbiol.*, **138**, 333–337.

Ritts, B.D., Hanson, A.D., Zinniker, D. and Moldowan, J.M. (1999) Lower-middle Jurassic nonmarine source rocks and petroleum systems of the northern Qaidam basin, northwest China. *AAPG Bull.*, **83**, 1980–2005.

Rosen, M.R., Turner, J.V., Coshell, L. and Gailitis, V. (1995) The effects of water temperature, stratification, and biological activity on the stable isotopic composition and timing of carbonate precipitation in a hypersaline lake. *Geochim. Cosmochim. Acta*, **59**, 979–990.

Rosen, M.R., Coshell, L., Turner, J.V. and Woodbury, R. J. (1996) Hydrochemistry and nutrient cycling in Yalgorup National Park, Western Australia. *J. Hydrol.*, **185**, 241–274.

Rouchy, J.M. (1988) Relations évaporites-hydrocarbures: l'association laminites-récifes-évaporites dans le Messinien de Mediterranée et ses enseignements. In: *Evaporites et Hydrocarbures* (Ed. G. Busson). *Mém. Mus. Natl. Hist. Natur.*, **55**, 15–18.

Rouchy, J.M., Noel, D., Wali, A.M.A. and Aref, M.A.M. (1995) Evaporitic and biosiliceous cyclic sedimentation in the Miocene of the Gulf of Suez; depositional and diagenetic aspects. *Sed. Geol.*, **94**, 277–297.

Rouchy, J.M., Taberner, C., Blanc-Valleron, M.M., Sprovieri, R., Russell, M., Pierre, C., di Stefano, E., Pueyo, J.J., Caruso, A., Dinares-Turell, J., Gomis-Coll, E., Wolff, G.A., Cespuglio, G., Ditchfield, P., Pestrea, S., Combourieu-Nebout, N., Santisteban, C. and Grimalt, J.O. (1998) Sedimentary and diagenetic markers of the restriction in a marine basin: the Lorca Basin (SE Spain) during the Messinian. *Sed. Geol.*, **121**, 25–55.

Ruble, T.E., Bakel, A.J. and Philp, R.P. (1994) Compound-Specific Isotopic Variability in Uinta Basin Native Bitumens – Paleoenvironmental Implications. *Org. Geochem.*, **21**, 661–671.

Russell, M., Grimalt, J.A., Hartgers, W.A., Taberner, C. and Rouchy, J.M. (1997) Bacterial and algal markers in sedimentary organic matter deposited under natural sulfurization conditions (Lorca Basin, Murcia, Spain). *Org. Geochem.*, **26**, 605–625.

Sassen, R. (1990) Geochemistry of carbonate source rocks and crude oils in Jurassic salt basins of the Gulf Coast. *Geol. Soc. London Spec. Publ.*, **50**, 265–277.

Schoell, M., McCaffrey, M.A., Fago, F.J. and Moldowan, J.M. (1992) Carbon isotopic compositions of 28, 30-bisnorhopanes and other biological markers in a Monterey crude oil. *Geochim. Cosmochim. Acta*, **56**, 1391–1399.

Schoellkopf, N.B. and Patterson, B.A. (2000) Petroleum Systems of Offshore Cabinda, Angola. In: *Petroleum Systems of South Atlantic Margins* (Eds M. R. Mello and B. J. Katz). *AAPG Mem.*, **73**, 361–376.

Schoenherr, J., Littke, R., Urai, J.L., Kukla, P.A. and Rawahi, Z. (2007) Polyphase thermal evolution in the Infra-Cambrian Ara Group (South Oman Salt Basin) as deduced by maturity of solid reservoir bitumen. *Org. Geochem.*, **38**, 1293–1318.

Schoenherr, J., Reuning, L., Kukla, P.A., Littke, R., Urai, J.L., Siemann, M.G. and Rawahi, Z. (2009) Halite cementation and carbonate diagenesis of intra-salt reservoirs from the Late Neoproterozoic to Early Cambrian Ara Group (South Oman Salt Basin). *Sedimentology*, **56**, 567–589.

Schouten, S., Hartgers, W.A., Lopez, J.F., Grimalt, J.O. and Damste, J.S.S. (2001) A molecular isotopic study of C-13-enriched organic matter in evaporitic deposits: recognition of CO_2-limited ecosystems: *Org. Geochem.*, **32**, 277–286.

Schreiber, B.C., Philp, R.P., Benali, S., Helman, M.L., de la Pena, J.A., Marfil, R., Landais, P., Cohen, A.D. and Kendall, C.G.St.C. (2001) Characterisation of organic matter formed in hypersaline carbonate/evaporite environments: Hydrocarbon potential and biomarkers obtained through artificial maturation studies. *J. Petrol. Geol.*, **24**, 309–338.

Schwark, L., Vliex, M. and Schaeffer, P. (1998) Geochemical characterization of Malm Zeta laminated carbonates from the Franconian Alb, SW-Germany (II). *Org. Geochem.*, **29**, 1921–1952.

Schnyder, J., Baudin, F. and Deconinck, J.F. (2009) Occurrence of organic-matter-rich beds in Early Cretaceous coastal evaporitic settings (Dorset, UK): a link to long-term palaeoclimate changes? *Cretaceous Res.*, **30**, 356–366.

Schoenherr, J., Littke, R., Urai, J.L., Kukla, P. and Ruwahi, Z. (2007) Polyphase thermal evolution in the Infra-Cambrian Ara Group (South Oman Salt Basin) as deduced by maturity of solid reservoir bitumen. *Org. Geochem.*, **38**, 1293–1318.

Schoenherr, J., Reuning, L., Kukla, P.A., Littke, R., Urai, J.L., Siemann, M.G. and Rawahi, Z. (2009) Halite cementation and carbonate diagenesis of intra-salt reservoirs from the Late Neoproterozoic to Early Cambrian Ara Group (South Oman Salt Basin). *Sedimentology*, **56**, 567–589.

Shiba, H. and **Horikoshi, K.** (1988) Isolation and characterization of novel anaerobic, halophilic eubacteria from hypersaline environments of western America and Kenya. In: *The Microbiology of Extreme Environments and its Biotechnological Potential. Proc. FEMS Symp.*, Portugal, 371–373.

Simmons, R.E. (1995) Population declines, viable breeding areas and management options for flamingos in southern Africa. *Conserv. Biol.*, **10**, 504–514.

Sinninghe Damste, J.S., Rijpstra, W.I.C., de Leeuw, J.W. and **Schenck, P.A.** (1989) The occurrence and identification of series of organic sulfur compounds in oils and sediment extracts: II. Their presence in samples from hypersaline and non-hypersaline palaeoenvironments and possible application as source, palaeoenvironmental and maturity indicators. *Geochim. Cosmochim. Acta*, **53**, 1323–1341.

Sinninghe Damste, J.S., Kenig, J., Koopmans, M.P., Koster, J., Schouten, S., Hayes, J.M. and **de Leeuw, J.W.** (1995) Evidence for gammacerane as an indicator of water column stratification. *Geochim. Cosmochim.*, **59**, 1895–1900.

Sloss, L.L. (1953) The significance of evaporites. *J. Sed. Petrol.*, **23**, 143–161.

Sonnenfeld, P. (1985) Evaporites as oil and gas source rocks. *J. Petrol. Geol.*, **8**, 253–271.

Sorokin, D.Y. and **Kuenen, J.G.** (2005) Chemolithotrophic haloalkaliphiles from soda lakes. *FEMS Microbiol. Ecol.*, **52**, 287–295.

Stasiuk, L.D. (1994) Oil-prone alginite macerals from organic-rich Mesozoic and Paleozoic strata, Saskatchewan, Canada. *Mar. Petrol. Geol.*, **11**, 208–218.

Stephens, N.P. and **Carroll, A.R.** (1999) Salinity stratification in the Permian Phosphoria sea; a proposed paleoceanographic model. *Geology*, **27**, 899–902.

Summons, R.E. and **Powell, T.G.** (1987) Identification of arylisoprenoids in a source rock and crude oils: Biological markers for the green sulfur bacteria. *Geochim. Cosmochim. Acta*, **51**, 557–566.

Svengren, H. (2002) *A Study of the Environmental Conditions in Lake Nakuru, Kenya, Using Isotope Dating and Heavy Metal Analysis of Sediments.* Unpubl. MSc thesis, Dept. Structural Chemistry, University of Stockholm, Sweden.

Tang, Z.H., Parnell, J. and **Longstaffe, F.J.** (1997) Diagenesis of analcime-bearing reservoir sandstones – the Upper Permian Pingdiquan Formation, Junggar Basin, Northwest China. *J. Sed. Res. Sect. A Sed. Petrol. Proc.*, **67**, 486–498.

ten Haven, H.L., de Leeuw, J.W. and **Schenk, P.A.** (1985) Organic geochemical studies of a Messinian evaporite basin, northern Appennines (Italy), part 1: Hydrocarbon biological markers for a hypersaline environment. *Geochim. Cosmochim. Acta*, **49**, 2181–2191.

ten Haven, H.L., de Leeuw, J.W., Peakman, J.W. and **Maxwell, T.M.** (1986) Anomalies in steroid and hopanoid maturity indices. *Geochim. Cosmochim. Acta*, **50**, 853–855.

ten Haven, H.L., de Leeuw, J.W., Rullkötter, J. and **Sinninghe Damste, J.S.** (1987) Restricted utility of the pristane/phytane ratio as a paleoenvironmental indicator. *Nature*, **330**, 641–643.

ten Haven, H.L., de Leeuw, J.W., Sinninghe Damsté, J.S., Schenk, P.A., Palmer, S.E. and **Zumberge, J.E.** (1988) Application of biological markers in the recognition of palaeo-hypersaline environments. *Geol. Soc. London Spec. Publ.*, **40**, 123–130.

Terken, J.M.J.,and **Frewin, N.L.** (2000) The Dhahaban petroleum system of Oman. *AAPG Bull.*, **84**, 523–544.

Terken, J.M.J., Frewin, N.L. and **Indrelid, S.L.** (2001) Petroleum systems of Oman: Charge timing and risks. *AAPG Bull.*, **85**, 1817–1845.

Trindade, L.A.F., Dias, J.L. and **Mello, M.R.** (1995) Sedimentological and geochemical characterisation of the Lagoa Feia Formation, rift phase of the Campos Basin, Brazil. In: *Petroleum Source Rocks* (Ed. B. Katz), pp. 149–165. Springer Verlag, Berlin.

van der Wielen, P.W.J.J., Corselli, C., Giuliano, L., D'Auria, G., de Lange, G.J., Huebner, A., Varnavas, S.P., Thomson, J., Tamburini, C., Marty, D., McGenity, T.J., Timmis, K.N., Bolhuis, H., Borin, S. and **Daffonchio, D.** (2005) The enigma of prokaryotic life in deep hypersaline anoxic basins. *Science*, **307**, 121–123.

Vareschi, E. (1978) The ecology of Lake Nakuru (Kenya). I. Abundance and feeding of the lesser flamingo. *Oecologia*, **32**, 11–35.

Vareschi, E. (1979) The ecology of Lake Nakuru (Kenya). II. Biomass and spatial distribution of fish. *Oecologia*, **37**, 321–325.

Vareschi, E. (1982) The ecology of Lake Nakuru (Kenya). III. Abiotic factors and primary production. *Oecologia*, **55**, 81–101.

Ventosa, A. (2006) Unusual micro-organisms from unusual habitats: hypersaline environments. In: *Prokaryotic Diversity: Mechanisms and Significance* (Eds N.A. Logan, H.M. Lappin-Scott and P.C.F. Oyston). Cambridge University Press.

Ventosa, A., Nieto, J.J. and **Oren, A.** (1998) Biology of moderately halophilic aerobic bacteria. *Microbiol. Molecul. Biol. Rev.*, **62**, 504–544.

Vincelette, R.R.,and **Chittum, W.E.** (1981) Exploration for oil accumulation in Entrada Sandstone, San Juan Basin, New Mexico. *AAPG Bull.*, **65**, 2546–2570.

Viohl, G. (1996) The paleoenvironment of the Late Jurassic fishes from the southern Franconian Alb (Bavaria, Germany). In: *Mesozoic Fishes, a Systematics and Paleoecology* (Eds G. Arratia and G. Viohl), pp. 513–528. Verlag Dr. Friedrich Pfeil, München, Germany.

Walsby, A.E. (1980) A square bacterium. *Nature*, **283**, 69–71.

Wang, R.L. (1998) Acyclic isoprenoids – molecular indicators of archaeal activity in contemporary and ancient Chinese saline/hypersaline environments. *Hydrobiologia*, **381**, 59–76.

Wang, R.L., Brassell, S.C., Fu, J.M. and **Sheng, G.Y.** (1998) Molecular indicators of microbial contributions to recent and Tertiary hypersaline lacustrine sediments in China. *Hydrobiologia*, **381**, 77–103.

Waples, D.W., Haug, P. and **Welte, D.H.** (1974) Occurrence of a regular C25 isoprenoid hydrocarbon in Tertiary sediments representing a lagoonal saline environment. *Geochim. Cosmochim. Acta*, **38**, 381–387.

Warren, J.K. (1986) Shallow water evaporitic environments and their source rock potential. *J. Sed. Petrol.*, **56**, 442–454.

Warren, J.K. (2000) Evaporites, brines and base metals: low-temperature ore emplacement controlled by evaporite diagenesis. *Aust. J. Earth Sci.*, **47**, 179–208.

Warren, J.K. (2003) Interpreting ancient evaporites: Quaternary sabkhas and salinas are not the "hole" story (abstr.): Proc. Geol. Soc. Am., Ann. Meeting, Seattle, Nov. 2–5.

Warren, J.K. (2006). *Evaporites: Sediments, Resources and Hydrocarbons. Springer*, Berlin, 1036 pp.

Warren, J.K. (2008) Salt as a sediment in the Central European Basin system as seen from a deep time perspective. In: *Dynamics of Complex Intracontinental Basins: The Central European Basin Systems* (Ed. R. Littke). Elsevier, Amsterdam, pp. 249–276.

Warren, J.K. (2010). Evaporites through time: Tectonic, climatic and eustatic controls in marine and nonmarine deposits. *Earth-Science Reviews*, **98**, 217–268.

Weeks, L.G. (1958) Habitat of Oil: A Symposium. AAPG, Tulsa, OK.

Weeks, L.G. (1961) Origin, migration and occurrence of petroleum. In: *Petroleum Exploration Handbook* (Ed. G. B. Moody), pp. 5–50. McGraw-Hill, New York.

Whittle, G.L. and **Alsharhan, A.S.** (1996) Diagenetic histroy and source rock potential of the Upper Jurassic Diyab Formation, offshore Abu Dhabi, United Arab Emirates. *Carbonates Evaporites*, **11**, 145–154.

Winsborough, B.M., **Seeler, J.S.**, **Golubic, S.**, **Folk, R.L.** and **Maguire, B. Jr.** (1994) Recent fresh-water lacustrine stromatolites, stromatolitic mats and oncoids from northeastern Mexico. In: *Phanerozoic Stromatolites II*

(Eds J. Bertrand-Sarfati and C. Monty), pp. 71–100. Kluwer Academic Publishers, Amsterdam.

Woese, C.R. (1993) Introduction. The archaea: their history and significance. In: *The Biochemistry of the Archaea* (Eds M. Kates, D.J. Kushner and A.T. Matheson), **26**, pp. vii–xxix. Elsevier, Amsterdam.

Wood, A.P. and **Kelly, D.P.** (1991) Isolation and characterisation of *Thiobacillus halophilus* sp. nov., a sulfur-oxidising autotrophic eubacterium from a Western Australian hypersaline lake. *Arch. Microbiol.*, **156**, 277–280.

Wood, W.W., **Sanford, W.E.** and **Al Habshi, A.R.S.** (2002) Source of solutes to the coastal sabkha of Abu Dhabi. *Geol. Soc. Am. Bull.*, **114**, 259–268.

Woolnough, W.G. (1937) Sedimentation in barred basins and source rocks of oil. *AAPG Bull.*, **29**, 1101–1157.

Zahran, H.H. (1997) Diversity, adaptation and activity of the bacterial flora in saline environments: *Biol. Fertil. Soils*, **25**, 211–223.

Zhang, S.C., **Zhu, G.Y.**, **Liang, Y.B.**, **Dai, J.X.**, **Liang, H.B.** and **Li, M.W.** (2005) Geochemical characteristics of the Zhaolanzhuang sour gas accumulation and thermochemical sulfate reduction in the Jixian Sag of Bohai Bay Basin. *Org. Geochem.*, **36**, 1717–1730.

Zhilina, T.N. and **Zavarzin, G.A.** (1994) Alkaliphilic anaerobic community at pH 10. *Curr. Microbiol.*, **29**, 109–112.

Zhilina, T.N., **Zavarzin, G.A.**, **Detkova, E.N.** and **Rainey, F.A.** (1996) *Natroniella acetigena* gen. nov. sp. nov., an extremely haloalkaliphilic, homoacetic bacterium: A new member of Haloanaerobiales. *Curr. Microbiol.*, **32**, 320–326.

Zumberge, J.E. (1987) Prediction of source rock characteristics based on terpane biomarkers in crude oils: A multivariate statistical approach. *Geochim. Cosmochim. Acta*, **51**, 1625–1637.

Int. Assoc. Sedimentol. Spec. Publ. (2011) **43**, 393–404

Coupled passive extension and compression on salt-based passive margins analyzed by physical models

ELISABETTA COSTA[*][1], C. CAVOZZI[*] and N. DOGLIONI[†]

[*]*Dipartimento Scienze della Terra, Università di Parma, Via G.p. Usberti 157/a, 43100 Parma, Italy (E-mail: cristian.cavozzi@unipr.it)*
[†]*Alpigeo Soc.Coop., Via Barozzi 45, 32100 Belluno, Italy (E-mail: nicolo.doglioni@alpigeo.it)*

ABSTRACT

This study analyzes the kinematics of thin-skinned, gravitational extension and coupled passive contraction affecting the salt cover on passive margins. Analogue models were designed to simulate the sliding and/or spreading of low viscosity horizons lying beneath frictional covers. No syn-kinematic sedimentation was reproduced, with the scope of analyzing the factors controlling the geometry and kinematics of active faulting. Our models, after deformation, are partitioned into domains whose cross-sectional lengths are independent of the areal extent of the basal viscous layer. However the cross-sectional length, its geometry, the distribution of deformed domains as well as the total length and the rate of extension are directly controlled by the margin's slope, the thickness of viscous décollement layers and its overlying frictional cover and the relative ratio of their respective thicknesses. This ratio also controls the fault throw and the grounding of the cover, forming welding surfaces. Grounding tends to stop extension and limit the cross length of the deformed domains. Due to the lack of sin-cinematic sediment supply, upslope extension results mainly in the form of symmetrical grabens that are concave basinward in the map view. When rafts form, their cross-sectional length commonly increases downslope.

Compression downslope takes place at the frontal termination of the viscous décollement, due to the increased basal friction. Compression is expressed by allochthonous tongues in the central part of the model, whereas folds and thrusts form at the lateral termination, enhanced by the lateral friction. These structures evolve as a breakback propagation sequence. Although analogue models are simplified replicas of natural systems, the results obtained by our models provide valuable insights concerning the evolution of these geodynamic systems.

Keywords: Salt tectonics, gravitational systems, physical modelling.

INTRODUCTION

Deformation of salt-based structures is unique due to the different rheology of the viscous salt and brittle overlying clastics during thin-skinned tectonics. Diverse geotectonic settings differ in their complex kinematics thus displaying different tectonic styles. In fact salt has a viscous behaviour even at very shallow conditions, where surrounding rocks deform in a brittle way (Weijermars *et al.*, 1993). As salt has negligible yield strength, it starts deforming under gravitational forces far before the overlying rocks begin to deform. For

these reasons, although salt is many times weaker than its cover (Costa & Vendeville, 2002 and references therein), it is here inferred that it leads the deformation, at least at early stages of gravitational contexts. We attempt to model the kinematics of thin-skinned, gravitational extension associated with salt spreading and/or sliding, and the kinematics of its contraction in the frontal gravitational belt of the passive margin. The latter may be controlled by either the salt pinchout and/or an abrupt change of basal slope.

Coupled extension and contraction are analyzed by analogue modelling designed to reproduce sliding and/or spreading of low viscosity horizons lying beneath frictional covers. First, we consider

[1]deceased

the factors that control structures and kinematics associated with upslope extension. Secondly, we focus on the geometry and kinematics of the contractional structures of the frontal part of the gravitational fold belt of a passive margin. As this study was aimed at investigating the factors controlling geometry and kinematics of active faulting induced by gravity, very simple models were designed involving few variables and no syn-kinematic sediment supply.

PASSIVE MARGINS AND GRAVITATIONAL STRUCTURES

Continental extension leads to rifted basins that initially form along passive margins. These are fundamental structures that influence the further evolution of these settings if either the salt predates or postdates rifting. Rifted basins are often repositories for sediments that may contain significant accumulation of hydrocarbons. Commercial grids typically do not display the Moho but selected deeper crustal transects show high lateral and vertical variability, related to rifted or volcanic genesis of passive margins, their basement and surface slope. Consequently these margins include a wide range of structural styles. The scenario is further complicated by the movement of salt, often present in rifted basins either predating or, mainly, postdating rifting deposited in post-rift sag sequences (Mohriak *et al.*, 1995; Cohen & Hardy, 1996; Koyi, 1996; Szatmari *et al.*, 1996). Salt-based passive margins display complex arrays of regional and counter-regional normal faults as well as linked contraction folds and faults located at the basinward pinch out of salt (Peel *et al.*, 1995; Spathopoulos, 1996; Rowan, 1997; Trudgill *et al.*, 1999; Rowan *et al.*, 2004). Often deformation in salt-based basins is influenced by lateral changes in salt thickness and geometry, karst, diapirism, massive sediment sliding, change in salt creep mechanism and/or differential sedimentary loading (Diegel *et al.*, 1995; Peel *et al.*, 1995; Costa *et al.*, 2002, 2004 and references therein). The tectonically and mechanically controlled submarine morphology may lead to the formation of ponded basins, which are subordinate when compared to the overall dimension of the whole systems. Thus morphology strongly controls the distribution and facies of sediments which in turn triggers more deformation. Many of these factors are still poorly understood but they all contribute to the complexity of deformation on salt-based passive margins. Therefore, insights gained from analogue modelling may be useful particularly if such experiments, as our do, allow testing the influence of a number of variables, one by one, on both the geometry and the kinematics of deformation affecting passive margins.

MODELLING PASSIVE MARGINS

In general this study attempts to balance the relative influence of the following controls: (1) thickness of rock salt; (2) thickness of the cover; and (3) steepness of slope on: (a) the geometry of deformation, (b) the localization and width of the deformed areas and (c) the sequence of nucleation and growth of deformational structures. To clarify the relative importance of these controlling factors and to balance their cause-effect relationship, we designed two different sets of models to respectively investigate the upslope extensional structures and the downslope contractional structures that form the frontal folded belts. To keep the experiments as simple as possible, in our models we did not simulate sin-tectonic sedimentation and did not change the basal slope through time.

In our models, the viscous décollement was reproduced by silicone gel and the brittle cover by dry sand (Table 1). The movies showing modification of the model's top surface during its deformation allowed complete reconstruction of the timing and kinematics of structure nucleation and growth. After deformation, models were wetted, cut and photographed along several cross sections, from which corresponding line drawings were produced. The first set of models was designed to eliminate the buttressing effect due to any kind of obstacle (i.e. the assumption is that the salt was deposited in post-rift sag sequences) allowing the forward propagation of the model's front, whereas in the second set of models the frontal buttressing effect (mainly due to primary salt pinchout) was reproduced by pinching out the basal viscous layer.

Table 1. Mechanical and physical properties of the materials used in the models

Materials	Density (g/cm³)	Grain size (μm)	Angle of internal friction φ	Dynamic shear viscosity η (Pa s)
Sand	1.75	300	30°	- - - -
Silicone	0.976	- - - -	- - - -	2×10^4

Table 2. Geometric and kinematic characteristics of the models

Model	Thickness of sand (mm)	Thickness of silicone (mm)	Slope (°)	Total time (h, m)	Final length of model (mm)	Total extension (%)
STA2	18	10	0°	05, 00	260	30
STA3	18	10	6°	04, 30	284	42
STA4	10	5	0°	08, 00	226	13
STA5	10	5	6°	07, 35	252	26
STA7	10	10	6°	05, 00	280	40
STA8	10	10	0°	06, 40	250	25
STA9	5	5	6°	13, 45	255	27,5
STA10	5	5	0°	22, 15	230	15

Upslope extension

Model set up and deformation

Here we show the variable geometries and kinematics of structures obtained from eight models that simulate extension of passive margin sediments underlain by salt (Table 2). These models had the same dimension (i.e. 200 × 200 mm) with an unconfined "seaward" side that allows the viscous silicone to spread freely and/or to slide under its own weight and that of its brittle overburden. These models independently test the following variables: (1) thickness of the viscous décollement layer, (2) thickness of the frictional cover, (3) thickness ratio between the brittle cover and the viscous décollement (B/V) and (4) the steepness of the margins' basal slope. All the models (see Table 2

for more details on model features) were free to deform until their forward extension stopped (i.e. when the movement rate <1 mm day^{-1}): in practice, every model was allowed to deform as much as possible. The total deformation time varies significantly for the various models (Table 2), depending on their total thickness as well as on the steepness of the basal slope.

Model results

Line drawings of cross sections cut at the centre of each model show a set of extensional structures bounded by normal faults that generated either symmetrical or asymmetrical grabens (Figs 1 and 2), which delimit tilted blocks. In all the models, most of the faults form during the first

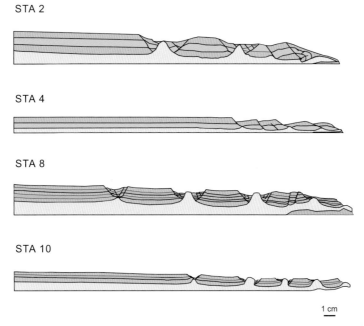

STA 2

STA 4

STA 8

STA 10

1 cm

Fig. 1. Line drawings obtained from the cross sections cut at the centre of the horizontal models reproducing passive extension.

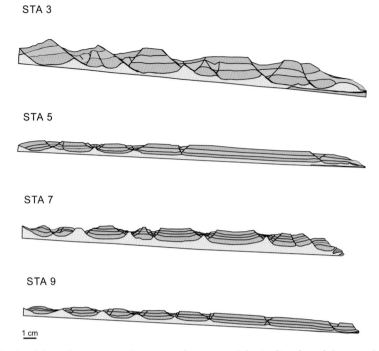

STA 3

STA 5

STA 7

STA 9

1 cm

Fig. 2. Line drawings obtained from the cross sections cut at the centre of the inclined models reproducing passive extension.

2.5 cm (i.e. 12.5%) of extension; these structures then evolve to form grabens and tilted blocks.

(A) Models that assume a flat base (Fig. 1):
  A-1: Never deformed at their rear;
  A-2: Show deformation propagating from the front, backwards;
  A-3: Show the cross length of the deformed frontal areas proportional to the thickness of the basal silicone layer but inversely proportional to the thickness of the cover.
(B) Models tilted toward the" deep sea" (Fig. 2):
  B-1: Deformed at both their front (with deformation propagating backwards) and rear (with deformation propagating forward) at the same time;
  B-2: The deformation structures form following a back-and-forth sequence of propagation (Fig. 3A, B);
  B-3: Deformation moving from both the front and the rear toward the centre of the system occurs by faults dipping downslope while deformation moving away from the centre toward both the front and the rear occurs by faults dipping upslope (i.e. the counter-regional faults of some areas);
  B-4: Show two different deformation bands with amplitudes proportional to the slope angle (as also shown in Mauduit *et al.*, 1997), but also proportional to the

thickness of the viscous layer, and inversely proportional to the thickness of the brittle cover;
  B-5: Have the two deformed sectors merging toward the centre of the models only when they are thicker whereas the thinner models have less or undeformed zones in their central part.

In all the models, the spreading velocity is directly related to the model thickness. The viscous layer pierces where the cover is weakened and thinned by normal faulting (Vendeville & Jackson, 1992), and in some models diapirs extrude. In all the gravitational models, a phase of almost instantaneous acceleration (that brings them to the maximum speed) is followed by progressive deceleration (Fig. 4). The maximum and minimal velocities are mostly due to the increased thickness of both the viscous and the brittle layers, rather than to the increased slope. The tilted models show a more uniform deceleration curve than the horizontal models (Fig. 4).

Coupled upslope extension and downslope compression

Model set up and deformation

Here we present different geometries and kinematics of two models that simulate sediment

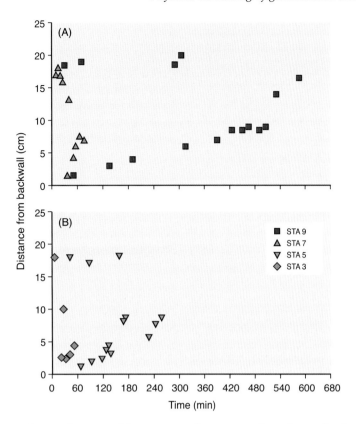

Fig. 3. Graphs showing the nucleation location of the extensional structures through time for the inclined models having a sand/silicone thickness ratio = 1 (A), and = 2 (B). The two graphs highlight the back-and-forth sequence of propagation converging toward the centre of the nucleating normal faults (centripetal back-and-forth sequence in Costa *et al.*, 2002; Costa & Vendeville, 2002).

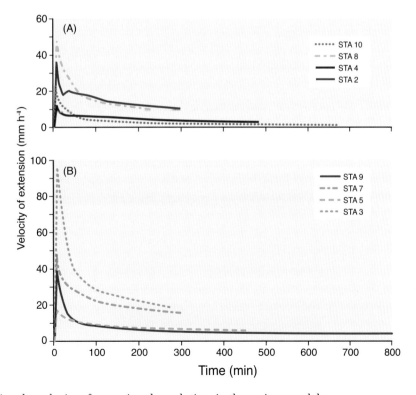

Fig. 4. Graph showing the velocity of extension through time in the various models.

Table 3. Characteristics of Models STA11 and STA12

Model	Thickness of sand (mm)	Thickness of silicone (mm)	Slope (°)	Model length (mm)	Model width (mm)
STA11	10	5	6°	900	600
STA12	10	5	6°	400	600

deformation on salt-based passive margins. Both models have the same initial conditions, except but they differ in length along the direction of extension (Table 3). The basal layer of silicone pinches out downslope toward the frontal part of the models, causing compression due to the buttressing effect of the sand directly overlying the base of the model. The models were designed to test the effect of changing the ratio between length and width of the initial setup while keeping the other variables equal. This helps to assess structures that are due to the overall shape of a salt basin, whether associated with gravitational sliding and/or whether associated with spreading. Furthermore, here the models were allowed to deform until their deformation stopped (i.e. every model continued to deform to its completion). Therefore, the total deformation time varies in the two models (Table 3), depending on the total thickness of the "sedimentary" load.

Model results

Both models aimed to reproduce salt-based sediments on passive margins. They linked thin-skinned extensional and compressional structures upslope and downslope, respectively. The two deformed sectors are separated by a non-deformed area; i.e. they are partitioned into three different deformation domains (Fig. 5). The proportional cross-sectional length of these domains, as measured along the central profiles, is the same for the two models (Fig. 6).

1. *Contraction.* In the shorter model STA12 (Table 3), frontal contraction is achieved by basinward overthrusting of the silicone-bearing sequence onto the sand layer lying beyond the silicone pinch out. In the longer model STA11 (Table 3), contraction is achieved by different structures depending on their position along strike of the system. At the centre of the model the structure is similar to that described for model STA12 (Fig. 7A), while at the lateral edges shortening is mainly achieved by box folds and thrusts-related-anticlines (Fig. 7B and C). This is due to frictional shear along the lateral walls, which is larger in the STA11 model. Along these margins, the shortened area is proportionally more extensive than the area measured at the centre of the models (Fig. 8). This is caused by the salt overriding the undeformed area beyond the pinchout. In the natural systems, the formation of fold-and-thrust belts at the frontal pinch out of the basal viscous layer located mainly at the lateral terminations of the salt basin occurs also in the northern Gulf of Mexico Basin, where fold-and-thrust belts mainly form at the westerly and easterly terminations of the

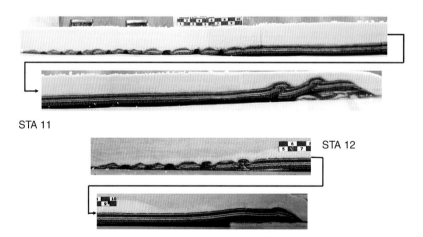

STA 11

STA 12

Fig. 5. Cross sections cut at the centre of the two models STA11 and STA12 showing their partition into three deformation domains: extended, undeformed and shortened domains.

STA 11

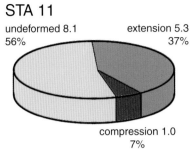

undeformed 8.1
56%

extension 5.3
37%

compression 1.0
7%

STA 12

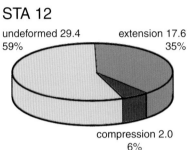

undeformed 29.4
59%

extension 17.6
35%

compression 2.0
6%

Fig. 6. Proportional cross-section length of the deformed domains in the Models STA11 and STA12.

salt basin (Perdido and Mississippi Fan Fold-belts, respectively; Wu *et al.*, 1990; Weimer & Buffler, 1992; Rowan, 1997; Trudgill *et al.*, 1999; Rowan *et al.*, 2004; Fig. 9). The kinematics of

contraction in the models (break-back sequence) reproduces nicely those of the Mississippi Fan Fold Belt and probably also that of the Perdido fold belt (Blickwede & Queffelec, 1988; Wu *et al.*, 1990; Weimer & Buffler, 1992; Rowan, 1997; Trudgill *et al.*, 1999). Even considering the differences existing between our models (autochthonous basal viscous layer and no sedimentary supply during deformation) and the northern Gulf of Mexico Basin system (mainly allochthonous salt layer and substantial sedimentary supply during deformation), the formation of break-back fold-and-thrust belts at the frontal sides of both models and natural systems mainly results from the shear strain occurring along the lateral termination of salt basins.

2. *Extension.* Extension upslope is here achieved by both regional and counter-regional dipping normal faults forming curved linked half-grabens showing, in the more internal part of the deformed zone, the same dip direction as that of the slope surface. More basin-ward, the grabens are concave forward in map view (Fig. 10). This is due to the flow pattern of the basal viscous layer, which becomes increasingly more concave updip or convex downdip with increasing flow. This spoon shape draws the silicone not only downslope, but also toward the centre of

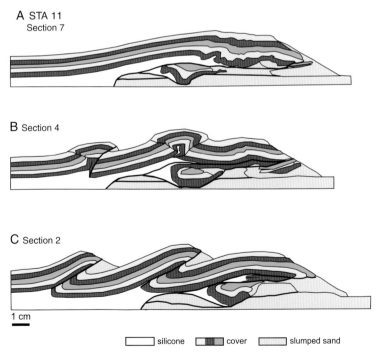

Fig. 7. Shortening structures formed at the frontal pinch out of the viscous layer in Model STA11: (A) at the centre of the model, (B) in between the centre and the lateral termination and (C) at the lateral termination.

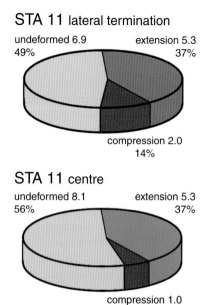

Fig. 8. Proportional cross-section length of the shortened domains at the centre and at the lateral termination of the model STA11.

the model. The combined effect of these two flow directions leads to oblique flow that causes the formation of oblique grabens at the lateral margins of the models. The cross sectional length and thickness of rafts increases forward (Fig. 11) in both models, whereas the wavelength of the intervening diapirs increases backward in the shorter model, STA12 and forward in the longer

model, STA11 (Fig. 10). In the STA12 model, the extensional structures seem more influenced by the coupled compression than in the STA11 model.

CONCLUSIONS

The models presented in this study were designed to be greatly simplified replicas of thin-skinned tectonics on salt-based passive margins. Simplicity was chosen to introduce only a small number of variables in the models. This permits to test the influence of each variable on the resulting structures. Of course additional important variables commonly act in natural settings (e.g. syn-tectonic sedimentation, thick-skinned tectonics and many others, even not yet adequately described), which furthermore influence the resulting overall structure of the deformed system. Thus, our models are only a first approximation to a model-based analysis of salt tectonics on passive margins.

Upslope extension

The location, spacing and kinematics of the structures depend on whether only spreading acts on the system, or else whether sliding is also involved. In this second case, the system follows a back-and-forth sequence of propagation converging toward the centre, which has been observed in many active fold-and-thrust belts detaching on salt (Costa *et al.*, 2002, 2004; Costa & Vendeville, 2002). The

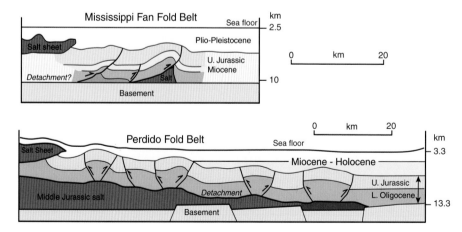

Fig. 9. Schematic cross sections across the Eastern Mississippi Fan Fold Belt and the Perdido fold belt, modified from Blickwede & Queffelec (1988).

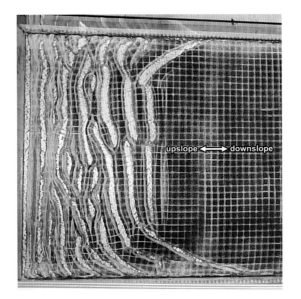

Fig. 10. Map view of the upslope extended domain in Model STA11, showing the concave-forward shape of the grabens, whose lightest, elongated cores show the walls of the basal silicone piercing the cover. The light-grey strips, flanking the silicone walls on both sides, mark the upper portions of the conjugated normal faults bordering the grabens.

of cover loading and cover grounding, the balance of which depends on the reciprocal thickness of cover and salt, as well as on the steepness of the margin (Mauduit *et al.*, 1997; Cramez & Jackson, 2000; Jackson *et al.*, 2000).

Coupled upslope extension and downslope compression

Passive margins that have experienced coupled upslope extension and downslope compression are partitioned into deformation domains (extended, undeformed and shortened domains) whose reciprocal length is independent from the cross length of the whole salt system. The models reproducing thin-skinned extension and coupled compression in salt-bearing passive margins show that both the total amount of extension and the extension rate, as well as the geometry, extent and distribution of deformed domains are directly tied (a) to both the thickness of viscous décollement layers and overlying frictional cover, (b) to their thickness ratio, as well as (c) to the margin's slope. Compression downslope is achieved by folds and thrusts at the frontal pinch out of salt, mainly at the lateral termination of the salt basin. These structures form following a break-back sequence of propagation. Extension upslope is achieved by half-graben and graben as well. When extension eventually leads to the formation of rafts, their cross length often increases downslope (Duval *et al.*, 1992). The formation of grabens (instead of half-graben, more widespread in passive margins) in our models is due to the absence of sediment supply during extension. In fact the first fault that nucleates is the sea-dipping one that is soon after followed by its conjugated pair when no or very

thickness of the cover directly influences the total amount of both sliding and spreading of the underlying salt layer and, consequently, the cross sectional length of the deformed areas and the fault throws. Thus, if the cover is thicker than the underlying salt, the hangingwall faulted blocks may ground even with very little or no sediment supply (fault welding of Jackson & Cramez, 1989). Fault welds tend to stop salt flow, which, in turn, limits the cross sectional length of the deformed areas. This depends, however, on the contrasting effects

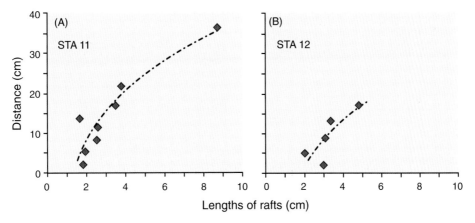

Fig. 11. Length of rafts versus distance from the back wall (in cross-section) of the models STA11 and STA12.

little sediment is supplied. On the contrary a rapid and abundant sediments infill tends to hamper the formation of the conjugated fault, so that the extension is completely achieved by sea-dipping faults. Results obtained by our simple modelling technique provide well circumscribed (described?) cases where known variables were assigned and changed under close management of the experiment. It is critical to have a foundation, even as simple as this, when interpreting structures that are further complicated by a wide array of variables such as inversion or, more important, when attempting to reconstruct the original areal extent of a salt basin as well as the salt thickness. A massive salt flow downslope, which is well-imaged may be related to extensive welding surfaces upslope. This may be most relevant when exploring for hydrocarbons, as the removal of salt associated with welding surface removes a seal, and so permits hydrocarbons sourced from the presalt sequence into the overlying sequences. The overall areal distribution of the extensional structures allows also reconstructing the angle of basal slope at the time of deformation, which may have changed through time due to thermal contraction of the oceanic crust and/or by tectonic or isostatic uplift of the continent.

ACKNOWLEDGEMENTS

The authors would like to express their gratitude to Albert Bally who helped to improve the manuscript by his helpful and useful comments.

REFERENCES

Blickwede, J.F. and Queffelec, T.A. (1988) Perdido Fold Belt: A New Deep-Water Frontier. In: *Western Gulf of Mexico* (abstr.) *AAPG Bull.*, **72**, 163.

Cohen, H.A. and Hardy, S. (1996) Numerical modelling of stratal architectures resulting from differential loading of a mobile substrate. In: *Salt Tectonics* (Eds G.I. Alsop, D.J. Blundell and I. Davison). *Geol. Soc. London Spec. Publ.*, **100**, 265–273.

Costa, E. and Vendeville, B.C. (2002) Experimental insights on the geometry and kinematics of fold-and-thrust belts above weak, viscous evaporitic décollement. *J. of Struct. Geol.*, **24**, 1729–1739.

Costa, E., Mosconi, A. and Cavozzi, C. (2002) Fold-And-Thrust Belts on Salt: Kinematics and 3d Geometry of Thrust Fronts in Natural Settings and in Physical Models. *Boll. Geofis. Teor. Appl.*, **42**, 157–161.

Costa, E., Camerlenghi, A., Polonia, A., Cooper, C., Fabretti, P., Mosconi, A., Murelli, P., Sormani, L. and Wardell, N. (2004) Modelling salt tectonics in the Mediterranean

Ridge Accretionary Wedge. *Geol. Soc. Am. Bull.*, **116**, 880–894.

Cramez, C. and Jackson, M.P.A. (2000) Superposed deformation straddling the continental-oceanic transition in deep-water Angola. *Mar. Petrol. Geol.*, **17**, 1095–1109.

Diegel, F.A., Karlo, J.F., Schuster, D.C., Shoup, R.C. and Tauvers, P.R. (1995) Cenozoic structural evolution and tectono-startigraphic framework of the northern Gulf Coast continental margin. In: *Salt Tectonics: a Global Perspective* (Eds M.P.A. Jackson, D.G. Roberts and S. Snelson). *AAPG Mem.*, **65**, 109–151.

Duval, B., Cramez, C. and Jackson, M.P.A. (1992) Raft tectonics in the Kwanza Basin, *Angola. Mar. Petrol. Geol.*, **9**, 389–404.

Jackson, M.P.A. and Cramez, C. (1989) Seismic recognition of salt welds in salt tectonic regimes. In: Gulf Coast Section SEPM Foundation 10th Annual Research Conference Program and Abstracts. Houston, Texas, pp. 66–89.

Jackson, M.P.A., Cramez, C. and J-M Fonck (2000) Role of subaerial volcanic rocks and mantle plumes in creation of South Atlantic margins: implications for salt tectonics and source rocks. *Mar. Petrol. Geol.*, **17**, 477–498.

Koyi, H. (1996) Salt flow by aggrading and prograding overburden. In: *Salt Tectonics* (Eds G.I. Alsop, D.J. Blundell and I. Davison). *Geol. Soc. London Spec. Publ.*, **100**, 243–258.

Mauduit, T., Guerin, G., Brun, J.P. and Lecanu, H. (1997) Raft tectonics: the effects of basal slope angle and sedimentation rate on progressive extension. *J. Struct. Geol.*, **19**, 1219–1230.

Mohriak, W.U., Macedo, J.M., Castellani, R.T., Rangel, H.D., Barros, A.Z.N., Latgé, M.A.L., Mizusaki, A.M.P., Szatmari, P., Demercian, L.S., Rizzo, J.G. and Aires, J.R. (1995) Salt tectonics and structural styles in deep water province of the Cabo Frio region, Rio de Janeiro, Brazil. In: *Salt Tectonics: a Global Perspective* (Eds M.P.A. Jackson, D.G. Roberts and S. Snelson). *AAPG Mem.*, **65**, 273–303.

Peel, F.J., Travis, C.J. and Hossack, J.R. (1995) Genetic structural provinces and salt tectonics of the Cenozoic offshore U.S. Gulf of Mexico: a preliminary analysis In: *Salt Tectonics: a Global Perspective* (Eds M.P.A. Jackson, D.G. Roberts and S. Snelson), *AAPG Mem.*, **65**, 153–175.

Rowan, M.G. (1997) Three dimensional geometry and evolution of segmented detachment folds, Mississippi Fan foldbelt, Gulf of Mexico. *J. Struct. Geol.*, **19**, 463–480.

Rowan, M.G., Peel, F.J. and Vendeville, B.C. (2004) Gravity-driven fold belts on Passive margins. In: *Thrust Tectonics and Hydrocarbons Systems* (Ed. K.R. McClay), *AAPG Mem.*, **82**, 157–183.

Spathopoulos, F. (1996) An insight on salt tectonics in the Angola basin, South Atlantic. In: *Salt Tectonics* (Eds G.I. Alsop, D.J. Blundell and I. Davison). *Geol. Soc. London Spec. Publ.*, **100**, 153–174.

Szatmari, P., Guerra, C.M. and Pequeno, M.A. (1996) Genesis of large counter-regional normal fault by flow of Cretaceous salt in the South Atlantic Santos Basin, Brazil. In: *Salt Tectonics* (Eds G.I. Alsop, D.J. Blundell and I. Davison). *Geol. Soc. London Spec. Publ.*, **100**, 259–264.

Trudgill, B.D., Rowan, M.G., Fiduk, J.C., Weimer, P., Gale, P.E., Korn, B.E., Phair, R.L., Gafford, W.T., Roberts, G.R. and **Dobbs, S.W.** (1999) The Perdido foldbelt, Northwest deep Gulf of Mexico. Part 1: Structural Geometry, Evolution and Regional Implications. *AAPG Bull.*, **83**, 88–113.

Vendeville, B.C. and **Jackson, M.P.A.** (1992) The rise of diapirs during thin-skinned extension. *Mar. Petrol. Geol.*, **9**, 331–353.

Weijermars, R., Jackson, M.P.A. and **Vendeville, B.C.** (1993) Rheological and tectonic modelling of salt provinces. *Tectonophysics*, **217**, 143–174.

Weimer, P. and **Buffler, R.T.** (1992) Structural geology and evolution of the Mississippi Fan fold belt, deep Gulf of Mexico. *AAPG Bull.*, **76/2** 225–251.

Wu, S., Bally, A.W. and **Cramez, C.** (1990) Allochthonous salt, structure, and stratigraphy of the northeastern Gulf of Mexico, Part II Structure. *Mar. Petrol. Geol.*, **7**, 334–369.

Int. Assoc. Sedimentol. Spec. Publ. (2011) **43**, 405–420

Salt tectonics and structural styles of the western High Atlas and the intersecting Essaouira-Cap Tafelney segments of the Moroccan Atlantic margin

MOHAMAD HAFID*, A.W. BALLY[†], A. AIT SALEM[‡] and E. TOTO*

*Equipe de Géophysique, Département de Géologie, Faculté des Sciences, BP 133, Kénitra, Morocco
(E-mail: hafidmo@yahoo.com)*
[†]*Rice University, Department of Geology and Geophysics, 6100 Main Street, TX 77005-1892, Houston, USA*
[‡]*Office National des Hydrocarbures et des Mines (ONHYM), 5 Avenue Moulay Hassan, BP99, 10050, Rabat, Morocco*

ABSTRACT

The Safi–Agadir Moroccan coastal region is the only area where the Moroccan Atlantic salt province makes an inland incursion. It is also an area where the High Atlas alpine fold-belt intersects at right angles the Atlantic passive margin. Regional reflection seismic transects have been constructed across this area and its offshore extension. These are used here to analyze the effects of the post-sedimentary mobility of salt on sedimentation and on structural styles, and to link these effects to the main stages of the geodynamic evolution of the Atlas and the Atlantic systems. Salt was deposited in Late Triassic–Early Jurassic times during the last stage of rifting prior to continental breakup. It played an important role in the genesis of most of the structures presently encountered in the area. A great variety of salt structures have been identified, including: gentle salt-cored folds and pillows; compressional diapirs; and salt withdrawal synclines. Completely allochthonous pluri-kilometric salt sheets and canopies are bounded to the west by toe-thrust salt structures that overlie oceanic crust in the deep offshore basin. This offers a great variety of favourable settings for possible hydrocarbon traps.

Keywords: Western High Atlas, Essaouira, Cap Tafelney, Moroccan Atlantic margin, salt tectonics, allochthonous salt, toe thrust, raft structures.

INTRODUCTION

Massive halite-rich evaporites were deposited during the initial opening of the North Atlantic Ocean. They form extensive diapiric salt provinces along both the eastern and western North Atlantic margins (Hinz *et al.*, 1982). The eastern diapiric province extends along the Moroccan Atlantic margin over some 900 km (Fig. 1A). In this paper, the objective is to illustrate with regional and local reflection seismic transects, how the mobile syn-rift salt influenced the post-rift sedimentary and structural evolution in this province. This area includes the onshore basins located between Safi and Agadir (Abda, Essaouira, Haha, and Agadir basins) and their offshore extension beneath the present-day Atlantic Ocean (Fig. 1B). It is only in

this zone that the Moroccan salt province has an onshore extension.

The geology of this large area is relatively well studied, but its salt tectonics has been analyzed only in a few previous publications (Heyman, 1989; Hafid, 2000, 2006; Mridek, 2000; Hafid *et al.*, 2000, 2005b; Tari *et al.*, 2000, 2001, 2003). Here, it will show be shown that salt tectonics strongly controlled the geological evolution of the area. Tectonic styles varied along-strike from coastal basins towards the deep oceanic basin, forming different tectonic domains characterized by different halokinetic features that often controlled depositional systems. This paper is an updated summary of a larger study (Hafid, 2006), and a synopsis that relates salt tectonics to the main stages of the geodynamic evolution of the area.

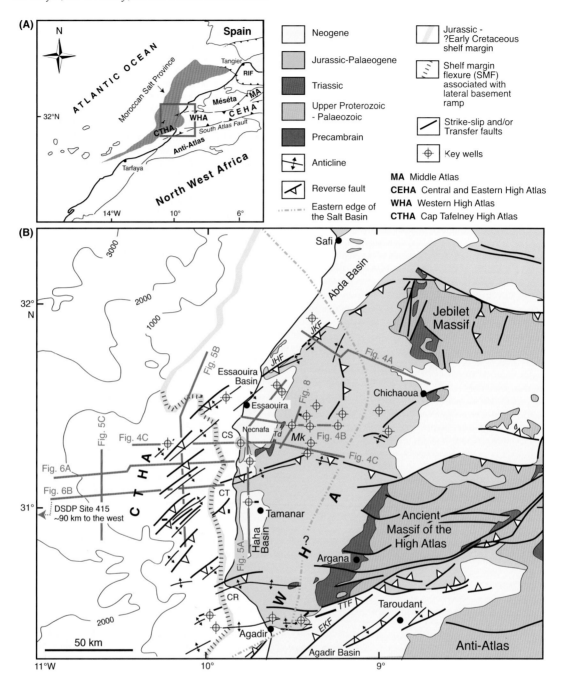

Fig. 1. The Moroccan Salt Province and its geological setting. (A) Location of the Moroccan Salt Province. (B) Simplified geological map of the western termination of the Jebilet–High Atlas system showing main Atlasic trends. Locations of seismic transects and key wells used in the present work are shown. CR = Cap Rhir; CS = Cap Sim; CT = Cap Tafelney; *EKF* = El Klea Fault; *JHF* = Jebel Hadid Fault; *JKF* = Jebel Kourati Fault; *Mk* = Meskala; *Td* = Tidsi; *TTF* = Tizi n'Test Fault.

REGIONAL AND STRATIGRAPHIC FRAMEWORK

The Essaouira-Agadir Basin and its corresponding Atlantic margin segment is located in an area where the High Atlas fold belt intercepts almost at right angles the Atlantic passive margin (Fig. 1). The geological evolution of this large area resulted from the interference of Atlantic passive margin rifting/drifting with the Atlas orogeny (Hinz *et al.*,

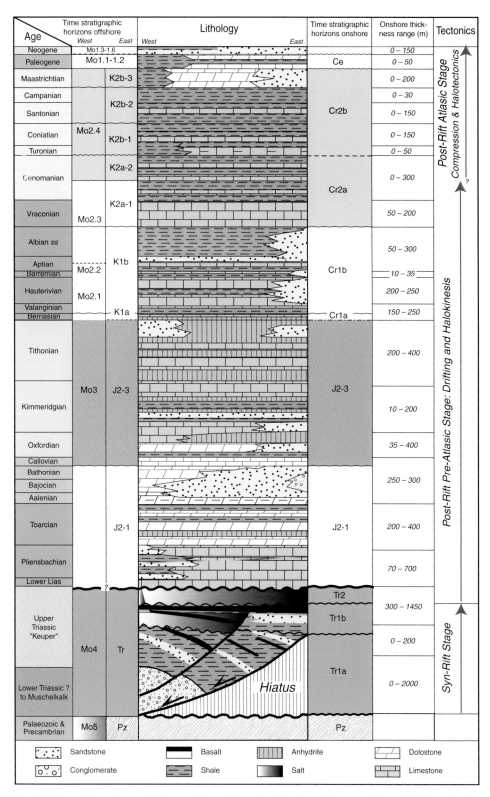

Fig. 2. Synthetic stratigraphic column of the onshore Essaouira Basin with age and correlation of seismostratigraphic subdivisions used in different zones of the study area.

1982; Heyman, 1989; Hafid, 2000, 2006). Overlying a Palaeozoic basement, three tectono-stratigraphic megasequences are differentiated (Fig. 2), from bottom to top, the: (1) syn-rift megasequence, from Lower to Upper Triassic; (2) post-rift pre-Atlasic megasequence, from Lower Jurassic to mid-Cretaceous; and (3) post-rift Atlasic megasequence, from mid-Cretaceous to the present day.

The structural and stratigraphic framework associated with each one of these megasequences is summarized below. For a detailed treatment of the general stratigraphy and structure of the area, the reader is referred to: Duffaud *et al.* (1966); Lancelot & Winterer (1980); von Rad (1982); Heyman (1989); Medina (1994); Amrhar (1995); Le Roy (1997); Hafid (2000, 2006); and Zühlke *et al.* (2004). Figure 2 shows the ages and lateral equivalents of the seismic units delimited in regional seismic cross-sections, and their stratigraphic formation equivalents as defined from surface geology by previous work (Duffaud *et al.*, 1966). Ages are based on industrial well data for the onshore and near-shore basins, and on DSDP Site 416 for the deep basin (Lancelot & Winterer, 1980).

Syn-rift megasequence

During the deposition of the syn-rift megasequence, the study area, like the rest of the Moroccan Atlantic margin, underwent regional crustal extension that opened, within a peneplained Palaeozoic basement, N–S to NNE–SSW-striking half-grabens and grabens laterally offset by E–W-striking transfer faults. Figure 3 shows these syn-rift structures in the onshore El Jadida-Agadir basins and in the near-shore platform margin. Farther offshore, the Triassic-Liassic section is deeper than can be resolved by available seismic sections. The extensional depocenters thus opened were filled with siliciclastic and evaporitic continental successions composed of at least three tectonostratigraphic sequences (Fig. 2; Olson *et al.*, 1997), that reflect the combined effect of tectonic pulses and climate changes on sedimentation.

In the subsurface of the coastal basins (Abda, Essaouira and Haha basins; Fig. 1), seismic profiles show that the uppermost of these sequences was deposited, probably during the transitional phase between rifting and drifting, in a continuous, considerably less faulted, salt-rich sag basin with extensive basalt flows (Fig. 4; Hafid, 2000, 2006). Most of the salt presently encountered at different structural levels within the studied

western Morocco basins was deposited in this shallow evaporitic sag basin which probably connected directly with the Tethys salt basin to the east.

Certain zones, such as the Necnafa depression in the western Essaouira onshore, and most of the Essaouira offshore (Fig. 1B), accumulated thick salt deposits that strongly influenced all subsequent sedimentary and tectonic events (see below). Triassic basalts have been encountered by drilling below and intercalated with salt in the Essaouira-Abda and the Doukkala basins (Hafid, 2000, 2006; Le Roy & Piqué, 2001; Echerfaoui *et al.*, 2002). Seismic evidence, however, shows that the most extensive basalt flows were emplaced just below the Triassic/Jurassic boundary (Tr2, Fig. 4), towards the end of salt deposition in the Moroccan Atlantic basin. These volcanics correspond to the spreading stage of the central Atlantic and constitute part of the widespread Central Atlantic Magmatic Province (CAMP; Marzoli *et al.*, 1999; Hames *et al.*, 2003; Olsen *et al.*, 2003; Youbi *et al.*, 2003; Gradstein *et al.*, 2004; Sahabi, 2004). In Morocco, CAMP magmatic events range in age from 197 to 203 Ma (Sebai *et al.*, 1991; Fietchner *et al.*, 1992; Knight *et al.*, 2003).

Post-rift pre-Atlasic megasequence

The post-rift pre-Atlasic stage began in the Early Jurassic, with the establishment of a widespread carbonate platform involving shallow-marine limestones and dolostones that were deposited near-shore and buried the syn-rift sequence (Fig. 2). Equivalent deeper water/basinal carbonates were deposited farther offshore. Overall, open-marine transgressive conditions prevailed until carbonate platform deposition terminated during the early Berriasian to Hauterivian. From the Barremian to the end of the Cretaceous, sedimentation was generally associated with a steadily rising sea level that resulted in the deposition of extensive marine shales and carbonates (Fig. 2; Heyman, 1989).

Post-rift Atlasic megasequence

During the Atlasic stage, the Essaouira and Cap Tafelney near-shore segments of the Moroccan Atlantic margin and their coastal basins underwent NNW–SSE compression that resulted from the Alpine collision between Africa and Iberia. In the western High Atlas, field and seismic evidence

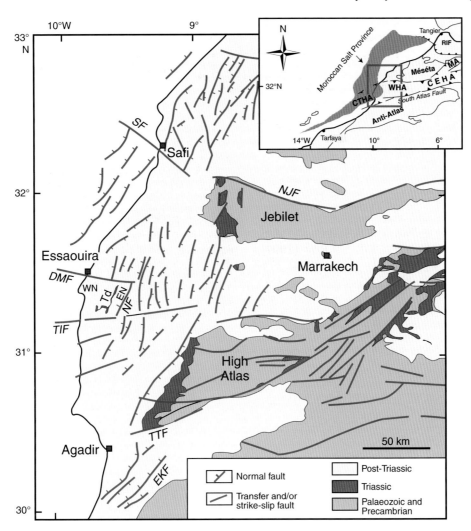

Fig. 3. Late Triassic normal and transfer faults of western Morocco. Mapping based on seismic and magnetic data for the El Jadida–Agadir coastal basins and the shallow Atlantic shelf, and existing surface maps for the Jebilet and the High Atlas (El Jadida lies on the coast, 30 km to the north of the map area). Named faults and localities: *DMF* = Diabet Meskala Fault; *EKF* = El Klea Fault; EN = East Necnafa depression; *TIF* = Tarhzout Ihchech Fault; *TTF* = Tizi n'Test Fault; *NF* = Necnafa Fault; *SF* = Safi Fault; *NJF* = North Jebilet Fault, Td = Tidsi salt diapir; WN = West Necnafa depression. After Hafid *et al.* (2005b). See Fig. 1 for additional abbreviations.

suggest that this compression started in the Late Cretaceous and reached its paroxysm in Paleogene times (Ambroggi, 1963; Fraissinet *et al.*, 1988; Froitzheim *et al.*, 1988; Frizon de Lamotte, 2000). This compression generated the onshore basins of Abda, Essaouira, Haha and Agadir, caused the inversion of selected Triassic-Liassic half-grabens, and initiated the formation of salt anticlines (Figs 1 and 3–6; Hafid, 2000, 2006). It also uplifted the Jebilet Massif and the ancient Palaeozoic massif of the High Atlas, between which the Essaouira basins evolved as a synclinorium (Hafid *et al.*, 2005a; Hafid, 2006).

In the offshore, the compression greatly disturbed the classic profile of the passive margin by uplifting its near-shore "platform" within the Cap Tafelney and the Essaouira segments. This near-shore platform appears on seismic sections very distinctly separated by a shelf margin flexure (SMF on Figs 1, 4C, 5B and 6B) from a southward-dipping post-Jurassic flexural basin: the Cap Tafelney Basin (Figs 4B and 5B). The Cap Tafelney Basin was mostly filled by thick northward-wedging Upper Cretaceous to Tertiary folded siliciclastic sediments (Fig. 5B). The folds thus generated strike NE–SW and form the Cap Tafelney High

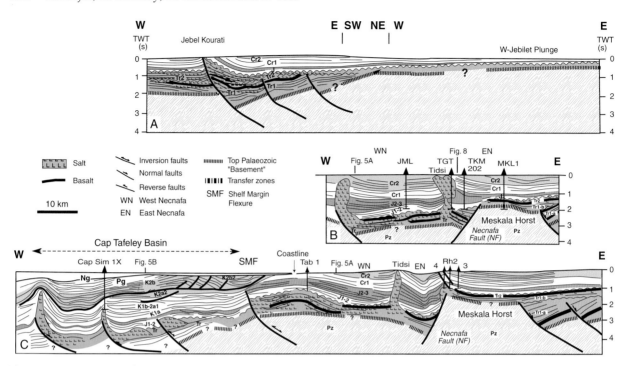

Fig. 4. Interpretations of east–west seismic sections through the study area. (A) Regional transect extending from the Chichaoua area in the east to the inverted Jebel Kourati near the coast. (B) Regional transect crossing, from east to west, the: Meskala horst; East Necnafa depression (EN); northern part of the northern branch of the Tidsi diapir; West Necnafa depression (WN). (C) Regional transect that crosses, from east to west, the: Meskala horst; East Necnafa depression (EN); southern part of the northern branch of the Tidsi diapir; West Necnafa depression (WN); present-day coastline; continental shelf; northern part of the Cap Tafelney Basin. SMF = shelf margin flexure. For location of sections, see Fig. 1.

Atlas (CTHA in Fig. 1). This underwater segment of the Atlas belt is interpreted as a system of lateral ramps that terminate the Atlas system at its intersection with the Atlantic margin, and mark the transition from a thick-skin deformation style in the onshore, to a thin-skin deformation style in the offshore (Hafid *et al.*, 2000, 2005b; Mridekh, 2002; Hafid, 2006). The basal décollement level of the Cap Tafelney folds is located above syn-rift salt and "ramps" down into the Palaeozoic continental basement underlying the coastal basins toward the east (Fig. 4C).

On a regional scale, the décollement level can be linked, in the north, to the Jebel Hadid-Jebel Kourati reverse fault which itself connects with the North Jebilet, a basement-involved reverse fault on the northern Atlas frontal ramp. In the south, it connects in the Agadir-Souss Basin to the South Atlas front (El Klea, and Tizi n'Test faults; Fig. 1B). All these major faults were interpreted by Hafid *et al.* (2000) and Hafid (2006) as merging together into a mid-crustal decoupling level revealed by previous geophysical studies (Giese & Jacobshagen, 1992).

Long regional composite seismic-profiles across the Abda-Essaouira and Haha coastal basins and their corresponding Atlantic Margin segments (Figs 4–6 and 8), illustrate the different salt-related structures that progressively formed in this large area during the three successive geodynamic stages described above. Figure 7 is a structural map of this area showing the main salt structures, as roughly mapped from the available seismic data.

Role of salt during the syn-rift stage

Structures due to an early mobility of salt are difficult to differentiate because of overprinting by more evolved subsequent structures. However, in certain zones of the onshore basins where Triassic layers are seismically resolved, and where the salt deposited was not very thick, there is some evidence of an early salt mobility that can still be seen. Figure 8 is an example from the Meskala zone where a Triassic–Liassic salt pillow (Tr1b and Tr2 sequences) is observed whose top was subsequently truncated by the base-Jurassic break-up

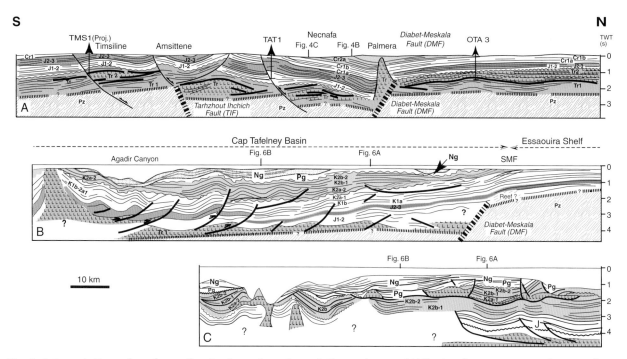

Fig. 5. Interpretations of north–south seismic sections through the study area. (A) Regional transect across the near-shore onshore Essaouira–Haha basins showing, from north to south, the: OTA platform; West Necnafa depression; outcroping Amsittene and Timsiline anticlinal structures. (B) Regional transect across the Essaouira-Haha offshore zone showing the Essaouira shelf to the north, with a shelf margin flexure (SMF) that separates it from the folded Cap Tafelney Basin, to the south. (C) Regional transect across the upper slope of the Essaouira offshore showing extensive allochthonous salt and an extruded diapir. For locations of sections see Fig. 1.

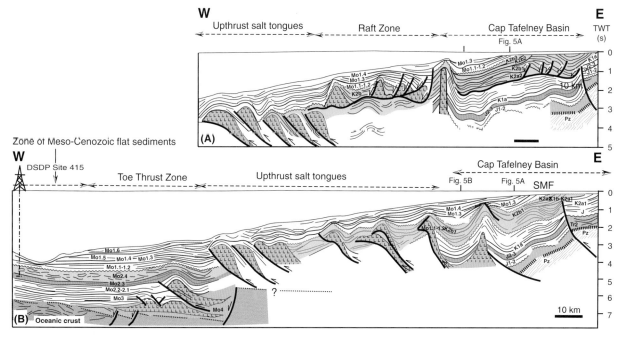

Fig. 6. Interpretations of east–west offshore seismic sections. (A) Regional transect extending from the Cap Tafelney shelf to the middle slope, showing the folded Cap Tafelney Basin, raft structures and upthrusted salt tongues. (B) Regional transect crossing the same zones as (A), but extending farther west to the deep flat-lying Mesozoic–Cenozoic sediments where DSDP Site 415 was drilled. For locations of sections see Fig. 1.

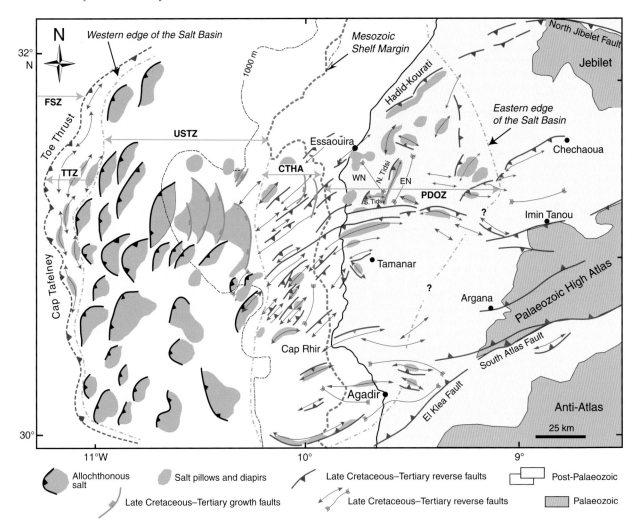

Fig. 7. Structural sketch and salt distribution map, based on available seismic data. The map shows the western termination of the Jebilet-High Atlas system and the Essaouira-Cap Tafelney segments of the Moroccan Atlantic margin. The toe-thrust zone structures are mapped after Tari *et al.* (2000). Salt tectonics related zones, from east to west, are: PDOZ = pillows and diapirs onshore zone; CTHA = Cap Tafelney High Atlas; USTZ = upthrusted salt tongues zone; TTZ = toe thrust zone; FSZ = flat sediments zone.

unconformity. Tr1b and Tr2 reflections clearly terminate against this unconformity. Salt began to move and influence sedimentation therefore, at least in the Meskala zone of the western onshore Essaouira Basin, towards the end of the rifting phase.

Role of salt during the post-rift pre-Atlasic stage

Salt continued to move within the basin during this stage and strongly controlled sedimentation especially in the Necnafa area, in western Essaouira onshore (Figs 1 and 3), as is clearly illustrated by the time isopach maps (Figs 9–11). In fact, this area contains the most evolved salt

structures of all the onshore Atlantic basins (Fig. 7). Two regional seismic transects that cross the area illustrate the character of this control (Fig. 4).

The strong thickness variations shown on the isopach maps (Figs 9–11) can be easily explained by salt withdrawal. Salt withdrawing towards rising diapirs and domes has created rapidly subsiding rim-synclinal depressions that accumulated very thick Jurassic and, to a lesser extent, Cretaceous successions. The areal/temporal migration of the axis of the rim synclines, thus created, reflects the areal/temporal distribution of the sedimentary load which diachronously drives salt towards different salt structures located in this

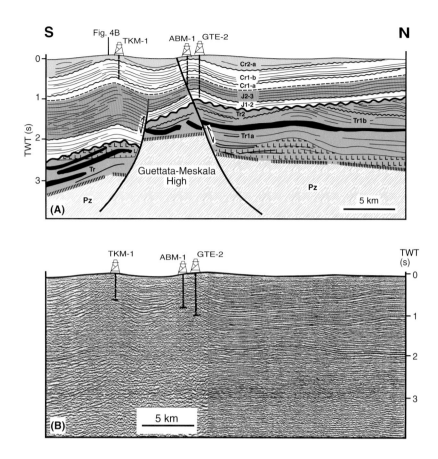

Fig. 8. North–south seismic section crossing the Guettata-Meskala Triassic high, central Essaouira onshore basin. (A) Interpretation of section illustrating a syn-rift early salt pillow. (B) Seismic profile.

area. Furthermore, the decrease with time of the importance and the number of the border synclines reflects progressive depletion of the syn-rift salt source layer at the base of the section, and welding of the top to the base of the salt.

There are two ways to account for the great expansion of the Jurassic and Lower Cretaceous successions within the Cap Sim-Necnafa depression. It may solely be due to: (a) halokinesis, that is, to salt withdrawal due to sediment loading; or to (b) halotectonics, that is, to the combined effect of gravity salt withdrawal enhanced by a regional crustal extension, at least during the deposition of Jurassic sequences J1-2 and J1-3 (Hafid, 2000, 2006; Hafid *et al.*, 2005a). The presence of an up-dip convergence of Jurassic (J1-2 and J2-3) reflections associated with the Tidsi salt dome, for example (Fig. 4B), corroborates the first option. On the other hand, the localization of these diapirs along the

strike of basement-involved extensional faults (Diabet-Meskala, Taghazout Ihchech, and Necnafa faults; Figs 3–5), and the massive thickening of the Jurassic against some of these faults (e.g. Tarzhout-Ihchech, Fig. 5A; Necnafa, Fig. 4), favours the second option. Similar fault-related thickening of the Upper Jurassic section has also been reported from the Haha and Agadir basins and may be explained by renewed extension at this time (Medina, 1984; Amrhar, 1995).

In any case, the spectacular divergence shown by the Upper Jurassic reflections towards both the Tidsi and the Palmera salt diapirs (Figs 4 and 5A), indicates that these two structures entered the diapiric stage as early as Late Jurassic. For the Cretaceous section, contrary to the conclusion of some previous surface studies (e.g. Taj-eddine, 1991), seismic profiles suggest that no faults are associated with the recorded thickness variations

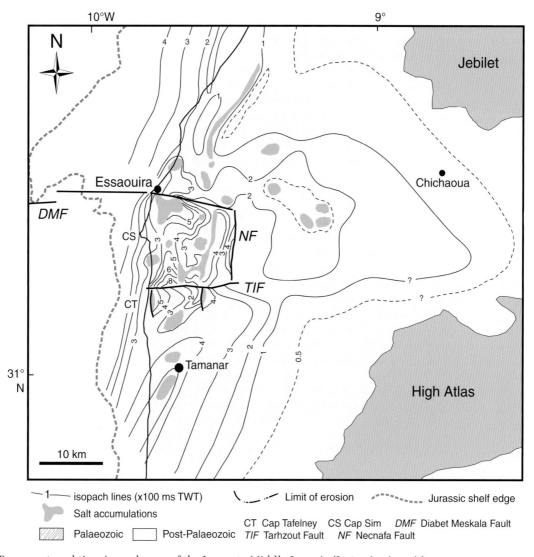

Fig. 9. Two-way travel time isopach map of the Lower to Middle Jurassic (J1-2 seismic unit).

during this period, and therefore these may be solely due to salt withdrawal (Hafid *et al.*, 2005a; Hafid, 2006).

In the offshore, gravity salt tectonics that locally characterize the present-day structure of the Essaouira slope (Fig. 6) may have started during mid-Late Jurassic times, when thermal subsidence of the Atlantic oceanic crust created a pronounced basinward slope (Tari *et al.*, 2003).

Role of salt during the post-rift Atlasic stage

The presence of a thick salt layer within the section of the study area controlled the structural styles of the Atlas compression in several important ways:

(1) In areas where the original salt layer was not very thick (east and NE of the Essaouira on-shore and Haha basins; e.g. Timsiline structure in Fig. 5A), salt was squeezed into the cores of the anticlines either from layered salt or from pre-existing pillows, similar to the one shown in Fig. 8.

(2) In the case of inverted Triassic-Liassic half-grabens, such as the Kourati-Hadid structure (Fig. 4A), salt was obliquely injected upward along the reverse fault planes (Hafid, 2000, 2006).

(3) In the case of well-developed pre-existing dia-pirs such as the Tidsi diapir in the western Essaouira onshore, and several other diapirs in the offshore Cap Tafelney Basin (Fig. 7),

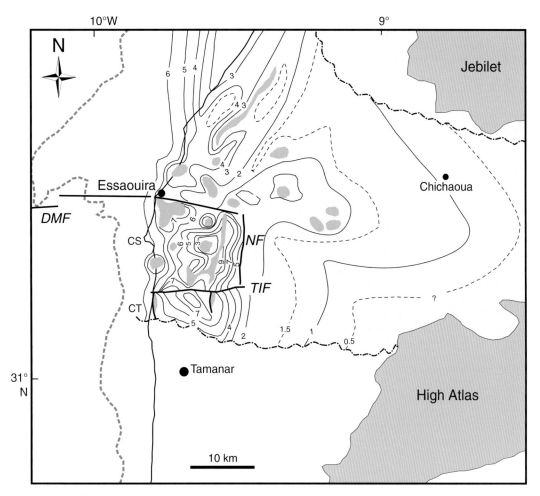

Fig. 10. Two-way travel time isopach map of the Upper Jurassic (J2-3 seismic unit). See Fig. 9 for legend.

compression simply reactivated them as compressional folds sometimes associated with a westerly verging reverse faults (Fig. 4B).

(4) In the Cap Tafelney offshore, the syn-rift salt acted as a detachment level above which the Cap Tafelney folds formed by thin-skin shortening (Hafid *et al.*, 2000; Figs 4A, 5B and 7). Salt was squeezed upwards along the reverse fault planes into inclined salt anticline cores, but except where pre-Atlasic well-formed diapirs exist (e.g. southern end of Fig. 11), it seems to stay in normal stratigraphic contact with the overlying strata.

(5) The previously formed slope structures were reactivated by Late Cretaceous to Tertiary Atlasic compression, which squeezed salt upwards into the section to form large west-verging salt tongues similar to those seen in Fig. 6. These salt tongues coalesce down-dip to form large salt canopies and salt sheets (Fig. 7).

(6) With compression starting in the Late Cretaceous (Hafid *et al.*, 2005b; Hafid, 2006), the progressively rising relief of the High Atlas to the east supplied plenty of siliciclastics that bypassed the shelf margin. Initiated locally by differential loading, Upper Cretaceous to Tertiary raft structures slid downdip either on shales (near-shore raft system in Fig. 6A) or on allochthonous salt (salt raft zone in Fig. 6A).

Figure 5C is a strike section along this later raft system. Its basal salt décollement surface extends over several tens of kilometres and its roof is dissected by gravity-driven listric growth faults that delimit supra-salt minibasins that have accumulated a thick Upper Cretaceous to Neogene siliciclastic section (Figs 5C and 6A). The inception of rafting was slightly earlier in the shale-based near-shore raft [mid-Cenomanian (K2a-2)/Santonian to Campanian (K2b2)]

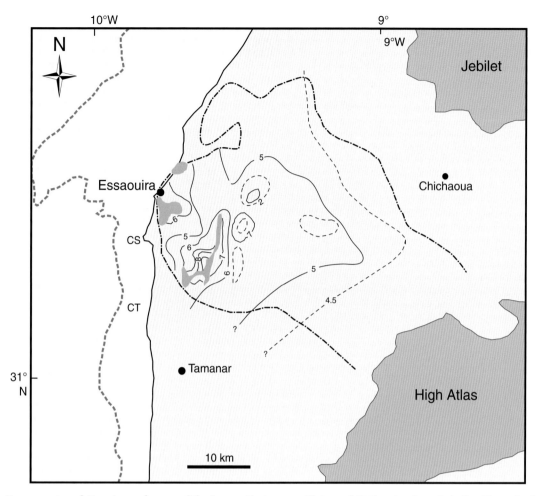

Fig. 11. Two-way travel time isopach map of the Lower Cretaceous (Cr1a and Cr1b seismic units). See Fig. 9 for legend.

than in its western neighbour, which outlived it and appears to be still active. Figure 7 shows a map view of these two raft systems. The allochthonous salt of the western raft system originated from the underlying salt tongues, from which it completely detached (Tari *et al.*, 2003), and from crestal spreading of mature diapirs that reach sea bottom in a slope position (Fig. 5C).

(7) On the lower slope of the Essaouira offshore, to the west of the salt tongues, lie prominent toe-thrust-related salt-cored folds (Figs 6B and 7) that presently represent the main exploration target in this area (Tari *et al.*, 2000, 2003). Widespread "filtering" due to salt, means that older vintage seismic lines do not image the basement underlying these folds as well as for much of the slope area. Newly acquired higher quality seismic data, however, show that these

salt-cored folds, which represent the western-most salt structures in the Essaouira offshore, are underlain by basement tilted blocks that are bounded by westerly dipping normal faults (Tari *et al.*, 2000; Tari & Molnar, 2005; Fig. 6B).

(8) To the west of the western termination of the offshore salt basin, there is a zone of non-deformed flat Mesozoic–Cenozoic sediments penetrated by DSDP Hole 415 (Fig. 6B).

DISCUSSION AND CONCLUSIONS

Salt tectonics played an important role in the tectono-sedimentary evolution of the western termination of the High Atlas and its intersecting Essaouira-Agadir segment at the Atlantic margin. This area is differentiated from the remainder of the Atlantic Moroccan salt basin by the fact that it corresponds to the only onshore extension of the Moroccan Atlantic salt basin, and that it is the only zone of this basin in which gravity-driven halokinesis was overprinted by the late Cretaceous to Tertiary compressive events associated with the High Atlas orogeny (Hafid *et al.*, 2000, 2005b; Hafid, 2006). The impact on salt tectonics of these compressive Atlasic events diminishes rapidly to the south and to the north of the Essaouira Basin (Tari *et al.*, 2003; Sahabi, 2004; Abouali, 2005; Abouali *et al.*, 2005).

There is fairly good seismic evidence in the onshore Essaouira Basin, that salt was mostly deposited in a widespread sag basin that is essentially free of normal faulting. In the offshore the data do not resolve the deep Triassic structures. Tari *et al.* (2003) suggested, based on new seismic data and on a comparison between the structural styles of the syn-rift Moroccan salt basin and the post-rift Angola salt basin, that the autochthonous syn-rift salt is characterized by its discontinuous patchy distribution that was controlled by the aerial location of the underlying half-grabens. The original thickness of the salt layer probably exceeded 1.5 km, judging by the size and frequency of the salt diapirs. This suggests that salt deposition was a very rapid process which probably averaged 1 mm yr^{-1} (Davidson, 2005). Individual salt structures, like the numerous salt tongues illustrated in Fig. 6, may have originated from autochthonous salt patches separated by tilted blocks. The timing of deformation of salt structures depends on the areal/temporal emplacement of the sedimentary load within individual sub-basins (Tari *et al.*, 2003).

Seismic evidence suggests that, in the Essaouira onshore, salt began to move in the Late Triassic and reached the diapiric stage during the Middle to Late Jurassic. Jurassic thickness variations control the style, distribution and size of these structures. The parallelism between salt structures that formed at this time in western Essaouira onshore and some faults that appear to be still active today in this area (Figs 9 and 10), suggests that halokinesis, at least in part, was coeval with crustal extension. This extension was terminated by the Early Cretaceous.

The seismic sections (Figs 4–6) and the map pattern of the salt structures at the western termination of the High Atlas and the adjacent Atlantic margin (Fig. 7), suggest that these structures show an areal/temporal distribution that is progressively more complex basinwards: salt-related tectonic styles thus varied along strike from the coastal basins towards the deep oceanic basin, forming different tectonic domains characterized by different halokinetic features (Tari *et al.*, 2000, Tari & Molnar, 2005; Hafid *et al.*, 2005a; Hafid, 2006).

Beginning at the onshore eastern edge of the salt basin and moving westward (Figs 6 and 7), the following salt-tectonic domains are encountered successively:

(1) gentle salt-cored folds and pillows and compressional diapirs associated with salt withdrawal synclines in the onshore basin (PDOZ in Fig. 7) and the Cap Tafelney basin (CTHA in Fig. 7);
(2) completely allochthonous pluri-kilometric upthrusted salt tongues and sheets that can coalesce downslope into large canopies (USTZ in Fig. 7);
(3) a toe-thrust zone characterized by salt-cored compressional anticlines that overthrust the oceanic crust and mark the western edge of the salt basin (TTZ in Fig. 7);
(4) the flat Mesozoic-Cenozoic zone (FSZ in Fig. 7), where there is no appreciable deformation and where DSDP Hole 415 was drilled.

Salt has been thrust westward onto the deep basin over some 15 to 20 km. This basinward gradient of salt-related structural complexity is due to the combined effect of the Atlantic margin geodynamics and the Atlas compressional events that spanned from Late Cretaceous to the present.

ACKNOWLEDGEMENTS

We would like to thank ONHYM, particularly its operating manager Ms Ben Khadra, for kindly making seismic profiles available for this study.

REFERENCES

Abouali, N. (2005) *Rapport de la sismique réflexion et de la sédimentologie à la reconstitution géodynamique du segment d'Ifni/Tan-Tan de la marge atlantique marocaine*. Unpubl. PhD thesis, University Ibn Zohr, Agadir, Morocco, 175 pp.
Abouali, N., Hafid, M. and **Chellai, E.** (2005) Structure de socle, sismostratigraphie et héritage structural au cours du rifting au niveau de la marge Ifni/Tan-Tan (Maroc sud-occidental). *CR Géosci.*, **337**, 1267–307.

Ambroggi, R. (1963) Étude géologique du versant méridional du Haut Atlas occidental et de la plaine du Souss. *Notes Mém. Serv. Géol. Maroc*, 157 pp.

Amrhar, M. (1995) *Téctonique et inversion géodynamiques post-rift dans le Haut Atlas occidental: Structures, instabilités tectoniques et magmatisme liés a l'ouverture de l'Atlantique central et la collision Afrique–Europe*. Unpubl. MSc thesis (Doctorat 3ème Cycle), Univ. Cadi Ayyad, Marrakech, Morocco, 253 pp.

Davison, I. (2005) Central Atlantic margin basins of North West Africa: Geology and hydrocarbon potential (Morocco to Guinea). *J. Afr. Earth Sci.*, **43**, 254–274.

Duffaud, F., Brun, L. and **Plauchut, B.** (1966). Le bassin Sud-Ouest Marocain. In: *Bassins Sédimentaires du Littoral Africain* (Ed. D. Reyre). *Publ. Assoc. Serv. Géol. Afr.*, **1**, 5–26.

Echarfaoui, H., Hafid, M., Aït Salem, A. and **Aït Fora, A.** (2002) Seismo-stratigraphic analysis of the Abda Basin (West Morocco): a case of reverse structures during the Atlantic rifting. *CR Géosci.*, **334**, 371–377.

Fietchner, L., Friedrichsen, H. and **Hammerschmidt, K.** (1992) Geochemistry and geochronology of Early Mesozoic tholeiites from Central Morocco. *Geol. Rundsch.*, **81**, 45–62.

Fraissinet, C., Zouine, E.M., Morel, J.-L., Poisson, A., Andrieux, J. and **Faure-Muret, A.** (1988) Structural evolution of the southern and northern Central High Atlas in Paleogene and Mio-Pliocene times. In: *The Atlas System of Morocco; Studies on its Geodynamic Evolution* (Ed. V. Jacobshagen). *Lecture Notes Earth Sci.*, **15**, 273–291. Springer, Heidelberg.

Frizon de Lamotte, D., Saint Besar, B. and **Bracène, R.** (2000) The two main steps of the Atlas building and geodynamics of the western Mediterranean. *Tectonics*, **19**, 740–761.

Froitzheim, N., Stets, J. and **Wurster, P.** (1988) Aspects of western High Atlas tectonics. In: *The Atlas System of Morocco; Studies on its Geodynamic Evolution* (Ed. V. Jacobshagen). Lecture Notes Earth Sci., **15**, 219–244. *Springer*, Heidelberg.

Giese, P. and **Jacobshagen, V.** (1992). Inversion tectonics of intracontinental ranges; High and Middle Atlas, Morocco. In: *Contributions to Moroccan Geology* (Eds V. Jacobshagen and M. Sarntheim), *Geol. Rundsch.*, **81**, 249–259.

Gradstein, F., Ogg, J. and **Smith, A.** (2004) *A Geologic Time Scale. Cambridge University Press*, Cambridge, 589 pp.

Hafid, M. (2000) Triassic–Early Liassic extensional systems and their Tertiary inversion, Essaouira Basin (Morocco). *Mar. Petrol. Geol.*, **17**, 409–429.

Hafid, M. (2006) Styles structuraux du Haut Atlas de Cap Tafelney et de la partie septentrionale du Haut Atlas Occidental: tectonique salifère et relation entre l'Atlas et l'Atlantique. *Notes Mém. Serv. Géol. Maroc*, **465**, 174 pp.

Hafid, M., Salem, A.A. and **Bally, A.W.** (2000) The western termination of the Jebilet-High Atlas system (offshore Essaouira Basin, Morocco). *Mar. Petrol. Geol.*, **17**, 431–443.

Hafid, M., Fedan, B., Toto, E. and **Mridekh, A.** (2005a) Structuration Jurassique post-rift à Crétacé inferieur du bassin d'Essaouira: distension ou halocinese? *Mem. N 1 MAPG* 289–307.

Hafid, M., Zizi, M, Bally, A.W. and **Ait Salem, A.** (2005b) Structural styles of the western onshore and offshore termination of the High Atlas, Morocco. In: *Geodynamics of the Magreb. CR Géosci.*, **338**, 50–64.

Hames, W., McHone, J.C., Renne, P.R. and **Ruppel, C.** (Eds) (2003) The Central Atlantic Magmatic Province: insights from fragments of Pangea. *Am. Geophys. Union Geophys. Monogr.*, **136**, 267 pp.

Heyman, M.A. (1989) Tectonic and depositional history of the Moroccan continental margin. In: *Extensional Tectonics and Stratigraphy of the North Atlantic Margin* (Eds A. Tankard and H. Balkwill). *AAPG Mem.*, **46**, 323–340.

Hinz, K., Dostmann, H. and **Fritsch, J.** (1982) The continental margin of Morocco: seismic sequences, structural elements and geological development. In: *Geology of the Northwest African Continental Margin* (Eds U. von Rad, K. Hinz, M. Sarthein, and E. Seibold,) pp. 34–60. *Springer*, Berlin.

Knight, K.B., Nomade, S., Renne, P., Marzoli, A. and **Youbi, N.** (2003) Magnetostratigraphy and ^{40}Ar/^{39}Ar dating of CAMP lava flows at the Triassic–Jurassic boundary, *Morocco. Geophys. Res. Abstr.*, **5**, 12482.

Lancelot, Y. and **Winterer, E.L.** (1980) Evolution of the Moroccan oceanic basin and adjacent continental margin: a synthesis. In: *Init. Rep. DSDP*, 50 (Eds Y. Lancelot and E.L. Winterer), pp. 801–821. US Govt Printing Office, Washington.

Le Roy, P. (1997) *Les bassins Ouest-Marocains; leur formation et leur évolution dans le cadre de l'ouverture et du développement de l'Atlantique central (marge africaine)*. Unpubl. PhD thesis, Université de Bretagne Occidental, Brest, 328 pp.

Le Roy, P. and **Piqué, A.** (2001) Triassic–Liassic western Moroccan synrift basins in relation to the Central Atlantic opening. *Mar. Geol.*, **172**, 359–381.

Marzoli, A., Renne, P.R., Piccirillo, E.M., Ernesto, M., Bellinei, G. and **de Min, A.** (1999). Extensive 200-million-year-old continental flood basalts of the Central Atlantic Magmatic Province. *Science*, **284**, 616–618.

Medina, F. (1994) *Evolution structurale du Haut-Atlas occidental et des régions voisines du Trias a l'actuel. Dans le cadre de l'ouverture de l'Atlantique Central et la collision Afrique–Europe*. Unpubl. MSc thesis (Doctorat de 3ème Cycle), *Univ. Mohammed V*, Rabat, Morocco, 271 pp.

Mridekh, A. (2002) *Géodynamique des bassins Méso-CénozoïMques de l'offshore d'Agadir (Maroc sud-occidental): Contribution à la connaissance de l'évolution Atlasique d'un segment de la marge Atlantique Marocaine*. Unpubl. PhD thesis, Univ. Ibn Tofaïl Kenitra, Morocco. 227 pp.

Olsen, P.E., Kent, D.V., Forwell, S.J., Schilische, R.W., Withjack, M.O. and **Le Tourneau, P.M.** (1997) Implications of a comparison of the stratigraphy and depositional environments of the Argana (Morocco) and Fundy (Nova Scotia, Canada) Permian-Jurassic basins. In: *Le Permien et le Trias du Maroc* (Eds M. Oujidi and **M. Et-Touhami**). *Actes de la Première et la Deuxième Réunion du Groupe Marocain du Permien et du Trias.* Oujda, Hilal, 28 pp.

Olsen, P.E., Kent, D.V., Et-Touhami, M. and **Puffer, J.** (2003) Cyclo-magneto and bio-stratigraphic constraints on the duration of the CAMP event and its relationship to the Jurassic/Triassic boundary. In: *The Central Atlantic Magmatic Province* (Eds W. Hames, J.G. McHone, P. Renne and C. Ruppel). *Am. Geophys. Union Geophys. Monogr.*, **136**, 7–32.

Sahabi, M. (2004) *Evolution cinematique Triasico-Jurassique de l'Atlantique Central: Implications sur l'évolution géodynamique des marges nord ouest africaine et est américaine.* Unpubl. PhD thesis, University Chaib Doukkali, Morroco/IFREMER, France, 2 vol.

Sebai, A., Feraud, G., Bertrand, H. and **Hanes, J.** (1991) ^{40}Ar/^{39}Ar dating and geochemistry of tholeiitic magmatism and related to opening of the Central Atlantic Rift. *Earth Planet. Sci. Lett.*, **104**, 455–472.

Taj-eddine, K. (1991) *Le Jurassique terminal et le Crétacé basal dans l'Atlas Atlantique (Maroc): Biostratigraphie, sédimentologie, stratigraphie séquentielle et géodynamique.* PhD thesis, Univ. Cadi Ayad, Morocco, Sér. 2, **16**, 289 pp. *Strata*, Toulouse, 1992.

Tari, G.C. and **Molnar, J.S.** (2005) Correlation of syn-rift structures between Morocco and Nova Scotia, Canada. Trans. Gulf Coast Section SEPM Foundation 25th Ann. Res. Conf., 132–150.

Tari, G.C., Molnar, J.S., Ashton, P. and **Hedley, R.** (2000a) Salt tectonics in the Atlantic margin of Morocco. *Leading Edge*, **19**, 1074–1078.

Tari, G.C., Ashton, P.R., Coterill, K.L., Molnar, J.S., Sorgenfrei, M.C., Thompson, P.W.A., Valasek, D.W. and **Fox, J.F.** (2001) Examples of deepwater salt tectonics from West Africa: are they analogs to the deepwater salt-cored foldbelts of the Gulf of Mexico? *Trans. Gulf Coast Section SEPM*, **21**, 251–269.

Tari, G.C., Molnar, J.S. and **Ashton, P.** (2003) Examples of salt tectonics from West Africa: a comparative approch. In: *Petroleum Geology of Africa: New Themes and Developing Technologies* (Eds T.J. Arthur, D.S. MacGregor and N.R Cameron). Geol. Soc. London Spec. Publ., **207**, 85–104.

Von Rad, U., Hinz, K., Sarnthein, M. and **Seibold, E.** (1982) *Geology of the Northwest Africa Continental Margin.* *Springer*, Heidelberg, **703** pp.

Youbi, N., Martins, L.T., Munhá, J.M., Ibouh, H., Madeira, J., Ait Chayed, E.H. and **El Boukhari, A.** (2003) The late Triassic-Early Jurassic volcanism of Morocco and Portugal in the framework of the Central Atlantic Magmatic Province: an overview. In: *The Central Atlantic Magmatic Province* (Eds W. Hames, J.G. McHone, P. Renne and C. Ruppel). Am. Geophys. Union Geophys. Monogr., **136**, 179–208.

Zühlke, R., Bouaouda, M.-S., Ouajhain, B., Bechstädt, T. and **Reinfelder, R.** (2004) Quantitative Meso-Cenozoic development of the eastern Central Atlantic continental shelf, western High Atlas, Morocco. *Mar. Petrol. Geology*, **21**, 225–276.

Int. Assoc. Sedimentol. Spec. Publ. (2011) **43**, 421–430

Carbonates and evaporites of the Upper Jurassic Arab Formation, Abu Dhabi: a petroleum exploration challenge

AHMAD S. AL SUWAIDI, MOHAMED EL HAMI, HIROSHI HAGIWARA, SABAH K. AZIZ and A.R. AL HABSHI

ADMA-OPCO, Abu Dhabi, United Arab Emirates (E-mail: asalsuwaidi@adma.ae)

ABSTRACT

The Upper Jurassic offshore Abu Dhabi consists of three formations: the Diyab (bottom), the Arab and the Hith (top). The Arab Formation contains large volumes of hydrocarbons in structural traps located in the western half of offshore Abu Dhabi. The Diyab and the Hith formations provide good source rocks and good cap rocks, respectively. Where the Hith Anhydrite is absent, as in the central and the eastern parts of offshore Abu Dhabi, no hydrocarbons are present in the Arab Formation. The Arab Formation consists predominantly of a mixed carbonate and evaporite succession, deposited on a carbonate shelf under open-marine to hypersaline conditions. Low occurrences and poor preservation of fossils in these rocks made determination of their exact age difficult: a Kimmeridgian–Tithonian age is assigned to the formation for offshore Abu Dhabi.

The Arab reservoirs show lateral and vertical facies and petrophysical property variations in a W–E direction. These variations are pronounced in central offshore Abu Dhabi, where the differentiation between Upper Arab A, B and C zone reservoirs becomes difficult. Well and seismic data show that the Hith Anhydrite and Upper Arab reservoirs (A, B and C), progressively onlap the underlying Lower Arab D reservoir towards central offshore Abu Dhabi. The onlap limit of the various intervals occurs along a NNW–SSE trend. An Arab D oolitic grainstone facies also seems to occur along the same trend. The lateral extent of the oolite facies belt is not well understood, and better well and seismic data are required for its delineation. Good potential for oil and gas accumulations is present in the Arab Formation reservoirs, both in structural and combined structural-stratigraphic traps along the onlapping edge. Proper definition of these onlapping traps represents a future challenge, and is considered to be vital for the success of future exploration in the area.

Keywords: Diyab, Arab, Hith, evaporite, petroleum reservoir, onlap, Kimmeridgian.

INTRODUCTION

The Upper Jurassic offshore Abu Dhabi consists of three formations, the Diyab (bottom), the Arab and the Hith (top). The Arab Formation represents the main reservoir interval in the region. The Diyab and the Hith formations provide good source rocks and good cap rocks, respectively (Hassan & Azer, 1983; Hassan, 1989). Where the Hith Anhydrite is absent, as in the central and eastern parts of offshore Abu Dhabi, no hydrocarbons remain in the Arab Formation.

The Arab Formation contains large volumes of hydrocarbons in structural traps located in the western half of offshore Abu Dhabi. New 3D-seismic interpretation of a giant central-offshore oil field has encouraged a search for potential new areas of hydrocarbon accumulation in regions where the Hith Anhydrite and the Upper Arab reservoirs onlap the Lower Arab reservoir, in the central offshore region. This paper incorporates results of the 3D-seismic study of a large offshore oilfield, and available well data.

STRATIGRAPHY

ADMA-OPCO's current stratigraphic nomenclature is shown in Fig. 1. The Upper Jurassic succession is provisionally named the Sila Group, which is believed to rest conformably on Middle Jurassic in western and central areas. Similarly, the

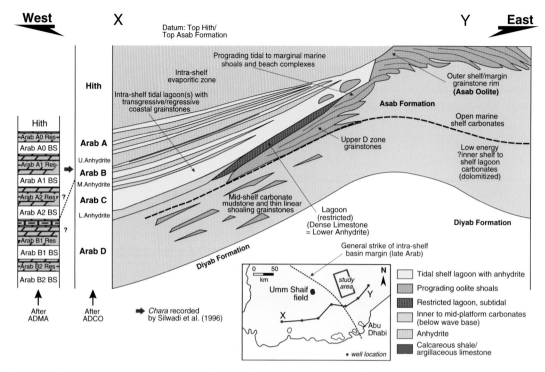

Fig. 1. Stratigraphy and schematic cross section through the Upper Jurassic of offshore Abu Dhabi. The locations of the giant Umm Shaif oilfield containing a major Upper Jurassic Arab Formation reservoir, and the present study area (red rectangle), are shown. X – Y indicates the line of section; red dots are well locations.

boundary between the Upper Jurassic and the Cretaceous is conformable, and coincides with a facies change (Habshan Formation) in most of offshore Abu Dhabi (British Petroleum, 1984; Alsharhan & Whittle, 1995).

The Arab Formation was first defined in the Dammam-7 Well of Saudi Arabia (Powers *et al.*, 1968) and was subdivided into four reservoir units, named (from top to bottom) the Arab A, B, C and D members. The Upper Arab of offshore Abu Dhabi approximately corresponds to the Arab A, B and C members of the type locality in terms of thickness and the occurrence of anhydrite intercalations; the Lower Arab comprises the Arab D Member.

The Arab Formation offshore Abu Dhabi conformably overlies the Diyab Formation, and its boundary is picked at a sharp contact from indented to flat clean gamma-ray response in the western area (Hassan, 1989). The Arab Formation is subdivided into four reservoir units: zones A, B, C, and D, in descending order (Fig. 1). Considerable lithofacies changes are recognized from the west to the east within the Upper Jurassic succession (Alsharhan & Magara, 1996; Silwadi *et al.*, 1996). As in the type area, Arab Formation zones A, B and C consist of interbedded dolostone and anhydrite

with minor limestones, representing a supratidal-intertidal-subtidal environment, whilst Arab zone D consists predominantly of limestones which were deposited in lagoonal to open-marine shelf environments. The top of the Arab Formation is picked at the base of the massive thick anhydrite of the overlying Hith Formation, which dominantly comprises thick anhydrites interbedded with thin dolostones, particularly in the western area. The proportion of anhydrite in Arab zones A, B and C decreases eastwards. Similarly, the Hith Anhydrite thins eastwards and disappears in the central offshore area (Fig. 1). As a result, the term "Hith/Arab equivalent" (Lekie, 1983) is used in the eastern area of offshore Abu Dhabi (i.e. east of the Hith Anhydrite limit) for age-equivalent beds.

ENVIRONMENT OF DEPOSITION

Following the deposition of the Diyab Formation, the offshore Abu Dhabi region was subject to a regressive cycle of deposition. In mid-Lower Arab time, the first appearance of sabkha conditions occurred in the northern area, while coral-stromatoporoid reef/oolite shoals frequently developed within lagoonal areas and prograded southwestwards (Fig. 2A and B). Oolite-coral facies representing offshore bars were formed in the southwestern area at this time (Japan Oil Development Company, 1989).

Following Lower Arab time, a sabkha-intertidal to partly subtidal complex developed over the western and central areas, represented by alternations of anhydrite and carbonate, a response to repeated fluctuations in sea level. These are represented by Arab zone A, B and C dolostones and their intercalated anhydrites (Figs 1, 3 and 4). An N–S trending shoreline developed in front of the sabkha facies (Fig. 2C), which prograded eastwards. Open-marine conditions prevailed to the east, where mudstones and wackestones of the Asab Formation were deposited (Hassan, 1989; Alsharhan & Whittle, 1995; Silawadi et al., 1996).

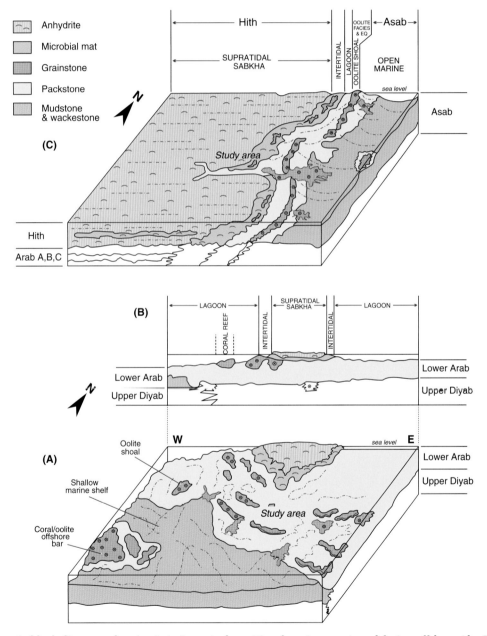

Fig. 2. Schematic block diagrams showing Late Jurassic depositional environments and facies, offshore Abu Dhabi. (A, B) Lower Arab (Arab zone D) Formation. (C) Upper Arab (Arab A, B and C zones) and Hith Anhydrite formations. The area shown represents that covered by cross-section X–Y in Fig. 1, a distance of around 230 km. Not to scale.

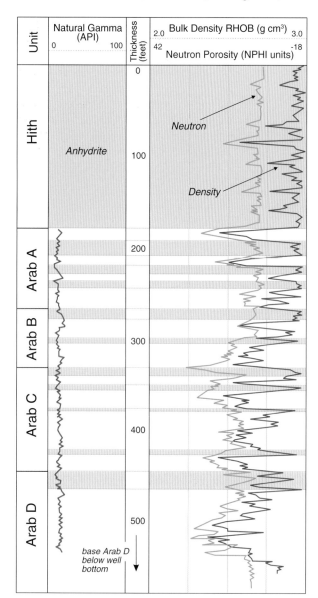

Fig. 3. Stratigraphy and wireline-log signatures of a typical well through the Hith and Arab formations, offshore Abu Dhabi. Pink bands are anhydrite.

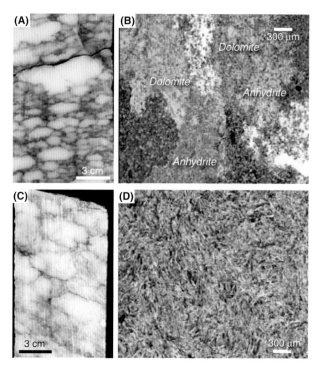

Fig. 4. Examples of anhydrite-rich samples from the Arab Formation, offshore Abu Dhabi. (A) Nodular anhydrite core. (B) Lath-shaped anhydrite crystals with dolomite in thin-section (crossed-polarized light, XPL). (C) Anhydrite nodular mosaic core. (D) Felted anhydrite in thin-section (XPL).

TECTONIC ACTIVITY

Tectonic activity in the offshore Abu Dhabi area is probably related predominantly to deep-seated basement tectonics and salt diapirism. The deep-seated basement tectonics caused the "Arabian Folds" system in this area (Fig. 5). This system generated the central Arabian Arch and the Qatar Arch, and also the dominant N–S aligned folds in the Abu Dhabi area (Marzouk & Sattar, 1993; Akbar & Sapru, 1994).

Salt diapirism is considered to be another major factor that has influenced the present-day structures in offshore areas (Marzouk & Sattar, 1993). Diapiric salt structures form the offshore islands of Das, Dalma and Sir Bani Yas, and it is reasonable to assume that many subsurface structures were also caused by salt diapirism. Salt diapirism is believed to originate in the Infra-Cambrian Hormuz salt basin, which underlies much of the Arabian Gulf (Marzouk & Sattar, 1993; Akbar & Sapru, 1994; Alsharhan & Salah, 1997). The timing of diapiric salt movement differs between individual structures but, in general, the present structures are

After mid-Upper Arab time, a mobile oolite shoal developed along the shoreline and continuously prograded eastward in a regressive sequence that probably continued to lower Habshan Formation (Early Cretaceous) time. This oolite facies, therefore, has diachronous contacts with the Upper Arab, the Hith Anhydrite, and the lower part of the Habshan Formation (Figs 1 and 2; Silwadi *et al.*, 1996).

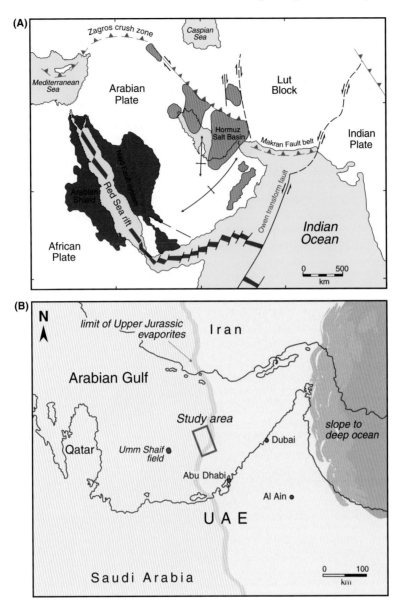

Fig. 5. Geographical and structural setting of the study area. (A) Major tectonic elements of the Arabian Peninsula (after Marzouk & Sattar, 1993). (B) Eastern Arabian Basin during Hith time (140 Ma; after ADNOC, 1990).

inferred to have grown intermittently from pre-Late Jurassic time. Besides the deep-seated basement tectonics and salt diapirism, the Zagros folding event during the Tertiary likely also had a large affect on the tectonic style in the study area (Marzouk & Sattar, 1993).

Predominant alignments of the present structures in the study area are: N–S and NNE–SSW (the dominant feature in offshore Abu Dhabi, and seen in western and central areas); E–W (related to Zagros folding during the Tertiary); NW–SE and NNW–SSE (recognized in the eastern part of offshore, and onshore Abu Dhabi (Fig. 5).

SEISMIC INTERPRETATION

A 3D-seismic interpretation study has been carried out on a large oil field in the central offshore area of Abu Dhabi (Fig. 1). The objective of the study was to identify the development and extent of Arab Formation reservoir horizons. The dataset used in the interpretation included 3D-seismic data covering an area of more than 1000 km^2 (Fig. 5B), and vertical seismic profile (VSP) and log data from five wells.

The regional geological data show that the Hith Anhydrite and the Upper Arab zones (A, B and C) exist only in the western part of the field, and they are missing over the crest and in the eastern part (Fig. 6). Normal and flattened seismic data, as well as a "coloured" inversion (Lancaster & Whitcombe, 2000) which was applied to the 3D-seismic data (Fig. 6), were used to delineate the seismic sequences of the Hith and the Arab formations. 'Coloured' inversion is a very popular and quick inversion technique that performs significantly better inversion than traditional fast-track routes such as recursive inversion, and benchmarks well against unconstrained sparse-spike inversion (Lancaster & Whitcombe, 2000).

The W–E seismic sections of the base case and the "coloured" inversion (Fig. 6) show that the interval from top Hith to top porous Arab D (i.e. Hith and Arab zones A, B and C) progressively onlap the top of the underlying Arab D near Well A, in the western part of the field. The interval seems to be missing in Wells B and C, located near the crest of the field. Indeed, the data from Well C show the presence of only 10 feet of Hith Anhydrite, which cannot be resolved from seismic data. It was noted that the onlapping edge of the Upper Arab interval is striking in a NNW–SSW direction in the western part of the field.

The exact limit of each Upper Arab zone cannot be exactly delineated due to the thin character of these units and their low impedance contrast (Fig. 6). An integrated amplitude map of "coloured" inversion data over the Arab D zone was developed for the western part of the structure, and showed some regular amplitude patterns in a belt that strikes almost N–S, passing through Well Z (Fig. 7). This belt is interpreted as a possible oolite shoal that may represent a tidal-shoal margin. A similar pattern of oolite shoals has been reported in onshore Abu Dhabi areas (Boekholt et al., 2002). This implies that the oolite belts of the Arab D are genuine and may be of a large scale, extending

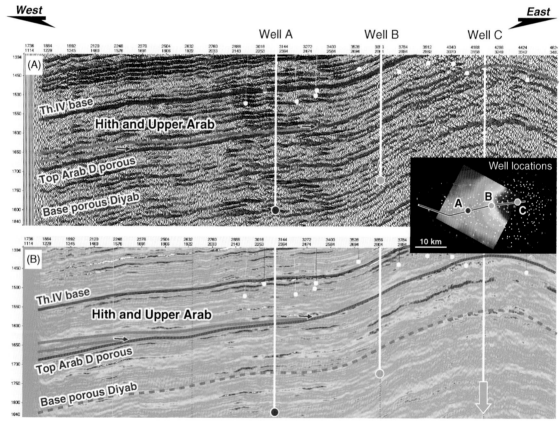

Fig. 6. Seismic cross-sections through the study area. (A) normal and (B) "coloured" inversion and flattened section, showing the Upper Arab (Arab zones A, B and C) onlapping (red arrows) the top Lower Arab (Arab zone D). Datums are: base unit Th.IV (above the study interval); top Arab D; and base porous Diyab Formation. Warm colours in (B) are higher porosity intervals. Locations of three study wells (A – C) are shown on the depth-contoured inset map (see Fig. 7 for enlarged well map).

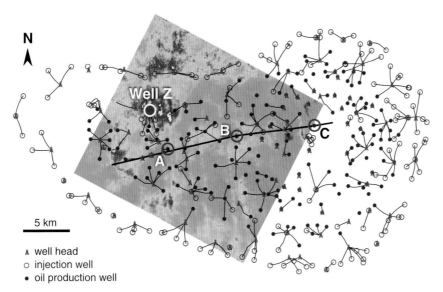

Fig. 7. Seismic amplitude attribute display over Arab zone D in the study area. Hot colours (e.g. in the vicinity of Well Z) are regions of negative amplitude, indicative of higher porosity and potential reservoir areas. Numerous exploration and production wells are located in the area.

over an area parallel to the edge of the platform margin and almost parallel to the onlapping edge of the Upper Arab.

The depth map to top Arab D generated for the western part of the structure exhibits a structural configuration similar to the overlying formation (Fig. 8). A small independent culmination 1.5 km × 3 km in width has been identified near Well A.

HYDROCARBON GENERATION AND ACCUMULATION

ADMA-OPCO carried out geochemical studies (1980–1984, unpublished data) and clarified the origin of Arab, Thamama, Mishrif and shallower oils in offshore Abu Dhabi, incorporating onshore data from ADCO studies. The Arab Formation oil

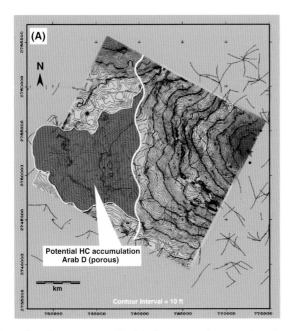

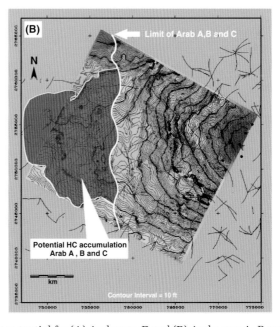

Fig. 8. Depth map to top Arab D for the study area, showing the potential for (A) Arab zone D and (B) Arab zone A, B and C hydrocarbon accumulations. Warm colours on the contour base map indicate structurally higher position (shallower depths); cools colours are lows. Thin lines are well orientations (see Fig. 7).

was generated from a Diyab source-rock in the western offshore area during Late Cretaceous times. Generated oil migrated updip towards structural highs to the north and accumulated in the present structures. The Hith Anhydrite acted as a cap rock to the Arab-reservoired oil in the western offshore area. In the eastern area, by contrast, oil migrated into the Lower Cretaceous Thamama Group due to the lack of the Hith Anhydrite seal. The Diyab source rock was deposited in an intra-shelf basin where euxinic conditions in the western offshore area were favourable for organic preservation during Lower Diyab time. Another source area for Thamama oil is presumed to exist in eastern and southern offshore areas.

The main structures present offshore are interpreted to have formed by diapirism of deep-seated Infra-Cambrian Hormuz salt and probably have grown intermittently from early Cretaceous time. Late Cretaceous palaeostructures, when oil generation started, have similar features to those of the present day. Oil migrated through structurally favourable paths to fill available reservoirs. Hydrocarbons found to date in Arab reservoirs are all located in the western and central areas of offshore Abu Dhabi; none have been confirmed in the eastern area.

POTENTIAL TRAPS

The northern offshore area of Abu Dhabi remained structurally in a higher position, probably from the Early Cretaceous to the present. In addition, an eastward structural dip was developed when the Ras Al Khaimah Trough was formed in the Tertiary (Hassan & Azer, 1985). Throughout this time, each structure had been growing independently. Taking migration paths into consideration, the possibility of hydrocarbon entrapment in the structural traps to the west of the Hith edge is higher than in the stratigraphic/diagenetic traps to the east of it.

In the western offshore areas, Arab zones A, B, C and D have been confirmed as the main oil and gas reservoirs (Hassan & Azer, 1985). The remaining unproven structural traps of the Arab Formation are located in the areas surrounded by the northern and central offshore producing fields. Optimal reservoir facies are developed in these areas. If structures are proved to exist here, the possibility of finding oil in the Arab Formation is very high.

Unconventional traps are expected to occur in areas where intense changes of depositional

environment take place or near the depositional limits of the main reservoir zones. The oolitic grainstone facies of the Arab D is a good reservoir unit that shifts its position diachronously and may develop in isolated bodies surrounded by less porous and less permeable facies near the platform margin. Furthermore, the depositional limit (onlapping edge) of the Arab A, B and C zones, aided by three-way dip-closure, may create potential traps for hydrocarbon accumulation (Fig. 8). The geological and seismic interpretation carried out for the present study area, has demonstrated that such traps may be present in the western part of the field. Log data for Well A show a 50-ft interval of good hydrocarbon saturation in the Arab Formation.

CONCLUSIONS

The Arab reservoir zones in central offshore Abu Dhabi are characterized by lateral and vertical lithofacies and reservoir variations. These are most pronounced near the Upper Arab onlapping edge and near the Hith Anhydrite pinch-out. The oolitic grainstones of the Upper Arab also show significant lateral variations in a W–E direction.

Well and 3D-seismic data have demonstrated that the onlapping edge of the Upper Arab can be delineated using a seismic stratigraphic approach, complemented by the "coloured" inversion technique. The interpreted data from the studied field and the surrounding areas have shown that the onlapping edge of Arab zones A, B and C and the oolitic facies of Arab D are aligned in a NNW–SSE direction, parallel to the Hith Anhydrite pinch-out.

Potential hydrocarbon accumulations in the Upper Arab zones A, B and C may occur near the onlapping edge, provided updip 3-way dip-closures exist. The oolitic Arab D zone also has reservoir potential along the NNW–SSE trending belt. Future exploration drilling in the western part of the field will provide vital data to consolidate the trapping concepts and the hydrocarbon potential of these productive reservoirs.

ACKNOWLEDGEMENTS

The authors gratefully acknowledge the management of Abu Dhabi National Oil Company (ADNOC) and Abu Dhabi Marine Operating Company (ADMA-OPCO) for permission to publish this paper.

REFERENCES

Abu Dhabi National Oil Company (ADNOC) (1990) Annual Company Report.

Akbar, M. and **Sapru, A.** (1994) In-situ stresses in the subsurface of "Arabian Peninsula" and their affect on fractures morphology and permeability. *6th Abu Dhabi Int. Petrol. Exhib. Conf. (ADIPEC) 16-19 October, ADSPE,* **99**, 162–180.

Alsharhan, A.S. and **Magara, K.** (1996) The Jurassic of the Arabian Basin: facies, depositional setting and hydrocarbon habitat. In: *Pangaea: Global Environment and Resources* (Eds A.F. Embry, B. Beauchamp and D.J. Glass), *Can. Soc. Petrol. Geol. Mem.,* **17**, 397–412.

Alsharhan, A.S. and **Salah, M.G.** (1997) Tectonic implications of diapirism on hydrocarbon accumulation in the United Arab Emirates. *Bull. Can. Petrol. Geol.,* **45**, 279–296.

Alsharhan, A.S. and **Whittle, G.L.** (1995) Carbonate-evaporite sequence of the Late Jurassic, southern and southwestern Arabian Gulf. *AAPG Bull.,* **79**, 1608–1630.

Boekholt, M.P. (2002) Direct hydrocarbon indicators in Cretaceous and Jurassic carbonate reservoirs, onshore Abu Dhabi. *GEO 2002 5th Middle East Geosci. Conf. Exhib. Abstracts,* 217–218.

British Petroleum (1984) *ADMA-OPCO Biostratigraphy Study (Phase II), Offshore Abu Dhabi.*

Hassan, T.H. (1989) The Lower and Middle Jurassic in offshore Abu Dhabi: stratigraphy and hydrocarbon occurrence. *Soc. Petrol. Eng., SPEM Conf. Bahrain,* SPE 18011-MS, 847-853, doi: 10.2118/18011-MS.

Hassan, T.H. and **Azer, S.** (1985) The occurrence and origin of oil in offshore Abu Dhabi. *Soc. Petrol. Eng., SPEM Conf. Bahrain,* SPE 13696-MS, 143-149, doi: 10.2118/13696-MS.

Japan Oil Development Company (1989) *Upper Jurassic Study, Offshore Abu Dhabi.*

Lancaster, S. and **Whitcombe, D.** (2000) Fast-track 'coloured' inversion. *SEG Expanded Abstracts,* **19**, 1572.

Lekie, G. (1983) *Regional Stratigraphy Review.* Jurassic Side-Group PDD/412.

Marzouk, I.M. and **Sattar, M.A.** (1993) Implication of wrench tectonics on hydrocarbon reservoirs, Abu Dhabi, U.A.E. *Soc. petrol. eng., middle east oil tech. conf. exhib.,* SPE 25608-MS, 119-130, doi: 10.2118/25608-MS.

Partyka, G. Gridley, J. and **Lopez, J.** (1999) Interpretational applications of spectral decomposition in reservoir characterization. *Leading Edge,* **18**, 353–360.

Powers, R.W. (1968) Saudi Arabia: Lexique Stratigraphique International, **3**. Centre National de la Recherche Scientifique, Paris, 177 pp.

Silwadi, S. Kirkham, A. Simmons, M. and **Twombley, B.** (1996) New insight into regional correlation and sedimentology, Arab Formation (Upper Jurassic), offshore Abu Dhabi. *GeoArabia,* **1**, 6–27.

Int. Assoc. Sedimentol. Spec. Publ. (2011) **43**, 431–464

Selenite facies in marine evaporites: a review

FEDERICO ORTÍ

*Departament de Geoquímica, Petrologia i Prospecció Geologica, Facultat de Geologia, Universitat de Barcelona,
C/Martí i Franqués s/n; 08028 Barcelona, Spain (E-mail: f.orti@ub.edu)*

ABSTRACT

The term "selenite facies" describes evaporitic, primary precipitates of coarse-crystalline gypsum formed on a depositional floor. The crystals comprising these autochthonous facies commonly display competitive growth patterns. Modern marine selenite facies have been studied in coastal salt works (evaporative salinas) and in some natural coastal lakes where gypsum is precipitating or has been precipitating in recent times. In ancient marine formations, selenite crystals display two major facies groups: bedded to massive selenites; and domal selenites. Good examples of selenite facies occur in the Miocene of the Mediterranean region (Messinian) and in the Carpathian Foredeep (Badenian). In ancient examples, selenite facies have been preserved as pseudomorphic anhydrite at depth and as pseudomorphic secondary gypsum at outcrop. In such cases, their identification may be difficult, owing to the intense diagenetic transformations undergone by many evaporites. Ancient occurrences of selenite facies cover a wide range of settings: (1) flanking selenite shelves; (2) narrow flanking selenite shelves; (3) selenite platforms covering salt-filled basins; and (4) selenite basins. All occurrences are sensitive to changes in sedimentological conditions and require specific environmental configurations. In particular, they depend on the existence of an outflow of heavy brines to prevent oxygen depletion in the water column, and the accumulation of chlorides on the platforms and in the basins.

Keywords: Evaporite, gypsum, selenite, evaporitic platform, evaporitic basin, diagenesis.

INTRODUCTION

The term "selenite facies" is used here for evaporitic, primary, coarse (>2 mm long), euhedral gypsum crystals, which form by subaqueous free-growth precipitation on a depositional floor. These crystals are characterized by a number of features: (1) they commonly display competitive growth (elongated upwards, but not exclusively); (2) they are arrayed in beds with variable thicknesses and fabrics; (3) they usually develop twins or intergrowths, which can be distinctive for each evaporite formation; and (4) they often exhibit zonation due to the presence of matrix inclusions. Although selenite facies are found both in marine and non-marine evaporites, the focus here will be on marine deposits.

The study of selenite facies of marine origin may be undertaken in many modern coastal salt works (evaporative salinas) and in some coastal saline lakes (natural salinas), where gypsum is currently precipitating or has been precipitating in recent times. The sedimentological observations derived from these modern settings are important for understanding the significance of the selenite facies in the geological past.

In ancient evaporitic platforms and basins, gypsum deposits are constituted by two major facies groups:

(1) Fine-grained gypsum facies. In deep basins, gypsum is characterized by fine-grained, basinal cumulates (laminae, thin beds) in association with fine- to coarse-grained slope clastics (gypsarenites, gypsum turbidites, gypsrudites, and gypsum breccias). On platforms, gypsum cumulates are not so finely laminated as in the basins, and the clastic facies are mainly composed of gypsarenites and minor gypsrudites.

(2) Selenite facies. This group involves autochthonous, bedded precipitates of relatively pure, coarse-crystalline gypsum.

In ancient evaporite successions, selenite facies are not as common as fine-grained gypsum facies. Moreover, identification of the former is sometimes difficult, given the intense diagenetic transformations undergone by many deposits. However, the term "selenite facies" refers not only to the facies preserved as primary gypsum, generally not older than Miocene, but also to the facies in ancient formations preserved as pseudomorphic anhydrite at depth or as pseudomorphic secondary gypsum at outcrop. Criteria for deciphering the selenitic origin of these pseudomorphic occurrences are fundamental for evaporite interpretation.

Research on the primary selenite facies has been carried out mainly in three types of evaporitic formations and settings: (1) Messinian of the Mediterranean region; (2) modern salinas, both in the evaporative salt works and in the Holocene salinas of South Australia; and (3) Badenian of Poland and Ukraine (Carpathian Foredeep).

Research on the Messinian (Upper Miocene) selenite facies has involved a large number of workers. In Sicily, pioneer observations made by Mottura (1871) were followed by Ogniben's reports (1954, 1955, 1957) and papers by Decima & Wezel (1971) and Hardie & Eugster (1971). Subsequently, a number of studies were carried out on Sicily, in the Apennines, on Cyprus and the Ionian Islands, and in other regions around the Mediterranean, which were published in two volumes by Drooger (1973) and Catalano et al. (1976), and Dronkert (1977) and Orti & Shearman (1977) also published articles on the Messinian selenite facies of SE Spain.

The study of selenite facies in coastal salt works (evaporative salinas) was initiated by Schreiber & Kinsman (1975), Schreiber et al. (1977) and Schreiber & Schreiber, 1977). This was continued with detailed studies carried out in southern France and SE Spain by, amongst others, Geisler (1982) and Orti & Busson (1984). Prior to the selenite studies in salt works, some papers made reference to the gypsum crystals of Marion Lake (Yorke Peninsula) in South Australia (Dickinson & King, 1951; Crawford, 1965; Goto, 1968; Hardie & Eugster, 1971; Schreiber & Kinsman, 1975; Eugster & Hardie, 1976; Shearman & Ortí, 1976). A paper by Warren (1982) on the assemblage of Holocene coastal salinas in South Australia was fundamental to understanding the hydrological conditions governing the growth of large selenites and to the application of actualism to the genesis of ancient selenites.

The study of the selenite facies in the Badenian (Middle Miocene) of the Carpathian Foredeep was initiated by Kwiatkowski (1972). During the following decades research was continued by a number of authors, including: Kubica (1992); Peryt 1996, 2001; Peryt & Jasionowski (1994); Kasprzyk (1993); Bąbel 1986, 1987, 1999b, 2002a, 2004a, 2005a, 2005b; and Petrichenko et al., (1977). More recently, other papers on the Badenian gypsum deposits in Ukraine and the Czech Republic (Peryt, 1996, 2001) have been published.

SELENITE CRYSTALS, FACIES AND GEOCHEMISTRY

Selenite crystals

A large number of crystalline forms, habits and twins have been described in gypsum sediments. However, only a few of them have generated selenite facies in marine successions. The main characteristics of these facies are as follows:

Crystals, twins, intergrowths, and fabrics

The crystals making up these selenite facies are single (individuals) or twins, and usually display free-growth fabrics and competitive growth patterns. The morphology and orientation of these gypsum crystals are shown in Fig. 1A. The most common gypsum twin in the Messinian formations is along the (100) plane (Fig. 1B). In this twin, the crystal grows by means of the (120) prism and either the (111) hemipyramid or a curved surface (Shearman & Ortí, 1976; see below). Less frequent is the twin along the ($\bar{1}$01) plane, which is present in Eocene deposits of the Paris Basin. The re-entrant angles are different in these contact twins (Fig. 1B).

Beside these twins, a crystalline intergrowth involving a composition surface occurs in the Badenian formations (Fig. 1C). This intergrowth is formed by the intersection of the (120) prism and the ($\bar{1}$03) face along the composition surface; on the two sides of this surface the cleavage plane is not coincident. This intergrowth has a very low cohesion at the composition surface (the intergrowth easily breaks down through it; Bąbel, 1987, 1990).

Both contact twins and intergrowths are commonly formed by the stack of pairs of subcrystals (also termed "segments", "inverted chevron-shaped twin blocks" or "aggregates"; Fig. 2A). These pairs of subcrystals are bounded by fine

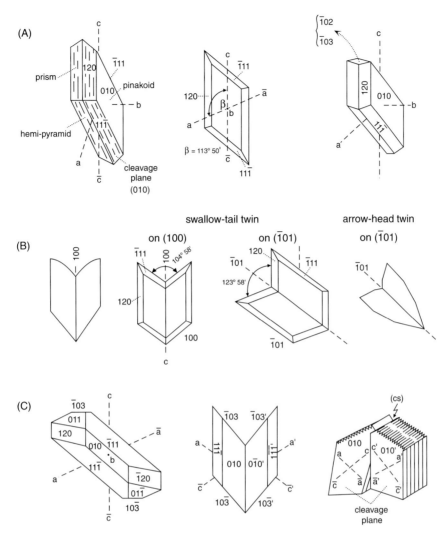

Fig. 1. Gypsum crystals, twins and intergrowths of most interest in marine evaporites. (A) Morphological orientation of the gypsum crystal corresponding to the lattice, selected by Palache *et al.* (1951). Adapted from several figures by Shearman (1971). (B) Most common contact twins in selenite facies. Swallow-tail twins: on (100), predominant in the Messinian (re-entrant angle: 104° 58′) and on (Ī01), known as Montmartre twin (re-entrant angle: 123° 58′). The Montmartre twin is commonly modified by other forms which give the crystal a tapering, arrow-head appearance. Adapted from several figures by Shearman (1971). (C) Left: single crystal of gypsum (after Palache *et al.*, 1951). Centre: the (Ī01) twin of gypsum with (Ī03) face. Right: the most frequent setting of the gypsum crystals in the Badenian intergrowths. Note the different orientations of the cleavage planes in the two halves of the intergrowth; (c.s.): composition surface of the intergrowth. Adapted from Bąbel (1987) and Bąbel (1990).

bands of sediment matrix. The stacking of subcrystal pairs apparently operates as an active nucleation on the re-entrant angles and as an epitaxial crystallization.

In the twinned selenite beds of the Messinian in Sicily, Mottura (1871) observed that the twin plane was perpendicular to bedding and the re-entrant angle invariably pointed upwards ("Mottura's Rule"). This rule is valid in all Messinian basins and also in many selenite basins of other ages, as well as in some twinned selenites of modern salt works. Although the rule was described for the (100) contact twin, as a general growth pattern, it seems to be valid for other twins and intergrowths.

Nevertheless, a non-marine occurrence has been cited in the Miocene lacustrine deposits of the Madrid Basin (Spain), where the geometry of the (100) twin is inverted: the re-entrant angle systematically points downward (Rodríguez Aranda *et al.*, 1995). These unusual crystals

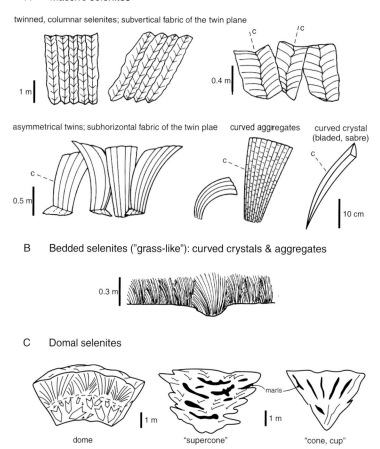

A   Massive selenites

twinned, columnar selenites; subvertical fabric of the twin plane

1 m

0.4 m

asymmetrical twins; subhorizontal fabric of the twin plae   curved aggregates   curved crystal (bladed, sabre)

0.5 m

10 cm

B   Bedded selenites ("grass-like"): curved crystals & aggregates

0.3 m

C   Domal selenites

1 m   marls   1 m

dome   "supercone"   "cone, cup"

Fig. 2. Selenite facies. (A) Massive selenites, symmetrical and asymmetrical twins; curved aggregates and single crystals. Adapted from Ortí & Shearman (1977). (B) Bedded selenites: "grass-like" variety. Adapted from Ortí & Shearman (1977). (C) Domal selenites: domes, cones and supercones.

("Christmas-tree gypsum") are developed horizontally due to a higher growth rate of the (010) face, and the (120) and (100) faces. The growth of the (111) face would be slow and accompanied by a vertical, discontinuous (seasonal) development.

Many twins and intergrowths in marine successions are symmetrical and are arranged in a subvertical fabric (in the sense of Mottura's Rule). In these twins, columnar or palisade selenites develop (Fig. 2A). However, asymmetrical twins and intergrowths, in which one half of the twin is well developed and the other is poorly developed ("aborted") are also common (Fig. 2A). In this case, the twin fabric is subhorizontal: both the twin plane and the c-axis of the crystals are at low angle (<45°, in general) on the bedding plane. Moreover, the subcrystals tend to be curved. Some selenite layers are formed by the juxtaposition of aggregates of subcrystals occurring in parallel clusters; when

seen in detail, each aggregate corresponds to one arm of a strongly assymetrical twin ("grass-like" selenites, Fig. 2B).

In the Messinian twinned crystals, the trace of the junction between the two arms is not always linear, but may zig-zag along the twin blocks (Fig. 2A, upper right drawing). The re-entrant angle varies from acute in some crystals to obtuse in others, and may even curve outwards until the sides come to extend almost at a right angle to the twin junction.

Selenites, single crystals, twins and intergrowths, commonly develop curved faces or surfaces, a fact that may strongly influence the morphology of the selenite beds. Some of the terms coined in the literature for these crystals are "bladed", "bent", "sabre", and "sabre-like". The curved surface is passive or has a slow growth rate. Lens-shaped crystals are also common in gypsum, which can be produced by various soluble organic compounds selectively absorbed on the (111) face; these compounds inhibit the growth parallel to the crystal c-axis (Cody, 1979; Cody & Cody, 1988).

Crystal zonation

Selenites commonly exhibit zoning due to the presence of thin films parallel to the crystal faces. These zones are made up of tiny solid inclusions, such as particles of the sedimentary matrix (clay, carbonate grains), diatoms, brine-shrimp eggs, or elongated faecal pellets. They may also include organic matter and "algal" filaments; this is because selenite growth is often associated with microbial mats.

The space between the zones is commonly less than one millimetre, suggesting seasonal growth. Active crystal faces, such as the (120) prism, are relatively clear (poor in solid inclusions), whereas passive faces (curved surfaces) such as the one dominating the re-entrant angle of the (100) twins, can be particularly dirty (rich in solid inclusions). Because of this, some selenites appear to be composed of a turbid "V-shaped core" and two transparent lateral zones, resembling syntaxial overgrowths.

Crystal growth interruptions

The interruption of selenite growth may be caused by a number of factors such as brine dilution, brine oversaturation, or the deposition of sediment on the crystal apices. In the case of brine dilution, the selenite apices are truncated by dissolution surfaces, and irregular horizons can crosscut the selenite layers. Moreover, sediment laminae or films parallel to the bedding plane and covering the dissolution surfaces may crosscut both crystals and zoning; these laminae become poikilitically enclosed within the selenites during the subsequent growth episode (Warren, 1982). In the case of brine oversaturation, halite precipitation (Schreiber & Schreiber, 1977) or accumulation of fine-grained gypsum laminae (see below) prevent the growth of the selenites, whose apices remain unaffected.

Selenite facies

Bedding and associated structures

Selenite layers have variable thicknesses, from a few centimetres to several metres, and may reach up to some tens of metres. Two main types of selenite facies, stratiform and domal, can be distinguished. The most common stratiform facies are (Fig. 2A and B): bedded or banded selenites (beds between a few cm and 0.5 m thick) and massive selenites (beds >0.5 m thick).

In the stratiform facies, rows of single crystals or twins commonly stand upright one beside the other ("palisade selenites"); in general, at any one level the crystals are all approximately of the same size (Fig. 3A and B). In this facies, however, the

Fig. 3. Stratiform selenite facies. (A) Bedded selenites. Beds of selenite crystals and small clusters (nucleation cones) overlying millimetre-sized laminae of fine-grained gypsum. Pencil for scale (circled). Messinian Upper Evaporite, Eraclea Minoa section (Sicily). (B) Massive selenites. Columnar, symmetrical twins. At the top of the picture the twins are progressively oriented toward the right side. Height of the picture is about 4 m. Quarry at San Miguel de Salinas (Alicante, SE Spain). Messinian, Lower Evaporite.

selenite crystals may vary considerably in size and twin morphology: (1) symmetrical twins can be elongated and narrow (columnar, in palisade) or short and wide (Figs 2A, upper part, and 3B); (2) twins can be strongly asymmetrical, with the developed arm forming adjacent aggregates of subcrystals that are curved to a variable degree (Figs 2A, lower part, and 4A and B); (3) selenites can be single crystals with variable orientations, from random to preferential; and (4) selenites can grade upward in the beds in size, fabric or morphology: they commonly show a fining trend (upward decreasing size) or may grade from symmetrical to asymmetrical twins. In some cases, however, the selenitic beds are formed by elongated crystals with horizontal or subhorizontal orientation ("non-palisade selenites").

Domal facies correspond to selenitic structures that can be interpreted as having had a positive relief during their growth on a depositional floor. These facies mainly belong to two types: domes and cones (Fig. 2C). Domes have convex upward boundaries both at the top and the base, although in some cases the base is flat. These structures may reach diameters up to >10 m. Moreover, they may display a hemi-cylindrical, elongated shape.

Cone facies, or cup facies, have flat or convex upward tops, but the bases are acute (conical) and are found sunk within the floor sediments (Fig. 5A). In the Messinian literature, cones of

Fig. 4. Domal selenites at San Miguel de Salinas (Alicante, SE Spain), Messinian Lower Evaporite. (A) Top of a layer of selenite domes. The domes are integrated by aggregates of slightly curved subcrystals. Each aggregate corresponds to the developed half of a strongly asymmetrical twin. Hammer shaft is 30 cm long. (B) Close view of the subcrystals forming an aggregate. These subcrystals are broken by the (010) cleavage plane; on the cleavage surfaces the traces of the (120) prism are seen indicating the orientation of the c-axis. Hammer shaft is 30 cm long. (C) Top view of the aggregate on a dissolution surface. The subcrystals composing the aggregate have the same orientation. The triangular shape of each subcrystal is composed of the traces of the prism (120) and the traces of the curved surface. The turbid appearance of the latter is due to abundant matrix *in situ* inclusions. Scale in cm.

Fig. 5. Domal selenites. (A) Selenite cones (supercones) totally surrounded by marine marls. On the left side of the picture other smaller cones are seen. San Miguel de Salinas (Alicante, SE Spain). Scientist is Dr İbrahim Gündoğan (March 1998). (B) Pseudomorphic structure of a selenite dome (note the flat base). The white material is alabastrine secondary gypsum. Evaporite unit (Chicamo Gypsum) in the Fortuna Basin (Murcia, SE Spain). Upper Tortonian–Lower Messinian.

relatively small size (<1 m, in general) were termed "cavoli" (or "cauliflower") by Richter-Bernburg (1973) and "nucleation cones" by Lo Cicero & Catalano (1976), and those of great size (>1 m, in general) were named "supercones" by Dronkert (1976, 1977). Very large domes are known also in the Badenian of Poland (Bąbel, 1999a). Both domes and cones are composed of single crystals or twins, with fabrics varying from random to radial (radial aggregates) or to preferentially oriented in one direction. In general, radial or preferentially oriented fabrics suggest competitive growth and faster growth rates of the selenites, whereas unoriented fabrics suggest non-competitive growth and slower growth rates.

Large domes and supercones can be made up of long, curved crystals (up to 1 m in length) either single or twinned. In the latter case, however, the twins are strongly asymmetrical. These curved crystals may display radial orientations or be preferentially oriented in one direction. All of these structures, which may be up to several metres in height and even greater in diameter, resemble certain types of biohermal or microbial constructions.

Vertical and lateral evolution of selenite facies

The presence at the base of many selenite layers of small domes (cavoli), followed upwards by columnar selenites, is a common feature in the Messinian. In other cases, however, bedded or massive selenite facies at the base of the layers are followed upwards by domes and supercones.

Facies gradation is observed also horizontally; for instance, columnar selenites can grade laterally into domal structures or into calcareous-gypsiferous stromatolites. Several factors could exert an influence on all these vertical and horizontal gradations.

Framework porosity

Selenite facies commonly display low porosity. In general, the crystals are arranged in a dense, interlocking packing. In some layers, however, some intercrystalline porosity can be observed without any evidence of diagenetic dissolution. In this case, apparently the depositional (framework) porosity is partly preserved. In the Badenian intergrowths, Babel (1987) assigned such an origin to the facies known as "skeletal gypsum". In some cases, early cementation of the selenites by halite may have favoured this preservation.

Selenite geochemistry

A number of studies have been carried out to determine the content of minor and trace elements in the primary gypsum crystals of the selenite facies. It was sought to establish some possible salinity indicators and to determine the evolution of the precipitating brines. These studies were based on results reported in some pioneer papers (e.g. Usdowski, 1973; Butler, 1973; Kushnir, 1980); one major conclusion of these studies, using data from the gypsum facies in coastal salt works, is that strontium is the most significant salinity indicator (Kushnir, 1982; Geisler, 1982; Ortí et al., 1984; Dronkert, 1985; Geisler-Cussey, 1985; Utrilla, 1985). Strontium data have been reported for the selenite facies in Messinian (Geisler, 1982; Dronkert, 1985; Michalzik et al., 1993; Rosell et al., 1994, 1998) and Badenian evaporites (Kasprzyk & Osmolski, 1989; Kasprzyk, 1994; Rosell et al., 1998). The Na, K and Mg contents of the gypsum facies have also been investigated in Messinian deposits, including the selenitic facies of the Betic basins (Playà, 1998; Lu et al., 2002) and the gypsum/anhydrite facies of the eastern Mediterranean region (Kushnir, 1982).

One of the major problems with the correct interpretation of the Sr data is the common presence of tiny celestite crystals associated with the gypsum. Given that the presence of this mineral will strongly influence the final Sr results, celestite should be eliminated prior to gypsum analysis

(Playà & Rosell, 2005). Moreover, the possible contribution of fluid inclusions to determinations of Na, Cl, Mg and K in gypsum also needs to be taken into account (Lu et al., 1997).

PRESERVATION OF SELENITES AS PSEUDOMORPHS

In ancient evaporite formations, the presence of selenite facies is recognized by means of pseudomorphs of crystals and structures (Fig. 5B). References to pseudomorphic features after gypsum selenites are common in the literature, but detailed studies of ancient pseudomorphic deposits are scarce. The pseudomorphs are composed of anhydrite when chlorides are absent in the evaporite succession, but halite and polyhalite may contribute to the preservation if chlorides are associated with the sulphates (Fig. 6A and B). Many anhydritization processes, however, are pervasive and hinder the preservation of pseudomorphic features. Indeed, some mosaic-nodular (chicken-wire) anhydrite facies can totally destroy a precursor selenitic fabric (Shearman, 1985).

In some Miocene and younger formations, where both selenitic gypsum and replacive anhydrite coexist, it can be observed that the anhydrite-to-gypsum replacement operates in a topotactic manner. Small anhydrite prisms penetrate the parent gypsum crystals along the cleavage planes; moreover, the anhydrite prisms are oriented parallel to each other and also parallel to the cleavage planes. Ghosts of this topotactic fabric are observed in some secondary gypsum rocks, where the gypsum fabric has mimetized the anhydrite prisms (Kasprzyk & Ortí, 1998). In the Messinian of SE Spain, Shearman & Ortí (1976) observed that anhydrite prisms do not occur at random in the selenites, but are concentrated in zones parallel to former crystals faces of the host gypsum. In many instances, even the length of the anhydrite crystals runs parallel to the c-axis of the host gypsum crystal.

Many published descriptions suggest that both the shape and the volume of the precursor selenite crystals were preserved during the replacive process. In this case, apparently the pseudomorphs behave as a rigid entity, the replacive process being isovolumetric. This fact suggests an external supply of calcium sulphate in order to compensate for the loss of volume in the gypsum-to-anhydrite conversion. In other cases, however, the pseudomorphs

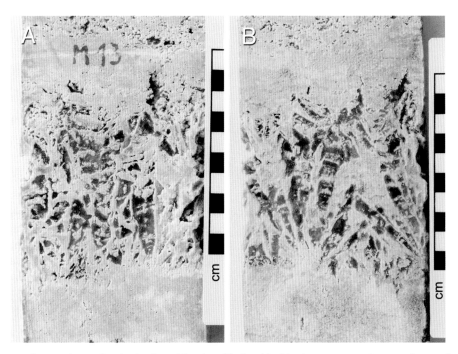

Fig. 6. Selenitic pseudomorphs made of anhydrite (clear) and halite (dark). The presence of twinned crystals can be deduced locally. View of the two sides of a core slab of about 1 cm thick. Core sample of the Upper Keuper evaporites (K5 unit) in the Carcelén borehole (Albacete, Spain). The sample is located at a depth of 1948 m.

are thin, irregular or strongly deformed, suggesting that they have behaved as a soft or plastic-like entity. In this case, a volume loss seems to occur during the replacement process (see below).

SELENITE FACIES IN MODERN ENVIRONMENTS

Salt works and coastal salinas

Selenite facies of salt works

A number of studies carried out in the water circuits of coastal salt works (evaporative salinas) have contributed to an understanding of the precipitation controls on the selenite facies (Schreiber & Kinsman, 1975; Geisler, 1982; Ortí *et al.*, 1984). The observations made in the salt works of the Mediterranean coast of Spain (Ortí *et al.*, 1984) can be summarized as follows (Fig. 7):

Precipitates. First precipitates in the pools of the water circuits are carbonate muds. Afterwards, laminated microbial mats of *Microcoleus* sp. colonize the bottoms of the next pools. In the subsequent pools, organic-rich sandy gypsum (chemical gypsarenite) and selenitic gypsum (selenitic crusts) form (Fig. 8A); these two gypsum facies are accompanied by non-laminated mats

of *Aphanothece* sp. Very fine-grained gypsum (acicular-prismatic crystals) may sporadically precipitate on the selenite apices if the salinity increases close to the beginning of halite formation (Fig. 8B). At the end of the circuits, halite precipitates in the crystallizers in summer.

Salinity range. Precipitation is controlled by the salinity gradient. Within the microbial mats, some fine laminae of gypsum microcrystals may sporadically precipitate to a salinity close to $140\,\mathrm{g\,l^{-1}}$. The gypsarenite facies, however, requires salinities between 140 and $230\,\mathrm{g\,l^{-1}}$, and the selenitic crusts form to salinities between 230 and $300\,\mathrm{g\,l^{-1}}$. These ranges, however, vary slightly from one circuit to another and from one salina to another.

Selenitic crusts. These crusts are composed of transparent, upward-directed, elongated gypsum crystals, up to 10 cm in length. These crystals grow by means of the (120) prism face and a slightly curved surface (Fig. 9A). The crystals may also be slightly bent (Fig. 9B). They are not twinned, in general, although the contact (100) twin, symmetrical or asymmetrical, is found locally (Fig. 9B).

Microtopography. The selenitic crusts are composed of irregularly spaced, coalescent rigid domes that occupy high zones (shoals) on the pool bottoms. In contrast, the low zones (the inter-dome

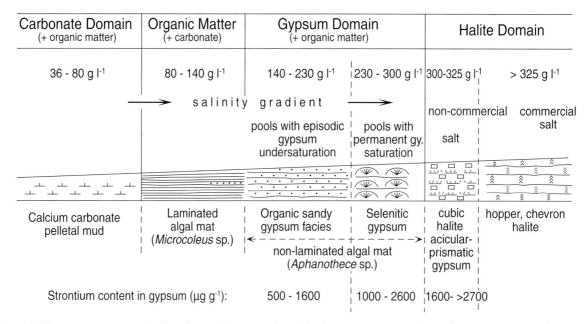

Carbonate Domain (+ organic matter)	Organic Matter (+ carbonate)	Gypsum Domain (+ organic matter)		Halite Domain	
36 - 80 g l⁻¹	80 - 140 g l⁻¹	140 - 230 g l⁻¹	230 - 300 g l⁻¹	300-325 g l⁻¹	> 325 g l⁻¹

salinity gradient

pools with episodic gypsum undersaturation | pools with permanent gy. saturation | non-commercial salt | commercial salt

| Calcium carbonate pelletal mud | Laminated algal mat (*Microcoleus* sp.) | Organic sandy gypsum facies | Selenitic gypsum | cubic halite | hopper, chevron halite |

non-laminated algal mat (*Aphanothece* sp.)

acicular-prismatic gypsum

| Strontium content in gypsum (µg g⁻¹): | | 500 - 1600 | 1000 - 2600 | 1600- >2700 | |

Fig. 7. Sedimentary spectrum in the salt work (evaporative salina) environment on the Mediterranean coast of SE Spain: facies continuum, salinity ranges, and Sr content in the gypsum facies. Adapted from Ortí *et al.* (1984).

depressions) are formed by soft sediment composed of fine (millimetre-sized) gypsum crystals (Fig. 9A).

Geochemical indicators. Strontium is the most significant marker of the gypsum facies in the salt works. A direct relationship exists between Sr contents in the crystals and the salinity of the precipitating brines (Fig. 7). However, Sr contents may vary significantly from one salt works to another (Rosell *et al.*, 1998).

Accumulation rate. Rates recorded in the selenitic crusts of the salt works in SE Spain were documented by Ortí *et al.* (1984). For a period of about 80 years, the maximum accumulation rate was close to 5 mm yr⁻¹. A similar value was recorded for the gypsarenite facies.

Selenite facies of the coastal salinas in South Australia

Along the coast of South Australia, a number of Holocene saline lakes record banded gypsum deposits (Warren, 1982, 1989). These deposits are formed by selenite facies and gypsarenite (autochthonous, fine-grained gypsum) facies. In these lakes, the depositional history started about 5000–6000 years ago when the last oceanic sealevel rise stabilized. In association with this rise,

most of the depressions in the coastal dune system were filled with marine water and resulted in perennial saline lakes, or salinas, up to 10 m deep. The gypsum facies precipitated in these salinas and their evolution was as follows (Fig. 10A and B):

Selenitic domes. On a depositional bottom covered by microbial mats, the initial gypsum deposit was made up of large, prismatic, unoriented crystals twinned along (100). These selenites formed coalescent domal structures and were associated with calcareous sediment rich in aragonitic peloids. These precipitates remained permanently below the halocline (Fig. 10B). The unoriented crystal fabric in the domes is thought to reflect the slow growth rate of the selenites and the soft character of the bottom.

Bedded selenites. When the water depth began to decrease, the perennial halocline became affected by some disturbances. Consequently, winter dilution slightly affected the deep water mass and some aragonite bands accumulated on the selenite domes. In summer, the salinity increased and accelerated the competitive growth of the selenites, which progressively oriented upwards. The depositional surface was dominated by the irregular (chevron-like) geometry of the selenite apices. The morphology of some single, zoned crystals in Marion Lake is shown in Fig. 11.

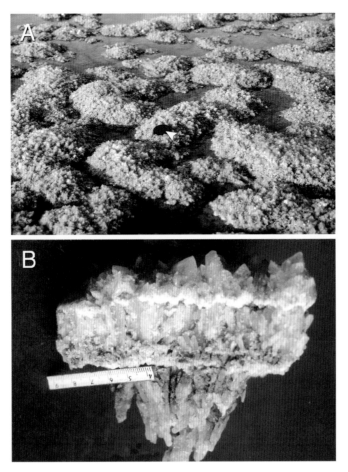

Fig. 8. Selenite facies in the coastal salt works of Santa Pola (Alicante, SE Spain). (A) Typical view of the bottom of a selenitic pool. Rigid selenite domes are distributed in high zones. Note the idiomorphic apices of the crystals in the domes. Scale: lens cap, arrowed (diameter = 6 cm). (B) Superposition of three selenitic episodes in a selenitic crust. In the central episode, the crystal size is finer at the base and extremely fine (white material) at the top. This white material records a moment of high concentration in the pool and includes tiny halite crystals mixed with the gypsum. Scale in cm.

Bedded selenites affected by dissolution. With the progressive infilling of the salina, winter dilution also affected the depositional surface. Then, a flat dissolution surface formed, in which the selenite apices appeared truncated; on this surface, laminae of aragonitic peloids accumulated. These laminae, however, were poikilitically enclosed within the selenites by growth reinitiating during the subsequent summer. The selenites grew to between 2 and 15 cm long, although some reached a length of up to 2 m.

Gypsarenite. In the shallowest salinas, where the salinity change in the brine was very rapid, selenite growth stopped, and a massive nucleation of microcrystals resulted in the accumulation of a gypsarenite lamina. However, in salinas where the selenites still precipitated together with gypsarenites, selenites always graded upwards into gypsarenites. Moreover, the gypsarenites precipitated toward the margins of the salinas, whereas the selenites predominated in the central depressions.

Discussion

The following aspects of the selenite precipitation in the coastal salt works and natural salinas can be highlighted:

Balance between freshwater influx and brine reflux

In the coastal salt works, the balance between freshwater influx and brine reflux is man-controlled. In the Mediterranean coast of SE Spain,

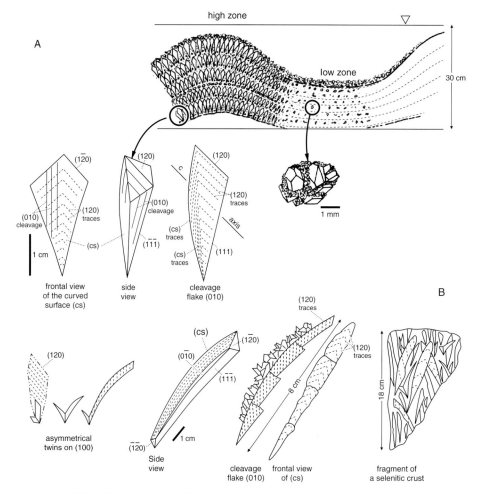

Fig. 9. Schematic diagram of the selenitic crust and selenite crystals developed in a salt works on the Mediterranean coast of SE Spain. Adapted from Ortí *et al.* (1984). (A) Selenitic crust: gypsum precipitates in a "shoal zone" and in a "low zone". (B) Asymmetrical twins and curved crystals found in other selenitic crusts.

the best developed selenite facies clearly appear in the salt works where water circulation is permanent during the whole year (even in winter). In contrast, in the salt works where the circulation stops each winter, with the associated dilution of the brines in the pools, the selenite facies are less developed or even totally absent. In the latter case, the gypsarenite facies occupy all the gypsum-precipitating pools and exhibit a number of uncommon facies (Ortí *et al.*, 1984).

In the Holocene selenitic salinas of South Australia, however, the influx of freshwater was balanced by the reflux of heavy brines into the sea; this reflux occurred interstitially through the highly porous dune system (seepage). It is worth noting that in some of these salinas selenite facies are totally absent and only gypsarenites have been recorded (Warren, 1982). In these shallow salinas,

Warren (1982) interpreted that the salinity oscillations were very rapid and the deep brines were unstable.

Presence of a permanent halocline versus constant gypsum supersaturation without halocline

The existence of perennial haloclines in the brines of the Holocene salinas of South Australia was noted by Warren (1982); this was regarded as a basic control on selenite development. This, however, does not mean that stagnant (anoxic) conditions develop in the deep water mass, where, apparently, bacterial sulphate-reduction processes are not intense. The case of the salt works in SE Spain is different. In the very shallow pools of these evaporative salinas, where haloclines are unimportant or absent, the protection of the

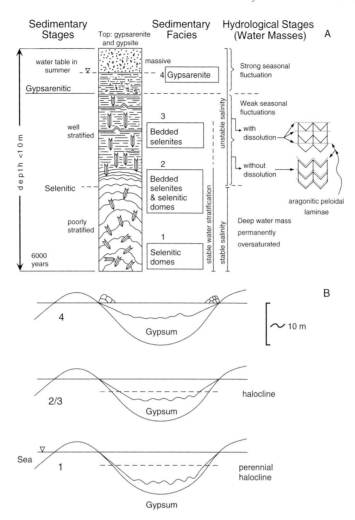

Fig. 10. Selenite facies in the Holocene coastal salinas of South Australia. Adapted from Warren (1982). (A) Sedimentary facies. (B) Infilling stages of the salinas. Numbers 1 to 4 correspond to the sedimentary facies in (A).

selenites against dissolution is achieved by means of a constant brine saturation in gypsum. This is because the selenite pools are fed with brines, already saturated in gypsum, coming from the gypsarenite pools (Fig. 7). In contrast, periodic undersaturation affects the brines in the pools of the gypsarenite facies: this happens each time they are fed with undersaturated waters coming from the pools colonized by the microbial mats. As this process is intermittent, the gypsarenite pools remain undersaturated in gypsum for long periods.

Crystal nucleation versus syntaxial overgrowth

In the gypsarenite facies of the salt works, crystal nucleation is important, whereas in the selenitic crusts, syntaxial overgrowth predominates. Only

in some water circuits, where operation is irregular (Ortí *et al.*, 1984), anomalous features in the selenitic crusts are commonly observed, such as truncation of the crystal apices due to dissolution or alternations between thin selenitic crusts and gypsarenite laminae.

Main conditions for selenite accumulation

It seems clear that selenite facies in marine environments mainly require stability in the brines. The wide range of Sr contents found in the selenites of the salt works also suggests that this is not a question of a high or a specific range of salinity. Moreover, the selenites in the Holocene salinas of South Australia seem to have formed at relatively low salinities. In both cases, fine-grained, autochthonous gypsum facies (gypsarenites)

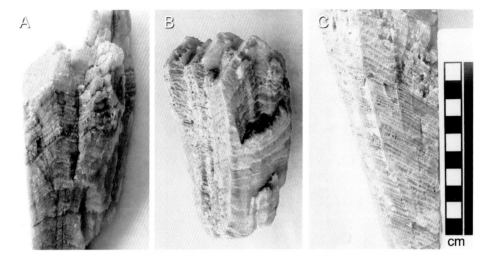

Fig. 11. Selenite crystals from Marion Lake (South Australia). Samples supplied to the author by Prof. Douglas J. Shearman in 1980. (A) Partial view of a specimen 22.5 cm long. At the top left, the (120) prism face is observed. The zonation produced by this prism face is seen throughout the crystal. In the lower part, the (010) cleavage plane (perpendicular to the picture plane) can be observed as vertical, dark lines. (B) Specimen 13 cm long. The specimen is composed of several subcrystals. At the top, short (120) prism faces are observed; on the right side, the curved surface of one subcrystal is shown; on the left side, several cleavage planes are seen. Zonation is present throughout all of the subcrystals. (C) Detail of the (010) cleavage planes of crystal in (A), showing the millimetric zonation. Long, straight zones correspond to the (120) prism, being parallel to the c-axis of the crystal. Zones corresponding to the curved surface are poorly developed, although they can be observed at the top of the picture.

instead of selenites form under unstable salinity conditions.

SELENITE FACIES IN ANCIENT ENVIRONMENTS

In this section, some examples of selenite units recognized in ancient evaporite formations will be discussed. Examples are taken from the Zechstein, Triassic, Eocene, Middle Miocene and Upper Miocene successions of Europe (Spain, Poland and Sicily). In the Miocene successions, the selenites still remain as primary gypsum, whereas in the older successions they have been preserved as anhydrite at depth.

Ancient marine selenite facies occur in both platform and basinal settings. Platform selenites can be classified as flanking selenite shelves, narrow flanking selenite shelves, and selenite-bearing platforms covering salt-filled basins (Fig. 12). Flanking selenite shelves involve selenite layers precipitated on coastal salinas, inner troughs, and shoals. These shelves, up to several hundred kilometres in width, are commonly adjacent to evaporitic basins (Fig 12Ai). However, they may also exist adjacent to non-evaporitic basins, as in the case of many Triassic platforms. In coastal salinas, the single selenite layers are usually thin (5–15 cm) and the Ca-sulphate deposits as a whole are not very thick (some tens of metres, except for cases of very high rate of subsidence). The association with microbial mats and nodular anhydrite is common in these occurrences. However, in large lagoons and inner troughs, single selenite layers may reach up to several metres in thickness, and the Ca-sulphate deposits as a whole may attain one hundred metres in thickness. In some cases, the selenite shelves are very narrow (few kilometres in width; Fig. 12Aii) and may display slump features. This type of narrow, flanking selenite shelf usually represents marginal gypsum wedges preceding falls of water-level in hypersaline basins.

The term "selenite platform covering a salt-filled basin" is used here for the occurrence of a selenite succession accumulated on a preexisting evaporitic basin (and adjacent shelves), after its total infilling with chlorides (Fig. 12Aiii). This platform has a rather flat topography, may cover broad areas, and may display well-developed evaporite cycles. However, the selenite facies only constitutes part of the whole evaporitic succession.

Selenite basins are basins filled mainly with selenite layers (Fig. 12B). These layers result from

A   Platform Selenites

(i)   Flanking shelf

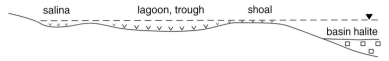

(ii)   Narrow flanking shelf

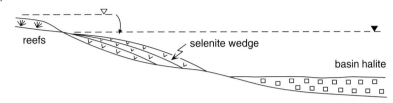

(iii)   selenite-bearing platform
on a salt-filled basin

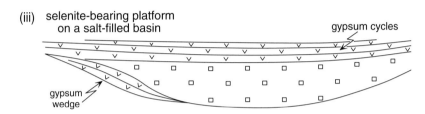

B   Basin Selenites

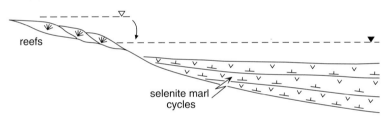

Fig. 12. Selenite occurrences in marine evaporites. (A) Platform selenites: (i) flanking selenite shelf; (ii) narrow, flanking selenite shelf (selenite wedge); (iii) selenite-bearing platform covering a salt-filled basin (with cycles bearing selenite facies). (B) Basin selenites (cycles of selenite layers and marl layers).

important sea-level oscillations in relatively deep basins, although these basins rarely or never reach the stage of "desiccated basin". The progressive infilling of these basins with selenite layers commonly records a cyclic alternation between deeper-water marls and shallower-water evaporites.

Platform selenites: some examples

Selenite facies associated with sabkha anhydrite in coastal salinas: the Anhydrite Zone (Triassic–Liassic boundary; NE Spain)

An example of pseudomorphic selenite facies of the coastal salina environment is found in the unit known as the "Anhydrite Zone" in Spain.

This unit, which is developed across the Triassic–Liassic boundary, has a Rhaeto–Hettangian age. The exploratory Alacón borehole, drilled near Alacón village (Teruel province) for hydrogeological purposes, cuts about 300 m of the anhydritic unit (Ortí & Salvany, 2004).

Some of the main facies forming this evaporitic succession are the following: (A) massive to banded carbonate mudstone; (B) alternation of carbonate laminae and anhydrite laminae; (C) banded anhydrite; (D) bedded pseudomorphs; (E) interstitial pseudomorphs; and (F) massive to nodular anhydrite (Fig. 13). These facies are cyclically repeated in the borehole log, where about 35 cycles are distinguished. The succession is interpreted as

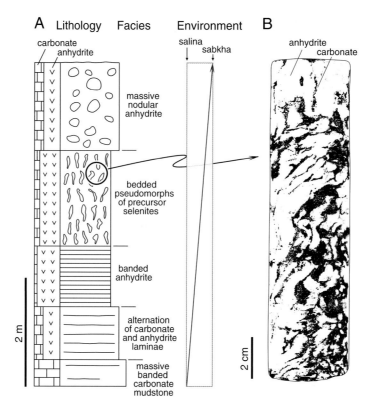

Fig. 13. Pseudomorphic selenite facies of the Anhydrite Zone (Rhaetian–Hettangian, NE Spain). Adapted from Ortí & Salvany (2004). (A) An example of a salina-sabkha cycle with carbonate and sulphate facies. (B) Detail of the anhydrite pseudomorphs after precursor, deformed selenites.

having been formed in a coastal salina-sabkha setting.

The banded anhydrite facies are interpreted as originally fine-grained, laminated to banded gypsum; the bedded pseudomorphs and interstitial pseudomorphs are interpreted as originally bedded (palisade) selenites and interstitial gypsum crystal, respectively. One of the cycles involving exposure conditions at the top (sabkha setting) is represented in Fig. 13A. The palisade pseudomorphs, without evidence of twinning, have a maximum length of 10 cm and exhibit truncation (dissolution surfaces) at the top. These pseudomorphs may display sutured contacts and a plastic-like deformation (Fig. 13B). This deformation was probably generated during the gypsum-to-anhydrite conversion and was favoured by several factors, such as mechanical compaction, progressive loss of strength in the selenites, and the presence of an unlithified carbonate matrix. This mineralogical transformation is thought to have occurred under shallow-to moderate-burial conditions (Ortí & Salvany, 2004).

Selenite coastal salinas: the Werra Anhydrite (Zechstein, Poland)

The Zechstein succession in the Polish Basin, with a total thickness of over 1500 m, is arranged in four major evaporite cycles. The first of these, the Werra Cycle, is up to 450 m thick. The lowest evaporite unit of the Werra Cycle is the Lower Werra Anhydrite. The general pattern of the Werra facies is a carbonate platform attached to the land and an evaporite platform in a more basinal position. The Werra evaporite platform comprises a complex of shoal zones with thick sulphate deposits, and topographic lows with thin sulphate and thick halite deposits (Peryt *et al.*, 1993).

Boreholes located in the shoal zones generally exhibit a succession of anhydrite pseudomorphs after bedded selenite facies. The pseudomorphic beds are relatively thin, less than 5 cm on average, and the maximum length of the individual pseudomorphs is up to 10 cm. The depositional setting of these selenites has been attributed to coastal salinas (Peryt *et al.*, 1993). In some samples, these pseudomorphs display a disorganized fabric due to mechanical compaction or to some gravitational instability (initial slumping). In other boreholes located in north Poland, Peryt (1994) illustrated the presence of twins in the Werra pseudomorphs. Similar pseudomorphic facies were reported by Richter-Bernburg (1985) in the Werra Anhydrite in Germany.

Flanking selenite shelf: Badenian gypsum (Miocene, Poland)

During the Badenian (Langhian and Serravalian p.p.), the Carpathian Foredeep of southern Poland was an area of evaporite sedimentation. Deposition took place on a platform made up of widespread lagoons (troughs) separated by elongated shoals (Kubica, 1992; Kasprzyk, 1993). Resulting evaporite deposits are up to 60 m thick and consist of calcium sulphates and microbial carbonates. At outcrop, the sulphate has been preserved as primary gypsum; in the subsurface, however, it changes progressively into anhydrite in a basinward direction.

The gypsum deposits form a constant, laterally extensive succession in which up to 18 layers have been distinguished (Fig. 14A). In this succession, a lower member and an upper member may be differentiated. The lower member is mainly

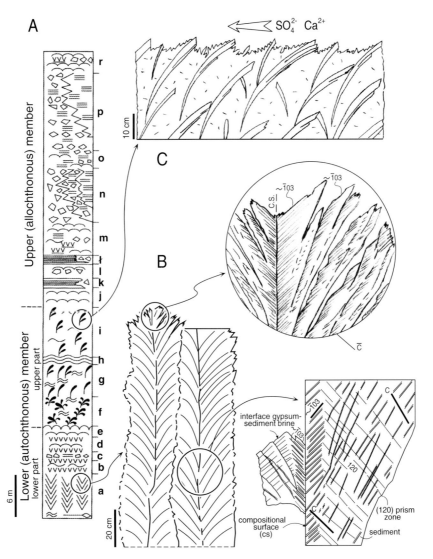

Fig. 14. Lower member of the platform succession of the Badenian gypsum in Poland. (A) Gypsum succession: lower and upper members; layers a–r. Adapted from Kasprzyk (1993). (B) Detail of the giant intergrowths in the basal layer, a; cs = composition surface of the intergrowth. Adapted from Bąbel (1987). (C) Oriented sabre-like selenites and deduced relation with brine currents on the platform. Adapted from Bąbel (2002a).

composed of autochthonous, selenite facies (layers a–i); the upper member is composed of allochthonous laminated and clastic gypsum facies (layers j–r; Kasprzyk, 1993). In the lower member, the main selenite facies are "giant gypsum intergrowths" (layer a, "glassy gypsum"), bedded selenites (layers b and d), "skeletal" gypsum (layer f), and "sabre-like" gypsum (layers g and i). The descriptions provided in some papers for the selenite facies of the lower member are summarized in Table 1. The layers composed of sabre-like crystals may form giant domes.

The largest selenite sizes, between ten centimetres and several metres in height, correspond to the giant intergrowths, whereas skeletal and sabre-like crystals only reach some tens of centimetres in length. The crystallography of the giant intergrowths (Fig. 14B) is dominated by a composition surface (Bąbel, 1987, 1990), and the two sides of the intergrowth have no common symmetry elements (Fig. 1C). All the crystals started to grow in pairs of subcrystals combined with the developing composition surface. This growth occurred perpendicularly to the c axis of the couplet crystals (Fig. 15A). As in the (100) contact twin, the intergrowth can also be asymmetrical. Moreover, the two sides can develop curved pair of subcrystals; the composition surface may also show a curved

Table 1. Selenite facies of the Badenian gypsum in Poland and Ukraine.

Peryt (1996)	**Giant gypsum intergrowths**
	Intergrowths are formed by large, vertically arranged gypsum crystals. Pairs of crystals are joined together along flat surfaces orientated normal to bedding. These are compositional surfaces, and neither a twin plane nor a twin axis is recognizable in them. Intergrowths may reach several metres high.
	Sabre gypsum
	Curved gypsum crystals resembling sabres. They grew upward and simultaneously curved laterally by advance of the (120) prism faces and parallel lens-shaped subcrystals or their aggregates. Sabre gypsum crystals may build thick units that are indistinctly bedded. Commonly the sabres are found orientated in the same direction.
	Skeletal gypsum
	Chaotically arranged intergrown crystals up to 20 cm long. The term was coined by Kwiatkowski (1972). Asymmetric twins are common, with the lower wing poorly developed. The intercrystalline space is filled by granular and microcrystalline gypsum with minor amounts of carbonate, clays and organic material.
	Bedded (selenite) (grass-like gypsum – stromatolite intercalations)
	Rows of upright gypsum crystals that are usually a few to 20 cm high; rows are separated by finely crystalline (alabastrine) gypsum layers, up to 30 cm thick, but often show crenulated patterns. The grass-like gypsum forms small nucleation cones or clusters.
Babel (1987)	**Giant (gypsum) intergrowths**
	Glassy gypsum built of vertical, juxtaposed giant crystals, up to 3.5 m in height, joined together in pairs along flat surfaces oriented perpendicularly to the depositional surface. Surfaces closely resemble the twin junctions, because the crystals on both sides of these surfaces show palmate structures which at first sight are symmetrical. Since crystallographic c axes are placed obliquely to this junction the crystal pair looks like the contact ($\bar{1}$01) twin. However, the planes of the (010) perfect cleavage belonging to both crystals are not strictly parallel. It seems that each crystal pair realizes its own orientation and there is no simple law of orientation common for all crystal pairs. There is no crystallographic symmetry in particular pairs, and neither a twin plane nor a twin axis is recognizable. The vertical surface that joints the crystals is herein named the "composition surface" as it is an interface of individuals of the oriented intergrowth of crystals.
	In a great number of outcrops the intergrowths are set up vertically with their composition surfaces perpendicular to the depositional surface. Such oriented intergrowths attain the greatest height. In another variety of glassy gypsum the composition surfaces are oriented obliquely to the depositional surfaces; the intergrowths are smaller, rarely attaining >1 m.
Bąbel (1991, 2002b)	Although similar to the contact gypsum twins ($\bar{1}$01), the intergrowths differ from twins in lacking any crystallographic symmetry between component crystals. Many features and data suggest that they are representative of a very peculiar type of organically-controlled mineral intergrowths so far recognized only in the Carpathian Foredeep (Babel, 1991). Due to low supersaturation, and/or organic compounds inhibiting crystallization, gypsum nucleation was sparse and crystal growth was mostly syntaxial, similar to growth in mineral druses. The protracted period of upward growth led to formation of extraordinarily large individuals.

Fig. 15. Badenian gypsum, Lower Gypsum Member (Poland). (A) Symmetrical, giant intergrowths with subvertical, almost straight composition surfaces. Some composition surfaces are seen. Hammer shaft is 30 cm long, layer a. (B) Asymmetrical giant intergrowth. Note that not only the subcrystals composing the intergrowth are curved, but also the composition surface is too. Hammer for scale, layer a. (C) Sabre-like gypsum facies. The view corresponds to a small part of a giant dome with a diameter of about 10 m. Scientist is Dr Maciej Bąbel (October 1985).

trend (Fig. 15B). According to Bąbel (1987, 2002b), these intergrowths probably formed under the control of some organic compounds.

Both skeletal and sabre-like selenites are mainly composed of single, curved crystals (Fig. 15C). The sabre-like gypsum crystals commonly display preferential orientations, which vary in different parts of the platform. Bąbel (1986) attributed this to the existence of brine currents during the development of the facies, with the result that selenites grew in an upstream direction and became progressively inclined (Fig. 14C). Detailed palaeocurrent analysis based on the orientation of the sabre-like crystals allowed Bąbel (2002a) and Bąbel & Becker (2006) to deduce a brine palaeoflow roughly parallel to the basin margin contour.

Rosell *et al.* (1998) studied a number of Sr profiles in the gypsum succession of the Badenian platform in Poland. In the lower member, a clear increase in Sr was demonstrated from the lower part (layers a–e) to the upper part (layers f–i). This was interpreted as the consequence of a salinity increase in the platform from the intergrowths at the base to the skeletal and sabre-like selenitic layers at the top.

In the Badenian deposits of west Ukraine, Peryt 1996, 2001 described the presence of significant lateral facies changes in the selenite facies of the lower member. These gradations involve giant intergrowths and stromatolitic gypsum (Fig. 16A), on the one hand, and sabre-like, skeletal selenites, stromatolitic gypsum and clastic gypsum, on the

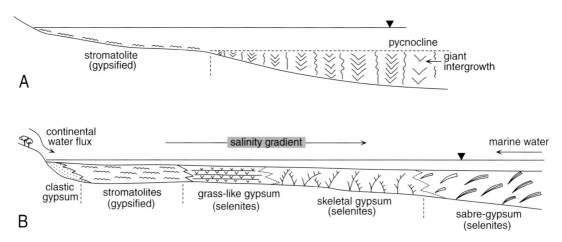

Fig. 16. Lateral facies changes in selenites and associated facies, Badenian gypsum of west Ukraine. Adapted from Peryt (1996). (A) Facies belts in basal layer a. (B) Facies belts in layers b–i.

other (Fig. 16B). According to Peryt (1996), the formation of the giant intergrowths in shallow lagoons and the stromatolitic gypsum on their margins was coeval (Fig. 16A), and the development of the various gypsum facies (shown in Fig. 16B) was mainly controlled by salinity variations (lateral salinity gradients).

More recently, the sedimentological relationships between the various gypsum facies exposed along the margins of the Carpathian Foredeep basin (Ukraine, Poland, Czech Republic, Moldava) were studied by Bąbel (2004a, 2005a, b). Detailed descriptions of the selenite facies were provided and their hydrological and palaeogeographic depositional patterns discussed, as well as the associations of the selenites with other facies, in particular, gypsum microbialites. Bąbel (2004a) interpreted all of these Badenian gypsum deposits as having formed in a giant, shallow basin of salina type.

Narrow, flanking selenite shelf: the Catalan Potash Basin (NE Spain)

At the end of the Eocene (Priabonian), the South-Pyrenean Potash Basin developed as an elongated marine trough, which extended over a distance of more than 300 km from Navarra, in the west, to Catalonia, in the east. In this trough, a wide shallow central zone with salt and sulphates may be distinguished from two marginal basins (the Navarrese and the Catalan potash basins), relatively deeper and more subsident, which reached the potash stage (Busquets *et al.*, 1985; Rosell & Pueyo, 1997).

In the Catalan Potash Basin (Fig. 17A), a calcium sulphate belt up to 30 m thick (Odena Gypsum Member) crops out overlying offshore marine marls (Igualada Formation). This belt occupies an intermediate position between a central chloride body (Cardona Formation, up to 300 m thick) and peripheral carbonates (La Tossa Formation). At outcrop, this calcium sulphate belt is made up of secondary gypsum and residual masses of anhydrite (not yet rehydrated), and is organized in a thick, massive lower unit, and a thinner, better stratified upper unit. At the base of the lower unit, a carbonate-sulphate stromatolite is present and the remainder of the unit is composed of pseudomorphic gypsum. The upper unit consists mainly of clastic gypsum facies.

The boundary between the two sulphate units is sharp and has been interpreted as an unconformity (Ortí *et al.*, 1985; Busquets *et al.*, 1985). According to Ortí *et al.* (1985), the development of this erosional surface between the two units could be related to the deposition of the chloride body in the basin (Fig. 17A). The gypsum belt thins out very fast towards the basin, where it is represented only by a thin anhydrite layer (Basal Anhydrite). A number of slump-related features are observed in the gypsum belt: intraformational gypsum breccias; folds affecting only particular gypsum layers; broken gypsum layers; and stacked portions of broken gypsum layers (Busquets *et al.*, 1985).

At outcrop, the lower unit contains centimetre- to decimetre-sized secondary gypsum pseudomorphs after selenites. Some (precursor) selenite facies are identified: bedded selenites, massive

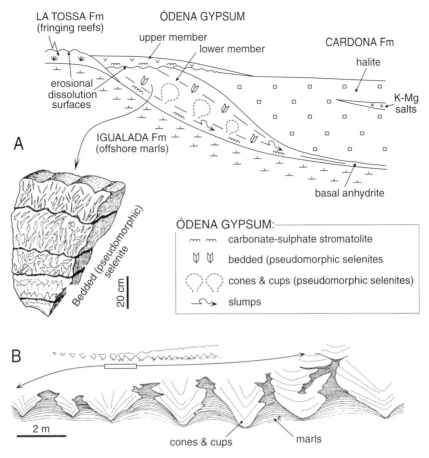

Fig. 17. Narrow flanking selenite shelf: the Eocene Odena Gypsum (NE Spain). (A) Schematic diagram, not to scale, of the facies distribution in the Catalan Potash Basin. The geometrical and sedimentological relationships between the lower and upper members of the Odena Gypsum is interpretative. Detail of the (pseudomorphic) bedded selenite facies. Adapted from Ortí *et al.* (1985) and Busquets *et al.* (1985). (B) Scheme of selenite cones and cups (or supercones) in the Odena Gypsum. Note the interfingering between marine marls and gypsum cones. Adapted from Busquets *et al.* (1985).

selenites, and large selenite cones and cups (Fig. 17A). The latter reach up to several metres high and may interfinger with the offshore marls (Fig. 17B).

Selenite platform covering a salt-filled basin: the Messinian Upper Evaporite (Central Sicily Basin)

The Messinian marine evaporites of the Mediterranean region comprise two major units: the Lower Evaporite and the Upper Evaporite (Rouchy, 1982). The Lower Evaporite consists of: (1) salt (Saline Unit), which occupies the centres of the large basins, located beneath the present abyssal plains of the Mediterranean Sea, and some marginal basins; and (2) sulphates (Lower Gypsum), which are developed in the marginal basins. The Upper Evaporite is made up of an alternation of marls and gypsum layers and extends more widely than the

lower unit, reaching the marginal basins in some peri-Mediterranean countries.

The Messinian selenite facies of Sicily and mainland Italy have been the subject of a number of works, including: Ogniben (1957); Nesteroff (1973); Richter-Bernburg (1973); Eugster & Hardie (1976); Lo Cicero & Catalano (1976); Schreiber *et al.* (1976); Vai & Ricci-Lucchi (1976); and Rouchy (1982). Some of these descriptions of the selenite facies are summarized in Table 2. Ogniben (1957) interpreted the selenites in Sicily as having formed by an early replacement of coarse-grained, primary anhydrite, but this secondary origin was rejected first by Hardie & Eugster (1971), and then by many other workers.

In the Central Sicily Basin, the Upper Evaporite covers a thick basinal chloride deposit (Saline Unit), which grades laterally into the Lower Gypsum. The geometrical relationships between these

Table 2. Selenite facies of the Messinian in Sicily and Italy

Some facies equivalences	*Bedded selenites* (Hardie & Eugster, 1971), *Grass-like selenites* (Richter-Bernburg, 1973), *banded selenites* (Vai &Ricci Lucchi, 1977), *bedded selenites* (Lo Cicero & Catalano, 1976)
	Wavy, anastomozing gypsum beds (Schreiber *et al.*, 1976), *flat lying selenite crystals* (Vai & Ricci Lucchi, 1977), *wavy, needle-like selenite layers* (Lo Cicero & Catalano, 1976)
	Massive selenites (Vai & Ricci Lucchi, 1977), *coarse twinned selenites* (Hardie & Eugster, 1971), *massive selenites* (Lo Cicero & Catalano, 1976)
Hardie & Eugster (1971)	**Very (coarse) selenite**
	These are single crystals up to one metre in length, in beds up to 30 m thick. The crystals are essentially pure gypsum with no discernible bedding.
	Bedded selenites
	As selenite grain size decreases bedding becomes more distinct, marked by discontinuous, undulating layers of carbonate interstitial to the selenite crystals; other selenite units are very distinctly bedded, with bedding defined by thin interbeds (0.5 cm) of gypsum sand or laminated gypsum-carbonate partings.
Schreiber & Decima (1976)	**Coarsely (crystalline) gypsum**
	Massive beds of selenitic gypsum, termed "spicchiolino", appear as orderly rows of oriented crystals. Richter-Bernburg (1973) divided the orderly crystalline gypsum into: grass-like variety; in this, the gypsum crystals are broad, well formed and are separated one from the other; the grass-like gypsum beds are usually separated from one another by irregular to semi-regular laminae of micritic carbonate and gypsum sands; the rows or banks of crystals may range in thickness from a few centimetres up to 3 m.
	"cavoli", cabbage or cauliflower form variety; this type is composed of crowded continuous clusters of narrower crystals; cone-like or stellate groupings are also common morphologies.
Lo Cicero & Catalano (1976)	**Coarsely (crystalline) selenites**
	Two different types of gypsum crystals are grouped in this lithofacies: individual crystals ranging in size from 50 cm to 1 m in length and from 2 to 4 cm in width; they develop perpendicular to the bedding along their long dimension, while their c-axis lie at low angle (about 20°) on the bedding plane; these crystals are not twinned and generally enclose impurities along (010); individual crystals ranging in size from 50 cm to 3 m in height and from 4 to 10 cm in width, occurring in parallel clusters, variously oriented; each crystal appears slightly curved, growing almost vertically to the bedding in the first part and then curving at the distal part; the crystals show their c-axis slightly inclined respect to the bedding plane, and appear twinned along an inclined twin-plane that lies near the base of the crystals.
	Selenite "(nucleation) cones"
	Conical clusters of selenitic crystals, irregularly spaced, growing on algal and gypsiferous laminations. The crystal clusters appear to bend and/or perforate the underlying laminae producing stellate pattern on the bottom of the bed; the crystals have a prismatic habit, 1 to 7 cm in length, are arranged in falt cone-like structures and are crossed by micritic films and algal filaments; the crystals have their c-axis parallel to the direction of major growth and are often twinned along (100).
	Massive selenites
	Coarse swallow tail selenites poorly interlayered with gypsum and carbonate mudstones; the couplets form units of 5–20 m thick separated by bituminous gypsiferous marls. At the base of each unit the crystals are arranged in conical structures with convex surfaces ("cavoli", according to Richter-Bernburg, 1973). The single crystals are elongated in the direction of c-axis which lies from perpendicular to slightly inclined regards the bedding; they are generally twinned and follow Mottura's rule. Each crystal shows a V shaped core crossed by more or less lined inclusions, which present an inverted chevron shape if seen along the (010) plane.
	Bedded selenites
	Selenitic crystal layers (2–4 cm thick) alternating with carbonate-clay-gypsum partings (0.5–30 mm thick). Selenites and partings are arranged in an upward-thinning sequence. The selenites, mostly thinly prismatic and twinned along (100), grew on the top of the partings with the c-axis parallel to their elongation and perpendicular to the bedding plane. The crystal apices are generally, either slightly rounded or sharply cut, and covered by thin micritic films. The bedded selenites differ from the massive selenites by the lack of enclosed algal filaments.

Table 2. (*Continued*)

Vai & Ricci Lucchi (1976, 1977)	***Wavy, needle-like (selenite) layers*** Flat or slightly wavy beds of needle-like gypsum crystals. The crystals (1–4 cm long) are generally oriented with their long axes parallel to or slightly inclined to the bedding plane. They appear in layers intercalated with 1–5 cm thick white, carbonate gypsiferous partings. Lineations and other sedimentary structures of flow patterns are present.

Stromatolitic (selenite)

Lenticular beds 5–30 cm thick composed of upstanding selenite crystals of length equal to bed thickness. The selenite encloses micritic algal laminae whose filaments can be seen with a hand-lens. The beds may be internally layered by dissolution surfaces and clastic calc-gypsum laminae interrupting the algal laminae and the selenite crystals.

Flat-lying (selenite) crystals
Patches of interconnected or loose selenite crystals lying with the long axes in the plane of bedding. The crystals have a prismatic-pseudohexagonal habit but contain a turbid core with a needle-like swallow-tail shape and clayey, organic and algal inclusions. The surrounding clear overgrowth has a greater volume than the core. This facies does not form well defined beds but a sort of pavement or crust some centimetres thick. Stellate aggregates or random orientations are visible locally.

Massive (selenite)
This facies forms parts or members of complex gypsum beds with a thickness ranging from 80 cm to more than 30 m. Typical swallow-tail twins show their maximum development (length: 10–100 cm) with a fining-upward trend through the member; they have a more or less pronounced orientation known as "Mottura rule" (elongation normal to bedding, tails pointing upwards). At the base of the member or bed, the twins are often arranged in conical clusters or "cavoli", with apices pointing downward and protruding into underlying laminated deposits which are bent and broken; when exposed, these basal surfaces show knobs known as "mamelloni". In other cases, the crystals have a more uniform vertical fabric and the base of the bed is flat.

Banded (selenite)
Subparallel bands, 1–20 cm thick, vertically repeated over a maximum thickness of 10 m with an upward-thinning trend. The bands look like palisades: upstanding selenite twins form their upper (and thicker) part and determine a more or less rough top surface, the lower part is composed of mechanically deposited crystals a few mm or less in size. A reverse grading is thus developed in each band. The growth was repeatedly interrupted by dissolution and or mechanical reworking, giving a sequence of minor cycles.

---

units are similar to that shown in Figure 12Aiii. In the Eraclea Minoa section (Fig. 18), up to six or seven cycles bearing selenite facies are developed with a total thickness of 300 m (Nesteroff, 1973; Rouchy, 1982). The gypsum facies in each cycle comprise the following sucession: (1) deep-water gypsarenites at the base; (2) fine-grained laminated to thin-bedded gypsum (called "balatino") in the middle; and (3) shallow-water selenite facies at the top. Rosell *et al.* (1998) studied the geochemistry (Sr content) of the primary gypsum in this profile, and interpreted an upward-increasing salinity trend for each cycle. This interpretation was consistent with former studies (Longinelli, 1979; Pierre, 1982; Rouchy, 1982). Thus, each selenite facies in these cycles seems to record a maximum fall of sea-level and an associated highest brine concentration on the platform.

Basin selenites: the Messinian basins of SE Spain and Italy

In the Eastern Betic Chain (SE Spain), a series of intermontane Neogene basins were formed during the Tortonian (Upper Miocene). Several of these record evaporite units that have been studied in a number of papers, including: Montenat (1973); Dronkert (1976, 1977, 1985); Rouchy (1976, 1982); Shearman & Ortí (1976); Ortí & Shearman (1977); Michalzik (1996); Rosell *et al.* (1998); Playà *et al.* (2000); and Fortuin & Krijgsman (2003).

Some of the basins, the so-called inner basins, are located near the present Mediterranean coast, the: Sorbas, Carboneras, Níjar, San Miguel de Salinas, and Palma de Mallorca basins. These basins exhibit cyclic selenite formations, which have been attributed to the Lower Evaporite of the

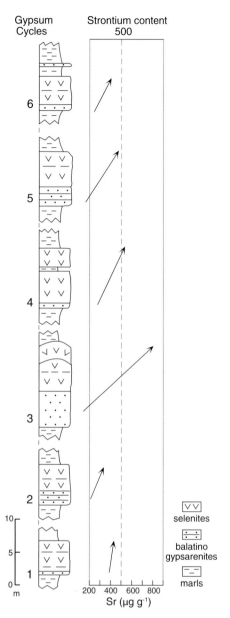

Fig. 18. Selenite cycles and Sr profile of the Messinian Upper Evaporite (Pasquasia Gypsum) in the Eraclea Minoa section, Sicily. Adapted from Rosell *et al.* (1998).

Messinian (Rosell *et al.*, 1998). In the selenite beds of these formations the crystals may be single, symmetrically twinned (Fig. 19A), or asymmetrically twinned (Fig. 19B), and their subcrystals are separated by sediment bands.

Details of the symmetrical twins in the San Miguel de Salinas Basin are shown in Fig. 20. In the selenite beds of this basin, a number of features are observed (Shearman & Ortí, 1976; Ortí & Shearman, 1977): (1) there is usually a relationship between the type of the re-entrant angle and the

crystal shape and orientation (Table 3); (2) the traces of the (120) prism face and the curved surface are easily seen on the cleavage surfaces and on the lateral view of the twins (Fig. 20); and (3) significant facies changes are recorded between the margins and the basin centre. Thus, a lateral gradation involving carbonate-gypsum stromatolites, disordered selenite domes, and columnar selenites is shown in Fig. 21. In this gradation, the different salinities assumed for the domes and the massive selenites are based on their patterns of Sr distribution: very homogeneous values in the columnar selenites *versus* a wide range of values in the selenite domes (Rosell *et al.*, 1998).

The selenite formations in the Neogene inner basins of the Eastern Betic Chain occupy the basinal centres, but not the adjacent shelves. These formations are constituted by 13 or 14 cycles, each cycle being composed of underlying marls and overlying selenite beds. The individual selenite beds have variable thicknesses, between a few centimetres and up to 25–30 m. The main selenite facies are bedded selenites, massive selenites, selenitic domes, and selenitic cones and supercones. In the selenite beds, the size of the crystals usually decreases upward (fining-upward trend) and the selenitic successions, as a whole, display thinning-upward trends (Rosell *et al.*, 1998).

The study of Sr profiles in these selenite successions enabled Rosell *et al.* (1998) to subdivide them into lower and upper parts. The three or four selenitic layers making up the lower parts of the successions are thicker, have massive columnar selenites, and display relatively homogeneous Sr profiles, suggesting that they were formed in stable, deeper water bodies. In contrast, the selenite layers comprising the upper parts of the successions are thinner, have variable selenite facies, and display irregular Sr profiles, suggesting that they were formed in unstable, shallower waters. As a whole, the successions seem to record shallowing-upward trends and are interpreted as the progressive evaporitic infilling of the Betic basins in a regressive situation during the Messinian. In the Sorbas Basin, however, Riding *et al.* (1998, 1999) interpreted a transgressive trend in the selenitic cycles; these authors also assumed that the evaporitic formation, the Yesares Member, belongs to the Upper Evaporite of the Messinian.

In the Vena del Gesso, a foreland basin adjacent to the emergent Apennines in mainland Italy, Vai & Ricci Lucchi (1976, 1977) studied a Messinian

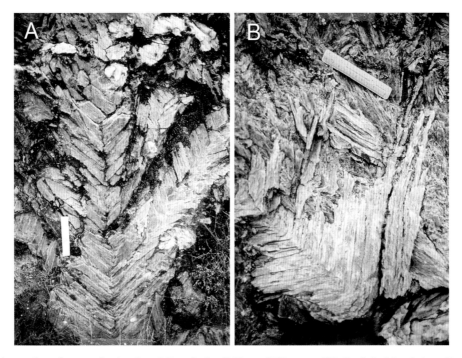

Fig. 19. Selenite twinned crystals in San Miguel de Salinas (Alicante, SE Spain). Messinian, Lower Evaporite. (A) Symmetrical twin. Note that the trace of the junction between the two arms of the twin is not linear. Scale bar = 11 cm. (B) Asymmetrical twin. The upward directed arm of the twin has been developed preferentially. Scale bar = 11 cm.

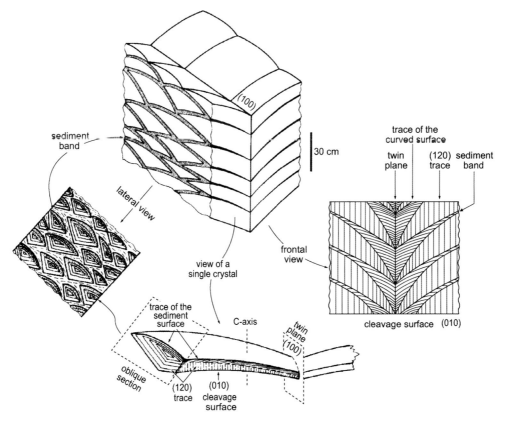

Fig. 20. Details of a symmetrically twinned selenite crystal in the Messinian of the San Miguel de Salinas Basin, Alicante, SE Spain. Adapted from Ortí & Shearman (1977).

gypsum succession belonging to the Lower Eva-porite. This succession, up to 150–170 m thick, is formed by 13–14 selenitic cycles; each cycle is made up of euxinic marls at the base and selenite beds at the top. The selenite beds were described as having a lower autochthonous part composed of vertically-grown selenites, and an upper part composed of nodular and lenticular selenites; the latter facies were interpreted as a mechanical accumulation of the selenitic gypsum forming the lower part in each cycle. The lowermost three cycles, which are the thickest and are made up of mainly auto-chthonous gypsum, form a lower unit; the over-lying cycles, which are thinner and have higher proportions of reworked gypsum at the tops, form an upper unit (Vai & Ricchi Lucchi, 1976; Manzi et al., 2005). This selenite deposition occurred in a lagoonal setting with an estimated depth of some tens of metres.

In a recent paper, Roveri et al. (2006) have rein-terpreted the clastic facies (nodular and lenticular selenites) in these cycles as in situ, interstitially-grown selenites grouped in branches, which project into a gypsiferous calcareous-marl matrix. The "branching selenites" represent an extreme evolution of the subaqueous selenitic supercone structures described by Dronkert (1976) in the Sorbas Basin (SE Spain). According to this new view, the selenitic succession of the Apennines is similar to that described by Rosell et al. (1998) in the Eastern Betic Chain, where also a distinction between some thick selenite cycles at the base (lower part), and the remaining thinner cycles at the top (upper part) may be distinguished in the Messinian successions of the Lower Evaporite.

Discussion

The following aspects of the selenite precipitation in the various examples of platform and basin settings can be highlighted:

Sea-level fluctuation

Selenite facies on the shelves appear to be very sensitive to sea-level oscillations and salinity changes. Sea-level fluctuations may result in the: (a) interruption of crystal growth and the development of erosional-dissolution surfaces, as observed in the narrow flanking shelf of the Catalan Potash Basin in NE Spain (Fig. 17A); and (b) cyclic repetition of selenite units, as seen in the platform deposits of the Messinian Upper Evaporite in Sicily (Fig. 18).

Role of salinity

The role played by salinity is more controversial. In some cases, the gradation of the gypsum facies seems to be clearly controlled by the salinity gradient. This idea was used by Peryt 1996, 2001 to explain the lateral change between the various gypsum facies in the Badenian platform in Ukraine (Fig. 16). In the Polish area of this platform, how-ever, the sharp transition from the selenite facies of the lower member, to the laminated and clastic gypsum facies of the upper member (Fig. 14A), has been assigned either to a structural control (Kasprzyk, 1993; Peryt, 1996; Rosell et al., 1998), i.e. to a general deepening of the platform and an associated salinity decrease, or to a salinity increase in this transition (Bąbel, 1991, 1999a). The last interpretation is based on the presence of halite casts at the base of the upper member (layers j, k and l).

Meaning of the oriented fabrics of selenites

In the Messinian Sorbas Basin (SE Spain), Dronkert (1976, p. 354) suggested that the common orienta-tion of the curved selenites may have been partly controlled by the surficial brine currents. In the Badenian platform deposits of Poland and west Ukraine, Bąbel (2002a) made systematic fabric measures in the sabre-like crystals and related them to the flow patterns of the brines circulating throughout the evaporitic shelf. These observa-tions suggest that evaporite platforms, and probably basins too, could have been affected by important systems of gypsum-precipitating bottom currents.

Maximum thickness of selenite accumulation in a single event

In the Messinian basins of the Eastern Betic Chain and mainland Italy, a permanent outflow of heavy brines prevented the accumulation of chloride brines and salt precipitation. Such an equilibrium situation, however, is difficult to maintain over long time periods. Thus, the thick (up to 150 m) successions of selenite facies recorded in these marginal basins are mainly the result of cyclic deposition: the infilling of the basins was

Table 3. Selenite fabrics and facies of the Messinian in SE Spain

Shearman & Ortí (1976); Ortí & Shearman (1977)	***Fabric with the c axis of the crystals vertical to subvertical** (with respect to bedding)*
	Narrow, elongated twins (palisade or columnar selenites; symmetrical twins with a high rate length/width (L/W) and also to wide, short twins (symmetrical twins with low rate L/W).
	In this fabric there is a relationship between the type of the re-entrant angle and the shape and orientation of the crystals:
	when the angle is acute, the crystals commonly have high length to breadth ratios and individuals may reach lengths of 3 m or more with widths of only 10 cm; in aggregate, the crystals are usually arranged with their long axes vertical or near vertical, and they form impressive parallel arrays that resemble palisades;
	when the re-entrant angle is obtuse, the crystals tend to be shorter and broader than in the previous type, and may achieve widths of up to 40 cm across a twin. These crystals tend to have a less ordered arrangement in the beds, but they nevertheless stand at a high angle to bedding.
	***Fabric with the caxis subhorizontal** (with respect to bedding)*
	Crystals with the c axis forming angles <45° (with respect to bedding) and generating asymmetrical twins. The developed half of the twin is formed by more or less curved (bladed) subcrystals reaching up to 70 cm in length. These subcrystals have the same crystallographic orientation and have a fan-like appearance.
	The crystals occur in parallel clusters in which the individuals are slightly but conspicuously curved, and the different orientations of the clusters imparts to the beds the appearance of a field of coarse grass waving in the wind. At first sight these crystals do not appear to be twinned but in suitable exposures a twin junction can be seen at the base of some of the clusters: it is evident that each vertical cluster of crystals represents one side of a twin. An interesting feature of this type of gypsum is the way in which the arms of a twin splits up laterally into a series of discrete curved blades.
	Fabric with radial caxis
	Upward convex, hemispherical structures, whose basal planes are coincident whit the bedding plane. The morphology of the crystals composing this structure is variable, and they may grade in size from the base to the top of the structure. Also, the spheroidal structure becomes elongated (hemi-cylindrical). Cones and supercone structures may also display this radial fabric.
	Fabric with the caxis randomly oriented
	Randomly oriented crystals; it is present in many selenitic layers and in some radial structures or cones.
Dronkert (1976, 1977)	***Nucleation cones***
	These usually form the base of the vertical tree-like columnar layers of selenites.
	Supercones
	Anastomosing clusters of selenite crystals. These crystals are mostly thinner, less twinned, and less well vertically oriented than in other associations. The larger selenite crystals (up to 1 m) are curved twins with one flank aborted and the other flank oriented subhorizontally.

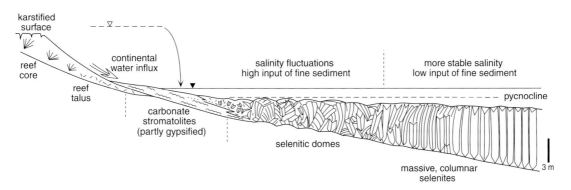

Fig. 21. Lateral facies change in a representative gypsum layer; Messinian of the San Miguel de Salinas Basin, Alicante, SE Spain. Adapted from Montenat *et al.* (1987) (left part of the figure) and from Rosell *et al.* (1998).

achieved by a progressive piling up of evaporative episodes that hardly ever or never reached the chloride stage. The maximum thickness of a single selenite bed recorded in any of these basins is close to 30 m.

Controls on upward gradations in the selenite facies

Vertical gradations between the selenite facies in basinal successions could theoretically be controlled by a number of factors, such as changes in depositional depth, salinity and matrix supply, as well as by fluctuations in the pycnocline level or in the type and amount of organic compounds. The crystal size gradation in a particular layer (fining-upward trend) could record a salinity increase associated with faster nucleation; in fact, the presence at the base of some layers of nucleation cones suggests initial low supersaturation conditions. This upward concentration trend, however, cannot be applied to the selenitic succession as a whole.

Mechanisms of supercone growth

The formation of large selenitic cones and domes dominated by asymmetrical twins is poorly understood. In the San Miguel de Salinas Basin of the Eastern Betic Chain, the large domes composed of dirty (matrix-rich) crystals are located in a marginal position, whereas the long palisade (columnar) fabrics, which are composed of relatively clear crystals, occupy the depocenter (Fig. 21). In the selenite layers of the Sorbas Basin, a "composite gypsum sequence" dominated by coarsening-upward gradation was described by Dronkert (1976). This gradation exhibits clear (matrix-poor) bedded or massive selenites in the lower half and supercones rich in marls in the upper half (Table 3). These supercones, integrated by large, curved crystals and strongly asymmetrical twins, are directly overlain by the marls of the next cycle (Fig. 22). Dronkert (1976) concluded that, probably, a slight salinity decrease combined with renewed input of fine sediment could account for this gradation. According to this, the large cones could represent a mechanism of the selenite structures to resist the progressive input of fine sediment: precipitation would be restricted to some "high" points on the depositional floor, where the domes would grow surrounded by marls. This could be also the case of the large cones and cups present in some marginal areas of the Messinian Betic basins (Fig. 5A) and in the narrow flanking shelf of the Catalan Potash Basin (Fig. 17B).

This mechanism, however, seems not to be valid for the giant domes made up of sabre crystals in the Badenian Gypsum, where an excess of sediment

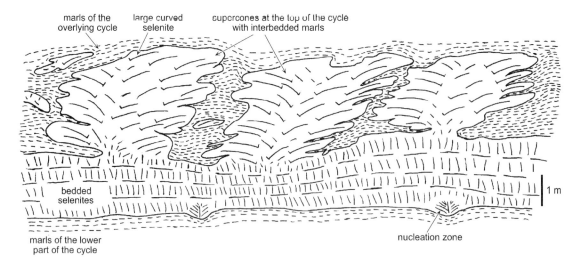

Fig. 22. Schematic representation of the selenite facies (nucleation cones, bedded selenites, supercones) in one of the Messinian gypsum-marl cycles of the Sorbas Basin (Almería, SE Spain). Note the interfingering between supercones in the upper part of the cycle and the overlying marls. Río de Aguas bridge section. Adapted from several figures by Dronkert (1976, 1977).

matrix is absent. For these domes, Bąbel (1999a) interpreted a simple accretionary, upward growth of layered sabre crystals during drowning of an elevated area.

Selenite versus chloride accumulation

The presence of autochthonous selenite layers in a basinal position can be assigned to the incomplete (interrupted) drawdown mechanism described by Tucker (1991), whereas basinal layers of salt and potash can be attributed to complete drawdown. Given the high solubility of chlorides, however, the thickness of salt and potash basinal facies can reach an order of magnitude higher than the basinal selenite facies. On the other hand, no alternations between selenite (or pseudomorphic selenitic) layers and chloride layers have been documented in the basin centres. This is also valid for the Zechstein cycles of central Europe, where the selenite formations are generally restricted to the marginal shelves or constitute marginal wedges (Tucker, 1991).

The geometric relationships between marginal selenites and the associated central chlorides remain enigmatic. In the Central Sicily Basin, the 13–14 selenite cycles comprising the Messinian Lower Evaporite surround a central, thick choride body. In the latter, however, cyclicity is not recorded. In the same basin, Rouchy (1982) and García-Veigas *et al.* (1995) assumed the existence of a lateral facies change between the cyclic selenite succession on the margins and the massive central chlorides, although the manner in which the deduced gradation happens, remains unknown.

Thus, it appears that the two facies groups (selenites and chlorides) are mutually exclusive in the depocentres of ancient basins. By contrast, alternations between fine-grained anhydrite laminae and chlorides are common in such settings.

CONCLUSION

As an important component of the facies continuum in marine evaporites, the study of selenitic gypsum can improve our understanding of palaeogeographic settings and the environmental conditions operating during sedimentation, and can also supply valuable information for sequence stratigraphy. For instance, it can help us to interpret shoaling- or deepening-upward trends. The lateral gradation of the selenitic facies into other facies such as fine-grained gypsum or stromatolites, suggests variations in water salinity, water depth, pattern of water circulation, or supply of fine clastic sediment to the platforms and basins. Furthermore, the ability of selenites to modify laterally or vertically their facies under changing environmental conditions is remarkable. In particular, the building up of large domes evokes the growth strategies displayed by some coral or microbial bioherms.

Selenite facies can develop in a variety of platform and basin settings with permanent outflow of heavy brines. In the majority of cases, these facies accumulate in regressive situations during lowstand or highstand positions of sea level, but selenite accumulation during transgressive episodes cannot be ruled out.

The identification of selenite facies in ancient anhydrite formations should not be a serious problem because the original geometry of the crystals very often can be observed as pseudomorphs or ghosts. Even if some deformation has affected the geometry of the precursor selenites, the pseudomorphic character can be identified under closer inspection. In this regard, further efforts are warranted to improve our knowledge of selenite formations in the stratigraphic record.

Possible controls (physical, chemical, biochemical, biological) on crystal habits and facies in selenite formations remain poorly understood. Thus, divergences commonly arise in the interpretation of sharp facies changes in the successions: salinity changes, bathymetry variations, or both? Observations carried out in modern gypsum-precipitating environments have supplied limited information on these controls. Factors such as the amount and the type of the associated organic matter require further investigation. Although some geochemical indicators, in particular Sr content, have been used to document salinity variations in marine gypsum deposits, further research would be desirable.

In the last years, renewed effort to understand selenite precipitation and the associated spectrum of gypsum facies was carried out by Bąbel (2004a,b, 2005a,b). This author applied systematically an integrated hydrological-sedimentological, quantitative model of the saline-type basin to the gypsum deposits of the Carpathian Foredeep. Among other significant results, some thin selenite layers have been shown to be valid isochronous markers that can be traced throughout the Badenian evaporite platform.

Finally, the broad accumulations of selenite facies in some marine formations have resulted in extensive mining operations, particularly for the plaster industry. For example, the exploitation of the thick selenitic cycles forming the lower part of the Messinian succession in the Sorbas Basin (SE Spain) constitutes the largest gypsum quarry in Europe (Regueiro & Lombardero, 1997).

ACKNOWLEDGEMENTS

This paper is dedicated to the memory of Prof. Douglas J. Shearman, who taught the author so much about evaporite petrography and sedimentology at Imperial College London in 1975, and during many years of fruitful co-operation and stimulating discussions. The author is indebted to Laura Rosell (*Universitat de Barcelona*) and Jean M. Rouchy (*Muséum National Histoire Naturelle*, Paris) for constructive comments on an earlier draft of this paper, to Prof. Abdulrahman Alsharhan (United Arab Emirates University) for reviewing the manuscript and making helpful suggestions to improve it, and to J. Agulló (*Servei de Dibuix i Disseny Gràfic, Universitat de Barcelona*) for drafting assistance.

REFERENCES

Bąbel, M. (1986) Growth of crystals and sedimentary structures in the sabre-like gypsum (Miocene, southern Poland). *Przegl. Geol.*, **4**, 204–207.

Bąbel, M. (1987) Giant gypsum intergrowths from the Middle Miocene evaporites of southern Poland. *Acta Geol. Pol.*, **37** (1–2), 1–19.

Bąbel, M. (1990) Crystallography and genesis of the giant intergrowths of gypsum from the Miocene evaporites of Poland. *Archiw. Mineral.*, **44**, 103–135.

Bąbel, M. (1991) Dissolution of halite within the Middle Miocene (Badenian) laminated gypsum of southern Poland. *Acta Geol. Pol.*, **41**, 168–182.

Bąbel, M. (1999a) Facies and depositional environments of the Nida Gypsum deposits (Middle Miocene, Carpathian Foredeep, southern Poland). *Geol. Quart.*, **43**, 405–428.

Bąbel, M. (1999b) History of sedimentation of the Nida Gypsum deposits (Middle Miocene, Carpathian Foredeep, southern Poland). *Geol. Quart.*, **43**, 249–447.

Bąbel, M. (2002a) Brine palaeocurrent analysis based on oriented selenite crystals in the Nida Gypsum deposits (Badenian, southern Poland). *Geol. Quart.*, **46**, 435–448.

Bąbel, M. (2002b) The largest natural crystal in Poland. *Acta Geol. Pol.*, **52**, 251–267.

Bąbel, M. (2004a) Badenian evaporite basin of the northern Carpathian Foredeep as a drawdown salina basin. *Acta Geol. Pol.*, **54**, 313–337.

Bąbel, M. (2004b) Models for evaporite, selenite and gypsum microbialite deposition in ancient saline basins. *Acta Geol. Pol.*, **54**, 219–249.

Bąbel, M. (2005a) Selenite-gypsum microbialite facies and sedimentary evolution of the Badenian evaporite basin of the northern Carpathian Foredeep. *Acta Geol. Pol.*, **55**, 187–210.

Bąbel, M. (2005b) Event stratigraphy of the Badenian selenite evaporites (Middle Miocene) of the northern Carpathian Foredeep. *Acta Geol. Pol.*, **55**, 9–29.

Bąbel, M. and Becker, A. (2006) Cyclonic brine-flow pattern recorded by oriented gypsum crystals in the Badenian evaporite basin of the northern Carpathian Foredeep. *J. Sed. Res.*, **76**, 996–1011.

Busquets, P., Ortí, F., Pueyo, J.J., Riba, O., Rosell, L., Sáez, A., Salas, R. and Taberner, C. (1985) Evaporite deposition and diagenesis in the Saline (Potash) Catalan Basin. In: *Excursion Guidebook* (Eds M.D. Milà and J. Rosell), pp. 13–59. Int. Assoc. Sedimentol., 6th European Regional Meeting, Lleida, April 1985,

Butler, G.P. (1973) Strontium geochemistry of modern and ancient calcium sulphate minerals. In: *The Persian Gulf: Holocene Sedimentation and Diagenesis in a Shallow Epicontinental Sea* (Ed. B.H. Purser), pp. 423–452. Springer-Verlag, Berlin.

Catalano, R. Ruggieri, G. and Sprovieri, R.(Eds) (1976) *Messinian Evaporites in the Mediterranean (Erice Seminare, October 1975).* Mem. Soc. Geol. It., **16**, 385 pp.

Cody, R.D. (1979) Lenticular gypsum: occurrences in nature, and experimental determination of effects of soluble green plant material on its formation. *J. Sed. Petrol.*, **49**, 1015–1028.

Cody, R.D. and Cody, A.M. (1988) Gypsum nucleation and crystal morphology in analog saline terrestrial environments. *J. Sed. Petrol.*, **58**, 247–255.

Crawford, A.R. (1965) The geology of the Yorke Peninsula. *S. Aust. Geol. Surv. Bull.*, **770**, 841 pp.

Decima, A. and Wezel, F. (1971) Osservazioni sulle evaporiti messiniane della Sicilia Centro-meridionale. *Riv. Min. Sic.*, **22**, 172–187.

Dickinson, S.B. and King, D. (1951) The Stenhouse Bay gypsum deposits. *S. Aust. Min. Rev.*, **91**, 95–113.

Dronkert, H. (1976) Late Miocene evaporites in the Sorbas Basin and adjoining areas. *Soc. Geol. It. Mem.*, **16**, 341–361.

Dronkert, H. (1977) The evaporites of the Sorbas Basin. *Rev. Inst. Inv. Geol. Diput. Prov. Barcelona*, **32**, 55–76.

Dronker, H. (1985) Evaporite models and sedimentology of Messinian and recent evaporites. *GUA Pap. Geol.*, **1**, 183 pp.

Drooger, C.W. (Ed.) (1973) *Messinian Events in the Mediterranean.* Geodynam. Sci. Rep., **7**. K. Ned. Akad. Wet., North Holland Publishing Company, Amsterdam, 272 pp.

Eugster, H.P. and Hardie, L.A. (1976) Some further thoughts on the depositional environment of the Solfifera Series of Sicily. *Soc. Geol. It. Mem.*, **16**, 29–38.

Fortuin, A.R. and Krijgsman, W. (2003) The Messinian of the Níjar Basin (SE Spain): sedimentation, depositional environments and paleogeographic evolution. *Sed. Geol.*, **341**, 1–30.

García Veigas, J., Ortí, F., Rosell, L., Ayora, C., Rouchy, J.M. and Lugli, S. (1995) The Messinian salt of the Mediterranean: geochemical study of the salt from the Central Sicily basin and comparison with the Lorca basin (Spain). *Bull. Soc. Géol. Fr.*, **166**, 699–710.

Geisler, D. (1982) De la mer au sel: les faciès superficiels des Marais Salants de Salin-de-Giraud (Sud de la France). *Géol. Mediterr.*, **9**, 521–550.

Geisler-Cussey, D. (1985) *Approche Sédimentologique et Géochimique des Mécanismes Génerateurs de Formations Evaporitiques Actuelles et Fossiles*. Unpubl. PhD thesis, Université de Nancy I, France, 298 pp.

Goto, M. (1968) Orientated growth of gypsum in Marion Lake gypsum deposits, South Australia. *J. Fac. Sci. Hokkaido Univ. Ser. IV, Geol. Mineral.*, **14**, 85–88.

Hardie, L.A. and Eugster, H.P. (1971) The depositional environment of marine evaporites: a case of shallow clastic accumulations. *Sedimentology*, **16**, 187–220.

Kasprzyk, A. (1993) Gypsum facies in the Badenian (Middle Miocene) of southern Poland. *Can. J. Earth Sci*, **30**, 1799–1814.

Kasprzyk, A. (1994) Distribution of strontium in the Badenian (Middle Miocene) gypsum deposits of the Nida area, southern Poland. *Geol. Quart.*, **38**, 497–512.

Kasprzyk, A. and Ortí, F. (1998) Palaeogeographic and burial controls on anhydrite genesis: the Badenian basin in the Carpathian Foredeep (southern Poland, western Ukraine). *Sedimentology*, **45**, 889–907.

Kasprzyk, A. and Osmolski, T. (1989) Strontium mineralization and its connection with the lithofacies of the Miocene chemical deposits in the regions of Solec Staszów and Zurawica (in Polish). *Biul. Państw. Inst. Geol.*, **362**, 97–118.

Kubica, B. (1992) Lithofacial development of the Badenian chemical sediments in the northern part of the Carpathian Foredeep (in Polish with English summary). *Pr. Państw. Inst. Geol.*, **133**, 1–64.

Kushnir, J. (1980) The coprecipitation of strontium, magnesium, sodium, potassium and chloride ions with gypsum. An experimental study. *Geochim. Cosmochim. Acta*, **44**, 1471–1482.

Kushnir, J. (1982) The composition and origin of brines during the Messinian desiccation event in the Mediterranean basin as deduced from concentrations of ions coprecipitated with gypsum and anhydrite. *Chem. Geol.*, **35**, 333–350.

Kwiatkowski, S. (1972) Sedimentation of gypsum in the Miocene of southern Poland (in Polish with English summary). *Pr. Muz. Ziemi*, **19**, 3–94.

Lo Cicero, G. and Catalano, R. (1976) Facies and petrography of some Messinian evaporites of the Ciminna Basin (Sicily). *Soc. Geol. It. Mem.*, **16**, 63–81.

Longinelli, A. (1979) Isotope geochemistry of some Messinian evaporites: paleo-environmental implications. *Palaeogeogr. Palaeoclimatol. Palaeoecol.*, **29**, 95–124.

Lu, F.H., Meyers, W.J. and Schoonen, M.A.A. (1997) Minor and trace element analyses on gypsum: an experimental study. *Chem. Geol.*, **142**, 1–10.

Lu, F.H., Meyers, W.J. and Hanson, G.N. (2002) Trace elements and environmental significance of Messinian gypsum deposits, the Níjar Basin, southeastern Spain. *Chem. Geol.*, **192**, 149–161.

Manzi, V., Lugli, S., Ricci Lucchi, F. and Roveri, M. (2005) Deep-water clastic evaporites deposition in the Messinian Adriatic foredeep (northern Apennines, Italy): did the Mediterranean ever dry out? *Sedimentology*, **52**, 875–902.

Michalzik, D. (1996) Lithofacies, diagenetic spectra and sedimentary cycles in Messinian (Late Miocene) evaporites in SE Spain. *Sed. Geol.*, **106**, 203–222.

Michalzik, D., Elbracht, J., Mauthe, J., Reinhold, C. and Schneider, B. (1993) Messinian facies relations in the San Miguel de Salinas Basin, SE Spain. *Z. Deut. Geol. Ges.*, **144**, 352–369.

Montenat, C. (1973) Le Miocène terminal des Chaines Bétiques (Espagne méridionale). Esquisse paléogéographique. In: *Messinian Events in the Mediterranean* (Ed. C.W. Drooger), pp. 180–187. North-Holland Publishing Company.

Montenat, C., Ott d'Estevou, P., Larouzière, F.D. and Bedu, P. (1987) Originalité géodynamique des bassins néogènes du domain Bétique Oriental (Espagne). *Notes Mem., Total Comp. Fr. Pétrol.*, **21**, 11–50.

Mottura, S. (1871) Sulla Formazione Terziaria della zona solfifera della Sicilia. *Mem. R. Coitm. Geol. It.*, **1**, 50–140.

Nesteroff, W.D. (1973) Un modèle pour les évaporites messiniennes en Mediterranée: des bassins peu profondes avec dépots d'évaporites lagunaires. In: *Messinian Events in the Mediterranean* (Ed. C.W. Drooger), pp. 69–81. North Holland Publishing Company, Amsterdam.

Ogniben, L. (1954) La "Regola di Mottura" di orientazione del gesso. *Period. Mineral. (Rome)*, **23**, 53–65.

Ogniben, L. (1955) Inverse graded bedding in primary gypsum of chemical deposition. *J. Sed. Petrol.*, **25**, 273–281.

Ogniben, L. (1957) Petrografia della Serie Solfifera Siciliana e considerazione geologiche relative. *Mem. Descr. Carta Geol. It.*, **33**, 275 p.

Orti, F. and Busson, G.(Eds) (1984) *Introducción a la Sedimentologiá de las Salinas Marítimas de Santa Pola (Alicante, España)*. *Rev. Inv. Geol. Diput. Barcelona*, **38/39**, 235 pp.

Ortí, F. and Salvany, J.M. (2004) Coastal salina evaporites of the Triassic-Liassic boundary in the Iberian Peninsula: the Alacón borehole. *Geol. Acta*, **2**, 291–304.

Ortí, F. and Shearman, D.J. (1977) Estructuras y fábricas deposicionales en las evaporitas del Mioceno superior (Messiniense) de San Miguel de Salinas (Alicante, España). *Rev. Inst. Inv. Geol. Diutp. Prov. Barcelona*, **32**, 5–53.

Ortí, F., Pueyo, J.J., Geisler-Cussey, D. and Dulau, N. (1984) Evaporitic sedimentation in the coastal salinas of Santa Pola (Alicante, Spain). *Rev. Inv. Geol. Diput. Prov. Barcelona*, **38/39**, 169–220.

Ortí, F., Pueyo, J.J. and Rosell, L. (1985) La halite du bassin potassique sud-pyrénéen (Eocène supérieur, Espagne). *Bull. Soc. Géol. Fr.*, **1**, 863–872.

Palache, C., Berman, H. and Frondel, C. (1951) *Dana, System of Mineralogy*. 7th Edn. John Wiley & Sons, New York, 1124 pp.

Peryt, T.M. (1994) The anatomy of a sulphate platform and adjacent basin system in the Leba sub-basin of the Lower Werra Anhydrite (Zechstein, Upper Permian), northern Poland. *Sedimentology*, **41**, 83–113.

Peryt, T.M. (1996) Sedimentology of Badenian (middle Miocene) gypsum in eastern Galicia, Podolia and Bukovina (West Ukraine). *Sedimentology*, **43**, 571–588.

Peryt, T.M. (2001) Gypsum facies transitions in basin-marginal evaporites, middle Miocene (Badenian) of west Ukraine. *Sedimentology*, **48**, 1103–1119.

Peryt, T.M. and **Jasionowski, M.** (1994) In situ formed and redeposited gypsum breccias in the Middle Miocene Badenian of southern Poland. *Sed. Geol.*, **94**, 153–163.

Peryt, T.M., Ortí, F. and **Rosell, L.** (1993) Sulfate platform-basin transition of the Lower Werra Anhydrite (Zechstein, Upper Permian), western Poland: facies and petrography. *J. Sed. Petrol.*, **63**, 646–658.

Petrichenko, O.I., Peryt, T.M. and **Poberengsky, A.V.** (1997) Peculiarities of gypsum sedimentation in the Middle Miocene Badenian evaporite basin of Carpathian Foredeep. *Slov. Geol. Mag.*, **3**, 91–104.

Pierre, C. (1982) *Teneurs en Iisotopes Stables ($^{18}O, ^{2}H, ^{13}C, ^{34}S$) et Conditions de Genèse des Evaporites Marines: Applications à Quelques Milieux Actuels et au Messinien de la Mediterranée.* Unpubl. PhD thesis, Université de Paris-Sud, Centre d'Orsay, 266 pp.

Playà, E. (1998) *Les Evaporites de les Conques Bètiques Marginals (Fortuna-Lorca, Miocè Superior): Comparació amb altres Conques Mediterrànies.* Unpubl. PhD thesis, Universitat de Barcelona, 248 pp.

Playà, E. and **Rosell, L.** (2005) The celestite problem in gypsum Sr geochemistry: An evaluation of purifying methods of gypsiferous samples. *Chem. Geol.*, **221**, 102–116.

Playà, E., Ortí, F. and **Rosell, L.** (2000) Marine to non-marine sedimentation in the upper Miocene evaporites of the Eastern Betics, SE Spain: sedimentological and geochemical evidence. *Sed. Geol.*, **133**, 135–166.

Regueiro, M. and **Lombardero, M.** (1997) *Innovaciones y Avances en el Sector de las Rocas y Minerales Industriales. Colegio Oficial de Geólogos de España*, Madrid, 78 pp.

Richter-Bernburg, G. (1973) Facies and palaeogeography of the Messinian evaporites in Sicily. In: *Messinian Events in the Mediterranean* (Ed. C.W. Drooger), pp. 124–141. North Holland Publishing Company, Amsterdam.

Richter-Bernburg, G. (1985) Zechstein-Anhydrite. Facies und Genese. *Geol. Jb., Reine A*, **85**, 85 pp.

Riding, R., Braga, J.C., Martín, J.M. and **Sánchez-Almazo, I.M.** (1998) Mediterranean Messinian Salinity Crisis: constraints from a coeval marginal basin, Sorbas, southern Spain. *Mar. Geol.*, **146**, 1–20.

Riding, R., Braga, J.C. and **Martin, J.M.** (1999) Late Miocene Mediterranean desiccation: topography and significance of the "Salinity Crisis" erosion surface on-land in southeastern Spain. *Sed. Geol.*, **123**, 1–7.

Rodríguez-Aranda, J.P., Rouchy, J.M., Calvo, J.P., Ordóñez, S. and **García del Cura, M.A.** (1995) Unusual twinning features in large primary gypsum crystals formed in salt lake conditions, Middle Miocene, Madrid Basin, Spain – palaeoenvironmental implications. *Sed. Geol.*, **95**, 123–132.

Rosell, L., Ortí, F. and **García-Veigas, J.** (1994) Geoquímica del estroncio en los yesos messinienses de San Miguel de Salinas (Alicante). *Geogaceta*, **15**, 82–85.

Rosell, L. and **Pueyo, J.J.** (1997) Second marine evaporitic phase in the South-Pyrenean Foredeep: the Priabonian Potash Basin (Late Eocene: Autochthonous-Allochthonous Zone). In: *Sedimentary Deposition in Rift and Foreland Basins in France and Spain* (Eds G. Busson, and C. Schreiber), pp. 358–387. Columbia University Press, New York.

Rosell, L., Ortí, F., Kasprzyk, A., Playà, E. and **Peryt, T.M.** (1998) Strontium geochemistry of Miocene primary gypsum: Messinian of southeastern Spain and Sicily and Badenian of Poland. *Jr. Sed. Res.*, **68**, 63–79.

Rouchy, J.M. (1976) Sur la genèse de deux principaux types de gypse (finement lité et en chevrons) du Miocène terminal de Sicile et d'Espagne meridionale. *Rev. Géogr. Phys. Géol. Dynam.*, **8**, 347–364.

Rouchy, J.M. (1982) *La Genèse des Évaporites vaporites Messiniennes de Méditerranée. Mus. Nat. Hist. Nat. Paris Mém., Nouv. Sér., Sér. C, Sci. Terre*, **50**, 267 pp.

Roveri, M., Lugli, S., Manzi, V., Gennari, R., Iaccarino, S.M., Grossi, F. and **Taviani, M.** (2006) *The record of Messinian events in the Northern Apennines Foredeep basins.* Pre-Congress Field Trip, R.C.M.N.S. Interim Colloquium "The Messinian salinity crisis revisited-II", Parma (Italy), 7–9 September 2006. *Acta Nat. Aten. Parm.*, **42**, 65 pp.

Schreiber, B.C. and **Decima, A.** (1976) Sedimentary facies produced under evaporitic environments: a review. *Mem. Soc. Geol. It.*, **16**, 111–126.

Schreiber, B.C. and **Kinsman, D.J.J.** (1975) New observations of the Pleistocene evaporites of Montallegro, Sicily and a modern analog. *J. Sed. Petrol.*, **45**, 469–479.

Schreiber, B.C. and **Schreiber, E.** (1977) The salt that was. *Geology*, **5**, 527–528.

Schreiber, B.C., Friedman, G.M., Decima, A. and **Schreiber, E.** (1976) Depositional environments of Upper Miocene (Messinian) evaporite deposits of the Sicilian Basin. *Sedimentology*, **23**, 729760.

Schreiber, B.C., Catalano, R. and **Schreiber, E.** (1977) An evaporitic lithofacies continuum: Latest Miocene (Messinian) deposits of Salemi Basin (Sicily) and a modern analog. In: *Reefs and Evaporites* (Ed. J.H. Fisher). *AAPG Stud. Geol.*, **5**, 169–180.

Shearman, D.J. (1971) *Calcium Sulphate Evaporitic Facies.* Short Course. Imperial College, London, 65 pp.

Shearman, D.J. (1985) Syndepositional and late diagenetic alteration of primary gypsum to anhydrite. In: *Sixth International Symposium of Salt* (Eds B.C. Schreiber and H.L. Harner), **1**, 41–50. Salt Institute, Alexandria, VA.

Shearman, D.J. and **Ortí, F.** (1976). Upper Miocene gypsum: San Miguel de Salinas, SE Spain. Mem. Soc. Geol. It., **16**, 327–340.

Tucker, M. (1991) Sequence stratigraphy of carbonate-evaporite basins: models and application to the Upper Permian (Zechstein) of northeast England and adjoining North Sea. *J. Geol. Soc. London*, **148**, 1019–1036.

Usdowski, E. (1973) Das geochemische Verhalten des Strontiums bei der Genese und Diagenese von Ca-Karbonat und Ca-sulfat Mineralen. *Contrib. Mineral. Petrol.*, **38**, 177–195.

Utrilla, R. (1985) *Estudi Sedimentologic i Geoquímic de les Salines de la Trinitat (Delta de l'Ebre) i San Pedro del*

Pinatar (Mar Menor). Unpubl. MSc thesis, Universitat de Barcelona, 124 pp.

Vai, G.B. and **Ricci-Lucchi, F.** (1976) The Vena del Gesso in Northern Apennines: growth and mechanical breakdown of gypsified algal crusts. *Mem. Soc. Geol. It.*, **16**, 217–249.

Vai, G.B. and **Ricci-Lucchi, F.** (1977) Algal crusts, autochthonous and clastic gypsum in a cannibalistic evaporite basin: a case history from the Messinian of the Northern Apennines. *Sedimentology*, **24**, 211–244.

Warren, J.W. (1982) The hydrological setting, occurrence and significance of gypsum in late Quaternaly salt lakes in South Australia. *Sedimentology*, **29**, 609–638.

Warren, J.W. (1989). *Evaporite Sedimentology. Importance in Hydrocarbon Accumulation*. Prentice Hall, New Jersey, 285 pp.

Index

Note: page numbers in *italics* refer to figures, those in **bold** refer to tables